ENVIRONMENTAL ENGINEERING IV

PROCEEDINGS OF THE CONFERENCE ON ENVIRONMENTAL ENGINEERING IV, LUBLIN, POLAND, 3–5 SEPTEMBER 2012

Environmental Engineering IV

Editors

Artur Pawłowski, Marzenna R. Dudzińska & Lucjan Pawłowski
Lublin University of Technology, Lublin, Poland

CRC Press
Taylor & Francis Group
Boca Raton   London   New York   Leiden

CRC Press is an imprint of the
Taylor & Francis Group, an **informa** business

A BALKEMA BOOK

Publication was supported by the Polish Ministry of Science and Higher Education

CRC Press/Balkema is an imprint of the Taylor & Francis Group, an informa business

© 2013 Taylor & Francis Group, London, UK

Typeset by V Publishing Solutions Pvt Ltd., Chennai, India
Printed and bound in Great Britain by CPI Group (UK) Ltd, Croydon, CR0 4YY

Published by: CRC Press/Balkema
P.O. Box 11320, 2301 EH Leiden, The Netherlands
e-mail: Pub.NL@taylorandfrancis.com
www.crcpress.com – www.taylorandfrancis.com

ISBN: 978-0-415-64338-2 (Hbk)
ISBN: 978-1-315-88748-7 (eBook)

Environmental Engineering IV – Pawłowski, Dudzińska & Pawłowski (eds)
© 2013 Taylor & Francis Group, London, ISBN 978-0-415-64338-2

Table of contents

Environmental Engineering IV – Pawłowski, Dudzińska & Pawłowski (eds)
© 2013 Taylor & Francis Group, London, ISBN 978-0-415-64338-2

Preface

The Persian king Xerxes lived at the turn of the 6th and 5th centuries B.C. During an expedition through today's Turkey, near the town of Kallatebus, he noticed a beautiful plane-tree. In order to preserve the tree, a special sign was hung on it and a guard stayed there to make sure that nobody will not cut it down. Perhaps it is the first documented human action for the protection of nature in history.

Probably the earliest formal decree on protection of the environment on wider scale was introduced in China, during the reign of Zhou Dynasty around 1122 B.C. It was based on the necessity to preserve the most valuable tree types, forests and green areas and to establish the Forester's office. The decree was continuously reintroduced and found its place among general rules of forest economy, which included even the financial support for afforestation of private property.

Unfortunately, it can be noticed that not only protection of nature dates back far into the past. We must mention that also the problems connected with human-caused destruction of the environment has also a long-lasting history, and actually it has been visible since the beginning of the human appearance on Earth. The earliest changes were purely local and caused no disturbance in the environment. The primitive nomad, in case of an ecological problem, usually moved elsewhere. With territorial expansion, the increase in the human population and obtaining new skills—the scale of our impact on the environment grew and the situation became much more complicated.

Underestimation of the environmental conditions led to the downfall of the first literate and highly advanced civilization in history, the Sumerians, who lived as early as 3000 B.C. The area, between the rivers of Euphrates and Tigris (Lower Mesopotamia), that they were occupying favored agricultural development. The yields were high due to the highly developed irrigating system. A rapid increase in population was observed, along with the increasing demand for food. The increase in yields slowed down and was depleting systematically, reaching one-third of the maximal yields about 1800 B.C. The signs of crisis were ignored, which led to a complete breakdown of the agriculture as well as the entire civilization. Among many causes of the yields decrease, two deserve a special attention, namely:

- Widespread irrigation favored the increase of soil salinity (one of the major causes of soil degradation).
- The growing demand for food, along with the increasing population entailed the expansion of cultivated area. After utilizing all available farming areas, forests were cut out and the land obtained was cultivated. This resulted in increasing erosion, which is another important form of soil degradation. Moreover, the depletion of plant cover and erosion contributed to the creation of large runoffs and the silting of rivers which, as a consequence, caused floods.

Modern technical powers of mankind are much bigger than those of the Sumerians. Our pressure on the environment has also increased. Not only can mankind cause its own extinction, but also can contribute to the destruction of the entire biosphere. Not so long ago it seemed that the environmental protection would bring rescue. The U'Thant report in 1969 was the breakthrough of its development. Media publicity, which accompanied the report, helped to shape the worldwide society awareness of the environmental threats. Alas, classic environmental protection was not able to stop the biosphere degradation. Therefore, the discussion was broadened in 1987, with the formulation of sustainable development concept merging various problematic groups including technology, ecology, economics, as well as politics, philosophy or even social backgrounds.

The central goals of the book Environmental Engineering IV are to summarize research carried out in Poland in the above mentioned area, and to improve technology transfer and scientific dialogue in the time of economic transformation from a planned to a free market economy, thereby leading to a better comprehension of solutions to a broad spectrum of environmentally related problems.

Ongoing political and social changes in Poland have caused some environmental improvements and at the same time it may pose some new problems, both expected and unpredicted. We have observed "ecological fashion". This "fashion" for environmental protection and "ecology" has resulted in a plethora of information in the media. This situation causes social pressure on pro-ecological behavior. However, there are also new conflicts, often associated with job losses that accompany the closing of polluting industries.

Money at the local level is now distributed by local and democratically elected councils. Because of the "ecological fashion" it is easier to make the decision of spending funds on the environment protection. Such decisions are popular among the local populace and this is a positive result of democracy. Democratic mechanisms are less satisfactory when considering the possibility of convincing people about the necessity of locating a landfill in their neighborhood or building a waste incinerator.

Increased use of motor vehicles is one of the most serious problems in Poland today. No incentives or economic stimulation for buying pro-ecological cars have yet been introduced. Nevertheless, due to EU pro-ecological programs, a lot of very important environmentally oriented projects are realized in Poland in which also international companies participate. Moreover, the number of multinational consortia with participation of Polish partners is steadily growing.

Therefore, a presentation of the scientific findings and technical solutions created by the Polish research community ought to be of the utmost interest not only for Polish institutions, but also for international specialists, searching for solutions for environmental problems in new emerging democracies, especially those who plan to participate in numerous projects sponsored by the European Union.

Finally, we would like to express my appreciation to all who have helped to prepare this book: Ms. Katarzyna Wszoła for improving the linguistic side of the papers. Anonymous reviewers who not only evaluated papers, but very often made valuable suggestion helping authors and editors to improve the scientific standard of this book. And finally, last but definitely not least Mrs. Katarzyna Wójcik Oliveira for her invaluable help in preparing a layout of all papers.

Lublin, February 2013
Artur Pawłowski
Marzenna R. Dudzińska
Lucjan Pawłowski

Environmental Engineering IV – Pawłowski, Dudzińska & Pawłowski (eds)
© 2013 Taylor & Francis Group, London, ISBN 978-0-415-64338-2

About the editors

ARTUR PAWŁOWSKI

Artur Pawłowski, Ph.D., D.Sc. (habilitation), was born in 1969 in Poland. He is a member of the European Academy of Science and Arts, and a member of the Environmental Engineering Committee of the Polish Academy of Science. In 1993 he received M.Sc. of the philosophy of nature and protection of the environment at the Catholic University of Lublin. Since that time he has been working in the Lublin University of Technology in the Faculty of Environmental Protection Engineering. In 1999 he defended Ph.D. thesis "Human's Responsibility for Nature" in the University of Card. Stefan Wyszyński in Warsaw. Also at this University in 2009 he defendend D.Sc. thesis "Sustainable Development—Idea, Philosophy and Practice". Now he works on problems connected with multidimensional nature of sustainable development. Member of European Academy of Science and Arts, Salzburg; Environmental Engineering Committee of the Polish Academy of Sciences, Warsaw; International Academy of Ecological Safety and Nature Management, Moscow and International Association for Environmental Philosophy, Philadelphia. Editor-in-chief of scientific journal "Problems of Sustainable Development". He has published over 100 papers (in Polish, English and Chinese) and 6 books.

MARZENNA R. DUDZIŃSKA

Marzenna R. Dudzińska received M.Sc. in physical chemistry in 1983 from Marie Curie-Skłodowska University in Lublin, Poland. She got a Fulbright Scholarship in 1989, and performed pre-doctoral research at University of Houston, USA. She received Ph.D. in environmental chemistry from Marie Curie-Skłodowska University (1992) and D.Sc. (habilitation) in 2004 from Warsaw University of Technology in Environmental Engineering. She is an associate professor at the Institute of Environmental Protection Engineering, Lublin University of Technology and the head of Division of Outdoor & Indoor Air Quality. She authored and co-authored 2 books and 135 papers and co-edited 12 books in the area of POPs in the environment, VOC and SVOC in indoor air. She is a member of Polish Chemical Society, and Committee of Environmental Engineering of Polish Academy of Sciences and International Society of Indoor Air Quality and Climate.

LUCJAN PAWŁOWSKI

 Lucjan Pawłowski, born in Poland in 1946, is the Member of the Polish Academy of Science and Deputy President of the Technical Science Division of the Polish Academy of Science, Member of the European Academy of Science and Arts, honorary professor of China Academy of Science. Director of the Institute of Environmental Protection Engineering of the Lublin University of Technology. He got his Ph.D. in 1976, and D.Sc. (habilitation in 1980 both at the Wrocław University of Technology). He started research on the application of ion exchange for water and wastewater treatment. As a result he together with B. Bolto from CSIRO Australia, has published a book "Wastewater Treatment by Ion Exchange" in which they summarized their own results and experience of the ion exchange area. In 1980 L. Pawłowski was elected President of International Committee "Chemistry for Protection of the Environment". He was Chairman of the Environmental Chemistry Division of the Polish Chemical Society from 1980–1984. In 1994 he was elected the Deputy President of the Polish Chemical Society and in the same year, the Deputy President of the Presidium Polish Academy of Science Committee "Men and Biosphere". In 1999 he was elected a President of the Committee "Environmental Engineering" of the Polish Academy of Science. In 1991 he was elected the Deputy Rector of the Lublin University of Technology, and this post he held for two terms (1991–1996). He has published 22 books, over 168 papers, and authored 98 patents, and is a member of the editorial board of numerous international and national scientific and technical journals.

Environmental Engineering IV – Pawłowski, Dudzińska & Pawłowski (eds)
© 2013 Taylor & Francis Group, London, ISBN 978-0-415-64338-2

Strategy for the security of energy resources in Poland—renewable energy sources

L. Gawlik & E. Mokrzycki
Mineral & Energy Economy Research Institute of the Polish Academy of Sciences, Poland

L. Pawłowski
Faculty of Environmental Engineering, Lublin University of Technology, Lublin, Poland

ABSTRACT: An evaluation of energy supply from renewable and nonrenewable sources has been discussed. It was stated that most of nonrenewable resources will last only for several decades. Therefore, an increase in the use of nonrenewable energy sources is of great importance for human civilization. What is more, potential supply of energy from biomass, hydroelectric, geothermal wind and solar installations has also been discussed. A special attention was given to the evaluation of European potential energy supply of these sources.

Keywords: sustainable development, energy security, renewable energy

1 INTRODUCTION

The objectives of environmental protection are increasingly derived from the paradigm of sustainable development. It is therefore advisable when characterizing the concept of sustainable development to put particular emphasis on environmental protection, linking the environmental objectives with this global trend. However, it should be noted that sustainable development is multidimensional. Setting the goals on a global scale (i.e. the reduction of the greenhouse effect and limiting the depletion of natural resources in order to ensure equal access to resources) creates the need to define appropriate behavior, which is primarily an ethical issue (Pawłowski 2009, 2011). Trends toward refuting the existence of phenomena occurring on a global scale do not result in an optimistic outlook for dealing with such problems. High consumption of fossil fuels leads to climate change but there are serious voices questioning the scale of this phenomenon. At the same time, the ever increasing consumption of fossil fuels ultimately threatens their exhaustion. According to Subramanian (2010), coal reserves should last for 122 years, while oil and natural gas reserves should last for 42 and 60 years respectively. Uranium reserves—with the technologies used today—should last for about 150 years but the introduction of breeder reactors would extend this time to hundreds of years. Breeder reactors are not used in power production, however, due to fears of uncontrolled access to weapons grade nuclear materials which can be produced by these reactors.

The basic problem our civilization now faces is ensuring supplies of energy (Udo et al. 2009). Faced with threats of exhausting fossil energy sources and increasing environmental degradation associated with the production of useful forms of energy, it is necessary to focus on the principles of sustainable development in energy policy. It means paying attention to the wider use of renewable energy sources combined with minimizing their negative impact on the environment. What is more, the technologies associated with the introduction and use of renewable energy sources contribute to an increase in employment levels. In 2010, the renewable energy sector in the 27 countries of the European Union generated jobs for 1.114 million people, which is 25 percent more than in 2009 (0.912 million people). Table 1 shows employment in the various branches of the renewable energy sector in 2010 (Euroobserver 2010). Taking into consideration that unemployment affects people as negatively as life in a degraded environment, combating unemployment by creating new jobs in the renewable energy sector is a very important and positive factor of pursuing sustainable development, especially from a social point of view. In 2010, 16.7% of the world energy consumption came from renewable sources, while in the 27 countries of the European Union this share was lower, amounting to 12.4% (see Table 2). The structure of energy generation was as follows: biomass—68.2%, hydroenergy—18%, wind energy—7.4%, geothermal energy—4.4%, solar energy—2%, and 0.03% from ocean tides.

Table 1. Employment in the renewable energy branches.

Industry	The total number of jobs	The total number of jobs created in selected countries
Biofuels	~1,500,000	Brazil 730,000 in the production of ethanol from sugar cane; China 150,000; Germany 100,000; United States 85,000
Wind energy	~630,000	Spain 40,000; Italy 28,000; Denmark 24,000; Brazil 14,000; India 10,000
Solar panels producing hot water	~300,000	China 250,000; Spain 7,000
Solar cells	~350,000*	China 120,000; Germany 120,000; Japan 26,000; United States 17,000; Spain 14,000
Energy from remaining biomass	~55,0000*	Germany 120,000; United States 66,000; Spain 5,000
Water power engineering	~150,000*	Europe 20,000—including: Spain 7,000; United States 8,000
Geothermal energy	~20,000*	Germany 13,000; United States 9,000
Total amount	~3,500,000	

*Estimated amounts.
Source: Euroobserver 2010.

Table 2. The share of renewable energy in EU energy consumption (%).

	2006	2007	2008	2009	2010	Objective to achieve
European Union (27)	9	9.9	10.5	11.7	12.5	20
Belgium	2.6	2.9	3.3	4.5	5.1	13
Bulgaria	9.6	9.3	9.8	11.9	13.8	16
Czech Republic	6.5	7.4	7.6	8.5	9.2	13
Denmark	16.5	18	18.8	20.2	22.2	30
Germany	6.9	9	9.1	9.5	11	18
Estonia	16.1	17.1	18.9	23	24.3	25
Ireland	2.9	3.3	3.9	5.1	5.5	16
Greece	7	8.1	8	8.1	9.2	18
Spain	9	9.5	10.6	12.8	13.8	20
France	9.6	10.2	11.3	12.3	12.9	23
Italy	5.8	5.7	7.1	8.9	10.1	17
Cyprus	2.5	3.1	4.1	4.6	4.8	13
Latvia	31.1	29.6	29.8	34.3	32.6	40
Lithuania	16.9	16.6	17.9	20	19.7	23
Luxembourg	1.4	2.7	2.8	2.8	2.8	11
Hungary	5.1	5.9	6.6	8.1	8.7	13
Malta	0.2	0.2	0.2	0.2	0.4	10
Netherlands	2.7	3.1	3.4	4.1	3.8	14
Austria	26.6	28.9	29.2	31	30.1	34
Poland	7	7	7.9	8.9	9.4	15
Portugal	20.8	22	23	24.6	24.6	31
Romania	17.1	18.3	20.3	22.4	23.4	24
Slovenia	15.5	15.6	15.1	18.9	19.8	25
Slovakia	6.6	8.2	8.4	10.4	9.8	14
Finland	29.9	29.5	31.1	31.1	32.2	38
Sweden	42.7	44.2	45.2	48.1	47.9	49
Great Britain	1.5	1.8	2.3	2.9	3.2	15

2 BIOMASS

It should be noted that obtaining energy from renewable energy sources will not eliminate conventional energy, though it can be an important element in supporting energy production.

Therefore, obtaining energy from renewable sources will not completely eliminate the negative environmental impact of the energy sector. However, an increase in the share of renewable energy resources in the energy balance is necessary because of the rapid depletion of fossil fuel reserves.

Unlike fossil fuels which play an important role in international trade (imports—exports), renewable energy sources are characterized by local acquisition and use. Each country implements its policy of diversifying energy sources based on local resources.

Biomass is the oldest and most widely used source of energy. It can exist as a solid, liquid, and gas. It is currently the third largest natural source of energy in the world. It typically comes in the form of wood and straw but also includes sewage sludge or waste containing waste paper. The main sources of biomass are agriculture (straw, biogas from manure), forestry (firewood), municipal economy (wastepaper, biogas from landfills and sewage treatment plants), and industry (waste from the pulp and paper industry, waste from the food industry).

Biomass can be used to produce heat in the combustion process. Biomass can be pre-processed before combustion, or highly processed wood briquettes or pellets are used.

2.1 Ability to cover the needs

There are several estimates of biomass potential which differ significantly from each other. Table 3 presents the evaluation of the potential of biomass according to the Institute for Renewable Energy (Wiśniewski (ed.) 2007).

Table 3 shows that the technical potential—the amount of energy that can be produced from domestic resources using the best technologies of energy conversion from renewable sources, taking into account spatial and environmental limitations—is almost 927 PJ. However, the economic potential—the part of the technical potential that can be used after taking into account economic conditions—is 600 PJ. The market potential until 2020—part of the economic potential that can be used during a stated period of time, with the optimal use of all available support mechanisms and funding—is 500 PJ. The use of the potential of biomass in 2006 was over 190 PJ. Renewable energy in the global energy balance amounts to 13.8%, while the share of energy extracted from biomass is about 11%; thus the share of biomass accounts for 80% of renewable energy.

2.2 Wood

Wood sources include forests, parks, orchards, and gardens. The production potential of forests depends on the structure and the surface of forests, forest cover, tree species composition, and the average age of the stand. A forestation conditions are characterized by the forest cover, which differs for different regions of Poland. The largest is the Lubuski Region (48.9% in 2009) and the smallest is the Łódź Region (21%). The market of wood used for energy purposes is difficult to estimate because of the high demand for wood by other industries.

In 2009, the combustion of biomass in boilers produced 500 GW·h of electricity from 15 units with a total capacity of 252 MW (Kamiński & Mirowski 2010). Both industrial waste and post-consumer waste are used for energy purposes. An important source of wood is the wood industry—waste pieces (coniferous, deciduous), sawdust, and bark. The sources include orchards, which annually provide about 1,000 dam^3 (1 dam^3 = 1000 m^3), including about 750 dam^3 as a result of tree thinning.

2.3 Straw

The amount of straw produced each year depends on a number of factors—the growing area, crops, plants, fertilizers, weather conditions, etc. The potential of straw is variable and depends on fluctuations in the harvest. In Poland, the share of cereals is about 60% of the total area of agricultural land. Straw yield of 1 hectare is about 2.5 Mg. The calorific value of the straw depends on its type, humidity, and storage conditions. In the case of gray straw, it is 16–18 MJ/kg.

It is estimated that the current production of straw is 25–28 million tons (Kaminski & Mirowski 2010). Fluctuations in the volume of straw are one of the barriers to its use outside of agriculture.

2.4 Energy use of liquid biofuels

Liquid substances considered to be biomass include pure vegetable oils (crude and refined), bio-ethanol, the esters (methyl or ethyl), biomethyloether, and glycerin. Currently available in the domestic

Table 3. The potential of biomass in Poland, TJ.

Biomass type	Technical potential	Economic potential	Utilization in 2006	Market potential up to 2020
Overall	926,950	600,168	192,097	533,118
Energy plantations	479,166	286,719	4,056	286,718
Dry solid waste	327,044	165,931	160,976	149,338
Wet waste (biogas)	175,809	123,066	2,613	72,609
Firewood from the forest	34,931	24,452	24,452	24,452

market and in the market of the European Union are vegetable oils, vegetable oil esters, and glycerin (Zuwała & Rejdak 2010). In Poland, a commonly grown oil crop is rapeseed. It is expected that cultivation for food will increase from 1 million tons (2007–2008) to 1.2 million tons by 2013. Rapeseed oil production for food will increase from 400,000 tons (2007–2008) to 480,000 tons (2013).

Low oil prices and strong competition among foreign producers (exporters) from Brazil, Germany, Switzerland, Czech Republic, and France are not favorable for the profitable production of methyl and ethyl esters and fatty acids. Since 1990, inland produced bioethanol has been used as an additive in gasoline. Directive 2009/28/EC of the European Parliament and of the Council of April 23, 2009 (effective from January 1, 2011) states that greenhouse gas emissions reductions from the use of biofuels and bioliquids shall be at least 35%; from the year 2017, at least 50%; and from 2018, at least 60%.

In 2010, the share of biofuels was to reach 5.3%, while in the coming years it will grow steadily to 10% in 2020 (IEA 2002). These requirements apply to the share of biofuels in diesel, consumed at levels nearly four times higher than gasoline in Poland.

2.5 Energy use of gaseous biofuels

Biogas is a gas mixture that results from the anaerobic digestion of various types of organic matter. Various types of biogas can be distinguished on the basis of how they are obtained, including the following: landfill gas, gas from sewage sludge and other biogas resulting from the anaerobic digestion of animal manure, slaughterhouse waste, waste from breweries, and other waste from the agri-food industry. Each tonne of municipal waste produces—after 20 years of storage—between 100 and 400 m³ of biogas containing 50% CH_4 and 10–17 MJ/m³ of calorific value.

Assessment of the energy potential of agriculture should be based on agro-climatic conditions, available production capacity of agriculture, crop structure, potential for yield formation, and the total of livestock. The energy potential of agriculture allows the acquisition of raw materials (substrates) necessary for the preparation of 5–6 billion m³ of biogas per year (Dasgupta et al. 2011, Shan et al. 2012, Piemental 2012).

2.6 Energy crops

Energy crops have a high calorific value, are characterized by rapid growth, and are rich in starch and oil. Selection of plant species should be adapted to existing climate and soil conditions for a particular country. Energy crops include: Energy willow,

Virginia mallow, Jerusalem artichoke, species of Miscanthus grasses, and many other plants (Spartina prairie, Japanese knotweed) characterized by rapid growth and easy assimilation of nutrients.

2.7 Covering the needs and the limitations

Biomass is one of the most versatile energy carriers among renewable energy sources. Conversion of biomass for energy can be physical, chemical, and biochemical. Biomass can be accomplished through direct combustion and indirect combustion acquisition of biogas. Power plants show great interest in the technology of coal and biomass co-firing (direct co-combustion, indirect co-combustion, parallel co-combustion). Biomass co-combustion tests have shown that the addition of biomass in an amount exceeding 5% of the primary fuel causes a number of problems associated with the preparation of the fuel, as well as adversely affects the stability of the boiler (Golec 2004; Gaj 2008; Kruczek et al. 2008).

Alkali metals contained in the biomass act as fluxes, lowering the melting point of slag, which creates serious problems for the stable operation of boilers. In addition, as a result of degassing, light particles of coal—entrained with the flue gas—are produced. They react with the water vapor separated from the biomass and form a mixture of CO and H_2. Therefore, during the co-firing of biomass, carbon dioxide emissions may easily exceed emission limits.

2.8 The main courses of action

The current draft of the new regulation (Project 2011) changes the required share of biomass during electricity production to recognize it as renewable energy.

2.9 Problems to be solved

Wood is similar to carbon, though its organic matter has a different chemical composition. Approximately 20% of wood pulp is of non-volatile compounds of carbon. The remaining 80% contains volatile compounds. Efficient combustion of these fuels requires special techniques and boilers which provide the necessary conditions for the complete combustion of volatile products from biomass thermolysis. An assessment of the gasification of biomass is necessary. It would avoid problems arising from melting slags. Furthermore, post-pyrolytic residues are suitable for soil fertilization. The resulting biocoal (biochar) can be used as a fertilizer and—to a certain extent— for the sequestration of carbon dioxide produced during gasification. Variation in the yields of straw

4

production is among the obstacles preventing its use outside of agriculture. There is a need for straw storage in case of crop failure, which affects the costs of 1 kW·h of electricity production and 1 GJ of heat production. Liquid fuels (bioethanol, biomethanol, vegetable oil, bio-oil biodiesel) are currently used in Poland and in the EU. There is a conflict between the food industry, energy industry, and transportation because of the cultivation of rapeseed and sunflower (resulting products: diesel and biodiesel). Concerns have been voiced that the existing malnutrition of millions of people makes the production of crops for fuel purposes immoral. Additionally, the cultivation of plants for the production of liquid biofuels poses a serious threat to the environment and can lead to monoculture, deforestation, and excessive water consumption. Most important, it does not necessarily lead to a reduction in CO_2 emissions when compared to fossil fuels. Using LCA analysis, Pimentel & Patzek (2005) have shown that taking into account the whole cycle of production and consumption of bioethanol produced from corn in the United States, the cumulative CO_2 emissions from ethanol are about 50% greater than the emissions from combustion of gasoline.

Biogas (landfill gas, sewage sludge gas and other biogas derived from the anaerobic digestion of animal waste, food waste, etc.) is obtained through different types of devices, as well as in various industrial processes. Selecting the types of energy plants deployed should be adapted to the existing climate and soil conditions of Poland. The diversity of the soil environment in different parts of Poland is high, which is important for growing plants.

It is commonly believed that all biofuels are low-carbon fuels. However, as has been shown by Pimentel (2008, 2012), CO_2 emissions from some biofuels may be higher than those of fossil fuels. Therefore, before making a final decision, life-cycle analysis is necessary.

3 WATER POWER ENGINEERING

3.1 Worldwide and European situation

Water power engineering is a mature technology.

It is relatively simple and of high reliability. Global, theoretical hydropower resources are estimated at about 40,700 TW·h/year, with exploitable resources estimated at about 14,400 TW·h/year. The largest hydropower resources are located in China, Russia, Canada, Congo, India, and the U.S. (Biedrzycka 2004). During the evaluation of hydroelectric resources, the term "potential" is used. It includes theoretical, technical, and economic potential. Theoretical potential describes the maximum volume of water resources available. The technical potential includes the current and an achievable state of technology as well as structural and ecological limitations. The economic potential includes the part of the technical potential that meets the requirements for commercial purposes in addition to the part which is used in a different way (Wiśniewski (ed.) 2007).

3.2 National needs and capabilities of their overage

Poland has limited hydropower resources. It is due to the low level of rainfall averaging 600 mm per year, lowland terrain, and the presence of land with high permeability. Potential energy resources of Polish rivers are mainly concentrated in the Vistula basin—more than 73%, including 25% (5.9 TW·h/year) in the lower section of the river. The rest is concentrated in the Oder river basin and coastal rivers. Polish theoretical hydropower potential is estimated at 23–25 TW·h/year. However, the possibility of technical use is estimated at a level of around 12–17 TW·h/year, while currently just over 16% is being used. Economic potential, after abandoning the Lower Vistula Cascade Project, is estimated at about 5 TW·h/year.

This potential is currently used at approximately 41% (Wiśniewski (ed.) 2007; Steller 2002). Experts estimate that the technical potential of 12 TW·h/year capacity should be supplemented by a potential of 1.7–2.0 TW·h/year associated with small rivers and other watercourses where only Small Hydropower Plants (SHP) can be built. Table 4 shows the energy potential of domestic rivers.

Table 4. The energy potential of domestic rivers.

Area or river	Potential	
	Theoretical [GW·h/year]	Technical [GW·h/year]
Vistula river basin	16457	9270
Including: vistula river	9305	6177
Left-bank tributaries	892	513
Right-bank tributaries	4914	2580
Other small rivers	1346	–
Oder river basin	5966	2400
Including: oder river	2802	1273
Left-bank tributaries	1615	619
Right-bank tributaries	1540	507
Other rivers	338	70
Coastal rivers	582	280
Overall	23005	11950

Source: Gołębiowski & Krzemień 1998.

5

3.3 Covering the needs and the limitations

Using the potential of small hydropower plants is related to the individual decisions of small investors which—because of procedural difficulties and high investment costs—slow the use of potential hydropower in Poland. It is estimated that about 65% of the economic potential would be used by 2020 (Wiśniewski (ed.) 2007).

3.4 The main courses of action

Information on the theoretical and technical potential is based on the work of the team led by Professor A. Hoffmann between 1953 and 1961 "Cadastre of water power of Poland" which includes all of the rivers or their sections where potential exceeds 100 kW/km. Due to technological advancements, it is necessary to update this data. The development of small hydroelectric plants is also environmentally friendly. Large hydropower plants have a significant impact on the ecosystem; while in the case of SHP, this problem virtually does not exist. The vast majority of small hydro power plants use the natural flow of water (run-of-river power plant) and are used to produce electricity for local customers.

About 2,000 locations to build new small hydroelectric plants have been selected in Poland. They are a good option for areas where the construction of power plants for a group of households or small villages is cheaper than supplying power from distant places.

3.5 Problems to be solved

Ongoing work on improving flow management should focus on increasing the efficiency of water flow energy conversion and reducing the costs of infrastructure. Another important issue is the production of the technical equipment to collect water energy. The main threats caused by the construction of a dam and its operation include (Wawręty & Żela-ziński 2007):

- reduction in the variation amplitude of the flows and water levels below the reservoir, leading to the degradation of alluvial forests,
- silting the reservoir with debris particles from the river bed,
- erosion of the river bed below the dam,
- limiting the migration of aquatic organisms,
- standing deep water of the reservoir may inhibit self-cleaning process associated with the turbulent flow of water, which increases the oxygen content,
- accumulation of contaminants from the river,
- reservoir eutrophication caused by a constant flow and accumulation of nitrogen and phosphorus,
- mass fish kill as a result of the expansive growth of algae receiving oxygen from the water,
- creation of backwater—the larger the backwater, the higher the dam, and the smaller the river fall,
- slowing the flow of the river,
- humidity change over a relatively large area.

4 WIND ENERGY

4.1 Worldwide and European situation

Awareness of the growing demand for electricity and the danger of devastation to the natural environment have led to a gradual shift towards new energy generation technologies. The leader among these technologies is wind power. Due to the high potential for development, it has become the primary direction of investment in the sector of renewable energy sources. Investing in wind energy has become a global trend. This technology is proven, as evidenced by many years of experience in Germany, Spain, and Denmark. Wind is the movement of air caused by the uneven heating of the Earth's surface due to solar radiation and the Earth's rotation. It results in global and local differences in temperature, pressure, and density. It is estimated that about 1–2% of the solar radiation reaching the Earth's surface is converted into wind energy. Therefore, wind potential is considerable. Global wind resources are estimated at about 53,000 TW·h/year, greatly surpassing the needs of humanity. However, wind energy potential—possible to use in a cost-effective manner—is estimated at about 5,000 TW·h/year. Wind energy resources are not abundant everywhere. About 90% of these resources are located in Europe and the USA. In Europe, the largest wind energy resources are in the United Kingdom and Ireland, as well as in the northern parts of the Netherlands and Germany. However, good wind conditions may occur locally in all European countries (Soliński et al. 2008). Worldwide, the installed capacity of wind power in 2009 was 157,932 MW. Europe accounted for 48.24% of the world's total wind power generation (European Union—98.18%), North America—24.36%, Asia—24.64%, the rest of the world—2.76%.

4.2 Ability to cover the needs

Investment prospects for wind energy in Poland over the next 10 years are optimistic. The Polish legislature has taken important steps to increase the attractiveness of this technology. Slowly improving economic conditions provide a solid basis for the exploitation of wind energy.

It is estimated that Polish theoretical wind energy potential is about 2,049 TW·h/year on land and about 374 TW·h/year at sea (assuming the use of wind power across the whole country, internal waters, and territorial sea). The Institute of Meteorology and Water Management (Instytut Meteorologii i Gospodarki Wodnej) claims that there are good wind conditions over 30% of the country, while 5% has very advantageous conditions (Skulimowska 2009).

Poland's most attractive areas in terms of wind conditions include (Boczar 2010) the Baltic Sea coast, Słowińskie Sea coast, and Kashubian Sea coast, the islands of Wolin and Usedom, Suwałki, most of the Polish lowlands including Mazovia and the central part of the Wielkopolskie Lakeland, the Silesian and Żywiec Beskids, the Bieszczady Mountains, Dynowskie Foothills, and San River valley from the borders of the country to Sandomierz.

According to the Energy Regulatory Office (U-rząd Regulacji Energetyki), the installed capacity of wind power in Poland as of January, 2011 amounted to a total of 1,180 MW, while by March, 2012 it increased to 1,968 MW. In 2010, wind farms in Poland produced 1.845 TW·h of electricity.

It is estimated that in 2020, installed wind power capacity in Poland will range from 11 to 14 GW, while in 2030 it will rise to 16 GW, which will increase electricity production from 24 TW·h in 2020 to over 35 TW·h in 2030 (Sztuba & Marcinkowski 2009). Wind turbines reach their rated power at wind speeds of 11 to 16 m/s.

4.3 Covering the needs and the limitations

A power system of wind turbines meeting 10% of annual electricity consumption requires the installation and maintenance (Paska & Kłos 2010) of operating reserves as follows: a second reserve of capacity equivalent to about 1% of the total installed capacity of the wind power plants, a minute reserve worth a several percent of the total installed capacity in wind turbines, and an hour reserve with a value of a few percent of the total installed capacity of wind power plants.

4.4 The main courses of action

The European Union supports research projects on wind energy. These studies focus on, *inter alia*, power production forecasting in wind farms and on increasing the capacity of power systems (Barzyk 2004). There is ongoing research on prototype designs for offshore wind turbines floating on the surface of water (they do not require foundations). Another option is multi-rotor turbines placed on a floating support structure.

The average capacity of offshore wind turbines will soon exceed 5 MW. Such technologies are already available but there is ongoing research into more powerful turbines (Wiśniewski et al. 2010).

4.5 Problems to be solved

Poland does not have well-documented wind energy resources. These resources have been documented only for selected areas.

Disadvantages of wind power include:

– low power factor of 20 to 40%,
– noise emission during operation,
– presence of so-called stroboscopic effect, when rotating blades reflect solar radiation,
– the risk of deterioration of landscape values and the negative impact on avifauna,
– possible influence on the microclimate (in case of large wind farms),
– television and radio interference significant "ripple", caused by the rotor blades passing through the tower shadow and the voltage changes in the distribution network due to the high volatility of the generated power,
– high dependence on climate,
– high investment costs,
– high maintenance costs.

5 GEOTHERMAL ENERGY

5.1 Worldwide and European situation

Geothermal energy is a renewable energy with great potential which could meet the energy needs of the world in the not too distant future. The thermal energy contained in the interior of the Earth consists of primary heat from the period of the planet's formation (planetary accretion process), the heat generated from the decay process of radioactive elements, and the heat from solar radiation.

Geothermal energy is the amount of heat energy compared to the average temperature of the Earth's surface (calculated for an average surface temperature equal to 15°C). It is about $12.6 \cdot 10^{24}$ MJ.

The energy accumulated in the earth's crust can occur in the following forms (Górecki et al. 2006; Węgrzyk 2010): hydrogeothermal—with warm, underground water as the energy carrier; petrogeothermal—with media (usually water) penetrating into hot rock formations through drillings as the energy carrier; geopressure energy—hydraulic energy accumulated in the Earth's crust; and the energy of magma—magma bodies and hot magma intrusions penetrating tectonic faults as the source of thermal anomalies.

Geothermal energy plays an important role in the energy balance of some countries. In 2001, the

Philippines share of electricity generated from geothermal energy was 27% of total energy, in Kenya it was 12.4%, Costa Rica—11.4%, Iceland—16%, and El Salvador—4.3% (Dickson & Fanelli 2004).

The global potential of geothermal reserves possible to use in the process of electricity generation in 2050 is estimated at a level of 140 GWe (Lako 2010).

The largest part of geothermal energy reserves is located in Hot Dry Rocks (HDR) which are present in almost all regions of the world at great depths (from 4,000 to 5,000 m). The most advanced research projects in the field of HDR are being conducted in Japan and Western Europe.

5.2 Ability to cover the needs

Poland has significant geothermal potential with energy reserves exceeding Polish energy needs but its vast quantity of geothermal water is characterized by low enthalpy and high mineralization.

The high potential of geothermal energy results from the occurrence of large, easily accessible heat reserves linked with three large geothermal sedimentary basins covering about 80% of the country.

5.3 Covering the needs and the limitations

The future of geothermal energy is determined both by the economics of local power generation and macroeconomic conditions.

High costs and high uncertainty associated with the identification of new geothermal resources and the assessment of their potential makes geothermal project risk relatively high when compared to other energy sources.

5.4 The main courses of action

The costs of generating electricity in geothermal power plants are much higher than in conventional power plants.

5.5 Problems to be clarified

Hot Dry Rock Geothermal Energy Technology, HDR, belongs to a family of experimental technologies, the results of which have not yet been verified. In practice, it creates a number of problems and is very expensive.

Disadvantages of geothermal energy include:

– possibility of a negative impact on the environment; open-cast exploitation releases radon and hydrogen sulfide,
– possibility to activate or increase the frequency of seismic events during the exploitation of geothermal fluids or water injection,
– potential to cause a gradual subsidence (ground fall) during the extraction of large quantities of geothermal fluids.

6 SOLAR ENERGY

6.1 Worldwide and European situation

The Sun releases an enormous amount of energy as a result of nuclear fusion, which combines light hydrogen nuclei and converts them into heavier helium nuclei. Annually, $5.6 \cdot 10^6$ EJ of energy reaches the Earth in the form of solar radiation. The intensity of solar radiation reaching the outer layer of the atmosphere is—on average—1,367 W/m^2, which is the, so-called, solar constant.

Solar energy has the largest reserves among all renewable energy sources, estimated at 788,000 EJ/a. This energy also has the highest technical potential, estimated at 600 EJ/a (Malej 2009), and even from 1,580 to 49,840 EJ/a (Michalski 2006).

Solar energy can be harnessed by three basic types of conversion—photothermal, photovoltaic, and photobiochemical. Photothermal systems use only direct radiation to produce heat, while photovoltaic systems also use scattered radiation to generate electricity.

6.2 Ability to cover the needs

The intensity of solar radiation in Poland (average) is 960–1,163 W/m^2, while the average number of hours of sunshine in a year is between 1,390 and 1,900. The largest solar radiation in our country takes place between April and September by the seaside and between October and March in the mountains. Due to the relatively long period of autumn and winter as well as frequent cloud cover, the annual solar energy resources in Poland per unit area are about 50% smaller than in such European countries as Italy, Greece, Turkey, France, Spain, or Portugal, while two times smaller than in African countries (Skoczek 2003).

6.3 Covering the needs and the limitations

Solar energy is seen as one of the most promising of all renewable energy sources. It is due to the widespread availability of this form of energy, as well as its vast resources. The progressive increase in the unit price of energy from conventional sources and a continuous decrease in the unit price of solar energy are the reasons behind steadily growing interest in this energy source (Malej 2009, Mokrzycki (ed.) 2011).

Table 5. The acquisition of renewable energy by type in the years 2001–2010.

	TJ			Structure of acquisition [%]		
	2001	2006	2010	2001	2006	2010
Biomass	160,406	181,108	245,543	94.1	90.8	85.4
Liquid fuels from biomass	9	6,965	19,123	0.0	3.5	6.6
Bioethanol	ND	3,542	4,538		1.8	1.6
Biodiesel	ND	3,423	14,584		1.7	5.1
Biogas	1,477	2,613	4,797	0.9	1.3	1.7
From waste	544	791	1,811	0.3	0.4	0.6
From wastewater treatment plants	933	1,803	2,652	0.5	0.9	0.9
Others	0	19	334	0.0	0.0	0.1
Municipal waste	22	27	123	0.0	0.0	0.0
Geothermal energy	120	535	563	0.1	0.3	0.2
Heat pumps	ND	33	888		0.0	0.3
Water	8,369	7,352	10,512	4.9	3.7	3.7
Wind	49	922	5,992	0.0	0.5	2.1
Solar radiation	0	11	100	0.0	0.0	0.0
Total	170,452	199,566	287,640	100.0	100.0	100.0

Source: GUS 2007, GUS 2011.

Table 6. Gross final energy consumption from renewable sources in 2010.

	Gross final energy consumption from RES		The share of energy from RES in gross final energy consumption
Sector	TJ	Ktoe	%
Heating and refrigeration	194,123	4,636	12.0
Electricity generation	31,475	894	6.7
Transport	27,732	887	5.9
Total	235,004	6,417	9.5

Source: GUS 2011.

6.4 *The main directions of research*

Main directions of research and development of photovoltaic technologies include (Malej 2009) silicon and amorphous solar cells, thin-film solar cells, solar cell modules, photovoltaic membrane systems, photovoltaic generators (both individual and co-working with energy systems), and hybrid systems. The idea of placing photovoltaic panels above the layer of clouds, which should ensure a constant stream of light energy throughout the day, is under investigation. Another concept employs the use of airships to generate electricity from solar thermal installations using ground-based steam turbines.

The concept of Space Solar Power, placing a collector in a geostationary orbit (35,000 km) to produce and transmit electricity to the Earth's surface by laser or microwave beam, is also being considered.

6.5 *Problems to be solved*

Space Solar Power (SSP) will be safer, more environmentally friendly, and more reliable than any other alternative energy source. Research has shown that it is possible to transfer energy via a microwave beam of density safe for all forms of life, though there are significant concerns about possible health effects resulting from the exposure to the beam. Moreover, interference in radio communications is likely to occur (Boyle 2004).

Disadvantages of solar energy include:

– construction of photovoltaic cells requires the use of toxic elements such as cadmium, arsenic, selenium, and tellurium,
– solar installations require large areas,
– the intensity of sunlight differs in different areas and depends on the climate.

7 CONCLUSIONS

Over the last decade in Poland, there has been a significant increase in obtaining energy from renewable sources. The share of renewable energy in total primary energy has increased from 5.1% in 2001 to 9.0% in 2009 (GUS 2011).

The development of individual renewable energy sources has occurred at different rates but in recent years progress with all types of renewable energy has been noted. Particularly dynamic development in the areas of liquid fuels (especially biodiesel), wind power, and heat pumps have fundamentally changed the structure of renewable energy. In 2001, biomass had a share of more than 94% and was basically supplemented only by energy produced from water (4.9% share). In 2010, the share of biomass decreased to 85.4% while the share of water decreased to 3.7%. Meanwhile, by 2010 the share of liquid fuels increased to 6.6%, while the share of wind energy increased to 2.1%.

To achieve the EU's renewable energy objective, further dynamic growth in the acquisition and the use of energy from renewable sources will be required. It is due to the objectives enshrined in the EU climate and energy package, and the implementation of the Directive 2009/28/EC, which establishes mandatory objectives for all member states in relation to the overall share of energy from renewable sources in gross final consumption of energy and in relation to the share of renewable energy in transport. Polish targets set for 2020 include 15% share of gross electricity consumption from renewable sources in gross final energy consumption, and a 10% share of renewable energy in transport.

Table 6 shows gross final energy consumption from renewable sources—total and by sector—as well as the increased share by sector in 2010. In the coming years, there will be a need for continued development in renewable energy use. Current conditions in Poland indicate further development of biomass co-firing technology, solar radiation and heat pumps for heating purposes, the development of wind power and small hydroelectric power plants, and increased use of biofuels in transport.

REFERENCES

Barzyk, G. 2004. Zastosowanie technologii czasu rzeczywistego w energetyce wiatrowej. (Application of real-time technology in wind energy), *Energetyka* 12: 815–817.

Boczar, T. 2010. Wykorzystanie energii wiatru (The use of wind energy). Wydanie PAK, Warszawa.

Boyle, G. 2004. Solar Photovoltaics. *Renewable Energy* (2nd Edition). Oxford University Press, Oxford: 66–104.

Biedrzycka, A. 2004. Energetyka wodna: nie jesteśmy potęgą... Małe hydroelektrownie dużymi producentami (Hydropower: we are not the power... Small hydroelectric large producers). *Gigawat Energia* 11.

Dasgupta, P. & Taneja, N. 2011. Low Carbon Growth: An Indian Perspective on Sustainability and Technology Transfer. Problemy Ekorozwoju/Problems of Sustainable Development 6(1): 65–74.

Dickson, M.H. & Fanelli, M. 2004. What is geothermal energy? International Geothermal Association, www.geothermal-energy.org.

Euroobserver 2010. The State of Renewable Energies in Europe.

Gaj, H. 2008. Wybrane aspekty systemowej efektywności wykorzystania biomasy—współspalanie (Selected aspects of the system efficiency of biomass—co-firing). *Czysta Energia* 12: 26–27.

Golec, T. 2004. Współspalanie biomasy w kotłach energetycznych (Co-firing of biomass in power boilers). *Energetyka* 7/8: 437–445.

Gołębiowski, S. & Krzemień, Z. 1998. Przewodnik inwestora małej elektrowni wodnej (Investors guide to small hydro). Fundacja Poszanowania Energii, Warszawa.

Górecki, W. et al. 2006. Atlas zasobów geotermalnych formacji paleozoicznej na Niżu Polskim (Atlas of geothermal resources in Paleozoic formations of the Polish Lowlands). Ministerstwo Środowiska, Narodowy Fundusz Ochrony Środowiska i Gospodarki Wodnej, Akademia Górniczo-Hutnicza, Państwowy Instytut Geologiczny, Kraków.

GUS 2007. *Energia ze źródeł odnawialnych w 2006 r. (Energy from renewable sources in 2006)*. Główny Urząd Statystyczny. Informacje i opracowania statyczne, Warszawa.

GUS 2011. *Energia ze źródeł odnawialnych w 2010 r. Energy from renewable sources in 2010)*. Główny Urząd Statystyczny. Informacje i opracowania statyczne, Warszawa.

IEA 2002. *Renewables in Global Energy Supply*. International Energy Agency. Paris.

Kamiński, J. & Mirowski, T. 2010. Rozwój energetyki odnawialnej w Polsce (Development of renewable energy in Poland). *Elektrownie:* 48–52.

Kruczek, S. et al. 2008. Kruczek, S., Skrzypczak, G. & Muraszkowski, R. Spalanie i współspalanie biomasy z paliwami kopalnymi (Combustion and co-combustion of biomass and fossil fuels). *Czysta Energia* 6: 32–35.

Lako, P. 2010. *Geothermal heat and power*. Energy Technology System Analysis Programme. IEA ETSAP, Technology Brief E07, www.etsap.org.

Malej, J. 2009. Bezpieczeństwo energetyczne świata a ochrona ekosfery. Technologie odnawialnych źródeł energii, technologie jądrowe, termojądrowe i wodorowe (Energy security and the protection of ecosphere. Renewable energy technologies, nuclear technologies, fusion and hydrogen). Wyd. Uczelniane Politechniki Koszalińskiej, Koszalin.

Michalski, M.Ł. 2006. Światowe zasoby energii słonecznej i kierunki ich wykorzystania (Global solar energy resources and their utilization trends). *Czysta Energia* 12: 16–18.

Mokrzycki, E. (ed.) 2011. Gawlik, L., Kryzia, D., Mokrzycki, E. & Uliasz-Bocheńczyk, A. Rozproszone zasoby energii w systemie elektroenergetycznym (Distributed energy resources in the power system). IGSMiE PAN. Kraków.

Paska, J. & Kłos, M. 2010. Elektrownie wiatrowe w systemie elektroenergetycznym—przyłączanie, wpływ na system i ekonomika (Wind power plants in the power system—connecting, the impact on the system and the economy). *Rynek Energii* 1: 3–10.

Pawłowski, A., 2009. The Sustainable Development Revolution, *Problemy Ekorozwoju/Problems of Sustainable Development* 4 (1), 65–76.

Pawłowski A., 2011. Sustainable Development as a Civilizational Revolution. Multidimensional Approach to the Challenges of the 21st century, CRC Press, Taylor & Francis Group, A Balkema Book, Boca Raton, Londyn, Nowy Jork, Leiden, 240 pages.

Pimentel, D. (ed) 2008. Biofuels, Solar and Wind as Renewable Energy Systems. Benefits and Risks. Springer.

Pimentel, D. & Patzek, T.W. 2005. Ethanol production using corn, switchgrass, and wood; biodiesel production using soybean and sunflower. *Nat Resour Res*, 14: 65–76.

Piementel D. 2012. Energy Production from Maize, *Problemy Ekorozwoju/Problems of Sustainable Development* vol. 7 no 2, 15–22.

Project 2011. Projekt rozporządzenia Ministra Gospodarki z dnia 17 lutego 2011 r. w sprawie szczegółowego zakresu obowiązków uzyskania i przedstawienia do umorzenia świadectw pochodzenia, uiszczenia opłaty zastępczej, zakupu energii (wersja 2.6) (Proposal for a Regulation of the Minister of Economy of 17 February 2011 on the detailed terms of reference for obtaining and present for redemption of certificates of origin, alternative payment, purchasing power (version 2.6)). Ministerstwo Gospodarki, Warszawa.

Shan, S. & Bi, X. 2012. Low Carbom Development of China's Yangtze River Delta Region. Problemy Ekorozwoju/Problems of Sustainable Development 7(2): 33–41.

Skoczek, A. 2003. Możliwości rozwoju fotowoltaiki w Polsce na tle programów rozwoju odnawialnych źródeł energii w Niemczech (Possibilities for the development of photovoltaics in Poland compared to programs for renewable energy sources in Germany). *Czasopismo Techniczne*, 94–97: 28–35.

Skulimowska, M. 2009. Energetyka wiatrowa—szanse i zagrożenia (Wind energy—opportunities and threats). Notatka nr 32, Bruksela.

Soliński, I. et al. 2008. Soliński, I., Soliński, B. & Solińska, M. Rola i znaczenie energetyki wiatrowej w sektorze energetyki odnawialnej (The role and importance of wind energy in the renewable energy sector). *Polityka Energetyczna* 11(1): 451–464.

Subramanian, M. 2010. *Global Energy Reserves*, CH 12002 Energy Management in Chemical Industries.

Steller, J. 2002. Wybrane problemy rozwoju energetyki wodnej w Polsce i na świecie (Some problems of the development of hydro energy in Poland and around the world). *Materiały VIII Konferencji Naukowo-Technicznej „Ogólnopolskie Forum Odnawialnych Źródeł Energii", Warszawa, 28−30 października*, Wyd. URM: 3–31.

Sztuba, W. & Marcinkowski, B. 2009. *Energetyka wiatrowa w Polsce (Wind energy in Poland)*. Raport TPA Horwath i Domański, Zakrzewski Palinka, Warszawa.

Udo, W. 2011. Human progress towards equitable sustainable development—part II: Empirical Exploration, *Problemy Ekorozwoju/Problems of Sustainable Development*. 6 (2), 33–62.

Wawręty R. & Żelaziński J. 2007. Środowiskowe skutki przedsięwzięć hydrotechnicznych współfinansowanych ze środków Unii Europejskiej (Environmental impacts of hydro projects co-financed by the European Union). Raport Towarzystwa na rzecz Ziemi i Polskiej Zielonej Sieci. Oświęcim—Kraków.

Węgrzyk, J. 2010. Elektrownie geotermalne—alternatywa w produkcji energii elektrycznej (Geothermal power plants—alternative power production). *Debata o Przyszłości Energetyki, 4−7 maja, Wysowa-Zdrój.*

Wiśniewski, G. (ed.) 2007. Możliwości wykorzystania odnawialnych źródeł energii w Polsce do roku 2020 (The possibilities of using renewable energy sources in Poland by 2020). Praca wykonana na zamówienie Ministerstwa Gospodarki w Instytucie Energetyki Odnawialnej EC BREC, Warszawa.

Wiśniewski, G. et al. 2010. Wiśniewski, G., Michałowska-Knap, P., Dziamski, P. & Regulski P. 2010. Gospodarcze i społeczne aspekty rozwoju morskiej energetyki wiatrowej w Polsce (Economic and social aspects of the development of offshore wind energy in Poland). Instytut Energetyki Odnawialnej, Warszawa.

Zuwała, J. & Rejdak M., 2010. Biomasa ciekła jako substytut ciężkiego oleju opałowego (Liquid biomass as a substitute for heavy oil). *Karbo* 1: 88–94.

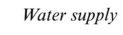

Water supply

Environmental Engineering IV – Pawłowski, Dudzińska & Pawłowski (eds)
© 2013 Taylor & Francis Group, London, ISBN 978-0-415-64338-2

Prediction of precipitation quantities for designing sewage system

B. Kaźmierczak & A. Kotowski
Institute of Environmental Protection Engineering, Wrocław University of Technology, Wrocław, Poland

ABSTRACT: The paper presents the methodology of model formation for description of the maximum rainfall depth in Wrocław for practical applications. The probabilistic model can be used for dimensioning and modeling effects of sewage or combined sewage systems, for the required EN 752 standard for drainage area. This paper attempts to develop a new probabilistic model for the maximum precipitation amounts of short-term rainfall, with the occurrence frequency of $C \in [1; 50]$ years and duration of $t \in [5; 180]$ minutes. Pluviographic measurement results from the meteorological station IMGW Wrocław-Strachowice from years 1960–2009 were the basis for this paper. Own criterion of precipitation amounts was assumed to isolate intensive rainfall from pluviograms, which made it possible to select a number of the most intensive rainfalls in each year.

Keywords: precipitation amount, probabilistic models, sewage system

1   INTRODUCTION

The designing as well as modernization of rainwater or combined sewage systems encounters a primary difficulty in Poland, which results from the lack of a reliable method of representative determination of rainfall intensity for dimensioning or verification of a sewage system flow capacity (Kaźmierczak & Kotowski 2012a). The precipitation model by Błaszczyk, recommended for dimensioning of drains and system features in Poland, significantly decreases measurement results of representative rain flows, which was shown in a number of comparative analyses (Kotowski 2006, 2011, Suligowski 2004). Błaszczyk's model is based on a statistical analysis of only 79 intense rainfalls (average height of the duration: $h > t^{0.5}$) registered in Warsaw in the years 1837–1891 and 1914–1925. Błaszczyk's model, which was created over a half century ago, also does not include existing climate change (De Toffol et al. 2009, Leonard et al. 2008, Willems et al. 2012).

It affects the dimensioning of the drainage areas in Poland according to the recommendations of the EU standard PN-EN 752 (2008). The lowered rainfall intensity directly influences the higher frequency of the sewage system outflows or the impossibility of rainwater collection (Kaźmierczak & Kotowski 2012b). The standard restricts the frequency of occurrence of the unfavourable phenomena to a rare "socially acceptable" repeatability: from once per 10 years in case of rural areas to once per 50 years for underground transportation facilities. However, the recommendations

concerning the frequency of the design rainfall amount between once a year to once per 10 years, depending on the land development type. The philosophy puts forward a new challenge of satisfying the recommendations by the sewage system designers.

Therefore, systematic research in precipitation patterns and statistic determination of the occurrence frequency of their maximum amounts are becoming so important nowadays, especially for such a rare rainfall repeatability. It is especially significant in order to match the requirements of the aforementioned standard (Baldassarre et al. 2006, Ben-Zvi 2009, Brath et al. 2003, Overeem et al. 2008). The appropriately long and uniform archival material from precipitation observation, not shorter than from the period of 30 years (Kossowska-Cezak 1999) is thus essential.

2   MATERIAL AND METHODS

Probabilistic model for maximum precipitation amount (h_{max} in mm) for Wroclaw for $C \in [1; 100]$ years and $t \in [5; 4320]$ min was proposed in (Kotowski et al. 2010):

Archival pluviograms from Wrocław-Strachowice IMGW station from years 1960–2009 constituted the research material to develop model (1). Own precipitation amount criterion $h \geq 0.75t^{0.5}$, related to the limit precipitation amount for strong rainfall was assumed to isolate intensive rainfall for statistical analyses. Precipitation amounts were determined for the following 16 intervals

Table 1. The recommended design frequencies of rainfall and permissible flooding occurrence frequencies.

Design rainfall frequency [1 per C years]	The area drainage standard category	Flooding occurrence frequency [1 per C years]
1 per 1 year	I. Out of town areas (rural)	1 per 10 years
1 per 2 years	II. Residential areas	1 per 20 years
1 per 5 years	III. City centres, service and industry areas	1 per 30 years
1 per 10 years	IV. Underground transportation facilities, underpasses, etc	1 per 50 years

Table 2. Selected time series of maximum precipitation amounts in Wrocław.

		Precipitation amounts h (in mm) in particular time intervals t (in minutes)															
No.	C	5	10	15	30	45	60	90	120	180	360	720	1080	1440	2160	2880	4320
1	50	13.1	18.7	24.7	32.9	34.7	35.3	42.7	57.7	61.9	63.1	64.2	72.9	80.1	92.6	103.9	116.9
2	25	11.6	18.0	22.8	30.3	34.7	35.3	37.7	41.5	42.8	50.4	64.2	71.5	77.9	92.5	103.2	111.6
5	10	9.9	15.7	20.1	28.2	32.1	34.7	35.4	36.2	38.4	43.9	54.2	69.1	72.2	85.4	94.5	101.9
10	5	9.3	13.8	17.7	26.7	28.8	30.5	33.9	35.4	35.7	38.7	49.2	57.4	65.0	73.1	76.2	87.5
25	2	8.0	11.0	13.9	17.9	19.9	20.2	24.2	25.6	27.3	35.2	40.8	45.3	48.3	55.2	60.6	63.4
50	1	6.4	8.9	10.1	13.7	14.8	15.3	16.3	17.9	20.0	26.2	32.0	36.5	39.9	45.2	48.1	49.0

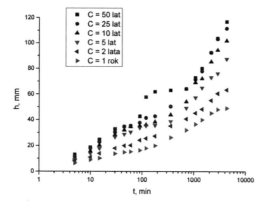

Figure 1. Measurement data from Wrocław.

of duration: 5, 10, 15, 30, 45, 60, 90 and 120 minutes and 3, 6, 12, 18, 24, 36, 48 and 72 hours, totalled by means of the moving sum method. In Table 2, interval precipitation amounts were arranged in a decreasing order (with durations ranging from 5 minutes to 72 hours) with $N = 50$ years of observation and the highest 50 time series of synthetic rainfalls were interpreted.

Describing the precipitation for the intervals of duration from 5 minutes to 3 days by means of the same model might be perceived as questionable as far as the accuracy of the description is concerned, especially for the short-term rainfalls. According to Kotowski et al. (2010) as much as 98% of tempestuous rainfalls constitute the ones that do not have the intervals of duration greater than 3.5 hours and they produce the largest flows in the sewage systems. The short-term rainfalls—with the intervals of duration up to 2 hours according to (Bogdanowicz & Stachý 1998, Soczyńska 1997), or up to 3 hours according to ATV A-121 (1985), are necessary to be described by means of different curves than the long-term rainfalls. The reason for this is a dissimilar mechanism of formation and duration of the precipitation occurrence. The graphical representation of the measurement data (from Table 2) was given in Figure 1, where the fact of changing the dependence of h on t around the 2nd and the 3rd hour of the interval of duration was presented.

Therefore, an attempt to develop a new practical for designing and modelling the operation of the sewage system probabilistic model of the maximum amount of short-term precipitation (from 5 to 180 min for $C \in [1; 50]$ years) was made.

3 RESULTS AND DISCUSSION

The new probabilistic model of the maximum amount of a short-term precipitation in Wrocław

$$h_{\max}(t,C) = -4.58 + 7.41t^{0.242}$$
$$+ \left(97.11t^{0.0222} - 98.68\right)\left(-\ln\frac{1}{C}\right)^{0.809} \quad (1)$$

was based on Fisher-Tippett type III_{min} probability distribution (Ciepielowski & Dąbkowski 2006, Ozga-Zielińska & Brzeziński 1997).

The density function, the credibility function logarithm and the quantile of a random variable for this distribution are represented by the following models:

$$f(x; \alpha, \lambda, \varepsilon) = \alpha \lambda^\alpha (x - \varepsilon)^{\alpha-1} e^{-\lambda^\alpha (x-\varepsilon)^\alpha} \qquad (2)$$

$$\ln L(\alpha, \lambda, \varepsilon) = N \ln \alpha + N \alpha \ln \lambda$$
$$+ (\alpha - 1) \sum_{i=1}^{N} \ln(x_i - \varepsilon) - \lambda^\alpha \sum_{i=1}^{N} (x_i - \varepsilon)^\alpha \qquad (3)$$

$$x_p(\alpha, \lambda, \varepsilon) = \varepsilon + \frac{1}{\lambda}(-\ln p)^{\frac{1}{\alpha}} \qquad (4)$$

where: α—the shape parameter, λ—the scale parameter, ε—the lower limit (mm), p—probability ($p = 1/C$).

The estimators of the density function parameters were calculated by the maximum likelihood method, through the equation maximization (3). In the calculation, 450 units of synthetic precipitation were used—50 series of the following intervals of duration: 5, 10, 15, 30, 45, 60, 90, 120 and 180 minutes. In the calculation, it was assumed that the family of curves of the precipitation amounts will be characterized by the constant shape parameter λ in the entire range of the model application, which is for $C \in [1; 50]$ years. Furthermore, the lower limit was assumed as the smallest analysed precipitation amount (Table 2, verse 50) reduced by 0.1 mm. The calculation results of the estimators of the parameters based on Fisher-Tippett type III_{min} probability distribution for the analysed data were presented in Table 3.

The graphical interpretation of empirical (resulting from measurements) and theoretical (resulting from the Fisher-Tippett type III_{min} probability distribution) were shown in Figure 2.

The dependence of the coefficient ε on the precipitation duration t was described (at R = 0.995) with the following function:

$$\varepsilon(t) = 3.76 \ln(t + 0.38) \qquad (5)$$

The dependence of the scale parameter λ on the precipitation duration t was described (at R = 0.996) with the function (Figure 4):

$$\lambda(t) = 0.104\left(1 - e^{-0.0313t}\right)^{-0.835} \qquad (6)$$

Finally, substituting equation (5) and (6) into equation (4), the new probabilistic short-term rainfall model was found with the following function (h_{max} in mm):

$$h_{max}(t, C) = 3.76 \ln(t + 0.38)$$
$$+ 9.61\left(1 - e^{-0.0313t}\right)^{0.835}\left(-\ln\frac{1}{C}\right)^{0.942} \qquad (7)$$

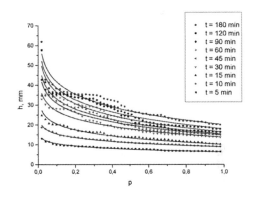

Figure 2.   Theoretical cumulative distribution functions for the distribution of Fisher-Tippett type III_{min}.

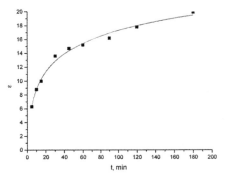

Figure 3.   Variability of empirical coefficient ε as a function of time t.

Table 3.   The calculation results of parameter values λ, α and ε.

t	λ	α	ε
5	0.520061	1.0615	6.3
10	0.330353	1.0615	8.8
15	0.210436	1.0615	10.0
30	0.163544	1.0615	13.6
45	0.135706	1.0615	14.7
60	0.123226	1.0615	15.2
90	0.108783	1.0615	16.2
120	0.106080	1.0615	17.8
180	0.100945	1.0615	19.9

The family of DDF (Depth-Duration Frequency) type curves for model (7) for $C \in [1; 50]$ years and $t \in [5; 180]$ min was shown in Figure 5.

The relative Residual Mean Square Error (rRMSE) was used for quantitative evaluation of the models (1) and (7):

$$rRMSE = \sqrt{\frac{1}{N} \sum_{i=1}^{N} \left(\frac{h_{c,i} - h_{m,i}}{h_{m,i}} \right)^2} \cdot 100\% \qquad (8)$$

where: h_c—calculated precipitation amount (mm), h_m—measured precipitation amount (mm), N—number of observations ($N = 450$).

For the analysed range: $t \in [5; 180]$ min and $C \in [1; 50]$ years, in case of model (1) the value of $rRMSE$ amounts to 8.3% but in case of model (7) it is only 6.5% (as a comparison, the value of $rRMSE$ for the aforementioned Błaszczyk's model amounts to 30.9%). Matching and partial residuals diagrams for models (1) and (7) were presented in Figures 6 and 7.

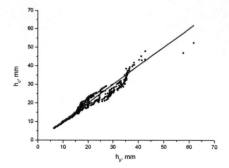

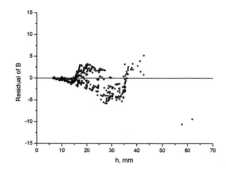

Figure 6. Matching and partial residuals diagrams for model (1), $rRMSE = 8.3\%$.

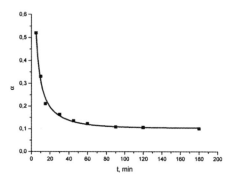

Figure 4. Variability of empirical coefficient λ as a function of time t.

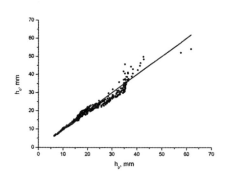

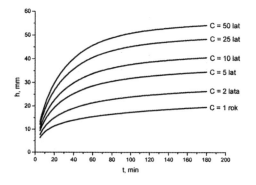

Figure 5. Depth-Duration-Frequency curves for the model (7).

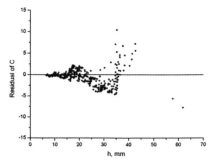

Figure 7. Matching and partial residuals diagrams for model (7), $rRMSE = 6.5\%$.

Table 4. The comparison of measurement data with calculation results for maximum rainfall intensities from models (1) and (7).

C	t	h_p	$h_p(1)$	$h_p(7)$	C	t	h_p	$h_p(1)$	$h_p(7)$
1	5	6.4	6.4	6.3	2	5	8.0	7.8	7.7
1	10	8.9	8.4	8.8	2	10	11.0	11.0	11.1
1	15	10.1	9.7	10.3	2	15	13.9	13.0	13.3
1	30	13.7	12.3	12.8	2	30	17.9	16.8	17.3
1	45	14.8	14.0	14.3	2	45	19.9	19.2	19.7
1	60	15.3	15.4	15.4	2	60	20.2	21.1	21.3
1	90	16.3	17.4	16.9	2	90	24.2	23.9	23.4
1	120	17.9	19.0	18.0	2	120	25.6	26.0	24.7
1	180	20.0	21.5	19.5	2	180	27.3	29.1	26.3
5	5	9.3	9.2	9.3	10	5	9.9	10.2	10.5
5	10	13.8	13.5	13.8	10	10	15.7	15.3	15.8
5	15	17.7	16.2	16.9	10	15	20.1	18.4	19.6
5	30	26.7	21.2	22.8	10	30	28.2	24.2	26.8
5	45	28.8	24.3	26.2	10	45	32.1	27.8	31.0
5	60	30.5	26.7	28.5	10	60	34.7	30.4	33.8
5	90	33.9	30.1	31.2	10	90	35.4	34.4	37.0
5	120	35.4	32.7	32.8	10	120	36.2	37.3	38.7
5	180	35.7	36.6	34.5	10	180	38.4	41.7	40.6

Table 4 presents the measurement data from Wrocław-Strachowice IMGW meteorological station from years 1960–2009 and the calculation results from the probabilistic models in formula (1) and (7). The analysis was carried for selected rainfall frequencies for dimensioning sewage systems. Generally, for $C \in [1; 10]$ years and $t \in [5; 180]$ min, model (7) has a high accuracy description of measurement data—to a few per cent deviation, while model (1) is less accurate—deviations of up to several percent.

The differences between models are small but short-term rainfall model (7) should be considered as better reflecting the results of the measurements. It can be a valuable complement to the probabilistic model (1), especially for designing rainwater or combined sewage systems in Wrocław.

4 CONCLUSION

The solution to the problem of obtaining the reliable data about the precipitation amounts, used to designing and modelling of the drainage areas in Wrocław conditions, is the maximum precipitation model (1), with the wide range of application $t \in [5; 4320]$ min and $C \in [1; 100]$ years. The provided new model of short-term precipitation (7) in Wrocław conditions increases the accuracy of the calculation results correlation with the empirical data in the range of $t \in [5; 180]$ min and $C \in [1; 50]$ years, in comparison to model (1), which has

a great importance in the engineering practice. For model (7) the relative residual mean square error $rRMSE$ amounts to 6.5%, and for model (1) $rRMSE$ amounts to 8.3%. The new probabilistic model (7) is especially recommended to designing the sewage systems in Wrocław in the range of $t \in [5; 180]$ min and $C \in [1; 10]$ years.

REFERENCES

ATV A-121, 1985. Niederschlag—Starkregenauswertung nach Wiederkehrzeit und Dauer Niederschlagsmessungen Auswertung. Hannef.

Baldassarre G. Di, Castellarin A., Brath A., 2006. Relationships between statistics of rainfall extremes and mean annual precipitation: an application for design-storm estimation in northern central Italy. *Hydrol. Earth Syst. Sci.* 10: 589–601.

Ben-Zvi A., 2009. Rainfall intensity–duration–frequency relationships derived from large partial duration series. *Journal of Hydrology* 367 (1–2): 104–114.

Bogdanowicz E., Stachý J., 1998. Maximum rainfall in Poland. Design characteristics. Series: *Hydrologia i Oceanologia* no. 23. The Publishing House of the Institute of Meteorology and Water Management (IMGW), Warsaw (in Polish).

Brath A., Castellarin A., Montanari A., 2003. Assessing the reliability of regional depth-duration-frequency equations for gaged and ungaged sites. *Water Resources Research* 39: 1367–1379.

Ciepielowski A., Dąbkowski S.L., 2006. Methods of maximum discharge calculations in small river catchments (with examples). Projprzem-EKO, Bydgoszcz (in Polish).

De Toffol S., Laghari A.N., Rauch W., 2009. Are extreme rainfall intensities more frequent? Analysis of trends in rainfall patterns relevant to urban drainage systems. *Water Science and Technology* 59 (9): 1769–1776.

Kaźmierczak B., Kotowski A., 2012a. DDF rainfall model for dimensioning and modelling of Wrocław drainage systems. *Environmental Protection Engineering* 38: 127–138.

Kaźmierczak B., Kotowski A., 2012b. Verification of storm water drainage capacity in hydrodynamic modeling. The Publishing House of Wrocław University of Technology, Wrocław (in Polish).

Kossowska-Cezak U. (Rec.), 1999. Climatological normals (CLINO) for the period 1961–1990. WMO No. 847, Geneva 1996. *Przegląd Geofizyczny* 44 (1–2).

Kotowski A., 2006. Of the need to adapt the dimensioning rules of sewage systems in Poland to the requirements of the standard PN-EN 752 and recommendations of the European Committee for Standardization. *Gaz, Woda i Technika Sanitarna* 80 (6): 20–26 (in Polish).

Kotowski A., 2011. The principles of safe dimensioning of sewage systems. Seidel-Przywecki, Warsaw (in Polish).

Kotowski A., Kaźmierczak B., Dancewicz A., 2010. The modelling of precipitations for the dimensioning of sewage systems. The Publishing House of the Civil Engineering Committee the Polish Academy of Sciences (PAN). *Studies in Engineering* no. 68, Warsaw (in Polish).

Leonard M., Metcalfe A., Lambert M., 2008. Frequency analysis of rainfall and streamflow extremes accounting for seasonal and climatic partitions. *Journal of Hydrology* 348 (1–2): 135–147.

Overeem A., Buishand A., Holleman I., 2008. Rainfall depth-duration-frequency curves and their uncertainties. *Journal of Hydrology* 348: 124–134.

Ozga-Zielińska M., Brzeziński J., 1997. Applied hydrology. Polish Scientific Publishers PWN, Warsaw.

PN-EN 752, 2008. Drain and sewer systems outside buildings.

Soczyńska U. (Red.), 1997. Prediction of the design storms and floods. Warsaw.

Suligowski Z., 2004. The management of precipitation waters. Particular problems. *Forum Eksploatatora* (3–4): 24–27 (in Polish).

Willems P., Arnbjerg-Nielsen K., Olsson J., Nguyen V., 2012. Climate change impact assessment on urban rainfall extremes and urban drainage: Methods and shortcomings. *Atmospheric Research* 103: 106–118.

Environmental Engineering IV – Pawłowski, Dudzińska & Pawłowski (eds)
© 2013 Taylor & Francis Group, London, ISBN 978-0-415-64338-2

Economical and reliability criterion for the optimization of the water supply pumping stations designs

J. Bajer

Faculty of Environmental Engineering, Cracow University of Technology, Krakow, Poland

ABSTRACT: In designing water supply pump stations, the optimal choice of technical solutions should be based on economic efficiency of the investment, which enables to treat quantitatively technological, economical and reliability criteria which will determine the proper work of these objects in the future. The article presents a proposal of such comprehensive criterion, including economical and reliability measures. The criterion is expressed as the effectiveness of extra investment expenditure in the increase of the reliability of a designed object. The proposed costs of unreliability related to the disturbance in the correct work, defined in a probabilistic way, play the fundamental role.

Keywords: pumping stations, reliability, cost of unreliability, optimization

1 INTRODUCTION

Modern design of Water Pumping Stations (WPS) as basic technical objects related to water supply should, in a measurable way, include reliability factor. First of all, it refers to the pumping stations of great economic significance (industrial or drainage pumping stations) or social significance (important municipal pump stations) work disturbances which cause great economic losses or drastically lower the comfort of human life. Problems of the reliability of water pump stations has been presented in professional literature for many years. Most of the authors, besides critics of not taking into account reliability at the designing stage (Cullinane 1985, Mays et al. 1986, Bajer 2004), concentrated mainly on the analysis of the impact of the results of failures on the work of water supply systems (Mays et al. 1986, Ilin 1987, Duan & Mays 1987, Dąbrowski & Bajer 2008, Tchórzewska-Cieślak 2009, Almeida & Ramos 2010, Mehzad et al. 2011, Zhuang et al. 2011, Mehzad et al. 2012), methods of the estimation of their reliability (Cullinane 1985, Hobbs 1985, Wieczysty & Rak 1985, Ilin 1987, Duan & Mays 1990, Bajer 2001a,b, Bajer & Głód 2003, Bajer 2006, Lindstedt & Sudakowski 2008, Liberacki 2010, Bajer 2011, Mehzad et al. 2011, Mehzad et al. 2012), and the determination of reliability indexes and their components by statistic analysis of operation records (Shultz & Parr 1981, Shamir & Howard 1981, Ilin 1987, Dawidowicz et al. 1993, OREDA 1997, Bajer & Iwanejko 2008, Liberacki 2010, Karmazinow et al. 2011).

Some authors pay particular attention to the need to abandon traditional approach which is based on intuitive and non-quantitative premises of the preliminary phase of designing WPS, connected with the adoption of a, so-called, technical structure (Bajer & Wieczysty 2000, Bajer 2001a, Bajer 2004, Bajer 2006, Liberacki 2010). Technical structure is defined by the number of working and spare pump aggregates as well as the way of the construction of suction parts and pumping parts including the number, location and links between the pipelines and their optimally distributed fittings. Although technical structure creates only a general concept of WPS designing, it is the most significant in allowing a proper solution for such an object.

The optimal designing of water pumping stations based on the balance of economical efficiency of the investment is so far less thoroughly discussed in literature, although the economic problems shown in the aspect of reliability were also presented (Shilston 1985, Ilin 1987, Duan & Mays 1987, Hashemi et al. 2011, Kurek & Ostfeld 2012). According to (Bajer 2002b, Bajer 2004), new possibilities in this area can be created by introduction of economic-reliability criterion, related to the costs resulting from failures of WPS (unreliability costs), and estimating the efficiency of spending financial resources for increasing of reliability, to such a measurable account. A proposal of such criterion is presented in this paper. It may be a starting point for the development of a detailed form of an optimization task, the solution of which allows to choose the best option of designing a water pumping station.

2 METHODS

2.1 Proposed optimization criterion

The most general approach looking for the optimal solution of the WPS designing of defined design parameters, i.e. desired nominal efficiency Q_n and required pressure in the consumer point H_r, taking into account loss costs due to unreliability (of costs), can be described by the following differential equation:

$$\frac{\partial C_{YR}}{\partial M_R} + \frac{\partial \overline{C}_{AR}}{\partial M_R} = 0 \qquad (1)$$

in which C_{YR} = annual costs including investment costs C_I and annual operation costs C_E for WPS, PLN/yr; $\overline{C}_{AR}$ = expected unreliability costs related to yearly operation of WPS, assumed to be fixed annual cost of operation, PLN/yr; M_R = reliability measure of a given pumping station (e.g. the probability of its correct work between the failures P_{WPS} or a stationary index of readiness A_{WPS} defining the probability of finding the station working at a given moment of time); measures or indexes of the technical object reliability (here WPS) is defined on the basis of the analysis of its reliability conducted with the use of different methods (Migdalski 1982, Rao 1992, Kwietniewski et al. 1993).

This equation (1) is derived from the minimization of the simple aim function presented as the equation (2). Often applied in the economical calculations it expresses unit annual costs of using the object (here WPS) including the costs of losses coming from the unreliability of the object. Due to the connection between unreliability costs and random factors (failures), the optimization was based on the expected value:

$$E[C] = \overline{C} = C_I + C_E + E[C_{AR}] = C_{YR} + \overline{C}_{AR} \qquad (2)$$

in which C, $E[C] = \overline{C}$ = total annual costs connected with the construction and operation of WPS, and their expected value, PLN/yr; C_I = investment costs of WPS at the beginning of the whole analyzed period, drawn into an annual cost PLN/yr; C_E = total annual costs of WPS operation, minus amortization of long-lasting resources and non-material investment, assumed to be constant in the consecutive years of operation, PLN/yr; C_{AR} = costs of unreliability connected with annual WPS operation assumed to be constant in the consecutive years of operation, PLN/yr; $E[C_{AR}] = \overline{C}_{AR}$ and C_{YR}—as in equation (1).

The solution of equation (1) demands the fulfillment of the assumption, in which all the occurring costs can be described by monotonic and continuous functions of the accepted measure of reliability. Because it is not fulfilled due to the fact that model of costs analysed in this paper, which can be attributed to the particular design options, is discrete, thus the application of optimization typical methods is not possible. Choosing the optimal option in this case demands the direct examination of all the options point by point, which may pose some difficulties. It is also problematic to assess the relationships between particular points (Peschel & Riedl 1979).

Taking all the above into account, it is proposed to make the final economic comparison of the WPS design solutions of varying in reliability, applying so-called internal efficiency indexes ε while coming to more and more expensive and reliable options. These indexes are determined on the basis of the difference criterion (Sozański 1990), assuming that more expensive option j is more profitable when decrease of unreliability costs $\overline{C}_{AR}$ covers the increase of annual costs C_{YR}.

Treating $\overline{C}_{ARi} - \overline{C}_{ARj}$ as the financial effect of the decrease of unreliability obtained by additional annual costs $C_{YRj} - C_{YRi}$, the differential criterion is defined by the following equation:

$$\overline{C}_{ARi} - \overline{C}_{ARj} \geq C_{YRj} - C_{YRi} \qquad (3)$$

In this case, the expression of the efficiency index ε_{ij}, illustrating the economic efficiency of the means spent on the decrease of the unreliability in the discussed options will be the following (Sozański 1990):

$$\varepsilon_{ij} = \frac{(\overline{C}_{ARi} - \overline{C}_{ARj}) - (C_{YRj} - C_{YRi})}{\overline{C}_{ARi} - \overline{C}_{ARj}} \qquad (4)$$

Relation (4) defines "the profitability" of the investment costs spent on the increasing reliability while starting from cheaper, but less reliable option i, to more expensive but less reliable option j, and makes ultimately accepted criterion of optimization (goal function). According to the general theory of making economic decisions (Kopecki 1960), in case of an approximate definition of unreliability costs and costs of the lack of confidence to possessed statistic materials, it is possible to apply the strategic model of making economic decisions. In this case, the costs of unreliability should have their weight assigned that would equal $1-v \leq 1$, where v is the measure of the lack of confidence to the obtained values C_{AR}. If the coefficient v equals zero, full confidence to expected unreliability costs is assumed: and if $0 < v < 1$ the confidence is respectively smaller. Thus, expected mean annual costs regarding unreliability costs and measures of the lack of confidence to the assessment of the amount of these costs can be expressed by the following equation:

$$\overline{C} = C_{YR} + \overline{C}_{AR} \cdot (1 - \nu) \tag{5}$$

A differential criterion (3) can be formulated in the following way:

$$(1 - \nu) \cdot (\overline{C}_{ARi} - \overline{C}_{ARj}) \ge (C_{YRj} - C_{YRi}) \tag{6}$$

and purpose function described by the coefficient of efficiency ε_{ij} coming from option i to option j, in a strategic model adopts the following shape:

$$\varepsilon_{ij} = \frac{(\overline{C}_{ARi} - \overline{C}_{ARj}) - (C_{YRj} - C_{YRi})}{\overline{C}_{ARi} - \overline{C}_{ARj}} \ge \nu \tag{7}$$

Using the relation (7) it is possible to define the required level of confidence β to the indicated costs of unreliability (Sozański 1990):

$$\nu \le \varepsilon_{ij} \tag{8}$$

thus

$$\beta = 1 - \nu \ge 1 - \varepsilon_{ij} \tag{9}$$

As it is confirmed by the aforementioned relationships, the bigger efficiency of going from one compared option to another is, the lower required confidence level to the results of calculation of costs of unreliability $\overline{C}_{AR}$ is and it, with a usual big error of their estimation, namely usually above 30% (Bojarski 1984), suggests that the optimal option is the one for which a low confidence level is required.

2.2 Components of the expected total annual costs

2.2.1 Costs of the unreliability of water supply pump stations

From the mathematical description of the proposed criterion of optimization (4), it can be stated that the value expressing the efficacy of choosing the more expensive and more reliable option of WPS design, shall depend on particular components of the total annual costs of each solution C i.e.: annual investment costs C_I, operation costs C_E and unreliability costs $\overline{C}_{AR}$. Undoubtedly, because of their probabilistic character, the most difficult to determine are unreliability costs. They are made

as a result of breaks or limitations in water supply and generally include:

1. losses incurred by the recipient due to:
   - lack of profit for the goods which could not be produced,
   - costs of destroyed resources and materials,
   - costs resulting from the deterioration of the product quality after the production is renewed,
   - indirect losses occurring in cooperating enterprises caused by the failure to supply the materials,
2. losses incurred by the supplier because of:
   - losses resulting from the lack of payments for the water that has not been supplied,
   - agreed fines paid to the customer,
   - costs of removing the failure.

Total cost of losses C_L, caused by a particular failure is a general function of the duration of the failure t_f, called a line of losses or temporal characteristic of losses (Bojarski 1978, Sozański 1990) with the equation:

$$C_L = f(t_f) \tag{10}$$

The knowledge of the shape of this function for the individual customer can be the base for the analysis of unreliability costs for the water supply pump station (Bajer 2000, Bajer 2002a). Annual costs of unreliability C_{AR} related to with the exploitation of the water pump station should reflect the economic consequences of its unreliability depending on breaks or limitations, randomly appearing in this period of time, in the water supply to the customers. General function of these costs can be described by the stochastic relation (Bajer 2002a):

$$C_{AR}(t, t_f) = c_L(t_f) \cdot Q^{YR}_0 \cdot \{1 - \Psi(t)\} \tag{11}$$

in which $\Psi(t)$ = random function taking value "1", when WPS is working and value "0", when WPS is not working (failure in which $Q < \alpha_f \cdot Q_n$); Q = efficiency of WPS in any moment of time; Q_n = nominal i.e. designed efficiency of WPS; α_f = coefficient of the acceptable decrease of efficiency during the failure; $c_L(t_f)$ = is the individual loss per one 1 m³ of water not delivered during the time of break t_f in water supply, PLN/yr.

$$c_L(t_f) = \frac{C_L(t_f)}{t_f \cdot \Delta Q} \tag{12}$$

in which $C_L(t_f)$ = functional dependence of the total cost of losses C_L caused by a given disturbance (failure) on the duration of the failure t_f, PLN/yr, t_f = value of the time of cessation in water supply

23

(duration of failure), hr; $\Delta Q = Q_0^{hr} - Q_f =$ the amount of water not supplied due to the failure of WPS, equal to the difference between the amount of normally pumped water (Q_0^{hr}) and the amount supplied during the failure (Q_f), m³/hr; $Q_0^{hr} =$ mean calculation efficiency of WPS per hour, m³/hr; $Q_0^{YR} =$ mean calculation efficiency of WPS per year, m³/yr; $Q_0^{YR} = Q_0^{hr} \cdot 24 \cdot 365$.

The expected value of annual costs of unreliability related to the exploitation of a water supply pump station, assuming that the result of a failure is a complete cessation of its work ($Q_f = 0$), and that value t_f is treated as a random variable, will become:

$$\overline{C}_{AR}(t, t_f) = E[C_{AR}(t, t_f)]$$
$$= E[c_L(t_f)] \cdot Q_0^{YR} \cdot E[1 - \Psi(t)] \qquad (13)$$

Marking $A(t)$ as the probability of the event that in the moment t the object is working, according to the definition it can be written (Rao 1992, Migdalski 1982):

$$E[\Psi(t)] = A(t) \qquad (14)$$

then:

$$\overline{C}_{AR}(t, t_f) = \frac{E[C_L(t_f)]}{E[t_f] \cdot Q_0^{hr}} \cdot Q_0^{YR} \cdot [1 - A(t)] \qquad (15)$$

because (Migdalski, 1982):

$$\lim_{t \to \infty} A(t) = A \qquad (16)$$

thus:

$$\lim_{t \to \infty} \overline{C}_{AR}(t, t_f) = \overline{C}_{AR}(t_f)$$
$$= \frac{E[C_L(t_f)]}{E[t_f] \cdot Q_0^{hr}} \cdot Q_0^{YR} \cdot (1 - A) \qquad (17)$$

Because $C_L(t_f)$ is a function of a random variable of the failure duration with the density of probability $f_f(t_f)$, thus the expected value of the costs of losses of a single switch off caused by a failure, can be determined by the formula (Sozański 1990):

$$C_L^1 = E[C_L(t_f)] = \int_0^\infty C_L(t_f) \cdot f_f(t_f) dt_f \qquad (18)$$

then the expected value of the duration of a single failure is:

$$E[t_f] = \overline{t}_f = \int_0^\infty t_f \cdot f_f(t_f) dt_f \qquad (19)$$

Then the relation showing the mean expected costs of unreliability connected with yearly operation of a pumping station is the following:

$$\overline{C}_{AR} = \frac{\int_0^\infty C_L(t_f) \cdot f_f(t_f) dt_f}{\overline{t}_f \cdot Q_0^{hr}} \cdot Q_0^{YR} \cdot (1 - A) \qquad (20)$$

Putting into the formula:

$$\overline{k}_S = \frac{\int_0^\infty C_L(t_f) \cdot f_f(t_f) dt_f}{\overline{t}_f \cdot Q_0^h} \qquad (21)$$

and regarding that $A = A_{WPS}$ finally we obtain:

$$\overline{C}_{AR} = \overline{c}_L \cdot Q_0^{YR} \cdot (1 - A_{WPS}) \qquad (22)$$

in which $\overline{C}_{AR} =$ expected costs of unreliability connected with yearly operation of WPS, assumed to be constant in the consecutive years of operation, PLN/yr; $\overline{c}_L =$ expected cost of losses per unit of the non-delivered volume of water, expressed in PLN/m³; $A_{WPS} =$ stationary index of the readiness of the designed pumping station; $Q_0^{YR} =$ as in formula (12).

In a physical sense an individual cost $\overline{c}_L$, is the equivalent to, often found in the calculations of the unreliability costs of electric energy supply (Bojarski 1978, Sozański 1990) a, so-called, economic equivalent of not supplied electric energy $E[c_A]$, expressed in PLN/kWhr and characterized by the sensibility of a given customer of electric energy per cessation in its supply. As the analogy to the above, it was accepted to define the value $\overline{c}_L$ as an economic equivalent of not supplied water.

In general the determination of $\overline{c}_L$ requires not only the knowledge of the temporal characteristic of losses but also the distribution of the periods of failures. For linear characteristics the relationship of $\overline{c}_L$ is significantly simplified, allowing a relatively easy determination of value $\overline{c}_L$ for a given industrial customer. In case of a municipal customer, the value can be only estimated, due to the difficulties in an accurate assessment of the consequences of a non-continuous water supply (Bajer 2000, Bajer 2002a).

2.2.2 Investment and operation costs

Apart from the costs of unreliability in the compared options of WPS design it is also required to know annual investment costs C_I and operation costs C_E, here reduced to costs of electric energy C_{EE} and costs of renovations C_{RE}. The remaining components of operation costs depend, only to small extent, on a technological solution of WPS and can be neglected.

Investment costs C_0 drawn to yearly costs C_I, paid at the beginning of the discussed period of time, are proposed to be defined from the formula:

$$C_I = W(C_0 \to C_I, p, Tr) \cdot C_0 = \frac{p \cdot (1+p)^{Tr}}{(1+p)^{Tr} - 1} \cdot C_0$$

(23)

in which C_I = investment costs paid at the beginning (C_0) of the whole period of time, drawn into the annual cost, PLN/yr; W = coefficient reducing all the costs paid once and paid continuously in a selected period of time that would equal a certain equivalent of the annual value (Stark & Nicholls 1979); p = rate of profit; Tr = period of the sustainability of the investment, years.

Assuming equally distributed yearly work of the WPS having identical pump aggregates, annual costs of electric energy C_{EE} (variable costs) for pumping water can, for the purpose of technical and economic analyses, be defined from the approximate formula:

$$C_{EE} = Q_0^{YR} \cdot H_m \cdot E_e \cdot c_e \cdot 10^{-3}$$

(24)

where Q_0^{YR} = yearly amount of water pumped, responding to mean yearly calculation efficiency of WPS (for the case of calculation in the most loaded—in terms of power use—period of the pumps work), m³/yr; H_m = manometric height of elevation for Q_0^{hr}, m of the water column; E_n = the amount of electric energy necessary to lift 1000 m³ of water through a given WPS to the height of 1 m, kWhr; c_e = stands for the price of 1 kWhr of electric energy used by the WPS, PLN/kWhr.

Annual costs of full and partial renovations C_{RE} (constant costs) can be calculated as the percentage of the amount of investment costs, e.g. from the following formula (Załutskij & Pietruchno 1987):

$$C_{RE} = 0.035 \cdot C_{0B} + 0.12 \cdot C_{0PU} + 0.01 \cdot C_{0PF}$$

(25)

In which C_{0B} = investment costs paid for the construction of the building of WPS, PLN; C_{0PU} = investment costs for the pump aggregates and other auxiliary equipment (vacuum pumps, compressors etc.), PLN; C_{0PF} = investment costs spent on the pipelines and the armament, PLN.

3 RESULTS AND DISCUSSION

3.1 Analysis of the criterial function

In order to initiate broader discussion on the efficiency of comparing and accepting different design options of WPS, on the basis of the values of the criteria function brought to extreme, its mathematical presentation was analyzed. To conduct such analysis, the relationships (22), (23), (24) and (25) were introduced to the equation (7), for the particular components of expected total annual costs, referring to the compared technical solutions in the WPS obtaining:

$$\varepsilon_{ij} = \frac{B - (D + E + F)}{B} \geq v$$

(26)

$$B = \bar{c}_L \cdot Q_0^{YR} \cdot [(1 - A_{WPS_i}) - (1 - A_{WPS_j})]$$
$$= \bar{c}_L \cdot Q_0^{YR} \cdot (A_{WPS_j} - A_{WPS_i})$$

(27)

$$D = \frac{p \cdot (1+p)^{Tr}}{(1+p)^{Tr} - 1} \cdot (C_{0_j} - C_{0_i})$$

(28)

$$E = Q_0^{YR} \cdot c_e \cdot (H_{mj} \cdot E_{ej} - H_{mi} \cdot E_{ei}) \cdot 10^{-3}$$

(29)

$$F = 0.035 \cdot (C_{0B_j} - C_{0B_i}) + 0.12 \cdot (C_{0PU_j} - C_{0PF_i})$$
$$+ 0.01 \cdot (C_{0PF_j} - C_{0PF_i})$$

(30)

Finally, after simple algebraic transformations, the relationship on the index of the internal efficiency of the transition from option i to option j shall have the following form:

$$\varepsilon_{ij} = -\frac{K_{ij}}{\bar{c}_L} + 1 \geq v$$

(31)

in which:

$$K_{ij} = \frac{D + E + F}{G}$$

(32)

$$G = Q_0^R \cdot (A_{WPS_j} - A_{WPS_i})$$

(33)

$$D + E + F = C_{YR_j} - C_{YR_i}$$

(34)

From the form of the criterial function written as the relationship (31), it can be stated that, for a definite pair of the compared options of WPS, the value K_{ij} is constant, thus the value of the efficiency index ε_{ij} will depend on the value of an economic equivalent of the not supplied water $\bar{c}_L$. This relationship, in a generalized graphic interpretation of the criteria function form is illustrated in Figure 1. It presents exemplified curves of changes in the value ε_{ij} in the growing function of the value $\bar{c}_L$ for two pairs of the compared options (options no. 1 and 2; options no. 2 and 3). Their more detailed analysis, combined with

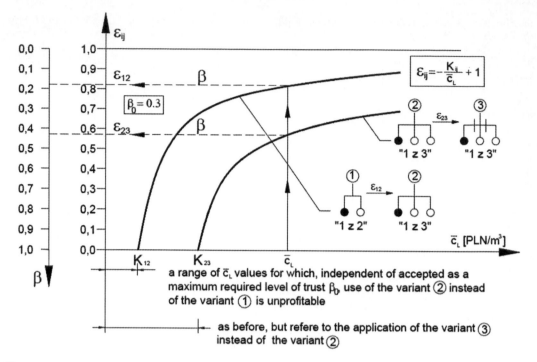

Figure 1. Relation between the inner efficiency index ε_{ij}, and an economic equivalent of unsupplied water $\overline{c}_L$ ($\beta = 1 - \varepsilon_{ij}$—required level of confession to the estimated cost of unreliability; β_0—the of β accepted as maximum).

the required confidence β level to the determined costs of unreliability, functionally combined with ε_{ij} relation (9), enable to notice certain rules:

- Efficiency of the transition from a cheaper or less reliable option i of PoW to more expensive, but at the same time more reliable option j, depends on the sensitivity of a customer to the shortage in water supply, expressed by the value $\overline{c}_L$.
- In a strategic model of making economic decisions, increase value of the economic equivalent of not supplied water $\overline{c}_L$ increases the efficiency of the transition from the compared options with the simultaneous decrease of the required confidence level β to the estimated reliability costs (see relation).
- For each pair of the compared options there is a limit value $\overline{c}_L$ (K_{12} for options no. 1 and 2, K_{23} for options no. 2 and 3), below which the value of the efficiency index $\varepsilon_{ij} < 0$. It can be directly drawn from the relation (31) after comparing it to zero:

$$\varepsilon_{ij} = -\frac{K_{ij}}{\overline{c}_L} + 1 \geq v = 1 - \beta = 0$$

then $\varepsilon_{ij} = 0$ dla $\overline{c}_L = K_{ij}$ i $\varepsilon_{ij} < 0$ dla $\overline{c}_L < K_{ij}$

then $v = 0$ a $\beta = 1$ a $\beta = 1$.

All the above mentioned justifies the statement that regardless of the defined maximum required confidence level β_0, the more expensive variant is completely unprofitable, even with the costs predicted with 100% accuracy.

For a given value $\overline{c}_L$ and a defined maximal required confidence level to the estimated costs of unreliability (β_0), the choice of the optimal option among the examined with the obtained efficiency indexes ε_{ij} of individual transitions and respective confidence levels β, depends on a mutual relation, in which values β_0 and β remain dependent on each other. If $\beta < \beta_0$ it is economically justified to choose the more expensive option. In Figure 1, such a situation refers to the first pair of the compared options of WPS, i.e. no. 1 and 2. The efficiency of the transition from option no. 1 to option no. 2 is high, while the required confidence level compared to the unreliability costs is lower than established ($\beta < \beta_0 = 0.3$). If $\beta \geq \beta_0$ it will be efficient to choose the cheaper option despite its smaller reliability. In Figure 1 such a situation takes place as the second among the compared pairs of technical solutions of WPS, i.e. in no. 2 and 3. The obtained efficiency index of the transition from option no. 2 to option no. 3 imposes the required confidence in the estimated costs of unreliability, exceeding the established level β_0, thus it is justified to choose option no. 2. Finally, among the discussed options of WPS, option no. 2, should be regarded as the optimal for the implementation.

4 CONCLUSIONS

Economic and reliability criterion, presented in this article, expresses the efficiency of extra investment to increase the reliability of the designed PoW. The presented in details in this article, so-called expected costs of unreliability played the crucial role. In the calculus of economic efficiency of the investment they make a measurable reflection of reliability of a designed object, so far established only by intuition. By making an analysis of the proposed criterion, great possibilities made by its use in optimization calculus were shown. It seems reasonable to continue the work on a bigger generalization of the analysis and make a mathematical form of the task of choosing the optimal option among the ones discussed on the stage of the WPS design. It can be particularly significant in the case of pump stations supplying water to economically important industrial objects, usually very sensitive to breaks in water supplies.

REFERENCES

Almeida, A.B. & Ramos, H.M. 2010. Water supply operation: diagnosis and reliability analysis in a Lisbon pumping system. *Journal of Water Supply: Research and Technology-AQUA* 59(1): 66–78.

Bajer, J. 2000. Metoda szacowania kosztów zawodności pompowni wodociągowych, Zeszyty Naukowe Politechniki Rzeszowskiej, No 180, Budownictwo i Inżynieria Środowiska, z. 32, Część 2—Inżynieria Środowiska; *Materiały V Konferencji Naukowej Rzeszowsko-Lwowsko-Koszyckiej nt. Aktualne problemy budownictwa i inżynierii środowiska, Rzeszów, 25–26 września 2000*, Rzeszów: Wydawnictwa Politechniki Rzeszowskiej.

Bajer, J. 2001a. Ocena niezawodności pompowni wodociągowych o różnych strukturach technicznych. In A. Wieczysty (ed.), *Metody oceny i podnoszenia niezawodności działania komunalnych systemów zaopatrzenia w wodę*, Monografie Komitetu Inżynierii Środowiska Polskiej Akademii Nauk, vol. 2, Kraków: Komitet Inżynierii Środowiska Polskiej Akademii Nauk.

Bajer, J. 2001b. Zastosowanie metody minimalnych przekrojów niesprawności do oceny niezawodności złożonych systemów technicznych, *Bezpieczeństwo, niezawodność, diagnostyka urządzeń i systemów gazowych, wodociągowych, kanalizacyjnych grzewczych; Materiały II Ogólnopolskiej Konferencji Naukowo-Technicznej, Zakopane Kościelisko, 21–23 listopada 2001, Nr 797/2001*, Kraków: Polskie Zrzeszenie Inżynierów i Techników Sanitarnych.

Bajer, J. 2002a. Teoretyczne podstawy obliczania kosztów strat produkcyjnych związanych z zawodnością funkcjonowania pompowni wodociągowych, *Czasopismo Techniczne z. 8-Ś*, Kraków: Wydawnictwo Politechniki Krakowskiej.

Bajer, J. 2002b, Kompleksowa metoda projektowania pompowni wodociągowych. In M.M. Sozański & J.A. Oleszkiewicz (eds), *Zaopatrzenie w wodę i jakość wód; Materiały XVII Krajowej Konferencji (V Międzynarodowej Konferencji), Poznań-Gdańsk, Polska, 26–28.06.2002*, Poznań: Polskie Zrzeszenie Inżynierów i Techników Sanitarnych Oddział Wielkopolski w Poznaniu.

Bajer, J. 2004. Zasady projektowania pompowni wodociągowych z uwzględnieniem ich niezawodności, Zeszyty Naukowe Politechniki Rzeszowskiej, Nr 211, Budownictwo i inżynieria środowiska z. 37; *Materiały IX Konferencji Naukowej Rzeszowsko-Lwowsko-Koszyckiej nt. Aktualne problemy budownictwa i inżynierii środowiska. Jakość–Niezawodność–Bezpieczeństwo, Rzeszów, 3–4 września 2004, Część 2—Inżynieria Środowiska*, Rzeszów: Wydawnictwa Politechniki Rzeszowskiej.

Bajer, J. 2006. Reliability analysis of variant solutions for water pumping stations. In L. Pawłowski, M. Dudzińska & A. Pawłowski (eds), *Environmental Engineering*, Taylor & Francis Group, New York, Singapore.

Bajer, J. 2011. Wpływ parametrów niezawodności elementów pompowni na jej wskaźniki niezawodności. *Instal* 11(323): 63–68.

Bajer, J. & Głód, K. 2003. Analiza niezawodności pompowni wodociągowych w celu doboru optymalnej liczby rezerwowych agregatów pompowych. *Instal* 11 (234): 6–10.

Bajer, J. & Iwanejko, R. 2008. Eksploatacyjne badania niezawodności podstawowych elementów uzbrojenia pompowni wodociągowych. *Instal* 10(288): 81–84.

Bajer, J. & Wieczysty, A. 2000. Badanie wpływu struktury technicznej pompowni wodociągowych na ich niezawodność. In M.M. Sozański (ed.), *Zaopatrzenie w wodę, jakość i ochrona wód; Materiały IV Międzynarodowej Konferencji (XVI Krajowej Konferencji), Kraków 11–13 września 2000*, Kraków: Polskie Zrzeszenie Inżynierów i Techników Sanitarnych o. Wielkopolski.

Bojarski, W.W. 1978. *Wytyczne metodyczne szacunku strat powodowanych w zakładach produkcyjnych ograniczeniami w dostawie czynników energetycznych*. Warszawa: Zakład Problemów Energetyki, Instytut Podstawowych Problemów Technicznych Polskiej Akademii Nauk.

Bojarski, W.W. 1984. *Przykładowe zastosowania analizy i inżynierii systemów*. Warszawa: Państwowe Wydawnictwo Naukowe.

Cullinane, M.J. Jr. 1985. Reliability evaluation of water distribution systems components, *U.S. Army Engineering, Waterways Experiment Station*, 353–358, New York: ASCE.

Dąbrowski, W. & Bajer, J. 2008. *Reliability of water supply systems in Central Europe using Poland as an example*, 13th World Water Congress, Montpellier—France.

Dawidowicz, L.A., Iwanejko, R., Bajer, J. & Wieczysty, A. 1993. Metody wyznaczania oczekiwanego czasu poprawnej pracy między uszkodzeniami wysoce niezawodnych elementów pompowni. *Gospodarka Wodna*4 (532): 92–94.

Duan, N. & Mays, L.W. 1987. *Reliability analysis of pumping stations and storage facilities*; Proc. of the National Conference on Hydraulic Engineering: ASCE.

Duan, N. & Mays, L.W. 1990. Reliability analysis of pumping systems. *Journal of Hydraulic Engineering* 116(2): 230–238.

Hashemi, S.S., Tabesh, M. & Ataee Kia, B. 2011. Ant-Colony Optimization of Energy Cost in Water Distribution Systems Using Variable Speed Pumps; *Proceedings of 4th ASCE-EWRI International Perspective on Water Resources and The Environment, January 4th–6th, National University of Singapore*, Singapore.

Hobbs, B. 1985. Reliability analysis of water system capasity. In W. Waldrob (ed.); *Proc, Speciality Conf. Hydr. and Hydro. in Small Computer Age*: ASCE.

Ilin, J.A. 1987. *Rascziot nadieżnosti podaczi wody*. Moskwa: Stroizdat.

Kopecki, K. 1960. *Rachunek awaryjności w energetyce i obliczanie rezerw*. Warszawa: Komitet Elektryfikacji Polski Polskiej Akademii Nauk.

Kurek, W. & Ostfeld, A. 2012. Multi-Objective Water Distribution Systems Control of Pumping Cost, Water Quality, and Storage-Reliability Constraints, *Journal of Water Resources Planning and Management*, doi: 10.1061/(ASCE)WR.1943-5452.0000309.

Kwietniewski, M., Roman, M. & Kłoss-Trębaczkiewicz, H. 1993. *Niezawodność wodociągów i kanalizacji*. Warszawa: Arkady.

Liberacki, R. 2010. Selected aspects of determining the reliability of the pump subsystems with redundancy, used in main engine auxiliary systems. *Journal of Polish CIMAC*, Vol. 5, No 2: 108–113.

Lindstedt, P. & Sudakowski, T. 2008. Method of prediction of reliability characteristics of a pumping station on the base of diagnostic information. *Journal of KONBiN* 2(5): 207–222.

Mays, L.W., Duan, N. & Su, Y-C. 1986. *Modeling reliability in water distribution network design*. University of Texas, Austin.

Mehzad, N., Tabesh, M., Hashemi, S.S. & Ataee Kia, B. 2011. Reliability of pumping station in water distribution networks considering VSP and SSP; *Proceedings of Computing and Control for the Water Industry, Urban Water Management—Challenges and* Opportunities, University of Exeter, United Kingdom, Vol. 1: 5–7.

Mehzad, N., Tabesh, M., Hashemi, S.S. & Ataee Kia, B. 2012. Reliability of water distribution networks due to pumps failure: comparison of VSP and SSP application, *Drinking Water Engineering and Science*, 5: 351–373.

Migdalski, J. 1982. *Poradnik niezawodności: Podstawy matematyczne*. Warszawa: Wydawnictwa Przemysłu Maszynowego "WEMA".

OREDA. 1997. *Offshore Reliability Data Handbook. 3rd Edition*. Trondheim: DNV.

Peschel, M. & Riedl, C. 1979. *Polioptymalizacja. Metody podejmowania decyzji kompromisowych w zagadnieniach inżynieryjno-technicznych*, Warszawa: Wydawnictwo Naukowo-Techniczne.

Rao, S.S. 1992. *Reliability-based design*. School of Mechanical Engineering Purdue University, U.S.A., New York, McGraw-Hill Inc.

Shamir, U. & Howard, C.D.D. 1981. Water supply reliability theory. *Journal of the American Water Works Association*, vol 73, 7: 379–384.

Shilston, P.B. 1985. Designing pump drive systems for reliability, *Technical Conference of the British Pump Manufacturers Association 9th*, Cranfield: BHRA.

Shultz, D.W. & Parr, V.B. 1981. *Evaluation and documentation of mechanical reliability of conventional wastewater treatment plant components (Draft Report)*, U.S. Environmental Protection Agency, Cincinnati—Ohio.

Sozański, J. 1990. *Niezawodność i jakość pracy systemu elektroenergetycznego*. Warszawa: Wydawnictwa Naukowo-Techniczne.

Stark, R.M. & Nicholls, R.L. 1979. *Matematyczne podstawy projektowania inżynierskiego*. Warszawa: Państwowe Wydawnictwo Naukowe.

Tchórzowska-Cieślak, B. 2009. Water supply system reliability management. *Environment Protection Engieneering*, vol. 35, 2: 29–35.

Wieczysty, A. & Rak, J. 1985. Metody oceny niezawodności działania pompowni wodociągowych. *Gaz, Woda i Technika Sanitarna* 5–6: 125–128.

Załuckij, J. E. W. & Pietruchno, A. I. 1987. *Nasosnyje stancji*. Moskwa: Izd. Wyższaja Szkoła.

Zhuang, B., Lansey, K. & Kang, D. 2011. Reliability/Availability analysis of water distribution systems considering adaptive pump operation, In R. Edward Beighley and Mark W. Killgore (eds), *World Environmental & Water Resources Congress, Palm Springs, California, United States,* 22–26 May 2011: ASCE.

Environmental Engineering IV – Pawłowski, Dudzińska & Pawłowski (eds)
© *2013 Taylor & Francis Group, London, ISBN 978-0-415-64338-2*

Multidimensional comparative analysis of water infrastructures differentiation

K. Pietrucha-Urbanik

Faculty of Civil and Environmental Engineering, Rzeszow University of Technology, Rzeszow, Poland

ABSTRACT: This paper presents the differentiation in the level of water supply infrastructure in the Polish provinces. The multidimensional comparative analysis (WAP) was used to illustrate the differentiation. Using the Cluster Analysis method (CA) observations similar to each other grouped in the so-called clusters are obtained. The indicators of water supply infrastructure for the years 1995–2010 were determined. On the basis of the method based on the arithmetic means, the parameters that have the greatest influence on the occurrence of each cluster, were defined. In order to assess grouping, and to compare and assess the results obtained using the classical method, the neural modelling—the Kohonen neural network (Self-Organizing Feature Maps), was used. The resulting topological map allows to determine the degree of similarity within and between the distinguished groups of analysed objects through dependencies visualisation.

Keywords: water supply infrastructure, cluster analysis, the Kohonen neural network

1 INTRODUCTION

During the period of the structural changes in the processes of the European integration and regionalisation, the issue of the differentiation in the level of water supply infrastructures in Polish provinces, which is associated with the social, economic and demographic development of the country, becomes more and more important. A properly functioning Collective Water Supply System (CWSS) should ensure a continuous water supply to the recipients, with a suitable quality and adequate quantity (required pressure), at a specific time (Rak & Pietrucha 2008a,b; Studziński & Pietrucha-Urbanik 2012). These parameters must be maintained at the required level, which in reality is not so simple. One of the reasons hindering the provision of water supply with the appropriate parameters is a continuous development of water supply infrastructure, caused mainly by the connection of the new customers or changes in water consumption by the users already connected (Pawłowski & Dudzińska 1994). Therefore, in order to know the current situation of water supply infrastructure it is necessary to conduct the periodic analyses that were proposed in this work. Such approach will help to establish methodology for reliability analysis and assessment system (Valis et al. 2010, 2012). Due to the importance of the discussed issue, the analysis of the differentiation of the Polish provinces regarding the level of water supply infrastructure development in the Polish provinces in the years 1995–2010, seems to be important. In order to know this differentiation the cluster analysis was performed and the selected grouping methods for the multivariate objects were compared, namely the Ward agglomeration method, the k-means and the Kohonen neural modelling.

2 MATERIAL AND METHODS

2.1 *Material*

The study included the following input characteristics: the number of water connections, the number of water connections per 1 km of distribution pipe, the average daily unit value of water consumption, the intensity of the network load, the indicators of the level of water supply facilities in settlements.

The primary source of data for this study was Central Statistical Office. The analysis was performed with the use of the neural network application contained in a StatSoft computer program STATISTICA.Pl 10.0. In order to carry out an initial classification procedure of the Polish provinces, the linear classification determining the synthetic indicator being the average member variable, according to which the provinces have been ordered, was used. Before the analysis the variables showing little variation and highly correlated with each other, were removed, then unitarisation of variables with different titers was made.

2.2 Using the WAP grouping methods to assess the differentiation in the level of water supply infrastructures in the Polish voivodships

The analysis of the differentiation in the level of services provided by water supply companies was performed using the tools of the multidimensional comparative analysis—WAP. Observations similar to each other are grouped in the so-called clusters of the Cluster Analysis (CA). Clusters created in such a way are then arranged into a clear structure on the basis of the analysis similarities in the area. These similarities may be suitable indicators specific to the group, similarities or distances (Jajuga et al. 2002). In this way we obtain such a representation of the objects that shows their common features and their differences. To estimate the distance between the clusters the analysis of variance can be used. This approach is presented in the Ward method. This method lying in minimizing the sum of squared deviations of any two clusters that can be formed at each stage. Although such approach generates the formation of clusters of very small distance, it is often used due to its effectiveness. In details this method is described in Ward (1963).

To determine the assessment of the level of infrastructure development of provinces the Euclidean distance matrix was determined. Its values are calculated using the formula:

$$d(x_i, x_k) = \sqrt{\sum_{j=1}^{m}(x_{ij} - x_{kj})^2} \qquad (1)$$

where x_{ij} = the value of the j-th variable for the i-th object; x_{kj} = the value of the j-th variable for the k-th object; and $d(x_i, x_k)$ = the distance between the i-th and the k-th object.

Then, the agglomeration method, lying in the determination of a hierarchical tree called dendrogram, was used. To verify belonging of the individual provinces to the dendrogram the authors used the method proposed by Z. Hellwig. When using this method, one should estimate a distance between the points belonging to two different subsets (Hellwig 1973). This distance is called the critical value, its value is determined by the following procedure:

– determination of the minimum values for each line in the distance matrix.
– determination of the arithmetic mean $\bar{x}$ and the standard deviation δ for the above values (new variables).
– calculation of the critical values from the formula: $W_k = \bar{x} + 2\delta$.
– After determining the critical value, the segregation of clusters in order to create a dendrogram that shows the hierarchy of the individual clusters, was made. To determine the individual

clusters and collective water supply systems belonging to them, the authors use the critical value determined according to the above algorithm.

– Another tool used for the grouping of objects is the k-means method. The significant difference between the k-means method and the Ward's agglomeration method (and other hierarchical methods) is that the latter method requires the formulation of a priori assumption about the number of clusters obtained by grouping (Greenacre, 1984). Grouping by the k-means method is an iterative procedure—in each iteration some objects are moved to other clusters, until the minimum differentiation within clusters and the maximum differentiation between clusters are obtained. The need to define the number of clusters can be regarded as a disadvantage of the k-means method if the structure of a set of objects is unknown and is not possible to formulate hypotheses regarding the expected number of clusters (Everitt et al. 2011).

– Clustering results depend, among others, on the sequence of observations in the set, the choice of initial cluster centres and the occurrence of unusual observations (although it is considered that a sensitivity to their occurrence is smaller than in the Ward method). The advantage of the k-means method is that it assures to get k-clusters that differ from each other to the most possible extent (Grabiński & Sokołowski 1980). The algorithm moves the objects into different clusters, leading to such an arrangement that will ensure the maximum homogeneity within the subgroups and, at the same time, the maximum heterogeneity between the subgroups. Another advantage of the method is the possibility to be applied for a large set of objects (Helsel & Hirsch 1992).

– Both methods presented above are the classical methods of the cluster analysis. Their common feature is the use of least squares criterion.

– Another method of carrying out the process of grouping of multivariate objects without external correction (without a pattern) is self-organizing feature maps neural network (SOFM—Self-Organizing Feature Maps), also known as the Kohonen network. The name comes from the name of the Finnish scientist Teuvo Kohonen, who first developed and studied the maps in the years 1979–1982 (Kohonen 1997). The SOFM is one of the most advanced neural network model that provides a topological mapping of a multi-dimensional space onto a two-dimensional map of neurons. For the Kohonen network learning an iterative algorithm, called "a winner takes all", is used. During the SOFM network training the neural weights are modelled in such a way that very similar cases are represented by one neuron and the less similar are represented by neighbouring neurons (Dreyfus 1995). The process of forming the connections weights between the individual neurons leads to mapping the input data on a plane. The advantages of the Kohonen network are: its non parametricity, no need for a preliminary assumptions, the maintenance of the nonlinear relationships between units, the processing of large sets of objects or data mining detecting relationships often invisible for other methods and unexpected for the analyst (Tadeusiewicz et al. 1999). Furthermore, the neural networks cope better with the lack of data, the presence of extreme values, the collinearity of variables or their excessive correlation.

3 RESULTS AND DISCUSSION

In Figure 1 the provinces classified in the *linear hierarchical order* are shown. The provinces that ranked highest in terms of the level of water supply infrastructure are: Masovian Voivodship (a synthetic indicator: 63.32%), Silesian Voivodship (the second in the ranking, a synthetic indicator: 52.12%), Greater Poland Voivodship (the third place in the ranking, a synthetic indicator: 44.08%), West Pomeranian Voivodship (the fourth place in the ranking, a synthetic indicator: 44.00%). The lowest level of infrastructure development has Subcarpathian Voivodship (a synthetic indicator: 29.69%). An analogous situation with regard to ranking of provinces occurs in rural areas of particular Polish provinces.

As a result of the clustering by the Ward method a dendrogram containing information about distances of bonds, shown in Figure 2, was obtained. To determine the individual clusters and provinces

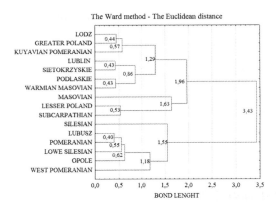

Figure 2. The dendrogram of the differentiation in the level of water supply infrastructures in the Polish provinces.

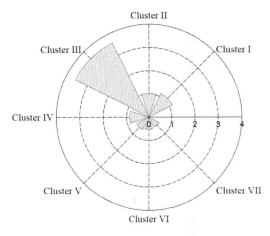

Figure 3. The quotient of the feature describing the increase of the network in [%].

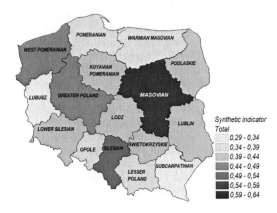

Figure 1. The ranking of Polish provinces in terms of the level of water supply infrastructure (total).

be longing to them, the critical value: $W_k = 0.96$, which divided the dendrite into seven clusters, was used.

The segregation analysis of the particular clusters was performed by calculating the quotient of the arithmetic mean of the successive clusters—X_i to the group mean—X, which, for the selected features, is shown graphically in Figures 3–8. The calculated ratio X_i/X higher than 1 indicates a dominance of a specific parameter in a given cluster.

The dominant role in the grouping played a feature describing % of water supply network users (clusters: I, V, VI, VII), the intensity of network load (clusters: III, V, VI, VII) and the length of network per unit area (clusters: I, III, IV, V). The increase of the network played an important role only in the third and second grouping, while the number of water supply connections per 1 km

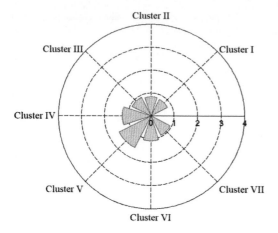

Figure 4. The number of water connections per 1 km of distribution pipe.

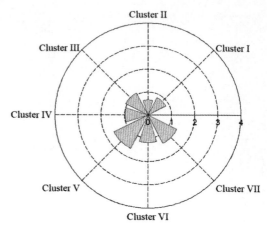

Figure 7. The intensity of the water supply network load.

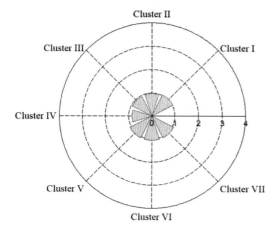

Figure 5. % of water supply network users.

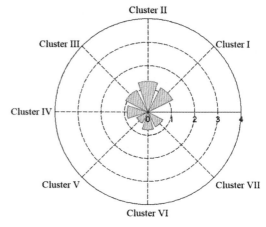

Figure 8. The length of water supply network per capita.

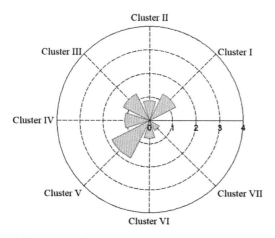

Figure 6. The network length per unit area.

of distribution pipe in creating clusters: V, VI, VII, the feature describing the investments in the clusters: II, III and VII, whereas the length of water supply network per capita dominated in the clusters I, II, III. In comparison to the Ward agglomeration method, using the k-means method we obtain similar grouping results, with the difference lying in the creation of separate clusters for the provinces: Lesser Poland and Subcarpathian, previously forming a common cluster, and including West Pomeranian Voivodship to the cluster with Lubusz Voivodship, Lower Silesian Voivodship, Opole Voivodship and Pomeranian Voivodship. By analysing the chart of means (Fig. 9), we can observe that the highest increase of the network occurred in Masovian Voivodship (Variable 1), therefore there are high investment expenditures on the construction of water supply

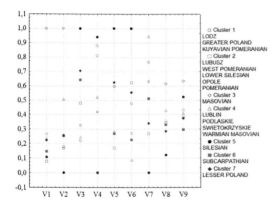

Figure 9.   Chart of the means for each cluster.

Cluster 1	•	Masovian
Cluster 2	•.•	Podlaskie, Kuyavian Pomeranian, Warmian Masovian
Cluster 3	••	Lublin, Swietokrzyskie
Cluster 4	••	Lesser Poland, Subcarpathian
Cluster 5	••	Lodz, Greater Poland
Cluster 6	•.•.	Lubusz, West Pomeranian, Lower Silesian, Opole, Pomeranian
Cluster 7	•	Silesian

Figure 10.   Topological map of obtained clusters.

network (Variable 9). This province is also characterized by a high indicator of water supply network load intensity (Variable 7) and an indicator of the level of water supply facilities (Variable 8). Silesian Voivodship is also characterized by large investments expenditures (Variable 9) and high values of indicators characterizing the number of water connections per 1 km of distribution pipe (Variable 3), the percentage of people using the water supply network (Variable 4), the intensity of the water supply network load (Variable 5) and the level of water supply facilities (Variable 6). High population density affects the indicator of a length of water supply network per capita (Variable 7). In the remaining clusters we can observe the average values of the indicators describing the water supply infrastructures.

Grouping of provinces with the use of the Kohonen network was carried out on the basis of several different sizes of the output layer: 1×2, 1×5, 1×7. After the completion of learning, the network in the form of a topological map was obtained, as shown in Figure 10.

For the clusters I, IV, VII the results of grouping obtained using the SOM are fully consistent with the results obtained by the Ward method.

The provinces included in the clusters I, VI, VII are identical with those obtained from the k-means method. A large convergence is also received in the clusters I, II and VI. Completely unlike the situation presented in the clusters II and VI.

4   CONCLUSIONS

The multivariate analysis method can be a helpful tool in making the ranking of the level of infrastructure development in different regions of the country. It is worth emphasizing the similarity between the results of grouping made by the classical methods and the differences in the results obtained using the SOM. The methods differ in the assumptions, conditions of using and presentation of results and therefore they should be used complementarily. The determination of the synthetic indicator allowed for an initial indication of the hierarchy of analysed regions. Then, the use of the Ward method allowed the separation of seven clusters of provinces, with different levels of infrastructure development, which is associated with the economic development of the country. The next step to deepen the analysis was grouping with the use of the SOFM neural network method, which provided additional conclusions, validated the results obtained from the earlier classifications and led to discovering new aspects of relationships between objects. The analysis separated seven clusters, the structure of the separated clusters, taking into account the dominant features, is the following: The cluster I is created by Lodz Voivodship, Greater Poland Voivodship, Kuyavian Pomeranian Voivodship, the particular indicators in these provinces are at the average level, with the exception of the indicator associated with the percentage of people using the network, whose average value for the above mentioned provinces is 90.9%, as well as the indicator of the level of water supply facilities. The cluster II contains Lublin Voivodship, Swietokrzyskie Voivodship, Podlaskie Voivodship, Warmian Masovian Voivodship. Noteworthy is the highest in the country indicator describing the length of water supply network per capita, which in abovementioned provinces is about 11 m/Mk, high values of the indicators related to the increase of network, which in comparison to 1995 was 210% and the high investment expenditures on the construction of water mains. In the cluster III (Masovian Voivodeship) we can see clearly distinguishing features, such as the increase of the network (300%) and inherent in this situation investments expenditures on water supply network. The cluster IV covers the provinces: Lesser Poland Voivodeship and Subcarpathian Voivodeship, for which the analysis of each indicator is unfavourably compared with other clusters. The formation of this cluster was caused by the following indicators: the number of water connections per 1 km of water pipe of approximately 23 pieces/km and the length of the network per unit area of about 8 m/ha. An unusual case among all the provinces is Silesian Voivodeship, assigned to one-element cluster VII, it is the most industrialized and urbanized province, resulting in the highest in the country indicator of the level of water supply facilities, associated with the network length per unit area (which is affected by the largest population density in the country, 376 persons/km^2), the highest number of water connections per 1 km of distribution pipe (about 29 pieces/km) and the highest intensity of network load amounting to 16.51 m^3/km·d, with, at the same time, the lowest investment expenditures on water supply. The cluster VI includes the following provinces: Lubusz Voivodeship, Pomeranian Voivodeship, Lower Silesian Voivodeship and Opole Voivodeship, whose most distinctive feature are the lowest investment expenditures on the construction of water supply which translates into its small increase (ca. 157%). The differentiators of the level of infrastructure development in West Pomeranian Voivodeship forming the cluster VII are: the intensity of network load amounting to 16.51 m^3/km·d, a very low level of network length per unit area, with low population density of 74 persons/km^2.

REFERENCES

Dreyfus, G. 2005. Neural Networks. Methodology and Applications. Berlin Heidelberg: Springer.

Everitt, B.S., Landau, S., Leese, M., Stahl, D. 2011. Cluster Analysis. Chichester: John Wiley & Sons, Ltd.

Grabiński, T. & Sokołowski A. 1980. The Effectiveness of Some Signal Identification Procedures; in: Signal Processing: Theories and Applications, ed. Kunt M., De Coulon F., Amsterdam: North-Holland Publishing Company, EURASIP.

Greenacre, M.J. 1984. Theory and applications of correspondence analysis. New York: Academic Press.

Hellwig, Z. 1973. Zarys ekonometrii. Warszawa: PWE.

Helsel, D.R. & Hirsch, R.M. 1992. Statistical Methods in Water Resources. New York: Elsevier Science Publishing Company Inc.

Jajuga, K. & Sokołowski, A., Bock, H.H. 2002. Classification, clustering and data analysis: recent advances and applications. Berlin Heidelberg: Springer.

Kohonen, T. 1995. Self-Organizing Maps. Berlin Heidelberg: Springer.

Pawłowski, L. & Dudzińska, M.R. 1994. Environmental problems of Poland during economic and political transformation. Ecological Engineering 3: 207–215.

Rak, J.R, Pietrucha, K. 2008a. Ryzyko w kontroli jakości wody do spożycia. Przemysł Chemiczny, 87(5): 554–556.

Rak, J., Pietrucha, K. 2008b. Some factors of crisis management in water supply system. Environment Protection Engineering, 34(2): 57–65.

Studziński, A., Pietrucha-Urbanik, K. 2012. Risk Indicators of Water Network Operation. Chemical Engineering Transactions, 26: 189–194. DOI: 10.3303/CET1226032.

Tadeusiewicz, R. & Lula, P. 1999. STATISTICA Neural Networks. Kraków: StatSoft.

Valis, D., Vintr, Z., Koucky, M. 2010. Contribution to highly reliable items' reliability assessment. Reliability, Risk and Safety: Theory and Applications, 1–3: 1321–1326.

Valis, D., Vintr, Z., Malach, J. 2012. Selected aspects of physical structures vulnerability—state-of-the-art. Eksploatacja i Niezawodność—Maintenance and Reliability, 14(3): 189–194.

Ward, J.H. 1963. Hierarchical grouping to optimize an objective function, Journal of the American Statistical Association 58(3): 236–244.

Environmental Engineering IV – Pawłowski, Dudzińska & Pawłowski (eds)
© 2013 Taylor & Francis Group, London, ISBN 978-0-415-64338-2

Economic effect of the liquid soil technology compared to traditional sewage system construction methods

D. Słyś & A. Stec
Department of Infrastructure and Sustainable Development, Rzeszow University of Technology, Rzeszow, Poland

M. Zeleňáková
Department of Environmental Engineering, Technical University of Košice, Košice, Slovakia

ABSTRACT: The paper presents results of cost analysis concerning sewage system construction with the use of traditional technology involving mechanical earth compacting versus the innovative liquid soil technology. The results of the investment cost simulation for both technologies are presented with special stress put on the best solution selection problem. The economical balance takes also into account certain additional costs and social cost related to construction work.

Keywords: liquid soil, sewage systems, innovative technologies, economic results

1 INTRODUCTION

One of the most important issues relating to construction of underground services infrastructure in urban areas is the necessity to close the streets to traffic for some time. It results in traffic impediments and measurable financial losses for users of buildings located in the construction work area. An additional problem, important especially in historic town quarters, consist in the necessity to use vibration-generating machinery for compacting earth around the pipelines. Such earth compacting method affects negatively technical condition of period buildings and is bothersome for their users. In many cases, the compaction of earth layers around the pipes cannot be performed correctly because of the scarcity of space required for such operation, careless workmanship, or high density of media network in the ground. Achieving proper earth compaction in the duct zone is the most important and decisive factor for strength and static parameters of the pipeline and thus for its defect-free service life.

Therefore, an optimum pipeline construction technology should be characterised with limited effect on the natural environment and minimum troublesomeness for residents and buildings, guaranteeing at the same time that proper parameters of earth around the constructed/repaired underground pipelines will be achieved. A technology meeting all these requirements, known as the liquid soil technology, is described in patent WO 2004/065330 (Stolzenburg 2004).

The central idea of the technology consists in fabrication of an earth material with temporary and reversible liquid consistency and using it as pipeline trench backfill. Liquid soil is produced from the excavated material or from aggregates and recycled materials with an addition of environmentally neutral mineral, and plant components, and water. The absence of negative effects on natural environment was confirmed in opinions issued by competent institutes, including the Hygienic Certificate of the Polish National Institute of Hygiene.

Technological and technical advantages of the innovative liquid soil technology resulted in the increasing number of cities in Germany (Wegweiser 2008, Schluter 2007), Poland (Dziopak et al. 2007, Nalaskowski 2008), and other countries that have already started or plan to start their media infrastructure construction works involving the use of this technology.

This paper presents the results of an analysis of the economic effect generated when the construction of a sewage system uses the innovative liquid soil technology, compared to the achieved when the traditional technology, is employed.

2 METHODS

Fluidisation technologies of the earth material become more and more universally utilised in engineering practice. Conditions and requirements, which should be met when using this process, have been established e.g. in German draft guidelines DWA-A 139. According to this document, liquid soil is fabricated by mixing loose or compact earth material or a recycled material with plasticisers and stabilisers, cement, water and, in some specific cases, substances increasing porosity or generating foam. Description of the liquid soil fabrication method is presented in the company standard (Werksnorm WN 04.02) developed by Logic, one of the firms offering the implementation of the new technology. The first step of production consist in the preparation of base material by means of its fragmentation until an average grain size is obtained, not exceeding 25 mm. In the next stage, additional substances are dosed including plasticiser, stabiliser and conditioner, as well as a mineral accelerator and water. All components in correct amounts and proportions are transferred to a mixing device agitating the content in the course of transportation to the construction site. The next step consists in pouring liquid soil into the trench in which the pipeline was laid earlier. After being poured into the trench, earth becomes the subject to spontaneous compaction filling in all possible voids. The compensation of the buoyancy forces acting on pipes in the course of flooding with liquid soil can be obtained by means of mechanical devices or locks made of earth mass with thick-plastic consistency. When work is carried out in a trench with reinforced walls, sheeting boards should be removed from the trench before the earth mass is hardened, i.e. when it is still in liquid or soft-plastic state, allowing the liquid soil to fill in precisely all voids remaining after removal of sheeting structures. The buoyancy force measurement and the resulting determination of the earth hardening point, followed by the determination of possibility to remove the buoyancy compensating equipment, must be consistent with recommendations contained in the technology guidelines. Liquid soil parameters after transition from fluid (liquid or thick-plastic) phase into solid one, are in most cases identical with or similar to those of the earth in its original undisturbed state. In the standard conditions as per the RSS technology (Werksnorm WN 04.02: 15°C, uncontaminated earth material such as sand, gravel, clays, loess) reconstruction of the earth solid structure takes from 4 to 8 hours. After that period, it is possible to resume traffic of construction machinery over the trench backfilled with liquid soil. The results of the detailed examination of mechanical parameters of liquid soil fabricated with the use of components supplied by different manufacturers have been presented by Triantafyllidis et al. (2006).

3 ECONOMICAL EFFECTS OF THE TECHNOLOGY

The aim of the presented analysis consisted in the demonstration of significant cost differences between the traditional technology and the innovative pipeline construction technology utilising the liquid soil properties.

The total costs of pipeline construction with the use of trenching technology can be determined by the following formula:

$$C = C_{iv} + C_{ad} + C_{sc}, \qquad (1)$$

where:

C_{iv} the total of investment costs incurred for investment realisation, including the costs of: excavation and sheeting, pipes purchase and lying; installation of fittings; earth backfilling and compacting; dewatering; pavement stripping and reconstruction; auxiliary materials and works, in €;

C_{ad} the total of additional costs including costs related to: alterations to traffic organisation; additional vehicle operation expenses; loss of time for passengers; increased number of road accidents; compensations to land owners and buildings users on account of damage and limited access; deterioration of street and sidewalk surface condition as a result of construction work; and waste material storage, in €;

C_{sc} social costs, including expenditure relating to: delays; increased traffic accident rate; damage to natural environment caused by noise, vibration, and contamination of environment elements, in €;

Investment costs represent the most important component of total construction expenses. The analysis takes into account these costs for three different sewage system construction technologies:

- traditional technology—consisting in using mechanically compacted loose earth for construction of both the pipeline zone and the higher layers;
- liquid soil technology—where the line and trench are filled with temporarily fluidised earth material;
- intermediate technology—consisting in filling the pipeline zone with liquid soil and backfilling the remaining portion of the trench with mechanically compacted loose earth.

Height of the pipeline embedding zone was assumed as equalling the pipe external diameter plus 0.6 m representing the minimum earth embedment that should be laid under and over the pipeline. Figure 1 shows a cross-section of earth layers for the technologies adopted for the analysis.

A very important feature of the liquid soil technology consists in the possibility to reduce the trench width normally required for correct earth compaction, as no such operation exists in the new

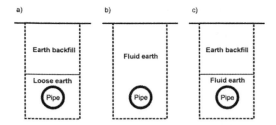

Figure 1. Section of soil layers, a) traditional technology, b) liquid soil technology, c) intermediate technology.

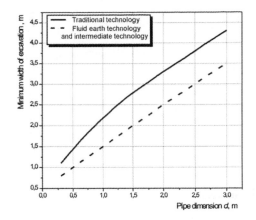

Figure 2. Minimum width of excavation for traditional technology, liquid soil technology and intermediate technology.

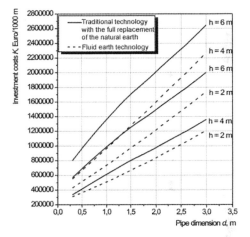

Figure 3. Investment costs of network construction by traditional technology and liquid soil technology, in dependence on pipe diameter and cover with the full replacement of the natural soil diameter for the traditional and liquid soil technology.

process. Minimum trench widths with sheeting as per European standard EN 1610 for the technologies for which the calculations were performed, are presented in Figure 2.

A very important component of sewage system construction expenses consists in earth replacement costs. Such a situation may result from both improper mechanical parameters of the earth borrowed from the excavation and its contamination, especially in the case of replacement of an old leaking sewer with a new one.

Possible necessity to fully replace earth is decisive for economic effect achieved as a result of application of the innovative liquid soil technology. It follows from our analyses that for any diameter and the depth of the pipeline, the earth fluidisation technology is less expensive than the traditional one when the whole earth needs to be replaced. The difference is from about 8% to 25% depending on the line lying depth and diameter. The results of calculations are presented in Figure 3.

On the other hand, Figure 4 shows a comparison of construction costs relating to ducts with different diameters covered with a 4 m-thick layer of earth in the conditions of total replacement of virgin soil. It can be noted that both technologies based on trench backfilling with liquid soil, either partially or entirely, are more price-effective than the traditional technology. For diameters less than about 1.2 m, the technology providing for full backfilling of the excavation with liquid soil is cheaper, while for larger diameters, less expenses will be incurred with the use of the intermediate technology. Similar relationship exists when comparing the investment costs for different depths of main backfill over the pipeline embedment zone. The results of example calculations performed for a pipe with diameter of 2.0 m are shown in Figure 5.

When the whole earth needs to be replaced, the economic balance must also include the cost of rehabilitation of the contaminated earth. Storage costs for such material are very high and in the years 2005–2007 amounted, as an average, to about 20–25 €/Mg in Poland, and in Germany, depending on the earth contamination level, from

about 15 €/Mg to 35 €/Mg. For example, storage cost for an earth layer 2 m thick, 2 m wide and 1,000 long will amount to 80,000–100,000 € in Poland, and at German prices, 60,000–140,000 €. One should also mention a disadvantageous effect resulting from displacement and useless storage of large earth quantities and necessity to look for new substitution land leading to further devastation of the natural environment.

The issue related to the disposal of contaminated earth that, in the case of the traditional technology, must be removed from the excavation, is solved in the liquid soil technology by means of adding an immobilising substance to the virgin

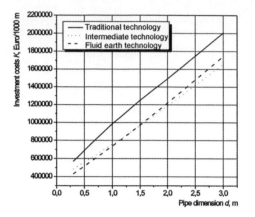

Figure 4. Investment costs of network construction by liquid soil technology, intermediate and traditional technology, in dependence on pipe diameter, cover width of 4 m, with natural soil exchange.

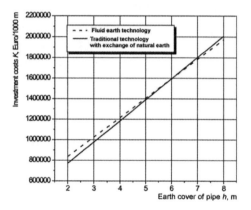

Figure 5. Investment costs of network construction by liquid soil technology and intermediate technology in dependence on the cover the pipe with a diameter of 2.0 m, with exchange of natural soil.

soil and its reuse in the trench. Some studies on the possibility of binding contaminants within the structure of temporarily fluidised earth material carried out to this day led to very promising results (Stolzenburg 2008). To sum up, it should be emphasised that the costs incurred by the investor on soil rehabilitation make the liquid soil technology even more economically viable. If there is a possibility to reuse the earth excavated from the trench for backfill, the balance of costs is slightly different. In such case, according to the analyses, the liquid soil technology applied to the whole excavation is more expensive than the traditional process. The lowest cost is generated in this case by the intermediate technology. It follows from the fact that the costs of fabrication of temporarily fluidised earth

are limited in the view of small amounts involved, and additionally, the technology allows to reduce the trench width. The results of the investment cost analysis for the technologies considered here, under assumption of no need to replace the virgin earth, are presented in Figure 6.

The analysis of the obtained results leads to the conclusion that in the case where the excavated earth must be replaced with new one showing appropriate mechanical and strength parameters, the most advantageous financial effects will be achieved by using the fluidised earth for backfilling the whole trench or the intermediate technology.

On the grounds of analysis of the obtained calculation results, it can be concluded that there are specific cost-effectiveness areas for both technologies depending on pipe diameter and lying depth. This key relationship is presented in Figure 7. For instance, the profitability threshold for full backfilling of the trench with liquid soil is 2.5 m for a pipeline with diameter of 0.3 m. For deeper pipelines, more advantageous solution will consist in limiting the application of temporarily fluidised earth only to the pipeline embedment and backfilling the rest of the trench with loose earth.

An important technical problem generating additional, and sometimes quite significant, investment costs consists in trench dewatering. In the liquid soil technology, the issue is limited only to the pipe lying or suspending stage. Shorter period of operation of dewatering devices results not only in significant reduction of the related cost but also allows to reduce amounts of discharged waters by 60–75%, with respect to the traditional technology.

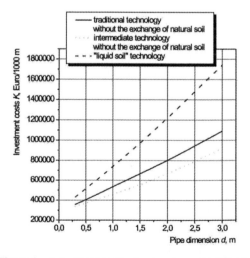

Figure 6. Investment costs of network construction by innovative liquid soil technology, intermediate and traditional technologies, in dependence on pipe diameter for 4 m of cover without the exchange of natural soil.

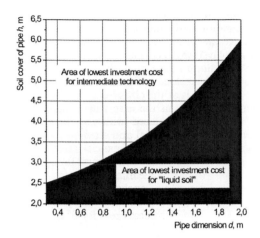

Figure 7. Area of lowest investment cost for liquid soil technology and intermediate technology in dependence of the pipe diameter and depth of its laying with exchange of natural soil.

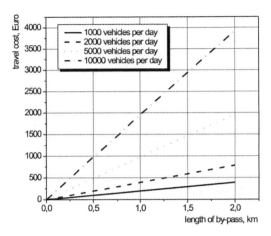

Figure 8. Vehicle operating costs resulting from the necessity of by-passes.

A portion of additional investment costs can be determined in an all-purpose manner. It is the case when costs related to additional mileage of vehicles on the created detours are considered. The costs for a single diversion can be determined from the following formula:

$$C_{tr} = \sum_{i=1}^{n}(L-R)\cdot n_i \cdot k \cdot t \qquad (2)$$

where:
L   diversion length, in km;
R   length of the road closed section, in km;
n   number of i-th category vehicles per day, in 1/days;
k   unit operational cost for i-th category vehicle, in €/km;
t   road closure time, in days.
A   specific feature of the new technology resulting, in particular, from technological properties of temporarily fluidised earth, consists in shorter total pipeline construction time and larger number of shorter work stages (Patent WO 2004/065330). These features of the new technology offer the possibility to shorten road closure periods and reduce diversion lengths. Example results of calculations concerning daily costs related to marking out detours for different numbers of vehicles using normally the closed road are presented in Figure 8.

In calculations, the average traffic structure in lower category roads in Poland has been adopted: motor cars 82.4%; lightweight trucks 8%; trucks without trailers 3.2%; trucks with trailers 3.5%;

buses 1.7%; motorcycles 0.6%; other vehicles 0.6%; and the corresponding average operational costs: motor cars 0.17 €/km; lightweight trucks 0.23 €/km; trucks without trailers 0.32 €/km; trucks with trailers 0.40 €/km; buses 0.45 €/km; motorcycles 0,09 €/km; other vehicles 0.29 €/km. On higher category roads and at higher traffic intensities, the costs would increase significantly.

Apart from vehicle operation expenses connected with the extension of road length, there are also costs relating to the loss of travellers' time depending on detour length and traffic speed limit on the created diversion, as well as, possible cost related to increased accident rate, which should be taken into account. Zwierzchowska (2003) has presented results of an analysis concerning costs caused by the extension of journey for travellers versus daily traffic intensity and diversion length for different speeds. For traffic intensity of 5000 vehicles per day and a detour 1 km long, the additional cost in Polish circumstances amounted to about 2200 €/day and 900 €/day for average drive speed of 20 km/h and 50 km/h, respectively.

Other expenses that should be taken into account in cost estimates relating to the traditional underground network service technology are connected with decreasing road service life and possible compensations paid for damage occurring in buildings as a result of vibrations generated in the course of earth compaction. The issue is important especially in city centres and areas with historic buildings. The technology consisting in spontaneous hardening of fluidised earth in the trench reduces definitely the negative effects of mechanical earth compaction. Achieving high mechanical strength parameters of road subsoil is very important for durability of road surfaces and, as a result, reduces costs relating to road repair and modernisation work (Stolzenburg 2008).

It is also possible to estimate other social costs related to carrying on the investment with the use of different technologies. However, it would be difficult to propose any universal figures in the view of individual nature of local conditions and their dependence on numerous parameters. Nevertheless, such difficulty must not be an excuse for excluding social costs from the analysis of the most advantageous investment variants. In general it can be assumed that because of specific features of the new technology and limited scope of earthwork all social costs will be similar to or lower than those incurred when traditional technology is used.

4 CONCLUSIONS

The above-described analysis of costs related to sewage system construction works performed with the use of traditional pipeline lying methods, the innovative liquid soil technology, and the intermediate technology representing a combination of these, allows to formulate the following utilitarian conclusions:

- one should strive after the maximum possible utilisation of excavated earth for pipeline trench backfilling in order to limit amount of waste as per European regulations on natural environment resources protection, which will have also an advantageous effect on pipeline construction costs;
- when full replacement of earth is necessary, the most cost-effective backfilling method consists in the utilisation of the liquid soil technology or the intermediate technology, depending on pipeline diameter and lying depth. The reduction of investment costs may vary from several percent to several dozen percent compared to traditional technology, while with soil rehabilitation costs taken into account, the financial effect is even more advantageous;
- where earth replacement is not necessary, the most advantageous solution is to apply the liquid soil technology to the pipe zone and backfilling the remaining part of the trench with virgin soil;
- in the case of technologies using temporarily liquified soil material, additional costs and social costs are lower than those incurred when traditional technology is employed;
- the economies achieved as a result of using the liquid soil technology compared to the traditional process depend mainly on pipeline lying depth and diameter. With increasing pipe diameter, opportunities related to reduce trench width by means of using the liquid soil technology increase, resulting in possibility to reduce total construction costs.

The selection of appropriate technology for pipeline construction works should be always performed on the grounds of thorough technical and economical analysis that, apart from the direct investment and any additional expenditure, would also take into account the, so-called, social costs. Some components of these costs are difficult to estimate and usually strongly dependent on specificity of the investment. However, even with the investment costs alone taken into account, it can be concluded that the liquid soil technology represents an interesting alternative to traditional pipeline construction methods.

REFERENCES

Arbeitsblatt Dwa-A 139. 2008. Einbau und Prüfung von Abwasserleitungen und –kanälen. German Association for Water, Wastewater and Waste.
Construction and Testing of Drains and Sewers. Standard EN 1610. 2002. Polish Committee of Standardization. Warsaw, Poland.
Dziopak, J., Słyś, D., Nalaskowski J., Świzdor, S. 2008. Conception of sewage collector reconstructing in Nisko city. Rzeszów, Poland.
Feickert, M.A. 2008. Vielfältige Vorteile In der Praxis. bi UmweltBau, 4: 33–36.
Hygienic Certificate. 2006. National Institute of Hygiene, Warsaw, Poland.
Nalaskowski, J. 2008. Technology of temporary soil liquefaction in sewerage works on the example of "Bobrek" sewage collector construction in Sosnowiec. Ochrona Środowiska BMP, 1: 60–62.
Schluter, M. 2007. Gut gebettet liegt länger [online]. eNewsletter April, IKT-Institut fur Unterirdische Infrastruktur, Gelsenkirchen, Germany. Available from http://www.ikt.de/iktnewsneu.php?doc=704.
Słyś, D., Dziopak, J., Nalaskowski, J. 2009. Liquid Soil—revolution in the underground infrastructure. Gaz, Woda i Technika Sanitarna, 3: 22–25.
Stolzenburg, O. 2004. Building material and method for production thereof. 2004. Patent WO 2004/065330.
Stolzenburg, O. 2008. Die Vorteile des Einsatzes von Flüssigboden für das Rohr-Boden-System. Proc. Rohrleitungen—Unternehmen im Umbruch, 22. Oldenburger Rohrleitungsforum 2008, Vulcan Verlag. Germany: 262–279.
Triantafyllidis T., Bosseler B., Arsic I., Liebscher M. 2006. Forschungsbericht, Einsatz von Bettungs -und Verfüllmaterialien im Rohrleitungsbau, Laboruntersuchungen und Versuche im Maßstab 1:1, Kurzfassung, IKT-Institut fur Unterirdische Infrastruktur, Gelsenkirchen, Germany. Available from http://www.ikt.de.
Wegweiser, S., 2008. Flüssigboden nach allen Regeln der Kunst. Steinbruch und Sandgrube, 10: 34–35.
Werksnorm WN 04.02. Für Herstellung und Lieferung von RSS®—Flüssigboden, Leipzig, Germany, 2002.
Zwierzchnowska A. 2003. Optimization of selection of non-excavation methods of underground pipelines construction. Kielce University of Technology, Kielce, Poland.

Environmental Engineering IV – Pawłowski, Dudzińska & Pawłowski (eds)
© 2013 Taylor & Francis Group, London, ISBN 978-0-415-64338-2

Numerical modeling of water flow through straight globe valve

J. Janowska, M.K. Widomski, M. Iwanek & A. Musz
Environmental Engineering, Lublin University of Technology, Lublin, Poland

ABSTRACT: This paper presents the modeling studies concerning the water flow through straight globe valve of the nominal diameter $d_n = 20$ mm. Numerical calculations were performed by the commercial CFD (Computational Fluid Dynamics) software Fluent, Ansys Inc. for three various degrees of the valve bore opening and for three different values of water flow velocity. The distributions of water flow velocity magnitude and turbulence intensity were presented in the paper for each studied variant. Our numerical calculations showed that there is a dependence, not only between the coefficient of local pressure loss and degree of valve opening, but also to the Reynolds number. The obtained results of local pressure loss coefficients calculations were compared to those suggested by standards, fittings manufactures and literature reports.

Keywords: local pressure losses, globe valve, computational fluid dynamics, flow modeling

1 INTRODUCTION

A globe valve is a type of fixture for opening, closing, diverting or regulating of flow in a pipeline (Whitehouse 1993). The composition of a typical globe valve is presented in Figure 1. It consists of a body containing a movable closure (plug or disc) connected to a stem operated by screw action, and a stationary ring seat. A bonnet is connected to the upper part of the body.

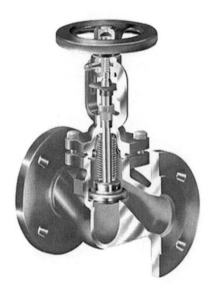

Figure 1. Typical globe valve (http://www.directindustry.com/industrial-manufacturer/steel-ball-75741-_4.html).

The design of globe valves restricts flow, so this kind of valves should not be used where full, unobstructed flow is required (Frankel 2002).

Globe valves are generally made of metal or metallic alloys, as well as some synthetic materials. A main element of a globe valve—plug—is available as grey iron, ductile iron, steel and more rarely—cast steel, brass or bronze construction. Rubber, graphite or PTFE is usually used as a material for packing. The choice of valve material depends on application of the valve, its dimension and properties of flowing media. There are three basic types of globe valve (Frankel 2002): a standard globe, an angle valve and a needle valve. Losses in valves can form a significant portion of the total energy drop in the relatively short pipes. These losses become much more relevant in laminar flow (Slatter 2006).

The head loss across the valve can be expressed in terms of the so-called resistance or pressure loss coefficient ζ and the velocity energy head, according to the following formula (Slatter 2006, Munson et al. 2009, EN 1267:2012):

$$\Delta H = \zeta \cdot \frac{v^2}{2 \cdot g} \qquad (1)$$

where ΔH is the head loss in the valve (m), ζ is the valve loss coefficient (dimensionless), v is the average velocity of the fluid flowing in the pipe (m/s) and g is the gravitational acceleration (m/s^2).

Many researchers report that pressure loss coefficient increases highly when Reynolds number decreases in laminar flow of Newtonian fluids.

This dependence does not occur so clearly for turbulent conditions (Kittredge and Rowley 1957; Edwards et al. 1995; Kimura et al. 1995; Gan 1997; Martínez-Padilla et al. 2001; Fester and Slatter 2009).

Another factor influencing the pressure loss coefficient ζ is the construction of a valve, especially a seat in a valve body (Nowakowski 2005). In accordance with this dependence, ζ can be determined as (Nowakowski 2005):

$$\zeta = \zeta_s \cdot \frac{ID^4}{SD^4} \qquad (2)$$

where ζ_s is the pressure loss coefficient in relation to a valve seat (dimensionless), ID is the internal diameter of a pipe in mm and SD is the diameter of a valve seat (mm). The loss coefficient ζ_s depends on SD according to the formula (Nowakowski 2005):

$$\zeta_s = C \cdot SD^n \qquad (3)$$

where C and n are empirical parameters.

Values of a pressure loss coefficient for valves reported in the PN-M-34034:1976 standard, technical literature and offered in valves producers' lists are different and might not be equal to the real values. The discrepancies are the result of using new materials and technologies in the processes of valves production. Moreover, it is difficult to take into account every factor influencing the loss coefficient in investigations. Values of ζ presented in the PN-M-34034:1976 standard, in several reports and numerous engineering handbooks depend on a nominal diameter of a valve only. In contrast, numerous science reports and engineering practice suggest dependence of pressure loss in valves not only on resistance dimensions, but also on degree of valve opening (Miller 1990, Perry 1997, Boele 2004, Jeon et al. 2010) and a Reynolds number (Turian 1998, Fester & Slatter 2009, Crane Company 2009, Munson et al. 2009). Thus, necessary simplifications make the obtained results more or less incorrect. The examples of published ζ coefficient for globe valve—20 mm in diameter, are presented in Table 1.

Computational Fluid Dynamics (CFD), as a method of calculations involving application of computers into fluid dynamics (Wesseling 2000, Dufresne et al. 2009) enables solving and analyzing problems of fluid flows, especially to develop spatial or temporal solutions of fluid pressure, temperature and velocity (Norton 2007). The results of simulation can be useful in designing, optimization and operating within the chemical, aerospace, hydrodynamic, food, agricultural,

Table 1. Pressure loss coefficient (ζ) for globe valve 20 mm in diameter reported in literature.

Source of information	Coefficient of local pressure loss ζ
PN-M-34034:1976	4.7
Kołodziejczyk and Płuciennik, 1995; Górecki et al. 2006; Western Dynamics, Inc. (2011); Turian et al. 1998 (for turbulent flow)	8.5
Nowakowski 2005 (depending on material and application of valve)	10.9–18.9
Nowakowski 2005 (calculated on the basis of PN-EN 1213: 2002)	17.1
Marketing materials of producers (Ari-Armaturen, Gestra)	4.5
Marketing materials of producer (2005) (Aquatherm)	9.5
Larock et al. 2000	6.4
U.S. Department of energy 2007 (for fully open valve)	3–8
Munson et al. 2009 (for fully open valve)	10

HVAC (heat, ventilation and cooling) and other industries (Xia & Sun 2002; Farmer et al. 2005; Asteriadou et al. 2006; Kondo et al. 2006; Norton et al. 2007, Shi et al., 2010, Kaushal et al. 2012). The governing equations of CFD cover mathematical formulations of the conservation laws of fluid mechanics based on the Naviere-Stokes equations. Taking into account a fluid continuum, these laws can be considered as the law of conservation of mass, the law of conservation of momentum and the law of conservation of energy (Wesseling 2000, Norton et al. 2006). CFD simulations of water flow in pipes systems can be performed with the use of the commercial software Fluent, Ansys Inc. (Ansys 2009) based on Finite Elements Method (Zienkiewicz et al. 2005, Madenci & Guven 2006, Eguchi et al. 2011). Considering water flow through a globe valve, ANSYS Fluent enables identification and analysis of changes in flow parameters, i.e. flow velocity magnitude and turbulence intensity distribution, as well as absolute and gauge static pressure required to determination of pressure losses for various Reynolds number and valve bore opening. Use of the program is a cheaper and less demanding alternative to laboratory experiments (Coppel 2008, Blel et al. 2009).

Therefore, the aim of the article is to determine the pressure losses coefficient for a globe valve (20 mm in diameter) for three different degrees of valve opening and three different values of water flow velocity on the basis of the CFD simulation.

The obtained results will be compared to literature data.

2 MATERIALS AND METHODS

The three dimensional modeling of water flow through the strait globe value of nominal diameter equal to 20 mm was performed by commercial CFD software Fluent, Ansys Inc. The modeled domain reflected water body inside the tested valve fitted to the straight pipe of 100 diameters length. Our model reflected three possible degrees of valve opening—100%, 66% and 33%. Figure 2 presents the exemplary solid reflecting the water body for 66% opened valve, while Figure 3 shows the developed mesh of finite elements for the same valve. All the developed modeling domains consisted of approx. 900000 elements and 190000 nodes.

Our numerical calculations were performed for the following assumptions: flow rate and properties of fluid were constant, fluid was incompressible, flow was turbulent (k-epsilon model of turbulent flow (Launder & Spalding 1974, Comini & Del Giudice 1985, Liu et al. 2012)), roughness of valve material was omitted while absolute roughness for pipe material was assumed as $k = 0.7$ mm. The following properties of water were assumed as input data to modeling: temperature $t = 10°C$, density $\rho = 999.69$ kg/m^3 and dynamic viscosity $\mu = 0.001308$ kg/(m·s). For each assumed degree of valve opening, the three variants of calculations were performed for three values of water flow velocity 0.5, 1.0 and 2.0 m/s (Reynolds number 7642.89, 15285.78 and 30571.59, respectively).

The accepted boundary conditions covered flow characteristics for inlet and outlet zone to the modeled domain as well as roughness of inner wall of the pipe. The inlet zone was characterized by mass flow rate related to the flow velocity, hydraulic diameter (20 mm) and turbulence intensity calculated by the following equation (Ansys 2009):

$$TI = 0.16 \cdot \mathrm{Re}^{-\frac{1}{8}} \qquad (4)$$

The calculated turbulence intensity for inlet zone of our modeling domain was in range

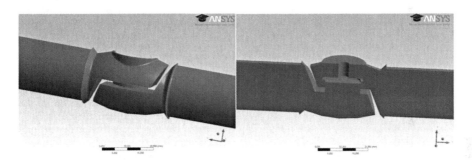

Figure 2.   Geometrical model of globe valve—33% of valve opening.

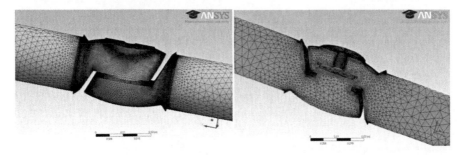

Figure 3.   Finite elements mesh for modeled globe valve—33% of valve opening.

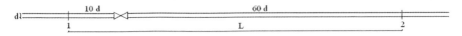

Figure 4.   Location of static pressure reading surfaces in the modeled domain.

4.8–5.2%, which values are typical for a fully developed turbulent flow in pipe under pressure (Zamora et al. 2008; Minkowycz et al. 2009). The outlet zone boundary condition was assumed as outflow coefficient equal of 1.0, which means that the same flow rate enters and leaves the domain. The initial conditions were set according to the boundary conditions for the inlet zone. Our steady state calculation were based on iteration procedure covering maximum 100 steps. Results of our modeling calculations covered distribution of magnitude of flow velocity, static, dynamic and absolute pressure as well as turbulence intensity.

Values of local pressure loss coefficients, characterizing minor losses generated by the modeled globe value for flows with specified flow velocity (Reynolds number), were based on static pressure readings conducted for surfaces presented in Figure 4. The reading surfaces were located inside the modeled domain according to literature suggestions (Cisowska & Kotowski 2004; Siwiec et al. 2002) and in accordance with the latest EN 1267:2012 standard considering tests of flow resistance for the industrial valves when water is a test fluid.

Values of local pressure loss coefficients, characterizing minor losses generated by the modeled globe value for flows with specified flow velocity (Reynolds number), were based on static pressure readings conducted for surfaces presented in Figure 4. The reading surfaces were located inside the modeled domain according to literature suggestions (Cisowska & Kotowski 2004, Siwiec et al. 2002) and in accordance with the latest EN 1267:2012 standard considering tests of flow resistance for the industrial valves when water is a test fluid.

Calculations of local/minor pressure loss coefficients were based on standard Bernoulli's equation for real, incompressible, viscous fluid.

$$z_1 + \frac{p_1}{\rho g} + \frac{v_1^2}{2g} = z_2 + \frac{p_2}{\rho g} + \frac{v_2^2}{2g} + \Delta h \qquad (5)$$

where: z_1, z_2 are the elevations of the point above a reference plane in m; p_1, p_2 are the pressure at the chosen point in Pa; v_1, v_2 are the fluid flow velocities in m/s and ρ is the density of the fluid in kg/m³ and Δh is total pressure head loss in m.

The above equation for horizontal orientation and constant diameter of pipe ($z_1 = z_2$ and $v_1 = v_2$) may be transformed to:

$$\frac{p_1}{\rho g} = \frac{p_2}{\rho g} + f \frac{l}{d} \frac{v^2}{2g} + \zeta \frac{v^2}{2g} \qquad (6)$$

which allows for the calculation of local pressure loss coefficient for tested globe valve.

$$\zeta = \frac{2(p_1 - p_2)}{\rho v^2} - f \frac{l}{d} \qquad (7)$$

Friction factor coefficient f was calculated according to Collebrook-White's formula:

$$\frac{1}{\sqrt{f}} = -2 lg \left(\frac{2,5}{\mathrm{Re}\sqrt{f}} + \frac{k}{3,7d} \right) \qquad (8)$$

where f is the dimensionless Darcy friction factor and k is absolute roughness in m, while d is pipe diameter in m.

3 RESULTS AND DISCUSSION

Results of our modeling calculations covered the distribution of flow velocity magnitude, static, dynamic and absolute pressure as well as turbulence intensity. Calculated distribution of flow velocity and turbulence intensity allowed to analyze factors influencing local pressure loss during flow of incompressible viscous fluid through the modeled globe valve while static pressure distribution allowed for determination of local/minor pressure loss coefficients.

The influence of valve geometry on streamline distributions for variable bore opening stages is clearly visible in Figures 5–7.

The sudden change of flow direction as well as occurrence of stagnation, reverse flow and contraction zones are visible for the fully opened globe valve (Fig. 5).

In case of the globe valve opened in 2/3 of its full bore opening the additional influence of the flat closure element, as an obstacle located on streams

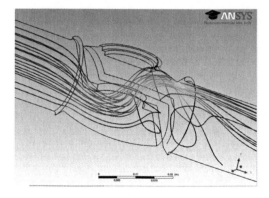

Figure 5. Streamlines for water flowing through fully opened globe valve.

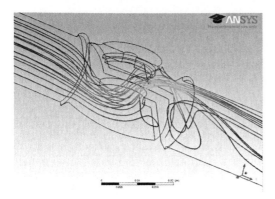

Figure 6. Streamlines for water flowing through 2/3 opened globe valve.

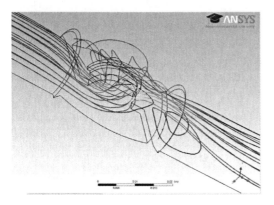

Figure 7. Streamlines for water flowing through 1/3 opened globe valve.

path, in causing flow redirection and stagnation zone (Fig. 6). The further closing of the valve, i.e. lowering the disc by rotating the spindle, decreases the flow area between the closure disk and valve body seats, so the flow streamlines distribution varies significantly than in case of fully opened globe valve. The streams surrounding closure disk and the spindle change their directions and cause the additional pressure loss (Fig. 7).

The influence of globe valve elements at various stage of closure disc spatial location on velocity and turbulent intensity distribution may be analyzed in Figures 8–10, all for the same inflow velocity equal to 1 m/s. The velocity magnitude and turbulent intensity distributions for the fully opened valve and 1 m/s inflow mean velocity are spatially developed according to streamlines layout presented in Figure 8. The highest turbulent intensity on the longitudinal section of the model, up to approx. 61%, may be observed on and after outflow area of the modeled value.

Turbulences visible on the cross section of the valve are mainly caused by valve body seats and closure plug in the upper position. The authors wish to underline that turbulence intensity for fully developed turbulent flow in a pipe without any obstacles causing local pressure/energy loss should not exceed 5% and turbulence intensity in the range of 5–20% is treated as very high (Zamora et al. 2008; Minkowycz et al. 2009). In all presented cases turbulent intensity in the straight section of pipe before the modeled globe valve was below 5%, so the inflowing streams should be treated as stable. Turbulent intensity distribution in longitudinal section of the modeled valve opened in 2/3 (Fig. 9), reaching the value of approx. 73%, have the similar distribution than in case of the fully opened valve but the appearance of turbulences caused by lowered closure disc is obvious. This situation is even better visible in case of distribution of turbulences on the cross section of the modeled valve.

The clear influence of closing the closure disk to the seats of valve body on flow turbulences is visible in Figure 10 for globe valve opened to 33%. The flow velocity magnitude and turbulent intensity (reaching the level of approx. 120%) visible on the longitudinal section of the model clearly repeats the pattern of streamlines presented in Figure 7. The analysis of the cross section distribution of turbulent intensity supports this observation—the highest rate of turbulences occurs in the narrow space between the valve body, closure disk and body seats.

Summarized results of the numerical calculations are presented in Table 2 where the inflow velocity, Reynolds number, maximum turbulence intensity and calculated coefficient of local pressure loss were compared for each studied variant.

It is clearly visible that construction features of the globe valve influence the velocity magnitude value and distribution as well as turbulence intensity. Generally, the higher inflow velocity for the same valve opening, the higher maximum observed increase of flow velocity and turbulence intensity. The increase of local pressure loss coefficients for studied openings of modeled globe valve was observed as it was expected but the results of our calculations showed that the obtained values of local/minor losses coefficient for the same bore opening shows dependence to Reynolds number. The calculated increase of pressure loss coefficient according to reduction of flow area dependent on stage of valve opening (from 100%, through 2/3 and 1/3) was equal from 8.5–9.7% for 66% of bore opening and 60.1–65.4% for 33% opening, if compared to fully opened globe valve (100%) opening.

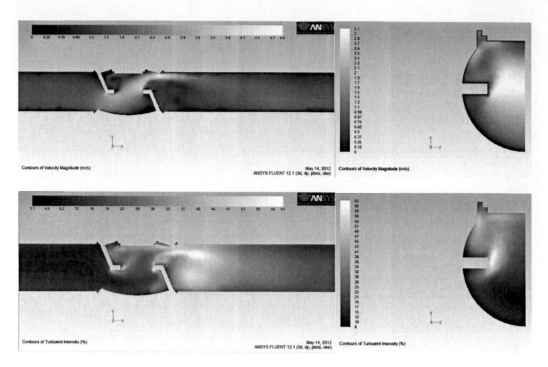

Figure 8. Contours of velocity magnitude and turbulent distribution for longitudinal and cross section of fully opened globe valve.

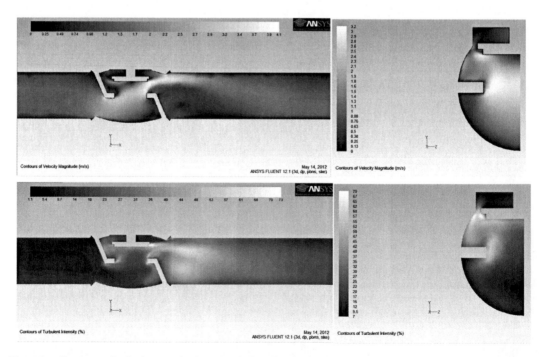

Figure 9. Contours of velocity magnitude and turbulent distribution for longitudinal and cross section of globe valve with closure disk opened in 2/3.

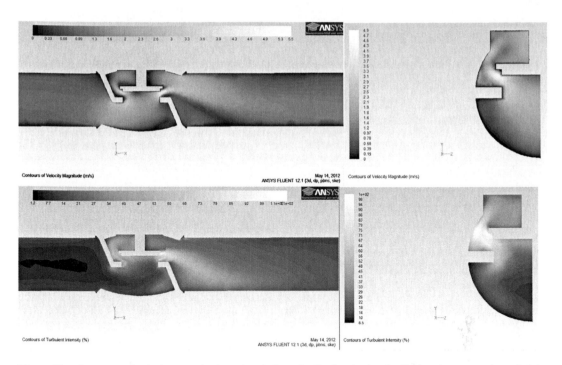

Figure 10. Contours of velocity magnitude and turbulent distribution for longitudinal and cross section of globe valve with closure disk opened in 1/3.

Table 2. Summarized results of numerical calculations.

Valve bore opening	Inflow velocity (m/s)	Reynolds number (–)	Maximum calculated local turbulence intensity (%)	Calculated coefficient of local pressure loss ζ (–)
100%	0.5	7642.89	41.20	15.97
	1.0	15285.78	72.84	16.36
	2.0	30571.59	135.85	16.68
66%	0.5	7642.89	41.17	17.52
	1.0	15285.78	76.72	17.88
	2.0	30571.59	144.73	18.11
33%	0.5	7642.89	61.61	25.56
	1.0	15285.78	115.66	26.81
	2.0	30571.59	218.90	27.59

The influence of increase of Reynolds number for value of local pressure loss coefficient for the constant bore opening is also visible. The maximum applied increase of Reynolds number equals 300% resulted in increase of calculated loss coefficient by 3.5–7.9% for tested variants.

The presented calculated values of local pressure loss coefficients for globe valve are higher than values available in common, including Polish, standards and handbooks. As it was mentioned before, according to Polish Standard PN-M-34034:1976 value of minor pressure loss for straight globe value $\zeta = 8,0$, while other sources e.g. fittings producers or hydraulics textbooks present lower or higher values 3.0–9.5 (marketing materials of ARI-Armaturen, Gestra, Aquatherm; Jeżowiecka-Kabsch & Szewczyk 2001). On the other hand, the results of our calculations are comparable to the reports presenting values of 12.0–17.0 for pressure loss coefficients of fully opened steel or brass straight globe valves (Nowakowski 2005).

4 CONCLUSION

The model tests of water flow through straight globe valve were performed using the software Fluent, Ansys Inc. Investigations allowed to determine the minor pressure losses coefficients for globe valve (20 mm in diameter) for three different opening degrees and three different velocities of water flow.

It was observed that a change of the opening degree of the valve caused changes of streamlines paths. In turn, these changes were the reason of changing a momentum of individual streams creating different pressure losses in the valve for different opening degree. However, changes of the inflow velocity for the same valve opening caused changes in the turbulent intensity, also creating different pressure losses for different velocities. Thus, the obtained results confirmed that local pressure losses appearing during flow of water through a globe valve depends not only on the valve dimension (like the former Polish Standard PN-M-34034:1976 publicized), but also on its opening degree and the Reynolds number, connected with water velocity. Some Polish and foreign authors present in their publications similar conclusions, so it can be claimed that the nowadays standards require further update.

Moreover, the computer simulations enabled to determine the distribution of velocity magnitude and turbulent intensity of water in longitudinal and cross section of a globe valve, as well as location of reverse flow appearance and local twirls of streamlines. This kind of information can be used by the valves producers to design geometry of valves component parts to receive the best possible hydraulic conditions of water flow through a globe valve.

REFERENCES

Ansys Fluent UDF Manual. Ansys Inc. 2009.
Aquatherm-Polska. 2005. System instalacyjny PP-R. Poradnik. Warszawa.
Ari-Armaturen. FABA—Katalog produktów (http://www.klimatech.net.pl/?n=prod&prod_id=1&id=1).
Asteriadou, K., Hasting, A.P.M., Bird, M.R. and Melrose, J. 2006. Computational fluid dynamics for the prediction of temperature profiles and hygienic design in the food industry. *Food and Bioproducts Processing*, 84(C2): 157–163.
Blel, W., Pierrat, D., Le Gentil, C., Legentilhomme, P., Legrand, J., Hermon, C., Faille, C., and Bénézech, T. 2009. Numerical and experimental investigations of the flow structures through a gradual expansion pipe, *Trends in Food Science and Technology* 20:70–76.
Boele A.M. 2004. Pressure Drop test H/V Pressure Vacuum. Relief Valve Test NEW-ISO-HV-80. Prepared for Pres-Vac Engineering A/S Denmark.

Cisowska I., Kotowski A. 2004. Straty ciśnienia w układach kształtek z polipropylenu. *Gaz, Woda i Technika Sanitarna* 10: 340–345.
Comini, G., Del Giudice, S. 1985. A (k-epsilon) model of turbulent flow. *Numerical Heat Transfer* 8(2): 133–147.
Coppel, A., Gardner, T., Caplan, N. et al. 2008. Numerical Modelling of the Flow Around Rowing Oar Blades (P71). In: M. Estivalet and P. Brisson (eds.), *The Engineering of Sport* 7: 353–361. Paris, Springer.
Crane Company 2009. Flow of Fluids Through Valves, Fittings and Pipe. Technical Paper No. 410, Crane Co.
De-Pan Shi D.P., Luo Z.H., Zheng Z.W. 2010. Numerical simulation of luquid—solid two—phase flow in tubular loop polymeryzation reactor, *Powder Technology* 198:135–143.
Dufresne, M., Vazquez, J., Terfous, A., Ghenaim, A., Poulet, J. 2009. Experimental investigation and CFD modelling of flow, sedimentation, and solids separation in a combined sewer detention tank, *Computers & Fluids* 38: 1042–1049.
Edwards, M.F., Jadallah, M.S.M. and Smith, R. 1985. Head losses in pipe fittings at low Reynolds numbers. *Chem Eng Res Des*, 63(1): 43–50.
Eguchi, Y., Murakami, T., Tanaka, M., Yamano, H. 2011. A finite element LES for high-Re flow in a short-elbow pipe with undisturbed inlet velocity, *Nuclear Engineering and Design* 241: 4368–4378.
EN 1267:2012. Industrial valves—Tests of flow resistance using water as test fluid.
Farmer, R., Pike, R., Cheng, G. 2005. CFD analyses of complex flows. *Computers and Chemical Engineering* 29: 2386–2403.
Fester, V.G. and Slatter, P.T. 2009. Dynamic similarity for Non-Newtonian fluids in globe valves, Trans IChemE, Part A, *Chemical Engineering Research and Design* 87: 291–297.
Frankel, M. 2002. *Facility Piping Systems* Handbook. McGraw-Hill.
Gan, G., Riffat, S.B. 1997. Pressure loss characteristics of orifice and perforated plates *Experimental Thermal and Fluid Science* 14 (2): 160–165.
Gestra, Karty katalogowe zaworów grzybkowych (http://www.saga.info.pl/Katalog/Armatura-przemyslowa/).
Górecki, A., Fedorczyk, Z., Płachta, J., Płuciennik, M., Rutkiewicz, A., Stefański, W., Zimmer, J. 2006. Instalacje wodociągowe, ogrzewcze i gazowe na paliwo gazowe, wykonane z rur miedzianych. *Wytyczne stosowania i projektowania*. Biblioteka Polskiego Centrum Promocji Miedzi. Warszawa.
Jeon, S.Y., Yoon, J.Y. and Shin, M.S. 2010. Flow characteristics and performance evaluation of butterfly valves using numerical analysis. *IOP Conf. Ser.: Earth Environ. Sci.* 12(1).
Jeżowiecka-Kabsch, K., Szewczyk, H. 2001. *Mechanika płynów*. Oficyna Wydawnicza Politechniki Wrocławskiej, Wrocław.
Jones, W.P, Launder, B.E. 1972. The prediction of laminarization with a two-equation model of turbulence. *International Journal of Heat and Mass Transfer* 15(2): 301–314.
Kaushal, D.R., Thinglas, T., Tomita, Y., Kuchii, S., Tsukamoto, H. 2012. CFD modeling for pipeline flow of fine particles at high concentration, *International Journal of Multiphase Flow* 43: 85–100.

Kimura T., Tanaka T., Fujimoto K. and Ogawa K. 1995. Hydrodynamic characteristics of a butterfly valve-Prediction of pressure loss characteristic, *ISA Transactions* 34: 319–326.

Kittredge, C.P. and Rowley, D.S. 1957. Resistance coefficients for laminar and turbulent flow through one-half-inch valves and fittings. *Trans. Am. Soc. Mech. E.* 79: 1759–1766.

Kołodziejczyk, W., Płuciennik, M. 1995. Wytyczne projektowania instalacji centralnego ogrzewania. COBRTI INSTAL.

Kondo H., Asahi K., Tomizuka T., Suzuki M. 2006. Numerical analysis of diffusion around a suspended expressway by multi-scale CFD model. *Atmospheric Environment* 40: 2852–2859.

Larock, B.E., Jeppson, R.W. and Watters, G.Z. 2000. *Hydraulics of Pipeline Systems*. CRC Press.

Launder, B.E., Spalding, D.B. 1974. The numerical computation of turbulent flows. *Computer Methods in Applied Mechanics and Engineering* 3(2): 269–289.

Liu, C.C., Ferng, Y.M., Shih, C.K., 2012. CFD evaluation of turbulence models for flow simulation of the fuel rod bundle with a spacer assembly, *Applied Thermal Engineering* 40: 389–396.

Madenci, E. and Guven, I. 2006. The Finite Elements Method and Applications in Engineering using ANSYS, Springer.

Martínez-Padilla, L.P. and Linares-García, J.A. 2001. Resistance Coefficients of Fittings for Power-law Fluids in Laminar Flow. *Journal of Food Process Engineering,* 24: 135–144.

Miller, D.S. 1990. Internal Flow Systems, 2nd edition, BHRA (Information Services) (ed.). *The Fluid Engineering Centre*, Cranfield, Bedford, UK.

Minkowycz, W.J., Abraham, J.P., Sparrow, E.M. 2009. Numerical simulation of laminar breakdown and subsequent intermittent and turbulent flow in parallel-plate channels: Effects of inlet velocity profile and turbulence intensity. *International Journal of Heat and Mass Transfer* 52: 4040–4046.

Munson, B.R., Young, D.F., Okiishi, T.H. and Huebsch, W.W. 2009. *Fundamentals of Fluid Mechanics*. John Wiley & Sons, Inc., Sixth Edition.

Norton, T. and Sun, D.W. 2006. Computational fluid dynamics (CFD)—an effective and efficient design and analysis tool for the food industry: A review. *Trends in Food Science & Technology* 17: 600–620.

Norton T., Sun D.-W., Grant J., Fallon R., Dodd V. 2007. Applications of computational fluid dynamics (CFD) in the modelling and design of ventilation systems in the agricultural industry: A review. *Bioresource Technology* 98: 2386–2414.

Nowakowski, E. 2005. Straty ciśnienia na zaworach prostych stosowanych w instalacjach sanitarnych. *Gaz, Woda i Technika Sanitarna* 4: 20–23.

Perry, R.H. 1997. *Perry's Chemical Engineering Handbook*, 7th edition McGraw-Hill Companies, NY, USA.

PN-EN 1213:2002 Armatura w budynkach. Zawory zaporowe ze stopów miedzi do instalacji wodociągowych w budynkach. Badania i wymagania.

PN-M-34034:1976 Rurociągi. Zasady obliczeń strat ciśnienia.

Siwiec, T., Morawski, D., Karaban, G. 2002. Eksperymentalne badania oporów hydraulicznych w zgrzewanych kształtkach z tworzyw sztucznych. *Gaz, Woda i Technika Sanitarna* 2: 49–50.

Slatter, P.T. 2006. Plant design for slurry handling. Journal of the Southern African Institute of Mining and Metallurgy, 106: 687–691.

Turian, R.M., Ma, T.W., Hsu, F.L.G., Sung, M.D.J. and Plackmann, G.W. 1998. Flow of concentrated slurries. 2. Friction losses in bends, fittings, valves and venturi meters. *Int J Multiphase Flow*, 24(2): 243–269.

U.S. Department of Energy. Energy Efficiency and Renewable Energy. Energy Tips—Pumping Systems. Industrial Technologies Program. Pumping Systems Tip Sheet #10. March 2007 (http://www1.eere.energy.gov/manufacturing/tech_deployment/pdfs/control_valves_pumping_ts10.pdf).

Wesseling, P. 2000. Principles of Computational Fluid Mechanics. Springer *Series in Computational Mathematics* 29.

Western Dynamics, Inc. Hydraulic System Design & Consulting 2011. (http://www.westerndynamics.com/Download/friclossfittings.pdf).

Whitehouse, R.C. 1993. The Valve and Actuator User's Manual (British Valve Manufacturers' Association). Mechanical Engineering Publications, London.

Xia, B. and Sun, D.W. 2002. Applications of computational fluid dynamics (CFD) in the food industry: a review. *Computers and Electronics in Agriculture* 34 (2002): 5–24.

Zamora, B., Kaiser, A.S., Viedma, A. 2008. On the effects of Rayleigh number and inlet turbulance intensity upon the buoyancy-induced mass flow rate in sloping and convergent channels. *International Journal of Heat and Mass Transfer* 51: 4985–5000.

Zienkiewicz, O.C., Taylor, R.L., Nithiarasu, P. 2005. *Finite Elements Method for Fluid Dynamics,* Butterwarth Heinemann, 6th Edition, Oxford & Burlington.

Environmental Engineering IV – Pawłowski, Dudzińska & Pawłowski (eds)
© 2013 Taylor & Francis Group, London, ISBN 978-0-415-64338-2

An evaluation of potential losses associated with the loss of vacuum sewerage system reliability

J. Królikowska & B. Dębowska
Faculty of Environmental Engineering, Technical University of Cracow, Cracow, Poland

A. Królikowski
Institute of Environmental Engineering, Podhale State Higher Vocational School, Nowy Targ, Poland

ABSTRACT: This work is devoted to the problems of operating a vacuum sewer system. The presented results are based on the authors' own research and data from scientific sources. The analysis included in this work relates to the failure of vacuum sewer system components and the operating costs of such systems, which are governed primarily by the electricity consumption.

Keywords: vacuum sewer system, failure, reliability, power consumption

1   INTRODUCTION

Vacuum Sewer Systems (VSS) are used wherever the construction of conventional gravity systems cannot be applied for technical and economic reasons. These systems, classified as unconventional, are the most frequently used where there is a flat terrain, a high groundwater table or dispersed residential developments. As the system is impermeable, its usage is recommended in areas protecting inland waters and groundwater, intakes, water sources and reservoirs, as well as in organic food production areas. VSS provides an alternative solution to the sewage disposal problem in relation to gravity and pressure systems but can also simultaneously cooperate with them. The flow of sewage is forced by the creation of a vacuum in the sewers, and its mixing with the air during flow, which has a dynamic, unstable, pulsating character. Important features of this system should be described as reliable and safe operation. Ongoing studies aim to identify the major operational problems and to identify the most unreliable components of the sewer system under discussion, and hence, to increase the future reliability of the VSS.

Reliability of sewage systems remains a very current topic, not only due to the economic reasons but most of all due to its ecological consequences. However, Polish and foreign research investigations focusing on reliability and safety of sewer systems are rather modest, whether measured with a number of publications or the area of interest, if compared with studies on water supply systems. The Russian scientists can be singled out as the leading group here (Ermolin 1983, Kuranov et al. 2009, Oren M 2001).

This article is based on the operational studies performed on the VSS operating in Stanisławice, in the administrative district of Bochnia in southern Poland, located in the protected zone of the Niepołomice Forest.

2   MATERIAL AND METHODS

2.1   *Characteristics of the vacuum sewer system under investigation*

Stanisławice is situated along the Bochnia—Niepołomice—Kraków road on a flat terrain covering about 1770 ha. In 2007, there were 1332 inhabitants (Dębowska 2012).

There is a partial sewerage system in Stanisławice which includes the drainage and treatment of domestic sewage from approximately 85% of the settlement area. It operates as a gravity-vacuum-pressure system and is divided into 20 catchment areas, from which the domestic sewage is drained by 20 independent gravity networks into 13 collection chambers. Then it is drained, through a network of vacuum and pressure sewers to the Vacuum Pump Station (VPS) adjacent to a BIOBLOK PS-100 sewage treatment plant. There are 298 households connected to the sewer system. Sewage from the remaining households is collected withcesspools, and then transported by tanker fleet to a Sewage Treatment Works (STW) in adjacent Damienice, or treated in domestic sewage treatment plants. Surface runoff is taken up by a system of open ditches. As already mentioned, the households' sewage flows by means of gravity into the collection chambers, equipped with a Vacuum Interface Valve (VIV) and a pipe sensor, and then into the VSS. Sewage from several to several dozens of households, but on average from about 30 households, is drained into one collection chamber. When the sewage in the collection chamber reaches a predetermined level the VIV is opened and the sewage is sucked into the vacuum main thanks to the partial vacuum produced by the VPS. Sewage from the system is then pumped to the STW, whilst the treated sewage is drained into the Raba river. The Stanisławice VSS network, 8.6 km in length, is constructed from PVC pressure pipes with diameters ranging from 110 mm to 160 mm, buried at a depth of 1.5 to 2.0 metres, effectively parallel to the ground surface. The network components are the inspection chambers and the connection chambers (18), which permit the proper functioning and operation of the system. Inspection chambers are separated by a distance of not more than 75 metres. The VPS at the STW is a freestanding building. It has an above-ground part (monitoring and control unit, staff room, workshop and storage) and an underground part (collection chamber, a 10.5 m³ capacity collection tank, two submersible sewage units, two vacuum pumps with vacuum systems and a control panel). The VPS is equipped with automatically controlled valve fittings. The vacuum units' operational parameters can be adjusted for various conditions. When a pumping unit fails the faulty pump is automatically switched off. Under normal operating conditions the pumps operate alternatively. The electricity supply to the VPS is dual sourced from a 220/380 V network. The installed capacity is about 9 kW. A backup source of electricity is a 13 kVA generator, in case of main supply failure to the vacuum pumps and for maintaining proper system operation at the required level of reliability.

3 RESULTS AND DISCUSSION

3.1 *VSS operational problems in Stanisławice— system failure*

The VSS turned out to be a critical system in the Stanisławice sewer network. The operational data for the 2007–2010 period was the basis for the failure assessment of the VSS which concerned the various network components i.e. VSS mains, collection chambers with a VIV and VPS. Table 1 summarises the most common types of system failures during this period. The collection chambers with their equipment and damage to vacuum mains play a dominant part. The number of failures (several dozen) over the years remained fairly constant. The graph in Figure 1 illustrates

Table 1, as a percentage, for each type of failure. The average downtime due to failures ranged from 1 to 12 hours; the longest for the collection chambers and VIVs. From Table 1 and Figure 1 below (its graphical representation) it also follows that the greatest operational problem, affecting the time between failures is the smooth operation of the collection chamber containing the VIV and the pipe sensor. They are characterized not only by the highest number of failures (71%), but cause a periodic loss of VSS reliability, and depending on the location, also of the whole system.

Particularly relevant here is the non-closure of the VIVs and the damage to the link connecting the pipe sensor to the valve (corrosion). It results in a loss of vacuum in the VSS network. Flaps protecting the inlet to the VIV can also be damaged, which if made of rubber can become deformed over the years. Figure 2 shows such a deformed component.

A common occurring malfunction within the collection chamber is "sump blockage". This problem arises in a situation where there is no automatic activation by the pipe sensor. One reason for this may be too high a vacuum, preventing the VIV from opening due to buoyancy (for collection chambers close to the VPS). Another cause for the blockage of the VIV is the accumulation of solids at the bottom of the sump. In this case the irresponsible behaviour of users, flushing various objects (refuse) e.g. rubbish, used clothing which blocks the vacuum sewers, preventing the closing or opening of the valve flap has a significant impact on the smooth operation of the system.

The loss of power (fortunately a rare occurrence in the analysed system) is very troublesome for the reliable operation of the system, as it is in the case with pump failures (both vacuum and sewage), which cause more adverse consequences for the entire sewage disposal system than the failure of a single collection chamber which can cause a temporary shutdown of an individual section of the sewer system, although in the event of a leak in one collection chamber its effect spreads to the entire VSS.

In the past year in Stanisławice, there were additional circumstances conducive to the possibility of failure in the sewer system, i.e. the construction of a dual carriageway running through the settlement. The construction of a flyover in the vicinity of the VPS has resulted in more damage to the vacuum mains, fittings or manhole covers for the collection chambers. This problem, however, is temporary and its impact on the sewer system will end once the project is completed. Currently, it contributes to a reduction in the operational reliability and security of the system and generates additional operational costs. Another factor causing sewer damage is the

Table 1. Summary of component failures in the Stanisławice VSS.

Item	Type of failure	Number per year 2007	2008	2009	2010	Total
1.	Loss of electricity	–	–	–	1	1
2.	Vacuum sewer failures	11	13	5	11	40
3.	Vacuum pump failures	–	–	1	–	1
4.	Other VPS failures	4	3	1	1	9
5.	Collection chamber and VIV failures	40	37	29	30	136
6.	Sewage pump failures	2	1	–	1	4
	Total	57	54	36	44	191

5% 0,5% 0,5% 2% 2,1% 71%

■ Loss of power
▥ Sewer network failure
■ Vacuum pump failure
▨ Other VPS failures
∴ Collection chamber and VIV failures
▨ Sewage pump failure

Figure 1. Summary of failures (by percentage) in the Stanisławice VSS for the period 2007–2010 (Dębowska 2012).

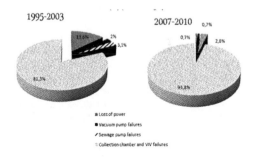

Figure 2. Deformed flap which closes the inlet to the VIV (Dębowska 2012).

Table 2. Summary of VSS component failures in Stanisławice during the 1995–2003 and 2007–2010 periods (Dębowska 2012).

Item	Type of failure	Number per period 1995–2003	2007–2010
1.	Loss of electricity	48	1
2.	Vacuum pump failures	7	1
3.	Sewage pump failures	11	4
4.	Collection chamber and VIV failures	285	136

1995-2003   2007-2010   0,7%

13,6% 2% 3,1% 81,3%   0,7% 2,8% 95,8%

■ Loss of power
■ Vacuum pump failures
⁄ Sewage pump failures
∴ Collection chamber and VIV failures

Figure 3. Summary (by percentage) of different failure types in the Stanisławice VSS during the periods 1995–2003 and 2007–2010 (Dębowska 2012).

aging of the network and its components. In contrast to the above-described problem, this problem will not disappear, but will affect the future failure rate. An analysis of the number of different failures in the Stanisławice VSS was performed comparing the 1995–2003 period with the 2007–2010 period (Dębowska 2012, Kapcia et al. 2005) based on the operational data from the "Dziennik Pracy Oczyszczalni—Stanisławice" (Sewage Treatment Works Journal—Stanisławice). For this purpose, the most common types of failure in the sewerage system were identified in each of the operational periods. Table 2 summarises the number of different network failures for both periods.

The percentage share for the different failures is illustrated in Figure 3. The analysis of

53

the operational data showed that the collection chambers with a VIV are consistently the most unreliable components in this system. The greatest percentage of failures was for the collection chamber in both periods. Over the analysed periods, this type of failure increased by almost 15%, which is associated with the aging of the sewer system components resulting in a higher failure rate. Sewage pump and vacuum pump failures within the VPSs grew similarly over the same periods. This type of failure was small and did not exceed 3.1%. The loss of power was a bigger problem during the earlier operational period. The percentage of events associated with the loss of power during the former period was 13.6% but only 0.7% during the latter, reflecting the increased reliability of the electricity supply. In addition, it should be noted that the VPS is equipped with an alternative electricity supply, as mentioned previously.

In the above analysis the number of failures in the vacuum sewer network has not been taken into account due to the lack of data in the earlier years of operation.

For a comparative assessment of the Stanisławice system with other VSS systems operating in Poland, Table 3 summarises the failure data for four VSSs, both quantitatively and by percentage.

These systems are vacuum systems in their entirety. Comparing the sewer system in Stanisławice one must bear in mind that it is a mixed sewer system with a much smaller number of collection chambers incorporating a VIV, as compared with conventional vacuum systems.

The summary confirms that the greatest failure is for collection chambers with their equipment, and points to a relatively high failure rate for vacuum pumps. The high failure rate of the vacuum pumps is a result of greater than anticipated daily pump activity due to faster loss of vacuum in vacuum sewers caused by vacuum leaks in collection chambers. The solution in Stanisławice results in

fewer VIV failures, due to their smaller number in the whole VSS, which obviously indicates that mixed system solutions are more reliable.

3.2 VSS operational problems in Stanisławice— energy consumption

Electricity in the VSS powers the vacuum and sewage pumps, the monitoring and control equipment, and is used to illuminate the building and the area around the station. The size of the vacuum pump has a significant impact on the energy consumption. Analysis of the electricity consumption was based on data from the 2001–2010, contained in (Dębowska 2012) which was concerned with the energy consumption of vacuum pumps. The monthly data for 2010 is shown as an example in Table 4.

Table 4. Summary of electricity consumption by vacuum pumps in 2010.

2010			
Month	Pump 1 (kWh)	Pump 2 (kWh)	Total
January	345.43	0.00	345.43
February	166.30	222.34	388.64
March	38.46	519.32	557.78
April	210.57	44.35	254.92
May	300.07	6.09	306.16
June	203.31	2.48	205.79
July	189.56	0.11	189.67
August	152.16	0.00	152.16
September	210.49	0.42	210.91
October	98.87	1.79	100.66
November	80.00	0.24	80.24
December	122.52	1.18	123.70
Annual consumption			2965.60

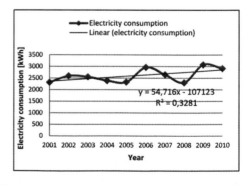

Figure 4. Changes in electricity consumption by vacuum pumping units in the Stanisławice VPS during the period 2001–2010 (Dębowska 2012).

Table 3. Summary of the most common VSS failures in the selected settlements (Błażejewski & Matz 2011).

Lp.	Type of failure	Number	Percentage
1.	Loss of electricity	19	3
2.	Vacuum sewer failures	13	2
3.	Vacuum pump failures	69	9
4.	Other VPS failures	15	2
5.	Collection chamber and VIV failures	611	83
6.	Sewage pump failures	5	1
	Total	732	100%

Table 5. Summary of monthly sewage volume flowing into the STW per day and per month (Dębowska 2012).

Month	2007		2008		2009		2010	
	Q_{ave}	Q_m	Q_{ave}	Q_m	Q_{ave}	Q_m	Q_{ave}	Q_m
January	72	2324	72	2258	72		2242	
February	76	2138	72	2107	73		2052	
March	76	2384	73	2283	77		2388	
April	72	2184	71	2150	71		2155	
May	74	2321	72	2247	71		2221	
June	75	2253	72	2187	89		2691	
July	71	2288	73	2284	71		2226	
August	71	2220	72	2234	72		2260	
September	98	2951	73	2202	71		2157	
October	80	2489	70	2295	72		2254	
November	74	2229	73	2177	76		2288	
December	71	2221	73	2260	72		2243	
Total	28,002		28,684		27,177		27,210	

The units of flow are: Q_{ave} (m³/day) and Q_m (m³/month).

The above summary shows that the greatest electricity consumption occurs during the winter months, while the least occurs during the autumn months. Consumption did not exceed 400 kWh apart from March. Comparing the energy consumption of vacuum pumps in different years, which is summarized in work (Dębowska 2012), it can be seen that over the years energy consumption remained more or less constant (below 3000 kWh), fluctuating between 2309 kWh (2008) to 2966 kWh (2006, 2009, and 2010). Thus, the electricity consumption trend of the vacuum pumps can be considered to be upward due to two factors—an increase in the number of system users producing a greater volume of sewage and the aging of the equipment resulting in more failures, especially in the vacuum sewers. This aging results in fewer sealed sewers and longer periods of vacuum pump activity in order to maintain the appropriate level of vacuum in the system. Intermittent surface runoff from illegal connections results in VIVs opening up to a dozen times more frequently (Kapcia et al. 2005) and consequently the vacuum pumps switching on more often.

3.3 VSS operational problems in Stanisławice-technological problems associated with the operation of a STW

The previously mentioned Stanisławice STW (for domestic sewage) is an integral part of the VSS. It has not been modernised for many years. In its current configuration it treats sewage to the required standard as regulated by the Environment Minister. The sludge formed during the treatment is discharged onto the open plot for dehydration.

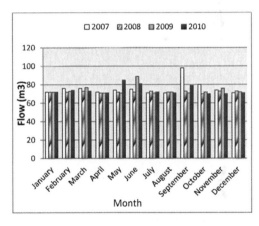

Figure 5. Summary of monthly average daily flow into the Stanisławice STW during the period 2007–2010 (Dębowska 2012).

A lack of roofing causes rehydration of the sludge when it rains. Since 2004, the STW does not accept sewage transported by tanker fleet. Table 5 shows the volume of sewage from the VSS flowing into the Stanisławice STW during the period of 2007–2010.

During 2001–2004, the volume of sewage transported to, and treated in the STW ranged from 80 to 346 m³/month, with an average of about 185 m³/month, while between January and May in 2004, this volume was only between 21 to 115 m³/month with an average of 45 m³/month.

The summary in Table 5 shows that the daily inflow of sewage from the VSS was steady during the 2007–2010 period, fluctuated between 72 to

76 m³/day, with an average of 74 m³/day. Similarly, the monthly inflow during this period was between 2273 m³/month (2008) to 2334 m³/month (2007), with an average of 2273 m³/month. The graph in Figure 5 illustrates these figures. The decline in the sewage inflow volume into the STW during the last decade (2001–2010) should be noted as it was due to a lack of sewage delivered by tanker fleet in recent years and an emerging trend in the reduction of household water consumption. A major problem for the smooth operation of the Stanisławice VSS is, as indicated above, the illegal connection of equipment and fittings in order to drain the surface runoff particularly troublesome during periods of intense rainfall. They increase the flow of sewage into the works causing, in addition to greater electricity consumption by the vacuum and sewage pumps, the leaching of active sludge in the reactor. Therefore, in recent years, measures have been taken to eliminate illegal surface runoff connections, with the only effective method being to impose financial penalties on the perpetrators resulting from checks carried out on sewer connections. Similar actions are carried out for flushing solids into the sewerage system, which adversely affect the operation of the system. Their overall effect has been to reduce the number of illegal connections and sewer pollution, which allows for a more reliable operation of the system and thus the possibility to limit potential losses.

The graph in Figure 5 shows a steady monthly inflow of sewage into the STW, currently only accepting sewage from the VSS, with favourable impact on the sewage treatment process.

3.4 VSS operational problems in Stanisławice— operating costs

The actual operating conditions for a small VSS are difficult to predict, which means that specifying the operating costs is not easy. The total operational cost of the VSS consists of a number of factors, e.g. the age of the sewer network, the materials used, the size and design of the network. The main constituent of these costs is the cost of electricity to the whole system. Table 6 summarises these costs for the period 2007–2010.

In assessing the annual electricity charges, it should be noted that they remain at a similar level and do not exceed 31,000 PZL per year.

Another important components of the VSS operating costs are the repair costs. They most commonly result from the need to repair or replace VIVs or their components, due to progressive corrosion. It significantly affects the system operational costs.

The VSS requires constant supervision and control, so two workers are employed full-time at the VPS to carry out this work. It is interesting to compare the operating costs of the components for the gravity sewer system with the VSS (Table 7).

Table 6. Summary of electricity charges for Stanisławice STW (Dębowska 2012).

	Gross electricity charges (PZL)			
Month	2007	2008	2009	2010
January/February	1,699.68*	7,552.98	6,290.61	4,865.93
March/April	3,723.79	2,882.53	5,746.75	3,998.51
May/June	2,574.74	5,187.09	4,590.02	3,432.89
July/August	3,034.20	4,219.58	4,241.65	4,401.73
September/October	3,067.21	3,837.75	4,034.30	5,750.68
November/December	2,894.13	6,630.64	4,672.21	–
Total	16,993.76	30,310.57	29,575.54	22,449.74

*Only February.

Table 7. Summary of elements affecting the operational costs of a gravity sewer system and a VSS(Dębowska 2012).

Sewer system type	
Gravity	Vacuum
Work related to sewage treatment	Energy consumption by the VPS
Recovery after failure	Recovery after failure
Employment of workers for maintenance and repairs	Permanent manning of the VPS
Hidden costs associated with the harmful effects on the environment in the event of a failure	Employment of workers who resolve abnormal system behaviour

Table 7 shows a summary of the similarities and differences between both systems. Without doubt the recovery after failure is common for both, with the comment that failures in the gravity sewers often require excavation work, generating additional costs resulting from the fact that these sewers are located at a much greater depth than the vacuum sewers. In the VSS, most failures occur in collection chambers which are more easily accessible for the maintenance crew.

Almost every fault and leak in the VSS network is quickly detected and localised. Another common operating cost is the work related with network maintenance and cleaning of the gravity network or the resolution of VSS malfunctions, although even here there are some differences arising from the need for the maintenance crew to have appropriate qualifications. Furthermore, for the VSS, the flow velocity in the pipes is high, which prevents them from becoming blocked. To the differences between both systems, so-called, hidden costs should be included in the case of a gravity sewer system. These are due to non-sealed sewers and for example sewage exfiltration into the surrounding ground, causing contamination of soil and water. Such damage is difficult to detect, and the cost of repairs difficult to estimate. However, in the case of a VSS it is the cost of electricity consumption by the VPS and the necessity of constant supervision by the maintenance crew (Dolecki 1984). Taking into account all the above-mentioned elements that constitute the operating costs of the compared systems, it should be noted that under certain circumstances, the construction of a VSS may also be economically justified. We must not forget that the use of the gravity sewer system with its local pumping of sewage raises the cost of operating such a system.

4 CONCLUSIONS

Non-conventional sewage disposal such as by a VSS is not yet popular in Poland. Its usefulness in a sewerage area flat terrain or high groundwater level is undisputed. Choosing a VSS should always be preceded by a technical, economic and reliability analysis of the solutions under consideration. The VSS performance must complyabsolutely with the requirements of the technical solution, e.g. the vacuum sewer configuration ensuring apportioned flow of sewage with great care.

An assessment of the exemplary VSS operational problems allowed us to present both positive and negative aspects of its operation. With respect to the settlement in which the VSS was implemented the following positive aspects can be included:

- It is well adapted to site and groundwater conditions.

- It avoided a large number of sewage pumping stations in the event of a conventional sewer system (gravitational) being chosen.
- It resulted in a reduction in the number of collection chambers through the use of an unusual solution, such as a gravity-vacuum-pressure system.

When it comes to negative aspects that were also found, they include:

- Vacuum Interface Valve (VIV) failures are most common, requiring them to be made of materials resistant to corrosion, along with the pipe sensor.
- A lack of awareness by people using the system.
- A dated Sewage Treatment Works (STW) not providing a high degree of purification.
- The building of a nearby dual carriageway causing damage to Vacuum Sewer System (VSS) components.
- As in other systems of this type, costs associated with electricity consumption and the need for permanent maintenance staff
- The discussed system has worked for many years. However, the newly emerging VSS are characterised by greater reliability and consume less electricity.

Characterised and evaluated in terms of gravity-vacuum-pressure reliability, the sewer system in Stanisławice can be regarded as properly functioning in accordance with the Polish regulations.

REFERENCES

Błażejewski, R. & Matz, R. 2011. Vacuumsewer system operational problems, *Gas, Water and Sanitation* No. 7–8,: 293–298.

Dębowska, B. 2012. The main operational problems of the vacuum sewer system using the sewer system in Stanisławice as anexample, *Diploma thesis in engineering at Cracow University of Technology, supervised by J.Krolikowski, 2012.*

Dolecki, J. 1984. Analysis of the economic effectiveness of a vacuum sewer system, *Environmental Protection,* No. 434/3–4,: 20–21.

Ermolin Fu. A. Optimal'noe upravlenie kanalizacionnoj set'ju po kriteriju minimum energozatrtat, *Stroitel'stvo i architektura,* No. 6,: 118–123.

Kalenik, M. 2011. *Vacuumsewer systems,* Warszawa, Publishers S.G.G.W.

Kapcia, J., Lubowiecka, T. & Mucha, Z. 2005. Evaluation of the sewage treatment system in Stanisławice, *Gas, Water and Sanitation,* No. 6,: 28–31.

Kuranov N.P. & Rozanov N.N., Timofeeva E.A. 2009. Rasćety riska avarij gidrotechnićeskich soorużenij *Vodosnabżenie I Sanitarnaja Technika* No 1.:41–44.

Orne M. 2001, Estimates of the energy impact of ventilation and associated financial expenditures, *Energy and Buildings,* No 33 (3),: 199–205.

Environmental Engineering IV – Pawłowski, Dudzińska & Pawłowski (eds)
© *2013 Taylor & Francis Group, London, ISBN 978-0-415-64338-2*

Analysis of water pipe breakage in Krosno, Poland

K. Pietrucha-Urbanik & A. Studziński
Faculty of Civil and Environmental Engineering, Rzeszow University of Technology, Rzeszow, Poland

ABSTRACT: The subject of this paper is a reliability analysis of water supply failure frequency in the town of Krosno. Operating data on the failure rate of water supply pipelines in the city were used. The analysis of water pipes failures, including pipes diameters, their age, the material they are made from and their function in the water supply subsystem in the city was made. There is a noticeable downward trend in both the number of damages as well as the intensity of damages of water pipes. The survival analysis was used to estimate the time to failure using the Kaplan-Meier estimator.

Keywords: water network, reliability, survival analysis

1 INTRODUCTION

Existing research determine the potential causes of failures, including both the factors changing over time and independence of time. Most of studies focused on identification and development of the relationship between the intensity of damages and the significant risk factors that contribute to failure in water supply network (Rak & Pietrucha 2008a; Studziński & Pietrucha-Urbanik 2012). The examples of such already analysed factors are the age (Shamir & Howard 1979; Vališ et al. 2012), diameter and material of pipe (Goulter et al. 1993; Le Gat & Eisenbeis 2000), ground conditions, operating pressure, temperature in the supply network (Tabesh et al. 2009), climatic change (Harrada 1988), possible external load (Rajani & Maka 2000) and the course of failure (Christodoulou & Deligianni 2010; Vališ et al. 2010, 2012). In Krosno, water is provided to consumers via water pipe network having radial-ring arrangement, which is beneficial to the reliability of water supply system. Currently, the water supply network has total length of about 604,7 kilometres and is made of the following materials: steel pipes (34%), cast iron pipes (26%), PVC pipes (22%), PE pipes (18%). The life of the water supply network is as follows: up to 5 years—6%, from 6 years to 10 years—12%, from 11 years to 20 years—25%, from 21 to 30 years—22%, from 31 to 50 years—33%, more than 50 years—3%. The water network supplies water to about 100000 recipients of the city of Krosno and neighbouring municipalities. The number of residential water supply connections in all served places is 5730, including 4675 terminals located in the Krosno municipality.

2 MATERIALS AND METHODS

2.1 Material

The purpose of this study is to characterize the unreliability of the Krosno water-pipe network. Detailed analysis of the water network failure should be main element of the managing system of the urban water networks, particularly in strategic modernization plans (Herz 1998). The calculations were made on the basis of the operational data on the water-pipe network in the town of Krosno, as well as on the failure protocols received from the Municipal Enterprise for Communal Economy in Krosno. The analysis was performed using the application contained in a ReliaSoft computer program Weibull++7.

2.2 Evaluation of water supply failure rate using the survival analysis

The survival analysis is a collection of statistical methods analysing the processes in which the examined variable is the time of specific event occurrence. Using this analysis we can predict the exploitation time of water pipes. In this type of analysis data censoring occurs, due to the fact that the observations are examined at a specific time (Cox 1972). If in the analysed period of time the event is not observed, then the right censoring occurs, while the left censoring occurs the given event was observed as a prior to the considered time interval, what is presented in Figure 1.

The survival function can be estimated using the Kaplan-Meier estimator, also known as the product limit estimator, which for the first time was proposed by E.L. Kaplan and P. Meier in 1958

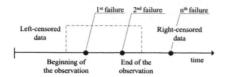

Figure 1. Types of data censoring in water supply systems.

(Kaplan & Meier 1958). The survival probability evaluation is the product of conditional probabilities of survival over the next periods of time. The estimator of the survival function is expressed by the formula:

$$\hat{S}(t) = \prod_{t_j \leq t}\left(1 - \frac{d_j}{r_j}\right) \qquad (1)$$

and the nonparametric estimator of the cumulative hazard function, the Nelson-Aalen estimator, is:

$$\tilde{H}(t) = \prod_{t_j \leq t}\left(\frac{d_j}{r_j}\right) \qquad (2)$$

where d_j = a number of failures occurring in time t_i; r_j = pipes prone to failure.

The hazard function (risk), contrary to the survival function, focuses on the occurrence of undesirable events, for example, failure, giving a negative supplement of the information carried by the survival function (Klein & Moeschberger 1997).

In the Kaplan-Meier method such parameters as number of previously observed water pipe failures, water pipe type, length and diameter of pipes, were taken into account. Test for many trials was checked using the Mantel's procedure, which assigns points to every survival time, then the value of chi-square statistics, based on the sums for each group of these points, was calculated. The p-value is 0.0267, which indicates a significant difference in survival of the particular groups differentiated by the above mentioned parameters.

2.3 Water network failure

The main criterion for assessing the state of water pipes is the failure rate index—λ_i. Failure rate index estimator, was determined, per year for particular type of water pipes (mains, distributional and water supply connections), from the following formula:

$$\lambda_i = \frac{k_i}{l_i \cdot \Delta t} \qquad (3)$$

where λ_i is the failure rate index estimator per year for particular type of water pipes per one year,

[km^{-1}a^{-1}]; k_i is the number of failures in one year for particular type of water pipes; l_i is the lenght of particular type of water pipes, on which failures appeared per one year, [km]; i is a type of water pipes (M—mains, R—distributional, P—water supply connections); Δt is the length of time that equals 1 year, [1 year].

3 RESULTS AND DISCUSSION

Figure 2 shows the unit values of the failure rate in the Krosno water-pipe network in the years 2005–2011.

The lowest failure rate have distribution pipelines ($\lambda_{Ravg} = 0.20$ km^{-1}a^{-1}) and the highest failure rate have main pipelines ($\lambda_{Mavg} = 0.74$ km^{-1}a^{-1}). The European criteria say that the pipeline needs repairing when the failure rate index exceeds 0.5 km^{-1}a^{-1} (Mays 1998).

However, one should seek the following values of the failure rate index: mains: $\lambda \leq 0.3$ km^{-1}a^{-1}; distributional: $\lambda \leq 0.5$ km^{-1}a^{-1}; water supply connections: $\lambda \leq 1.0$ km^{-1}a^{-1}.

Taking into consideration the percent of failures in the main water pipes, the distribution pipes and the water supply connections, depending on the material from which they were made, 63% of the failures occurring in the main water pipes happened in the cast iron pipes, which results from a significant share of this material in the construction of these pipelines and their significant age. The distribution pipelines are characterized by the high failure rate of iron pipes (52% of all failures), and PVC pipes (34%). Most failures in the water supply connections, as many as 68%, occur in steel pipes, this is due to their poor technical condition. The lower number of failures in pipelines made of PVC and PE is caused by the fact that they are part of younger sections of water network and that they are resistant to corrosion. The detailed analysis of the failure rate in water pipes in the years 2005–2011 showed that:

- Pipes made of PE are characterized by the lowest failure rate. In the water supply connections the

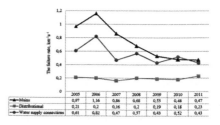

Figure 2. The failure rate for particular type of water pipes in the years 2005–2011.

60

failure rate index does not exceed 0.12 km^{-1}a^{-1} and in the distribution pipelines 0.05 km^{-1}a^{-1}.

- Iron pipelines in the main pipelines show the highest failure rate. The average failure rate is 1.59 km^{-1}a^{-1}. In the distribution pipelines the average failure rate is 0.39 km^{-1}a^{-1}.
- The highest failure rate in PVC pipes is seen in the water connections, from 0.8 to 1.31 km^{-1}a^{-1}, and the lowest in the distribution pipelines, from 0.22 to maximum of 0.42 km^{-1}a^{-1}. In the main water pipes the average failure rate was 0.74 km^{-1}a^{-1}.
- In the steel pipelines the highest failure rate is in the water supply connection, it ranges from 0.75 to over 1.39 km^{-1}a^{-1}, the main pipelines have the average failure rate up to 0.3 km^{-1}a^{-1}. The distribution pipelines have the lowest failure rate, which amounted to 0.08 km^{-1}a^{-1}.

Comparing the determined average failure rate in the water supply system made of different materials with the required values, it is concluded that:

- In the case of the main pipelines made of cast iron and PVC the average failure rate is 1.51 and 0.75 km^{-1}a^{-1}, respectively, while the required value of failure rate index in the main water network is $\lambda_{Mreq} = 0.3$ km^{-1}a^{-1}, the steel pipes, however, have the average failure rate 0.30 km^{-1}a^{-1}.
- In the case of the distribution pipelines, the average failure rate for pipelines made of steel (0.08 km^{-1}a^{-1}), made of iron (0.39 km^{-1}a^{-1}), made of PE (0.01 km^{-1}a^{-1}) and PVC (0.36 km^{-1}a^{-1}) is lower than the required value $\lambda_{Rreq} = 0.5$ km^{-1}a^{-1}.
- In case of the water supply connections, the average failure rate in steel pipelines and PVC pipelines (values respectively 1.07 and 1.01 km^{-1}a^{-1}) is higher than the required value $\lambda_{Preq} = 1.0$ km^{-1}a^{-1}.

Taking into account the percentage of failures depending on the diameter and material of water pipes in the years, it is clearly visible that the most often the failures occur in the water connections with a diameter of 32 mm, made of steel (14.5% of all failures). The next frequent are the failures in pipes made of PVC and steel, with a diameter of 40 mm, respectively 12% and 9.6% of all failures.

Figures 3 and 4 show the hazard function determining the occurrence of a failure for data concerning the material and type of water pipes.

Considering the type of water pipes, the highest values of the hazard function occur for the main pipelines, then the water supply connections and the distribution pipelines. The hazard function at the end of the range of observation is about 0.04, increasing from the beginning of the observation from the value 0.001. The hazard function for the main pipelines and the water connections is at a similar level, while the function describing the distribution pipelines shows the lowest hazard rate.

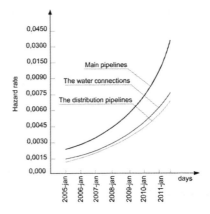

Figure 3. The hazard function for data concerning the type of water pipes.

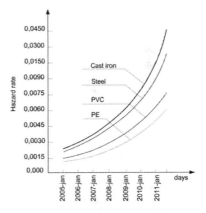

Figure 4. The hazard function for data concerning the material of water pipes.

Considering the material of water pipes, the highest risk occurs for pipes made of cast iron, which are characterized by the highest values of the hazard function. During the observed time interval the hazard rate associated with pipes made of PE increased from 0.0009 to 0.006, in comparison with the hazard rate of pipes made of steel is increasing to 0.024. It means that pipes made of cast iron and steel should be under special supervision, including the continuous monitoring and the replacement program.

Based on the performed analysis of the failure protocols the recovery time and the total time of repair of damaged main pipelines were determined, these times are greater than the times for the distribution pipelines. The range of changes of the recovery time for the distribution pipelines was 1–16.5 h (average of 4 hours) and for the main pipelines 1–18 h (average of 5 hours). The total time of repair of damaged water connections 8.7 h is longer than the time of repair of damaged distribution pipelines 7.5 h, while the recovery time (related to the lack of water supply) for the water supply connections 4 h is similar to the recovery time of the distribution pipelines.

4 CONCLUSIONS

On the basis of the present analysis of the research results of the, the following conclusions and statements were made:

- the average failure rate for the main pipelines is $\lambda_{Mavg} = 0.74$ km^{-1}a^{-1}, the distribution pipelines $\lambda_{Ravg} = 0.20$ km^{-1}a^{-1} and the water connections—$\lambda_{Pavg} = 0.55$ km^{-1}a^{-1},
- the survival analysis based on material and kind of waterworks pipe play a very important role in the procedures of risk analysis and lead to investment-modernization undertakings, also strategic modernization plans,
- the majority of failures (55.22%) happen in the water connections (25 to 63 mm diameter), representing about 44% of the total length of the pipelines,
- the lowest failure rate is in the distribution pipelines made of PE—$\lambda_{PEavg} = 0.01$ km^{-1}a^{-1}, which results from their use since the 1990s,
- for the water connections made of steel the average failure rate is $\lambda_{steelavg} = 1.07$ km^{-1}a^{-1}, the highest average failure rate is found in the cast iron main water pipelines $\lambda_{castironavg} = 1.52$ km^{-1}a^{-1}, which is caused by the fact that steel and iron are the oldest materials used to build that water supply network,
- the value of failure rate indexes corresponds to national trends (Hotloś, 2003; Zimoch, 2010; Rak and Pietrucha 2008b), the declining trend in both the number of failures and failure rate in water pipelines is seen.

ACKNOWLEDGEMENTS

Scientific work was financed from the measures of National Center of Research and Development as a development research project No N R14 0006 10: "Development of comprehensive methodology for the assessment of the reliability and safety of water supply to consumers" in the years 2010–2013.

REFERENCES

Christodoulou, S. & Deligianni, A. 2010. A neurofuzzy decision framework for the management of water distribution networks. *Water Resour Manag* 24(1): 139–156.

Cox, D.R. 1972. Regression models and life-tables. *J R Stat Soc B* 34(2): 187–220.

Goulter, I.C., Davidson J., Jacobs P. 1993. Predicting water-main breakage rates. *Journal of Water Resources Planning and Management* 119(4): 419–436.

Harada, H. 1988. Statistics of the cold wave in the temperate region and prediction of the number of damaged service pipes. *Journal of AWWA* 57(8): 12–15.

Herz, R.K. 1998. Exploring rehabilitation needs and strategies for water distribution networks. *Journal of Water Supply Research and Technology-Aqua* 47(6): 275–283.

Hotloś, H. 2003. Reliability level of municipal water pipe networks. *Environmental Protection Engineering* 32(32): 141–151.

Kaplan, E.L & Meier, P. 1958. Nonparametric estimation from incomplete observations. *J Am Stat Assoc* 53: 457–481.

Klein, J.P. & Moeschberger, M.L. 1997. *Survival analysis techniques for censored and truncated data*. New York: Springer.

Le Gat, Y. & Eisenbeis, P. 2000. Using maintenance records to forecast failures in water networks. *Urban Water* 2(3): 173–181.

Mays, W.L. 1998. *Reliability Analysis of Water Distribution Systems*, New York: American Society of Civil Engineers.

Rajani, B. & Makar, J. 2000. A methodology to estimate remaining service life of grey cast iron water mains. *Can. J. Civ. Eng.* 27: 1259–1272.

Rak, J.R. & Pietrucha, K. 2008a. Ryzyko w kontroli jakości wody do spożycia. *Przemysł Chemiczny* 87(5): 554–556.

Rak, J. & Pietrucha, K. 2008b. Some factors of crisis management in water supply system. *Environment Protection Engineering* 34(2): 57–65.

Shamir, U. & Howard, C.D.D. 1979. An analytical approach scheduling pipe replacement. *Journal of AWWA* 71(5): 248–258.

Studzinski, A. & Pietrucha-Urbanik, K. 2012. Risk Indicators of Water Network Operation. *Chemical Engineering Transactions* 26: 189–194. DOI: 10.3303/CET1226032.

Tabesh, M., Soltani, J., Farmani, R. & Savic, D. 2009. Assessing pipe failure rate and mechanical reliability of water distribution networks using data-driven modeling. *Journal of Hydroinformatics* 11(1): 1–17.

Valis, D., Vintr, Z. & Koucky, M. 2010. Contribution to highly reliable items' reliability assessment. *Proceedings of the European Safety and Reliability Conference ESREL, Prague, Czech Republic. Reliability, Risk and Safety: Theory and Applications.* Taylor & Francis 1–3: 1321–1326.

Vališ, D., Vintr, Z. & Malach, J. 2012. Selected aspects of physical structures vulnerability—state-of-the-art. *Eksploatacja i Niezawodnosc—Maintenance and Reliability* 14(3): 189–194.

Vališ, D., Koucký, M. & Žák, L. 2012. On approaches for non-direct determination of system deterioration. *Eksploatacja i Niezawodnosc—Maintenance and Reliability* 14(1): 33–41.

Zimoch, I. 2010. Reliability analysis of water-pipe networks in Cracow, Poland; in: *Environmental Engineering III*, Pawłowski L., Dudzińska M.R. (eds.), London: Taylor and Francis.

Environmental Engineering IV – Pawłowski, Dudzińska & Pawłowski (eds)
© 2013 Taylor & Francis Group, London, ISBN 978-0-415-64338-2

Analysis and assessment of the risk of lack of water supply using the EPANET program

K. Boryczko & A. Tchorzewska-Cieślak
Department of Water Supply and Sewage Systems, Rzeszow University of Technology, Rzeszow, Poland

ABSTRACT: Water supply system belongs to the, so called, critical infrastructure of cities, and a priority task for waterworks should be to ensure the suitable level of its safety. The aim of this study is to propose the method to analyse risk of interruptions in water supply to consumers using the EPANET simulation program. The paper presents an application example for the water supply system supplying 63,000 inhabitants, for whom the risk maps of water supply interruptions were drawn.

Keywords: water supply, water distribution system, risk maps

1 INTRODUCTION

The management of safety of Collective Water Supply System (CWSS) was carried out through risk analysis and assessment, in particular risk identification, including identification of threats and consequences for possible emergency scenarios, risk assessment and its treatment.

Risk identification is based mainly on the analysis of risk factors, their sources, the definition of the, so-called, weak points (vulnerabilities) and consequences (effects) of their occurrence. The most often it concerns undesirable events that may occur with a specified probability "P", lead to a specified losses "C" in the system with a specified degree of protection "V".

The analysis of the results, in terms of the possibility of a crisis, can be performed on the basis on the risk maps, that is the distribution of tolerant, controlled and unacceptable risks. Technical risk maps allow a global and comprehensive assessment of any technical system (Boryczko 2010). Collective water supply system is one of the elements of critical infrastructure. It provides residents with drinking water, in the required quantity and quality, at the required pressure, at time convenient for a customer and for an acceptable price (Tchorzewska-Cieslak & Rak 2010). The most common cause of water supply interruptions are failures in Water Distribution Subsystem (WDS). The failure of a main or a transit line may cover a large area and cause reduction of water supplies to thousands of residents. Due to that fact, the identification of the areas which are the most threatened by risk of water supply interruption is an important process (Farley & Mounce et al. 2010;

Liserra & Artina et al. 2010; Tchorzewska-Cieslak 2010; Tchorzewska-Cieslak 2012; Weber 2012).

Bursts and leaks are the major components of water loss from distribution systems (Lambert 2003). Water losses have several associated costs, namely the direct cost of the water lost, the cost of interrupting the supply, the cost of repairing the system and the cost for society associated with the interruption of supply (Farley & Mounce et al. 2010).

The aim of this study is to propose the method to analyse risk of interruptions in water supply to consumers using the EPANET simulation program. The paper presents an application example for the water supply system supplying 63,000 inhabitants, for whom the risk maps of water supply interruptions were drawn.

2 MATERIAL AND METHODS

Proposed methodology takes into consideration basic triplet definition of risk. The three following parametric definitions were assumed for the risk

$$r_C = P_i \cdot C_j \cdot V_k \tag{1}$$

where P_i = probability of a single event that may cause the risk; C_j = losses caused by a single undesirable event that may cause the risk; V_k = vulnerability associated with the occurrence of a single undesirable event that may cause the risk.

Tables 1 and 2 presents the criteria of point and descriptive scale for particular risk parameters. The criteria presented below were created through own research and were consulted with literature (Ferwtrell 2001; Hrudey 2001; Pollard 2004).

Table 1. Criteria of point and descriptive scale for the aparameter P.

Point weight P	Description of the parameter P	Ranges of probability of undesirable event occurence
1	Very low probability	Once in 10 years
2	Low probability	Once in 5 years
3	Medium probability	Once in 2 years
4	High probability	Once in 0.5 Years
5	Very high probability	Once a month and more often

Table 2. Criteria of point and descriptive scale for the parameter V.

Point weight	Description of the parameter V
1	Very low vulnerability to failure (very high resistance): – The network in the closed system, the ability to cut off the damaged section of the network (in order to repair it), – The ability to avoid interruptions in water supply to customers, full monitoring of water-pipe network (continuous measurements of pressure and flow rate at strategic points of the network) covering the entire area of water supply, using Supervisory Control And Data Acquisition (SCADA) and Geographic Information System (GIS) software, the possibility to remote control of network hydraulic parameters, – Emergency reserve in network water tanks covering the needs of the city for at least 24 h, (Q_{dmax} or $Q_{d.avg}$—daily average water production), – Comprehensive system of emergency warning and response, – Full use of alternative water sources.
2	Low vulnerability to failure (high resistance): – The network in the closed or mixed system, the ability to cut off the damaged section of the network (in order to repair it), – Standard monitoring of water-pipe network (continuous measurements of pressure and flow rate), – Early warning system, – Use of alternative water sources.
3	Medium vulnerability to failure (medium resistance): – The network in the mixed system, the ability to cut off the damaged section of the network by means of gates, (water supply to customers is limited because of the network capacity), – Water-pipe network standard monitoring, measurements of pressure and flow rate, – Delayed emergency response system, – Alternative water sources do not cover the needs completely.
4	High vulnerability to failure (low resistance): – The network in the open system, the inability to cut off the damaged section of the network by means of gates without interrupting water supply to customers, – Limited water-pipe network monitoring, – Delayed emergency response system, – Limited access to alternative water sources.
5	Very high vulnerability to failure (very low resistance): – The network in the open system, the inability to cut off the damaged section of the network by means of gates without interrupting water supply to customers, – Lack of water-pipe network monitoring, – Lack of emergency warning and response system – Very limited access to alternative water sources.

A five-point scale related to the percentage of people at risk C was proposed. For the percentage of threatened population, the following weights were assumed:

– (0%÷12,5%)—weight 1,
– (12,5%÷25%)—weight 2,
– (25%÷50%)—weight 3,
– (50%÷75%)—weight 4,
– Higher than 75%—weight 5.

The criteria for risk assessment are as follows:

– Negligible (1÷9),
– Tolerable (9÷20),

– Controlled (20÷45),
– Unacceptable (45÷60),
– Inadmissible (60÷125).

3 CASE STUDY

3.1 *Characteristics of water distribution subsystem*

Figure 1 shows a diagram of the main pipelines in the analysed WDS.

The vast majority of the WDS works in the general supply pumping system without expansion tanks. Most of the city is covered with a peripheral network. Only a small part of the network works as a branched one. Table 3 presents the lengths of the mains and distribution network in km for the years 2003–2011.

Figure 2 shows the material structure of the water supply network.

As can be seen in Figure 2, most of the WDS is built of steel and PVC. Figure 3 shows the age structure of the WDS.

Figure 1. Scheme of the analysed water supply network with the highlighted Water Treatment Plant (WTP) and the second stage Pumping Station (PS II).

As presented in Figure 3, more than 40% of water supply pipeline is more than 20 years old. At present, urban water supply system serves around 63,000 residents. The state of drinking water metering is high, which equals 100% for the mass recipients and 98% for individual customers. An uneven hourly intake was assumed as $N_h = 1.44$. The second stage of pumping station was operating at a constant pressure at discharge (40 kPa).

An average daily water production is about 8200 m³. Water supply network is connected with a pipeline DN 400 providing water from the underground water intake in the north-west of the city. In case when one party finds itself in a situation when water supply provided to the network has to be stopped or reduced because of the lack of water in water intake, water pollution dangerous to health, the need to increase water supply to fire hydrants, the necessity to carry out necessary repairs of water supply facilities, breaks in electric power supply for water supply facilities; the other party will provide water to the network of the party affected by the aforementioned events.

Each party is obliged to start the supply within 4 hours of telephone notification made by the other party. The city was divided into 41 zones in order to conduct the risk analysis (Fig. 4).

3.2 *Analysis and evaluation of the risk of lack of water supply in case of the pipeline 405 failure*

The pipeline 405 failure, near to the second stage pumping station in the zone D3, was simulated. This pipeline failure causes the pressure drop across a large area of the city. Failures of other piping lines that create ring mains did not cause the pressure drop, which would result in water supply interruptions. The pipeline 405 supplies water to the south-eastern area of the city. The location of the pipeline 405 is marked in Figure 5.

In order to determine the weight associated with the percentage of the population of the housing estate without water supply, the simulation of the WDS operation was made with the use of Epanet 2.0. According to the previous assumptions, a 12-hour failure, lasting from 8.00 am to 8.00 pm, was simulated in a day of medium intake.

Table 3. The lengths of the mains and distribution network in km for the years 2003–2011.

The lengths in km	Years								
	2003	2004	2005	2006	2007	2008	2009	2010	2011
Mains	28,1	28,1	29,4	29,4	29,6	29,6	29,6	29,6	29,6
Distribution	142,9	143,13	143,6	145,7	147,8	148,8	148,8	149,1	149,3

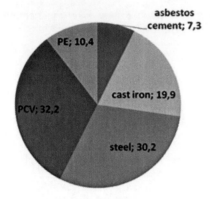

Figure 2.   The material structure of the WDS.

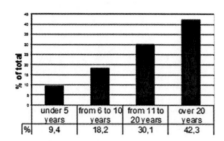

Figure 3.   The age structure of the WDS.

%	under 5 years	from 6 to 10 years	from 11 to 20 years	over 20 years
%	9,4	18,2	30,1	42,3

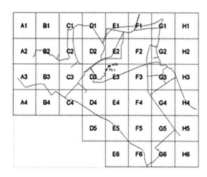

Figure 4.   Distribution of the WDS into zones.

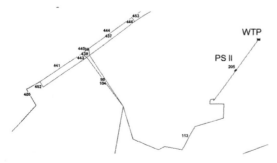

Figure 5.   Location of the pipeline 405.

Table 4.   Number and percentage of people deprived of water supply (failure of the main pipeline 405).

District	NR*	Number of residents without water supply	Percentage of residents of district without water supply	Point weight related to the percentage of residents of given district
A1	500	0	0%	1
A2	500	38	8%	1
A3	700	401	57%	4
A4	1400	1006	72%	4
B1	600	0	0%	1
B2	1000	131	13%	2
B3	1400	772	55%	4
B4	1500	1176	78%	5
C1	2900	0	0%	1
C2	2500	110	4%	1
C3	800	283	35%	3
C4	1300	0	0%	1
D1	2500	0	0%	1
D2	2500	0	0%	1
D3	1800	189	11%	1
D4	1600	783	49%	3
D5	100	53	53%	4
E1	600	0	0%	1
E2	900	0	0%	1
E3	1800	12	1%	1
E4	2500	592	24%	2
E5	1000	155	15%	2
E6	1000	94	9%	1
F1	400	0	0%	1
F2	600	0	0%	1
F3	1600	0	0%	1
F4	2000	0	0%	1
F5	500	0	0%	1
F6	1000	181	18%	2
G1	1500	0	0%	1
G2	2000	0	0%	1
G3	3500	0	0%	1
G4	3000	0	0%	1
G5	1000	0	0%	1
G6	700	49	7%	1
H1	2000	0	0%	1
H2	1500	0	0%	1
H3	5000	0	0%	1
H4	4000	0	0%	1
H5	700	0	0%	1
H6	600	0	0%	1

*NR—number of residents in district.

The model simulated the pipeline failure resulting from the closure of this pipeline. For the network operating during the failure of mains, the layered charts of pressure were made for the last hour of failure. To determine the number of

66

residents deprived of water supplies, it was assumed that water supply interruption occurs:

- for the top floor, if the network pressure drops below 30 mH$_2$O,
- for the last but one floor, if the network pressure drops below 25 mH$_2$O,
- to the third floor from the top, if the network pressure drops below 20 mH$_2$O.

In Figure 6 the layered pressure system in the water supply network, during the failure in the pipeline 405, was presented.

On the basis of the results of the pipeline failure simulation, made with the use of the EPANET program, the percentage of the inhabitants of the area without water supply, was determined.

On the basis of the operating data for the analysed WDS it was found that:

- point weight associated with the likelihood of failure P is 4, high probability (once in 0.5 year),

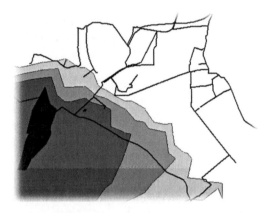

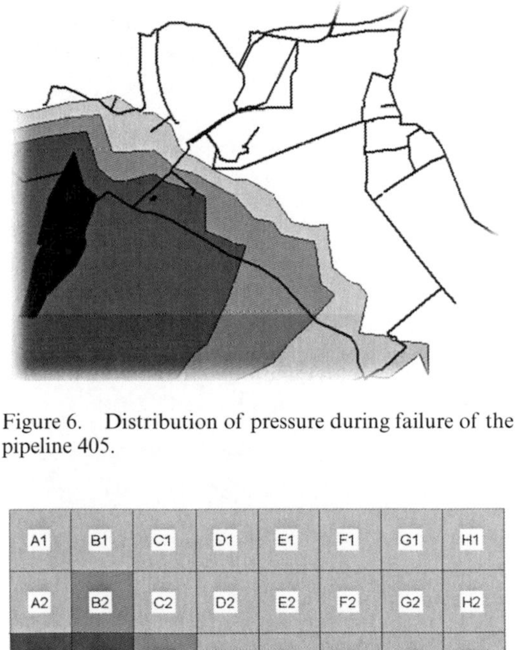

Figure 6. Distribution of pressure during failure of the pipeline 405.

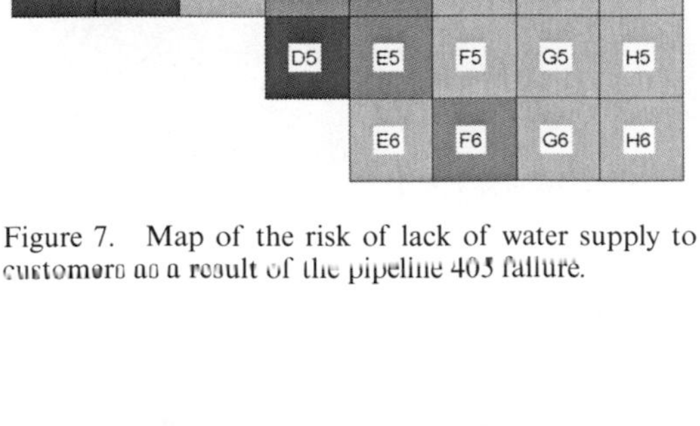

Figure 7. Map of the risk of lack of water supply to customers as a result of the pipeline 405 failure.

- point weight associated with the vulnerability of V is 3; medium vulnerability to failure.

Based on the above point weights the map of the risk of lack of water supply for the analysed city was made.

4 CONCLUSION

The developed risk maps allow the identification of threatened zones and taking appropriate decisions about modernization to improve water consumers safety, zones where on the level of risk that should be lowered. The model was adapted for an exemplary WDS; and a simulation of the potentially exposed residents was carried out using EPANET software. During the failure in the main pipeline 405 the most threatened zones are A3, B3, A4, B4, D5 (unacceptable risk), and also B2, C3, D4, E4, E5, F6 (controlled risk). The condition for the application is to have a hydraulic model of water distribution subsystem, data about network failure rate and about vulnerability to failure.

Hydraulic models of water distribution subsystems and risk maps represents new elements in the WDSs risk analysis and can be used for the WDSs complex modelling and managing in the context of water consumers safety.

REFERENCES

Boryczko, K., 2010, Analysis of risk of interruption in water supply to consumers, in Journal of KONBiN, *Safety and Reliability Systems*, 1: 79–90.

Farley, B., Mounce S.R., et al., 2010, Field testing of an optimal sensor placement methodology for event detection in an urban water distribution network, *Urban Water Journal*, 7: 345–356.

Ferwtrell, L., Bartram, J., 2001, Water Quality: Guidelines Standards Health. Assessment of Risk Management for Water Related Infection Disease. World Health Organization Series, London, IWA Publishing.

Hrudey, S., E., 2001, Drinking water quality—a risk management approach, *Water*, 26: 29–32.

Lambert, A., 2003, Assessing non-revenue water and its components: a practical approach, *Water21*, 8: 50–51.

Liserra, T., Artina, S., Bragalli, C., Lenzi., 2010, Water loss dynamic control by Automatic Meter Readings in water distribution network, Boca Raton, Crc Press-Taylor & Francis Group.

Pollard, S., J., T., Strutt, J., E., Macgillivray, B., H., Hamilton, P., D., Hrudey, S., E., 2004, Risk analysis and management in the water utility sector, *Process Safety and Environmental Protection*. 82: 453–462.

Tchórzewska-Cieślak, B., 2010, Water consumer safety in water distribution system, *Environmental Engineering III*, D.M.R. Pawłowski, I., Pawłowski, A., (eds), London, Taylor & Francis Group.

Tchórzewska-Cieślak, B., Rak, J., 2010, Method of identification of operational states of water supply system, *Environmental Engineering III,* D.M.R. Pawłowski, L., Pawłowski, A. (eds), London, Taylor & Francis Group.

Tchórzewska-Cieślak, B., Boryczko, K., Eid, M., 2012, Failure scenarios in water supply system by means of fault tree analysis, Advances in Safety, *Reliability and Risk Management.* G.A. Bérenguer Ch., Guedes Soares C. (eds), London, Taylor & Francis Group.

Weber, P., Simon, C., Theilliol, D., Puig, V., 2012, Control allocation of k-out-of-n systems based on Bayesian Network Reliability model: Application to a drinking water network, Advances in Safety, *Reliability and Risk Management*, G.A. Bérenguer Ch., Guedes Soares C. (eds), London, Taylor & Francis Group.

Water and wastewater treatment

Environmental Engineering IV – Pawłowski, Dudzińska & Pawłowski (eds)
© 2013 Taylor & Francis Group, London, ISBN 978-0-415-64338-2

Application of hybrid system ultrafiltration reverse osmosis in geothermal water desalination

M. Bodzek
Institute of Water and Wastewater Engineering, Silesian University of Technology, Gliwice, Poland
Institute of Environmental Engineering of the Polish Academy of Sciences, Zabrze, Poland

B. Tomaszewska
Mineral and Energy Economy Research Institute of the Polish Academy of Science, Cracow, Poland

ABSTRACT: The study assessed the potential of RO system to reduce total dissolved solids and enhance removal of microelements, such as boron, fluoride, arsenic, to make geothermal waters suitable for reuse in various purposes. Preliminary treatment involved an iron-removal system and UF modules. The RO system was equipped with spiral wound polyamide thin-film composite membranes. Waters from three different geothermal areas were tested, i.e., the Podhale basin, Polish Lowlands and Western Carpathian Mountains. Efficient and stable performance of the BWRO-membrane equipped desalination system was achieved with geothermal waters containing 7 g/L TDS and a boron concentration of up to 10 mg/L.

Keywords: Desalination, geothermal water, membrane, reverse osmosis, boron

1 INTRODUCTION

The use of geothermal energy is of significant importance in the context of the goals and provisions of Directive 2009/28/EC of the European Parliament and of the Council on the promotion of the use of energy from renewable sources (OJ L 140/16 of 5.6.2009), Poland's National Energy Strategy until 2030 and the National Action Plan on support for developing the utilization of renewable energy sources (RES) (Kępińska & Tomaszewska 2010, Tomaszewska & Hołojuch, 2012). Therefore, the interest of Polish municipal local government bodies and businesses in the utilisation of geothermal waters for heating, and also for balneological and leisure purposes, has been growing in recent years (Bujakowski et al., 2008, Bujakowski 2010, Bujakowski et al., 2010, Tomaszewska et al., 2010, Kasztelewicz & Pajak 2010). Elevated salinity levels and the presence of microelements such as boron, fluoride, barium, strontium, bromides and heavy metals often lead to difficulties related to the disposal of spent (cooled) water. Spent (cooled) geothermal water may be discharged, e.g., into surface waters, or reinjected into the reservoir. In both cases, it is treated as a waste product of the energy extraction process. Water reinjection prevents its excessive extraction, which could result in a decrease of the formation pressure. However, the injection process poses numerous technical challenges and requires additional consumption of energy to drive the pumps. A partial solution to this problem may lie in the deployment of water treatment technologies, which will enable its further use (Bujakowski et al., 2012, Tomaszewska, 2011). Among the technological enhancements related to the desalination of seawater and saline underground waters which are continuously being developed and implemented, separation processes using pressure-driven membrane processes and hybrid methods that combine the advantages of various technologies play a significant role (Bodzek & Konieczny 2011a, 2011b, Mezher et al., 2011, Sauvet-Goichon 2007). Membrane-based water desalination processes and hybrid technologies that combine membrane processes are widely used to produce drinking water in many regions of the world. They are also considered a technologically and economically viable alternative for desalinating water (mainly seawater), often with the use of renewable (solar, wind, geothermal, photovoltaic) energy (Bodzek et al., 2011, Bodzek & Konieczny 2005). In these processes, the membrane can be viewed as a barrier between contaminated and purified water streams. The separation of these two streams often allows for operation with no or minimal chemical water pre-treatment, which otherwise can form deleterious by-products (Bodzek & Konieczny 2011b).

The study assessed the potential of UF-RO system to reduce total dissolved solids (TDS) and enhance the removal of microelements to make geothermal waters suitable for discharge into surface waters, reuse for drinking purposes or use to replenish network water losses in the geothermal heating system.

The study was performed at the Geothermal Laboratory of the Mineral and Energy Economy Research Institute, Polish Academy of Sciences (PAS MEERI), using a dual hybrid process combining ultrafiltration and two independent stages (RO-1 and RO-2) connected in series (Tomaszewska & Bodzek 2012). The membrane separation performance was assessed in short and long-term tests, at a semi-production scale (ca. 1 m³/h of desalinated water production).

2 MATERIALS AND METHODS

2.1 Geothermal waters

Waters from three different geothermal areas were tested, i.e. the Podhale basin (GT-1), Polish Lowlands (GT-2) and Western Carpathian Mountains (GT-3). The total dissolved solids (TDS) ranged from 2.5 to 24.4 g/L and they have high concentrations of iron, boron, fluoride, arsenic, strontium, silica and other microelements. The physical properties and chemical composition of the tested waters are shown in Table 1.

The feedwater had a temperature of 30°C at wells GT-1 and GT-2, and 22°C at GT-3 and the pH ranged from weakly acidic to alkaline.

2.2 Apparatus used

The water desalination facility include the following components (Fig. 1):

- a water pretreatment facility: mechanical filter, iron removal stage and ultrafiltration module (UFC M5, X-Flow, with hydrophilic capillary polyethersulfone membranes),
- a two-stage reverse osmosis setup with NaOH dosing before stage two (spiral wound DOW FILMTEC BW30HR–440i reverse osmosis membranes, designed for brackish water with increased silica, boron content, Table 2),
- final treatment to achieve drinking water parameters (mineralization, disinfection).

To bring the boron concentration below its maximum level for drinking water (1 mg/L) the desalination was carried out in a two-step process. The system was fitted with typical industrial plant components.

2.3 Analytical measurements

The quality of both the feed water and the treated (permeate) water was assessed using a continuous

Table 1. Physical properties and chemical composition of the geothermal water tested.

Element	GT-1	GT-2	GT-3
TDS, mg/L	2561.86	556.02	4447.0
pH	6.41	7.88	7.81
Total hardness, mg CaCO$_3$/L	645.4	474.2	328.2
Carbonate hardness, mg CaCO$_3$/L	213.9	184.1	328.2
Conductivity, mS/cm	3.550	10.960	35.500
Na, mg/L	466.8	2297	9492
K, mg/L	45.2	27.2	83.1
Ca, mg/L	196	146.8	71.24
Mg, mg/L	42.7	26.2	36.56
Cl, mg/L	536.0	3574	12815
SO$_4$, mg/L	938.2	193.7	<3
As, mg/L	0.026	<0.005	0.094
B, mg/L	8.98	2.53	96.73
F, mg/L	2.230	0.696	1.050
Cr, mg/L	0.012	0.044	0.758
Cd, mg/L	<0.0003	<0.0005	<0.0005
Ni, mg/L	0.013	<0.005	0.005
Pb, mg/L	0.001	0.0005	<0.0005
PO$_4$, mg/L	0.03	0.289	0.315
Hg, mg/L	<0.0001	<0.0001	<0.0001
Al, mg/L	0.015	0.01	0.049
Mn, mg/L	0.041	0.181	0.051
Fe, mg/L	3.89	1.93	1.378
Sr, mg/L	6.0	5.5	41.34
SiO$_2$, mg/L	42.73	35.78	11.63

on-line measurement of its unstable physical parameters, temperature and specific electrolytic conductivity. Alkalinity was measured immediately after sampling the water from the installation using the electrometric method. Inorganic components were measured using the inductively coupled plasma mass spectroscopy method (ICP-MS) and, for the fluoride concentration, the spectrophotometric method (Cary UV-vis). Chloride ion concentration and alkalinity were determined by titration following accredited testing procedures. Tests concerning the total count of microorganisms at 22°C after 68 hours were conducted as per PN-EN ISO 6222:2004.

2.4 UF-RO tests

The membrane separation performance was assessed at a semi-production scale (ca. 1 m³/h of desalinated water production). The effectiveness of a membrane water treatment process was evaluated on the basis of permeate flux and relevant rejection ratios. The performance of the filtration process was determined by measuring

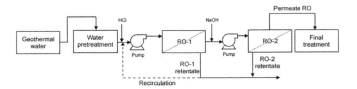

Figure 1. Water desalination facility diagram.

Table 2. Manufacturer's specifications of the DOW FILMTEC BW30HR-440i membrane.

Membrane type	Polyamide thin-film composite
Active area (m^2)	41
Maximum operating pressure (bar)	41
Maximum pressure drop (bar)	1.0
Maximum operating temperature (°C)	45
Maximum temperature for continuous operation above pH 10 (°C)	35
pH range, continuous operation	2–11
pH range, short-term cleaning (30 min.)	1–13
Maximum feed flow (m^3/h)	19
Maximum feed Silt Density Index (SDI)	5
Salt rejection minimum (%)*	99.4
SiO$_2$ rejection stabilized (%)*	99.9
Boron rejection stabilized (%)*	83

*Salt rejection and specific solute stabilized rejections based on the following standard test conditions: 2,000 ppm NaCl, 15.5 bar, 25°C, pH 7 and 15% recovery.

the permeate flux J_v (Bodzek & Konieczny 2005):

$$J_v = \frac{V}{F \cdot t} \tag{1}$$

where: V—volume (L), F—membrane surface area (m^2), t—filtration time (s).

Relevant rejection ratios (R, %) were calculated with the following formula:

$$R = \left(1 - \frac{C_p}{C_f}\right) \cdot 100\% \tag{2}$$

where: C—concentration (mg/L), p—permeate, f—feedwater.

3 RESULTS AND DISCUSSION

The GT-1 geothermal well is situated within the PAS MEERI Geothermal Laboratory. It has enable to conduct on-site studies with the use of its water on a semi-industrial scale for almost a year (and for 8 months on a continuous basis). At the other two wells (GT-2 and GT-3), 30 m^3 of water was tapped directly from the well into an insulated road tanker and sent immediately for desalination. The peak flow rate into the desalination system amounted to ca. 4–5 m^3/h (average 2.4 m^3/h) for GT-1 feedwater and 2–3 m^3/h (average 1.5 m^3/h) for GT-2 and GT-3.

Cooled water was fed to the desalination facility via a sealed pipeline, retaining its reduction potential. The water was only aerated in the iron removal system. The reduction of 99% in iron concentration was obtained by using an iron removal system. Water pressurised to around 0.3–0.5 MPa passes an MTM catalyst bed layer. Oxidised iron hydroxides, which precipitate in the form of flocks that settle easily, are retained on the surface of the catalyst bed. After iron removal system, the water was fed to the UF module. UF membranes were used to remove microsuspensions (<0.03 µm), colloids, bacteria and viruses. After UF, water electrolytic conductivity and hardness decreased by around 10%. The SDI measured in raw geothermal water was from 4.6 to 5 and after pre-treatment was from 2 to 2.8 (average 2.4). It is advantageous from the point of view of protecting RO membranes from fouling. After pre-treatment, the water was fed to the reverse osmosis stage. Due to the high carbonate hardness of the geothermal water studied, the feedwater was acidic with a pH ca. 5 and the process was carried out using module RO-1 at a transmembrane pressure of 1.1 MPa and feedwater recovery level of 75–78%. The feedwater had a temperature of 30°C at wells GT-1 and GT-2, and 22°C at GT-3. To bring the boron concentration below its maximum level for drinking water (1 mg/L) the pH of the permeate after RO-1 was corrected to about 10 and put to further filtration at a transmembrane pressure of 1.0 MPa (Tomaszewska & Bodzek, 2012). The process yielded a water recovery level of about 75%. The removal of boron compounds from waters are of special importance for its desalination by reverse osmosis. The rejection of boron for RO membranes under low or neutral pH varies from 40 to 60%, what is not sufficient to obtain the permissible level for drinking water or water disposed to the environment (Bodzek & Konieczny 2011b). Thus, the RO permeate is alkalized to pH ca. 9.5–10 and once more treated by RO (Redondo et al., 2003, Dydo et al., 2005, Kabay et al., 2010, Koseoglu et al., 2010, Bodzek & Konieczny 2011b, Tomaszewska & Bodzek 2012). Test results indicate that, in general, apart from the boron content, the chemical composition of the water obtained after the RO-1 stage (from the GT-1 intake) met the requirements for drinking water. Taking into account a feed pH of ca. 5 before the RO-1 stage, the boron content in water from the GT-1 intake was reduced from 8.98 to 4.51 mg/L, and the rejection ratio was 48%. The rejection ratio in the case of mineralisation (TDS) was 93%; for SiO_2, it was 94%. Boron rejection following RO-2 was 97% (Tab. 3). High rejection ratios were also achieved for other elements—97% after RO-1 and 99% after RO-2 for chlorides and 99% for sulphates already being achieved after RO-1 (Tab. 3). Rejection ratios for key ingredients of water from the GT-2 well following RO-1 were in general lower than for the water tested from the GT-1 well but sufficient with respect to the requirements for drinking water. The one microelement the retention of which in this regard was insufficient from the point of view of the study objectives, was the boron

ion. Although its concentration in the feed was relatively low (2.53 mg/L), the permeate after RO-1 exhibited a boron content slightly above the limit for drinking water (1 mg/L). Taking into account new guidelines adopted by the WHO (2011) which proposed to revise the maximum allowable boron concentration in drinking water to 2.4 mg/L, a single-step desalination system would prove sufficient in this particular case. The best stable performance of the system (i.e. an average flux (J_v) of 6.25×10^{-6} m³/m²s at RO-1 and 9.03×10^{-6} m³/m²s at RO-2) was recorded for the feedwater with the lowest total dissolved solids (GT-1). Permeability of the membranes for feedwater with TDS at ca. 6.5 g/L ranged from 2.95×10^{-6} m³/m²s to 3.64×10^{-6} m³/m²s at RO-1 and from 4.86×10^{-6} m³/m²s to 8.68×10^{-6} m³/m²s at RO-2 (Tomaszewska & Bodzek, 2012). Desalination attempts failed on waters with a TDS of 24.4 g/L. Permeability of the BW30HR-440i membrane (with a transmembrane pressure of 1.1 MPa and initial flux of 5×10^{6} m³/m²s) dropped at RO-1 to 0.35×10^{-6} m³/m²s within one hour (Fig. 2). Concentrations of cations and anions of permeate after RO-2 were compared with national drinking waters standards (Tab. 4). According to Table 4, the permeate quality from GT-1 was much close to deionized water characteristics. Permeate reaction has to be adjusted following the second stage of RO in order to meet the requirements for drinking water in this respect (according to national regulations, the admissible pH for drinking water lies in the range 6.5–9.5), and the need to increase carbonate hardness during the remineralisation process. Bacteriological tests of desalinated water demonstrated that the total number of microorganisms after 68 hours for a water temperature of 22°C amounted to 9 units per 1 ml in product water from GT-1, 21 units per 1 ml in product water from GT-2 and 10 units from GT-3. The admissible number of microorganisms in water for potable water is included in standards.

Table 3. Rejection ratios (R, %) for key compounds: macro- and microelements after RO-1 (including iron removal and ultrafiltration process) and RO-2 in relation to feedwater concentration.

Element	RO-1/RO-2 [%]		
	GT-1	GT-2	GT-3
TDS	93/97	91/96	33/89
Total hardness	99/100	88/100	30/92
Carbonate hardness	99/100	68/100	86/92
Conductivity	90/97	92/93	26/87
Na	92/96	93/93	32/94
K	91/98	81/94	24/90
Ca	95/99	88/99	20/90
Mg	99/99	88/99	42/95
Cl	97/99	94/99	27/90
SO_4	99/99	99/99	–
As	95/96	–	77/94
B	48/97	56/94	12/66
F	92/94	89/90	88/90
Cr	99/99	89/99	59/87
Cd	–	–	–
Ni	88/93	–	99/99
Pb	99/99	–	–
Hg	–	–	–
Al	99/99	99/99	59/59
Mn	99/99	58/99	25/75
Fe	99/99	87/98	97/99
Sr	99/99	90/96	63/98
SiO_2	94/99	81/99	0/84

"–" Concentration in feedwater and in permeate lower than the laboratory quantification limit.

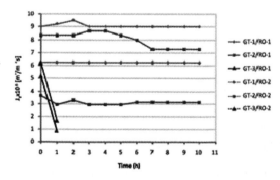

Figure 2. Changes in the volume water flux (J_v) during the filtering of geothermal waters from the GT-1, GT-2 and GT-3 intakes through RO-1 and RO-2 membranes.

Table 4. Comparison of permeate after RO-2 with standards.

Element	GT-1	GT-2	GT-3	Standard
TDS, mg/L	83.1	291.6	2588.0	–
pH	9.91	9.5	10.28	6.5–9.5
Total hardness, mg CaCO$_3$/L	0	0.6	25.6	60–500
Carbonate hardness, mg CaCO$_3$/L	0	0.6	25.6	–
Conduct., mS/cm	0.086	1.887	4.500	2.500
Na, mg/L	18.6	151.8	575.1	200
K, mg/L	0.83	1.76	8.19	–
Ca, mg/L	<10	0.24	7.11	–
Mg, mg/L	<0.1	0.1	1.905	30–125
Cl, mg/L	7.6	11.2	1294.0	250
SO$_4$, mg/L	6.4	<3.0	<3	250
As, mg/L	0.001	<0.005	0.006	0.010
B, mg/L	0.24	0.1593	2.98	1.0
F, mg/L	0.137	0.078	0.104	1.5
Cr, mg/L	<0.005	0.005	0.095	0.050
Cd, mg/L	<0.0005	<0.0005	<0.0005	0.005
Ni, mg/L	<0.005	<0.005	<0.005	0.02
Pb, mg/L	<0.0005	<0.0005	<0.0005	0.025
PO4, mg/L	<0.006	<0.006	0.06	–
Hg, mg/L	<0.0001	<0.0001	<0.0001	0.001
Al, mg/L	0.005	<0.01	0.02	0.200
Mn, mg/L	<0.003	<0.005	0.013	0.050
Fe, mg/L	0.013	0.03	0.037	0.200
Sr, mg/L	0.006	0.2	0.907	–
SiO$_2$, mg/L	0.198	0.31	2.15	–

Table 5. Results of microbiological tests on desalinated water.

Element	GT-1	GT-2	GT-3	Standard
Coliform bacteria count [units/100 mL]	0	0	0	0
Escherichia coli count [units/100 mL]	0	0	0	0
Total number of microorganisms after 68 hours for a water temperature of 22°C	9	21	10	100

It should be noted that 100 microorganism units per 1 ml of water are admissible (Tab. 5).

The tests and analyses conducted have demonstrated that geothermal water from GT-1 well, treated using membrane processes has a secondary application, namely it can be used to replenish water losses in heating circuits and to fill such circuits. In Poland, the requirements for distribution water and the water used to replenish heating circuits are set forth in Polish Standard PN-85/C-04601 Water for Power Engineering Purposes. Water Quality Requirements and Tests for Water Boilers and Closed Heating Circuits. The standard sets separate requirements for water in installations that are replenished to a limited extent (up to 5 m^3/h) and with respect to those where losses exceed 5 m^3/h. Requirements concerning the quality of boiler water for filling and replenishing heating circuits are listed in Table 5. Water distribution that circulates in heating installations must meet certain parameters depending on the technical requirements set by the manufacturers of the equipment through which it flows. It should not cause scaling of boilers, should not corrode parts of the installation and should not foam. The permeate obtained may be used for filling and replenishing heating circuits, an anti-corrosion adjustment is required, in particular pH stabilisation and degassing. Adjustment of pH to a value around 10 (through NaOH dosage) before the second RO stage permits a high boron retention coefficient (96%) to be obtained. However, reducing the content of this ion is not required, and therefore pH adjustment may be introduced after water treatment.

The PN-85/C-04601 Polish Standard does not specify any requirements concerning the admissible concentrations of, *inter alia*, chloride and sulphate ions, i.e. anions that affect water corrosivity, mostly pitting corrosivity. These ions form soluble compounds with metals and thereby inhibit the formation and precipitation of metal oxides. The use of the membrane technology discussed here in the desalination of geothermal waters made it possible to achieve high retention coefficients—97% after RO-2 for chlorides and 99% for sulphates already after RO-1. During the desalination process, the water becomes oxygenated, and therefore it must be degassed by chemical means or in a vacuum degasifier before it is finally fed into the heating circuit. As a result of the tests conducted, it was found that treated water from GT-1 well met the requirements set forth in the standard and it can be used to replenish network water losses after degassing.

4 CONCLUSIONS

Each water intake has its specific features, and solutions for desalinating water, and wastewater disposal must be developed on a case-by-case basis. When engaging in theoretical considerations, one must therefore take into account the physical properties and chemical composition of water, environmental conditions, the manner in which the retentate will be utilized and the availability of suitable technologies.

Membrane-based water desalination technologies and hybrid technologies are widely used to produce drinking water in many regions all over the world. They are also considered a technologically and economically viable alternative for desalinating water (mainly seawater), often with the use of renewable (solar, wind, geothermal) energy. The tests conducted using the UF-RO system fitted with low-pressure BWRO membranes (ca. 1.1 MPa) showed that favorable retention ratios for key ingredients were obtained from geothermal waters with a TDS of up to 7 g/L and boron concentrations of up to ca. 10 mg/L. Desalinated geothermal waters exhibited high quality that allow them to be used for drinking, technological or household purposes.

Geothermal energy is being used for heating purposes to an ever greater extent. In many cases, using cooled waters for drinking purposes may be considered an alternative method of disposing of them. The tests and analyses conducted have demonstrated that geothermal water treated (with TDS of up to 3 g/L) using membrane processes can also be used to replenish water losses in heating circuits and to fill such circuits. It may be of particular significance in areas where there is a deficit of fresh water. One of such areas is the Podhale Basin where the facility discussed in the article is situated. It uses geothermal water sourced by the largest Polish geothermal heating installation.

REFERENCES

Bodzek, M. & Konieczny, K. 2011a. Usuwanie zanieczyszczeń nieorganicznych ze środowiska wodnego metodami membranowymi. Warszawa, Wyd. Seidel-Przywecki.

Bodzek, M. & Konieczny, K. 2011b. Membrane techniques in the removal of inorganic anionic micropollutants from water environment—state of the art, Archives of Environmental Protection, 37(2): 15–29.

Bodzek, M., Konieczny, K. & Kwicińska A. 2011. Application of membrane processes in drinking water treatment—state of art. Desalination and Water Treatment, 35: 164–184.

Bodzek, M. & Konieczny, K. 2005. Wykorzystanie procesów membranowych w uzdatnianiu wody. Bydgoszcz, Oficyna Wydawnicza Projprzem-EKO.

Bujakowski, W. 2010. The use of geothermal waters in Poland (state in 2009), Przegląd Geologiczny. 58(7): 580–588.

Bujakowski, W., Barbacki, A., Czerwińska, B., Pająk, L., Pussak, M., Stefaniuk, M. & Trześniowski, Z. 2010. Integrated seismic and magnetotelluric exploration of the Skierniewice, Poland, geothermal test site. Geothermics, 39(1):, 78–93.

Bujakowski, W., Pająk, L. & Tomaszewska, B. 2008. Renewable energy resources in the Silesian Voivodship (Southern Poland) and potential utilization, Gospodarka Surowcami Mineralnymi, 24(2): 409–426.

Bujakowski, W., Tomaszewska, B. & Bodzek M. 2012. Geothermal water treatment—preliminary experiences from Poland with a global overview of membrane and hybrid desalination technologies. In: J. Hoinkis & J. Bundschuh (ed.), Renewable Energy Applications for Freshwater Production (Sustainable Energy Developments), London, CRC Press INC.

Directive 2009/28/EC Of The European Parliament And Of The Council of 23 April 2009 on the promotion of the use of energy from renewable sources and amending and subsequently repealing Directives 2001/77/EC and 2003/30/EC (Official Journal of the European Union, 5.6.2009, L 140/16).

Dydo, P., Turek, M., Ciba, J., Trojanowska, J. & Kluczka J. 2005. Boron removal from landfill leachate by means of nanofiltration and reverse osmosis, Desalination, 185: 131–137.

Kabay, N., Güler, E. & Bryjak, M. 2010. Boron in seawater and methods for its separation—A review, Desalination, 261: 212–217.

Kasztelewicz, A. & Pająk. L. 2010. GEOCOM Project of the 7th Framework Programme of the European Union, Przegląd Geologiczny, 58(7): 631.

Kępińska B. & Tomaszewska, B. 2010. Główne bariery rozwoju wykorzystania energii geotermalnej w Polsce. Propozycje zmian. Przegląd Geologiczny, 58(7): 594–598.

Koseoglu, H., Harman, B.I., Yigit N.O., Guler, E., Kabay, N. & Kitis, M. 2010. The effects of operating conditions on boron removal from geothermal waters by membrane processes, Desalination, 258: 72–78.

Mezher, T., Fath, H., Abbas, Z. & Khaled, A. 2011. Techno-economic assessment and environmental impacts of desalination technologies, Desalination, 266: 263–723.

National Action Plan—Krajowy Plan Działania W Zakresie Energii Ze Źródeł Odnawialnych, 2010, (http://www.mg.gov.pl/files/upload/12326/KPD_RM.pdf)

National Drinking Water Standard—Rozporządzenie Ministra Środowiska z dnia 24 lipca 2006 r. w sprawie warunków, jakie należy spełnić przy wprowadzaniu ścieków do wód lub do ziemi, oraz w sprawie substancji szczególnie szkodliwych dla środowiska wodnego (Dz. U. z 2006 r. Nr 137, poz. 984 z późn. zm.).

PN-85/C-04601 Woda do celów energetycznych. Wymagania i badania jakości wody dla kotłów wodnych i zamkniętych obiegów ciepłowniczych.

Poland's National Energy Strategy until 2030—Krajowa Strategia Energetyczna Polski do 2030 r., 2009, (http://www.mg.gov.pl/files/upload/8134/Polityka%20energetyczna%20ost.pdf)

Redondo, J., Busch, M. & De Witte, J.P. 2003. Boron removal from seawater using FILMTECTM high rejection SWRO membranes, Desalination, 156: 229–238.

Sauvet-Goichon, B. 2007. Ashkelon desalination plant—A successful chellenge, Desalination, 203: 75–81.

Tomaszewska, B. 2011. The use of ultrafiltration and reverse osmosis in the desalination of low mineralized geothermal waters, Archives Of Environmental Protection, 37(3): 63–77.

Tomaszewska, B. & Bodzek M. 2012. Desalination of geothermal waters using a hybrid UF-RO process. Part I: Boron. removal in pilot-scale tests. Desalination, doi:10.1016/j.desal.2012.05.029.

Tomaszewska, B. & Hołojuch, G. 2012. Pozyskanie energii geotermalnej w świetle nowych uwarunkowań prawnych. Biul. PIG 448(2): 281–284.

Tomaszewska, B., Bujakowski, W., Barbacki, A.P. & Olewiński, R. 2010. Zbiornik geotermalny jury dolnej w rejonie Kleszczowa, Przegląd Geologiczny, 58(7): 603–608.

World Health Organization (WHO), Guidelines for Drinking-water Quality. Fourth edition. (www.who.int), 2011, ISBN 978 92 4 154815 1.

Environmental Engineering IV – Pawłowski, Dudzińska & Pawłowski (eds)
© 2013 Taylor & Francis Group, London, ISBN 978-0-415-64338-2

The removal of anionic contaminants from water by means of MIEX®DOC process enhanced with membrane filtration

M. Rajca

Faculty of Energy and Environmental Engineering, Silesian University of Technology, Gliwice, Poland

ABSTRACT: The impact of the water matrix (synthetic and surface water) on the effectiveness of the organic and inorganic anionic contaminants removal of using single stage MIEX®DOC process, and its hybrid combinations with low pressure membrane processes (micro-MF and ultrafiltration UF) were investigated. The ion exchange process with the use of strongly basic MIEX® resin was performed. The MIEX®DOC process is an alternative for classical ion exchange carried out on ionic columns as the anion exchange constantly takes place on the resin suspended in the treated water. The synthetic water containing fulvic and humic acids as well as anions (F^-, Br^-, NO_3^-) and surface water collected from Paprocanskie Lake (Tychy, Poland) were treated. The results of the study adjusted both, the applicability of MIEX®DOC to remove anionic contaminants from water as well as its combination with low pressure membrane technique.

Keywords: MIEX®DOC, ultrafiltration, microfiltration, fulvic and humic acids, anions

1 INTRODUCTION

The significant instability of surface water composition has a great impact on the classical treatment methods, as they do not always guarantee the production of water quality required for drinking water. MIEX®DOC system is an interesting ion exchange process, which significantly differs from classical ion exchange generally used on the water treatment plants, as the ion exchange takes place on the resin suspended in the treated water. The granulation of MIEX® resin is almost five times smaller than one of traditionally used resins, what increases the specific surface of the process and provides better performance of the ion exchange.

MIEX®DOC process can be applied to treat both, surface and ground water, which is characterized by significant color level and high concentration of natural organic matter (Kabsch-Korbutowicz et al. 2008). MIEX® resin removes from water various groups of contaminants i.e. anionic organic compounds (DOC—Dissolved Organic Carbon), inorganic anions (sulphates, sulphides, bromides, fluorides, nitrates, arsenites and arsenates), but also ions responsible for water hardness (calcium and magnesium) (Apell et al. 2010, Boyer et al. 2006). In Table 1 the origin of contaminants removed during MIEX®DOC process, their permissible concentration in drinking water as well as hazardous effect on human health are shown.

The integrated systems, in which MIEX® resin can cooperate with other water treatment processes, are of great interest. The ion exchange with the use of MIEX®DOC process can be an alternative for a classical coagulation. The application of MIEX®DOC process to water treatment becomes the very promising method which enables the elimination of disinfection byproducts (THM, bromates). The use of the process also allows for the reduction of required coagulant and disinfecting agents doses (Hsu et al. 2010, Rajca 2012).

MIEX®DOC process can also be combined with low pressure membrane techniques (microfiltration, ultrafiltration). These combinations can be made within two pathways:

- Integrated system, in which water is treated with the MIEX®DOC and the supernatant is directed to the membrane unit
- Hybrid system, in which the resin is introduced to the tank equipped with the immersed membrane (Kabsch-Korbutowicz et al. 2008, Jung et al. 2009).

It has already been found that there is a great possibility of wide range of contaminants removal from water using MIEX®DOC process in combination with membrane filtration. It is caused by the removal of low molecular weight compound during MIEX®DOC operation, while high-molecular weight compounds are rejected by the membrane in the extent that depends on the membrane

Table 1. The characteristic of ions removed during MIEX®DOC process.

Ion	Main source of contamination	Standard (acc. to the regulation 2007)	Hazards
DOC^-	Natural source	5.0 mg/L	THM precursors
F^-	Natural sources aluminum production, artificial fertilizers	1.5 mg/L	Teeth diseases, bones fluorosis
BrO_3^-	Byproducts ozonation in the presence of bromides	0.01 mg/L	Cancerogenic precursors of N-nitroso
NO_3^-	Eutrophication, wastes, artificial fertilizers	50 mg/L	Of cancerogenic properties

properties (i.e. size of pores). The additional advantage of such a system is the limitation of fouling phenomenon, which is responsible for the decrease of membrane capacity in time (Jung et al. 2009).

The aim of the study was to determine the effectiveness of anionic organic and inorganic compounds removal from water using single stage MIEX®DOC process and its combination with membrane filtrations.

2 MATERIALS AND METHODS

The synthetic water was made on the deionized water matrix in which fulvic (HF), humic (HA) acids and anions (F^-, Br^-, NO_3^-) were dissolved, while surface water was collected at Paprocanskie Lake in Tychy. The powdered humic acid was supplied by Sigma-Aldrich, powdered fulvic acid by Beijing Multigrass Formulation Co. Ltd., while anions in the form of KBr, NaF and $Mg(NO_3)_2$ by POCH. The characteristic of water used is shown in Table 2. In the ion exchange process the strongly basic macroporous resin MIEX® by Orica Watercareof granules size 150 μm was used. The resin was dosed to water (5 ml/L) in the form of suspension and after the process it was regenerated with 10% NaCl solution. The single stage ion exchange MIEX®DOC process was carried out for 30 minutes including time of the contact of the resin with water and sedimentation, while in the hybrid process configuration MIEX®DOC/UF (MF) the resin was dosed directly to the filtration chamber (in-line process).

The membrane filtration was carried out with the use of flat sheet PVDF membranes by Ge Infrastructure Water & Process Technologies, UF one and of cut off 30 kDa and MF of pore size 0,2 μm. Millipore CDS10 installation with Amicon 8400 chamber operated in the dead end mode (Fig. 1) was also used. The effectiveness of membranes separation (the retention coefficient R, %) was determined at the transmembrane

Table 2. The characteristic of water used.

Parameter	Synthetic water	Surface water
Temperature, °C	22	22
pH	6.92	7.02
Conductivity, mS/cm	0.28	0.32
Turbidity, NTU	1.99	2.48
Absorbance*UV_{254}, 1/m	45.3	24.0
Color*, mg Pt/L	73	22
TOC, mg/L	8.89	14.9
DOC*, mg/L	8.70	13.5
F^-, mg/L	10.1	0.53
Br^-, mg/L	0.86	Non measurable
NO_3^-, mg/L	98.9	0.88
SO_4^-, mg/L	Not added	46.4
SUVA**, m^3/gC·m	5.20	1.78

TOC—total organic carbon. DOC—dissolved organic carbon.
*Samples filtered via 0.45 μm filter. **Specific ultraviolet absorbance UV_{254}/DOC.

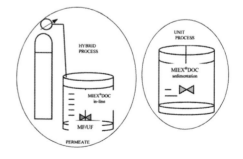

Figure 1. Schematic of research state.

pressure 0.1 MPa. The analysis of parameters of raw and treated water streams obtained in single stage MIEX®DOC process and in the hybrid MIEX®DOC/UF (MF) system included: Total and Dissolved Organic Carbon (TOC, DOC) content using HiperTOC analyzer by Thermo Corporation, absorbance at 254 nm (UV_{254}) with the use of UV/VIS CE 1021 spectrophotometer by Instruments, color using Spectroquant NOVA 400 photometer by Merck, anions (F^-, Br^-, NO_3^-) concentration using DX-120 ion chromatograph by Dionex.

3 RESULTS AND DISCUSSION

Results of the synthetic and surface water treatment using single stage and hybrid systems are shown in Tables 3 and 4.

The studies revealed that MIEX® resin was highly effective in the removal of DOC (theoretically, in the form of low and medium molecular weights up to 10 kDa) (Kabsch-Korbutowicz et al. 2008, Matilainen et al. 2010), nitrates, bromides and, in a lower extent, fluorides. In case of synthetic water treatment via MIEX®DOC process the concentrations of DOC and NO_3^- were in the range of permissible values for drinking water, established in the regulation (Table 1). The removal rate of bromides was also high, while for fluorides it was not as satisfactory. The high removal rates of DOC, UV_{254}, color and anions were also obtained for ultrafiltration and hybrid systems MIEX®DOC/UF (MF), while in case of single stage microfiltration the values of DOC, color and UV_{254} were ca. 10% lower. It confirmed the usability of

microfiltration to remove suspension and turbidity from water, and the efficiency of the process could be increased by its combination with other processes e.g. MIEX®DOC. Such a configuration of hybrid system is fully adjusted as MF membrane rejects resin granules in the filtration chamber, what enables the constant run of the ion exchange. As the number of active sites of the resin is limited, part of it is taken to the regeneration and the loses are completed with regenerated resin.

In case of surface water the treatment via hybrid MIEX®DOC/UF system enabled to produce water of quality that corresponded to drinking water parameters. The application of MIEX®DOC ion exchange decreased DOC value by 40%, while the removal rate was obtained for sulphate anions.

The effectiveness of DOC removal during ion exchange depends on many factors e.g. type and origin of raw water, pH, temperature and presence of competitive anions (especially sulphates), which are able to occupy active sites of the resin. It was found that DOC removal rates were higher for waters with higher SUVA value (Humpert et al. 2005). The results of this study confirmed such a relation.

One of the most important parameter which has a great impact on the MIEX®DOC process is the number of times of resin volume exchange. In Figure 2 the results of the experiments on the number of resin volumes exchange in the hybrid

Table 3. Values of anionic organic and inorganic contaminants in synthetic water treated using single stage and hybrid systems.

Treatment method	Parameter					
	UV_{254}, 1/m	DOC, mg/L	Color, mgPt/L	F^-, mg/L	Br^-, mg/L	NO_3^-, mg/L
Synthetic water	45.3	8.70	73	10.1	0.86	98.9
Microfiltration	40.9	7.12	62	–	–	–
Ultrafiltration	13.5	4.29	14	–	–	–
MIEX®DOC	7.10	2.55	18	6.90	0.20	57.0
MIEX®DOC/MF	12.9	3.89	20	8.85	0.36	58.6
MIEX®DOC/UF	1.90	1.55	2	8.55	0.37	43.0

Table 4. Values of anionic organic and inorganic contaminants in surface water treated using single stage and hybrid systems.

Treatment method	Parameter					
	UV_{254}, 1/m	DOC, mg/L	Color, mgPt/L	F^-, mg/L	SO_4^-, mg/L	NO_3^-, mg/L
Surface water	24.0	13.5	22	0.53	46.4	0.88
Microfiltration	20.0	13.2	19	–	–	–
Ultrafiltration	14.5	10.9	11	–	–	–
MIEX®DOC	4.10	7.50	5	0.33	5.03	0.20
MIEX®DOC/UF	2.30	2.58	2	–	–	–

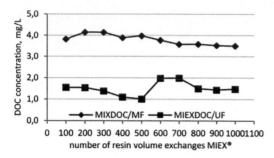

Figure 2. The dependence of DOC concentration on number of MIEX® resin volume exchange.

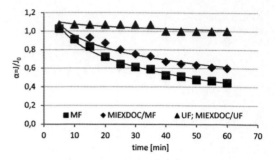

Figure 3. The dependence of relative membrane permeability of micro and ultrafiltration membranes on time during synthetic water treatment.

systems MIEX®DOC/UF and MIEX®DOC/MF are shown.

The obtained results indicated on the stable and efficient operations of MIEX®DOC/UF and MIEX®DOC/MF systems which decreased DOC concentration in water even after 1000-times resin volume exchange. It showed that the application of the resin to water highly contaminated with organic compounds before membrane filtration would improve the membrane operation, without the significant affection of the water.

In Figures 3 and 4 results of the study on the relative permeability of micro and ultrafiltration membranes used for synthetic (Fig. 3) and surface (Fig. 4) water treatment via single stage and hybrid systems are shown. The relative permeability of membranes allowed to indicate the intensity of membrane blocking (fouling) and was calculated as the ratio $\alpha = J/J_0$, where J—volumetric permeate flux, J_0—volumetric deionized water flux.

The experiments showed that the use of hybrid systems improved the relative permeability of both membranes in case of surface water treatment, while for synthetic water the constant and high capacity of ultrafiltration membrane, both as single stage or hybrid system operation, was observed (Fig. 3). The UF membrane was not affected by fouling, while the treatment of the synthetic water via single stage MF caused the decrease of α to ca. 0.4, and in the hybrid combination 10% improvement of the capacity was observed.

The comparison of effectiveness of synthetic and surface water treatment processes showed the impact of raw water composition on the cleaned water quality and membranes capacity. In case of synthetic water (which was less diversified in composition) the higher contaminants removal rate and less severe fouling were observed. The surface water was more diversified in composition, thus the higher competiveness of ions to resin active sites occurred. Moreover, the presence of low-molecular weight hydrophilic organic compounds (Table 1,

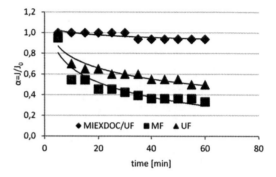

Figure 4. The dependence of relative membrane permeability of micro and ultrafiltration membranes on time during surface water treatment.

low SUVA—1.78 m³/gC·m) probably resulted in more severe blocking of both used membranes.

4 CONCLUSIONS

The studies revealed that the ion exchange process with the use of MIEX® resin enabled the efficient removal of both, anionic organic (DOC) and inorganic (fluorides, bromides, nitrates, and, in the higher extent, sulphates) compounds. The effectiveness of the process depended on the water matrix.

The combination of MIEX®DOC process with ultrafiltration or microfiltration improved significantly the quality of water in comparison with single stage operations. Moreover, the application of the hybrid system limited the intensity of membrane blocking, especially microfiltration ones. The hybrid systems, in which the MIEX® resin is directly introduced to the membrane filtration chambers can be successfully applied at water treatment plants, even those small ones, as such a combination does not require much space.

Additionally, membranes are a great barrier for resin granules, thus their loses during the process are eliminated.

ACKNOWLEDGEMENTS

This work was performed with the financial support from The Polish Ministry of Education and Science under the Grant No. N N523 61 5839.

Orica Watercare and Beijing Multigrass Formulation Co. Ltd. are acknowledged for supply of their products used in that work.

REFERENCES

Apell, J.N. & Boyer, T.H. 2010. Combined ion exchange treatment for removal of dissolved organic matter and hardness. *Water Research* 44: 2419–2430.

Boyer, T.H. & Singer, P.C. 2006. A pilot-scale evaluation of magnetic ion exchange treatment for removal of natural organic material and inorganic anions. *Water Research* 40 (15): 2865–2876.

Hsu, S. & Singer, P.C. 2010. Removal of bromide and natural organic matter by anion exchange. *Water Research* 44: 2133–2140.

Humbert, H. & Gallart, H. & Suty, H. & Crouc, J.P. 2005. Performance of selected anion exchange resins for the treatment of high DOC content surface water. *Water Research* 39 (9): 1699–1708.

Jung, C.W. & Son, H.J. 2009. Evaluation of membrane fouling mechanism in various membrane pretreatment processes. *Desalination and Water Treatment* 2: 195–202.

Kabsch-Korbutowicz, M. & Kozak, A. & Krupinska, B. 2008. Ion Exchange-ultrafiltration integrated process as a useful method in removing natural organic matter from water. *Environment Protection Engineering* 34 (2): 79–93.

Matilainen, A. & Vepsalainen, M. & Sillanpaa, M. 2010. Natural organic matter removal by coagulation during drinking water treatment: A review. *Advances in Colloid and Interface Science* 159: 189–197.

Rajca, M. 2012. The influence of selected factors on the removal of anionic contaminants from water by means of ion exchange MIEX®DOC process. *Archives of Environmental Protection* 38 (1): 115–121.

Rozporządzenie Ministra Zdrowia z dnia 29 marca 2007 r. w sprawie jakości wody przeznaczonej do spożycia przez ludzi (Dz.U.07.61.417) i z dnia 20 kwietnia 2010 r. zmieniające rozporządzenie w sprawie jakości wody przeznaczonej do spożycia przez ludzi (Dz.U.07.61.417).

Singer, P.C. & Boyer, T.H. & Holmquist, A. & Morran, J. & Bourke, M. 2009. Integrated analysis of NOM removal by magnetic ion exchange. *Journal American Water Works Association* 101 (1): 65–73.

Environmental Engineering IV – Pawłowski, Dudzińska & Pawłowski (eds)
© 2013 Taylor & Francis Group, London, ISBN 978-0-415-64338-2

Possibilities of using treated water analyses to monitor groundwater chemistry evolution

P. Kondratiuk

Department of Environmental Protection and Management, Technical University of Bialystok, Bialystok, Poland

ABSTRACT: Recently we have proposed a method of monitoring groundwater chemistry in an aquifer by the use of treated water analyses. This new method can be helpful in cases when only simple treatment of water is used (removing Fe and Mn ions during aeration and filtrations). Using the Holt's exponential smoothing method, the author extracted trends of physical-chemical parameters changes of the treated water in order to make predictions regarding the groundwater chemistry evolution. The research confirms that the presented method is a useful tool which may help with the identification of the processes occurring in the aquifer and for preparing water chemical composition changes forecasts in the regions without a groundwater monitoring system.

Keywords: groundwater, monitoring, trend, quality of water

1 INTRODUCTION

A good system of monitoring guarantees early recognition of unfavourable groundwater chemical composition changes and enables to undertake proper actions early enough to reverse such trends. Eventually, good condition of groundwater may be achieved, which is the basic aim formulated in the EU water framework directive (Directive 2000/60/EC). In the literature one can find examples of monitoring application for the implementation of the directive, as well as examples confirming the effectiveness of groundwater chemistry monitoring (Giedraitiene et al. 2002, Onorati et al. 2006). Unfortunately, the groundwater monitoring system organized within the confines of the state programme of monitoring is very limited, compared with the number of existing potential pollution sources. For example, the currently executed programme in the whole Podlaskie Province (NE Poland) involves only about 80 points (wells) for both shallow and deep groundwater. Unfortunately, the cases of organizing local monitoring systems, which would control local pollution sources and observe the trends of groundwater quality changes, are very rare in Poland. However, it should be noted that practically for every groundwater intake vast amounts of data are collected, which characterize the chemical composition of the water, not only the raw but also the treated one, which is distributed to the water-supply system. The data is produced by the sanitary-epidemiological stations, which regularly control the drinking water. Every year such stations make from a few dozen to a few hundred water analyses for every water intake. In general, the results remain unused apart from the up-to-date drinking water control and quality assessment.

Very often the water from groundwater intakes undergoes only a short treatment. Therefore, it can be assumed that in many cases a large part of the water chemical parameters does not change significantly during the process of treatment. Consequently, the results of the regular controls of the drinking water from the water-supply system can be used for the indirect observation of the raw water, i.e. to monitor the processes occurring in the aquifer.

The method of using the results of treated water analyses for the monitoring of the groundwater chemistry changes was introduced for the first time in 2009. The method was tested on the data from the town intake in Hajnówka, NE Poland (Kondratiuk 2006). In the current paper, we present a slightly modified method of extracting trends from the data and creating forecasts of the water chemistry changes. Another important aim of this paper was to assess the usefulness and reliability of the above-mentioned method. For this reason, the author analysed and compared the data from two intakes located in two different places, collected during the same period.

2 MATERIALS AND METHODS

Possibilities of using sanitary stations investigations for monitoring changes of various ions concentrations in groundwater were checked on the basis of the results of physical-chemical analyses of pipe water done for pipe water from water-supply system in two towns, Hajnówka and Bielsk Podlaski, both located in Podlaskie Province, north-east Poland (Fig. 1). According to Kondracki's physicgeographical division (2002), the area of the cities belongs to the Bielsk Upland mesoregion (the Północnopodlaska Lowland macroregion). Both towns have the population of about 25,000 inhabitants and serve as administrative, food and building industry centres of the counties.

The main relief features of the Bielsk Upland surface were formed in the Quaternary, during the Warta glaciation—the last one which covered this region. In principle, the relief is rolling-plain, but spatially quite differentiated. The elevation of the surface varies from 130 to 150 m above the sea level. There are only Quaternary sediments within the town areas. In general, they are glacial (tills) and fluvioglacial sediments (sands), whose thickness

varies significantly—from 107 to 206 m. In small river valleys peat can also be found.

Bielsk Podlaski and Hajnówka are located in the Podlachian climate region, in the Bialystok subregion (Górniak 2000). The climate is transitional, with distinct features of the continental climate. The meteorological data (years: 1961–1995 for Białystok) suggests that the mean annual air temperature is equal to 6.8°C, the warmest month is July (monthly mean temperature: 17.3°C) and the coldest one is January (monthly mean temperature: –4.3°C). The mean annual precipitation is equal to 593 mm. The months from May to August are characterized by highest rainfalls, with the maximum in July. The precipitation is the lowest during the period from January to March. Snowfall make up approximately 21% of the annual precipitation.

For the purpose of the research, 491 drinking water analyses were used, made by sanitary stations during the period 1991–2000. The determined parameters in these control tests were: the reaction (pH),

Figure 1. Location of the study areas.

84

total hardness, ammonia, nitrate and nitrite nitrogen, chloride, iron, manganese, turbidity, colour as well as the oxidability of water. These results came from typical, routine investigations, whose performing belongs to the statutory tasks of the stations. Controlling quality of piped water was based on testing samples taken from various, usually random places in the towns. In both towns, the water supplied to the pipe system was treated by aeration and filtrations only, which is a method used at most water intakes in Podlaskie Province. The next step was calculating the characteristic values (in this case—the averages) of every parameter for each quarter of a year. In fact, if more data had been available, the average value of each parameter for a shorter period of time (for instance a month) could have been calculated. After preparing the time series the following part of work was extracting the development tendencies (the trends).

The applied smoothing method was Holt's trend corrected exponential smoothing (Gardner 2006). It is a procedure eliminating accidental data from the time series and providing formal trend model used for forecasting. The Holt's trend corrected exponential smoothing is appropriate when both the level and the growth rate of a time series vary.

The formulae for the Holt's trend corrected exponential smoothing are the following (Eq. 1).

In Hajnówka, the water supplied to the pipe system is exploited from Tertiary fine sand aquifer from the depth of about 110 m. In Bielsk Podlaski, the water is pumped from the depth of about 60–70 m from medium sandy aquifer. Both the exploited aquifers are well isolated from the surface of the ground but remain in hydraulic connection with another Quaternary aquifers, which lays above.

At the first stage of the research, the author checked whether statistically significant differences exist between the parameters of raw and treated water, sampled for the analysis at the same time. For this reason, the results of nearly thirty analyses were used, which satisfied the above mentioned condition. For the comparison, the t-Student test was used, with the significance level set to 0.05.

The second part of the work consisted of setting together all the results of the water chemical analyses and exposing them to the, so called, crossing tests (Rao 1997). They involved all the preliminary work done in order to identify the nature of the data, detect measurement errors, find false observations and observations not fitting to the rest of the data. Among others, "box-and-whisker" graph drew by means of the Gnumeric 1.10 spreadsheet program were used for that reason. Those charts were used with the initial matrix of data, to among others, determine the extreme observations. In some probes, the extreme value of a particular parameter did not match to the distribution of the other ones. In such cases, the outlying result was regarded as wrong or presenting, so called, "emergency" or atypical situation (for example: a test performed after receiving by the sanitary station some information about a rapid decrease of water quality, or testing a probe sampled from place where water had not been taken from for a long time) and removed from the dataset. Afterwards, the data was sorted by time.

The next step was calculating the characteristic values (in this case—the averages) of every parameter for each quarter of a year. In fact, if more data had been available, the average value of each parameter for a shorter period of time (for instance a month) could have been calculated. After preparing the time series the following part of work was extracting the development tendencies (the trends).

The applied smoothing method was Holt's trend corrected exponential smoothing (Gardner 2006). It is a procedure eliminating accidental data from the time series and providing formal trend model used for forecasting. The Holt's trend corrected exponential smoothing is appropriate when both the level and the growth rate of a time series vary.

The formulae for the Holt's trend corrected exponential smoothing are the following (Eq. 1):

$$l_t = \alpha y_t + (1 - \alpha)(l_{t-1} + b_{t-1})$$
$$b_t = \gamma(l_t - l_{t-1}) + (1 - \gamma)b_{t-1} \tag{1}$$

where: y_t—stands for the true value at time t; l_t—stands for the estimated level at time t; b_t—stands for the estimated growth rate at time t; α—the "damping factor" or the "smoothing constant"; γ—stands for the "growth damping factor".

The standard error formula for the Holt's trend corrected exponential smoothing is shown in Eq. (2).

$$s_t = \sqrt{\frac{\sum_{i=1}^{t}[(y_i - (l_{i-1} + b_{i-1})]^2}{t - 2}} \tag{2}$$

Using the available in the Gnumeric 1.10 spreadsheet program exponential smoothing algorithm, a few variants of the trend were calculated for each tested parameter of water. The models with various values of the smoothing constant α from the range <0.1; 0.3> were checked. The final choice of the trend model illustrating best the development tendency was made on the basis of the trend line shape on the graph and the comparison of the fitting measures (errors) calculated for each variant. In the last stage of the research, we performed statistical tests (t-Student and the parametric correlations) in order to compare and determine relations between the extracted trends for the individual parameters of water taken from the two examined aquifers. The statistical significance level was taken as 0.05.

3 RESULTS

The results of the statistical tests revealed that the only significant impacts of simple treatment process (aeration and filtration) are the reduction of the iron content in water, its turbidity and (to a smaller extent) of the manganese content. No other statistically significant changes of the water chemical composition were observed. Those phenomena are clearly illustrated in the "box-and-whisker" graphs (Figs. 2 and 3), which show the influence of treatment on the iron (the values strongly depend on the treatment) and ammonium (the values are practically independent from the treatment) content in the water.

The analysis of the trends illustrating the changes of the parameters which turned out to be practically independent from the process of treatment, brought a series of interesting results. Some part of these parameters did not undergo any significant changes during the whole examined period

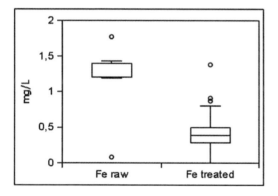

Figure 2. Box-and-whisker graph for iron content in water.

and remained almost constant. This phenomenon was observed for the water from both towns and concerns, among others, ions which may indicate human pressure, i.e. nitrate and nitrite. Both ions appeared in very low concentrations in the water—the first one usually amounted to 0.1–0.3 mgN/L, and the second one—usually a few thousandth of mgN/L. Also, the chloride concentration in water, which is one of the best human pressure indices, was very low all the time in both aquifers no more than a few mg/L. The water reaction was similarly stable and the most varied from 7.1 to 7.2 pH.

A distinct increase (Fig. 4) was noticed while monitoring the ammonia in the Tertiary aquifer in Hajnówka. At the beginning of the nineties, the average content of ammonia in water amounted to 0.2 mgN/L, while in the second half of the decade—to about 0.6 mg N/L. The changes of ammonia content clearly corresponded with the observed changes of the water hardness (Fig. 5), which increased from 240 to 280 mg $CaCO_3$/L. This relation was confirmed by calculating the Pearson correlation coefficient—its value turned out to be as high as 0.95! As for the calculations, it was taken into consideration that the changes of the water hardness were delayed by about 4 quartiles in relation to the ammonia concentration changes (as can be seen in the figures).

The same procedure of extracting trends was applied to the above-mentioned parameters of water taken from the Quaternary aquifer in Bielsk. This analysis brought similar results. Similar changes of ammonia concentration (values: 0.3 to 0.6 mg N/L), as well as the water hardness (values: 265 to 305 mg $CaCO_3$/L) were noted (Figs. 6 and 7). The correlation coefficient for these trends was as high as 0.80! Also in Bielsk Podlaski, the water hardness were delayed by about 4 quartiles in relation to the ammonia concentration changes too.

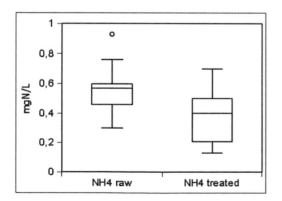

Figure 3. Box-and-whisker graph for ammonia content in water.

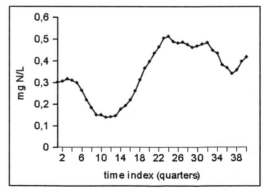

Figure 4. Trend line for ammonia in water in Hajnówka (1991–2000).

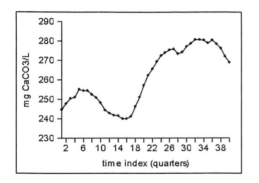

Figure 5. Trend line for water hardness in Hajnówka (1991–2000).

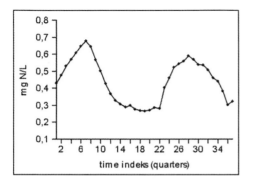

Figure 6. Trend line for ammonia in water in Bielsk-Podlaski (1991–1999).

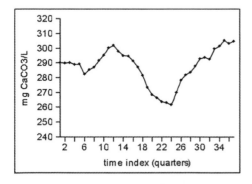

Figure 7. Trend line for water hardness in BielskPodlaski (1991–1999).

4 DISCUSSION AND CONCLUSION

As literature suggests, the lack of statistically significant differences between the values of some water parameters before and after the treatment is typical and connected with the chemical properties of substances occurring in water, as well as with the process of treatment. Chloride, which does not participate in the adsorption-exchange and oxidation-reduction processes (Macioszczyk & Dobrzyński 2007), the water hardness, but also the nitrogen compounds do not undergo significant changes during treatment of this type. As a result of traditional aeration and filtrations, the concentration of the ammonia can be decreased by not more than about 0.1–0.3 mgN/L (Kowal & Świderska-Bróż 1997). The reason of this phenomenon is that iron is always oxidized before NH_4^+, whose oxidization may be additionally disturbed by the simultaneous process of the manganese oxidization. The nitrate is not susceptible to traditional methods either, and in order to remove it from water, more complicated and expensive methods, such as electrodialysis, reverse osmosis, or ion exchange, have to be used (Kowal & Świderska-Bróż 2003). These facts indicate that in many cases, if the water is treated by simple methods of aeration and filtrations, it is possible to use water samples analyses results to conclude about the chemical composition of raw water and to identify processes occurring in the aquifer (Kondratiuk 2001). In our research, the data describing such parameters as: the reaction (pH), the total hardness, chloride, ammonia, nitrate and nitrite, was used for concluding about the groundwater chemistry evolution in aquifer.

The extracted trends let us state that neither of the analysed aquifers was subject to human pressure, as low nitrate concentrations suggest (Hansen et al. 2011, Nagarajan et al. 2010, Rupert 2008, Rosen & Lapman 2008) and the observations of chloride confirm this fact. The noticed trends of water hardness and ammonia concentration increase should not be associated with people's negative impact on the groundwater. They are rather related to the decrease of groundwater exploitation (a distinct decrease in both towns in the second half of the nineties). It had to cause a decrease of the recharging pace of the exploited aquifer by water from higher aquifers, prolonging the contact of the water with rock. Consequently, the water hardness and the content of NH_4^+ ions increased. Higher concentrations of NH_4^+ ion, especially in deeper aquifers, are typical for the local environment and are observed in north-east Poland very often (Małecka 1987).

In two analysed aquifers the same phenomena could be observed, which confirms the authors' reasoning and supports their new method. In general, our results confirm the advantages of using statistical methods for water chemistry analyses (Kumar et al. 2009, Dar et al. 2011).

The applied procedure of the time series analysis turned out to be a tool sensitive to even small changes of groundwater chemistry, such as those caused by normal exploitation. Changes of physical-chemical parameters of water caused by infiltration of pollution are usually much larger and would be stressed in the appropriate trends incomparably stronger. Thus, long-term observations of treated water may be successfully used as a basic tool for the analysis of processes and phenomena occurring in groundwater. As it is important from the practical point of view, it is definitely cheaper than the traditional methods of groundwater analysis (Dhar et al. 2008, Davies et al. 2010). The trend analysis performed in the above-mentioned way can be also used for planning the investments in the water-treating devices.

ACKNOWLEDGEMENTS

Financial support for research was provided by project S/WBiIS/1/2011.

REFERENCES

Dar, M.A.; Sanakar, K.; Dar, I.A. 2011. Major ion chemistry and hydrochemical studies of groundwater of parts of Palar river basin, Tamil Nadu, India, *Environmental Monitoring and Assessment* 176 (1–4): 621–636.

Davis, A.; Heatwole, K.; Greer, B.; Ditmars, R.; Clarke, R. 2010. Discriminating between background and mine-impacted groundwater at the Phoenix mine, Nevada USA, *Applied Geochemistry* 25 (3): 400–417.

Dhar, R.K.; Zheng, Y.; Stute, M.; van Geen, A.; Cheng, Z.; Shanewaz, M.; Shamsudduha, M.; Hoque, M.A.; Rahman, M.W.; Ahmed, K.M. 2008. Temporal variability of groundwater chemistry in shallow and deep aquifers of Araihazar, Bangladesh, *Journal of Contaminant Hydrology* 99 (1–4): 97–111.

Directive 2000/60/EC of the European Parliament and of the Council of 23 October 2000 establishing a framework for Community action in the field of water policy.

Gardner, E. 2006. Exponential smoothing: the state of the art—part II, *International Journal of Forecasting* 22: 637–666.

Giedraitiene, J.; Satkunas, J.; Graniczny, M.; Doktor, S. 2002. The chemistry of groundwater: A geoindicator of environmental change across the Polish-Lithuanian border, *Environmental Geology* 42 (7): 743–749.

Górniak, A. 2000. *Klimat województwa podlaskiego*. Instytut Meteorologii i Gospodarki Wodnej, Białystok. [In Polish]

Hansen, B.; Thorling, L.; Dalgaard, T.; Erlandsen, M. 2011. Trend Reversal of Nitrate in Danish Groundwater—a Reflection of Agricultural Practices and Nitrogen Surpluses since 1950, *Environmental Science & Technology* 45 (1): 228–234.

Kondracki, J. 2002. *Geografia regionalna Polski*. Wydawnictwo Naukowe PWN, Warszawa. [In Polish]

Kondratiuk, P. 2001. Ocena środowiska hydrogeologicznego z wykorzystaniem analizy statystycznej. *Zeszyty Naukowe Politechniki Białostockiej* 135: 215–224. [In Polish]

Kondratiuk, P. 2006. Possibilities of using results of investigations done by sanitary-epidemiological stations for monitoring changes of groundwater chemical composition. *Polish Journal of Environmental Studies* 15 (5D): 438–440.

Kowal, A.L. & Świderska-Bróż, M. 2003. *Oczyszczanie wody*, Wyd. Naukowe PWN, Warszawa. [In Polish]

Kumar, M.; Ramanathan, A.; Keshari, A.K. 2009. Understanding the extent of interactions between groundwater and surface water through major ion chemistry and multivariate statistical techniques, *Hydrological Processes* 23 (2): 297–310.

Macioszczyk, A. & Dobrzyński, D. 2007. *Hydrogeochemia strefy aktywnej wymiany wód podziemnych*. Wyd. Naukowe PWN, Warszawa. [In Polish]

Małecka, D. 1987. Aktualne tło hydrochemiczne wód podziemnych zlewni Supraśli i Horodnianki jako poziom odniesienia ich antropogenicznych przekształceń. in: *Ochrona wód podziemnych na obszarach zurbanizowanych. Materiały VII sympozjum – Częstochowa* 74–95. [In Polish]

Nagarajan, R.; Rajmohan, N.; Mahendran, U.; Senthamilkumar, S. 2010. Evaluation of groundwater quality and its suitability for drinking and agricultural use in Thanjavur city, Tamil Nadu, India, *Environmental Monitoring and Assessment* 171 (1–4): 289–308.

Onorati, G.; Di Meo, T.; Bussettini, M.; Fabiani, C.; Farrace, M.G.; Fava, A.; Ferronato, A.; Mion, F.; Marchetti, G.; Martinelli, A.; Mazzoni, M. 2006. Groundwater quality monitoring in Italy for the implementation of the EU water framework directive. *Physics and Chemistry of the Earth* 31 (17): 1004–1014.

Rao, C.R. 1997. *Statistics and Truth*, World Scientific Publishing, Singapore.

Rosen, M.R. & Lapham, W.W. 2008. Introduction to the U.S. Geological Survey National Water-Quality Assessment (NAWQA) of Ground-Water Quality Trends and Comparison to Other National Programs, *Journal of Environmental Quality* 37 (5) Supplement: 190–198.

Rupert, M.G. 2008. Decadal-Scale Changes of Nitrate in Ground Water of the United States, 1988–2004, *Journal of Environmental Quality* 37 (5) Supplement: 240–248.

Environmental Engineering IV – Pawłowski, Dudzińska & Pawłowski (eds)
© *2013 Taylor & Francis Group, London, ISBN 978-0-415-64338-2*

The influence of natural organic matter particle size on the haloacetic acids formation potential

A. Włodyka-Bergier, M. Łągiewka & T. Bergier
AGH University of Science and Technology, Faculty of Mining Surveying and Environmental Engineering, Department of Environmental Protection and Management, Kraków, Poland

ABSTRACT: The paper presents the results of experiments on the influence of particle size of natural organic matter on the haloacetic acids formation potential. The experiments were conducted on water samples taken from rivers nearby Krakow. Organic matter of each sample was separated with membrane filters into fractions of a different particle size: <0.1 μm, 0.1–0.22 μm, 0.22–0.45 μm and >0.45 μm. The samples from this fractionation were chlorinated with sodium hypochlorite. Its dose was adjusted to obtain a residual free chlorine concentration between 3 and 5 mg/dm³ after 24 hours. After this time, the water chlorination by-products were analyzed with a gas chromatography. The result of the research, namely the fractions, which have the highest potential to form the particular organic water chlorination by-products, have been defined.

Keywords: disinfection by-products, haloacetic acids, natural organic matter, particle size

1 INTRODUCTION

Natural Organic Matter (NOM) is present in each reservoir of surface water and also in groundwater. NMO is a diverse mixture of organic compounds (large and small molecules), and its predominant part in water is constituted by dissolved humic substances (Świetlik & Sikorska, 2005, Teixeira et al. 2011). These are mostly products of living organisms metabolism and compounds resulting from the decomposition of dead organisms. Other compounds that form the NOM are the following: carbohydrates, proteins, porphyrins, plant pigments, amino acids, phenols, organic acids, hydrocarbons and others (Wei et al. 2008). This abundant variety of NOM causes a lack of a specific research methodology, which could ultimately determine the structure of the individual fractions and the nature of NOM. The presence of NOM in water is a reason of specific odor, discoloration and taste change (Teixeira et al., 2011), as well as a need for disinfection. NOM is also characterized by high reactivity with chlorine, what makes it a precursor of by-products during water chlorination (Bo et al. 2008, Lu et al. 2009, Teixeira et al. 2011). NOM by reacting with hypochlorous acid or sodium hypochlorite contributes to the formation of halogenated organic compounds (including chlorinated and brominated species) which are deemed harmful or dangerous (Richardson et al. 2007). The main chlorine disinfection by-products are Trihalomethanes (THM), Haloacetic Acids (HAA),

haloacetonitriles, haloaldehydes (including chloral hydrate), chlorophenols, halonitromethanes (including chloropicrin) (Bo et al. 2008, Lu et al. 2009, Nikolau et al. 2004). THM are the best examined and the dominant chlorination by-products group in terms of quantity. HAA are the second largest group of chlorination by-products, formed in reaction of chlorine mainly with humic and fulvic acids (Richardson et al., 2007). The most common of HAA are Monochloroacetic Acid (MCAA), Dichloroacetic Acid (DCAA), Trichloroacetic Acid (TCAA), Monobromoacetic Acid (MBAA), Dibromoacetic Acid (DBAA) and Bromochloroacetic Acid (BCAA) (Agus et al. 2009; Sérodes et al. 2003; Villanueva et al., 2003). HAA are not well investigated, although they generate much interest due to their negative impact on health. Toxicology studies have shown that HAA are potentially carcinogenic. However, dihaloacetic acids cause the greater threat to human health in comparison with mono- and trichloroacetic acids. Mutagenicity of the brominated HAA is higher than chlorinated species. Medical studies have shown that DCAA, DBAA, BCAA and TCAA are carcinogenic in laboratory tests on animals (Agus et al. 2009; Qi et al. 2004). In addition, DCAA and TCAA have been classified as probable human carcinogens. TCAA can also cause detrimental effects on liver, kidney, spleen and in general developmental defects (Włodyka-Bergier & Bergier, 2011). The HAA limit for drinking water is regulated by WHO (2008) and US EPA (2011) but it has not been defined in Polish regulations (the Regulation of the Minister of Health 2007). HAA formation is a subject of numerous research investigations and experiments worldwide. Their main challenge is to develop the methods of removing HAA from disinfected water. However, the improvement of the chlorination techniques to minimize these by-products formation is even more important (Chang et al. 2001, Lin et al. 2007, Lu et al. 2009, Nikolau et al. 2004, Włodyka-Bergier & Bergier 2011).

The studies on the influence of NOM particle size on the chlorination by-products focus on the analysis of particles whose sizes enclose in the very low range of several tens of nm (Amy et al. 1990, Chang et al. 2001, Huang et al. 2008, Wei et al. 2008). Amy et al. (1990) has affirmed that the organic matter fraction <1 kDa is the major precursor of THM formation. Other researchers, analyzing DOM molecular weight distribution, have shown that 64–87% of DBPs precursors have a molecular weight of 1–30 kDa (Chang et al. 2001). In another research (Wei et al., 2008) samples of raw water were taken from inlet to four large water treatment plants in China, and the stirred ultra-filtration cell with disc membranes were used to fraction them with following cut-offs: 1, 3, 10 and 30 kDa. This study has shown that fractions of DOM <1 kDa and 10–30 kDa are the major THM precursors. Generally research studies on DOM focus mainly on the particles below 30 kDa (0.007 μm) (Huang et al. 2008, Chalatip et al. 2009).

To extend their knowledge and complement up-to-date studies in this area, the authors of this article have decided to examine the HAA formation potential of NOM particles larger than those previously analyzed. In this research, water samples were fractionated into four groups of the NOM particle sizes: <0.1 μm, 0.1–0.22 μm, 0.22–0.45 μm, >0.45 μm, which were chlorinated, and the most reactive fraction regarding HAA formation has been identified.

2 MATERIALS AND METHODS

2.1 Samples

The water samples for this study were collected from the rivers nearby Krakow, that is Dlubnia, Pradnik and Rudawa. All samples were taken in the spring season (May 2011), during the period when the level of organic matter is high in surface water. The samples were stored in a dark room at 4°C until fractionation and chlorination.

2.2 Fractionation procedure

To divide NOM particles into above-mentioned fractions, water samples were fractionated under vacuum with nitrocellulose membrane filters by Millipore, with pore-sizes: 0.1 μm, 0.22 μm and 0.45 μm. The concentration of organic carbon (C-org) was measured in samples prior and after the fractionation. The reason of this operation was to determine the share of each fraction in each water sample. Each water sample was used to obtain 4 further samples: raw (unfiltered) water, water with particle size <0.45 μm, water with particle size <0.22 μm, water with particle size <0.1 μm. Each of these samples was chlorinated and HAA concentration was measured. From the difference of C-org concentrations in sequent filtered samples, a percentage share of each fraction (<0.1 μm, 0.1–0.22 μm, 0.22–0.45 μm, >0.45 μm) has been calculated. Based on HAA concentration in filtrates, an analogical method was used to calculate HAA concentrations formed by the fractions: <0.1 μm, 0.1–0.22 μm, 0.22–0.45 μm and >0.45 μm.

2.3 Chlorination

A 24-hour chlorination test was carried out to examine disinfection by-products formation potential. The water samples were chlorinated using chlorine water (NaClO), with a free chlorine dosage that would result in a residual free chlorine from 3–5 mg/dm³ after 24 h. All samples were adjusted to pH 7 by adding sulphuric acid or sodium hydroxide and a phosphate buffer. A chlorinated water samples were incubated at 25 ± 2°C in amber bottles with PTFE liners.

2.4 Analytical methods

The six HAA were analyzed, that is MCAA, MBAA, DCAA, TCAA, BCAA and DBAA. HAAs concentration were analyzed using acidic methanol esterification method (Nikolau et al. 2002) and GC-MS (Trace Ultra DSQII, Thermo Scientific). Helium was used as the carrier gas. The Rxi™-5 ms capillary column (Restek) was also used (film thickness 0.5 μm, column length 30 m, column diameter 0.25 mm). The HAAs were extracted using the liquid-liquid extraction method with MTBE (methyl tert-butyl ether). 0.9 cm³ of the extract was transferred into a 15-cm³ amber vial, then 2 cm³ of a solution of sulphuric acid in methanol (10%) was added and the vial was placed in water bath at 50°C for 1 hour. After this time, the vial was cooled in 4°C for 10 min and 5 cm³ copper (II) sulphate pentahydrate and anhydrous sodium sulphate solution (50 g/dm³ and 100 g/dm³ respectively) was added. The vial was shaken for 2 min and allowed to stand for about 5 min. The upper layer was used for an injection into GC. The column was heated from 40°C (0 min) to 100°C (5 min) with the temperature increase rate of 40°C/min, then to 200°C (0 min) with the temperature increase rate of 8°C/min. The method detection limit was 0.5 µg/dm³ for MCAA and MBAA, 0.01 µg/dm³ for other HAA. Free chlorine was analyzed using the DPD (N,N-diethylphenylendiamine) method (according to Polish Standard PN-ISO 7393-2). The free chlorine concentration was measured using the Aurius 2021 UV-VIS spectrophotometer (Cecil Instruments). The detection limit of this method was 0.03 mg/dm³. C-org was analyzed following Polish Standard PN-EN 1484. To oxidize organic matter the method of the chemical oxidation in fluid phase has been applied (sodium persulfate/100°C). CO_2, which was released in the process, has been analyzed with the use of the gas chromatograph with the Trace Ultra DSQII GC-MS mass spectrometer (Thermo Scientific). The method detection limit was 0.03 mg/dm³.

3 RESULTS AND DISCUSSION

Table 1 presents the concentration of HAA and C-org in each particle size of NOM fractions. C-org concentration was measured after fractionation but before chlorination. HAA concentration was measured after 24 hours of chlorination. All values of concentrations are given for all considered rivers, that is Dlubnia, Pradnik and Rudawa.

3.1 Raw water

Figure 1 presents the C-org concentration in each particle size fraction before chlorination for samples from rivers: Dlubnia, Pradnik and Rudawa.

Taking into consideration C-org in the organic matter fractions, in all analyzed samples the highest share of <0.1 µm fraction was observed (89% for Dlubnia, 68% for Pradnik and 71% for Rudawa River), than >0.45 µm (9% for Dlubnia, 28% for Pradnik and 15% for Rudawa water sample), than 0.22–0.45 µm (1% for Dlubnia, 13% for Rudawa and <1% for Pradnik). The lowest share in organic matter had 0.1–0.22 µm fraction (1% for Dlubnia, 4% for Pradnik and 1% for Rudawa water sample).

3.2 MCAA formation potential

The results of experiments on MCAA formation potential for NOM fractions of water from Dlubnia, Pradnik and Rudawa Rivers have been presented in Figure 2.

The greatest potential to form MCAA was presented with the 0.22–0.45 µm fraction. Its MCAA formation potential was 222.4 µg/mg C-org (Pradnik) and 166.9 µg/mg C-org (Dlubnia). The smallest MCAA formation potential was observed for organic matter particle size >0.45 µm (0.5 µg/mg C-org for Pradnik, 1.0 µg/mg C-org

Table 1. Organic compounds concentration in NOM fractions.

| | River | Concentration in NOM fraction, µg/dm³ | | | |
		<0.1 µm	0.1–0.22 µm	0.22–0.45 µm	>0.45 µm
MCAA	Pradnik	36.2	0.9	2.5	0.9
	Dlubnia	23.8	1.4	4.7	5.4
	Rudawa	25.3	nd	2.2	2.2
MBAA	Pradnik	4.2	3.9	2.3	1.1
	Dlubnia	6.0	7.6	0.5	1.6
	Rudawa	4.4	0.9	7.0	nd
DCAA	Pradnik	23.68	0.46	2.28	35.48
	Dlubnia	32.68	6.79	1.69	23.88
	Rudawa	65.37	0.86	5.44	30.43
TCAA	Pradnik	26.31	5.66	0.56	55.34
	Dlubnia	0.94	0.14	1.05	6.06
	Rudawa	1.68	0.30	2.97	2.56
BCAA	Pradnik	1.68	0.08	0.19	0.10
	Dlubnia	0.08	0.03	0.01	0.01
	Rudawa	0.08	0.01	0.04	0.01
DBAA	Pradnik	0.08	nd	0.04	0.01
	Dlubnia	0.02	0.02	0.01	nd
	Rudawa	0.01	0.01	nd	nd
C-org	Pradnik	4430	250	10	1780
	Dlubnia	3710	30	30	390
	Rudawa	4100	60	750	890

nd—not detected.

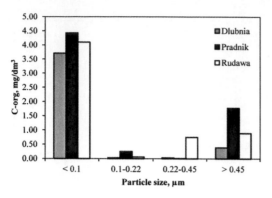

Figure 1. C-org concentration in NOM fractions of water from Dlubnia, Pradnik and Rudawa.

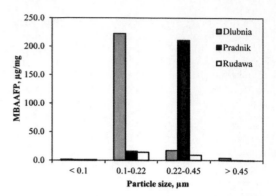

Figure 3. MBAA formation potential in NOM fractions of water from Dlubnia, Pradnik and Rudawa Rivers.

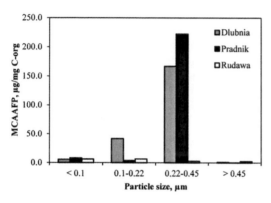

Figure 2. MCAA formation potential in NOM fractions of water from Dlubnia, Pradnik and Rudawa Rivers.

for Dlubnia. MCAA has not been detected in the sample from Rudawa river in the 0.1–0.22 μm fraction.

3.3 *MBAA formation potential*

The results of experiments on MBAA Formation Potential (MBAAFP) for NOM fractions of water from Dlubnia, Pradnik and Rudawa rivers have been presented in Figure 3. The 0.1–0.22 μm fraction from the Dlubnia River has acquired the highest value of MBAAFP: 222.1 μg/mg C-org. In the Pradnik River, in turn, most MBAA was formed by the fraction of 0.22–0.45 μm (210.6 μg/mg C-org). Whereas, the lowest MBAAFP characterized the fraction >0.45 μm from Pradnik (0.6 μg/mg C-org), in Rudawa MBAA was not detected for this fraction. The <0.1 μm fraction also reached low MBAAFP: 1 μg/mg C-org in Pradnik and slightly more than 1 μg/mg C-org both in Dlubnia and Rudawa.

3.4 *DCAA formation potential*

The results of experiments on DCAA formation potential (DCAAFP) for NOM fractions of water from Dlubnia, Pradnik and Rudawa Rivers have been presented Dlubnia, Pradnik and Rudawa Rivers have been presented in Figure 4.

The highest values of DCAAFP were observed for the 0.22–0.45 μm fraction in Pradnik (206.91 μg/mg C-org) and for the 0.1–0.22 μm fraction in Dlubnia (199.74 μg/mg C-org). The <0.1 μm fraction has shown the smallest DCAAFP, which was 5.35 μg/mg C-org in Pradnik.

3.5 *TCAA formation potential*

The results of experiments on TCAA Formation Potential (TCAAFP) for NOM fractions of water from Dlubnia, Pradnik and Rudawa Rivers have been presented in Figure 5.

TCAA was formed by all fractions of organic matter. However, the highest values of TCAAFP were observed for the 0.22–0.45 μm fraction, there were: 51.09 μg/mg C-org for Pradnik and 37.61 μg/mg C-org for Dlubnia. For the 0.1–0.22 μm fraction, the relatively high values of TCAAFP was also observed in Pradnik (31.05 μg/mg C-org), slightly lower for >0.45 μm fraction: 22.72 μg/mg C-org in Pradnik. The lowest TCAAFP was observed for the <0.1 μm fraction: it was 0.22 μg/mg C-org Dlubnia.

3.6 *BCAA formation potential*

The results of experiments on BCAA Formation Potential (BCAAFP) for NOM fractions of water from Dlubnia, Pradnik and Rudawa Rivers have been presented in Figure 6.

The highest value of BCAAFP was observed for 0.22–0.45 μm fraction from Pradnik (17.00 μg/mg C-org). The other values were significantly lower.

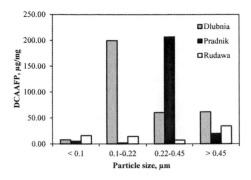

Figure 4.   DCAA formation potential in NOM fractions of water from Dlubnia, Pradnik and Rudawa Rivers.

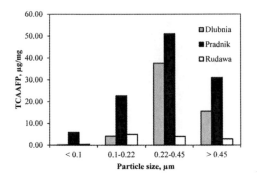

Figure 5.   TCAA formation potential in NOM fractions of water from Dlubnia, Pradnik and Rudawa Rivers.

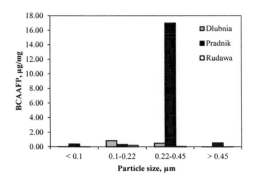

Figure 6.   BCAA formation potential in NOM fractions of water from Dlubnia, Pradnik and Rudawa Rivers.

For the same fraction from remaining rivers, it was 0.50 µg/mg C-org and 0.06 µg/mg C-org for Dlubnia and Rudawa respectively. The smallest amount of BCAA was formed by the >0.45 µm fraction— BCAAFP was 0.01 µg/mg C-org both for Pradnik and Dlubnia. Similarly low value of BCAAFP occurred for the <0.1 µm fraction: 0.02 µg/mg C org for both Dlubnia and Rudawa.

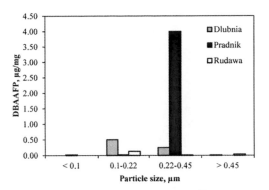

Figure 7.   DBAA formation potential in NOM fractions of water from Dlubnia, Pradnik and Rudawa Rivers.

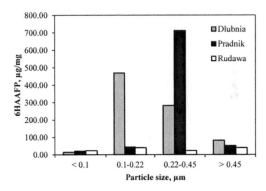

Figure 8.   6HAA formation potential in NOM fractions of water from Dlubnia, Pradnik and Rudawa Rivers.

3.7   *DBAA formation potential*

The results of experiments on DBAA Formation Potential (DBAAFP) for NOM fractions of water from Dlubnia, Pradnik and Rudawa Rivers have been presented in Figure 7.

The highest DBAAFP value (4 µg/mg C-org) was measured in the 0.22–0.45 µm fraction in Pradnik. The values for other rivers and fractions were noticeably lower. The lowest value (0.01 µg/mg C-org) was observed simultaneously for the 0.22–0.45 µm fraction for Rudawa and for the >0.45 µm fraction from Dlubnia. DBAA was not found in the >0.45 µm fraction from Dlubnia, 0.1–0.22 µm from Pradnik and 0.22–0.45 µm from Rudawa.

3.8   *6HAA formation potential*

The results of the sum of six considered haloacetic acids (6HAA) formation potential (6HAAFP) for NOM fractions of water from Dlubnia, Pradnik and Rudawa Rivers have been presented in Figure 8.

The highest value of 6HAAFP was observed in the 0.22–0.45 µm fraction from Pradnik: 712.0 µg/mg C-org. The high values of 6HAAFP were also observed in Dlubnia water for the 0.1–0.22 µm (469.15 µg/mg C-org) and for the 0.22–0.45 µm fraction (282.5 µg/mg C-org). In comparison with other rivers, the lowest 6HAAFP was observed for all NOM fractions from Rudawa.

4 CONCLUSIONS

The research results presented in the article, especially an analysis of HAAFP values for NOM fractions, have presented the relations between the NOM particle size and HAA concentrations resulted from chlorination of water from Dlubnia, Pradnik and Rudawa. The NOM fraction of particle size 0.22–0.45 µm has showed the highest formation potentials for MCAA, DCAA, TCAA, BCAA, DBAA (five of six analyzed HAA), as well as for 6HAA. The lowest formation potential was observed for NOM particle size <0.1 µm. For all three examined rivers, the lowest values of 6HAAFP were observed for this fraction. The relatively low HAA formation potentials were also noticed for >0.45 µm fraction.

Among all analyzed HAA, the higher formation potential was obtained for MCAA: 222.36 µg/mg C-org—it was observed in the 0.22–0.45 µm fraction from Pradnik. The high values of formation potential were also observed for acids MBAA (222.09 µg/mg C-org for the 0.1–0.22 µm fraction from Dlubnia and 210.64 µg/mg C-org for the 0.22–0.45 µm fraction from Pradnik) and DCAA (206.91 µg/mg C-org in the 0.22–0.45 µm fraction from Pradnik and 199.74 µg/mg C-org in the 0.1–0.22 µm fraction from Dlubnia). The lowest formation potentials among all 6HAAs were observed for BCAA (0.01 µg/mg C-org for the >0.45 µm fraction both from Dlubnia and Rudawa) and DBAA (0.01 µg/mg C-org both for the >0.45 µm fraction from Dlubnia and for the 0.22–0.45 µm fraction from Rudawa).

World Health Organization (WHO, 2011) defines the potable water standards for: MCAA, DCAA and TCAA, which is an evidence of the higher toxicity of these acids in a comparison to other HAA. MCAA and DCAA have showed the highest formation potential in presented studies. The Environmental Protection Agency (US EPA, 2011) establishes standards for the sum of 5 HAA: MCAA, DCAA, TCAA, MBAA, DBAA. For three of them (MCAA, MBAA, DCAA) the highest formation potential have been observed in conducted research. Defining the precursors of HAA formation is the important step to minimize the formation of these compounds in water

treatment utilities and can be employed to reform water treatment technology, and generally potable water management.

ACKNOWLEDGEMENTS

The work was completed under AGH University statutory research No. 11.11.150.008 for the Department of Management and Protection of Environment.

REFERENCES

Agus, E., Voutchkov, N., Sedlak D. 2009. Disinfection by-products and their potential impact on the quality of water produced by desalination systems: A literature review. *Desalination* 237: 214–237.

Amy, G.L., Alleman, B.C., Cluff, C.B. 1990. Removal of dissolved organic-matter by nanofiltration. *Journal of Environmental Engineering* 116 (1): 200–205.

Bo, L., Jiuhui, Q.U., Huijuan, L., Xu, Z. 2008. Formation and distribution of disinfection by-products during chlorine disinfection in the presence of bromide ion. *Chinese Science Bulletin* 53 (17): 2717–2723.

Chalatip, R., Chawalit, R., Nopawan, R. 2009. Removal of haloacetic acids by nanofiltration. *Journal of Environmental Sciences* 21 (1): 96–100.

Chang, E.E., Chiang, P., Ko, Y., Lan, W. 2001. Characteristics of organic precursors and their relationship with disinfection by-products. *Chemosphere* 44 (5): 1231–1236.

Huang, H., Charles, R.O. 2008. Direct-flow microfiltration of aquasols: II. On the role of colloidal natural organic matter. *Journal of Membrane Science* 325 (2): 903–913.

Lin, Y., Chiang, P., Chang, E.E. 2007. Removal of small trihalomethane precursors from aqueous solution by nanofiltration. *Journal of Hazardous Materials* 146 (1–2): 20–29.

Lu, J., Zhang, T., Ma, J., Chen, Z. 2009. Evaluation of disinfection by-products formation during chlorination and chloramination of dissolved natural organic matter fractions isolated from a filtered river water. *Journal of Hazardous Materials* 162 (1): 140–145.

Nikolaou, A., Golfinopoulos, S., Kostopoulou, M., Lekkas T. 2002. Determination of haloacetic acids in water by acidic methanol esterification–GC–ECD method. *Water Research* 36: 1089–1094.

Nikolau, A.D., Golfinopoulos, S.K., Lekkas, T.D., Arhonditsis, G.B. 2004. Factors affecting the formation of organic by-products during water chlorination: A Bench-Scale Study. *Water, Air & Soil Pollution* 159 (1): 357–371.

Qi, Y., Shang, C., Lo, I. 2004. Formation of haloacetic acids during monochloramination. *Water Research* 38: 2375–2383.

Richardson, S., Plewa, M., Wagner, E., Schoeny, R., Demarini, D. 2007. Occurrence, genotoxicity, and carcinogenicity of regulated and emerging disinfection by-products in drinking water: A review and roadmap for research. *Mutation Research* 636: 178–242.

Rozp. Min. Zdrowia (Rozporządzenie Ministra Zdrowia) 2007. *Rozporządzenie w sprawie jakości wody przeznaczonej do spożycia przez ludzi*, Dz. U. nr 61 poz. 417.

Sérodes, J., Rodriguez, M., Li, H., Bouchard, C. 2003. Occurrence of THMs and HAAs in experimental chlorinated waters of the Qubec City area (Canada). *Chemosphere* (51), 253–263.

Świetlik, J., Sikorska, E. 2005. Characterization of Natural Organic Matter Fractions by High Pressure Size-Exclusion Chromatography, Specific UV Absorbance and Total Luminescence Spectroscopy. *Polish Journal of Environmental Studies* 15 (1): 145–153.

Teixeira, M., Rosa, S.M., Sousa, U. 2011. Natural Organic Matter and Disinfection By-products Formation Potential in Water Treatment. *Water Resources Management* 25 (12): 3005–3015.

US EPA 2011. 2011 Edition of the Drinking Water Standards and Health Advisories. Washington DC: US Environmental Protection Agency.

Villanueva, C., Kogevinas, M., Grimalt J. 2003. Haloacetic acids and trihalomethanes in finished drinking waters from heterogeneous sources. *Water Research* 37: 953–958.

Von Gunten, U., Driedger, A., Gallard, H., Salhi, E. 2001. By-products formation during drinking water disinfection: a tool to assess disinfection efficiency? *Water Research* 35 (8): 2095–2099.

Wei, Q., Wang, D., Qiao, C., Shi, B., Tang, H. 2008. Size and resin fractionations of dissolved organic matter and trihalomethane precursors from four typical source waters in China. *Environmental Monitoring Assessment* 141 (1–3): 347–357.

WHO 2011. Guidelines for Drinking-water Quality. Fourth edition. Geneva: World Health Organization.

Włodyka-Bergier, A., Bergier, T. 2011. The occurrence of haloacetic acids in Krakow water distribution system. *Archives of Environmental Protection* 37 (3): 21–29.

Environmental Engineering IV – Pawłowski, Dudzińska & Pawłowski (eds)
© 2013 Taylor & Francis Group, London, ISBN 978-0-415-64338-2

Adsorption of phenol on clinoptilolite modified by cobaltions

R. Świderska-Dąbrowska
Faculty of Civil Engineering, Environmental and Geodetic Sciences, Koszalin University of Technology, Koszalin, Poland

R. Schmidt
Inspectorate Health in Koszalin, Koszalin, Poland

ABSTRACT: The paper presents the physicochemical properties of natural zeolite modified by cobalt (II) ions, the effect of the solution pH, and the dose of zeolite on the effectiveness of phenol removal from aqueous solution. The experiment has confirmed the high effectiveness of phenol removal on Co zeolite. Co zeolite may be used in purification of water with increased phenol concentration. The phenol removal process is the most effective in an acidic environment (pH 2–4), which is caused by electrostatic interaction of the phenol-cobalt complex with the negative charge of the zeolite surface.

Keywords: adsorption, clinoptilolite, phenol, modification of zeolite, impact of pH

1 INTRODUCTION

Phenols and phenol derivatives which occur in an aquatic environment are extremely toxic to humans (Calace et al. 2002). A small concentration of such compounds in drinking water causes an unpleasant taste and smell to it. Drinking water containing phenols causes degeneration of proteins, paralysis of the central nervous system and damage to the kidneys, liver and pancreas. Major sources of phenolic compounds in natural waters are oil refineries and petrochemical plants. Due to the harmful effect they have on the environment, phenol-containing wastewater has to be treated before being discharged to a receiving waters. Phenolic wastewater is treated in physicochemical processes, including advanced oxidation (Esplugas et al. 2002), membrane filtration (Rzeszutek et al. 1998), electrochemical oxidation (Rodgers et al. 1999) and adsorption (Dąbrowski et al. 2005; Nevskaia et al. 1999). Adsorption on activated carbon is the most frequently used method.

Due to high investment and operation costs generated by the treatment of phenolic wastewater with activated carbon, new technologies are being sought which would use cheaper materials. It is also important that they should be cheap, simple, and environmentally justified. These criteria are met by natural zeolites, which are used as adsorbents. Adsorptive properties of natural and synthetic zeolites are different (Anielak 2006). The catalytic and adsorptive properties of natural zeolites can be enhanced by modifying their surface.

A different degree of oxidation of silicon (IV) and aluminium (III) imparts an electronegative character to zeolites. The amount of negative charge is determined by the Al/Si ratio, owing to which they can be modified, for example, with cationic surfactants (Li 1999; Roy et al. 1998). Surfactant-modified zeolites can be used in the adsorption of organic substances and oxyanions from waters and industrial wastewater. Adsorption of oxyanions on a surface-modified zeolite takes place on its positively charged surface, whereas adsorption of hydrophobic organic contaminants takes place due to adsorption of the apolar part of their molecules on the apolar active sites of the zeolite surface, formed by the hydrocarbon chains of the surfactant (Li et al. 2000). Chlorophenols have been found to be removed more efficiently than phenols. Adsorption of phenols on surfactant-modified zeolites increases within the range of 0.1–0.6 mmol/g with the molar ratio of Si/Al increasing from 5 to 500 (Khalid et al. 2004). The efficiency of phenol removal from aqueous solutions is determined by the Si/Al ratio rather than by the size of pores on the zeolite surface (Okolo et al. 2000; Yousef et al. 2011). Adsorption of phenols on a synthetic Na-Y zeolite is equal to a mere 0.8 mmol/g (Beutel et al. 2001). Beutel et al. (2001) have shown that alkalinity of zeolite, interaction of hydrogen bonds in phenols with Lewis' basic centres as well as interactions of phenol aromatic rings with Na$^+$ cations of Na-Y zeolite are important factors which affect phenol adsorption. Application of modified zeolites as catalysts allows for the use of the unique properties of those minerals, which—apart from their catalytic properties—show remarkable adsorption, exchange capacity and molecular sieve capacity. The effectiveness of adsorption is significantly affected by the conditions in which Na zeolites are modified, e.g. pre-heating (calcination), acid, alkali or steam treatment, indirect drying of zeolites and modification with complexing agents. These methods also increase the thermal and chemical resistance of a zeolite and enable obtaining purer zeolite as a result of dissolution or destruction of other, non-zeolitic materials. It is necessary to find cheap and easily regenerated adsorptive material which shall effectively remove phenols. Therefore, it was the aim of this study to examine phenol removal from aqueous solutions on clinoptilolite modified by cobalt ions. The degree of oxidation of cobalt on modified zeolites depends on a number of parameters; most of all on the calcination temperature and the type of zeolite. A high temperature of calcination entails formation of molecules of Co_3O_4 inside the crystalline structure and on the zeolite surface, whose catalytic reactivity is very low (Khemthong et al. 2010; Salavati-Niasari 2009), especially if the calcination process was conducted after zeolite modification in the ion exchange process (Khemthong et al. 2010). However, the experiment conducted by the authors (Shukla et al. 2011) showed a high effectiveness of oxidation of phenols in aqueous solutions on nanoparticles of Co_3O_4/SiO_2.

This paper presents the physicochemical properties of natural zeolite modified by cobalt (II) ions and the effect of the solution pH and the dose of zeolite on the effectiveness of phenol removal.

2 MATERIALS AND METHODS

Natural zeolite from Slovakia was used in the experiment. It consisted of 84% clinoptilolite, 8% cristobalite, 4% of feldspar and illite as well as traces of silica and carbonate minerals. Natural zeolite with a grain size ranging from 0.40 to 0.75 mm was transformed into the hydrogenous form with 5% HCl, with which the zeolite was poured over three times at the ratio of 3:1 and shaken for 2 hours. The hydrogenous zeolite was subsequently modified by co-precipitation with cobalt (II) ions. In order to achieve that, zeolite was stirred with 0.05 mol/L $Co(NO_3)_2$ for 4 hours at the temperature of 50°C, and subsequently, pH was increased up to 11.0 with NaOH and the solution was stirred for another hour. The zeolite was washed with deionised water and dried at the temperature of 105°C. The procedure was repeated three times to increase the degree of exchange into cobalt (II) ions. After the exchange procedure, the zeolite was calcined for 2 hours in a muffle furnace, with the calcination temperature set at 450°C. Determination of ions: Co, Fe, Mn, Ca, Mg, Al, Si, K and Na was performed with a Varian Spectr AA 20 plus and Solar S4 atomic absorption spectrometer, manufactured by Thermo Jarrell Franklin, USA. The electrokinetic potential and size of the zeolite particles (in suspension of finely ground zeolite with the concentration of 0.1 g/L) was measured by the method of phase analysis of dispersed laser beam in a ZetaPals zetameter, manufactured by Brookhaven. Photographs of zeolite grains were taken with a JSM 5500 LV scanning electron microscope. The washout capacity was determined by the following procedure: deionised water at the ratio of 1:10 was added to a zeolite sample and shaken for 24 hours. The pH was adjusted with NaOH 0.1 mol/L or HCl 0.1 mol/L. Resulting variables included pH, specific conductivity and redox potential, as well as the amount of cobalt in decanted solution after a 24-hour shaking. The process analysis covered adsorption of phenol on clinoptilolite modified by cobalt ions (Z-Co). The phenol (from POCH company) solution was prepared 24 hours before the experiment. The adsorption process was examined by the static method in conical flasks with the capacity of V = 250 mL; 50 mL of phenol solution at the concentration of 100 mg/L was poured into each of them and variable doses of the adsorbent were added. The control sample of the adsorbate without the adsorbent was also prepared. Subsequently, the flasks were closed tightly and shaken on a laboratory shaker with the frequency of 220 1/minute and amplitude of 20 mm. After the contact time of t = 120 minutes, at which the adsorption equilibrium state between the zeolite and the solution was established, the sample was taken for the analysis and the phenol content was determined. The dose of zeolite ranging from 20 to 200 g/L and the pH of the model solution were the independent variables. Phenol concentration in the samples that were and were not filtered through a medium grade quantitative filter, as well as the amount of cobalt eluted from zeolite to the solution, were resulting variables. The effect of the solution pH on the adsorption of phenol at a constant concentration of 100 mg/L was examined at pH values ranging from 2 to 12, adjusted with 0.1 m solutions of HCl and NaOH and for a constant dose of zeolite of 100 g/L. Phenol concentration was determined by reverse-phase HPLC with spectrophotometric detection.

3 RESULTS AND DISCUSSION

The experiment was conducted with zeolite Z-Co, which contains high concentrations of silicon and aluminium compounds as well as alkaline earth metals (sodium, calcium, iron and magnesium)—Figure 1. Cobalt content after the modification process amounts to 3%, which is equivalent to 28 mg/g, and the molar ratio of Si/Al is equal to 5.2. Therefore, the zeolite is belongs to group of a medium level of silicon content.

Examination of the washout capacity of Z-Co has shown that an increase in duration of zeolite contact with water from 0 to 1440 minutes results in an increase in pH—Figure 2. After a contact time of 1440 minutes and with an initial pH = 3, pH increases to 7. Extending the time of contact between zeolite and distilled water small changes of pH were observed. Modification of Z-Co considerably increased Co content in the crystal intra- and extra-lattice. It is the reason for cobalt compounds migrating to the solution—Figure 3.

Their content in the solution whose initial pH was 3—after 24 hours of contact with zeolite—is equal to 1.5% of the total amount of Co in the zeolite. Increasing the solution pH resulted in more cobalt ions being eluted to it, to reach as much as 2.5% after t_k = 1440 minutes and at the initial pH of 9.0. The cobalt content in the solution (Fig. 3) is the function of zeolite abrasion, initial pH value and the time of Z-Co contact with water. This also explains the increased turbidity, which is equal to 60 NTU after t_k = 1440 minutes and at pH = 9—Figure 4.

The experiment results shown in Figure 5 indicate differences in the zeta potential of Z-Co. The ζ potential values ranged from –15 to –28 mV when pH changed from 2.5 to 10.5. It can be concluded, on the basis of the experiment results, that the Co zeolite has an excess of negative charge within the range of pH change under study.

The surface of zeolite modified by Co (II) ions considerably affects the adsorption of phenol from aqueous solutions. The study results have shown

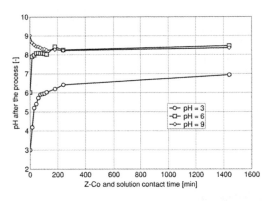

Figure 2. Impact of washout capacity of Z-Co on pH after the process.

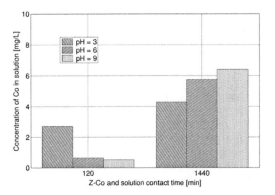

Figure 3. Impact of Z-Co and aqueous solution contact time on migration of Co from zeolite structure.

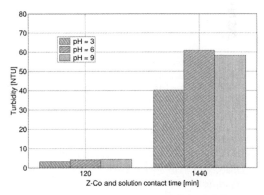

Figure 4. Impact of Z-Co and aqueous solution contact time on solution's turbidity after the process.

that the zeolite surface after the modification has an excess of negative charge and is more acidic. The electrokinetic potential at a pH of about 3 is equal to –18 mV, and the amount of washed-out cobalt—about 1.5%. It is the pH value that may

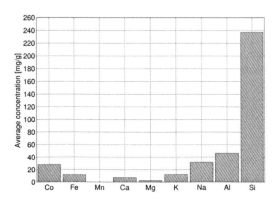

Figure 1. Chemical composition of zeolite modified with Co(II) ions.

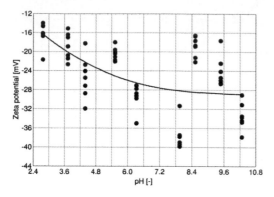

Figure 5. Electrokinetic potential of zeolite modified with Co(II) ions.

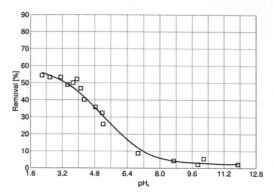

Figure 6. Impact of initial solution pH on amount of adsorbed phenol. Zeolite dose 100 g/L, phenol concentration 100 mg/L.

have a decisive effect on the effectiveness of phenol adsorption.

Phenols are weak acids (pK_a for phenol is equal to 9.9) and, therefore, they dissociate in water showing an alkaline reaction. The acidity of phenol is largely affected by the functional groups in the ring. They may have a mesomeric or inductive effect. The mesomeric effect is greater from the orto- and/or para-positions and the strength of the inductive effect decreases with increasing distance from the phenolic group. The electronodonor groups that show +M effect (EDG) decrease phenol acidity, whereas those with the −M effect (EWG) increase the acid strength of phenol. Therefore, the effect of initial pH on the amount of removed phenol has been examined. The results of examination of phenol adsorption on Co zeolite are shown in Figure 6.

The analysis of the experiment results indicates that the largest rate of removal of phenol is observed at pH values from 2 to 4, i.e. when non-dissociated forms dominate. The Co zeolite surface is always electronegative at this pH value. Therefore, it may be assumed that electrostatic adsorption is not a dominant process. It is possible that phenol adsorption takes place due to the interaction of both the acidic sites of the zeolite and as a result of the action of hydrogen bonds, it is about 50% within the range of pH changes from 2 to 4. Adsorption decreases with increasing pH. This was observed within the pH range between 5.0 and 12.0. Decreasing phenol adsorption at higher pH values is caused by an increase in its solubility; phenol acidity decreases its adsorption at the acidic sites of zeolite. At the same time, neutral phenol-Co complexes may interact with the surface of Co zeolite. The highest phenol removal rate is observed at a pH of about 2, and the Z-Co surface is the least electronegative at this pH. This explains why, despite favourable conditions at higher pH values (increase in cobalt concentration in the washout

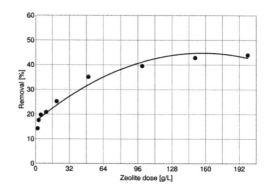

Figure 7. Impact of Z-Co dose on amount of removed phenol.

process), adsorption by electrostatic forces does not take place—the zeta potential of the Z-Co surface is about −30 mV at pH of 10.0. Adsorption on Co zeolite takes place quantitatively depending on the dose and initial pH—Figures 6 and 7.

It was observed that the amount of removed phenol was close to 50% when the zeolite dose was 200 g/L. When the process is complete, the pH value characteristically decreases to about 4. On the other hand, when the zeolite dose was 100 g/L, the final pH was equal to 4.31, and the amount of adsorbed substance was close to 40%, which is consistent with the study results presented in Figure 6. It can be concluded by comparing Figures 4 and 5 that phenol adsorption depends on pH and on the surface charge. The structure of zeolite formed from cyclic sets of tetrahedrons contains open tunnels, which impart the zeolites with the character of a molecular sieve.

In modified zeolites, the specific capacity can be reduced by as much as 40%. As a consequence, phenol adsorption is considerably hindered, which has been shown by the findings presented in Figures 5–7. Interaction of phenol molecules with the zeolite structure can depend on several factors, namely zeolite grain size, pH and concentration of complexing ions (Co(II)) and ligands (phenol). Adsorption of phenol-cobalt complexes depends on the acidic sites on the zeolite surface, mainly those of Brönsted type, on zeolite pore size and related polarisation and separation of the phenol-Co complex with zeolite. The sieve adsorption of phenol-Co is small due to the dimensions of the phenol molecule, which—according to WINMOPAC—are 0.80 nm × 0.67 × 0.15 nm. The molecule is too large to infiltrate the zeolite structure.

Phenol concentration was determined independently in filtered samples, taken after the sorption process. The differences between the amounts of adsorbed molecules for filtered samples as compared to non-filtered ones were very small, which can be attributed to the solubility of both the phenol-Co complexes and those complexes bonded to the active sites of zeolite. An increase in turbidity caused by zeolite abrasion (Fig. 4) after the process of shaking can additionally intensify complex solubility.

4 CONCLUSIONS

The experiment has confirmed the high effectiveness of phenol removal on Co zeolite; it can be as high as 50% at a concentration of C = 100 mg/L and as the function of pH.

What is more, the considerably high sorptive capacity of Co zeolite with respect to phenol molecules shows that the material can be used in purification of water with increased phenol concentration.

When it comes to the phenol removal process, it is the most effective in an acidic environment (pH 2–4), which is caused by electrostatic interaction of the phenol-cobalt complex with the negative charge of the zeolite surface.

ACKNOWLEDGEMENTS

The scientific research was financed from the resources of the National Science Centre in the years 2010–2013, as the research project no. N523 559138.

REFERENCES

Anielak A.M. 2006. Właściwości fizykochemiczne klinoptylolitu modyfikowanego ditlenkiem manganu. *Przemysł Chemiczny* 85/7; 487–491.

Beutel T., Peltre M.J. & Su B.L. 2001. Interaction of phenol with NaX zeolite as studied by 1H MAS NMR, 29Si MAS NMR and 29Si CP MAS NMR spectroscopy. Colloids and Suraces. A: *Physicochem. Eng. Aspects.* 187–188: 319–325.

Calace N., Nardi E., Petronio B.M. & Pietroletti M. 2002. Adsorption of phenols by papermill sludges. *Environ. Pollut.* 118: 315–319.

Dąbrowski A., Podkościelny P., Hubicki Z. & Barczak M. 2005. Adsorption of phenolic compounds by activated carbon—a critical review. *Chemosphere.* 58: 1049–1070.

Esplugas S., Gimenez J., Contreras S., Pascual E. & Rodriguez M. 2002. Comparison of different advanced oxidation processes for phenol degradation. *Water Res.* 36: 1034–1042.

Khalid M., Joly G., Renaud A. & Magnoux P. 2004. Removal of phenol from water by adsorption using zeolites. Ind. Eng. *Chem. Res.* 43: 5275–5280.

Khemthong P., Klysubun W., Prayoonpokarach S. & Wittayakun J. 2010. Reducibility of cobalt species impregnated on NaY and HY zeolites. *Materials Chemistry and Physics* 121: 131–137.

Li Z. 1999. Sorption kinetics of hexadecyltrimethylammonium on natural clinoptilolite. *Langmuir.* 15: 6438–6445.

Li Z., Burt T. & Bowman R.S. 2000. Sorption of ionisable organic solutes by surfactant modified zeolite. *Environ. Sci. Technol.* 34: 3756–3760.

Nevskaia D.M., Santianes A., Munoz V. & Guerrero-Ruiz A. 1999. Interaction of aqueous solutions of phenol with commercial activated carbons: an adsorption and kinetic study. *Carbon.* 37: 1065–1074.

Okolo B., Park C. & Keane M.A. 2000. Interaction of phenol and chlorophenols with activated carbon and synthetic zeolites in aqueous media. J. *Colloid Interface Sci.* 226(2): 308–317.

Rodgers J.D., Jedral W. & Bunce N.J. 1999. Electro-chemical oxidation of chlorinated phenols. *Environ. Sci. Technol.* 33(9): 1453–1457.

Roy Z. Li, S., Zou Y. & Bowman R. 1998. Long-term chemical and biological stability of surfactant modified zeolite. *Environ. Sci. Technol.* 32: 2628–2632.

Rzeszutek K. & Chow A. 1998. Extraction of phenols using polyurethane membrane. *Talanta.* 46(4): 507–519.

Salavati-Niasari M. 2009. Template synthesis and characterization of cobalt(II) complex nanoparticles entrapped in the zeolite-Y. J *Incl. Phenom. Macrocycl. Chem.* 65(1–2): 317–327.

Shukla P.K., Sun H.Q., Wang S.B., Ang H.M. & Tade M.O. 2011. Nanosized Co3O4/SiO2 for heterogeneous oxidation of phenolic contaminants in wastewater. *Separation and Purification Technology.* 77(2): 230–236.

Yousef R.I., El-Eswed B. & Al-Muhtaseb A.H. 2011. Adsorption characteristics of natural zeolites as solid adsorbents for phenol removal from aqueous solutions: Kinetics, mechanism, and thermodynamics studies. *Chemical Engineering Journal.* 171(3): 1143–1149.

Environmental Engineering IV – Pawłowski, Dudzińska & Pawłowski (eds)
© 2013 Taylor & Francis Group, London, ISBN 978-0-415-64338-2

Treatment of dairy wastewater by ozone and biological process

D. Krzemińska & E. Neczaj

Institute of Environmental Engineering, Czestochowa University of Technology, Częstochowa, Poland

ABSTRACT: The present study was aimed to treat the dairy wastewater by ozonation process and an aerobic sequencing batch reactor. The first part of this study examined the effect of operating conditions on ozonation pretreatment of dairy wastewater. The effectiveness of the AOP pretreatment was assessed by evaluating wastewater biodegradability enhancement (BOD$_5$/COD), as well as monitoring major pollutant concentrations (COD) with reaction time. The optimum time and ozone concentration was found to be 5 and 11,1 mg/L/h, respectively. In a single biological treatment the average removal efficiencies of COD, and BOD$_5$ were 70%, and 91%, respectively. Integration of ozonation and biological treatment resulted in 92% removal of COD and 97% BOD$_5$ from the dairy wastewater. The results indicated that the combined process would be a promising alternative for the treatment of dairy wastewater.

Keywords: Sequencing batch reactor (SBR), ozonation process, dairy wastewater, advanced oxidation processes

1 INTRODUCTION

The dairy industry is one of the main lines of agricultural production in Europe. In 2009, milk production in the European Union amounted to 147924 million liters, with 8.2% share of Poland (12084 million liters) (Anielak 2008, Struk-Sokołowska 2011). In dairy industries, water has been a key processing medium. Water is used throughout all steps of the dairy industry including cleaning, sanitization, heating, cooling, in production and floor washing—water requirements is huge (Bylund 2003, Malińska 2005a, Sarkar et al. 2006). A typical dairy industry in Poland creates 450–600 m^3/d effluents with an average Biochemical Oxygen Demand (BOD$_5$) value of about 1.167 mgO$_2$/L and Chemical Oxygen Demand (COD) 2.077 mgO$_2$/L. The milk processing plants, there are drainage facilities over 5.000 m^3/d plants, where the value of BOD$_5$ reaches 6.000 mgO$_2$/L and COD 9.000 mgO$_2$/L (Struk-Sokołowska 2011).

The dairy industry is one of the most polluting of industries, not only in terms of the volume of effluent generated, but also in terms of its characteristics (Banu et al. 2008, Kushwaha et al. 2011). The composition of wastewater produced in the milk processing plants depends primarily on the type of production (such as fluid milk, butter, cheese, buttermilk, whey, yogurt, condensed milk, flavored milk, milk powder, ice cream, etc.) (Bylund 2003, Sarkar et al. 2006, Göblös et al. 2008, Wojnicz 2009, Struk-Sokołowska 2011). The factors influencing the composition and charge of waste water

are the raw materials used, level of technology plant, cleaning and disinfection processes and the amount of water used (Wojnicz 2009, Kushwaha et al. 2011, Dasgupta & Taneja 2011, Venkatesh 2012, Shan & Bi 2012, Michalowski 2012).

Dairy wastewater contains high level of organic contents (proteins, fats, and carbohydrates in the form of lactose) and inorganic ions. The detergents and their additives are also present in small quantities (Mohan et al. 2007, Seesuriyachan et al. 2009, Kushwaha et al. 2011). Dairy wastewater is distinguished by the high BOD$_5$ and COD contents, high levels of dissolved or suspended solids including fats, oils and grease, nutrients such as ammonia or minerals and phosphates and therefore require proper attention before disposal (Sarkar et al. 2006, Banu et al. 2008). Dairy wastewaters are generally treated using biological methods such as activated sludge process, aerated lagoons, trickling filters, anaerobic sludge blanket (UASB) reactor, anaerobic filters, etc. (Demirel et al. 2005, Neczaj et al. 2008, Kushwaha et al. 2011). Among biological treatment processes SBR have been applied for treating dairy wastewater (Li & Zang 2002, Sirianuntapiboon et al. 2005, Mohan et al. 2007, Wojnicz 2009, Kushwaha et al. 2011, Struk-Sokołowska 2011).

Sequencing Batch Reactor (SBR) is the name given to wastewater treatment systems based on activated sludge, operated on a sequence changes of anaerobic and aerobic conditions (Lu et al. 2006, Anielak 2008, Grosser et al. 2009). The difference between the conventional activated sludge systems and SBR is that, in conventional systems, these two processes take place in two different tanks whereas, in SBR systems, they occur sequentially in the same tank (Casellas et al. 2006, Mahavi 2008). The advantages of this technology can include high flexibility and ease of adaptation of operating parameters (Mace & Mata-Alvarez 2002, Kamizela 2011, Elmolla & Chaudhuri 2012). The operation cycle is divided into five phases: filling, aeration-reaction, settling, decantation and idle (Neczaj et al. 2007, Figueroa et al. 2008, Mahavi 2008, Seesuriyachan et al. 2009, Kamizela 2011). Only in theory (Mace & Mata-Alvarez 2002, Casellas et al. 2006, Lu et al. 2006), the alternation of aerated and anoxic phases, followed by a settling period, leads to the high removed of suspended solids, carbon, phosphorus and nitrate ions produced during aerobic nitrification. Still increased pollution, combined with increased industrial activity (Casellas et al. 2006, Li et al. 2010, Kushwaha et al. 2011) and increasingly restrictive laws concerning discharges (Malińska 2005b, Sarkar et al. 2006), focuses on the problem of optimal industry wastewater treatment. High concentration of organic matter in dairy wastewater causes problems with their removal in biological treatments (Banu et al. 2008, Neczaj et al. 2008, Seesuriyachan et al. 2009).

Combining Advanced Oxidation Process (AOP) and biological process has received attention in recent years as a promising alternative. Using AOP pretreatment is important to improve the biodegradability and produce an effluent that can be treated biologically (Neczaj et al. 2007, Casellas et al. 2006, Li et al. 2010, Elmolla & Chaudhuri 2012). These processes involve the generation of highly free radicals, mainly hydroxyl radical (HO•) via chemical, photochemical and photocatalytic reactions. Their application is unavoidable in the treatment of refractory organic pollutants (Balcıoğlu & Ötker 2003, Pera-Titus et al. 2004, Oller et al. 2010, Tarek et al. 2011). Ozonation has been successfully applied to the treatment of winery and distillery wastewater (Lucas et al. 2010), olive mile wastewater (Oller et al. 2010), meat industry wastewater (Wu & Doan 2005), molasses wastewater (Coca et al. 2005) ect.

Ozonation is an oxidative process in which the oxidizing agent used is ozone (O_3). Although, the cost of ozone production still is high, the interest in the use of ozone in wastewater treatment has increased considerably in recent years due to the numerous advantages of this process. Amongst them there are the high oxidation potential of ozone, even at low concentrations, its high efficiency in the decomposition of organic matter, the addition of oxygen to water, and its low sensitivity to changes in temperature. Wastewater characteristics (i.e. pH, concentration of initiators, promoters and scavengers) play an important role in process efficiency (Beltrán et al. 2001, Balcıoğlu & Ötker 2003, Gogate & Pandit 2004, Pera-Titus et al. 2004, Meriç et al. 2005, Ulson de Souza et al. 2010, Turhan & Turgut 2011).

This arises from the fact that pH affects the double action of ozone on the organic matter, that may be a direct or an indirect (free radical) oxidation pathway. At low pH, ozone exclusively reacts with compounds with specific functional groups through selective reactions such as electrophilic, nucleophilic or dipolar addition reactions (i.e., direct pathway). At basic pH, ozone decomposes yielding hydroxyl radicals, a highly oxidizing species which reacts nonselectively with a wide range of organic and inorganic compounds in wastewater (i.e., indirect ozonation pathway) (Beltrán et al. 2001, Pera-Titus et al. 2004, Meriç et al. 2005, Turhan & Turgut 2011).

The aim of this study was to investigate the effectiveness of Ozonation-Sequencing Batch Reactor (SBR) coupled process for the treatment of the dairy wastewater. The first part of this study examined the effect of operating conditions on ozonation pretreatment of a dairy wastewater. The second part of this study examined the feasibility of using combined O_3-SBR treatment of the dairy wastewater.

2 MATERIAL AND METHODS

2.1 Materials

Raw wastewater was collected from Dairy Factory (Silesian Region, Poland) and stored in the refrigerator, if required to be used for long periods. Table 1 shows the average characteristic of the wastewater composition during the experimental period.

2.2 SBR reactor

Two laboratory scale-reactors were used for the examination of dairy treatment efficiency. The reactors were constructed from plexi glass; each

Table 1. Composition of wastewater from dairy factory.

Parametres	Units	Value
COD	mgO_2/L	2300–2600
BOD_5	mgO_2/L	820–980
pH	–	6,6–6,8
Alkalinity	$mgCaCO_3/L$	280–300
VFA	mgCHCOOH/L	870–920
$N-NH_4^+$	mg/L	200–240
TOC	mg/L	760–820

reactor with 12 cm diameter, and 30 cm height had a total volume of 2,5 L. Magnetic stirrers were used for mixing.

A set of two peristaltic pumps was used to feed and discharge the effluent in both reactors, respectively. The reactors were supplied with oxygen by fine bubble air diffuser to keep dissolved oxygen concentration above 3 mg/L in the aerobic phase (Fig. 1). The reactors operated at room temperature (20–23°C). The cycle time of the reactors was 24 h and consisted of five distinct modes: aerobic fill, aerobic react, anoxic react, settle and draw. The initial volume of the culture in the tank was 200 ml which was 10% v/v total volume.

2.3 Ozonation

Ozone gas was generated from air using an ozone generator (CH-KTB-3G with a power on 0,075 KW). The gas mixture of air and ozone was fed into the glass tank containing wastewater via a porous diffuser. All experiments were performed at room temperature and at original pH of wastewater. The applied ozone doses were controlled by adjusting the flow rate of ozone in the feed gas.

2.4 Analytical methods

Samples were withdrawn from the reactor at the beginning and at the end of each cycle for analysis. The following parameters were analyzed: BOD_5, pH, ammonia nitrogen, Volatile Fatty Acids (VFA) and Total Organic Carbon (TOC). All analyses were carried out according to Standard Methods (APHA, 1999). COD was determined by the dichromate method using DR/4000 spectrophotometer (Hach Company, USA). Alkalinity was determined by titration.

2.5 Ozonation as pretreatment of dairy wastewater

The first phase of the study was based on choosing the most favorable dose of ozone and ozonation time (Table 2).

Batch experiments were conducted with the use of 1 L wastewater with natural pH (6,7) and temperature room (20–23°C). A 2–3,5 mg/L of ozone dose was applied for ozonation experiments with an air flow rate of 0,18 m³/h. The ozone residue was destroyed by two sequential washing bottles containing 250 mL of acidified 2% KI solution. Samples were withdrawn and analyzed for the changes in the COD and BOD_5/COD ratio.

2.6 Start up of SBR

In the first step of the experiment both reactors (SBR1, SBR2) were in a start-up mode to allow the biological populations to adapt to the dairy wastewater. The SBRs was inoculated with 200 mL of sludge from the aeration tank of a municipal wastewater treatment plant. Concentration of Mixed Liquor Suspended Solids (MLSS) in the reactors after inoculation was 2300 mg/L. For the acclimation of the biomass, a Hydraulic Retention Time (HRT) of 2 days was applied and the dairy wastewater was mixed with domestic wastewater at ratios of 25:75, 50:50, 75:25 and 100:0, and the acclimation period was extended to 8 days. After acclimatization process two different operating modes has been studied.

In mode I: SBR1 reactor where the raw dairy wastewater was treated, the reactor operated at organic loading rate of 1,23 kg COD/m³ d, hydraulic retention time of 2 d and Sludge Retention Time (SRT) of 10 days.

In mode II: SBR2 reactor treated dairy wastewater after ozonation. Reactor operated at

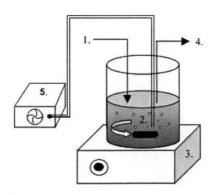

1. Input wastewater
2. SBR reactor
3. Magnetic stirrers
4. Treated wastewater
5. Air diffuser

Figure 1. Schematic of the experimental setup.

Table 2. Parameters of ozonation.

Time [min]	Dose [mg O₃/L/h]	
1	11,11	19,5
3		
5		
7		
10		
15		
20		
25		

organic loading rate of 0.97 kg COD/m³ d, hydraulic retention time of 2 d and SRT of 10 days.

The cycle time of the reactors was 24 h and consisted of five distinct phase:

– aerobic fill (2 h),
– aerobic react (18 h),
– anoxic react (2 h),
– settle (1,5 h),
– draw and idle (0,5 h).

This approximate period starting from filling to supernatant removal until the last 30 min of idleness was defined as a 'cycle'. The next cycle started with restoring the reactor to 2.5 L with fresh dairy wastewater, and the operation was then repeated during 21 days. Supernatant samples were taken on a daily basis and assayed for COD, BOD_5, pH, ammonia nitrogen, alkalinity, VFA and TOC.

3 RESULT AND DISCUSSION

3.1 Pre-treatment of dairy wastewater by ozonation process

Two series of experiments were performed at room temperature and at original pH of wastewater. The first one was carried out using 11,1 mg/L/h ozone and 8 different time amounts for treating dairy wastewater (Table 2). Table 3 shows the normalized COD, BOD_5, TOC, VFA, ammonia nitrogen, pH and alkalinity. With the increment of the time ozonation slight pH increased and nearly three-fold increased alkalinity. After processes VFA, concentration was on the same level for all the time around 410 mg/L CH_3COOH. The BOD_5 value in the processes rose and was maintained during the whole time within the range of 1500–1600 mgO_2/L. Figure 2a presents the effect of the biodegrability change as a ratio BOD_5 to COD. This value increased only in the first 5 minutes, after this time ratio BOD/COD went down. The same situation

was observed for COD removal. Reduction COD increased to 22% in 5 min and drop down to −17% in 25 min. Figure 2b shows the influence of ozonation on ammonia nitrogen removal and TOC reductions. In the second series, the raw effluents of dairy industry were treated by ozone dose of 19,5 mg/L/h and the same 8 ozonation time. We can see that the increase in ozone dose resulted in similar changes in pH, VFA and alkalinity (Table 4). While the value of BOD_5 during the first 3 min increased, and then began to decline in 5 min. The opposite situation was observed for the COD.

Figure 3a and b show the efficiency removal of particular parameters. Ammonia nitrogen removal efficiency significantly has changed over time and reached a lower value compared to the previous dose of ozone. Moreover TOC removal efficiency decrease with increasing ozonation time.

There is a tendency towards an increase in COD with increasing ozonation time, although in some cases the values decrease with increasing ozonation time. When COD removal decreasing, the ratio BOD/COD also decreases in the results and consequently decreased susceptibility to biodegradation. Among the investigated dose and times ozonation higher BOD/COD and COD removal were achieved for 11,1 5 $mgO_3/L/h$ and reaction time of 5 min.

3.2 Effect of dairy wastewater treatment in SBR and combined processes (O_3/SBR)

After acclimatization process in SBR I the 24 h cycle were applied to reactor operated at organic loading rate of 1.23 kg COD/m³/d, hydraulic retention time of 2 d and SRT of 10 days. The process was monitored by 21 days and dairy raw wastewater were characterized by the presence COD, BOD_5, TOC, VFA, ammonia nitrogen, alkalinity and pH of 2367 mgO_3/L, 850 mgO_3/L, 778,9 mg/L, 909 mg/L, 229,6 mg/L CH_3COOH, 280 mg $CaCO_3/L$ and 6,61

Table 3. Effluent quality and removal efficiency of ozonation for dose 11,1 $mgO_3/L/h$ under various times.

Time [min]	COD [mgO₂/L]	VFA [mg/L CH₃COOH]	BOD₅ [mgO₂/L]	pH [–]	Alkalinity [mgCaCO₃/L]	Ammonia nitrogen [mg/L]	TOC [mg/L]
0	2516 ± 56	891 ± 12	958	6,7	280 ± 20	227 ± 3	804,3
1	2382 ± 76	394 ± 27	1513	7,2	940 ± 10	90 ± 7	743,7
3	2153 ± 63	411 ± 9	1543	7,5	960 ± 20	87 ± 1	752,8
5	1969 ± 28	411 ± 18	1578	7,6	940 ± 10	81 ± 3	755,1
7	2314 ± 36	429 ± 11	1641	7,7	880 ± 10	81 ± 2	769,7
10	2513 ± 43	446 ± 21	1567	7,3	740 ± 20	81 ± 3	731,4
15	2697 ± 33	411 ± 12	1458	7,5	700 ± 10	73 ± 1	739,3
20	2854 ± 40	429 ± 18	1551	7,8	680 ± 20	62 ± 3	721,9
25	2940 ± 34	446 ± 6	1529	7,8	680 ± 10	56 ± 2	719,7

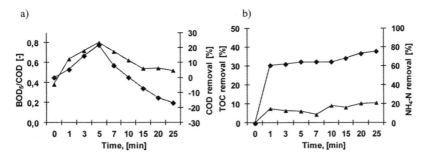

Figure 2. Performance of the initial step of ozonation process for dose 11,1 mgO₃/L/h under various times: (a) ▲BOD/COD, ◆COD removal, (b) ▲TOC removal, ◆Ammonia nitrogen removal.

Table 4. Effluent quality and removal efficiency of ozonation for dose 19,5 mgO₃/L under various.

Time [min]	COD [mgO₂/L]	VFA [mg/L CH₃COOH]	BOD₅ [mgO₂/L]	pH [–]	Alkalinity [mgCaOC₃/L]	Ammonia nitrogen [mg/L]	TOC [mg/L]
0	2516 ± 56	891 ± 12	958	6,7	280 ± 20	227 ± 3	804,3
1	2164 ± 42	411 ± 6	1570	7,2	780 ± 10	76 ± 1	767,6
3	2266 ± 51	429 ± 9	1511	7,5	720 ± 20	70 ± 3	764,9
5	2507 ± 53	429 ± 18	1434	7,6	680 ± 10	56 ± 2	754,8
7	2679 ± 45	463 ± 12	1409	7,7	620 ± 20	56 ± 1	770,1
10	2875 ± 36	480 ± 18	1321	7,3	560 ± 10	48 ± 5	789,1
15	3037 ± 48	497 ± 9	1332	7,5	520 ± 10	140 ± 9	792,2
20	3098 ± 46	463 ± 3	1298	7,8	540 ± 20	179 ± 6	798,7
25	3158 ± 18	514 ± 6	1265	7,8	560 ± 10	210 ± 6	754,8

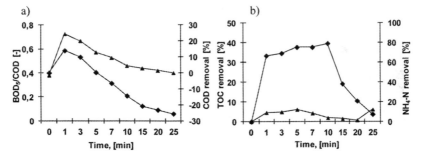

Figure 3. Performance of the initial step of ozonation process for dose 19,5 mgO₃/L/h under various times: (a) ▲BOD/COD, ◆COD removal, (b) ▲TOC removal, ◆Ammonia nitrogen removal.

respectively. Summary of changes in particular parameters value during the process are shown in Table 5. From day 1 to 15 an effluent COD decreased up and after this time remained at similar levels. The COD removal in the medium varied from 60 to 70%. The maximum COD removal of 70% was observed in 21 days and COD had value 704 mg/L. A good BOD, TOC, VFA and ammonia nitrogen removal efficiencies were achieved, while the biological treatment process caused an increase in alkalinity of original dairy effluent, the alkalinity in the treatment plant effluent rose to 420 mg CaCO₃/L. Concentrations of BOD₅, VFA, ammonia nitrogen and TOC in the effluent were 73–75, 120–137, 72,8–81,2 and 27,8–29,9 mg/L, respectively, corresponding to removal efficiencies of 83%–87%, 65%–68% and 96%, respectively.

The same dairy raw wastewater was subjected to ozonation process and then purified by SBR II reactor. Effluent after ozonation process characterized by a greater susceptibility to biodegradation (the ratio of BOD/COD doubled) and lower content of organic compounds (about 1900 mg COD/L).

Table 5. Summary of changes parameters during the dairy wastewater treatment in SBR1.

Parameters [units]	Time [d]							
	1	3	6	9	12	15	18	21
pH [–]	7,62	7,72	7,8	7,6	7,73	7,9	8,12	7,97
COD [mgO₂/L]	1412	1393	1309	1280	1190	818	771	704
BOD [mgO₂/L]	71,4	70,1	81,2	81,3	72,4	75,4	74,6	73,9
BOD/COD	0,05	0,05	0,06	0,06	0,06	0,09	0,10	0,10
Alkalinity [mgCaCO₃/L]	320	320	400	420	400	430	420	400
VFA [mg/L CH₃COOH]	137	120	154	86	103	120	154	137
TOC [g/L]	77,3	75,8	61,1	52,8	33,4	29,9	28,5	27,8
Ammonia nitrogen [mg/L]	112	106,4	86,8	75,6	72,8	81,2	72,8	75,6
COD removal [%]	40	41	45	46	50	65	67	70

Table 6. Summary of changes parameters dairy wastewater treatment in SBR II.

Parameters [units]	Time [d]							
	1	3	6	9	12	15	18	21
pH [–]	7,93	7,21	8,14	7,89	8,26	8,13	7,9	8,04
COD [mgO₂/L]	257	249	276	261	199	189	205	192
BOD [mgO₂/L]	56,7	34,2	38,3	31,1	24,6	28,5	27,7	26,6
BOD/COD	0,22	0,14	0,14	0,12	0,12	0,15	0,14	0,14
Alkalinity [mgCaCO₃/L]	520	480	540	500	540	5200	5200	500
VFA [mg/L CH₃COOH]	137	103	120	103	103	120	86	120
TOC [g/L]	55,1	65,8	48,3	48,7	52,3	44,8	41,7	43,8
Ammonia nitrogen [mg/L]	72,8	72,8	81,2	75,6	81,2	70	70	72,8
COD removal [%]	89	89	88	89	92	92	91	92

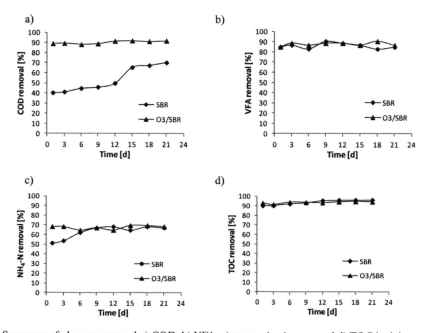

Figure 4. Summary of changes removal a) COD, b) VFA, c) ammonia nitrogen and d) TOC in dairy wastewater.

In this case, organic loading rate in SBR was 0,97 kg COD/m^3 d. HRT and SRT was unchanged of 2 d and 10 days, respectively. Summary of changes in particular value during the combined process is presented in Table 6.

From day 1 to the last day removal value of COD, VFA, TOC, BOD_5 and ammonia nitrogen was on the same high level 89%–92%, 85%–91%, 92%–95%, 93%–97% and 65%–70%, respectively. As in the SBR I, a significant increase in alkalinity was observed, and its value oscillated between 480–540 mg $CaCO_3$/L. It can be noted (Fig. 4) that the O_3/SBRII system reached steady state within 1 d, while in the SBRI optimum COD removal was reached in 15th day. For combined process COD removal was higher of about 12% and as a consequence of the treatment process the effluent COD was on the level of 200 mgO_2/L. The rest of removal parameters like TOC, VFA and ammonia nitrogen in both study was almost on the same level.

4 CONCLUSION

Single Biological (SBR) and coupled O_3/SBR treatment methods were evaluated for the treatment of dairy wastewater. On the based of the results of the present study several conclusions can be drawn:

– Use of ozonation process as a pretreatment needs to be optimized in terms of the treatment time as well as the dose of the oxidant.
– In this study the optimum dose and time ozonation pretreatment of an dairy wastewater were 11.1 mgO_3/L and 5 minutes. Ozonation increased twice, BOD_5/COD ratio to 0.78 and had positive effect on the SBR performance.
– SBR showed maximum COD removal 70% in 21 d treatment, whereas in combination treatment, the percentage of COD removal has increased up to 90% in first day.
– This indicates that the pretreatment with O_3 is stimulating biodegradability and reducing the treatment time by SBR.

ACKNOWLEDGEMENTS

This work was supported by the Faculty of Environmental Protection and Engineering (Czestochowa University of Technology) BS/MN-401-315/11 and BS/PB-401-303/11.

REFERENCES

Anielak, A.M. 2008. Gospodarka wodno-sciekowa przemysłu mleczarskiego. *Agro Przem.*, 2:57–59.

APHA-AWWA-WEF 1999. Standard methods for the examination of water and wastewater, 20th edition, Washington, DC.

Balcioğlu, I.A. & Ötker, M. 2003. Treatment of pharmaceutical wastewater containing antibiotics by O_3 and O_3/H_2O_2 processes, *Chemosphere*, 50 (1): 85–95.

Banu, J.R., Anandan, S., Kaliappan, S., Yeom, I-T. 2008. Treatment of dairy wastewater using anaerobic and solar photocatalytic methods. *Sol. Energy*, 82: 812–819.

Beltrán, F.J., García-Araya, J.F., Álvarez, P.M. 2001. pH sequential ozonation of domestic and wine-distillery wastewaters. *Wat. Res.*, 35 (4): 929–936.

Bylund G. 2003. *Dairy processing handbook.* Sweden Tetra Pak Processing Systems AB: 415–423.

Casellas, M., Dagot, C., Baudu, M. 2006. Set up and assessment of a control strategy in a SBR in order to enhance nitrogen and phosphorus removal. *Process Biochem.*, 41: 1994–2001.

Coca, M., Pena, M., Gonzalez, G. 2005. Variables affecting efficiency of molasses fermentation wastewater ozonation, *Chemosphere*, 60: 1408–1415.

Dasgupta, P. & Taneja, N. 2011. Low Carbon Growth: An Indian Perspective on Sustainability and Technology Transfer. *Problemy Ekorozwoju—Problems of Sustainable Development*, 6(1): 65–74.

Dąbrowski, W. 2011. Determination of pollutants concentration changes during dairy wastewater treatment in 'Mlekovita' Wysokie Mazowieckie. *Inżynieria Ekologiczna*, 24: 236–142.

Demirel, B., Yenigun, O., Onay, T.T. 2005. Anaerobic treatment of dairy wastewaters: A review. *Process Biochem.*, 40: 2583–2595.

Elmolla, E.S. & Chaudhuri, M. 2012. The feasibility of using combined Fenton-SBR for antibiotic wastewater treatment. *Desalination*, 285: 14–21.

Figueroa, M., Mosquera-Corral, A., Campos, J.L., Mendez, R. 2008. Treatment of saline wastewater in SBR aerobic granular reactors. *Water Sci. Technol.*, 58 (2): 479–485.

Göblös, S., Portörő, P., Bordás, D., Kálmán, M., Kiss, I. 2008. Comparison of the effectivities of two-phase and single-phase anaerobic sequencing batch reactors during dairy wastewater treatment. *Renew. Energ.*, 33 (5): 960–5.

Gogate, P.R. & Pandit, A.B. 2004. A review of imperative Technologies for wastewater treatment I: Oxidation technologies at ambient conditions. *Adv. Environ. Res.*, 8 (3–4): 501–551.

Grosser, A., Kamizela, T., Neczaj, E. 2009. Treatment of wastewater from the fiberboard production enhanced with ultrasound sonification in the SBR reactor. *Inżynieria i Ochrona Środowiska*, 12 (4): 295–305.

Kamizela, T. 2011. Co-treatment of Wastewater from Fiberboard Production in the SBR Reactors. *Inżynieria i Ochrona Środowiska*, 14 (2): 157–166.

Kushwaha, J.P., Srivastava, V.Ch., Mall, I.D. 2011. An overview of various technologies for the treatment of dairy wastewaters. *CRC Cr. Rev. Food Sci.*, 51: 442–452.

Li, B., Xu, X-Y., Zhu, L. 2010. Catalytic ozonation-biological coupled processes for the treatment of industrial wastewater containing refractory chlorinated nitroaromatic compounds. *J. Zhejiang Univ-Sci. B*, 11(3): 177–189.

Li, X. & Zhang, R. 2002. Aerobic treatment of dairy wastewater with sequencing batch reactor systems. *Bioproc. Biosyst. Eng.*, 25 (2): 103–109.

Lu, S., Park, M., Ro, H-S., Lee, D.S., Park, W., Jeon, Ch.O. 2006. Analysis of microbial communities using culture-dependent and culture-independent approaches in an anaerobic/aerobic SBR reactor. *J. Microbiol.*, 44 (2): 155–161.

Lucas, M.S., Peres, J.A., Puma, G.L. 2010. Treatment of winery wastewater by ozone-based advanced oxidation processes (O_3, O_3/UV and O_3/UV/H_2O_2) in a pilot-scale bubble column reactor and process economics. *Sep. Purif. Technol.*, 72: 235–241.

Mace, S. & Mata-Alvarez, J. 2002. Utilization of SBR technology for wastewater treatment: an overview. *Ind. Eng. Chem. Res.*, 41: 5539–5553.

Mahvi, A.H. 2008. Sequencing batch reactor: a promising technology in wastewater treatment. *Iran. J. Environ. Healt.*, 5 (2): 79–90.

Malińska, K. *2005a* . Problemy ochrony środowiska w przedsiębiorstwach przemysłu spożywczego. *Środowisko a Zdrowie—2005 VII Ogólnopolska Sesja Popularnonaukowa*, Częstochowa.

Malińska, K. 2005b. Organic waste management in agri-food industry in Poland. In M.P. Bernal, R. Moral, R. Clemente, C. Paredes, EAO & CSIC (eds.), *Sustainable Organic Waste Management for Enviromental Protection and Food Safety*, Spain.

Meriç, S., Selçuk, H., Belgiornov. 2005. Acute toxicity removal in textile finishing wastewater by Fenton's oxidation, ozone and coagulation–flocculation processes. *Wat. Res.*, 39 (6): 1147–1153.

Michałowski, A. 2012. Ecosystem services in the light of a sustainable knowledge-based economy. *Problemy Ekorozwoju—Problems of Sustainable Development*, 7(2): 97–106.

Mohan, S.V., Babu, V.S., Sarma, P.N. 2007. Anaerobic biohydrogen production from dairy wastewater treatment in sequencing batch reactor (AnSBR): Effect of organic loading rate. *Enzyme Microb. Tech.*, 41: 506–515.

Neczaj, E., Kacprzak, M., Kamizela, T., Lach, J., Okoniewska, E. 2008. Sequencing batch reactor system for the co-treatment of landfill leachate and dairy wastewater. *Desalination*, 222: 404–409.

Neczaj, E., Kacprzak, M., Lach, J., Okoniewska, E. 2007. Effect of sonication on combined treatment of landfill leachate and domestic sewage in SBR reactor. *Desalination*, 204: 227–233.

Oller, I., Malato, S., Sánchez-Pérez, J.A. 2010. Combination of advanced oxidation processes and biological treatments for wastewater decontamination—a review. *Sci. Total. Environ.*, 409 (20): 4141–4166.

Pera-Titus, M., García-Molinam., Baños, M.A., Giménez, J., Esplugas, S. 2004. Degradation of chlorophenols by means of advanced oxidation processes: a general review. *Appl. Catal. B-Environ.*, 47(4): 219–256.

Sarkar, B., Chakrabarti, P.P., Vijaykumar, A., Kale V. 2006. Wastewater treatment in dairy industries—possibility of reuse, *Desalination*, 195: 141–152.

Seesuriyachan, P., Kuntiya, A., Sasaki, K., Techapun, Ch. 2009. Biocoagulation of dairy wastewater by Lactobacillus casei TISTR 1500 for protein recovery using micro-aerobic sequencing batch reactor (micro-aerobic SBR). *Process Biochem.*, 44: 406–411.

Shan, S. & Bi, X. 2012. Low Carbon Development of China's Yangtze River Delta Region. *Problemy Ekorozwoju—Problems of Sustainable Development*, 7 (2): 33–41.

Sirianuntapiboon, S., Jeeyachok, N., Larplai, R. 2005. Sequencing batch reactor biofilm system for treatment of milk industry wastewater. *J. Environ. Manage.*, 76: 177–183.

Struk-Sokołowska, J. 2011. The influence of dairy wastewater on COD fractions in municipal wastewater. *Inżynieria Ekologiczna*, 24: 130–144.

Tarek, S.J., Ghaly, M.Y., El-Seesy, I.E., Souaya, E.R., Nasr, R.A. 2011. A comparative study among different photochemical oxidation processes to enhance the biodegradability of paper mill wastewater. *J. Hazard. Mater.*, 185 (1): 353–358.

Turhan, K. & Turgut, Z. 2009. Decolorization of direct dye in textile wastewater by ozonization in a semi-batch bubble column reactor. *Desalination*, 242: 256–263.

Ulson De Souza, S.M., Santos Bonilla, K.A., Ulson De Souza, A.A. 2010. Removal of COD and color from hydrolyzed textile azo dye by combined ozonation and biological treatment. *J. Hazard. Mater.*, 179 (1–3): 35–42.

Venkatesh, G. 2012. Future prospects of Industrial Ecology as a Set of Tools for Sustainable Development. *Problemy Ekorozwoju—Problems of Sustainable Development*, 7(1): 77–80.

Wojnicz, M. 2009. Ifluence of the process phases modifications on the effectiveness of the dairy wastewater treatment in the SBR reactor. *Monografie Polskiej Akademii Nauk*, Komitetu Inżynierii Środowiska, 59 (2): 281–292, Lublin.

Wu, J. & Doan, H. 2005. Disinfection of recycled red-meat-processing wastewater by ozone. *J. Chem. Technol. Biot.*, 80: 828–833.

Environmental Engineering IV – Pawłowski, Dudzińska & Pawłowski (eds)
© *2013 Taylor & Francis Group, London, ISBN 978-0-415-64338-2*

Application of the computer simulations to determine an operation strategy for intensification of denitrification

P. Beńko & W. Styka

Faculty of Environmental Engineering, Cracow Technical University, Cracow, Poland

ABSTRACT: The paper presents the results of computer simulations performed for the Kujawy WWTP in Kraków (PE = 260000). The main objective of simulations was to modify the existing treatment process and to increase denitrification efficiency without any major modifications of the biological system. The simulations were conducted with the BioWin program, for a continuous flow model of biological treatment comprising a three-phase biological reactor and secondary clarifiers. The model was calibrated and validated using a set of real operational data. Simulations were carried out for 6 different operation strategies. The most promising results for intensification of denitrification were obtained for a strategy assuming an increase of the COD/N_{tot} ratio in the influent to the biological system.

Keywords: biological nitrogen removal, denitrification, computer simulation

1 INTRODUCTION

Polish biological reactors, used for an integrated removal of organic compounds, nitrogen and phosphorus, were designed at times when less stringent environmental standards concerning effluent quality were in force. Now, they face the challenge to comply with not only the current requirements but also an upcoming perspective of more stringent regulations. It imposes an obligation to look for methods of intensification of nutrients removal technology, mostly nitrogen removal, on the operators.

Many years of experience have confirmed that most of the wastewater treatment plants maintain high nitrification efficiency, and only the denitrification process requires some optimization strategies. An attempt to find operational strategies/process modification that would assure the required nitrogen removal without enlargement of reactor volume and major alterations in the biological system itself, constitutes a unique challenge for operators. Mathematical modeling seems to be a useful tool in solving such research and operational problems. Modeling was introduced into both scientific/research field and engineering practice by the IWA task group. IWA has its roots in the International Association on Water Pollution Research (IAWPR) and the International Water Quality Association (IAWQ). The IWA task group worked on the application of mathematical modelling for design and operation of biological wastewater treatment. In 1987 they presented a general

kinetic model for organics and nitrogen removal from wastewater (ASM1) (Henze et al. 1987), and then its further modifications (Henze et al. 2000). Since that time, mathematical modelling of biological systems has become a widely accepted tool in research and design, additionally supported by a quickly growing computing power of computers. Now, mathematical models are widely used, mostly for biological process optimization (Oleszkiewicz et al. 2004; Mąkinia et al. 2002; Cinar et al. 1998), but also for optimization of entire sewer system (Gujer 2009, Sochacki et al. 2010). Simulations offer a broader range of analyzed technological solutions, which otherwise could not have been tested in the real life, due to technical, technological and economic constrains. This argument has been considered while undertaking the research on intensification of denitrification at the Kujawy WWTP in Kraków.

2 MATERIALS AND METHODS

2.1 *Kujawy WWTP in Kraków*

The Kujawy wastewater treatment plant treats municipal wastewater from Nowa Huta, a district of Kraków. The plant provides mechanical and biological wastewater treatment with additional phosphorus precipitation. Currently, a hydraulic capacity of the plant is Q = 51000 m³/d, PE = 260000.

Mechanical treatment comprises the following units: coarse screens, aerated longitudinal grit chambers and fine step screens. Then wastewater enters three biological treatment lines; each one consists of: circular primary clarifier, biological reactor (with pre-denitrification chamber, anaerobic chamber, 3 anoxic chambers and 5 aerobic chambers equipped with submerged aerators) and 2 circular secondary clarifiers. The schematic diagram of the biological reactors is presented in Figure 1.

The sludge from primary clarifiers is thickened in 2 gravity thickeners, while the excess sludge feeds 2 belt thickeners. The thickened sludge, together with fats separated in grit chambers, is digested in 4 anaerobic digesters. Then the digested sludge is dewatered in 2 belt presses and finally incinerated in the sludge incinerator, located at the PŁASZÓW II, another Kraków's WWTP. Supernatant from gravity thickeners enters the treatment line in front of the biological reactors, while other streams are diverted to the head of the treatment plant. During the research study, the plant complied with the effluent standards, imposed by the water legal permit, although it was not ready to meet the perspective effluent requirements concerning total nitrogen (TN ≤ 10 g N/m³). Poor biological treatment efficiency, especially denitrification of nitrates, seemed to be the main source of the problem

2.2 Sampling procedures and analysis

Model calibration and simulations were based on the actual measurements and qualitative analysis of sludge samples, both daily composite and grab samples. All samples were collected at the inlet of the biological stage and at the secondary clarifiers' outlet, with an appropriate delay considering the actual detention time. The analysis of average daily wastewater samples included: COD, $COD_{filtred}$, BOD_5, $BOD_{5filtred}$, TKN, $TKN_{filtred}$, NH_4-N, NO_3-N, NO_2-N, P_{tot}, PO_4-P, Z_{tot}, pH, Alkalinity, TS_{inorg}, Mg and Ca. In grab samples, collected every 2 h for model dynamic calibration, the following analysis were performed: COD, BOD_5, TKN, NH_4-N, NO_3-N, P_{tot}, TS_{tot}, pH and Alkalinity.

All parameters were determined according to Polish Standard Methods, except for COD, NO_2-N, P_{tot}, and PO_4-P; which were performed with Hach's Test'N Tube. A biodegradable fraction of COD (S_S and X_S) was estimated with OUR batch tests, following the procedures presented by Ekama et al. (1986); the coefficient Y_H was determined during the OUR test and simultaneous measurements of oxygen used and COD removed (Orhon et al. 1995). A concentration of dissolved oxygen was measured with the oxygen probe Cellox 325 and the meter pH/Oxi 340i, manufactured by WTW. A dissolved non-biodegradable fraction of COD (S_i) was determined on the basis of the corrected COD value in filtered effluent.

2.3 Configuration, calibration and verification of the model

During the research project many simulation runs were performed using the BioWin General Model (Activated Sludge/Anaerobic Digestion—AS/AD). The model is a part of simulation software BioWin, developed by the Canadian group EnviroSim Associates Ltd. Prior simulations were necessary to configure the technological system, and then, to calibrate and validate the model using an independent data set. The model configuration was limited to a single technological line that included:

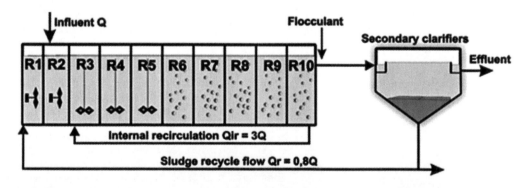

	R1	R2	R3	R4	R5	R6	R7	R8	R9	R10	sec. clarif.
Volume (m³)	612	1510	1780	1870	1870	1730	1730	1870	1870	1780	2 * 4110

Figure 1. Flow diagram and biological process operating conditions at the Kujawy WWTP.

biological reactor, two secondary clarifiers, Internal Recycle (IR) system and Sludge Recycle (SR) system. The model is presented in Figure 1.

A static calibration procedure was based on the quality parameters and a close monitoring of the process performance over 13 sampling days. The monitoring period covered the period of time with no equipment failures or weather anomalies that could influence the treatment process in any way. The stoichiometric and kinetic constants were assumed, as proposed in the BioWin General Model, except for the aerobic biomass yield coefficient Y_H, which was determined using analytical methods ($Y_H = 0,63$ g COD/g COD_{rem}). The calibration procedure was completed when the measured and computed average concentration values for TS, COD, N_{tot}, $N-NO_3$, $N-NH_4$ and P_{tot} had matched.

After model's calibration in static conditions, the authors started to calibrate it in dynamic conditions, introducing daily changes of parameters. A comparison of calculated values and actually measured effluent parameters was performed only for the key parameters, which were considered the most important for the evaluation of the problem i.e.: COD, $N-NH_4$ and $N-NO_3$. A goodness of fit of dynamic model reactions and the actual system was measured with a correlation coefficient for the variables defined as the time series. Then, the model was positively validated using the independent data set, obtained from the treatment plant during its stable operation.

2.4 Operation strategy

A proposal for technological changes focused mainly on looking for such process solutions, which would intensify denitrification without any substantial investment costs and change of the biological reactors volumes; an addition of an external carbon source was not considered. Simulation was performed for 6 different operation strategies including changes of:

- Internal Recycle (IR) rate,
- Impact of supernatant on the quality of the biological reactor influent,
- Organic substrate/nitrogen ratio in the biological reactor influent,
- Oxygen concentration in the aerobic chamber,
- Anoxic chamber/reactor volume ratio,
- Biological stage configuration (introduction of a final endogenous denitrification zone).

In each series only one studied parameter was changed, whereas other model constituents remained constant. The strategy had been considered as "adequate" if the total nitrogen concentration dropped below 10 g N/m³ during denitrification.

The concentration of dissolved oxygen in the aerobic chamber was estimated at 2 g O_2/m^3, except for the run when aeration conditions were examined. Also constant temperature of T = 15 °C was maintained in all reactor chambers; such temperature is close to the lowest temperature level observed at the Kujawy WWTP over the year.

3 RESULTS AND DISCUSSION

The input data, used as the base for simulation calculations, comprised the characteristic of wastewater entering the biological reactor (Table 1) and biological process parameters (Table 2).

3.1 Internal recycle (IR) rate

The analyzed biological system was operated at the 300% IR rate. The rate was calculated on the basis of the daily average flow Q_{dav}. There were over ten simulation runs in which internal recirculation rates varied from 200% to 700% Q_{dav} (Fig. 2). Such an increase of the recycle rate in real life would require pumping mixers of higher capacity.

Operation strategy employing the IR rate had little effect on denitrification efficiency in the analyzed system. An increase of the IR rate over 350% would have positive impact on denitrification, though it had to be accompanied by other operational strategies.

3.2 COD/N ratio and a structure of an organic substrate

In the following experimental run, possible increase of COD in the wastewater entering the biological reactor was expected; the organic material would be generated from the internal sources. The calculations were performed for a broad range of COD/N_{tot} ratios, ranging from 6,6 to 12; only total COD was altered while total nitrogen remained constant (Fig. 3). Moreover, some independent series of simulations were conducted for the constant COD/N_{tot} ratio, while the biodegradable fraction of COD varied from 10 to 40% of total COD.

An increase of the COD/N_{tot} ratio within the anticipated range resulted in a gradual decrease of nitrates in the effluent, with no adverse effect on nitrification. As a result, an almost linear decrease of total nitrogen was observed. The required value ($N_{tot} \leq 10$ g N/m³) was reported for the COD/N_{tot} ratios above 8,8. At the same time, no substantial changes in denitrification efficiency were observed as a result of a biodegradable COD fraction. The best denitrification results were observed for the S_S fraction ranging from 18 to 22%. The structure of organic substrate would have much more

Table 1. Characteristic of wastewater entering the biological reactor.

COD (g O$_2$/m^3)	S$_s$ (g O$_2$/m^3)	S$_I$ (g O$_2$/m^3)	X$_s$ (g O$_2$/m^3)	N$_{tot}$ (g N/m^3)	NH$_4$-N (g N/m^3)	NO$_3$-N (g N/m^3)	P$_{tot}$ (g P/m^3)	Akalinity (val/m^3)	pH
448,3	71,7	13,5	286,9	67,4	47,2	0,0	8,0	9,0	7,7

Table 2. Operational parameters of the "Kujawy" WWTP.

Q (m^3/d)	X (kg$_{dry\ solids}$/m^3)	Sludge age (d)	SR (%)	IR (%)	T (°C)	Flocculant (g/m^3)
51000	2,65	15,3	80	300	19	45

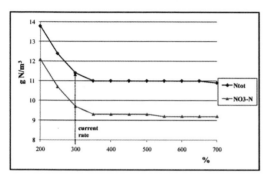

Figure 2. N$_{tot}$ and N-NO$_3$ concentrations in the secondary effluent as a function of the IR rate.

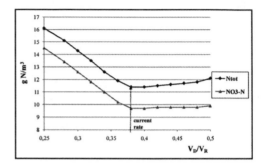

Figure 4. N$_{tot}$ and N-NO$_3$ concentrations in the secondary effluent as a function of the V$_D$/V$_R$ ratio.

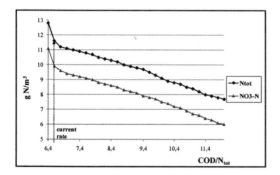

Figure 3. N$_{tot}$ and N-NO$_3$ concentrations in the secondary effluent a function of the COD/N$_{tot}$ ratio.

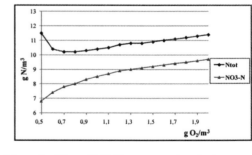

Figure 5. N$_{tot}$ and N-NO$_3$ concentrations in the secondary effluent as a function of oxygen concentrations in the aerobic zones.

significance, if COD in the influent had reached the higher values.

3.3 Volume of the anoxic chamber in the reactor

Possible changes of the anoxic zone volume in the biological reactor are determined by its actual construction; each reactor is divided into 8 sections of almost equal volumes. Without any substantial constructional adjustments (shift of walls) the only possible to obtain ratios of the anoxic volume to the total reactor volume (V$_D$/V$_R$) are 0,25; 0,38 (currently in operation) and 0,50. The calculations were performed for over ten V$_D$/V$_R$ ratios, ranging from 25 to 50% (Fig. 4). The results of calculations showed that the current V$_D$/V$_R$ ratio, which equals 0,38, provides the best conditions for the maximum nitrogen removal, assuming that other constituents of the system remain at the initial level.

3.4 Oxygen concentration in the aerobic zone

The simulations were conducted for the dissolved oxygen concentrations varying from 0,5 to 2,0 g O_2/m^3; the concentration remained constant in all aerobic reactor zones (Fig. 5).

During the calculations some promising results were observed. They were related to the fact that some volume of the aerobic zones was used for simultaneous denitrification. Reducing the oxygen concentration to the range of 0,7–0,8 g O_2/m^3 resulted in a substantial decrease of total nitrogen. At the same time though, the concentration of ammonia nitrogen in the effluent increased (up to 0,7–1,0 g N/m^3) suggesting a small nitrification instability. It should be stated that in the real life conditions, the maintaintenance of complete nitrification conditions would withhold an oxygen drop to such a low level.

Table 3. Simulation results for the strategy with supernatant treatment.

	Effluent			
	COD ($g\ O_2\ m^{-3}$)	N_{tot} ($g\ N/m^3$)	NH_4-N ($g\ N/m^3$)	NO_3-N ($g\ N/m^3$)
No supernatant treatment	22,9	11,4	0,2	9,7
Supernatant treatment	19,7	9,2	0,2	7,5

Table 4. Simulation results for the strategy with a new biological reactor configuration.

	Effluent			
	COD ($g\ O_2\ m^{-3}$)	N_{tot} ($g\ N/m^3$)	NH_4-N ($g\ N/m^3$)	NO_3-N ($g\ N/m^3$)
Option A	21,5	11,7	1,0	9,7
Option B	21,9	9,5	1,6	6,9

3.5 Supernatant

Supernatant coming from excess sludge thickening and dewatering of digested sludge shows the adverse effect on the treatment process. To minimize this effect the authors analyzed the possibility of supernatant treatment in a separate treatment line. It was supposed that supernatant, after pretreatment, would be directed to the head of the treatment line, as it is being practiced now. Therefore, the expected efficiency of the primary treatment would remain at the same level. The quality of wastewater entering the biological reactor, calculated with such mass balance assumptions, was acknowledged to be the fundamental element for new simulation calculations. Other parameters of the models remained at the same initial level. Table 3 illustrates the revised values of the main parameters in the influent to the biological reactor, as well as the simulation results.

The obtained results showed that elimination of the supernatant impact would result in much lower concentrations of total nitrogen in the effluent. On the other hand, such strategy would require large investment costs associated with the construction of the reactor and purchasing the new equipment. It should be also stated, that such option would lead to higher operational costs and make the operation more complex, due to the presence of two separate treatment trains. A reasonable management of supernatant and its controlled dosing to the treatment line, as a way of supporting of other technological actions, seems to be a much better solution.

3.6 Final denitrification

The next scenario tested by the authors was a change of the biological reactor configuration. As a result of the change, the final endogenous denitrification was introduced in the last reactor's chamber and the IR pumping mixers were relocated to the chamber next to the last one. Simulation was performed (Table 4) for $V_D/V_R = 0,38$

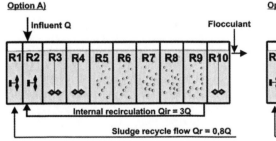 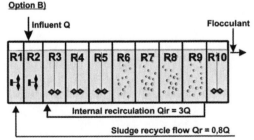

Figure 6. Diagrams for the strategy with a new biological reactor configuration.

(option A—Fig. 6) and $V_D/V_R = 0,50$ (option B—Figure 6).

The results of calculations (Table 4) indicate that the introduction of another anoxic zone for final denitrification may significantly improve the process efficiency. However, the positive effect was observed only when the higher anoxic volumes (primary and final) were implemented and the V_D/V_R ratio reached 0,5 (option B); such action resulted in higher ammonia concentrations in the effluent. The implementation of such configuration may destabilize nitrification and eventually reduce total nitrogen removal, especially at low temperatures.

4 CONCLUSION

The main objective of the computer simulation was to provide information necessary to select an appropriate operation strategy for the Kujawy WWTP. The strategy was to assure an efficient removal of total nitrogen ($N_{tot} \leq 10$ g N/m^3) without major investment projects. The goal was achieved using the different strategies considering: increase of an organic substrate supply, modification of supernatant management and introduction of final denitrification. In the last scenario, the assumed effect was feasible only with a simultaneous increase of all anoxic volumes in the reactor up to $V_D/V_R = 0,5$; such action resulted in a gradual deterioration of the nitrification process. However, treatment of supernatant as a side stream would require substantial investment costs, and afterwards operation of two separate treatment lines of different process parameters. The most promising option of the intensification of denitrification turned out to be a strategy offering an increase of the COD/N$_{tot}$ ratio in the influent to the biological reactor. Thus, the computer simulations confirmed that supply of adequate organic material remained a crucial element of nitrogen removal at this plant.

To sum up, an application of simultaneous hydrolysis of wastewater and sludge during preliminary treatment would be the best solution for the plant. It would not only enhance COD generation from sludge but also improve the structure of the organic substrate in wastewater.

Implementation of other proposed operation strategies did not improve denitrification in a significant way, as long as they were introduced separately.

REFERENCES

Cinar, Ö. Daigger, G.T. Graef S.P. 1998. Evaluation of IAWQ Activated Sludge Model No. 2 using steady-state data from four full-scale wastewater treatment plants, *Water Environ. Res.* 70(6): 1216–1224.

Ekama, G.A. Dold, P.L., Marais, G.v.R. 1986. Procedures for determining influent COD fractions and the maximum specific growth rate of heterotrophs in activated sludge systems. *Water Science & Technology* 18(6): 91–114.

Gujer, W. 2006. Activated sludge modelling: past, present and future. *Water Science & Technology* 53(3): 111–119.

Henze, M. Grady, C.P.L. Jr. Gujer, W. Marais, G.v.R. Matsuo, T. 1987. Activated sludge model no. 1. IAWPRC *Scientific and Technical Report No. 1*, London, IAWPRC.

Henze, M. Gujer, W. Mino, T. van Loosdrecht M. 2000. Activated sludge models ASM1, ASM2, ASM2D and ASM3. *IWA Scientific and Technical Report No. 9*, London, IWA.

Mąkinia, J. Swinarski, M. Dobiegała, E. 2002. Experiences with computer simulation at two large wastewater treatment plants in northern Poland. *Water Science & Technology* 45(6): 209–218.

Oleszkiewicz, J.A. Kalinowska, E. Dold, P. Barnard, J.L. Bieniowski, M. Erenc, Z.F. Ones, R.J. Rypina, A. Udol, J.S. 2004. Feasibility Studies and Pre-design Simulation of Warsaw's New Wastewater Treatment Plant. *Environmental Technology* 25(12):1405–1411.

Orhon, D. Yildiz, G. Çokgör, E. Sözen, S. 1995. Respirometric evaluation of the biodegradability of confectionary wastewaters. *Water Science & Technology* 32(12):11–19.

Sochacki, A. Płonka, L. Miksch, K. 2010. Kilka refleksji o wykorzystaniu modeli matematycznych w symulacji procesów oczyszczania ścieków metodą osadu czynnego. *Materiały III Kongresu Inżynierii Środowiska*, 1: 289–298, Lublin.

Environmental Engineering IV – Pawłowski, Dudzińska & Pawłowski (eds)
© *2013 Taylor & Francis Group, London, ISBN 978-0-415-64338-2*

Organic matter removal efficiency in treatment wetlands simulation by COD fractions

M.H. Gajewska, M.J. Marcinkowski & H. Obarska-Pempkowiak
Department of Water and Wastewater Technology, Faculty of Civil and Environmental Engineering, Gdansk University of Technology, Gdansk, Poland

ABSTRACT: The objective of this paper was to investigate the organic matter fraction (colloidal, particulate and dissolved) in sewage during treatment in treatment wetland. Performance of TWs was simulated well with a simple first order model k—C* based on monitoring data. In this investigation background or residual concentration (C*) was assumed to be equal concentration of input influent organic matter in particulate (CODp). Calculated reaction rate constants (kCOD) for each organic matter fraction showed that fraction present in particulate (CODp) undergoes a multiple-step removal process than it was originally estimated. Mass removal rates computed for COD fractions directly show a relationship between removal efficiency of organic matter and background concentration (C*), resulting in higher efficiency of the second and third treatment step.

Keywords: colloidal, dissolved and particulate organic matter, sewage treatment, treatment wetlands

1   INTRODUCTION

Treatment Wetland (TW) (as also known as Constructed Wetland—CW) is human design and build (constructed) facility where wastewater is discharged to be treated. Thus, it is an engineered marsh system where a variety of biochemical and physical processes are incorporated leading to effective removal of pollutants, including persistent pollutants (Kadlec & Wallace 2008). Recent developments e.g. unconventional pathways of nitrogen removal indicating the mechanisms of pollutants removal in TWs have more versatile properties than it was expected and are not fully understood due to lack of appropriate models. The traditional first order kinetic equation is the most widely employed model (Kadlec & Knight 1996). First-order models may be either area specific determining the necessary wetland area, most frequently used for Vertical Subsurface Flow (VSSF) beds, or volume specific determining the wetland water volume (usually for Horizontal Subsurface Flow HSSF beds) (Kadlec 2000). These models are frequently based on inlet/outlet pollutant concentration, less frequently on internal wetland sampling data. The presumption is often made that the parameters of this model, i.e. the rate constants (for BOD$_5$ or COD), are true constants and do not depend on factors such as hydraulic loading rate and inlet concentration. Another common presumption is that plug flow is a reasonable approximation to the hydraulic conditions in the wetland. However, available literature presents

several studies that show inaccuracy of such a model due to hydraulic conditions inside wetland filter bed (flow characteristics) and influent wastewater composition (Kadlec & Knight 1996, Rousseau et al. 2004, Kadlec 2005, Garcia et al. 2005). Recognizing the shortcomings of a one-parameter model and to better fit available data the modified first-order model, often called the k—C* model, has been proposed (Kadlec 2000, Stein et al. 2006). This model allows for a background or residual concentration (C*) of input, which accounts for the generation of organic matter within the wetland, from atmospheric or ground sources and existence of a recalcitrant fraction of influent organic matter (Stein et al. 2006, Kadlec 2000). Many authors used numerical models to fit the parameters and used definition of empirical or apparent background concentration, which has nothing in common with the treatment processes in TWs (Kadlec 2000). Important for further consideration is to note that during the process of data fitting the parameter C* was observed close to zero for pollutants like ammonium, nitrate or phosphorous (dissolved forms), and often not for organic nitrogen, TSS or organic matter. It could lead to a conclusion that background concentration, as it was mention, mostly is connected with wastewater inlet composition. Very little information exists on suitable values for C*, e.g. for BOD$_5$ suggested values are in the range of 1.7 to 18,2 mg L^{-1} and for COD reasonable is to expect a higher values of C*, since COD is the measure of organic matter that would be considered recalcitrant in a BOD$_5$

tests (Stein et al. 2006). According to Shepherd et al. (2001) the C* for COD in winery wastewater could vary from 23 to 450 mg L^{-1}. Until now, research carried out by several authors demonstrates that a part of organic matter is trapped in the TW filter bed and does not undergo rapid transformation processes. This observation leads to a thesis that background concentration could be easily estimated by quantification of organic matter, e.g. COD in particulate form. Thus, the increasing interest in application of TWs combined with more restrictive water quality standards force a development of better design tools. These tools should be both simple (first order kinetic equation) and easily fitted to different characteristic of wastewater (basic measurements methods).

The aim of this paper is to present how suspended organic matter fractions (COD) can be used in prediction of effluent organic matter concentrations to minimize the inadequacy of first order models.

2 MATERIAL AND METHODS

The research and sampling were carried out at two Multistage Treatment Wetlands in Darżlubie and Wikilno (Northern Poland) in the period 2007–2009. The samples of wastewater were collected according to hydraulic retention time for each stage of treatment. The characteristics of analyzed MTWs is given in Table 1.

At this stage of the investigation, standard analyses of wastewater chemical properties (COD, BOD$_5$, TN, NH$_4$-N, NO$_3$-N, NO$_2$-N, TKN, TSS and VSS) were carried out in accordance with the Polish Standard Methods and the procedures adopted by Hach Chemical Company (Gajewska 2011). The quantification (dissolved, colloidal, and particulate) of organic matter and N was based on the filtration of the influent and subsequent stage effluents through a series of filters (0.1 and 1.2 µm pore

size Millipore nitrocellulose filters). The dissolved fraction was achieved by the filtration of a wastewater sample through a 0.1 µm pore size filter. The filtration of the wastewater sample through a 1.2 µm pore size filter allowed for achieving a non-dissolved fraction (particulate) and the sum of the colloidal and dissolved ones. Having the value of the dissolved fraction, it was possible to calculate the colloidal one. Then the filtrates were analyzed for COD, NH$_4$-N and TKN (Gajewska 2011).

This paper is focused only on fractions of organic matter expressed by COD and their transformation in Horizontal Subsurface Flow Beds (HSSF).

Removal efficiency was calculated as a quotient of pollutants load difference in influent (L$_{inf}$) and effluent (L$_{out}$) after subsequent stages of treatment in MTW and load in influent (L$_{inf}$):

$$\eta = (L_{inf} - L_{out})/L_{inf} \qquad (1)$$

Mass Removal Rate (MRR) was calculated on the basis of the following equation:

$$MRR = [(C_{in} \cdot Q_{inf}) - (C_{out} \cdot Q_{out})]/A \; [g \cdot m^{-2} \cdot day^{-1}] \quad (2)$$

where:
A—area of TW, [m^2];
Q$_{inf}$ and Q$_{out}$—average quantities of wastewater discharged with inflow and outflow, [m^3 day^{-1}];
C$_{in}$ and C$_{out}$—average concentration of pollutant in inflow and outflow, [mg l^{-1}].

First order kinetics of analyzed HSSF was calculated according to the modified first-order model (Kadlec & Knight 1996, Kadlec 2000):

$$\frac{C_{out} - C^*}{C_{in} - C^*} = e^{-k_v \cdot t} \qquad (3)$$

where:
t—is hydraulic retention time [d], k$_v$—is volumetric rate constant [d^{-1}];
C*—background concentration [mg·l^{-1}].

Table 1. Characteristics of the multistage treatment wetland systems.

Plant	Q, [m^3 day^{-1}]	Configuration	Area, [m^2]	Depth, [m]	Hydraulic load, [mm day^{-1}]	HRT, [d]
Wikilno	18.6	HF I	1050	0.6	17.7	12,3
		VF	624	0.4	46.9	–
		HF II	540	0.6	34.4	6,3
			Σ 2 214			
Darżlubie	56.7	HF I	1 200	0.6	47.3	5,1
		Cascade filter	400	0.6	141.2	–
		HF II	500	1.0	113.4	2,1
		VF	250	0.6	226.8	–
		HF III	1 000		56.7	4,2
			Σ 3 350			

118

3 RESULTS AND DISCUSSION

3.1 *Organic matter concentration and efficiency removal*

The characteristics of organic matter expressed in COD and fractions: CODp (in particulate), CODc (colloidal) and CODd (dissolved) concentrations of the influent and effluent after each HSSF bed in multistage treatment wetlands, including their mean values with standard deviation, are presented in Table 2. In the case of this study, the mean and median values did not vary significantly. Additionally, the standard deviation was less than 30% of mean values, which suggests that the data were normally distributed. Thus, the mean values were taken under further consideration and calculation of efficiency removal as well as mass removal rate. The concentrations of COD in the inflow of analyzed MTWs were much higher in comparison with the values given in the literature for domestic wastewater discharged to TWs (Puigagut et al. 2007, Vymazal 2005, Kuschk et al. 2003).

The reasons for higher pollutant concentrations could be the lower water consumption by person

equivalent (about 100 l day^{-1}), the lack of rain water infiltration to the sewer system, and what is most possible—the septic tank failure resulting in high concentrations of TSS (and COD) discharged to MTW.

As presented in Table 2, COD concentrations at Wikilno and Darżlubie TWs are consecutively lower for the individual treatment beds. However, values of CODp show different trend. It was observed that while the first treatment bed at Wikilno and Darżlubie TW systems removed only 49,40% and 48,14% of CODp load, respectively. The second treatment bed of Wikilno TW system presented significant increase in removal of organic matter in particulate up to 81,39%, and third filter bed of Darżlubie TW system up to 79,63%. Such functioning of the wetland system may be explained by several factors. The most possible phenomenon responsible for such a conversion of organic matter is that anoxic process responsible for removal of COD has a longer decomposition period, and the fact that it does not take place in the first treatment bed, where most of COD associated with particulate

Table 2. COD fractions after each stage of treatment at Wikilno and Darżlubie MTWs [mg·l^{-1}].

Parameter	Mean ± SD	Median	Mean ± SD	Median	Removal %
Wiklino					
	Influent		After HSSF I		
COD	714,63 ± 110	720,40	324,69 ± 85	309,40	54,57
CODp	501,70 ± 62	520,60	213,04 ± 72	201,90	49,35
CODc	322,23 ± 52	322,60	137,85 ± 36	139,40	57,22
CODd	166,74 ± 38	163,60	97,01 ± 22	99,00	41,82
	Influent to HSSF II		After HSSF II (effluent)		
COD	156,94 ± 16	158,70	84,45 ± 15	78,60	46,19
CODp	103,84 ± 9	104,80	19,32 ± 15	10,20	81,39
CODc	39,56 ± 8	40	36,69 ± 7	37,80	7,26
CODd	21,13 ± 6	19,80	28,05 ± 5	26,80	−32,71
Darżlubie					
	Influent		After HSSF I		
COD	843,76 ± 40	839,05	408,34 ± 20	405,35	51,60
CODp	607,29 ± 20	602,37	285,71 ± 24	294,30	48,14
CODc	393,52 ± 14	398,35	194,83 ± 19	193,80	50,49
CODd	253,68 ± 24	253,10	132,44 ± 16	132,40	47,79
	Influent to HSSF II		After HSSF II		
COD	281,25 ± 18	283,70	223,31 ± 19	22,05	20,60
CODp	156,43 ± 12	156,65	178,88 ± 17	181,40	64,40
CODc	133,32 ± 8	133,10	122,24 ± 11	126,35	8,31
CODd	106,74 ± 7	106,30	71,34 ± 4	70,95	33,16
	Influent to HSSF III		After HSSF III (effluent)		
COD	172,17 ± 12	172,85	71,28 ± 3	70,80	58,60
CODp	122,24 ± 11	126,35	61,11 ± 2	60,40	79,60
CODc	79,52 ± 8	80,15	38,02 ± 4	37,15	52,19
CODd	57,41 ± 6	59,30	32,90 ± 2	32,95	42,69

matter (CODp) is trapped. It suggests that influent with dominating dissolved organic matter (CODd) creates favorable conditions for removal in the second (Wikilno MTW), and the second and third (Darżlubie MTW) filter bed and thus results in higher MRR values (Fig. 1). It can be associated with similar phenomenon taking place in activated sludge treatment systems, where dissolved organic nitrogen undergoes ammonification process, while organic nitrogen in suspension or colloidal fraction is removed in either one- or two-step hydrolysis process (Czerwionka et al. 2009).

3.2 Organic matter fractions deposition simulation

Volumetric rate constants for mean values of COD fractions (Table 3) calculated from formula (3) present significant differences in processes of each treatment step of the analyzed MTWs. Values for the first treatment step (in this case the first filter bed—HSSF I) are lower in comparison with those calculated for the next treatment steps. It confirms the previous measurements for these MTWs carried out by Gajewska & Obarska-Pempkowiak (2007), and Gajewska & Obarska-Pempkowiak (2011).

Considering rate constant for COD in particulate fraction (kCODp), the difference between the first and second HSSF is significant: for Wikilno MTW it is 0,06 and 0,33; for Darżlubie MTW it is equal to 0,14 and 0,65. These results could suggest that reaction rate constant is not only dependant on temperature but also on how organic matter present in wastewater is easily reached for decomposition. First treatment beds (HSSF I) are responsible here for sedimentation process, which was described earlier in this paper. Next, the trapped organic matter undergoes hydrolysis process and is transported to next treatment beds in colloidal and dissolved form. This process can be observed

Table 3. Volumetric rate constants calculated for mean values of COD fractions after each treatment step of Wikilno and Darżlubie MTWs [d^{-1}].

	kCOD	kCODp	kCODc	kCODd
Wikilno				
HSSF I	0,08	0,06	0,10	0,05
HSSF II	0,12	0,33	−0,04	−0,05
Darżlubie				
HSSF I	0,15	0,14	0,17	0,13
HSSF II	0,15	0,65	−0,49	0,25
HSSF III	0,23	0,42	0,22	0,15

Where: kCOD—rate constant for COD; kCODp—rate constant for COD in particulate fraction; kCODc—rate constant for COD in colloidal fraction; kCODd—rate constant for COD in dissolved fraction.

as negative values of MRR, kCODc and kCODd parallel with decrease of COD and COD fractions concentrations, connected with possible internal release of organic matter (Tables 2 and 3, Fig. 1) (Kadlec 2000).

New effluent concentrations were calculated (C_{calc}) for COD based on previously computed reaction rate constants (kCOD). The values of C_{calc} were lower than concentrations of effluent measurements for HSSF I. If dimensioning process of a treatment wetland will be based on these new COD concentrations and computed kCOD values, the newly designed treatment wetland system will be too small (in terms of available area), and in a result the potential removal efficiency of this system will be not sufficient to reach the design criteria of effluent standard.

Trend lines for C_{out} and $C_{out(calc)}$ presented in Figures 2 and 3 are almost parallel, suggesting a possible constant background concentration of organic matter (C^*). It can be observed that presented approach, where CODp concentration is the same as C^*, is not the best solution for calculating effluent concentrations ($k—C^*$) as presented in Figures 2 and 3. However, it causes the forecasted effluent concentrations to be higher than real concentrations and in fact results in dimensioning of a larger wetland system. A treatment wetland designed in such a way is capable of having a better effluent standard with lower pollutant concentrations. Following this scheme, the proposed approach to dimensioning of a treatment wetland can be considered as safety factor that directly increases total area of TW filter beds. This factor takes into account individual processes of pollutants removal in different form (e.g. COD fractions), and with different availability for biological decomposition in wastewater.

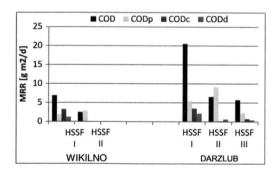

Figure 1. Mass Removal Rates (MRR) of organic matter expressed as COD fractions for Wikilno and Darżlubie MTWs.

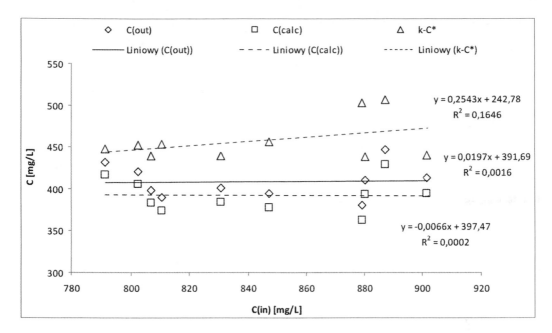

Figure 2. Measured and calculated effluent COD concentrations for Darżlubie MTW [mg/L].

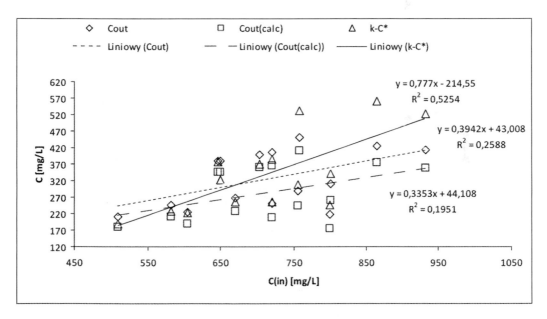

Figure 3. Measured and calculated effluent COD concentrations for Wikilno MTW [mg/L].

4 CONCLUSIONS

The paper presented an alternative approach for simulation of MTWs effluent pollutants concentration. Methodology given here describes clearly that organic matter removal is a complex process, and COD fractions (particulate, colloidal and dissolved) must be considered during modelling and dimensioning of treatment wetlands.

Performance of TWs was simulated well with a simple first order model k—C* based on monitoring data. Calculated reaction rate constants (kCOD) for each organic matter fraction showed that fraction present in particulate (CODp) undergoes a multiple-step removal process than it was originally estimated. Mass removal rates computed for COD fractions directly show a relationship between removal efficiency of organic matter and background concentration (C*), resulting in higher efficiency of the second and third treatment step.

REFERENCES

Czerwionka K., Makinia J., Drewnowski J., 2009. Przemiany azotu organicznego w komorach osadu czynnego z biologicznym usuwaniem związków biogennych, W: Ogólnopolska Konferencja Naukowa: Inżynieria Ekologiczna: (red) H. Obarska-Pempkowiak., Komitet Inżynierii Środowiska PAN., Monografie Komitetu Inżynierii Środowiska Polskiej Akademii Nauk 56: 84–93.

Diederik P.L., Rousseau Peter A., Vanrolleghem, Niels De Pauw, 2004. Model-based design of horizontal subsurface flow constructed treatment wetlands: a review, *Water Research*, 38: 1484–1493.

Gajewska M., Obarska-Pempkowiak H., 2011. Efficiency of pollutant removal by five multistage constructed wetlands in a temperate climate, *Environment Protection Engineering*, 37 (3): 27–36.

Gajewska M., 2011. Fluctuation of nitrogen fraction during wastewater treatment in a multistage treatment wetland, *Environment Protection Engineering*, 37 (3): 119–128.

Gajewska M., Obarska-Pempkowiak H., 2007. Nitrogen pathways during sewage treatment in constructed wetlands: seasonal effects, Environmental Engineering— Pawlowski, Dudzinska & Pawlowski (eds), Taylor & Francis Group, London, 71–85.

Garcia J., Aguirre P., Barragána J., Mujeriegoa R., Matamorosb V., Bayonab J.M., 2005. Effect of key design parameters on the efficiency of horizontal subsurface flow constructed wetlands, *Ecological Engineering*, 25: 405–418.

Kadlec R.H., Knight R.L., 1996. Treatment wetlands. Boca Ration, FL: CRC Pres: pp. 893.

Kadlec R.H., Tanner C.C., Hally V.M., Gibbs M.M., 2005. Nitrogen spiraling in subsurface-flow constructed wetlands: implication for treatment response, *Ecological Engineering*, 25: 365–381.

Kadlec RH., 2000. The inadequacy of first-order treatment wetland models, *Ecological Engineering*, 15: 15–119.

Kadlec R., Wallace S. Treatment Wetlands (2nd ed.), 2009. CRC Press Taylor and Francis Group, Boca Raton, FL 33487-2742: 1016.

Kuschk P., Wießner A., Kappelmeyer U., Weißbrodt E., Kästner M., Stottmeister U., 2003. Annual cycle of nitrogen removal by a pilot-scale subsurface horizontal flow in a constructed wetland in moderate climate, *Water Res.*, 37 (17): 4236–4242.

Puigagut J., Villaseñor J., Salas J.J., Becares E., Garcia J., 2007. Subsurface-flow constructed wetlands in Spain for the sanitation of small communities: A comparative study, *Ecological Engineering*, 30: 312–319.

Shepherd H.L., Tchobanoglous G., Grismer M.E., 2001. Time dependent retardation model for chemical oxygen demand removal in a subsurface-flow constructed wetland for winery wastewater treatment, *Water Environ. Res.*, 73 (5): 597–606.

Stein O.R, Biederman J.A., Hook P.B., Allen W.C., 2006. Plant species and temperature effects on the k—C* first-order model for COD removal in batch-loaded SSF wetlands, *Ecological Engineering*, 26: 100–112.

Vymazal J., 2005. Horizontal sub-surface flow and hybrid constructed wetland systems for wastewater treatment, *Ecological Engineering*, 25: 478–490.

Environmental Engineering IV – Pawłowski, Dudzińska & Pawłowski (eds)
© *2013 Taylor & Francis Group, London, ISBN 978-0-415-64338-2*

Removal of nitrogen and phosphorous compounds by zeolites and algae

M. Zabochnicka-Świątek & K. Malińska
Institute of Environmental Engineering, Czestochowa University of Technology, Czestochowa, Poland

ABSTRACT: The goal of the presented research was to investigate the effects of natural clinoptilolite and algae on removal of nitrogen and phosphorous compounds from synthetic wastewater. The treatments included: synthetic wastewater treated with clinoptilolite and synthetic wastewater treated with clinoptilolite and algae. The degree of removal of $N-NH_4$, $N-NO_3$, N_t, $P-PO_4$ and P_t was determined for both treatments. The results showed that nitrogen and phosphorus compounds were removed in high levels (above 90%). Only the total nitrogen was removed in 74% for both treatments. Algae and clinoptilolite are very good sorbents for removal of phosphorus and nitrogen compounds. Clinoptilolite used in the experiment increased the sorption efficiency.

Keywords: nitrogen and phosphorous compounds, removal of nitrogen and phosphorous, algae, zeolites, synthetic wastewater, wastewater treatment

1 INTRODUCTION

Nitrogen and phosphorous compounds are contaminants typically present in industrial and municipal wastewater. Currently, the technologies applied for removal of these contaminants from wastewater are not sufficient. Therefore, there is a need for more advanced and efficient solutions for removal of phosphorous and nitrogen compounds from wastewater and/or contaminated drinking water.

A potential solution for removal of phosphorous and nitrogen compounds could be the application of zeolites, algae or combinations of algae biomass and zeolites.

Zeolites are naturally occurring aluminosilicate minerals with a diversified chemical composition and wide range of properties and crystal forms. This large group of minerals includes fibrous, lamellar and block zeolites, and analcym. Due to high cation exchange and adsorption capacity zeolites are suitable media for selective adsorption of e.g. heavy metals, nitrogen and phosphorous compounds from soil, water and air (Zabochnicka-Świątek 2007, Chmielewska 2010, Zabochnicka-Świątek & Malińska 2010). Among different types of zeolites clinoptilolite is of great interest due to relatively low price, and sorption and ion exchange properties. Clinoptilolite is a natural aluminosilicate that can occur in sedimentary rocks, volcanic rocks and metamorphic compositions. In Poland, these mineral deposits are found mainly in sedimentary rocks in the vicinity of Rzeszów (Kaleta 2001). Specific density of clinoptilolite varies within 2.02–2.25 g/cm3 and hardness on the Mohs scale ranges from 3.5 to 5.5.

Clinoptilolite also shows significant chemical resistance to acids (Gottardi & Galli 1985).

Clinoptilolite can be used as a sorption or ion exchange medium by different industries (e.g. to water filtration). What is more, it can also be used in environmental engineering (e.g. wastewater treatment, removal of heavy metals from water, etc.) and agriculture (Groffman et al. 1992, Kaleta 2001, Anielak & Piaskowski 2005). Natural clinoptilolite could be used as a sorption medium for removal of heavy metals and ammonium nitrogen from contaminated water and wastewater. Natural clinoptilolite could be employed as a substrate for further modifications to obtain products with improved sorption properties. Due to a wide range of properties and availability clinoptilolite can be applied for wastewater treatment and water purification.

Algae use nitrate, ammonia and urea for growth, also phosphorous is a very important element for almost all cellular processes in algae. Therefore, nitrogen and phosphorous compounds present in wastewater can be utilized, and thus removed, by algae for growing biomass that could be used for various purposes, e.g. production of biofuels (Mallick 2002, Shi et al. 2007, Kim et al. 2010, Chinnasamy et al. 2010, Li et al. 2011, Pittman et al. 2011).

The overall goal of the presented research was to investigate the effects of natural clinoptilolite and algae on removal of nitrogen and phosphorous compounds from synthetic wastewater. It has been hypothesized that the addition of clinoptilolite, algae and combination of clinoptilolite and algae would improve the removal of nitrogen and phosphorous compounds from synthetic wastewater.

2 MATERIALS AND METHODS

Synthetic wastewater, algae biomass and natural zeolite were used in the experiment. Synthetic wastewater was prepared according to the PN-72/C-04550 standard. However, the concentrations of nitrogen and phosphorous compounds were increased for the purpose of this investigation. The concentrations of NH_4Cl, KH_2PO_4 and K_2HPO_2 were 100 mg/l, 64 mg/l and 160 mg/l, whereas according to the PN-72/C-04550 standard, they equalled 20 mg/l, 16 mg/l and 40 mg/l, respectively. 1 L of synthetic wastewater consisted of casein peptone (226 mg/l), dry nutrient broth (152 mg/l), NH_4Cl (100 mg/l), NaCl (7 mg/l), $CaCl_2$ (7.5 mg/l), $MgSO_4$ (2 mg/l), KH_2PO_4 (64 mg/l), K_2HPO_2 (160 mg/l) and $NaNO_3$ (40 mg/l). Algae biomass (*Chlorella vulgaris*) was obtained through densification of 1 L suspension of algae culture (pH = 7.55, density = 111.6 NTU). Clinoptilolite was used as natural zeolite. Treatments included: (1) synthetic wastewater (1 L) + clinoptilolite (10 g), (2) synthetic wastewater (1 L) + clinoptilolite (10 g) + algae biomass (6 g). The 1 L glass bottles were filled with synthetic wastewater and clinoptilolite and clinoptilolite with algae biomass were added. The experiment was conducted within the period of 3 days.

Samples of synthetic wastewater with treatments were taken on each day at 8 am., 2 pm., and 8 pm. The following measurements were conducted: Total Nitrogen (Nt), Total Phosphorous (Pt), Chemical Oxygen Demand (COD), total organic carbon (TOC), Total Inorganic Carbon (TIC), total alkalinity, pH, density, density in the Thom's chamber, $N-NH_4$, chlorophyll a, $N-NO_3$ and $P-PO_4$. Biological Oxygen Demand (BOD) was measured on the first and last day of the experiment at 8 am.

$N-NH_3$, $N-NO_3$, $P-PO_4$, total N, total P measurements were carried out in clear supernatant. $N-NO_3$, $P-PO_4$, total P concentrations were measured by spectrophotometric methods. The total N and $N-NH_4$ contents were measured by titration. The pH was measured using pH-meter. Turbidity was determined with a densitometer. The COD in wastewater was determined using the spectrophotometric method, and the BOD was determined using the manometric method. TOC, IC, TC were determined with a Multi N/C 2100 analyser.

3 RESULTS AND DISCUSSION

3.1 pH

From the beginning of the experiment to its termination pH was above 6.99 but did not exceed 7.6 (Fig. 1). pH of wastewater treated with algae

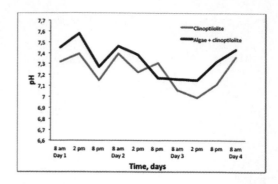

Figure 1. pH of wastewater treated with clinoptilolite and with algae and clinoptilolite during the 3-day experiment.

and clinoptilolite decreased insignificantly over the time. The maximum pH was 7.58 whereas the minimum was 7.15. As for wastewater treated with clinoptilolite the pH ranged from the maximum value of 7.39 to 6.99. It was observed that algae contributed to the increase in pH at 2 pm. on Day 1 (pH = 7.58). In wastewater treated with clinoptilolite there was a decrease in pH on Day 3 (pH = 6.99 at 2 pm.). During the 3-day experiment wastewater treated with algae and clinoptilolite showed higher pH than wastewater treated with clinoptilolite with the exception on Day 2 from 8 pm. to 8 am. when pH of wastewater treated with clinptilolite was higher (pH = 7.31).

3.2 Total Organic Carbon (TOC), Inorganic Carbon (IC), Total Carbon (TC)

During the experiment TOC for both treatments decreased over time. The highest values were observed on Day 1. TOC for wastewater treated with algae and clinoptilolite was 112.0 mg/l, whereas for wastewater treated with clinoptilolite it was 124.0 mg/l. The lowest values of TOC were observed on Day 3. For wastewater treated with clinoptilolite and algae TOC was 27.66 mg/l, and for wastewater treated with clinoptilolite TOC was 41.41 mg/l. Total organic carbon is the sum of carbon contained in the organic compounds. Some algal species can shift from autotrophy to heterotrophy by modifying the source of carbon. Inorganic and organic carbon substrates can be consumed by photoautotrophic algae. In the light *C. vulgaris* and some other algae can grow on CO_2 and in the darkness different forms of organic carbon are used. The limitations in the permeability of the cell-wall causes the lack of versatility in using some sugars and different organic carbon substrates. In the dark and in the light-limiting conditions the growth-stimulating effect occurs. However, *Dunaliella* is not able to grow with the use of organic carbon as the only source (Becker 2008).

Over the 3-day experiment the decrease in TOC was observed. TOC of wastewater treated with clinoptilolite was 124.0 mg/l at the beginning and 41.41 mg/l on Day 3. It was due to the fact that clinoptilolite adsorbed the organic carbon. TOC of wastewater treated with algae and clinoptilolite was 112.0 mg/l at the beginning and 27.66 mg/l on Day 3. The values of TOC were lower in comparison to the treatment with algae. In this case algae consumed some amount of organic carbon. The measurements were taken in the morning and consumption of organic carbon by algae occurred during the night. On the last day both treatment showed a slight increase in TOC (for wastewater treated with clinoptilolite—42.69 mg/l and with algae and clinoptilolite—44.93 mg/l). At the beginning and the end of the experiment the concentrations of organic carbon in the investigated wastewater were above 30 mg/l that is the upper limit for organic carbon imposed by the Regulation of the Minister of the Environment.

During the experiment the inorganic carbon increased in both treatments. IC in wastewater treated with algae and clinoptilolite was 2.05 mg/l, whereas for wastewater treated with clinoptilolite was 2.43 mg/l. The highest values of IC were observed on Day 3–18.19 mg/l for wastewater with algae and clinoptilolite and 25.72 mg/l for wastewater with clinoptilolite. Algae need CO_2 for growth. Most algae species are autotrophs and need CO_2 in the process of photosynthesis. CO_2 is present in water in the following forms (Becker 2008):

$$CO_2 + H_2O \leftrightarrow H_2CO_3 \leftrightarrow H^+ + HCO_3^-$$
$$\leftrightarrow 2H^+ + CO_3^{2-}$$

These forms depend on the pH, the concentration of the nutrients and temperature of the solution. For the pH in the range from 6.99 to 7.58—observed during the 3-day experiment—HCO_3^- forms predominated (Becker 2008). CO_2 from the atmosphere causes the increase in inorganic carbon concentration (Travieso et al. 1996). This growth occurred from the beginning (2.43 mg/l) to the end (25.72 mg/l) of the experiment for wastewater treated with clinoptilolite. For wastewater treated with algae and clinoptilolite the increase in IC was observed from the beginning (2.05 mg/l) to Day 3 (18.19 mg/l) of the experiment. The amount of inorganic carbon compounds was higher than the algae needed for consumption. After that the inorganic carbon decreased to the value of 15.18 mg/l, which indicated that algae started consuming CO_2. It was confirmed by total alkalinity.

Total carbon decreased in both treatments. The highest values were observed at the beginning of the experiment. For wastewater treated with algae and clinoptilolite TC was 114.1 mg/l whereas for wastewater treated with clinoptilolite it was 127.0 mg/l. The lowest values of TC were observed on Day 3. For wastewater treated with algae and clinoptilolite TC was 45.86 mg/l and for wastewater treated with clinoptilolite it was 58.70 mg/l. Total carbon is the sum of carbon contained in organic and inorganic compounds including elemental carbon. The decrease in the values of TC from the beginning of the experiment to Day 3 and slight increase in the last day indicated that organic carbon was the main source of carbon for both treatments. The trends were similar for organic and total carbon values. The increase in the concentration of inorganic carbon did not affect the total carbon concentrations.

3.3 Removal of nitrogen compounds

Nitrate and ammonia assimilation is regulated by pH of a medium. Nitrogen absorption changes pH value. pH of a medium can decrease to 3.0 when ammonia is used as the sole nitrogen source. The growth of some algae can be inhibited by high ammonia concentrations (about 1 mM). The pH increases when the assimilation of nitrate ions occurs. Ammonia tends to volatilize from the medium. However, this is the case when the luxury consumption by some algae occurs (ammonia is taken up in excess of the immediate metabolic needs). Thus, the adjustment of the supply to e.g. pH and temperature appears (Becker 2008). Nitrate, ammonia and urea can be utilised by algae. Nitrate can be also used, however, higher concentrations of this compound is toxic. Elemental nitrogen from the atmosphere can be assimilated by certain cyanobacteria. Ammonia or urea are absorbed by algae in the first order than nitrate or nitrite. It is due to the fact that before consumption, the nitrate or nitrite must be reduced, which requires higher metabolic energy. For Dunaliella at pH above 8 the concentration of ammonia higher than 5 mM is toxic (Becker 2008). Approximately 5–50 mM or 5–10% of the dry weight is the average nitrogen requirement for most algae species. Nitrogen source does not affect the growth rate that is the same for both sources (Becker 2008).

The value of $N-NH_4$ (Fig. 2) significantly increased at 8 pm. on Day 1 of the experiment and reached its highest value of 42.00 mg/l and 39.20 mg/l for wastewater treated with algae and clinoptilolite and with clinoptilolite, respectively. The lowest values were observed at 8 am. of the last day (8.40 mg/l for wastewater treated with algae and clinoptilolite and 8.40 mg/l for wastewater treated with clinoptilolite). During the experiment the values of $N-NH_4$ for both treatments were almost the same.

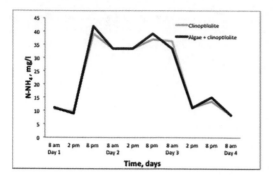

Figure 2. N-NH$_4$ for wastewater treated with clinoptilolite and with algae and clinoptilolite during the 3-day experiment.

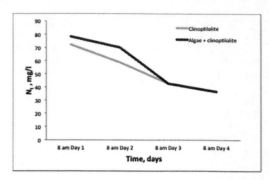

Figure 4. N$_t$ for wastewater treated with clinoptilolite and with algae and clinoptilolite during the 3-day experiment.

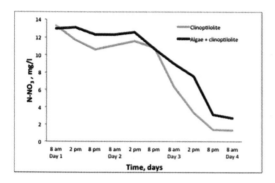

Figure 3. N-NO$_3$ for wastewater treated with clinoptilolite and with algae and clinoptilolite during the 3-day experiment.

In the pH range between 6 and 9 NH$_4^+$ ion dominates, and in the range between 9 and 12 NH$_3$ is the main form. From Day 1 to Day 3 of the experiment the forms of NH$_4^+$ constituted over 95% and the forms of NH$_3$ were less than 5%. On Day 3 at 2 pm. in case of wastewater treated with clinoptilolite at pH = 6.99 ammonium nitrogen occurred only in the form of NH$_4^+$. NH$_4^+$ is the form that can be absorbed by algae. Ammonium nitrogen is absorbed in the first place. Then, N-NO$_3$ is consumed (Becker 2008). The increase in N-NH$_4$ value is connected with transformations in the solution. Prior to assimilation, nitrate-N is reduced to ammonium-N. At the same time ammonium-N is sorbed. It is indicated by the decrease in N$_t$ value.

The decrease in the amount of N-NO$_3$ was observed for both treatments (with and without the algae) (Fig. 3). The highest values of N-NO$_3$ concentration were at the beginning of the experiment. For wastewater treated with algae and clinoptilolite the concentration of N-NO$_3$ was 13.0 mg/l and for wastewater treated with clinoptilolite it

was 13.3 mg/l. The lowest values were observed at the end of the experiment. For wastewater treated with algae and clinoptilolite and with clinoptilolite it was 2.7 mg/l and 1.3 mg/l, respectively.

The values of N$_t$ for both treatments decreased during the experiment (Fig. 4). The highest values were observed at the beginning, i.e. 78.40 mg/l for wastewater treated with algae and clinoptilolite and 77.00 mg/l for wastewater treated with clinoptilolite. The lowest values were the same for both treatments, i.e. 36.40 mg/l. Algae were in the adaptive stress on Day 1 of the experiment, and thus insignificant absorption was observed.

3.4 Removal of phosphorous compounds

Phosphorus occurs in different forms in wastewater. It depends on the pH of the medium. Organic phosphorus can be metabolized only after decomposition to PO$_4$. For algae it is available in the form of phosphate ions (H$_2$PO$_4^-$, HPO$_4^-$) (Gardolinski et al. 2004; Becker 2008). In the pH range typical for natural water (pH 4–9) HPO$_4^{-2}$ and H$_2$PO$_4^-$ predominate. Inorganic orthophosphates in water and wastewater below pH = 6 occur almost exclusively in the form of H$_2$PO$_4^-$—, at pH higher than 7 in form of HPO$_4^{-2}$ and at pH above 9 PO$_4^{-3}$ ions are formed (Kowal et al. 1998). In the investigated wastewater HPO$_4^{-2}$ forms dominated in over 60% over the H$_2$PO$_4^-$—forms. For wastewater treated with clinoptilolite at 2 pm. on Day 3 of the experiment, it was observed that H$_2$PO$_4^-$—and HPO$_4^{-2}$ forms occurred in the same amount. For algae both of these forms are available (Gardolinski et al. 2004). Insignificant fluctuations in the P-PO$_4$ concentration from the first measurement to the end of the experiment were caused by the blockage of assimilation of phosphorus compounds. It was due to the fact that nitrogen compounds were adsorbed in the first place (Becker 2008).

The initial concentration of P-PO$_4$ at the beginning of the experiment was 224.00 mg/l (Fig. 5). After the first measurement of the investigated wastewater treated with clinoptilolite this value decreased to 19.89 mg/l and for wastewater treated with algae and clinoptilolite to 20.54 mg/l. It indicated that very strong sorption occurred immediately after the addition of additives. During the next days P-PO$_4$ was removed to a very small extent. At the end of the experiment the concentration of P-PO$_4$ was 16.30 mg/l for wastewater treated with clinoptilolite and 18.26 mg/l for wastewater treated with algae and clinoptilolite. The removal degrees of ions were 92.72% and 91.85% for these treatments, respectively. There was no significant difference between the removal of P-PO$_4$ by clinoptilolite and the mixture of clinoptilolite with algae.

The initial concentration of P$_t$ at the beginning of the experiment was 224.00 mg/l (Fig. 6). After the first measurement this value decreased to 20.86 mg/l for both treatments. It indicated that very strong sorption occurred immediately after the addition of treatments. During the next

days, P$_t$ was removed to a very small extent. At the end of the experiment the concentration of P$_t$ was 16.95 mg/l for wastewater treated with clinoptilolite and 18.91 mg/l for wastewater with algae and clinoptilolite. The removal degrees of ions were 92.43% and 91.56% for these treatments, respectively. There was no significant difference between the removal of P$_t$ by clinoptilolite and the mixture of clinoptilolite and algae.

The similar values of P-PO$_4$ and P$_t$ indicated that phosphorus occurred mainly in the form of H$_2$PO$_4^-$ and HPO$_4^-$ (Gardolinski et al. 2004; Becker 2008).

The removal degrees of N-NH$_4$, N-NO$_3$, N$_t$, P-PO$_4$ and P$_t$ for wastewater treated with cilnoptilolite and with algae and clinoptilolite are presented in Figure 7.

During the experiment nitrogen and phosphorus compounds were removed at high levels (above 90%). Only the total nitrogen was removed in 74% for both treatments. The removal was the highest in the first hour after the addition of treatments. Hernandez et al. (2006) removed P-PO$_4$ using *C. vulgaris* from the synthetic wastewater in 69%. These researchers also observed that with the increase in algae biomass the removal degree also increased. In the experiment P-PO$_4$ the removal was 92.72% for wastewater treated with clinoptilolite and 91.85% for wastewater treated with algae and clinoptilolite. The clinoptilolite used in the experiment increased the sorption efficiency (Groffman et al. 1992; Kaleta 2001). Tam and Wong (1996) used *Chlorella vulgaris* to remove N-NH$_3$. The removal degree for N-NH$_3$ was 53.89%, and it was the highest removal efficiency these researchers obtained. This value depended on the initial N-NH$_3$ concentration, chlorophyll content and the amount of algae added (Tam & Wong 1996). In the experiment the efficiencies of removal of nitrogen compounds were very high for N-NH$_4$ and N-NO$_3$.

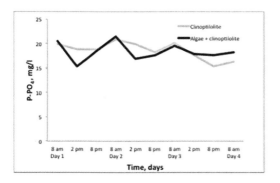

Figure 5. P-PO$_4$ for wastewater treated with clinoptilolite and with algae and clinoptilolite during the 3-day experiment.

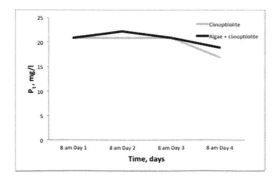

Figure 6. P$_t$ for wastewater treated with clinoptilolite and with algae and clinoptilolite during the 3-day experiment.

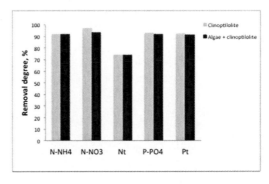

Figure 7. Degree of removal of N-NH$_4$, N-NO$_3$, N$_t$, P-PO$_4$ and P$_t$ for wastewater treated with cilnoptilolite and with algae and clinoptilolite after 3 days.

For wastewater treated with clinoptilolite and with algae and clinoptilolite the N-NH$_4$ removal degree was 91.60%. In case of the NO$_3$ removal for wastewater treated with clinoptilolite the efficiency was 96.75% and for wastewater treated with algae and clinoptilolite it was 93.25%.

The clinoptilolite used in the experiment increased the sorption efficiency (Groffman et al. 1992; Kaleta 2001). The Nt removal was lower (74%) which was caused by less favourable uptake of other nitrogen compounds.

4 CONCLUSIONS

With reference to the obtained results the following conclusions can be formulated:

- algae and clinoptilolite were very suitable sorbents of phosphorous and nitrogen compounds,
- a significant difference in sorption of the investigated nitrogen and phosphorous compounds was not observed due to the fact that the concentrations of algae and clinoptilolite additives were too low,
- the most efficient sorption was observed shortly after clinoptilolite and algae were added.

ACKNOWLEDGEMENTS

This scientific work was supported by the BS/ PB-401-301/11.

REFERENCES

Anielak A.M., Piaskowski K. 2005. Oczyszczanie ścieków zeolitami w SBR przy różnych układach faz. *Inżynieria i Ochrona Środowiska* 8 (1), 73–86.

Becker E.W. 2008. Microalgae: biotechnology and microbiology. Cambridge University Press.

Chinnasamy S., Bhatnagar A., Hunt R.W., Das K.C. 2010. Microalgae cultivation in wastewater dominated by carpet mill effluents for biofuels applications. *Bioresource Technology* 101, 3097–3105.

Chmielewska E. 2010. Zeolites—materials of sustainable significance (short retrospective and outlook). *Environmental Protection Engineering* 36 (4), 127–135.

Gardolinski P.C.F.C., Worfold P.J., McKelvie I.D. 2004. Seawater induced release and transformation of organic and inorganic phosphorus from river sediments. *Water Research* 38, 688–692.

Gottardi G., Galli E. 1985. Natural zeolites. Springer-Verlang.

Groffman A., Peterson S., Brookins D. 1992. Removing lead from wastewater using zeolite. *Water Environment and Technology* 5, 54–59.

Hernandez J-P., De-Bashan L.E., Bashan Y. 2006. Stravation enhances phosphorus removal from wastewater by the microalga Chlorella spp. Co-immobilized with Azospirillum brasilense. *Enzyme and Microbial Technology* 38, 190–198.

Kaleta J. 2001. Zastosowanie klinoptylolitu do usuwania wybranych zanieczyszczeń organicznych z roztworów wodnych. *Gaz, Woda i Technika Sanitarna*, Wyd. SIGMA NOT.

Kim J., Lingaraju B.P., Rheaume R., Lee J-Y., Siddiqui K.F. 2010. Removal of ammonia from wastewater effluent by Chlorella Vulgaris. *Tsinguha Science and Technology* 15 (4), 391–396.

Kowal A.L., Świderska-Bróż M. 1998. Oczyszczanie wody. Wydawnictwo Naukowe PWN, Warszawa-Wrocław.

Li Y., Chen Y-F., Chen P., Min M., Zhou W., Martinez B., Zhu J., Ruan R. 2011. Characterization of a microalga Chlorella sp. well adapted to highly concentrated municipal wastewater for nutrient removal and biodiesel production. *Bioresource Technology* 102, 5139–5144.

Mallick N. 2002. Biotechnological potential of immobilized algae for wastewater N, P and metal removal. A review. *BioMetals* 15, 377–390.

Pittman J.K., Dean A.P., Osundeko O. 2011. The potential of sustainable algal biofuel production using wastewater resources. *Bioresource Technology* 102, 17–25.

Shi J., Podola B., Melkonian M. 2007. Removal of nitrogen and phosphorous from wastewater using microalgae immobilized on twin layers: an experimental study. *Journal of Applied Phycology* 19, 417–423.

Tam N.F.Y., Wong Y.S., 1996. Effect of ammonia concentrations on growth of Chlorella vulgaris and nitrogen removal from media. *Bioresource Technology* 57 (1), 45–50.

Travieso L., Benitez F., Welland P., Sanchez E., Dupeyron R., Dominguez A.R. 1996. Experiments on immobilization of microalgae for nutrient removal in wastewater treatments. *Bioresource Technology* 55 (3), 181–186.

Rozporządzenie Ministra Środowiska z dnia 28 stycznia 2009r. zmieniające rozporządzenie w sprawie warunków, jakie należy spełnić przy wprowadzaniu ścieków do wód lub do ziemi, oraz w sprawie substancji szczególnie szkodliwych dla środowiska wodnego.

Zabochnicka-Świątek M. 2007. Czynniki wpływające na pojemność adsorpcyjną i selektywność jonowymienną klinoptylolitu wobec kationów metali ciężkich. *Inżynieria i Ochrona Środowiska* 10 (1), 27–43.

Zabochnicka-Świątek M., Malińska K. 2010. Removal of ammonia by clinoptilolite. *Global NEST Journal* 12 (3), 256–261.

Environmental Engineering IV – Pawłowski, Dudzińska & Pawłowski (eds)
© *2013 Taylor & Francis Group, London, ISBN 978-0-415-64338-2*

Influence of nitrogen and organic matter fractions on wastewater treatment in treatment wetland

M. Gajewska

Gdansk University of Technology, Civil & Environmental Engineering, Gdansk, Poland

ABSTRACT: Municipal wastewater contains both biodegradable and non-degradable pollutants. Moreover, they are present in dissolved, colloidal and particulate forms. These fractions are characterised by various decomposition rates, hence their bioavailability during microbial respiration. It was confirm that biological degradation rate in terms of COD reduction is influenced by particle size speciation. Therefore, the objective of the study was to investigate organic matter (COD) and N fractions during wastewater treatment in treatment wetland with the special respect to vegetative and non-vegetative period. The analyses of samples filtered in the standard way through a 0.45 μm pore size filter does not provide reliable information about the dissolved (soluble) and particulate fractions due to the colloids content. Based on content of dissolved, colloidal and particulate forms of organic matter and nitrogen it is possible to estimate their potential for degradations and removal in multistage treatment wetland.

Keywords: wastewater treatment, treatment wetlands, organic matter and nitrogen fraction, biodegradability

1 INTRODUCTION

At present, the evaluation of WWTP effectiveness is based upon organic matter expressed by BOD and COD as well as the removal of nutrients (N and P). In order to obtain efficient nitrogen removal in the denitrification process, sufficient amounts of bioavailable carbon source should be guaranteed. In the case of Treatment Wetlands (TWs), the preferred approach is to use the internal carbon source already present in wastewater. According to Tanner et al (2002), nitrogen removal in many TWs occurs together with organic matter removal due to limited carbon source. Another problem which occurs in the case of wastewater treatment in TWs is filter bed clogging due to organic matter accumulation. Although organic matter accumulation is a typical feature of both natural and treatment wetlands, the over net inputs of external organic matter present in wastewater and outputs from decomposition potentially contribute to clogging in the pore space in sub-surface TWs flow (Kadlec and Knight, 1996, Nguyen, 2000, Knowels et al., 2011). Such conditions lead to decrease in HRT, and in a consequence, to reduction of the efficiency of pollutant removal (Kayser et al., 2001, Gajewska et al, 2004, Langegraber et al, 2006).

Municipal wastewater contains both biodegradable and non-degradable pollutants. The amount of biodegradable compounds is relatively high and constitutes up to 70–75% of total organic matter (COD). Approximately 10–30% of influent COD is in an easily biodegradable and 40–60% hardly degradable form (Henze 2002). According to ATV-A 131 (1995), organic matter in municipal wastewater is present in dissolved (easily and non-degradable) and suspended (hardly and non-degradable) fractions. In literature, these COD fractions are regarded as a good indicator to describe organic matter in wastewater. According to Pagilla et al. (2006), pollutants in wastewater (both influent and effluent) are present in three physical states and they should be fractionated. The dissolved, colloidal and particulate forms are characterised by various decomposition rates, hence their bioavailability during microbial respiration. Many authors confirm that biological degradation rate in terms of COD reduction is influenced by particle size distribution (Tiehm et al., 1999, Pagilla et al., 2008, IUPAC Recommendations 2000). Most of the initial Total Nitrogen (TN) can be successfully removed by nitrification and denitrification in a wastewater treatment plant. In contrast, a portion of Dissolved Organic Nitrogen (DON) which has not been converted into inorganic N forms is very difficult to remove from wastewater (Pagilla et al. 2006). Recently it has been found out that wastewater treatment leads to formation of dissolved organic compounds resistant to biochemical degradation, similar to humic acids (Pempkowiak et al. 2009). The colloidal material also constitutes a portion of the organic matter characterized by very slow biodegradation (Pagilla et al. 2008). Thus, the bioavailability and environmental fate of organic matter and N, in treated effluent released into the receiving waters, is not only related to the residual concentration of pollutants, but also to their physical state (fractions).

Since significant variation of influent wastewater quality and relatively stable efficiency of pollutants removal was observed for analysed MTW, it was interesting to investigate the composition and biodegradability of pollutants in subsequent stages of treatment in this facility. Therefore, the goal of this study was to investigate the organic matter (COD) and N fractions during wastewater treatment with the special respect to vegetative and non-vegetative period. Moreover, the results of conventional fractionation (two fractions) and of "true" with three fractions (dissolved, colloidal and particulate) were compared.

2 MATERIAL AND METHODS

2.1 Characteristics of the study site

The study was carried out in a full-scale Multi stage Treatment Wetland (MTW) in Darżlubie near Puck (at the Puck Bay in the northern part of Poland). The facility was designed for 600 PE. Wastewater after mechanical treatment in an Imhoff tank (primary settling) was pumped into the first horizontal subsurface flow bed (HSSF I). The Darżlubie facility consists of five subsurface beds with both horizontal (HSSF) and vertical (VSSF) flows of wastewater and their characteristic is given in Table 1. With different intervals, long term monitoring of wastewater quality has been carried out since 1996 in Darżlubie MTW. Significant fluctuation in pollutant concentration of influent wastewater was confirmed for this MTW. In vegetation period, the facility was loaded with wastewater more rich in pollutants than in the non-vegetation period. No significant differences for organic matter removal in both periods were observed. In the case of total nitrogen, about 10% higher removal efficiency was observed during the

vegetation season. Consequently, the quality of the effluent in vegetation period was much worse and did not meet the requirements given in the Regulation of the Ministry of the Environment of 24 July 2006 (Obarska-Pempkowiak & Gajewska 2005, Gajewska & Obarska-Pempkowiak, 2007).

2.2 Sampling and anlytical methods

The average samples of the influent and effluent, and samples after subsequent treatment stages were collected during twenty sampling events between February 2008 and September 2009. At this stage of the investigation, standard analyses of chemical properties (COD, BOD, TN, NH_4-N, NO_3-N, NO_2-N, TKN, TSS and VSS) were carried out according to the procedures adopted by Hach Chemical Company and Polish Standard Methods. Organic matter and nitrogen speciation were conducted additionally during sampling events from June 2008 to August 2009 (nine samplings).

Additionally, COD_f was analysed after filtration through a membrane filter with pore size of 0.45 μm (Millipore nitrocellulose filters), in aqueous phase. Furthermore, the content of volatile suspended solids in the total suspended solids was determined as loss on ignition.

The quantification (dissolved, colloidal and particulate) of organic matter (DCOD, CCOD and PCOD) and Org-N (DON, CON and PON) was based on filtration of the influent and subsequent stages effluent through a series of filters (0.1, and 1.2 μm pore size Millipore nitrocellulose filters). The dissolved fraction was achieved by filtration of wastewater sample through a 0.1 μm pore size filter. The filtration of the wastewater sample through a 1.2 μm pore size filter allowed for measurement of non-dissolved fraction (particulate) and the sum of colloidal and dissolved fractions. After measurement of the dissolved fractions it was possible to calculate the colloidal ones. Then the filtrates were analysed for COD and TKN. The COD/BOD and BOD/N ratios provide information about the biodegradability of microbiological transformations. Additionally, COD_f/BOD was presented as an indicator of easy degradable, dissolved organic

Table 1. The characteristics of the multistage treatment wetland.

Plant	Q, m³ day⁻¹	Configuration	Area, m²	Depth, m	Hydraulic load, mmday⁻¹	Unit area, m² pe⁻¹
Darżlubie	56.7	HSSF I	1200	0.6	47.3	2.0
		Cascade filter	400		141.2	0.67
		HSSF II	500	0.6	113.4	0.8
		VSSF	250	1.0	226.8	0.4
		HSSF III	1000	0.6	56.7	1.7
		Σ 3350			16.9	Σ 5.6

matter (COD_f—filtration through a 0.45 μm pore size filter). Organic nitrogen (Org-N) concentration was estimated on the basis of the difference between TN and the summation of NH_4-N, NO_3-N, NO_2-N concentrations.

3 RESULTS AND DISCUSSION

The average values for vegetative and non-vegetative period are presented in Table 2. The statistical analysis of the results showed that coefficient of variation in the wastewater influent varied from 18% (COD) to 31% (Org-N). This values confirmed significant fluctuation of pollutant concentration occurring during the vegetative (April–October) and the non-vegetative period. In the vegetative period, the concentrations of organic matter were almost double than the concentrations of organic matter in the non-vegetative one. This variation was mainly associated with the contribution from small agro-industrial activities. During the monitoring period no considerable rainfalls occurred and the flow was relatively constant, which caused almost double loads of pollutants discharged to the analyzed facility during the vegetative periods. Despite such a big fluctuation in the discharged loadings, the facility showed a stable and high efficiency of pollutant removal. There was no difference in TSS removal (72.0%). For BOD and COD, the effectiveness was 80.0% and 72.0% in the vegetative period, and 86.6% and 56.8% in the non-vegetative period. In the case of nitrogen compounds, the results indicated very small difference in efficiency removal for non-vegetative and vegetative periods: 67.4 and 73.0% for TN, 64.0 and 78.1% for NH_4^+-N and 84.0 and 78.2% for Org-N.

Although the efficiency removal was similar to the data reported in the literature by Albuquerque et al., (2009), Vymazal, (2005), Langergraber et al., (2006) as well as by Gajewska and Obarska-Pempkowiak, (2007), the effluent concentration of pollutants often exceeded the limits imposed by the Polish Regulation of the Ministry of the Environment of 24 July 2006. The achieved results seem to indicate a relationship between the form of physical state of organic matter, its biodegradability, and the efficiency removal.

The biodegradability for microbiological transformations was estimated on the basis of COD/BOD and BOD/N ratios (Fig. 1) supported by VSS and MSS (mineral suspended solids) profiles along the treatment stages at MTW in Darżlubie (Fig. 2) for vegetative and non-vegetative periods.

In this study, COD_f/BOD is considered as an indicator of easy degradable, dissolved organic matter (Paggila et al., 2008). The analysed wastewater was characterised by a typical COD/BOD ratio of about 2.1 and BOD/N of about 2.8 in the influent. Furthermore, the COD_f/BOD ratio was near 1.0, and therefore it suggested that organic matter was mostly present in the easily biodegradable dissolved fraction (e.g ATV-131, 1995). Compared with the high content of VSS in total solids the rest was probably present in easily decomposable suspended organic matter (Paggila et al., 2008, Gajewska & Obarska-Pempkowiak, 2007). For non-vegetative period the analysed ratios were characterised by smaller changes in comparison with adequate values for vegetative periods indicating more stable operation of the MTW. In the effluent, the COD/BOD ratio was over 6.0, and the BOD/N ratio was below 0.4, whereas the COD_f/BOD ratio reached

Table 2. Average operating conditions in vegetative and non-vegetative periods at MTW in Darżlubie.

Parameters	Unit	Influent $\Sigma X/N \pm \delta$ Vegetative	Non-vegetative	Min	Max	Effluent $\Sigma X/N \pm \delta$ Vegetative	Non-vegetative	Min	Max
TSS	mg l⁻¹	401.0 ± 87.9	283.5 ± 56.8	245.6	1123.5	110.0 ± 27.3	82.0 ± 21.4	30.0	139.4
VSS	mg l⁻¹	342.8 ± 62.1	213.6 ± 38.5	112.8	674.3	15.6 ± 3.2	8.6 ± 1.2	5.6	25.2
COD	mg O_2 l⁻¹	895.2 ± 156.3	420 ± 74.2	530.0	1450.9	250.0 ± 67.8	183.5 ± 12.8	135.8	268.2
BOD_5	mg O_2 l⁻¹	450.0 ± 61.3	197.8 ± 41.4	235.0	780.5	90.0 ± 21.4	27.8 ± 6.3	43.6	88.2
TN	mg l⁻¹	180.2 ± 35.6	120.4 ± 26.3	68.9	201.0	58.2 ± 12.9	32.0 ± 7.4	28.8	62.5
NH_4^+ N	mg l⁻¹	82.7 ± 23.4	60.8 ± 14.2	52.6	190.5	29.3 ± 7.5	13.2 ± 3.1	18.4	56.2
$N-NO_3^-$	mg l⁻¹	0.1 ± 0.03	0.1 ± 0.02	0.0	2.5	8.0 ± 1.4	4.4 ± 0.7	2.8	9.2
Org-N	mg l⁻¹	88.4 ± 26.8	45.8 ± 10.9	42.3	126.2	12.4 ± 2.8	9.8 ± 1.7	7.8	29.8

Where: ΣX—sum of measurements; N—number of sampling events: for vegetative N = 11 and non-vegetative N = 9 ± 2.8 periods; δ—standard deviation, min and max—bottom and top value of 95% confidence level, significance level equal to 0.05 for two sides test.

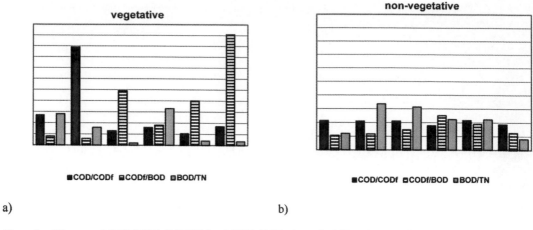

Figure 1. Changes of COD/BOD, BOD/TN and COD$_f$/BOD along the MTW a) vegetative and b) non-vegetative.

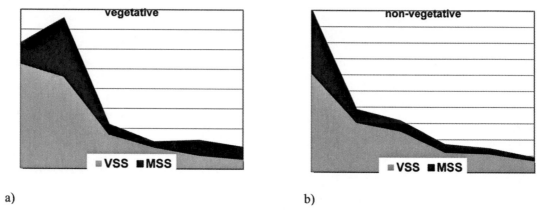

Figure 2. Concentration gradients VSS and MSS along the MTW a) vegetative b) non-vegetative.

values over 10.2 in vegetative period. These values could suggest that organic matter and nitrogen were present in the effluent mainly in dissolved, hardly degradable forms like refractory compounds (Pempkowiak et al., 2009, Nyuyen, 2000, Aquino and Stuckey, 2003; Ramesh et al., 2006). Table 3 presents a detailed fractionation of organic matter and N, including dissolved (<0.1 μm), colloidal (0.1–1.2 μm), and particulate (>1.2 μm).

The conventionally filtered effluent in a standard way through a 0.45 μm pore size filter does not provide reliable information about the dissolved (soluble) and particulate fraction due to the colloids content. The results were from 21.0 to 65.0% lower than the results achieved with "true" fractionation and were very significant for vegetative period (Table 3).

The analysis of different species of organic matter (COD) and organic nitrogen showed the highest

contribution of particulate forms of both COD (63.1%) and TON (61.3%), while 16.3% COD and 9.8% TON in the dissolved fraction were present in the influent during vegetative period. These results are similar to the results achieved by Tuszyńska and Obarska-Pempkowiak (2006), who estimated the fraction according to the guidelines given in ATV-131 (1995), proved that organic suspended solids (PCOD) were the dominant form of COD discharged to the TWs, and therefore responsible for the clogging process. During the subsequent treatment stages at MCW, the forms changed significantly, and finally COD and TON were present in the effluent mainly in dissolved and colloidal fractions. These two fractions could be products of treatment processes as in the case of organic matter produced during cell decay, previously recognized as soluble microbial products (including cell debris), and could be potentially hardly

Table 3. COD and organic N fractions (dissolved, colloidal, particulate) in the analysed MCW Darzlubie.

Stage of treatment	% of organic matter (COD)			% of total organic nitrogen (TON)		
	Dissolved (DCOD)	Colloidal (CCOD)	Particulate (PCOD)	Dissolved (DON)	Colloidal (CON)	Particulate (PON)
Vegetative						
Influent	16.3 (48.1)	20.6	63.1 (51.9)	9.8 (63.9)	28.9	61.3 (36.1)
After HSSF I	17.1 (11.9)	33.8	49.1 (88.1)	11.2 (70.2)	41.0	47.8 (29.8)
After cascade filter	23.8 (74.7)	34.6	41.6 (25.3)	10.3 (85.2)	38.2	51.5 (14.8)
After HSSF II	29.6 (60.9)	38.6	31.8 (39.1)	19.6 (51.9)	40.6	39.8 (48.1)
After VSSF	32.7 (93.4)	37.2	30.1 (7.6)	23.7 (87.4)	55.7	20.6 (12.6)
Effluent	42.1 (49.4)	37.5	20.4 (50.6)	43.5 (28.5)	41.9	14.6 (71.5)
Non-vegetative						
Influent	28,0 (47,5)	41,9	30,1 (52,5)	49,5 (69,7)	24,0	26,5 (30,6)
After HSSF I	30,0 (32,6)	37,5	32,4 (67,4)	31,7 (59,6)	36,8	31,5 (40,4)
After cascade filter	44,4 (46,2)	17,7	38,0 (53,8)	30,7 (51,3)	33,2	36,1 (48,7)
After HSSF II	19,2 (56,5)	48,2	32,7 (43,5)	21,2 (42,9)	43,3	35,4 (57,1)
After VSSF	29,0 (46,4)	37,7	33,3 (53,6)	37,5 (48,7)	16,7	45,8 (51,3)
Effluent	14,3 (52,5)	39,6	46,2 (47,5)	20,6 (35,1)	25,3	54,1 (64,9)

* ( )—data for conventional fraction.

degradable (Aquino and Stuckey, 2003; Ramesh et al., 2006). Such refractory compounds are similar to humic acids (fluvic and humin), and according to Nguyen (2000) they are the predominate stable organic matter fraction accounting for 63 to 96% of deposit in sub-surface TWs. According to Tuszynska and Obarska-Pempkowiak, (2006), the dominant form of COD in the treated effluent is the dissolved non-decomposable one, and it constitutes almost 80% of the total COD. It was confirmed in this study by the results achieved for the conventional fractionation. However, it is in contrast to the data given by Nguyen (2000), Pagilla et al., (2008). Also Aluquerque et al., (2009) indicate the relationship between the low removal of COD and TSS at Caphina TW in Portugal. His investigation suggested that a considerable amount of slowly biodegradable amount of organic matter was not removed in the sub-surface horizontal TW. He estimated that the VSS/TSS rate in the effluent was 0.53, which he assumed as a low degree of effluent mineralization, and as the presence of considerable organic matter content. In the light of the investigation presented in this paper, it could be the consequence of the improper procedure of COD fractionation, where the colloidal fraction was not included (including decay sub-products), which constitutes approximately 20% in the treated wastewater according to Korkusuz (2005). The achieved results are also in accordance with the results described in the literature for a conventional WWTP. According to Pagilla et al (2008), TOC in the effluent from a conventional WWTP in the USA consists of 10% CON and 85% of DON.

Such results may suggest that DON and CON as well as CCOD and DCOD can originate from the microbiological decomposition of raw wastewater, and can be both humic acids and microorganisms. Whereas during non-vegetative period the predominant form of both organic substance and nitrogen were in form of colloidal and particulate fractions, what could be attributed to slower microbial activity in lower temperature and lack of vegetation supportive activity.

4 CONCLUSIONS

Despite the significant difference between the pollutant loading discharged during the vegetative and non-vegetative periods, the Multistage Treatment Wetland in Darżlubie ensures a very effective and stable removal of organic matter and total nitrogen compounds (BOD:80.0% and TN:67.3% respectively). The effluent filtered in the standard way through a 0.45 μm pore size filter does not provide reliable information about the dissolved (soluble) and particulate fractions due to the colloids content. The results range from 21.0 to 65.0% and are lower than the results achieved in the case of "true" fractionation. In the analysed MCW Darżlubie, COD in the inflow was present in the same proportions for suspended (SCOD), colloidal (CCOD) and dissolved (DCOD) fractions. Fractions of both organic matter and nitrogen dominating in the effluent are present in colloidal and dissolved forms. They could be products of treatment processes (cell debris and decay sub-products) and may be potentially hardly degradable. While the high content of particulate fractions of organic matter and nitrogen in the effluent could be the result of presence of algae in well operated TWs.

There is a need for further investigation of the COD and TON fractions and their bioavailability for decomposition.

REFERENCES

Albuquerque A., Arendacz M., Gajewska M., Obarska-Pempkowiak H., Anderson P., Kowalik P., 2009 Water Science & Technology, 60.7:1677–1682.

Arbeitsblatt ATV-A 131., 1995. Bemessung und Betrieb von einstufigen Belebungsanlagen ab 5000 Einwonhnergleichwerten. St. Augustin.

Aquino S.F, Stuckey D.C 2003, Production of soluble microbial products (SMP) in anaerobic chemostats under nutrient deficiency. Jurnal of Envir. Engrg., 129(11), 1007–1014.

Environment Ministry Regulation according limits for discharged sewage and environmental protection from 24 July 2006 (Dz.U.no 137 item 984).

Gajewska M, Obarska–Pempkowiak H 2007, Experience of hybrid constructed wetlands operation in Gdansk region, Polish Journal of Environmental Studies.— Vol. 16, nr 2A,. 437–433.

Gajewska M. Tuszyńska A. Obarska-Pempkowiak H., 2004 Influence Of Configuration Of The Beds On Contaminations Removal In Hybrid Constructed Wetlands, Polish Journal of Environmental Studies. Vol 13, Ss 149–153.

Henze M, 2002, Weastwater Treatment, biological and chemical processes, third edition, Springer:430.

Kadlec R.H., Knight R.L., 1996, Treatment Wetland. CRC Press, Boca Raton, FL, pp 893.

Kayser K. Kunst S. Fehr G. Voermanek H, 2001, Nitrification in reed beds-capacity and potential control methods, World water congress, published at the IWA, Berlin, Germany, October 2001: 126–138.

Knowles P.R., Dotro G., Nivala J., Garcia J. 2011. Clogging in subsurface-flow treatment wetlands: Occurrence and contributing factors. Ecol. Eng. 37(2): 99–112.

Korkusuz E., 2005 Manual of practice on constructed wetlands for westwater treatment and reuse in Mediterraniean countries. Report AVKR 5, MED-REUNET, Crete, Greece.

Langergraber G. Prandtstetten C. Pressl A., Rohrhofer R., Harbel R., 2006 Removal efficiency of subsurface vertical flow constructed wetland for different organic loads In: 10th International Conference on Wetland Systems for Water Pollution Control: 587–599.

Nguyen L.M., 2000, Organic matter composition, microbal biomass and microbal activity in gravel-bed constructed wetlands treating farm dairy wastewater, Ecological Eng.,16:199–221.

Obarska-Pempkowiak H., Gajewska M., 2005 Recent development in wastewater treatment in constructed wetlands in Poland, In Modern Tools and Methods of Water Treatment for Improving Living Standards (ed) A. Omelchenko, The Netherlands, Springer, Series IV Earth and Environmental Series-Vil. 48: 279–307.

Pagilla K.R., Urgun-Demirtas M., Ramani R., (2006) Low effluent nutrient technologies for wastewater treatment. Water Science & Technology, 53 (3), 165–172.

Pagilla K.R., Czerwinka K., Urguj-Demirtas M., Makinia J., 2008 Nitrogen Speciation In Wastewater Treatment Plant Influent and Effluent—the US and Polish Case Studies, Water Science & Technology, 57 (10), 1511–1517.

Pempkowik J., Obarska-Pempkowiak, H, Gajewska M., Wojciechowska E., 2009, Influence of humic substances in treated sewage on the quality of surface water. Polish Journal of Env. Studies, Vol 3:27–33.

Ramesh A., Lee D.J., Hong S.G 2006, Soluble microbial products (SMP) and Soluble extracellular polymeric substances (EPS) from wastewater sludge, Appl. Microbiol. Biotechnol. 73(1), 219–225.

Tanner Ch.C., Kadlec H.R., Gibbs M.M., Sukias J.P.S., Nguyen M.L., 2002, Nitrogen processing gradients in subsurface-flow treatment wetlands-influence of wastewater characteristics, Ecol. Eng.18: 499–520.

Templeton D.M., Ariese F., Cornelis R., Danielsson L.G., Muntau H., van Leeuwen H.P., Łobiński R., 2000 Guidelines for trems releted to chemical speciation and fractionation of elements. Definitions, structural aspects, and methodological approaches IUPAC Recommentations, Pure Appl. Chem., Vol 72, No 8, pp 1453–1470.

Tiehm A., Herwig V., Neis U., 1999, Particle size analysis for improved sedimentation and filtration in waste water treatment, Wat. Sci. & Tech Vol 39, No 8 pp 99–106.

Tuszyńska A., Obarska-Pempkowiak H. 2006. Wpływ substancji organicznej na natlenienie i efektywność usuwania zanieczyszczeń w złożach hydrofitowych. Wydawnictwo Politechniki Gdańskiej, 74 s.

Vymazal J., 2005 Horizontal sub-surface flow and hybrid constructed wetland systems for wastewater treatment Ecol. Eng. 25: 478–490.

Environmental Engineering IV – Pawłowski, Dudzińska & Pawłowski (eds)
© 2013 Taylor & Francis Group, London, ISBN 978-0-415-64338-2

Specific impact of enhanced biological phosphorus removal from wastewater on fermentation gas generation

S. Rybicki

Faculty of Environmental Engineering, Cracow University of Technology, Poland

ABSTRACT: The paper summarizes results of observation of routine wastewater treatment plant operation, followed by laboratory and full scale experiments performed to discover a reason of a biogas production decrease during sludge digestion. Long-term observations showed that a lower-than-calculated gas yield from the decomposed biomass may be a considerable problem for operators. The values lower by 1/3rd than the calculated unit gas production were reported. Laboratory tests exposed some specific process control problems, which might occur in wastewater treatment plants due to acidic generation of short chained fatty acids, used in enhanced biological phosphorus removal. A relatively high consumption of biodegradable carbon was found due to unfavorable effects. A significant contradiction between needs for acidic fermentation products and biogas production was the main obstacle in the proper operation; both in laboratory tests and in full-scale operation it led to a decrease in fermentation gas production.

Keywords: sludge processing, sludge digestion, combustible biogas, sustainable development, EBPR, interaction between processes

1 INTRODUCTION

Contemporary wastewater treatment plants are usually sized with respect to ensure a proper amount of readily biodegradable carbon compounds ahead of an anaerobic zone (Barnard 1994, 2000). In case of municipal wastewater a need for electron donors led initially to addition of external carbon sources such as ethanol, methanol or acetic acid, and later an 'internal source' of carbonaceous matter was incorporated into WWTPs flow schemes. The 'internal source' of carbon means usually a mixture of short chained fatty acids produced from sludge during hydrolysis and acidic fermentation. The energy, which is available from hydrolysis (break of polyphosphate chains) is being used for storage of organic material within the cell. Under aerobic conditions material stored in this way is utilized when energy for growth and energy for phosphate storage is obtained via respiration (Carlsson et al. 1996, Rybicki 1997, Mino et al. 1995, Ekama et al. 1999, Barnard 2000). During these processes a conversion of an organic substrate originated from primary sludge into fermentable soluble carbon compounds and then the fermentation of products takes place (Mino et al. 1998; Carucci et al. 1999, Pijuan et al. 2005, Song 2005, Jönsson et al. 2007). Sludge hydrolysis— based on the same phenomena—is also a backbone of methanogenic fermentation of wastewater

sludge, customized for sludge stabilization and a biogas production (Jardin & Pöpel 1996, Rybicki 2009).

The volume of biogas produced during a mesophilic digestion phase can vary depending on the nature of the substrates delivered with raw unstabilized sludge. Operational check of the plant practice showed that it might be caused by an intensive use of pre-fermenters, which convert a part of an organic matter contained in raw sludge to less complex compounds, mostly Short Chain Fatty Acids (SCFA). These observations led to the initial conclusion that some part of organics load was missing during previous stages of operation, thus this specific problem was related to generation of SCFAs from primary sludge. The pre-fermentation was performed in side-stream pre-fermenters and an increase of the sludge retention in the pre-fermentation unit was observed. The average design hydraulic retention time in the pre-fermenter was 1,5 days but during a start-up phase operators found that such mode of operation did not ensure a proper SCFAs load for EBPR. An increase of a Sludge Retention Time (SRT) in pre-fermenters resulted in a higher volume of supernatant (rich with SCFA) that was added to a wastewater treatment train. Due to a relatively longer response time, two effects were observed simultaneously: a higher than expected dry mass concentration in a supernatant and a significant load of a SCFA rich mass of hydrolyzed sludge passing through a sludge thickener and being discharged to the treatment line ahead of the reactor.

2 MATERIALS AND METHODS

The investigations focused on the question whether, and to what extent, a primary fermentation might adversely impact biogas production in digestion chambers. Tests were performed in a respirometric test unit. The scope of tests were as follows:

- Measurement of possible SCFA production under various operational conditions;
- Tests on a phosphorus release rate in such a mixture of SCFA;
- Batch tests on a possible gas yield with and without pre-fermentation in laboratory conditions.

Tests were conducted with the use of a respirometry method, being described by the author in previous papers (Rybicki 2011). The tests stand was located at the Cracow University of Technology Research Lab. The AER-208 (Challenge Systems, USA) respirometer consisted of two basic parts apparatus to measure the gas volume and interface that transmits the obtained values to a computer for data registration. Sludge samples were placed in small air tight bottles (reaction chambers) of 0.5 litre volume, covered with rubber plugs and additionally secured with metal cups. (Cimochowicz-Rybicka, 2009).

2.1 Experimental protocol

Tests were performed in consecutive steps. First step was addressed towards SCFA generation in a pre-fermentation step (acidic fermentation). Next stage i.e. batch tests on Phosphorus Release Rates (PRR) were performed to check the process intensity when SCFAs were added with supernatant liquid from the first step investigation. A gas production test was the final step of investigation, that is thickened sludge was added to a testing vessel of the respirometer, which from that moment was switched to a mode reflecting mesophilic fermentation. Both first and third steps were performed in a respirometric test stand as described above. PRR lab batch test was performed routinely on the activated sludge coming from the same WWTP as the WAS. Basic experiments conditions were the following:

Temperature: during the acidic pre-fermentation tests the temperature was kept on a constant level of 18°C (equal to the average real temperature); the incubation temperature during third stage was 35°C.

pH: with regard to the sensitivity of methanogenic bacteria to pH, this parameter was restrictedly controlled in third stage and ranged from 7.0 to 7.4; in first (acidic) stage of tests the pH was uncontrolled and varied between 5.0 and 5.9.

Experimental system: A mechanically stirred system was applied to ensure a full mixing of samples; The total amount of fatty acids was measured by titration at pH 5 and pH 4 with 0.1 M HCl (Wild 1997). Specific single SCFAs were determined by gas chromatography. Phosphates were measured with a spectrophotometer (molybdenovanate method). The spectrophotometer was used in all COD measurements.

After the pre-fermentation stage, the sludge from the pre-fermenter stage was exposed to mesophilic fermentation in the test stand, while a supernatant liquid was used for PRR testing. The sludge was acclimatized in eight 1-litre fully mixed reactors for more than two weeks at 35°C. Samples were composed by mixing pre-fermented sludge (raw in a reference sample) and addition of 'inoculating sludge' from the WWTP. Applied solid retention times were 2, 3, 5 and 7 days long pre-fermentation of primary sludge, with non-fermented sludge as a reference sample. Each sample was tested twice in a respirometric unit.

3 RESULTS AND DISCUSSION

3.1 *Short chain fatty acids generation*

Figure 1 summarizes results of pre-fermentation as a proportion between dissolved and total COD measured in sludge samples versus SRT at 18°C. Total COD ranged between 3600 and 4200 mg/L. It was stated that intensity of fermentation (expressed as a ratio of COD soluble to COD total) reached 20% value at SRT = 2,5 days. The

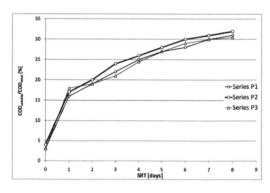

Figure 1. Soluble COD vs SRT in the pre-fermenter. T = 18°C, uncontrolled pH. Primary sludge fermentation.

Table 1. Phosphate release rate tests after pre-fermentation.

Sample number	Grab samples (Rybicki, 2009)							Extended SRT samples				
	1	2	3	7	8	9	10	14	21	22	13	25
SRT [days]	1.6	3.3	3.4	3.5	3.2	2.4	3.2	5.6	5.3	6.5	4.1	4.2
PRR [gP/ gMLSS/h]	2.4	3.87	4.12	4.60	3.90	3.10	4.50	3.45	3.87	4.12	4.60	4.35

specific composition of supernatant with respect to the SCFA showed that almost the same fraction of soluble COD was acetate (30%) and propionate (29%) as it had been expected (Moser-Engeler et al. 1998). This stage of investigation showed that there is no practical gain in operating a pre-fermenter at SRT longer than 4 days.

3.2 Phosphate release tests

A consecutive stage of investigation focused on kinetics of the phosphate uptake phase. As it has been known, performance of release also differs with respect to a type of fatty acids (Moser-Engeler et al. 1998). Both stages of experiment described in this paper and literature references confirm that linear fatty acids were initially degraded but the decrease of branched fatty acids was very slow. When linear fatty acids were consumed, the substrate uptake rates for an iso-butyrate and also 2-methyl butyrate increased. Table 1 summarizes results of experimental series, where SRT was equal or lower than the value assumed.

3.3 Impact of SRT on fermentation gas production

The specific goal of the research, performed along with laboratory experiments, was to determine a possible impact of the extended sludge retention time in pre-fermenters on SCFAs generation and a phosphates release. The effluent phosphorus concentration remained below 1 mgP/l throughout the study, thus the results obtained turned out to be satisfactory, from the wastewater quality perspective. The final results have been presented as gas production cumulative curves; see Figure 2 for SRT = 2 days and Figure 3 for SRT = 7 days. Summary of the results has been presented in Table 2.

The results confirmed that a decrease of unit gas production in WWTPs (being operated with pre-fermentation units) might be credited to the fact, that extended SCFA production with a limited possibility of sludge discharge control adversely impacted utilization of a prospective substrate in a wastewater treatment train rather than in mesophilic fermentation. A shorter SRT

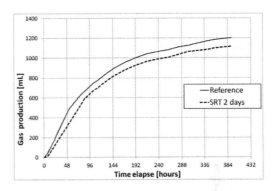

Figure 2. Gas production in batch tests—prefermentation SRT = 2 days. T = 35°C.

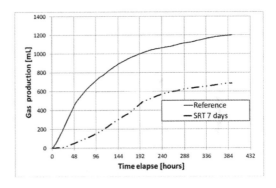

Figure 3. Gas production in batch tests—prefermentation SRT = 7 days. T = 35°C.

(2 days) caused a negligible decrease in overall gas production (approx. 5%, while 3 days SRT led to a 10% decrease in gas production). 7 days SRT resulted in a 38% loss of fermentation gas. All samples had a similar methane content of about 67% (by mass), although due to a limited number of series this conclusion should not be generalized. Table 2 summarizes the results re-calculated for unit gas production after 16 days; the expected range was 0,75 to 0,95 std m^3/kg VSS$_{removed}$ (Rybicki 2009), so at SRT higher than 3 days, unit gas production was lower than the lowest expected value.

Table 2. Unit gas production test; Initial VSS = 5.2 g/L; Temperature 35°C; Mesophilic fermentation.

Pre-fermentation SRT	Gas production [mL]	VSS_A [g/L]	VSS_M [g/L]	Unit gas production [std m^3/g $VSS_{removed}$]
No prefermentation	1200		3.87	0.91
2	1120	4.42	3.07	0.80
3	920	4.49	2.70	0.83
5	800	4.16	2.80	0.59
7	710	4.08	2.58	0.51

Where: Gas prod—gas production.
VSS_A—VSS before methanogenic tests.
VSS_M—VSS after methanogenic tests.

It can be stated that an extension of an acidic ('pre-fermentation') phase over 3 days resulted in a significant decrease of unit gas production. Then, responsibility for a decrease in overall gas production should be 'shared' between two main mechanisms, that is removal of a portion of hydrolysable mater (drop of VSS after pre-fermentation), which results in a lower than expected load of substrates to methanogenic phase, and a decrease of unit gas production, probably due to the fact that the process is based (in this phase) on a 'slowly' hydrolysable matter.

4 CONCLUSIONS

Generation of short chain fatty acids from a primary sludge provides a required amount of electron donors allowing a multi-phase reactor to be successfully operated without chemical addition.

In the described specific case, a visible contradiction between needs for acidic fermentation products for EBPR processes and biogas production was the main obstacle in system operation; both in laboratory tests and in full-scale operation it led to a significant decrease of fermentation gas production (over 30% decrease was found at 7 days SRT). A decrease in overall gas production (in WWTP with a pre-fermentation units) might be due to the consumption of some hydrolysable matter but also to a decrease of unit gas production.

Tests described in this paper were based on batch performance of acidic fermentation and gas production and as presented in the paper they may be applied for design purposes (assessment of the proper SRT) as well as for operational optimization. Proposed testing procedures led to accurate calculations of the SRT value for acidic and mesophilic fermentation, which are considered the main parameter for dimensioning of the sludge processing line at modern WWTPs.

ACKNOWLEDGEMENTS

Experiments presented in this paper were financially supported by the Ministry of Science, Republic of Poland; research grant No. NN 523 3773 33 (3775/B/TO2/2007/33).

REFERENCES

Barnard, J.L. 1994. Alternative Prefermentation Systems. *Proc. 67th Annual WEF Conference*, Chicago, Water Environment Federation.

Barnard, J.L. 2000. Design of pre-fermentation process. In J. Kurbiel (ed.) *Proceedings of seminar on design and operation of wastewater plants* (In Polish), 61–77.

Carlson, A. 1996. Interactions between wastewater quality and phosphorus release in the anaerobic reactor of the EBPR process, *Wat. Res,*. vol. 30 (6):1517–1527.

Carucci, A. et al. 1999. Microbial competition for the organic substrates and its impact on EBPR systems under condition of changing carbon feed, *Wat. Sci. Tech.*, 39(1):.75–85.

Cimochowicz-Rybicka, M. 2009. Application of Methanogenic Activity as a Monitoring Tool for Sludge Disintegration Control in Methane Gas Generation, *Polish Journ. Of Environmental Studies*, vol 18 (3A):52–59.

Cimochowicz-Rybicka, M. 2011. Application the Sludge methanogenic activity tests to improve overall methane production using sludge sonication; *Proc 4th IWA ASPIRE conference*, Tokyo, 22–52.

Ekama, G.A. &, Wentzel, M.C. 1999. Denitrification kinetics in biological N and P removal activated sludge systems treating municipal wastewaters, *Wat. Sci. Tech.* 39(6): 69–77.

Jardin, N.&Pöpel, H.J. 1996. Influence of the enhanced biological phosphorus removal on the waste activated sludge production, *Wat. Sci. Tech.* 34(1–2):17–23.

Jönsson, K et al. 2007. Utilising laboratory experiments as a first step to introduce primary sludge hydrolysis in full scale, *Proc. 10th IWA Specialised Conference LWTP*, IWA, 129–137.

Mino, T., et al. 1995. Estimation of the rate of slowly biodegradable COD (SBCOD) hydrolysis under anaerobic, anoxic and aerobic conditions using starch as model substrate; *Wat. Sci. Tech.*, 31(2):95–103.

Mino, T., et al. 1998, Microbiology and biochemistry of the enhanced biological phosphate removal process, *Wat. Res.* 32(11):3193–3207.

Moser-Engeler, R. et al. 1998, Products from primary sludge fermentation and their suitability for nutrient removal, *Wat. Sci. Tech.*, 38(1):265–273.

Pijuan, M. et al. 2005. Anerobic release linked to acetate uptake: influence of PAO intracellular storage compounds, *Biochemical Engineering Journal*, 26:184–190.

Rybicki, S.M. 1997. Monograph: Phosphorus Removal From Wastewater. A Literature Review; Stockholm, Royal Institute of Technology.

Rybicki, S.M. 2009. Interactions Between Advanced Wastewater Treatment and Sludge Handling Processes. *Polish Journal of Environmental Studies*, 18(3 A):396–403.

Rybicki, S.M., Cimochowicz-Rybicka, M. 2011. Selected Interactions Between Methane Recovery From Wastewater Sludge and Phosphorus Removal. *Proc 4th IWA ASPIRE conference*, Tokyo,

Seviour, R.J. et al. 2003. The microbiology of biological phosphorus removal in activated sludge systems, *FEMS Microbiology Reviews*, 27:99–127.

Song, C. 2005. Contributions of phosphate and microbial activity to internal phosphorus loading, *Science in China*, 49:102–113.

Wild, D. et al. 1997. Prediction of recycle phosphorus loads from anaerobic digestion, *Wat. Res.*, 31(9):2300–2308.

Environmental Engineering IV – Pawłowski, Dudzińska & Pawłowski (eds)
© 2013 Taylor & Francis Group, London, ISBN 978-0-415-64338-2

Detection of *Cryptosporidium*, *Giardia* and *Toxoplasma* in samples of surface water and treated water by multiplex-PCR

M. Polus & R. Kocwa-Haluch
Faculty of Environmental Engineering, Cracow University of Technology, Cracow, Poland

A. Polus
Medical College, Jagiellonian University, Cracow, Poland

ABSTRACT: Studies on the application of PCR to detect waterborne pathogens have been conducted for more than 15 years but there are still many obstacles in the routine use of molecular biology methods in environmental engineering. The extraction of DNA from environmental materials is one of the most important problem that may occur. While the isolation of DNA from microorganisms found in clean samples is usually successful and the detection by PCR is relatively easy, the variability of the composition of raw water, surface water effectively inhibits PCR. Immunomagnetic Separation (IMS) of cysts/oocysts of protozoa, prior to the DNA extraction phase is the solution which considerably improves the possibility of its detection. Unfortunately, it is possible to implement IMS only against a few pathogens. In the present study the authors demonstrated the possibility of simultaneous detection of the presence of three protozoa in the water without the need of IMS.

Keywords: *Cryptosporidium*, *Giardia*, *Toxoplasma*, PCR, multiplex-PCR, surface water, raw water, treated water

1 INTRODUCTION

According to the guidelines of the World Health Organization, water intended for human consumption, in general, should not contain organisms pathogenic to man (WHO 2004), water source requires special protection and technology of treatment should not prevent the emergence of parasites in the water treated (Marshal et al. 1997). Among the indicators of microbiological water quality, the bacterium *Clostridium perfringens* is of particular importance since their spores show considerable similarity to protozoan cysts/oocyst, which are similar in size, have comparable resistance to disinfectants and are equally difficult to eliminate in the water treatment process (Kocwa-Haluch & Polus 2004). Therefore, it is often claimed that the high costs and difficulty of detection and determination of cysts/oocysts of protozoan parasites, the bacterial content analysis of *C. perfringens* may be an indirect indicator of the presence of intestinal protozoa in the water—a negative result for *C. perfringens* implies the absence of protozoa of the genera *Giardia*, *Cryptosporidium* and *Toxoplasma* in a water sample.

Owing to the fact that there is no complete correlation between the prevalence of *C. perfringens* and protozoan parasites there is a need for specific determination of their cysts/oocyst, also in waters where there is no fecal bacteria. Currently, there are methods to detect cysts/oocysts of *Giardia* and *Cryptosporidium* (EPA 2001, DiGiovani et al. 1999, Jenkins et al. 1997, Robertson et al. 1998), but their common characteristics are: low and widely variable sensitivity, high dependence of the human factor, the limited capacity to assess the viability of cysts/oocysts and very high prices in detection, which are hardly acceptable in the Polish realities. PCR enables detection of DNA of any of the listed protozoa, ensuring high sensitivity but, unfortunately, the success of the PCR reaction depends, in great measure, on the purity of the DNA isolated from a given material. The DNA extraction from a clinical sample is relatively simple (as easy to isolate DNA from the samples of treated water) and there are many effective, widely used protocols. Whereas, the DNA extraction from environmental samples such as surface water, poses enormous challenges because removal of all factors present in the water, inhibiting the PCR, is necessary (Jiang et al. 2005, Isaac-Renton et al. 1998). Since the quantity and kind of contamination affecting the course of PCR can radically vary between different samples of water, most of the previously published work has revealed a large and unpredictable variation in the sensitivity of the detection (Yang et al. 2008,

Clancy 2001) compensated, to some extent, by nested-PCR (Mayer & Palmer 1996).

In short: the use of the PCR in environmental engineering is effectively hindered by the difficulties of extracting pure DNA from water samples (Jiang et al. 2005). Although there are many reports showing the advantages of PCR, most authors encounter difficulties at the stage of the material concentration (e.g. surface water) and the isolation and purification of DNA (Rochelle et al. 1997, Nichols et al. 2003, Johnson et al. 1995). Despite the existence of numerous DNA extraction protocols most of them become almost unusable for surface/raw water (Yu et al. 2009, Jiang et al. 2005). Due to the fact that the most optimal solution to these problems is Immunomagnetic Separation (IMS) of cysts/oocysts of waterborne parasites (Hsu et al. 2001), available only for very few genera (*Cryptosporidium*, *Giardia*, *Escherichia*), the aim of this study was to test the DNA purification protocol, which will provide a template for multiplex-PCR, and thus enable the detection of specific pathogens without IMS.

2 METHODS

Sample preparation. 4 litres of water samples were subjected to filtration through a MCE membrane filter (type RAWP, Millipore) with a diameter of 90 mm and porosity of 3,0 μm. In the case of very large water turbidity filtration was done on either a few membranes and then the collected pellets were combined or by initial sample pre-filtration using fiberglass (type APFD, Millipore) in order to remove suspended solids. Samples of the treated water prepared to control seeds were preliminary filtered through a membrane with a porosity of 1,2 μm. Cysts and oocysts used to examination came from Waterborne (USA).

Concentration. After filtration the membrane filter was transferred to a Petri dish, flooded with 5 ml of 0,01% Tween-20 (in PBS) and gently scraped. The collected suspension was overlaid on an aqueous sucrose solution (d = 1,15 g/ml) and centrifuged (1250xg, 10 min, 4°C). The entire top fraction together with the interphase was transferred to a new test tube, diluted to 50 ml with cold PBS and centrifuged (4500xg, 15 min, 4°C). The pellet was washed twice with 25 ml PBS and centrifuged (4500xg, 15 min, 4°C). The pellet was finally suspended in 1 ml PBS and transferred to a standard 1,5 ml test tube (Villena et al. 2004, Polus & Kocwa-Haluch 2010).

Immunomagnetic Separation (IMS). IMS was adopted from the manufacturer's protocol (Life Technologies, Dynabeads® GC-Combo Kit), as the scale of the original procedure was too large for PCR needs (moreover there is no need to remove paramagnetic beads before isolation of DNA), therefore some modification was made. To simplify the further steps of separation the most important modification was the abandonment of standard MPC magnets (Dynal) in favour of smaller DynaMag-2 magnets (Invitrogen), adapted for standard 1,5 ml test tubes. In brief, pellets gathered in the earlier steps were suspended in 1 ml of 1x SLA buffer (100 μl of 10x SLA and 100 μl of SLB in 1 ml of water) with the subsequent addition of 10 μl of Dynabeads®. The tubes were then placed on a rotary shaker and subjected to gently stirring/rotation (15 rpm) for 2 hours. Then, the tubes were placed on a magnet stand and gentle swinging was performed for 2 minutes until the suspension of magnetic beads accumulated onto the walls of the tubes. Buffer was discarded and the remaining beads were rinsed twice in 1 ml of 1x SLA buffer and subsequently in cold PBS.

DNA extraction. The material collected in the preceding step was centrifuged (14000xg, 10 min, 4°C), and the supernatant was discarded. The pellet was suspended in the 600 μl of lysis buffer (10 mM Tris, 0,5% SDS, pH 8,0). The use of 1 mM EDTA in the lysis solution was abandoned as a potential agent of PCR inhibition. The lysate was incubated at 37°C for 30 minutes after which it was subjected to 5 freeze/thaw cycles (5 min in liquid nitrogen and 5 min in 60°C per cycle) (Nichols et al. 2003). Samples were then incubated for 15 hours (overnight) with proteinase K (200 μg/ml, 52C) while being vigorously shaken in a thermomixer. Then, tubes were incubated for 15 min at 95°C to denaturate the proteinase K and single extraction with 500 μl of phenol:chloroform:isoamyl alcohol mixture (25:24:1 v/v, pH 8,0) was made. Lysate was centrifuged (7500xg, 15 min, 4°C) and the upper, water phase was transferred to a new 1,5 ml test tube. DNA was precipitated by addition of 100 μl of 10 M ammonium acetate and 800 μl of isopropanol. The samples were then centrifuged (20000xg, 30 min, 4°C). Pellets were washed once with 1 ml cold 70% ethanol, centrifuged (20000xg, 15 min, 4°C), air-dried (1 hour) and suspended in 50 μl of molecular biology grade water. Purification of DNA was performed using High Pure PCR Template Preparation Kit (Roche), according to the manufacturer's guidelines. The only modification was reducing to 50 μl the volume of buffer, used for elution of the DNA from the column.

In this the three methods of DNA isolation were compared, namely direct isolation from pellets collected from membrane filters, direct isolation from pellets and subsequent purification of DNA from columns (Roche), and immunomagnetic separation of cysts/oocysts (Life Technologies) that preceded the isolation of DNA. PCR. The reactions were performed in a final volume of 25 µl in thermal cycler Mastercycler Personal (Eppendorf) using 2,5 µl of DNA under the following conditions: 0,7 Unit of DNA polymerase DNA (Biotools, Spain), 2 mM Mg^{+2}, 0,2 mM dNTPs, 250 nM of each primer (Genomed). The temperature of primer annealing was established in the DNA Engine Opticon (MJ Research, USA). In the case of the nested-PCR, 0,5 µl of outer-PCR was used as a template.

Genomic DNA of *Cryptosporidium parvum* (ATCC PRA-67D), *Giardia intestinalis* (ATCC 30888D) and *Toxoplasma gondii* (ATCC 50174D) were used as positive controls and quantitative standards in the PCR reactions. The concentration of standard solutions of genomic DNA was determined on the basis of the measurements in the Qubit fluorometer (Invitrogen). The set of primers used in this study (Table 1, CP1Wf/r CP2Wf/r GI1Wf/r, GI2Wf/r, TGWf/r) allowed for the accurate detection of 0,1 pg of genomic DNA of each of the three protozoa in a single-step PCR (Grigg & Boothroyd 2001, Magpie et al. 2009, Rochelle et al. 1997). Each mentioned pair of starters can be combined with pair flanking slightly wider area (CP1Zf/r CP2Zf/GIZf r/r TGZf/r), and therefore suitable for the first step of the nested-PCR. All reactions were performed using one and the same thermal profile: 94°C/3 min., 40 cycles involving the phase of 94°C/45 s: 55°C/60 s: 72°C/60 s, and final extension 72°C/10 min. Electrophoresis was carried out on 1,5% standard agarose containing SYBR Green (Molecular Probes). In cases of weak fluorescence, gels were additionally stained in SYBR Gold (Molecular Probes).

3 RESULTS AND DISCUSSION

The implementation of a multiplex-PCR method means introducing a mixture of numerous pairs of primers into the reaction, thereby increasing the concentration of oligonucleotides (as per usual, each of the primers is used at a similar concentration as in a classical PCR with one pair of primers), thus increasing the risk of the interaction between them and therefore the emergence of unplanned products, and the exclusion of the expected synthesis. The initial action was to check the sensitivity of the PCR in a mixture with three pairs of primers. It appears that the sensitivity of the detection of genomic DNA of each of the three protozoa using multiplex-PCR is slightly lower than that where detection was carried out with the use of single pairs of starters. It is possible to detect 1 pg DNA of *C. parvum*, but only 10 pg DNA of *T. gondii* and *G. intestinalis* (Fig. 1/I). For a comparison of the multiplex-PCR *Cryptosporidium/Toxoplasma*, two pairs of primers were used, which detected 50fg DNA of *T. gondii*, and the presence of several hundred fold more DNA *C. parvum* (Fig. 2). However, if one performs the multiplex-PCR in two stages, as a sequence of external reaction and internal (nested) reaction, then it is possible to detect even 0,1 pg DNA of

Table 1. Primer sets used in the study. CP1Z, CP1W, CP2Z, CP2W: primers for the determination of *C. parvum*, GIZ, GI1W, GI2W: primers for the determination of *G. intestinalis*, TGW: primers for the determination of *T. gondii*.

Name	Sequence	Comments
CP1Z	f: 5′-GTATCAATTGGAGGGCAAGT r: 5′-CTATGTCTGGACCTGGTGAG	External
CP1W	f: 5′-AAGCTCGTAGTTGGATTTCTG r: 5′-TAAGGTGCTGAAGGAGTAAGG	Internal (435 bp)
CP2Z	f: 5′-AGTGCTTAAAGCAGGCAACTG r: 5′-CGTTAACGGAATTAACCAGAC	External
CP2W	f: 5′-TAGAGATTGGAGGTTGTTCCT r: 5′-CTCCACCAACTAAGAACGGCC	Internal (256 bp)
GIZ	f: 5′-CATAACGACGCCATCGCGGCTCTCAGGAA r: 5′-TTAGTGCTTTGTGACCATCGA	External
GI1W	f: 5′-CATAACGACGCCATCGCGGCTCTCAGGAA r: 5′-TTTGTGAGCGCTTCTGTCGTGGCAGCGCTAA	Internal (218 bp)
GI2W	f: 5′-AAGTGCGTCAACGAGCAGCTC r: 5′-TTAGTGCTTTGTGACCATCGA	Internal (171 bp)
TGZ	f: 5′-TGTTCTGTCCTATCGCAACG r: 5′-ACGGATGCAGTTCCTTTCTG	External
TGW	f: 5′-TCTTCCCAGACGTGGATTTC r: 5′-CTCGACAATACGCTGCTTGA	Internal (531 bp)

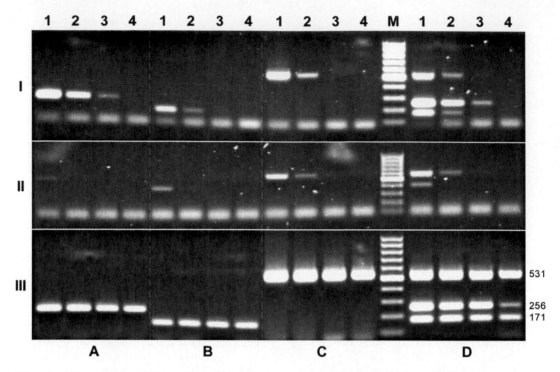

Figure 1. The sensitivity of the multiplex-PCR (D) compared to the classical PCR for each of the three pathogens (A: detection of *Cryptosporidium*; B: detection of *Giardia*; C: detection of *Toxoplasma*). I—classic, single-step PCR (primers as for internal PCR: CP2Wf/r GI2Wf/r TGWf/r); II—external/outer PCR (primers: CP2Zf/r GIZf/r TGZf/r); III—internal PCR (nested-PCR, primers: CP2Wf/r—256 bp, GIWf/r—171 bp, TGWf/r—531 bp), re-amplification of 0,5 µl products of external reaction. Quantity of genomic DNA of each protozoa used for PCR: 1—100 pg; 2—10 pg; 3—1 pg; 4—0,1 pg. M—ladder 100 bp. 1,5% agarose stained with SYBR Green.

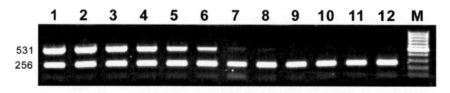

Figure 2. Multiplex-PCR with two pairs of primers (CP2Wf/r—256 bp, and TGWf/r—531 bp) on a combination of genomic DNA of *Cryptosporidium parvum* and with *Toxoplasma gondii*. Each lane contains the 10 pg DNA of *C. parvum* and subsequent amount of DNA *T. gondii*: 1—100 pg; 2—50 pg; 3—10 pg; 4—5 pg; 5—1 pg; 6—500 fg; 7—100 fg; 8—50 fg; 9—10 fg; 10—5 fg; 11—1 fg; 12—no DNA *T. gondii*. M—ladder 100 bp. 1,5% agarose stained with SYBR Green.

each of the three examined protozoa (Fig. 1/II, III). This is an acceptable sensitivity.

One of the reasons for the criticism of multiplex-PCR is the different length of the resulting products necessary for their separation in gel electrophoresis. The differences in the lengths of the products mean the difference in the pace of their synthesis, which means that in particularly adverse circumstances (e.g. if there is a significant difference in the number of matrices for each pair of primers in the sample material) only one product may be formed in

amounts detectable on a gel. Thus, the result will be a false negative, although the test material contains the investigated DNA. That is why a series of reactions were performed with material containing different amounts of genomic DNA of individual protozoa. They ranged from 0,1 pg to 100 pg DNA per reaction. As it has been shown, in some concentrations the sensitivity of detection was slightly reduced but never worse than 1 pg DNA (Fig. 3).

The next step was to determine the sensitivity as an expression of the number of cysts/oocysts

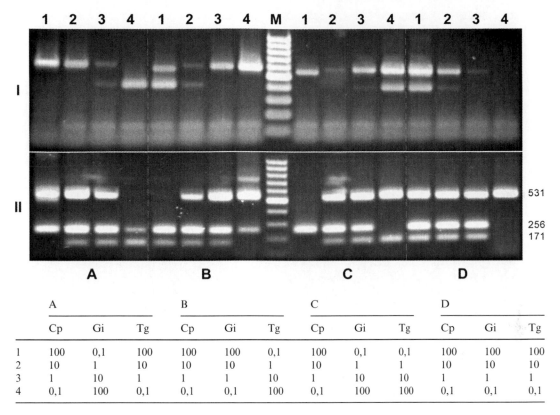

Figure 3. Sensitivity of multiplex-PCR for various combinations of genomic DNA of individual protozoa. I—external/outer PCR (primers: CP2Zf/r GIZf/r TGZf/r); II—internal PCR (nested-PCR, primers: CP2Wf/r—256 bp, GIWf/r—171 bp, TGWf/r—531 bp). Amounts of pg of DNA are given in the table below (Cp—*Cryptosporidium parvum*, Gi—with *Giardia intestinalis*, Tg—*Toxoplasma gondii*). M—ladder 100 bp. 1,5% agarose stained with SYBR Gold (I) or SYBR Green (II).

	A			B			C			D		
	Cp	Gi	Tg	Cp	Gi	Tg	Cp	Gi	Tg	Cp	Gi	Tg
1	100	0,1	100	100	100	0,1	100	0,1	0,1	100	100	100
2	10	1	10	10	10	1	10	1	1	10	10	10
3	1	10	1	1	1	10	1	10	10	1	1	1
4	0,1	100	0,1	0,1	0,1	100	0,1	100	100	0,1	0,1	0,1

detected in the various types of water. It is widely known that the type of water, turbidity, chemistry and composition of the biocoenosis impacts on several stages of the analysis. These significantly affect: the ease of filtration and therefore the volume of the sample that can be filtered; the recovery of cysts/oocysts from membrane filters, and the quantity of DNA and the degree of its possible pollution with PCR inhibitors. The concept of PCR inhibitors covers any chemical substances presented in the test material (surface water, raw water), which may affect the PCR. In surface waters there are numerous substances that are able to impede the course of the polymerase chain reaction/PCR. We prepared a number of seeds of *G. intestinalis* cysts and *C. parvum* oocysts in tap water (negative) and surface water (Wisła, Kraków, at the height of Tyniec) from 0 to 5000 cysts/oocysts per 4 litres of water (equal amounts of both protozoa). Then complete filtration of samples was performed together with concentration, IMS and the isolation of DNA. Two-step

PCR (external/internal) was performed for all the samples. PCR showed positive results even for the lowest seeded cultures in treated water (50 cysts/oocyst in 4 liters) (Fig. 4/IA, IB). The multiplex-PCR also allowed the detection of the smallest quantity of cysts/oocysts (Fig. 4/IC). Significantly lower sensitivity was shown, however, in the case of isolation from seeded surface water, the sensitivity of the multiplex-PCR decreased (Fig. 4/II). The figures illustrate only the results of the internal reaction (nested-PCR).

Similar multiplex-PCR results were obtained after the inoculation with variable concentration of cysts/oocysts: 5/5000, 50/500, 500/50 and 5/5 cysts/oocysts of *G. intestinalis* and *C. parvum* (Fig. 5). The achieved sensitivity of the entire procedure (filtration/insulation/PCR) was mainly due to the possibility of separation of cysts/oocysts from the rest of the pellets collected from the filters using IMS. Unfortunately, it is not always possible because currently there are only a few IMS separation kits to separate aquatic pathogens

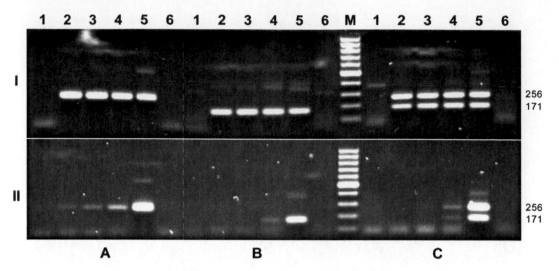

Figure 4. Sensitivity of detection of control seeds of *C. parvum* oocyst and *G. intestinalis* cysts in negative samples of 4 liters of treated water (I) and surface water; (II) using nested-PCR. A—*C. parvum*, primers CP2Wf/r—256 bp; B—*G. intestinalis*, primers GI2Wf/r—171 bp; C—multiplex-PCR with both pairs of primers. The following lanes are: 1—sample without the seeds of cysts/oocysts; 2—50 cysts/oocysts; 3—500 cysts/oocysts; 4—5000 cysts/oocysts; 5—positive control; 6—negative control. The photo shows the only results of the internal reaction. M—ladder 100 bp. 1,5% agarose stained with SYBR Green.

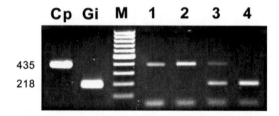

Figure 5. Sensitivity of detection of control seeds of *C. parvum* oocyst and *G. intestinalis* cysts in negative 4-litre samples of treated water by multiplex-PCR with pairs of primers: CP1Wf/r—435 bp and GI1Wf/r—218 bp. Quantity of oocyst/cysts on the following lanes: 1—5000/5; 2—500/50; 3—50/500; 4—5/5000. PCR positive controls: Cp—10 pg *C. parvum* genomic DNA, Gi—10 pg *G. intestinalis* genomic DNA. M—ladder 100 bp. 1,5% agarose stained with SYBR Green.

(*Cryptosporidium sp.*, with *Giardia sp.*, some serotypes of *E. coli*). Therefore, there is a need to direct the isolation of DNA in such a way that makes it effective for PCR. We therefore assessed the ability to detect the DNA of *C. parvum* extracted from the different types of water by direct isolation, supplemented by purification of the DNA on columns (Roche) and compared it to the effects of the isolation of DNA from material obtained by IMS. Four types of water were used as material for isolation: treated (tap) water free from parasites, surface water with low turbidity (Wisła, Kraków-Tyniec),

surface water with signifycant turbidity and pollution (Serafa, Kraków-Brzegi) and as a control sterile PBS. Three procedures were used for the isolation of DNA from 4 litres of each of the listed materials inoculated with 100 *C. parvum* oocyst: A) IMS separation and isolation of DNA, B) only direct DNA isolation, C) direct isolation of DNA and additional cleanup on columns (Roche). On the isolated DNA classic PCR was performed (with internal CP1Wf/r primers) and the results are shown in Fig. 6. As can be seen the use of IMS allowed for the detection of all cultures/seeds resulting in clean PCR products (Fig. 6/A). Direct isolation of DNA caused considerable reduction of sensitivity (Fig. 6/B1). Moreover, PCR performed on samples of surface water, showed numerous unspecific bands (Fig. 6/B2, 3) and did not give any positive results (Fig. 6/B4). Purification of DNA after the direct isolation on the columns (Roche), whilst not reducing the problem of the formation of unspecific bands, slightly improved the sensitivity of detection. For instance, 100 oocysts seeded into highly polluted water samples (Fig. 6, tracks 4, 5) were was not detectable by direct DNA extraction, although after purification on columns there was the generation of a weak but noticeable product (Fig. 6/C5).

Similar results were achieved by multiplex-PCR (*Cryptosporidium* and with *Giardia*) on seeds of oocysts/cysts in identical samples of waters as above (Fig. 7). What is interesting, DNA

146

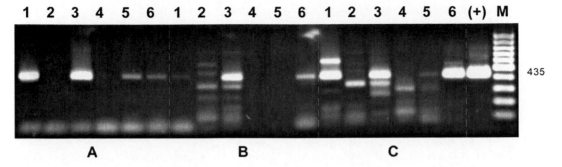

Figure 6.   Impact of water quality and DNA isolation techniques on the effectiveness of detection of 100 *C. parvum* oocysts by single-step, classic PCR (primers CP1Wf/r—435 bp). A—isolation of DNA preceded by IMS; B—direct isolation of DNA; C—direct isolation of DNA combined with purification on columns (Roche). 1—100 oocyst in treated water; 2—surface water of low turbidity; 3—100 oocyst in surface water of low turbidity; 4—surface water of high turbidity; 5—100 oocyst in surface water of high turbidity; 6—100 oocyst speeded in PBS; (+)—positive control. M—ladder 100 bp. 1,5% agarose stained with SYBR Gold.

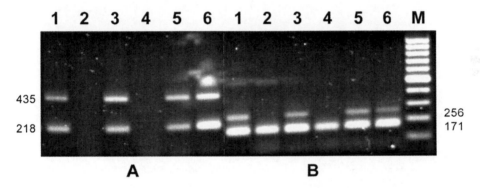

Figure 7.   Impact of water quality on the effectiveness of the detection of 100 cysts/oocysts (*Giardia/Cryptosporidium*) by multiplex-PCR. A—isolation of DNA preceded by IMS, primers CP1Wf/r—435 bp and GI1Wf/r—218 bp; B— direct isolation of DNA combined with purification on columns (Roche), primers CP2Wf/r—256 bp and GI2Wf/r— 171 bp. 1—100 cysts/oocyst in treated water; 2—surface water of low turbidity; 3—100 cysts/oocyst in surface water of low turbidity; 4—surface water of high turbidity; 5—100 cysts/oocyst in surface water of high turbidity; 6—100 cysts/ oocyst speeded in PBS; M—ladder 100 bp. 1,5% agarose stained with SYBR Green.

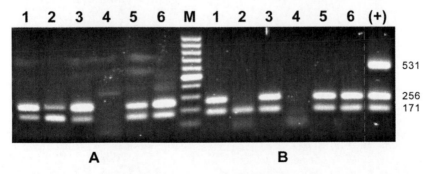

Figure 8.   Impact of water quality and DNA isolation techniques on the effectiveness of detection of 100 cysts/ oocysts (*Giardia/Cryptosporidium*) by multiplex-PCR (three pairs of primers: CP2Wf/r GI2Wf/r TGWf/r). A—isolation of DNA preceded by IMS; B—direct isolation of DNA combined with purification on columns (Roche). 1—100 cysts/oocysts in treated water; 2—surface water of low turbidity; 3—100 cysts/oocyst in surface water of low turbidity; 4—surface water of high turbidity; 5—100 cysts/oocyst in surface water of high turbidity; 6—100 cysts/oocyst speeded in PBS; M—ladder 100 bp. 1,5% agarose stained with SYBR Green.

147

purification on columns allowed for the detection of the presence of *Giardia sp.* in samples of unseeded surface water (Fig. 7/2.4) indicating the presence of these cysts in the tested rivers. Similar results were obtained with a triple multiplex-PCR (with primers for all three protozoa: CP2Wf/r GI2Wf/r and TGWf/r, as seen on Fig. 8). PCR performed on material isolated with or without IMS (in this case the material was purified on columns) produced similar effects. If a comparison of Figures 6 and 8 is made, it can be seen that a two-phase PCR (nested-PCR) allows to not only increase the sensitivity but also eliminate unspecific bands in the internal reaction.

4 CONCLUSION

To summarize, it can be noted that, first of all, it is possible to detect cysts/oocysts of parasitic protozoa present in different water samples. The strongly recommended method of choice is the isolation of DNA from the material obtained by IMS, but in those situations where it is not possible to use dedicated magnetic beads, the possibility exists for direct DNA isolation and its purification. The centrifugation on a sucrose gradient (and careful washing of the collected fraction) and final treatment of isolated DNA on the columns are the key issues in this case. In this work we used a commercial kit (Roche), providing suitable purification of DNA. It is not perfect, however, as can be seen when one compares the results of the PCR reaction when performed using different volumes of DNA as a template for PCR. For example, the use of 0,5 μl of DNA isolated from samples of surface water, compared to 5 μl (total volume of reaction mixture: 25 μl) results in much better amplification and fewer unspecific bands (data not shown in this work). In some publications there are protocols that presuppose the dilution of DNA pellets in 100–200 μl of water, after which for PCR only 1–2 μl is used in order to "dilute" the number of potential PCR inhibitors. This is a legitimate activity, however, it automatically reduces the sensitivity about two orders of magnitude. In this work, it was decided to dissolve the isolated DNA in a volume of 40–50 μl of nuclease-free water and use 2,5–5 μl for PCR (25 μl reaction volume). Even using 5 μl of DNA did not cause major difficulties in performing a PCR using the described DNA purification methods and executing the two-step nested-PCR. Additionally, internal reaction PCR prolonged the entire procedure by 2–3 hours, but this is not a significant change, as the complete protocol from the delivery of the water samples to the laboratory to producing a photo of the agarose gels, requires two-three working days.

Secondly, it demonstrated the possibility of the detection of several pathogens in a multiplex-PCR reaction. It appears that the risk for PCR from the possible interactions between multiple pairs of primers, significant enhancing of the concentrations of oligonucleotides in the reaction mixture and differences in the rate of synthesis of fragments of different lengths does not have the impact on reducing the sensitivity. It has been demonstrated that the multiplex-PCR enables detection of 0,1 pg of genomic DNA (which means the sensitivity of the order of a few copies of the genome) and only in extremely adverse circumstances (when the differences in the number of copies of each of the matrices was at four orders of magnitude, Fig. 3) sensitivity reached 1 pg.

In conclusion, we can say that the multiplex-PCR is a good, sensitive, inexpensive and convenient way to detect given protozoan parasites in water samples. A crucial difficulty is the isolation of the DNA in such a way as to remove the presence of PCR inhibitors in surface waters. In this work, the protocol for isolation and purification of DNA to obtain a clean template, suitable to use samples of surface water of significant pollution were tested. The effectiveness of this protocol is particularly important, given that for the most interesting waterborne parasites, there are no tools for their separation by the IMS.

ACKNOWLEDGEMENTS

The work was created with funding from a grant MNiSW N N523 748940.

REFERENCES

Clancy J.L., 2001, Lessons from the 1998 Sydney water crisis, *Water*, 28(1), 33–36.

DiGiovanni G.D., Hashemi F.H., Shaw N.J., Abrams F.A., LeChevalier M.W. & Abbaszadegan M., 1999, Detection of Infectious Cryptosporidium parvum Oocyst in Surface and Backfill Water Samples by Immunomagnetic Separation and Integrated Cell Culture-PCR, *Appl. Environ. Microbiol.*, 65(8), 3427–3432.

EPA Method 1623: Cryptosporidium and Giardia in Water by Filtration/IMS/FA. 2001, EPA-821-R-01-025, Washington, USA.

Grigg M.E. & Boothryod J.C., 2001, Rapid identification of virulent type I strains of the protozoan pathogen Toxoplasma gondii by PCR restriction fragment length polymorphism analysis at the B1 gene, *J. Clin. Microbiol.*, 39, 398–400.

Hsu B.-M., Huang C., Lai Y.-C., Tai H.-S. & Chung Y.-C., 2001, Evaluation of immunomagnetic separation method for detection of Giardia for different reaction times and reaction volumes, *Parasitol. Res.*, 87, 472–474.

Isaac-Renton J., Bowie W.R., King A., Irwin G.S., Ong C.S., Fung C.P., Shokeir M.O. & Dubey J.P., 1998, *Appl. Environ. Microbiol.*, 64(6), 2278–2280.

Jenkins M.B., Anguish L.J., Bowman D.D., Walker M.J. & Ghiorse W.C., 1997, Assessment of a Dye Permeability Assay for Determination of Inactivation Rates of Cryptosporidium parvum Oocyst, *Appl. Environ. Microbiol.*, 63(10), 3844–3850.

Jiang J., Alderisio K.A., Singh A. & Xiao L., 2005, Development of Procedures for Direct Eextraction of Cryptosporidium DNA from Water Concentrates and for Relief of PCR Inhibitors, *Appl. Environ. Microbiol.*, 71(3), 1135–1141.

Johnson D.W., Pieniazek N.J., Griffin D.W., Misener L. & Rose J.B., 1995, Development of a PCR Protocol for Sensitive Detection of Cryptosporidium Oocyst in Water Samples, *Appl. Environ. Microbiol.*, 61(11), 3849–3855.

Kocwa-Haluch R. & Polus M., 2004, Występowanie patogennych pierwotniaków jelitowych w wodzie wodociągowej. Część II: Usuwanie cyst i oocyst patogennych pierwotniaków jelitowych z wody przeznaczonej do picia, Czasopismo Techniczne Politechniki Krakowskiej, 8Ś, 113–124.

Marshall M.M., Naumowitz D., Ortega Y. & Sterling C.R., 1997, "Waterborne Protozoan Pathogens", *Clin. Microbiol. Rev.*, 10(1), 67–85.

Mayer C.L. & Palmer C.J., 1996, Evaluation of PCR, Nested PCR, and Fluorescent Antibodies for Detection of Giardia and Cryptosporidium Species in Wastewater, *Appl. Environ. Microbiol.*, 62(6), 2081–2085.

Nichols R.A.B., Campbell B.M. & Smith H.V., 2003, Identification of Cryptosporidium spp. Oocyst in United Kingdom Noncarbonated Natural Mineral Waters and Drinking Waters by Using Modified Nested PCR-Restriction Fragment Length Polymorphism Assay, *Appl. Environ. Microbiol.*, 69(7), 4183–4189.

Polus M. & Kocwa-Haluch R., 2010, Detection of Cryptosporidium so, in surface water: assessment of direct DNA isolation and nested-PCR, *Monografia PAN*, 2010, 64, 171–181.

Robertson L.J., Campbell A.T. & Smith H.V., 1998, Viability of Cryptosporidium parvum Oocysts: Assessment by the Dye Permeability Assay; Letter to the Editor. *Appl. Environ. Microbiol.*, 64(9), 3544–3545.

Rochelle P.A., Deleon R., Steward M.H. & Wolfe R.L., 1997, Comparison of Primers and Optimalization of PCR Conditions for Detection of Cryptosporidium parvum and Giardia lamblia in Water, *Appl. Environ. Microbiol.*, 63(1), 106–114.

Sroka J., Szymańska J. & Wójcik-Fatla A., 2009, The Occurence of Toxoplasma gondii and Borrelia burgdorferi sensu lato in Ixodes ricinus Ticks from Eastern Poland with the Use of PCR, *Ann. Agric. Environ. Med.*, 16, 313–319.

Villena I., Aubert D., Gomis P., Ferte H., Inglard J.-C., Denis-Bisiaux H., Dondon J.-M., Pisano E., Ortis N. & Pinon J.-M., 2004, Evaluation of a Strategy for Toxoplasma gondii Oocyst Detection in Water, *Appl. Environ. Microbiol.*, 70(7), 4035–4039.

WHO, 2004, Guidelines for Drinking-water Quality, 3rd Edition, Geneva.

Yang W., Chen P., Villegas E.N., Landy R.B., Kanetzky C., Cama V., Dearen T., Schultz C.L., Orndorff K.G., Prelewicz G.J., Grown M.H., Young K.R. & Xiao L., 2008, Cryptosporidium Source Tracking in the Potomac River Watershed, *Appl. Environ. Microbiol.*, 74(21), 6495–6504.

Yu X., Van Dyke M.I., Portt A. & Huck P.M., 2009, Development of a direct DNA extraction protocol for real-time PCR detection of Giardia lamblia from surface water, *Ecotoxicology*, 18, 661–668.

Environmental Engineering IV – Pawłowski, Dudzińska & Pawłowski (eds)
© *2013 Taylor & Francis Group, London, ISBN 978-0-415-64338-2*

Fecal indicators resistance to antimicrobial agents present in municipal wastewater

A. Łuczkiewicz & K. Olańczuk-Neyman
Faculty of Civil and Environmental Engineering, Gdansk University of Technology, Gdansk, Poland

E. Felis, A. Ziembińska, A. Gnida & J. Surmacz-Górska
Faculty of Energy and Environmental Engineering, The Silesian University of Technology, Gliwice, Poland

ABSTRACT: In this study removal rate of six antimicrobial compounds (belonging to folate pathway inhibitors and macrolides class) was analyzed in processes of local Wastewater Treatment Plant (WWTP) together with susceptibility patterns of Erythromycin (E) resistant *Enterococcus* spp. and trimethoprim/ Sulfamethoxazole (SXT) resistant *Escherichia coli*. According to the obtained data, six of tested antimicrobial agents, namely erythromycin, clarithromycin, trimethoprim, roxithromycin, sulfamethoxazole and N-acetyl-sulfamethoxazole were detected in both raw and treated wastewater samples. Among tested *E. coli* resistance to SXT was highly correlated with resistance to penicillins as well as to tetracycline (88% and 75%, respectively) while E-resistant *E. faecalis* and *E. faecium* showed significant differences in resistance patterns. Among tested isolates of wastewater origin resistance phenotype of clinical importance were detected. Thus, further study is needed to evaluate dissemination of antimicrobial resistance via treated wastewater.

Keywords: wastewater, fecal indicators, antimicrobial agents, antimicrobial resistance

1 INTRODUCTION

There are some evidences that antimicrobial agents, used in human and animal therapy, can create a selective pressure on microorganisms and may favor mutant strains, even in concentrations significantly lower than therapeutic dosages (Ohlsen et al. 1998). Outside hospital units wastewater is regarded as a significant reservoir of both antimicrobial agents and antimicrobial-resistance determinants (Kim et al. 2007, Al-Ahmad et al. 1999; Szczepanowski et al. 2009). Thus, in terms of resistance dissemination, wastewater impact on receivers should be considered.

Currently, wastewater processes focus on nutrients, organic matter and suspensions removal while other micropollutants as well as microorganisms are eliminated unintentionally. In case of antimicrobial agents, sorption and/or biodegradation are considered main removal processes (Kümmerer 2001, Golet et al. 2003, Lindberg et al. 2006). However, chemical properties of drug molecules as well as WWTP operating system cannot also be ignored (Golet et al. 2003, Batt et al. 2007). In particular, activated sludge processes are suspected to select some resistance patterns, especially associated with Multidrug Resistance (MDR) (Reinthaler et al. 2003, Luczkiewicz et al. 2010) as well as to promote spread of

antimicrobial resistance by horizontal gene transfer (Soda et al. 2008, Marcinek et al. 1998). Additionally WWTP effluents were found to be an important source of bacteria with resistance patterns of clinical significance, such as Extended-Spectrum Beta-Lactamase (ESBL) producing *Escherichia coli* (Dolejska et al. 2011) or Vancomycin (VRE) and High Level Aminoglycoside Resistance (HLAR) enterococci (Talebi et et al. 2008, Tejedor Junco et al. 2001). In general, prevalence of Erythromycin (E) resistant enterococci (with resistance rate even up to 82%) as well as common resistance to trimethoprim/sulfamethoxazole (SXT) among *E. coli* (reaching 11%) has been widely reported in wastewater compartments (Blanch et al. 2003; Ferreira da Silva et al. 2006; Martins da Costa et al. 2006, Luczkiewicz et al. 2010).

Therefore, the aim of the current study was to determine concentrations of folate pathway inhibitors (trimethoprim, sulfamethoxazole, N-acetyl-sulfamethoxazole) and macrolides (erythromycin, clarithromycin, roxithromycin) in wastewater samples. At the same time, the resistance patterns carried by SXT-resistant *E. coli* as well as E-resistant *Enterococcus* spp. were analyzed. Samples were taken from WWTP Gdansk—Wschod, which discharges into the recreationally important coastal waters of Gdansk Bay.

2 MATERIALS AND METHODS

2.1 Samples collection

The twenty-four hour flow-proportional wastewater samples (5 L) were taken from WWTP Gdansk-Wschod. The Raw (RW) and Treated (TW) wastewater was tested together with Activated Sludge (AS) samples in the three following days of January and June 2009. The 24-hour retention time in WWTP was considered during samples collection. The main characteristic of the WWTP Gdansk-Wschod are given in Table 1. During the study, the average values of treated wastewater parameters were lower than those legally established for the examined WWTP (COD < 125 mg O_2 L^{-1}, BOD5 < 15 mg O_2 L^{-1}, TN < 10 mg N L^{-1}, TP < mg P L^{-1}, TSS < 35 mg L^{-1}). Moreover, the reduction of fecal indicators was significant (up to 99.9%) and similar to results obtained previously for this system (Olanczuk-Neyman et al. 2003) as well as for other wastewater systems working on activated sludge (Ferreira da Silva et al. 2006, Blanch et al. 2003). It should be mentioned, however, that due to the occasional high number of bacterial cells detected in raw wastewater their number may reach even 3.7×10^4 CFU per 100 mL for enterococci and up to 3.2×10^4 CFU per 100 mL for E. coli in treated wastewater.

2.2 Determination of antimicrobial agents

Presence of antimicrobial agents was determined in RW and TW. All samples of the volume of 1 L were frozen and stored in dark-glass bottles at—20 °C prior to analysis. After defrosting, samples were filtered through the glass fiber filters with diameter 55 mm and a pore size of <1 μm (Schleicher & Schuell, Dassel, Germany). The 100 mL of influent (RW) and 200 mL of effluent (TW) were added to the mixture of Internal Standards (IS). The pH of all samples was adjusted to 7.5.

As the internal standards the following substances were used: sulfamerazine-d4 (IS for sulfonamides and trimethoprim), sulfamethoxazole-d4 (IS for sulfamethoxazole), N-acetyl-sulfamethoxazole-d5 (IS for N-acetyl-sulfamethoxazole) and (E)-9-[O-(2-methyloxime)]-erythromycin (IS for macrolides). The Solid Phase Extraction (SPE) of the wastewater samples was performed by means of Oasis HLB (200 mg, 6 mL) cartridges (Waters, Milford, MA, USA). The details concerning chromatographic analysis performance are presented in the publication of Hijosa-Valsero et al. (2011).

2.3 Antimicrobial susceptibility tests (AST)

Susceptibility of SXT-resistant E. coli (n = 16) as well as E-resistant E. feacium (n = 50) and E. faecalis (n = 28) were tested (Luczkiewicz et al. 2010). In brief, identification and susceptibility tests were carried out using the Phoenix Automated Microbiology System (Phoenix AMS, BD) using commercially available Phoenix panels and manufacturer's recommended procedure. In this study the confidence level of 90% was required as the lowest limit of acceptability for ID results. The bacterial susceptibility data was analyzed using standards of the Clinical and Laboratory Standards Institutefor antimicrobial susceptibility testing (CLSI, 2011). In characterization of antibiotic resistance only two categories were used sensitive and resistant (as an 'intermediate resistant' and 'resistant' behavior). Phenotype of Multidrug Resistance (MDR) was defined as resistance to 1 or more agents in 3 or more categories of antimicrobial agents, according to Magiorakos et al. (2011). The antimicrobial agents used in this study to test susceptibility of E. coli and Enterococcus spp., categories to define MDR as well as MIC Interpretive Standards according to CLSI (CLSI, 2011) are given in Table 2.

Table 1. Basic characteristic of WWTP Gdansk-Wschod as well as Raw (RW) and Treated (TW) wastewater samples (mean/σ).

Population equivalent/ Population		System		$Q_{aver.}$ (m^3 per day)	
700000/570000		Modified University of Cape Town (UCT) system		96000 Local industry (5%) Hospital wastewater (0.17%)[*]	
	mg L^{-1}			mg O_2 L^{-1}	
	TSS	TN	TP	BOD$_5$	COD
RW	520/91	72.8/5.5	10.6/0.7	370/51	1136/122
TW	5.9/0.4	8.2/0.3	0.11/0.06	5.6/0.3	44.6/3.9

Table 2. The MIC Interpretive Standards (μg mL^{-1}) used to define the susceptibility of *E. coli* and *Enterococcus* spp. to tested antimicrobial agents (acc. to CLSI, 2011) as well as the antimicrobial categories used to define MDR (acc. to Magiorakos et al. 2011).

MIC (μg mL^{-1})		Antimicrobial						MIC (μg mL^{-1})	
S	R[*]	code	agent	Category		Agent	code	S	R
E. coli						*Enterococcus* spp.			
Antimicrobial categories used to define MDR									
≤ 16	≥ 32	AN	Amikacin	Aminoglycosides [**]	Gentamicin-Synergy		GMS	≤ 500	> 500
≤ 4	≥ 8	GM	Gentamicin						
≤ 4	≥ 8	NN	Tobramycin						
-	-	-	-	Streptomycin [**]	Streptomycin-Synergy		STS	≤ 1000	> 1000
≤ 1	≥ 2	IPM	Imipenem	Carbapenems	-		-	-	-
≤ 1	≥ 2	MEM	Meropenem					-	-
≤ 4 c)	≥ 8 c)	CXM	Cefuroxime	Cephalosporins 1st and 2nd generation non-extended spectrum	-		-	-	-
≤ 4	≥ 8	CAZ	Ceftazidime	Cephalosporins 3th and 4th generation extended spectrum	-		-	-	-
≤ 1	≥ 2	CTX	Cefotaxime						
≤ 8	≥ 16	FEP	Cefepime						
≤ 4	≥ 8	ATM	Aztreonam	Monobactam	-		-	-	-
≤ 8	≥ 16	AM	Ampicillin	Penicillins	Ampicillin		AM	≤ 8	≥ 16
≤ 8/4	≥ 16/8	AMC	Amoxicillin/Clavulanate	Penicillins/β-lactamase inhibitors	-		-	-	-
≤ 16/4	≥ 32/4	TZP	Piperacillin/Tazobactam	Antipseudomonal penicillins/β-lactamase inhibitors	-		-	-	-
≤ 2/38	≥ 4/76	SXT	Trimethoprim/Sulphamethoxazole	Folate pathway inhibitors	-		-	-	-
≤ 1	≥ 2	CIP	Ciprofloxacin	Fluoroquinolone	Ciprofloxacin		CIP	≤ 1	≥ 2
≤ 2	≥ 4	LVX	Levofloxacin		Levofloxacin		LVX	≤ 2	≥ 4
-	-	-	-		Moxifloxacin		MXF	≤ 4	≥ 8
-	-	-	-	Glycopeptides	Teicoplanin		TEC	≤ 4	≥ 8
					Vancomycin		VA	≤ 8	≥ 16
-	-	-	-	Lipopeptides	Daptomycin		DAP	≤ 4	[***]
-	-	-	-	Macrolides	Erythromycin		E	≤ 0,5	≥ 1
-	-	-	-	Oxazolidinones	Linezolid		LZD	≤ 2	≥ 4
≤ 4	≥ 8	TE	Tetracycline	Tetracyclines	Tetracycline		TE	≤ 4	≥ 16
Antimicrobial categories not used to define MDR									
≤ 16	≥ 32	PIP	Piperacillin	Antipseudomonal penicillins	-		-	-	-
Others									
-	-	ESR	ESBL	test for the detection of extended- spectrum β-lactamases (ESBLs)	-		-	-	-

[*]—'intermediate resistant' and 'resistant' behavior, [**]—to define MDR of enterococci, streptomycin was given to the separate category, [***]—MICs value to define daptomycin resistant enterococci was not determined by CLSI.

2.4 *Data analyses*

Significant differences between antibiotic resistance rate in Raw (RW) and Treated (TW) wastewater were determined for fecal indicators using a two-sided z-test for two proportions ($p < 0.05$). For statistical computations 'Scilab environment' was used.

3 RESULTS AND DISCUSSION

Sulfonamides, despite their possible side effects, have been extensively used in many different clinical indications since 1935 (Sköld 2000). In medical practice sulfamethoxazole is commonly used together with trimethoprim, due to their suspected synergistic effect (Sköld 2000). Both components affect the following steps of bacterial folic acid synthesis and are active against many *Enterobacteriaceae*, as well as against important respiratory tract pathogens as *Streptococcus pneumoniae* and *Haemophilusinfluenzae*. Several advantages make trimethoprim/sulfamethoxazole (SXT, commercial name—Bactrim) accounted for 10%–15% of the total number of antimicrobial agents used worldwide. In Poland, SXT is commonly applied to treat uncomplicated Urinary Tract (UTIs) caused by gram-negative bacteria (Hryniewicz et al. 2001). From both components, which are taken orally in standard ratio 1:5, about 30% of sulfamethoxazole and up to 90% of trimethoprim are excreted unchanged in the urine and reach the wastewater system.

Therefore, in this study raw and treated wastewater was tested for presence of trimethoprim and sulfamethoxazole. Simultaneously the concentration of N-acetyl-sulfamethoxazole, generated in the human body in reaction of sulfamethoxazole acetylation, was analyzed. According to the obtained data trimethoprim was present in raw wastewater at the average concentrations of 482 ± 115 ng L^{-1} and 441 ± 81 ng L^{-1} in winter and summer periods, respectively (Table 3). Observed concentration of N-acetyl-sulfamethoxazole in raw wastewater was even slightly higher than observed for sulfamethoxazole (Table 3). Since cleaving back of N-acetyl-sulfamethoxazole to sulfamethoxazole was observed under environmental conditions (Kümmerer 2009; Göbel et al. 2005) such high concentration of this retransformable metabolite in wastewater suggests its importance in sulfamethoxazole mass balance. Together with folic acid antagonists presence of erythromycin, roxithromycin and clarithromycin, the most widely used macrolide antibiotics, were analyzed in this study. Erythromycin was the first macrolide antibiotic discovered and for over 60 years used for treatment of upper respiratory tract as well as skin and soft-tissue infections. Several drawbacks of erythromycin, including frequent gastrointestinal intolerance, have been excluded by its second generation derivatives roxithromycin and clarithromycin approved in early 1990s for clinical use. Thus, during last years in Poland, and other European countries, the increase of second generation macrolides consumption have been observed at the expense of erythromycin (Ferech et al. 2006). It was reflected also in tested raw wastewater, where

Table 3. Occurrence and removal of selected antimicrobial agents from WWTP Gdansk—Wschod during the wastewater treatment process.

Compound	Temperature	Raw wastewater (ng L^{-1})	Treated wastewater (ng L^{-1})	Removal (%)
Clarithromycin	T = 10°C	1416 ± 401	761± 106	43
	T = 20°C	904 ± 433	185 ± 32	76
Roxithromycin	T = 10°C	161 ± 0	132 ± 14	18
	T = 20°C	105 ± 17	55 ± 6	47
Erythromycin	T = 10°C	n.d.[a]	17	n.r.[b]
	T = 20°C	n.d.[a]	14	n.r.[b]
Sulfamethoxazole	T = 10°C	1464 ± 203	508 ± 25	65
	T = 20°C	1225 ± 173	642 ± 114	47
N-acetyl-sulfamethoxazole	T = 10°C	1763 ± 470	16	98
	T = 20°C	1358 ± 224	0	100
Trimethoprim	T = 10°C	482 ± 116	445 ± 72	7
	T = 20°C	441 ± 81	269 ± 14	38

[a]n.d.—not detected; [b]n.r.—not removed.

Table 4. Resistance patterns obtained for SXT-resistant *E. coli* isolated from Raw (RW) and Treated Wastewater (TW) as well as Activated Sludge (AS) samples.

Isolate		AN	GM	NN	IPM	MEM	CZ	CXM	CAZ	CTX	FEP	ATM	AM	PIP	AMC	TZP	STX	CIP	LVX	TE	MDR*
E. coli	RW-6	-	-	-	-	-	-	-	-	-	-	-	+	+	+	-	+	-	-	+	+
E. coli	AS-27	-	-	-	-	-	-	-	-	-	-	-	+	+	-	-	+	-	-	+	+
E. coli	AS-152	-	-	-	-	-	-	-	-	-	-	-	+	+	-	-	+	-	-	-	-
E. coli	AS-151	-	-	-	-	-	+	-	-	-	-	-	+	+	+	+	+	+	+	+	+
E. coli	AS-150	-	-	-	-	-	-	-	-	-	-	-	+	+	+	-	+	-	-	+	+
E. coli	AS-149	-	-	-	-	-	+	+	-	-	-	-	+	+	-	-	+	-	-	-	+
E. coli	AS-145	-	-	-	-	-	-	-	-	-	-	-	+	+	-	-	+	-	-	-	-
E. coli	AS-143	-	-	-	-	-	+	+	+	+	+	+	+	+	+	-	+	-	-	+	+
E. coli	AS-141	-	-	-	-	-	-	-	-	-	-	-	+	+	-	-	+	-	-	+	+
E. coli	TW-54	-	-	-	-	-	-	-	-	-	-	-	+	+	+	-	+	-	-	-	+
E. coli	TW-57	-	-	-	-	-	-	-	-	-	-	-	-	-	-	-	+	-	-	+	-
E. coli	TW-164	-	+	-	-	-	-	-	-	-	-	-	+	+	-	-	+	+	+	+	+
E. coli	TW-169	-	-	-	-	-	-	-	-	-	-	-	+	+	+	-	+	+	+	+	+
E. coli	TW-175	-	-	-	-	-	-	-	-	-	-	-	+	+	-	-	+	-	-	+	+
E. coli	TW-181	-	-	-	-	-	-	-	-	-	-	-	-	-	-	-	+	-	-	+	-
E. coli	TW-183	-	-	-	-	-	-	-	-	-	-	-	+	+	-	-	+	+	+	+	+
n		0	1	0	0	0	3	2	1	1	1	1	14	14	6	1	16	4	4	12	12
%		0	6	0	0	0	19	13	6	6	6	6	88	88	38	6	100	25	25	75	75

the highest concentration was obtained for clarithromycin (up to 1817 ng L^{-1}). Erythromycin was not detected in the raw wastewater samples but its occurrence was confirmed in treated wastewater analysis. The seasonal fluctuations of macrolides concentration observed in raw wastewater (Table 3) may be related to their intended use to treat respiratory tract infections. Since the presence of antibiotics in the wastewater reflects their usage and excretion aspect, it has to be mentioned that macrolides are excreted mainly in the bile (up to 90% of erythromycin and about 50% of clarithromycin and roxithromycin). However, about 40% of clarithromycin dose is also effectively excreted in urine (for roxithromycin 12%).

Important aspect of antimicrobial compounds presence in wastewater is their susceptibility for biological degradation and possible interactions with activated sludge particles (Heise et al. 2006, Junker et al. 2006). However, it has been suggested that other parameters, not related to the chemical structure of antibiotics, like activated sludge age, retention time in the system should be also investigated (Göbel et al. 2007). The complexity of this issue together with lack of analytical techniques standardization may explain the variety in results obtained for presence and removal of antimicrobial agents from wastewater (Batt et al 2007, Göbel et al. 2007, Junker et al. 2006, Perez

et al. 2005). In this study, the removal of tested macrolides and folic acid antagonists was found to be incomplete, except N-acetyl-sulfamethoxazole. This metabolite was removed from wastewater with efficiency reaching 100%. The removal efficiency observed for sulfamethoxazole and clarithromycin was also relatively high (up to 65% and 76%, respectively), while for trimethoprim it did not exceeded 38%.

Similar high resistance of trimethoprim and susceptibility of sulfamethoxazole to biodegradation was reported by Perez et al. (2005) in wastewater treatment based on activated sludge processes. Furthermore, according to the obtained data the removal efficiency depended on the temperature—increased (except sulfamethoxazole) during the summer season (Table 3). Simultaneously with detection of macrolides and folic acid antagonists, resistance patterns of E-resistant *Enterococcus* spp. and SXT-resistant *E. coli* were determined in tested wastewater. It has been already mentioned that trimethoprim-sulfamethoxazole is commonly used as a first-line therapy for acute uncomplicated UTIs. In Poland, as well as worldwide, the most prevalent uropathogen, is *E. coli*. It is accounted for 84% of hospital and 38% community–acquired infection (Hryniewicz et al. 2001). High prevalence of SXT-resistant *E. coli* of clinical (up to 31%) and

Table 5. Resistance patterns obtained for E-resistant *E. faecalis* isolated from Raw (RW) and Treated Waste-water (TW) as well as Activated Sludge (AS) samples.

Isolate		GMS	STS	AM	DAP	TEC	VA	E	C	LZD	FM	CIP	LVX	MXF	TE	MDR*
E. faecalis	RW-1	-	-	-	-	-	-	+	-	-	-	-	-	-	-	-
E. faecalis	RW-23	-	-	-	-	-	-	+	-	-	-	-	-	-	-	-
E. faecalis	AS-140	-	-	-	-	-	-	+	-	-	+	-	-	-	-	-
E. faecalis	AS-154	-	-	-	-	-	-	+	-	-	-	-	-	-	-	-
E. faecalis	AS-159	-	-	-	-	-	-	+	-	-	-	-	-	-	-	-
E. faecalis	AS-39	-	-	-	-	-	-	+	-	-	-	-	-	-	+	-
E. faecalis	AS-40	-	-	-	-	-	-	+	-	-	+	-	-	-	+	+
E. faecalis	AS-46	-	-	-	-	-	-	+	-	-	-	-	-	-	-	-
E. faecalis	AS-57	+	+	-	-	-	-	+	+	-	+	+	+	+	+	+
E. faecalis	TW-61	-	-	-	-	-	-	+	-	-	-	-	-	-	-	-
E. faecalis	TW-62	-	-	-	-	-	-	+	-	-	-	-	-	-	-	-
E. faecalis	TW-63	-	-	-	-	-	-	+	-	-	+	+	-	-	-	+
E. faecalis	TW-71	-	-	-	-	-	-	+	-	-	+	-	-	-	-	-
E. faecalis	TW-72	-	-	-	-	-	-	+	-	-	-	-	-	-	-	-
E. faecalis	TW-82	-	-	-	-	-	-	+	-	-	-	-	-	-	+	-
E. faecalis	TW-83	-	-	-	-	-	-	+	-	-	-	-	-	-	+	-
E. faecalis	TW-165	-	-	-	-	-	-	+	-	-	-	+	+	+	+	+
E. faecalis	TW-167	-	-	-	-	-	-	+	-	-	+	-	-	-	-	-
E. faecalis	TW-172	-	-	-	-	-	-	+	-	-	-	+	+	+	+	+
E. faecalis	TW-173	-	-	-	-	-	-	+	-	-	-	-	-	-	-	-
E. faecalis	TW-174	-	-	-	-	-	-	+	-	-	-	-	-	-	-	-
E. faecalis	TW-187	-	-	-	-	-	-	+	-	-	-	-	-	-	-	-
E. faecalis	TW-188	-	-	-	-	-	-	+	-	-	-	-	-	-	+	-
E. faecalis	TW-321	-	-	-	-	-	-	+	-	-	-	+	-	-	+	+
*E. faecali**	TW-401	-	+	+	-	-	-	+	-	-	+	+	+	+	-	+
E. faecalis	TW-511	-	-	-	-	-	-	+	-	-	-	-	-	-	+	-
E. faecalis	TW-621	-	-	-	-	-	+	+	-	-	+	+	-	-	-	+
E. faecalis	TW-625	-	-	-	-	+	+	+	-	+	-	-	-	-	-	-
	n	1	2	1	0	1	2	28	1	1	8	7	4	4	10	8
	%	4	7	4	0	4	7	100	4	4	29	25	14	14	36	29

wastewater (up to 11%) origin has been previously reported (Hryniewicz et al. 2001, Luczkiewicz et al. 2010). Thus, the resistance phenotype carried by SXT-resistant *E. coli* seem to be important from both clinical and environmental point of view. In this study resistance to SXT was highly correlated with resistance to penicillins (AM and PIP) as well as to tetracycline (88% and 75%, respectively).

Additionally, 25% of SXT-resistant *E. coli* were simultaneously resistant to fluoroquinolones (CIP and LVX), antimicrobial agents frequently prescribed for UTIs. What is also important, among SXT-resistant *E. coli* 75% of tested isolates showed clinically relevant Multidrug Resistance (MDR) phenotypes. As it was previously mentioned, genes conferring resistance to sulfonamides in MDR profile were more often connected with presence of *sul2* then *sul1* gene (75% and 50% of STX-MDR isolates respectively) (Luczkiewicz et al. 2011). Moreover, in activated sludge one strain producing Extended Spectrum Beta-Lactamase (ESBL) was detected (Table 4). This phenotype connected with plasmid mediated enzymes capable of hydrolyzing penicillins and broad spectrum-cephalosporins is regarded to be limited to intensive care units. Presence of ESBL producing isolates in treated waste-water suggest a dissemination of these enzymes in the environment.

Resistance to erythromycin is commonly noted among enterococci of wastewater and environmental origin (up to 80%) (Blanch et al. 2003; Ferreira da Silva et al. 2006; da Costa et al. 2006,

Table 6. Resistance patterns obtained for E-resistant *E. faecium* isolated from Raw (RW) and Treated Wastewater (TW) as well as Activated Sludge (AS) samples.

Isolate		GMS	STS	AM	DAP	TEC	VA	E	C	LZD	FM	CIP	LVX	MXF	TE	MAR*
E. faecium	RW-5	-	+	+	-	-	-	+	-	-	+	+	-	-	+	+
E. faecium	RW-7	-	+	+	-	-	-	+	-	-	+	-	-	-	+	+
E. faecium	RW-11	-	-	-	-	-	-	+	-	-	-	-	-	-	-	-
E. faecium	RW-14	-	+	+	-	-	-	+	-	-	+	-	-	-	+	+
E. faecium	RW-112	-	-	-	-	-	-	+	-	-	+	+	+	+	-	+
E. faecium	AS-43	-	-	-	-	-	-	+	-	-	+	+	+	+	-	+
E. faecium	AS-44	-	-	+	-	-	-	+	-	+	+	+	-	-	-	+
E. faecium	AS-45	-	-	-	>4**	-	-	+	-	-	+	-	-	-	-	-
E. faecium	AS-58	-	-	+	-	-	-	+	-	-	+	-	-	-	-	+
E. faecium	AS-135	-	-	+	-	-	-	+	-	-	+	+	+	+	-	+
E. faecium	AS-136	-	-	-	-	-	-	+	-	-	+	-	-	-	-	-
E. faecium	AS-143	-	-	-	-	-	-	+	-	-	+	-	-	-	-	-
E. faecium	AS-148	-	-	-	-	-	-	+	-	-	-	-	-	-	-	-
E. faecium	AS-152	-	-	-	-	-	-	+	-	-	+	-	-	-	-	-
E. faecium	AS-158	-	-	-	-	-	-	+	-	-	+	-	-	-	-	-
E. faecium	TW-68	-	-	-	-	-	-	+	-	-	+	+	-	-	-	+
E. faecium	TW-69	-	-	-	-	-	-	+	-	-	+	+	-	+	-	+
E. faecium	TW-73	-	-	-	-	-	-	+	-	-	+	+	+	+	-	+
E. faecium	TW-74	+	-	+	-	-	-	+	-	-	+	+	+	+	+	+
E. faecium	TW-84	-	-	-	-	-	-	+	-	-	+	-	-	-	-	-
E. faecium	TW-87	+	-	+	-	-	-	+	+	-	+	+	+	+	-	+
E. faecium	TW-88	-	-	-	-	-	-	+	-	-	+	+	+	+	-	+
E. faecium	TW-125	+	-	-	-	-	-	+	-	-	-	+	-	-	-	+
E. faecium	TW-164	-	-	-	>4	-	-	+	-	+	+	-	+	-	-	+
E. faecium	TW-166	-	-	-	-	-	-	+	-	-	+	+	+	+	-	+
E. faecium	TW-178	-	-	-	-	-	-	+	-	-	+	+	-	+	-	+
E. faecium	TW-180	-	-	-	-	-	-	+	-	-	+	+	-	-	-	+
E. faecium	TW-183	-	-	-	-	-	-	+	+	-	+	+	+	+	-	+
E. faecium	TW-184	-	-	-	-	-	-	+	-	-	+	-	-	-	-	-
E. faecium	TW-189	-	-	-	-	-	-	+	-	-	+	+	-	-	+	+
E. faecium	TW-190	-	-	-	-	-	-	+	-	-	+	+	-	+	-	+
E. faecium	TW-232	-	-	-	-	-	-	+	-	-	-	+	-	-	-	-
E. faecium	TW-301	-	-	-	-	-	-	+	-	-	+	+	-	-	-	+
E. faecium	TW-303	-	-	-	-	-	-	+	-	-	-	-	-	-	-	-
E. faecium	TW-323	-	-	-	-	-	-	+	-	-	-	-	-	-	-	-
E. faecium	TW-401	-	-	-	-	-	-	+	-	-	-	-	-	-	+	-
E. faecium	TW-402	+	-	-	-	-	-	+	-	-	-	-	-	-	+	+
E. faecium	TW-403	-	-	-	-	-	-	+	-	-	+	+	+	+	-	+
E. faecium	TW-505	-	-	-	-	-	-	+	-	-	-	-	-	-	-	-
E. faecium	TW-603	-	-	-	>4	+	+	+	-	-	+	+	-	+	-	+
E. faecium	TW-605	-	-	-	-	-	-	+	-	-	-	-	-	-	-	-
E. faecium	TW-606	-	-	-	-	-	-	+	-	-	+	-	-	+	-	+
E. faecium	TW-607	-	-	-	-	-	-	+	-	-	+	+	-	+	-	+
E. faecium	TW-608	-	-	-	-	-	-	+	-	-	+	-	-	-	-	+
E. faecium	TW-609	+	+	-	>4	-	-	+	-	-	-	-	-	-	-	+
E. faecium	TW-622	-	-	-	-	-	-	+	-	-	+	+	+	+	-	+
E. faecium	TW-623	-	-	-	-	-	-	+	-	-	-	-	-	-	-	-
E. faecium	TW-624	-	+	-	>4	+	+	+	-	+	+	+	+	-	+	+
E. faecium	TW-627	-	-	-	-	-	-	+	-	-	+	+	-	-	-	+
E. faecium	TW-629	-	+	+	-	-	-	+	-	-	-	-	-	-	+	+
	n	5	6	9	0	2	2	50	2	3	37	27	13	17	9	34
	%	10	12	18	0	4	4	100	4	6	74	54	26	34	18	68

*—categories used to define presence of MDR were distinguished by colour.

**—rare in clinic, elevated resistance to daptomycin (MIC > 4 µg mL⁻¹) was distinguished.

Luczkiewicz et al. 2010). However, macrolides are not used to treat enterococcal infections. Therefore, it is postulated that this resistance is developed by cross-resistance, since macrolides share mostly the same resistance Mechanisms with Lincosamides and Streptogramins B (MLS) (Ferreira da Silva et al. 2006, Blanch et al. 2003). In this study susceptibility of E-resistant *E. faecalis* (n = 28) and *E. faecium* (n = 50) was also tested against antimicrobial agents listed in Table 2. According to the obtained data, both species showed significant differences in resistance patterns (Tables 5 and 6), with general higher resistance rate noted for *E. faecium*. For instance, the most prevalent resistance to nitrofurantoin for *E. faecalis* did not exceed 30%, while for *E. faecium* reached 74%. Also resistance rate to fluoroquinolonesas well as clinically relevanthigh-level aminoglycosides were reportedtwo times higher for E-resistant *E. faecium* then E-resistant *E. faecalis*. In case of vancomycin, resistance to this important glycopeptide was detected in two isolates of *E. faecium* and two isolates of *E. faecalis*. It should also be noted that elevated resistance to daptomycin (MIC > 4 μgmL^{-1}) was observed among 5 isolates of E-resistant *E. faecium*, including all VRE strains (Table 5). Daptomycin is a cyclic lipopeptide with potent antimicrobial activity against vancomycin-resistant enterococci (Vouillamoz et al. 2006). Resistance to daptomycin remains rare in clinic (Vouillamoz et al. 2006).

4 CONCLUSION

This study provides the evidence that conventional wastewater processes do not prevent the receiver from dissemination of fecal indicators with clinically relevant resistance patterns. In the environmental risk assessment, the importance of continuous input of resistant bacteria versus continuous input of sub-inhibitory concentration of antimicrobial agents present in treated wastewater should be taken into consideration.

ACKNOWLEDGEMENTS

The authors wish to thank Prof. T.A. Ternes, Dr. M. Schlüsener, G. Fink from Federal Institute of Hydrology (Germany) for the opportunity and assistance during the analyzes of selected antimicrobial substances in the environmental samples. Resistance phenotypes to antimicrobial agents were partly reported by Luczkiewicz et al. (2010). This research was supported by the Polish Ministry of Science and Higher Education, grant no. N N523 493134.

REFERENCES

Al-Ahmad A., Daschner F.D., Kümmerer K. 1999. Biodegradability of cefotiam, ciprofloxacin, meropenem, penicillin G, and sulfamethoxazole and inhibition of waste water bacteria. *Arch. Environ. Contam. Toxicol.* 37, 158–63.

Batt A.L., Kim S., Aga D.S. 2007. Comparison of the occurrence of antibiotics in four full-scale wastewater treatment plants with varying designs and operations. *Chemosphere.* 68 (3): 428–35.

Blanch A.R., Caplin J.L., Iversen A., Kühn I., Manero A., Taylor H.D., Vilanova X. 2003. Comparison of enterococcal populations related to urban and hospital wastewater in various climatic and geographic European regions. *J. Appl. Microbiol.* 94 (6): 994–1002.

CLSI 2011, Clinical and Laboratory Standards Institute, Performance standards for antimicrobial susceptibility testing, in: *Twenty First Informational Supplement*, M100-S21, Vol. 21, No 1.

Ferech M., Coenen S., Malhotra-Kumar S., Dvorakova K., Hendrickx E., Suetens C., Goossens H., ESAC Project Group. 2006. European Surveillance of Antimicrobial Consumption (ESAC): outpatient macrolide, lincosamide and streptogramin (MLS) use in Europe. *J. Antimicrob. Chemother.* 58 (2): 401–7.

Ferreira da Silva M., Tiago I., Veríssimo A., Boaventura R.A., Nunes O.C., Manaia C.M. 2006. Antibiotic resistance of enterococci and related bacteria in an urban wastewater treatment plant: *FEMS Microbiol. Ecol.* 55: 322–9.

Göbel A., McArdell C.S., Joss A., Siegrist H., Giger W. 2007. Fate of sulfonamides, macrolides, and trimethoprim in different wastewater treatment technologies. *Sci. Total Environ.* 372 (2–3): 361–71.

Göbel A., Thomsen A., McArdell C.S., Joss A., Giger W. 2005. Occurrence and sorption behavior of sulfonamides, macrolides, and trimethoprim in activated sludge treatment. *Environ. Sci. Technol.* 39 (11): 3981–89.

Golet E.M., Xifra I., Siegrist H., Alder A.C., Giger W. 2003. Environmental exposure assessment of fluoroquinolone antibacterial agents from sewage to soil. *Environ Sci Technol.* 37 (15): 3243–9.

Heise J., Höltge S., Schrader S., Kreuzig R. 2006. Chemical and biological characterization of non-extractable sulfonamide residues in soil. *Chemosphere.* 65 (11): 2352–7.

Hijosa-Valsero M., Fink G., Schluesener M.P., Sidrach-Cardona R., Martín-Villacorta J., Ternes T., Bécares E. 2011. Removal of antibiotics from urban wastewater by constructed wetland optimization. *Chemosphere.* 83 (5): 713–19.

Hryniewicz K., Szczypa K., Sulilowska A., Jankowski K., Betlejewska K., Hryniewicz W. 2001. Antibiotic susceptibility of bacterial strains isolated from urinary tract infections in Poland. *J. Antimicrob. Chemother.* 47 (6): 773–80.

Junker T., Alexy R., Knacker T., Kümmerer K. 2006. Biodegradability of 14C-labeled antibiotics in a modified laboratory scale sewage treatment plant at environmentally relevant concentrations. *Environ. Sci. Technol.* 40 (1): 318–24.

Kim S., Aga D.S. 2007. Potential ecological and human health impacts of antibiotics and antibiotic-resistant bacteria from wastewater treatment plants. *J. Toxicol. Environ. Health. B. Crit. Rev.* 10 (8): 559–73.

Kümmerer K. 2001. Drugs in the environment: emission of drugs, diagnostic aids, and disinfectants into wastewater by hospitals in relation to other sources—a review. *Chemosphere.* 45: 957–69.

Kümmerer K. 2009. Antibiotics in the aquatic environment—A review—Part I. *Chemosphere.* 75 (4): 417–34.

Lindberg R.H., Olofsson U., Rendahl P., Johansson M.I., Tysklind M., Andersson B.A. 2006. Behavior of fluoroquinolones and trimethoprim during mechanical, chemical, and active sludge treatment of sewage water and digestion of sludge. *Environ. Sci. Technol.* 40 (3): 1042–8.

Luczkiewicz A., Fudala-Książek S., Jankowska K., Olanczuk-Neyman K. 2010. Antimicrobial resistance of fecal indicators in municipal wastewater treatment plant. *Wat. Res.* 44 (17): 5089–97.

Luczkiewicz A., Kotlarska E., Jankowska K., Pazdro K., Olanczuk-Neyman K. 2011. Tetracycline and sulfonamide resistance genes carried by Escherichia coli isolates from surface water and wastewater. *Proceedings of 4th Congress of European Microbiologists* FEMS 2011 "Advancing Knowledge on Microbes", 26–30.06.2011, Geneva, Switzerland.

Magiorakos A.P., Srinivasan A., Carey R.B., Carmeli Y., Falagas M.E., Giske C.G., Harbarth S., Hindler J.F., Kahlmeter G., Olsson-Liljequist B., Paterson D.L., Rice L.B., Stelling J., Struelens M.J., Vatopoulos A., Weber J.T., Monnet D.L. 2012. Multidrug-resistant, extensively drug-resistant and pandrug-resistant bacteria: an international expert proposal for interim standard definitions for acquired resistance. *Clin. Microbiol. Infect.* 18 (3): 268–81.

Marcinek H., Wirth R., Muscholl-Silberhorn A. 1998. Enterococcus faecalis gene transfer under natural conditions in municipal sewage water treatment plants. *Appl. Environ. Microbiol.* 64 (2): 626–32.

Martins da Costa, M.P., Vaz-Pires, P., Bernardo, F. 2006. Antimicrobial resistance in Enterococcus spp. isolated in inflow, effluent and sludge from municipal sewage water treatment plants. *Wat. Res.* 40, 1735–40.

Ohlsen K., Ziebuhr W., Koller K.P., Hell W., Wichelhaus T.A., Hacker J. 1998. Effects of subinhibitory concentrations of antibiotics on alpha-toxin (hla) gene expression of methicillin-sensitive and methicillin-resistant Staphylococcus aureus isolates. *Antimicrob. Agents Chemother.* 42 (11): 2817–23.

Olanczuk-Neyman K., Geneja M., Quant B., Dembinska M., Kruczalak K., Kulbat E., Kulik-Kuziemska I., Mikołajski S., Gielert M. 2003. Microbiological and biological aspects of the wastewater treatment plant "Wschod" in Gdansk. *Pol. J. Environ. Stud.* 12 (6): 747–57.

Pérez S., Eichhorn P., Aga D . 2005. Evaluating the biodegradability of sulfamethazine, sulfamethoxazole, sulfathiazole, and trimethoprim at different stages of sewage treatment. *Environ. Toxicol. Chem.* 24 (6): 1361–67.

Reinthaler F.F., Posch J., Feierl G., Wust G., Haas D., Ruckenbauer G., Mascher F., Marth E.; 2003, Antibiotic resistance of E. coli in sewage and sludge. *Wat. Res.* 37 (8): 1685–90.

Sköld O. 2000. Sulfonamide resistance: mechanisms and trends. *Drug Resistance Updates.* 3 (3):155–60.

Soda S., Otsuki H., Inoue D., Tsutsui H., Sei K., Ike M.; 2008, Transfer of antibiotic multiresistant plasmid RP4 from Escherichia coli to activated sludge bacteria. *J. Biosci. Bioeng.* 106 (3): 292–6.

Szczepanowski R., Linke B., Krahn I., Gartemann K.H., Gützkow T., Eichler W., Pühler A., Schlüter A. 2009. Detection of 140 clinically relevant antibiotic-resistance genes in the plasmid metagenome of wastewater treatment plant bacteria showing reduced susceptibility to selected antibiotics. *Microbiology.* 155 (7): 2306–19.

Talebi M., Rahimi F., Katouli M., Möllby R., Pourshafie M.R. 2008. Epidemiological link between wastewater and human vancomycin-resistant Enterococcus faecium isolates. *Curr Microbiol.* 56 (5): 468–73.

Tejedor Junco M.T., González Martín M., Pita Toledo M.L., Lupiola Gómez P., Martín Barrasa J.L. 2001. Identification and antibiotic resistance of faecal enterococci isolated from water samples. *Int J Hyg Environ Health.* 203 (4): 363–8.

Vouillamoz, J., P. Moreillon, M. Giddey, and J.M. Entenza. 2–6. Efficacy of daptomycin in the treatment of experimental endocarditis due to susceptible and multidrug-resistant enterococci. *J. Antimicrob. Chemother.* 58:1208–14.

Environmental Engineering IV – Pawłowski, Dudzińska & Pawłowski (eds)
© 2013 Taylor & Francis Group, London, ISBN 978-0-415-64338-2

Rational method of grit chamber efficiency determination

M.A. Gronowska & J.M. Sawicki
Faculty of Civil and Environmental Engineering, Gdansk University of Technology, Gdansk, Poland

ABSTRACT: Grit chambers constitute a crucial element in technological systems for waste water treatment. Although designed and manufactured with utmost care, grit chambers efficiency of operation still remains difficult to control. As concentration of grit in transit channels and inside the device itself is non-uniform, representative measurement of suspension concentration at inlet and outlet device cross-sections is a very complex issue. In the present paper, the results of measurements performed at waste water treatment plants in Gdansk and Gdynia, Poland, were analyzed. The proposed method to determine grit chambers efficiency is based on comparison of suspension flow in the inlet cross-section and flux of sediment removed from the device bottom.

Keywords: concentration, sand removal efficiency, grit chamber, suspension, waste water treatment

1 INTRODUCTION

The need to monitor operation efficiency of objects in environmental engineering is uncontested, both at start-up (when results of investigation are the basis for object acceptance or indicate a need for design correction), and during system operation. This applies to all objects and devices, but especially to reactors and installations for water and waste water treatment, that operate in harsh conditions—intensive changes in quality and quantity of feed liquid stream followed by very limited possibilities to control these parameters in case of waste water.

Efficiency "ε" is defined as relative change of state of substance pumped through the reactor (water or waste water). When concentration of considered substance (or a number of substances) is taken as the measure of change of state, then:

$$\varepsilon = \frac{C_{in} - C_{out}}{C_{in}} \qquad (1)$$

where C_{in} = initial concentration; C_{out} = final concentration.

With device efficiency described by Eq. 1, proper determination of both characteristic concentrations is the key element, which depends on:

– representativeness of samples to be analyzed;
– accuracy of the analysis.

During concentration measurements, the key issue is the degree of substance dispersion. In case of waste water, four categories of matter can be distinguished:

– dissolved substances;
– colloids;
– settling flocks (e.g. activated sludge, coagulated colloids);
– settling grits (commonly called sand).

The higher the degree of fineness, the more uniform the substance concentration in measurement cross-section (and the easier to acquire a representative sample). The fact applies especially to solutions and colloids, and also to easy-settling suspended flocks. On the contrary, non-uniform distribution of sand concentration (removed from waste water in grit chambers and separators) along cross-section causes sampling, and efficiency determination, to be troublesome.

Classical principles and procedures of determining efficiency of objects for water and waste water treatment have been widely discussed in specialist literature (e.g. Imhoff 1982, Sawicki 2000; Sawicki 2001). However, their application for mineral suspension removal yields some errors that have been outlined further in the paper. As an answer, the authors propose to update wide-known methodology with a well-justified change, in which incoming liquid stream is compared with the flow of sand removed from the object, instead of the outflowing stream.

2 MATERIAL AND METHODS

2.1 *Amount of sand in waste water*

According to literature sources (Badowski et al. 1991, Imhoff 1982), domestic waste water contains

40 g/PEd settling solids on average (including 10 g/PEd mineral solids and 30 g/PEd organic solids). When waste water unit discharge equals 0.2 m³/PEd, concentration of settling solids reaches:

– total settling solids 200 g/m³;
– mineral solids 50 g/m³;
– organic solids 150 g/m³.

Taking into account the specifics of system under the consideration, mineral suspension is determined by mass of sample after roasting, and organic suspension—by loss of sample mass during roasting. The approach results from the fact that waste water sample can be totally devoid of suspension mass (e.g. by means of sedimentation or filtration), however, it is almost impossible to separate particles of both fractions (grit particles may stick to flocculent particles of organic suspension). Besides fine sand particles, the resulting sample contains some amount of ash from incineration of organic substances. As such, the two constituents cannot be practically separated. With suspension limits for grit chambers set to:

– $d > 0.2$ mm removed in 100%;
– $0.1 < d > 0.2$ mm removed in 65–75%;
– total of $d < 0.1$ mm left in waste water stream.

The abovementioned fact does not create a problem.

Amounts of settling solids present in domestic sewage provided by literature were confirmed by the measurements performed on aerated grit chambers that operate in Gdansk and Gdynia waste water treatment plants (Ignatowska & Kowalska 2000). Average concentration of mineral solids after roasting in Gdynia amounted to 58.3 g/m³, while in Gdansk—29.5 g/m³. In both cases the finest fraction ($d < 0.1$ mm) was the dominating one with concentrations: 40.4 g/m³ in Gdynia and 19.6 g/m³ in Gdansk.

2.2 Change of sand concentration in time and space

In rectilinear liquid flow, vertical distribution of suspension concentration is described by the relation:

$$C_i(z) = C_{io}(z/z_o)^{-1.67 w_i/v_*}$$
(2)

where i = index of suspended fraction with free sedimentation velocity w_i; C_{io} = fraction concentration on depth Z_o; and v_* = shearing velocity given by:

$$v_* = v\sqrt{g}/C$$
(3)

where v = mean velocity of waste water flow; $g = 9{,}81$ m/s²; and C = Chezy's coefficient.

Relation (Eq. 2) results from the classical theory of turbulent transport (Prandtl 1958). Exemplary concentration distributions determined by the two characteristic fractions at inlet and outlet cross-sections in Gdansk waste water treatment plant (Ignatowska & Kowalska 2000) are presented in Figure 1.

At the inflow cross-section curves for both fractions are very similar, in contrast to the outflow. Taking into account the non-linear course of function (Eq. 2), as well as the occurrence of transversal concentration variations due to changing channel shape, sampling at the outflow cross-section proves to be pointless. Additionally, time variations in feed sand concentration make the situation more complex. For instance, sand concentrations in five consecutive samples obtained from Gdynia waste water treatment plant (sampled at 10:00 am every second day in November) were: 11.6; 97.8; 22.1; 16.4; 21.3 g/m³.

2.3 Granulometric composition of suspension

Exemplary composition (mass M and volume V for bulk density 1.7 kg/dm³) of suspension sample obtained by mixing single samples taken every hour during one day at Gdansk waste water treatment plant, are presented in Tables 1 and 2.

Organic fraction is the dominating one in samples obtained from grit chambers in plants under consideration. Incineration of the samples yield relatively high amounts (with respect to sample volume) of tiniest fraction $d < 0.1$ mm, that cannot be distinguished from fine mineral solids. Granulometric composition of sample from Gdansk, together with the analogical one from Gdynia, is presented in Figure 2. Particle size distribution graphs contain additional curves from sieve analysis performed on samples obtained from grit separators.

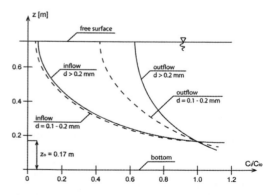

Figure 1. Exemplary concentration distributions of samples from plant in Gdansk.

Table 1. Composition of exemplary sample from inlet cross-section of grit chamber in Gdansk plant.

Sample material	Fraction	Mass	Volume
Dry sediment	–	137.5 g/m³	–
Mineral solids	–	29.5 g/m³	–
in which:	$d > 0.2$ mm	5.25 g	3.1 cm³
	0.1 mm $< d > 0.1$ mm	4.64 g	2.7 cm³
	$d < 0.1$ mm	19.61 g	11.5 cm³
Organic solids	–	108.0 g/m³	–
Total sample	–	–	17.3 cm³

Table 2. Composition of exemplary sample from outlet cross-section of grit chamber in Gdansk plant.

Sample material	Fraction	Mass	Volume
Dry sediment	–	119.9 g/m³	–
Mineral solids	–	22.6 g/m³	–
in which:	$d > 0.2$ mm	0.93 g	0.6 cm³
	0.1 mm $< d > 0.1$ mm	1.94 g	1.1 cm³
	$d < 0.1$ mm	19.73 g	11.6 cm³
Organic solids	–	97.3 g/m³	–
Total sample	–	–	13.3 cm³

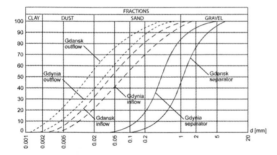

Figure 2. Granulometric composition of samples from plants in Gdansk and Gdynia.

The analysis in terms of sieve fractions indicate that the curves fulfil the requirements stated in subchapter 2.1—most particles at the outlet cross-section are finer than at the inlet, and the separator contains bigger ones.

Comparison of operation of Gdansk and Gdynia waste water treatment plants show that feed waste water stream in Gdansk contains more finer fractions than in Gdynia (the fact resulting probably from catchment and sewerage system specifics), while the reverse is true for sediments removed from grit separators. The situation may originate from different sediment washing systems applied in the plants, however, their evaluation is not a subject of the paper.

3 RESULTS AND DISCUSSION

The abovementioned characteristics of suspended solids underline the complexity of suspension distribution in transversal cross-section of sewage channels. The concentration is low and dominated by organic solids. Spatial distribution, as well as time variation of suspension, make designation of characteristic sampling points difficult. Moreover, after incineration, finer mineral fractions were mixed with fine ash fractions formed from organic suspension.

Despite all the difficulties, determination of suspension concentration is necessary, especially in case of large and complex installations (mainly aerated grit chambers). Practical examples confirm the determination at the inlet to be complex, but yielding quite repeatable results. Measurements at the outlet carry significant errors due to very low suspension concentrations at this cross-section. In case of measurements of limited range, concentration of finer fractions at the outflow may be even higher than at the inflow. According to data presented in subchapter 2.3 apparent volume of sand obtained from 1.0 cm³ of sample (a relatively high volume as for a sample) does not exceed 2.0 cm³ and is separated from a several times larger fraction of diameter smaller than 0.1 mm. All the factors call for a search for other possibilities of grit chamber efficiency determination.

163

4 CONCLUSIONS

Relation (Eq. 1) includes the relationship between inflow of substance mass M_{in} to the reactor and outflow of mass M_{out} from the reactor, that is difficult to be determined. The flow balance is closed by the flow of substance mass M_r removed from waste water stream in the reactor:

$$M_{in} - M_{out} = M_r \qquad (4)$$

In case of mechanical methods for sediment removal, it is not only possible to evaluate mass M_r but this mass is actually being determined (suspended solids are disposed on landfills as waste), the fact impossible for chemical and biological means of contaminants removal.

The value of M_r is being used so far, among others, by the owner of waste water treatment plant to control proper determination of masses M_{in} and M_{out}. However, it is more purposeful to apply mass M_r to calculate efficiency of grit chamber operation by setting it only with mass M_{in}, which is easier to determine. By denoting daily-averaged concentration of suspension at the inlet cross-section as C_d and daily-averaged waste water discharge as Q_d, then:

$$\varepsilon = \frac{M_r}{M_{in}} = \frac{M_r}{C_d \, Q_d} \qquad (5)$$

Application of the described possibility requires the amounts of sand removed from feed waste water to be regularly controlled (discharge measurements are already performed with utmost care).

ACKNOWLEDGEMENTS

Scientific research has been carried out as a part of the Project "Innovative resources and effective methods of safety improvement and durability of buildings and transport infrastructure in the sustainable development" financed by the European Union from the European Fund of Regional Development based on the Operational Program of the Innovative Economy.

REFERENCES

Badowski, M. (et al.) 1991. Wodociagi i kanalizacja: podstawy projektowania i eksploatacja: poradnik. Roman M. (ed.), Warszawa: Arkady.

Imhoff, K. & K.R. 1982. Kanalizacja miast i oczyszczanie sciekow: Poradnik. Warszawa: Arkady.

Ignatowska, A., Kowalska A. 2000. Hydrauliczna analiza piaskownika napowietrzanego na G.O.S. Debogorze. Master's Thesis, Politechnika Gdanska, Gdansk.

Prandtl, L. 1958. Fuehrer durch die Stroemungslehre. Braunschweig: Vieweg und Sohn.

Sawicki, J.M. 2000. Hydrauliczne aspekty poboru i analizy probek zawiesiny. Proc. Polish Conf. on Charakterystyka i Zagospodarowanie Osadow Sciekowych, Gdansk, Poland, 80–86.

Sawicki, J.M. 2001. Wyznaczanie sprawnosci piaskownikow. In: *Gaz, woda i technika sanitarna* 5: 164–168.

Neutralisation of solid wastes and sludge

Environmental Engineering IV – Pawłowski, Dudzińska & Pawłowski (eds)
© 2013 Taylor & Francis Group, London, ISBN 978-0-415-64338-2

Sewage sludge mass minimization technology—from legislation to application

M. Cimochowicz-Rybicka
Institute of Water Supply and Environmental Protection, Cracow University of Technology, Cracow, Poland

ABSTRACT: The paper presents analysis of new wastewater treatment and sludge disposal technologies that have been brought to attention together with new legal regulations in Europe. Agriculture application and composting processes are the leading methods of sludge disposal in the EU countries, where the most sludge is produced. Poland declares the highest amount of sludge (58% of dry mass) as being processed using "other methods of disposal". The most popular methods applied at a full scale in Europe include: mechanical disintegration, ultrasound disintegration and ozonation. The authors discussed results of their research studies carried out at the Cracow Technical University on dynamic of biogas production during ultrasonic disintegration. The methanogenic potential of sludge samples, defined as methanogenic activity and expressed as $gCOD_{CH_4}/gVSS*day$ was investigated. The parameter has been proposed as one of the indicators of the process operation control

Keywords: sewage sludge, minimization, disintegration

1   INTRODUCTION

The EU directives related to wastewater and sludge issues provide grounds for the development of legal acts for particular European countries. Following the directives, Poland has been obliged to undertake the following actions: a) limit possibilities of sewage sludge deposition; b) process more municipal sludge before its introduction to the environment, including thermal processing; c) utilize more municipal sewage sludge in biogas stations and as an energy source; d) increase the amount of municipal sewage sludge that is thermally processed; e) utilize nutrients present in sludge. The analysis of the Eurostat data (EUROSTAT 2012), related to the sludge generated by municipal sewage treatment plants in Europe, points at the diversity of approach to these issues by individual countries, even though a common policy is formally applied. The authors completed a detailed analysis of data from 10 selected EU countries (and Switzerland) producing the highest dry mass of municipal sludge. Table 1 contains a summary of data on the total sludge weight, the mass of sediment per capita and basic methods of sludge disposal.

The average unit sludge production in municipal Wastewater Treatment Plants (WWTP) was 23 kg/cap*year, with the highest value observed in Austria (30,28 kg/cap*year). The lowest unit value for Poland (14,76 kg/cap*year) can be explained by a low percentage of population connected with WWTP. The prevailing final sludge

handling methods are agricultural applications and composting or thermal incineration. It should be noted that the largest sludge 'producer' i.e. Germany and largest unit sludge 'producer' (Austria) apply agriculture/composting and incineration methods in similar proportions. In other countries, the predominance of one of these two methods may be observed e.g. in the United Kingdom, France, Spain, the Czech Republic and Hungary where agricultural methods dominate while in the Netherlands and Switzerland at least 90% of sludge is incinerated. Italy is the only country where landfilling remains the leading method of disposal, while Poland declares that over 50% of the sludge is subjected to the "other methods of disposal". The total annual mass of sewage sludge, generated at the Polish WWTP increased in years 2000–2008. However, currently it tends to decrease slightly. It may be credited to major investments in construction of new treatment plants that took place since 1990s until the end of last decade. The funds were followed by the operational optimization. An improvement of plants operation resulted in a decrease of amount of sewage sludge stored at the WWTP.

The analysis of statistic data, concerning sludge management at Polish municipal wastewater treatment plants, has showed that a dynamic growth of a number and capacity of WWTPs has been recently supported by improvements in sludge management. Another illustrative example for improvement of sludge handling is a formal

Table 1. Sludge produced at municipal sewage treatment plants and the ways of its management in the European countries.

Country	Total Thou. ton/year	Sludge dry mass per capita kg/cap*year	Basic method of disposal		
			Agriculture +Composting	Incineration	Other
			Percent of total mass		
Germany	2049	24,99	53%	47%	0%
United Kingdom	1771	28,63	70%	16%	9%
France	1087	17,37	73%	19%	0%
Spain	1065	23,12	65%	4%	16%
Italy	1056	17,51	44%	3%	11%
Poland	563	14,76	26%	2%	58%
Netherlands	353	21,37	0%	95%	0%
Hungary	260	25,95	60%	1%	10%
Austria	254	30,28	38%	36%	17%
Czech Republic	220	20,94	78%	1%	8%
Switzerland	210	27,14	10%	90%	0%

Table 2. Trends in sludge production and utilization at the Polish municipal WWTPs.

Year		2000	2005	2008	2009	2010
Total sludge	Thou. ton/year	359,8	486,1	567,3	563,1	526,7
Waste disposal methods						
Agriculture	Thou. ton/year		66	112	123,1	109,3
Land reclamation	Thou. ton/year		120,6	105,8	77,8	54,3
Composting	Thou. ton/year	25,5	27,4	27,5	23,5	30,9
Incineration	Thou. ton/year	5,9	6,2	6	8,9	19,8
Landfilled	Thou. ton/year	151,6	150,7	91,6	81,6	58,9
Other	Thou. ton/year	176,8	115,2	224,4	248,2	253,5

and economical requirement for an increase of a renewable energy share in the total country energy balance. At present, renewable energy from biomass and waste incineration equals 8.5% of the annual production of primary energy, while the average value for 27 member countries of the European Union is 12.5%. Development of specific technologies resulted in a significant reduction of the amount of residual substances that are released to the environment and remain in its circulation. This approach is of crucial importance for modern sustainable plants. Apart from sludge stabilization, the main task of contemporary sludge processing technology is to produce an additional source of easy biodegradable carbon compounds to intensify nutrient removal (Boehler & Sigrist 2006, Ødegaard 2004, Nickel & Neis 2007, Zielewicz 2007). The key mechanisms of sludge reduction, incorporated in wastewater treatment or sludge disposal units, can be described in the following way (Foladori et al. 2010): cell lysis and cryptic growth, uncoupled metabolism, endogenous metabolism, microbial predation, hydrothermal oxidation.

A conventional requirement, i.e. reduction of dry mass accompanied with an increase of biogas production in anaerobic digestion, remains still valid in a routine operations (Wei et al. 2003, Winter 2002, Valo et al. 2004, Zhang et al. 2008). Sludge disintegration appears to be an ecologically reasonable and economically feasible solution as it leads to a decrease of dry mass of processed sludge and intensifies generation of a methane-rich gas simultaneously (Müller 2000, Bień & Szparkowska, 2004, Rybicki 2004, Pilli et al. 2010, Cimochowicz-Rybicka & Rybicki, 2010).

2 MATERIALS AND METHODS

Disintegration of sludge is a relatively new area of sewage treatment, and therefore there are still no unified standards/procedures that would allow for a clear assessment of the process efficiency (Zielewicz 2007, Cimochowicz-Rybicka et al. 2009). Process parameters may be divided into three basic groups:

a. quantitative parameters—that evaluate sludge cell rupture as a direct effect of the sewage sludge disintegration process (analysis of particle size, disintegration degree—DD, change of physic-chemical properties),
b. process parameters of the disintegration unit and other parameters associated with the sludge characteristics (ultrasound frequency, intensity and density; specific energy),
c. parameters defining the sludge potential within the technological system, after its disintegration (proposed methanogenic activity).

The volume of biogas produced and the production rate itself are the most important technological parameters that describe intensity of anaerobic processes. A relationship between biodegradation of organic compounds and the methane production defines the ability of anaerobic biomass to anaerobic digestion. The parameter known as methanogenic activity fulfills all the mentioned requirements of the process description in the real life conditions. In case of biomass coming from anaerobic digesters, determination of its methanogenic activity is considered to be very difficult. There is still a shortage of literature data that would offer results obtained at the full scale treatment applications. Methanogenic activity of biomass, as the measure of methane production rate, depends on the presence of the methane formers and the availability of the substrate that is used for methane production. The authors have been involved in scientific, design and consulting projects dealing with anaerobic sludge decomposition processes. In recent years these investigations have focused on a methanogenic activity of sludge (Cimochowicz-Rybicka & Rybicki 2009, 2011).

The scope of tests was the following:

– assumptions of sonication parameters
– fermentation tests performed at a respirometric test stand with the WAS, after its disintegration at assumed parameters
– methanogenic activity tests—batch tests with the WAS.

2.1 Assumptions for ultrasonic disintegration

Constant ultrasound intensity $24*10^3$ W/m²
Frequency   22,5 kHz
Changing ultrasound density: from 2,9 to 17,5 kWatt*hours/m³ (adjustment of the ultrasound density was performed through change of

a sludge sample volume: 130 ml, 86 ml, 65 ml, respectively). Sonication time: 3, 5, 7, 9 min.

2.2 Assumptions for fermentation and methanogenic activity tests

The waste activated sludge was collected at the Kujawy WWTP, the large municipal treatment plant (200000 p.e.) treating the outskirts of Krakow, southern Poland. Temperature: 35°C; pH adjustment: with regard to the sensitivity of methanogenic bacteria; experimental system: a mechanically stirred; substrate and sludge concentration: the volatile fatty acids (VFA) mixture with concentrations ranging from 2 to 5 g/L; the concentration of sludge ranging from 3 to 5 gVSS/L (Lettinga & Hulshoff, 1991; Cimochowicz—Rybicka 2009, 2011); time of research: fermentation test—ca. 20 days, methanogenic test—ca. 10–14 days; repeatability of measurements: double (fermentation), triple (activity test). The test-stand consisted of:

Ultrasound disintegration: disintegrator UD 11, nominal frequency 22,5 kHz.
Fermentation and methanogenic tests: water bath with a magnetic stirrer, anaerobic respirometer AER-208, computer system with data recording programs.

2.3 Experimental protocols—methanogenic activity

The methanogenic activity of sludge was evaluated on the basis of the measurements of the amount of methane produced during the fermentation process in the reaction vessels filled with anaerobic sludge, VFA substrates and mineral nutrients. The sludge methanogenic activity was determined by the amount of fatty acids (g/L COD), which was consumed during the fermentation process and converted into methane (mL CH_4). The COD was referred to 1 gram of sludge VSS (volatile suspended solid) per 1 day.

Based on the amount of the gas produced during the batch methanogenic test, cumulative gas production curves were plotted. Using those curves, the maximum gas production (mL CH_4) was calculated. From the maximum slope of the curves, the gas production rate R (mL CH_4/h), which determined the amount of gas produced per hour, was obtained. The methanogenic activity in g COD_{CH_4}/g VSS*d, was calculated in the following way (Lettinga & Hulshoff 1991):

$$ACT = (R * 24)/(CF * V * VSS) \tag{1}$$

where:
ACT—methanogenic sludge activity, $gCOD_{CH_4}$/gVSS*d

Table 3. Waste activated sludge after exposition to ultrasound disintegration (average values).

Sonication time [min]	0	3	5	7	9
Series 1: ultrasound density [kWatt*h/m³]	0	2,90	4,80	6,80	8,70
VSS [g/l]	4,63	4,33	3,89	4,21	4,45
R	3,80	5,10	5,03	5,70	5,90
ACT	0,09	0,14	0,15	0,17	0,18
Series 2: ultrasound density [kWatt*h/m³]	0	4,40	7,33	10,27	13,20
VSS [g/l]	4,03	4,12	3,99	3,77	3,8
R	4,25	5,60	5,90	6,50	6,80
ACT	0,11	0,15	0,18	0,20	0,18
Series 3: ultrasound density [kWatt*h/m³]	0	5,80	9,70	13,50	17,40
VSS [g/l]	4,23	4,24	4,03	3,90	3,80
R	4,25	5,60	5,90	6,50	6,80
ACT	0,11	0,16	0,18	0,21	0,20

R—gas production rate, mL CH_4/h.
CF—conversion factor, mL CH_4/g COD.
V—effective liquid volume of reactor vessels, L.
VSS—sludge sample concentration, g VSS/L.

3 RESULTS AND DISCUSSION

Results obtained are summarized in Table 3.

A decrease of the VSS concentration was observed at entire ranges in VSS reduction of sonication time and an ultrasound density. Higher effect was obtained rather with increasing the density than adjusting a sonication time. However a decrease of the VSS concentrations was observed for sonication times: 3, 5 and 7 min, while after 9 minutes an increase of the VSS value was stabilized. On the contrary, a methanogenic activity increased with the applied ultrasonic density over 10 kWatthours/m³, despite a sonication time. Increasing of the ultrasound density up to approx. 13 kWatthours/m³ did not resulted in changes in an activity while further increase of an ultrasound density adversely impacted a methanogenic activity, which tended to decrease. These proved the methanogenic activity to be more reliable parameter to asses and ultrasound disintegration than conventional measurement of a VSS decrease.

4 CONCLUSIONS

Agriculture application and composting are the leading methods of sludge disposal in the EU countries where the most of sludge is produced. The six countries (i.e. Germany, the United Kingdom, France, Spain, Hungary, the Czech Republic) dispose over 50% of produced sludge. The Netherlands and Switzerland perform sludge incineration of 95% and 90% of total sludge dry mass respectively. Poland is declared to have the highest amount of sludge (58% of dry mass) as being processed using "other methods of disposal".

In Poland the mass of sludge landfilled and stored has significantly decreased since 2005. Recently more emphasis is put on the increase of biogas production and sludge incineration.

The studies on sludge disintegration, conducted by the authors, lead to the conclusion that methanogenic activity mesurements can serve as an operational guideline. A simultaneous increase of sonication time and ultrasonic density may result in a decrease of methanogenic activity beyond the limits of ultrasound energy that was applied. Therefore, control of the ultrasonic density at the constant sonication time, during anaerobic digestion of excess sludge at the WWTPs, seems to be the most suitable option from the perspective of higher gas production.

REFERENCES

Bień, J.B. & Szparkowska, I. 2004. Wpływ dezintegracji ultradźwiękowej osadów ściekowych na przebieg procesu stabilizacji beztlenowej (In Polish), *Inżynieria i Ochrona Środowiska* (in Polish), 7, 3–4: 341–352.

Boehler, M. & Siegrist H. 2006. Potential of activated sludge disintegration, *Wat. Sc. & Tech.*, 53, 12: 207–216.

Cimochowicz-Rybicka, M., Tal-Figiel, B., Rybicki, S. 2009. The effect of sonication parameters on methanogenic sludge activity. Proceedings of 12th Int. Conf. IWA "*Sustainable Management of Water & Wastewater Sludges*", Harbin, China: 329–336.

Cimochowicz-Rybicka, M. & Rybicki, S.M. 2010. Disintegration of fermented sludge—possibilities and potential gains. In Pawłowska M., Pawłowski L. (eds.), *Environment Engineering III;* London, UK: Taylor & Francis.

Cimochowicz-Rybicka, M. & Rybicki, S.M. 2011. Application of the sludge methanogenic activity tests to improve overall methane production using sludge sonication. Proceedings of 4th IWA-ASPIRE, *Sludge Management and Resources Recovery,* Tokyo, Japan: IWA.

Eurostat. 2012. http://epp.eurostat.ec.europa.eu/.

Foladori, P., Andreottola, G., Ziglio, G. 2010. Sludge Reduction Technologies in Wastewater Treatment Plants. IWA Publishing, London, UK.

Lettinga, G., Hulshoff, P., Pol, L.W. 1991. Anaerobic Reactor Technology. Proc. of AWWT International Course, IHE Delft, Wageningen Agricultural University, The Netherlands.

Müller, J. 2000. Disintegration as a key-step in sewage sludge treatment; *Water. Sci. Tech.* 41, 8:123–130.

Nickel, K. & Neis. U. 2007. Ultrasonic disintegration of biosolids for improved biodegradation. *Ultrasonic Sonochemistry*, Elsevier, 14:450–455.

Ødegaard, H. 2004. Sludge minimization Technologies—an overview, *Water Science and Technology*, 49, 10:31–40.

Pilli, S., Bhunia, P., Yan, S., Le Blanc, R.J., Tyagi, R.D., Surampalli, R.Y. 2010. Ultrasonic pretreatment of sludge: A review, *Ultrasonics Sonochemistry*, doi:10.1016/j. ultsonoch. 2010.02.014.

Rybicki, S.M. 2004. Modeling of phosphorus release in continuous presence on nitrates and its influence on an energy recovery, In: E. Plaza, E. Levlin, B. Hultman, (eds.), Proc. of a Polish-Swedish seminar, Integration and optimisation of urban sanitation systems. Vol. 13:39–48.

Valo, A., Carrere, H., Delgenes, J. 2004. Thermal, chemical and thermo-chemical pretreatment of waste activated sludge for anaerobic digestion, *Journal. Chem. Technol. Biotechnol.* 79:1197–1203.

Wei, Y., van Houten, R.T., Borger, A.R., Eikelboom, D.H., Fany, Y. 2003. Minimization of excess sludge production for biological wastewater treatment. *Wat. Res.*, 37:4453–4467.

Winter, A. 2002. Minimization of costs by using disintegration at a full-scale anaerobic digestion plant, *Wat. Sci. Tech.*, 46, 4–5:405–412.

Zielewicz, E. 2007. Dezintegracja ultradźwiękowa osadu nadmiernego w pozyskiwaniu lotnych kwasów tłuszczowych (In Polish), *Zeszyty Naukowe Politechniki Śląskiej Gliwice,* Poland.

Zhang, G., Zhang, P., Yang, J., Liu, H. 2008. Energy-efficient sludge sonication: Power and sludge characteristics. *Bioresource Technology,* 99: 9029–9031.

Environmental Engineering IV – Pawłowski, Dudzińska & Pawłowski (eds)
© *2013 Taylor & Francis Group, London, ISBN 978-0-415-64338-2*

Dewatering of sewage sludge conditioned by means of the combined method of using ultrasound field, Fenton's reaction and gypsum

K. Parkitna, M. Kowalczyk, T. Kamizela & M. Milczarek
Czestochowa University of Technology, Institute of Environmental Engineering, Czestochowa, Poland

ABSTRACT: The presented publication provides an analysis of studies results conducted in order to determine the influence of sewage sludge conditioning by ultrasonic field, the Fenton's reaction and chemical substances in the form of gypsum on efficiency of mechanical dewatering with the using pressure filtration. Sludge was prepared by mentioned factors, whether each individual, as well as a combined method using sonification, Fenton's reaction and gypsum at the same time, in various combinations. On the basis of changes in final hydration of the sludge pressure filtration process, efficiency of these methods in relation to values of the parameter discussed, obtained for sludge that have not been treated, was determined. Using only ultrasonic field, Fenton's reaction, or gypsum for sludge conditioning, impacted on reduction of the final hydration after dewatering process but is less effective than combined methods, which were used for the preparation of two or three of the conditioning factors used in various combinations.

Keywords: sewage sludge, dewatering, Fenton's reaction, sonification

1 INTRODUCTION

To protect natural environment it is necessary to limit the negative effect of human activities to the lowest possible degree. It can be achieved by e.g. improvement in efficiency of wastewater treatment and neutralization or limitation of the volume of sewage sludge generated during these processes. Since contemporary technologies do not offer non-sludge methods and despite a variety of popular methods of sewage sludge preparation, seeking new solutions that lead to improvement in sludge de-watering efficiency remains very current. The major part of sewage sludge does not meet the requirements of the standards that allow them to be used in nature or agriculture while the National Waste Management Plan 2014 emphasizes limitation of storage and the increase of sewage sludge in thermal conversion (Bień et al. 2010, Law of 27/04/2001, National Waste Managment Plan 2014, Bień et al. 2008). Therefore, it seems legitimate to search for new methods and combinations of methods that would increase the effectiveness of division of liquid and solid phases. Flocculent structure which forms sludge particles allows for binding water mechanically, which makes it difficult to remove in the future (Hrynkiewicz 1998, Bień et al. 2008). Application of such processes as e.g. pressure filtration improves the effectiveness of dewatering through changes in shape, dimensions and concentration of sludge particles (Szwabowska 1986).

A variety of methods of sludge conditioning are used in order to improve the effectiveness of dewatering (Kowalczyk et al. 2011). Among these methods, ultrasound field is regarded to be a very effective mean of preparation which causes intensification of initial process of sludge preparation and increased effectiveness of mechanical dewatering and improvement in biodegradation with insignificant effect on the environment (Bień et al. 2008, Tiehm et al. 2001, Bourgier et al. 2006). According to Bień (1989), active effect of ultrasound field causes an increase in ability of better spatial packing in the sonicated sludge, particularly in the sludge with a more rigid structure of molecules. In the practice of waste management, the effects of ultrasound field on e.g. specific resistance, changes in sludge structure, changes in capillary suction time, destruction of microorganisms during hygienization, acceleration of the process of stabilization and changes in rheological parameters have been thoroughly researched (Bień 2003). Due to the expected benefits, it can be assumed that industrial application of ultrasounds will gain in popularity, particularly if combined with other methods (Boruszko 2001).

Physical treatment is also used: it consists in addition of such materials as ashes from furnaces, smoke-box dust from electrofilters in power plants, sawdust, cement or gypsum (Bień et al. 1999). The conditions of dewatering for conditioned sludge are affected by the type and size of grains of the auxiliary filtration layer (Bień 2007). These substances do not cause changes in physical and chemical properties of the prepared sludge and positively affect dewatering process (Bień et al. 2007).

Recently, more attention has been attracted to increasingly wide application of one of the methods of advanced oxidation, which is Fenton's reaction (Kowalczyk et al. 2007). This process is catalyzed by hydrogen peroxide and iron (II) ions (Świderska et al. 2008, Wąsowski et al. 2002, Krzemieniewski et al. 2005). In practice the highest effect on oxidation of waste during Fenton's reaction is from such parameters as: iron dose, hydrogen peroxide dose, Fe^{2+}/H_2O_2 mass ratio, pH, process temperature and reaction time (Ledakowicz et al. 2005, Zhou et al. 2001, Krzemieniewski et al. 2003). Analysis of results presented in Kowalczyk et al. (2011), Krzemieniewski et al. (2003), Lu et al. (2001), Barbusiński et al. (2000) confirms that the introduction of Fenton's reagent into a technological system has an impact on the improvement of sewage sludge dewatering parameters through the reduction in specific filtration resistance, shortening of dewatering time during pressure filtration, reduction in final hydration of sludge cake as well as considerable improvement in CST results obtained for sewage sludge and reduction in concentration of the organic matter which is difficult to decompose. Apart from classical Fenton's reaction, the processes of advanced oxidation use more and more often its modifications in order to generate additional hydroxyl radicals and thus to intensify the process (Chang et al. 2005). According to the analysis results presented in publications Sun et al. (2007), Ioan et al. (2007), Biń (1998), Ratanatamskul et al. (2005), the combination of ultrasound field and Fenton's reaction have positive influence on the pollutants degradation. It turned out that this method is more effective in comparison to Fenton's reaction which is carried out independently or in comparison to ultrasound preparation. Due to the simplicity of the process, lack of necessity of having specialized apparatus and low costs of reagents, combination of ultrasound and Fenton's reaction is becoming a promising method of sewage sludge conditioning today (Torres et al. 2007, Pham et al. 2010, Lu and Ting et al. 2010, Lu and Hung et al. 2010).

The present publication shows the results of the examinations which give a picture of changes in final hydration of sewage sludge conditioned by means of a method combining propagation of ultrasound wave, Fenton's reaction and gypsum, obtained after the process of mechanical dewatering by means of pressure filtration.

2 MATERIAL AND METHODS

The examinations discussed in the present study were carried out in order to examine the mixture of excess thickened sludge and initial sludge from the Warta Central Sewage Treatment Plant in Często-chowa, Poland, where industrial and municipal waste is processed. The samples were characterized by hydration at the level 97% and dry matter of 30 [g/dm³].

The scope of the study included dewatering of sewage sludge in the process of pressure filtration, conditioned with selected preparation reagents such as ultrasound field, Fenton's reaction and gypsum, used independently or by means of combined methods that used all the factors at the same time. An ultrasound field with an amplitude of 30.5 and 61 μm was used for 2, 4 and 6 minutes by means of a *VCX 134* laboratory ultrasound disintegrator. Fenton's reaction was conducted in the following way:

- sewage sludge samples in the amount of 1 dm³ were acidified until its pH value reached 3.0 with the use of 50% solution of H_2SO_4,
- added 1 g of iron (II) ions—the source of iron ions was iron (II) sulphate—$FeSO_4$*7 hydrate,
- next added 4 g of hydrogen peroxide—the source of hydrogen peroxide was obtained from 30% solution of Perhydrol,
- the samples were then mixed at the speed of 100rpm with the use of Kemwater Flocculator 2000 for one hour,
- next, the samples were neutralized to pH 7.0 using 50% solution of NaOH;

A chemical substance in the form of gypsum was also used for conditioning with the amount of 2 and 4 g per 1 dm³ of the sewage sludge studied.

Figure 1 presents the methods used for sewage sludge conditioning. After the process of

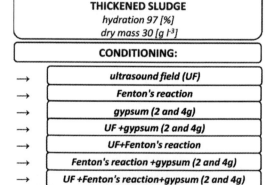

Figure 1. The methods of sewage sludge preparation used in the study.

conditioning the sludge samples were dewatered mechanically by means of pressure filtration. Sludge samples, which were previously conditioned and mixed, were taken with the volume of 100 cm³ each. Pressure filtration was carried out by means of a laboratory stand which was comprised of a pressure filter with a cotton filtration fabric inside (BT), a compressed air tank, cut-off valves, manometer, stopper and measuring cylinder for the filtrate.

The pressure of 0.5 MPa was used in the experiment. The values of final hydration presented for sewage sludge samples after the process of hydration are arithmetic means obtained after 3 repetitions.

3 RESULTS AND DISCUSSION

Figures 2 and 3 present changes in final hydration of sewage sludge conditioned with ultrasound field and gypsum with the volume of 2 and 4 g obtained after the process of pressure filtration.

The non-prepared sludge samples dewatered in the process of pressure filtration were characterized by final hydration of 83.5%. Regardless of the amplitude used (30.5 or 61 μm), ultrasound field caused a reduction in the value of the parameter discussed. In the case of the amplitude of 30.5 μm (Fig. 2), the lowest final hydration was obtained as a result of sonication for 4 minutes and it amounted to 81.8%. The use of higher amplitudes (61 μm)

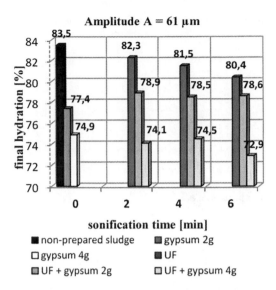

Figure 3. Changes in final hydration of sewage sludge conditioned with ultrasound field with amplitude of 61 μm and the time of 2, 4 and 6 min. and gypsum (2 or 4 g) obtained after the process of pressure filtration.

was found to be more beneficial. Final hydration of 80.4% was obtained after 6-minute exposure to ultrasound field, which equaled the value by 3.1% lower in comparison to the level obtained for raw sludge.

Conditioning of sewage sludge using chemical substance (gypsum) considerably affected the reduction in final hydration after the process of pressure filtration. After dosing 2 g of gypsum per 1 dm³ of sludge, a 6.1% reduction in the parameter discussed was obtained compared to non-prepared sludge. An increase in the dose of preparation substance to 4 g turned out to be more efficient. A final hydration of 74.9% was obtained in this case, which was by 8.6% lower compared to the level obtained for the raw sludge samples.

Sludge conditioning by means of the combined method, which used ultrasound field and gypsum at the same time, caused a more effective reduction in final hydration compared to the levels obtained for the same preparation agents used separately. Regardless of the amplitude (30.5 or 61 μm) and the gypsum dose (2 or 4 g), final hydration decreased proportionally to the time of exposure to ultrasound field. A reduction in the level of the parameter discussed was more noticeable when sonication and gypsum was used with higher doses (4 g). The most beneficial preparation was found for ultrasound field with the amplitude of 61 μm (6 minutes) and gypsum with volume of 4 g. The obtained final hydration was 72.9%, which was the level by 10.6% lower compared to the value

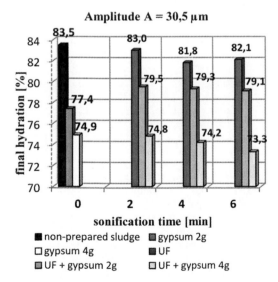

Figure 2. Changes in final hydration of sewage sludge conditioned with ultrasound field with amplitude of 30.5 μm and the time of 2, 4 and 6 min. and gypsum (2 or 4 g) obtained after the process of pressure filtration.

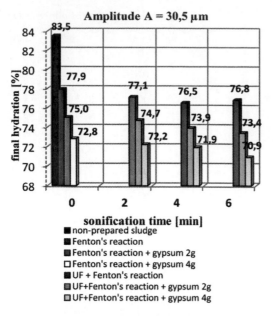

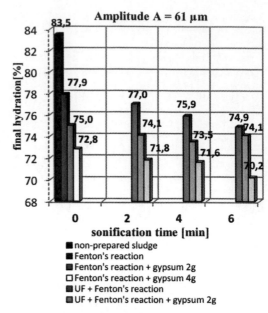

Figure 4. Changes in final hydration of sewage sludge conditioned using the combined method of ultrasound field with amplitude of 30.5 μm and the time of 2, 4 and 6 min, Fenton's reaction and gypsum (2 or 4 g) obtained after the process of pressure filtration.

Figure 5. Changes in final hydration of sewage sludge conditioned using the combined method of ultrasound field with amplitude of 61 μm and the time of 2, 4 and 6 min, Fenton's reaction and gypsum (2 or 4 g) obtained after the process of pressure filtration.

obtained for the non-conditioned sludge. Figures 4 and 5 present changes in final hydration of sewage sludge conditioned with Fenton's reaction separately and combined with other preparation agents such as ultrasound field with the amount of 2 and 4 g obtained after the process of dewatering using pressure filtration.

Preparation of sewage sludge with the use of Fenton's reaction caused improvement in its dewatering efficiency. The obtained final hydration was 77.9%, which was the level by 5.6% lower in comparison with the value obtained for the non-conditioned sludge. Regardless of the doses used, adding a chemical substance in the form of gypsum causes further reduction in the value of the parameter discussed. A more effective reduction in final hydration of sewage sludge was found when gypsum was dosed at the amount of 4 g. Final hydration of 72.8% was obtained. As a result of dewatering of the conditioned sludge by means of Fenton's reaction and gypsum (2 g), the value of the parameter discussed was reduced to 75%. Preparation of sewage sludge by means of the combined method of ultrasound field with the amplitude of 30,5 μm and Fenton's reaction (Fig. 4) caused further but insignificant reduction in final hydration compared to the hydration obtained for the sludge conditioned with Fenton's reaction alone.

Longer times of exposure to ultrasound field corresponded to lower values of the parameter discussed. Application of Fenton's reaction and ultrasound field with higher amplitude (61 μm) was found to be more effective (Fig. 5). The lowest value of final hydration was obtained for the longest sonication time (6 minutes) and amounted to 74.9%. It yielded the result by 8.6% lower compared to raw sludge and by 3% lower compared to the result obtained for sludge prepared with Fenton's reaction alone.

However, the highest efficiency was found for the sludge conditioned by the combined method that used all the three preparation agents at the same time i.e. ultrasound field, followed by Fenton's reaction and gypsum. Regardless of the amplitude used, the value of final hydration decreased in proportion to the elongation of the time of exposure to ultrasound field. More effective dewatering was observed for the sludge prepared with the use of the combination method with gypsum (4 g). The lowest value of the parameter discussed, obtained for the combined methods with gypsum at the amount of 2 g was 73.4%, and was obtained after the preparation with ultrasound field with the amplitude of 30.5 μm within 6 minutes using Fenton's reaction and then through gypsum with the above mentioned dose. This value was 10.1%

lower in comparison with the level obtained for raw sludge and 4.5% lower than the level found after Fenton's reaction alone.

The combined method of ultrasound field, Fenton's reaction and gypsum (4 g) was found to be the most efficient in terms of reduction in final hydration of sewage sludge. Regardless of the amplitude used for sonication, elongation of the time of exposure to ultrasound field caused that the value of the parameter analyzed in the study decreased.

The lowest value of 70.9% was obtained after 6 minutes of exposure for the method combined with sonication with the amplitude of 30.5 μm (Fig. 4). This yielded a reduction in final hydration by 12.6% compared to the non-prepared sludge. The use of ultrasound field with the amplitude of 61μm (Fig. 5) combined with Fenton's reaction and gypsum (4 g) was found to be the most efficient method of sludge conditioning in terms of the improvement in their susceptibility to dewatering in the process of pressure filtration. Final hydration decreased in proportion to sonication time and the lowest value was obtained after the time of 6 minutes. The obtained final hydration of the sludge studied was 70.2%, which was the level 13.3% lower in comparison with the value obtained for the non-prepared sludge.

4 CONCLUSION

The analysis of the results of the present study reveals that conditioning of sewage sludge using ultrasound wave propagation, Fenton's reaction and chemical substance in the form of gypsum improve the effectiveness of sludge dewatering during the process of pressure filtration. Changes in final hydration of the sludge prepared with the above conditioning factors lead to the following conclusions:

1. Application of ultrasound field alone for conditioning of sewage sludge causes a reduction in final hydration. The value of the parameter discussed decreased in proportion to exposure time both for the amplitude of 30.5 and 61 μm. The most efficient method was preparation with ultrasound field with amplitude of 61 μm and time of 6 minutes. The obtained final hydration was 80.4%, which was a value 3.1% lower in comparison with the value obtained for the non-conditioned sludge.
2. The use of Fenton's reaction considerably improves susceptibility of sewage sludge to dewatering. Higher reduction in final hydration was found in this case when compared to

application of UD field. The obtained final hydration was 77.9%, which was the level 5.6% lower in comparsion with the value obtained for the non-prepared sludge.
3. Conditioning of the sewage sludge samples analyzed in the study by means of a chemical substance of gypsum caused greater reduction in final hydration after the process of hydration compared to preparation with ultrasound field or Fenton's reaction, regardless of the dose used (2 or 4 g). Dosing gypsum with 4 g turned out to be more beneficial: the final hydration was 74.9%, which was by 8.6% lower in comparison with raw sludge.
4. The combined methods which used two or three of the preparation agents in different combinations were found to be more effective in terms of improvement in susceptibility of sludge to dewatering compared to the sludge conditioned with these agents separately.
5. The combined methods that used the effect of ultrasound field, both for the amplitude of 30.5 and 61 μm, Fenton's reaction and gypsum (4 g) improved the effectiveness of dewatering of the sewage sludge to the highest degree. The most effective conditioning was found when using the exposure to ultrasound field with the amplitude of 61 μm and the time of 6 minutes, Fenton's reaction and gypsum (4 g). The obtained final hydration was 70.2%, which was the level by 13.3% lower compared to the value obtained for the non-prepared sludge.

REFERENCES

Barbusiński K., Filipek K., 2000, Aerobic Sludge Digestion in the Presence of Chemical Oxidizing Agents Part II. Fenton's Reagent, *Polish Journal of Environmental Studies* 9, 3, 145:149.

Bień J., 1989, Ultrasound in preparation for mechanical sludge dewatering, nr 2, GWiTS.

Bień J. 2003, Ultrasound in the economy of the sludge, Conference Materials "New insights into sludge", Częstochowa.

Bień J., 2007, Sewage sludge, Theory and Practice, Second Edition revised and supplemented, published by Technical University of Czestochowa, Czestochowa.

Bień J., Kowalczyk M., Kamizela T., 2008, Effectiveness of sludge dewatering conditioned chemicals in combined methods with ultrasound, Publisher Częstochowa University of Technology, *Engineering and Environment*, 11, 1, 65–72.

Bień J., Matysiak B., Wystalska K., 1999, Stabilization and dewatering of sewage sludge, Publisher Czestochowa University of Technology, Czestochowa.

Bień J., Neczaj E., Worwąg M., Grosser A., Nowak D., Milczarek M., Janik J., 2010, Guidelines for management of sludge in Poland after 2013, *Engineering and Environmental Protection*, 14, (4) 375–384.

Bień J., Wystalska K., 2008, Problems of sludge management, Conference Materials. *Engineering and Environmental Protection*, Częstochowa, 11 (1) 5–11.

Biń A., 1998, Application of advanced oxidation processes for water treatment, *Environmental Protection*, 1, 68.

Boruszko D., 2001, Processing and utilization of sewage sludge—laboratory exercises, Bialystok University of Technology, Department of publishing and printing, Bialystok.

Bourgier C., Albasi C., Delgenes J.P., Carrere H., 2006, Effect of ultrasonics, thermal and ozone pretreatments on waste activated sludge solubilisation and anaerobic biodegradability, *Chem Eng Process*, 45, 711–718.

Chang C.Y., Hsieh Y.H., Eyao K.S., Wei M.C., Hsieh L.L, 2005, The effect of pH on hydroksyl free radical reaction rate in the Fenton process. 4th International Conference on Oxidation Technologies for Water and Wastewater Treatment, 343–348.

Hrynkiewicz Z., 1998, Sewage sludge in practice, Sludge dewatering filter presses and belt, Czestochowa-Ustroń.

Ioan I., Wilson S., Lundanes E., Neculai A., 2007, Comparison of Fenton sono-Fenton bisphenol A degradation, *Journal of Hazardous Materials* 142, 559–563.

Kowalczyk M., Kamizela T., Parkitna K., Milczarek M., 2011, Application the Fenton's reaction in the technology of sewage sludge, Scientific Papers of the University of Zielona Gora, Environmental Engineering—21, Zielona Gora, 141, 98–112.

Kowalczyk M., Parkitna K., Kamizela T., 2011, Influence of sewage sludge conditioning field ultrasonic-assisted chemical substances the efficiency dewatering process, *Engineering and Environment,* 14 (1) 87–94.

Krzemieniewski M., Dębowski M., Sikora J., 2005, Possibility of using the Fenton reaction in the process of conditioning and stabilization of sludge from the centers of intensive fish farming, University of Warmia and Mazury, Olsztyn.

Krzemieniewski M., Janczukowicz J., Pesta J., Dębowski M., 2003, Influence of Fenton reaction, and the phenomena occurring during the flow of electric current in the conditioning of sewage sludge, Scientific Papers of the Koszalin University of Technology, *Environmental Engineering*, 21, 195–214.

Law of. 27/04/2001 Waste (Journal of Laws 62, pos. 628).

Ledakowicz S., Olejnik D., Perkowski J., Segota H., 2001, The use of advanced oxidation processes for degradation of nonionic surfactant Triton X-114, *Chemical Industry*, 80 (10) 453–459.

Lu M.C., Lin C.J., Liao C.H., Huang R.Y., and Ting W.P., 2010, Deweatering of activatedsludge by Fenton's reagent, Chia Nan University of Pharmacy and science, Taiwan 717, ROC.

Lu M.C., Lin C.J., Liao C.H., Ting W.P., Huang R.Y., 2001, Influence of pH on the dewatering of activated sludge by Fenton's reagent. *Wat. Sci. Technol.* 44, 327–332.

Lu M.C., Lin C.J., Liao C.H., Ting W.P., and Huang R.Y., 2010, Influence of pH on the dewatering of activated sludge by Fenton's reagent, Chia Nan University of Pharmacy and science, Taiwan 717, ROC; *National Waste Management Plan* 2014.

Pham T.T.H., Bra S.K., Tyagi R.D., Surampalli R.Y., 2010, Influence of ultrasonication and Fenton oxidation pretreatment on rheological characteristics of wastewater sludge. *Ultrasonics Sonochemistry*, 17, 38–45.

Ratanatamskul C., Chintitanun S., Lu M-C., 2005, Catalytic degradation of aniline by iron oxide in the presence of hydrogen peroxide, 4th International Conference on Oxidation *Technologies for Water and Wastewater Treatment*, 374–377.

Sun J.H., Sun S.P., Sun J.Y., Sun R.X., Qiao L.P., Guo H.Q., Fan M.H., 2007, Degradation of azo dye Acid black 1 using low concentration iron of Fenton process facilitated by ultrasonic irradiation. Ultrasonics Sonochemistry 14, 761–766.

Świderska R., Czerwińska M., Kutz R., 2008, Oxidation of organic pollutants using Fenton's reagent, Department of Water and Wastewater Technology, Koszalin University of Technology, VII National Conference.

Szwabowska E., 1986, Design of sludge dewatering Processes, Gliwice.

Tiehm A., Nickel K., Zellhorn M., Neis U., 2001, Ultrasonics waste activated sludge dsintegration for Improving anaerobic stbilization, *Water Res.* 35, 2003–2009.

Torres R.A., Abdelmalek F., Combet E., P´etrirer C., Pulgarin C., 2007, A comparative study of ultrasonic cavitation and Fenton's reagent for bisphenol A degradation in deionised and natural Walter,. *Journal of Hazardous Materials* 146, 546–551.

Wąsowski J., Piotrkowska A., 2002, Distribution of organic water pollutants advanced oxidation processes, *Environmental Protection*, vol. 2, No. 85, 27–32.

Zhou T., Lim T.T., Lu X., Li Y., Wong F.S., 2009, Simultaneous degradation of 4CP and EDTA in a heterogeneous Ultrasound/Fenton like system at ambient circumstance, Separation and Purification Technology, 68, 367–374.

Environmental Engineering IV – Pawłowski, Dudzińska & Pawłowski (eds)
© 2013 Taylor & Francis Group, London, ISBN 978-0-415-64338-2

The use of combined methods in sewage sludge conditioning before centrifuging process

M. Kowalczyk, T. Kamizela & K. Parkitna
Czestochowa University of Technology, Institute of Environmental Engineering, Czestochowa, Poland

ABSTRACT: The main aim of the applied methods for sewage sludge processing is reduction in volume of sewage sludge which is obtained by removal of water. In order to improve the process of concentration and dewatering, sewage sludge is subjected to preprocessing referred to as conditioning. Too high compressibility of sewage sludge impairs the process of dewatering. Therefore conditioning of sewage sludge prior to dewatering is justified not only to increase the particular resistance but also to reduce compressibility of sewage sludge. The presented work provides the analysis of the efficiency of dewatering performed by: centrifugation of selected sewage sludge subjected to conditioning with selected chemical substances (i.e. ashes, gypsum), and combination of ultrasonic field with selected chemical substances and advances oxidation processes. The overall goal of this work was to test the hypothesis that the energy of ultrasounds and the effect of chemical substances can change the properties of sewage sludge by reducing the size and compaction of particles and at the same time increase the quantity of free water present in sewage sludge which results in improved efficiency of sewage sludge dewatering.

Keywords: mechanical dewatering, disintegration, sonication, Fenton's reaction

1 INTRODUCTION

Conditioning of sewage sludge can be carried out by means of ultrasound field (Bien et al. 2001). The coagulation effect of ultrasound waves depends primarily on the size of particles in the sonicated suspension and sound frequency while a particular range of frequencies exists for each size of the suspended particles at which colliding particles are coagulated. Active effect of ultrasound field causes an increase in ability of better spatial packing in the sonicated sludge, particularly in the sludge with a more rigid structure of molecules. The ultrasound field significantly affects the improvement in conditions of fermentation process (Zawieja et al. 2009, Bien et al. 2004) and dewatering of sludge in the process of filtration (Wolny et al. 2006). Using ultrasound field in the process of conditioning was aimed at intensification of the dewatering effect and limitation in consumption of chemical reactants (Wolny et al. 2006, Kowalczyk et al. 2011). The processes of sonication of sewage sludge can be supported with advanced oxidation. Fenton's reaction is one of many methods of advanced oxidation. This method is used mainly due to the easier access to the reactants and the simplicity of the process. Fenton's reaction is catalysed by hydrogen peroxide and iron (II) ions. The reaction leads to catalytic decomposition of hydrogen peroxide in the presence of Fe^{2+} ions, followed by generation

of reactive hydroxyl radicals OH* which are capable of reacting with virtually all contaminants of both organic and inorganic origin (Kowalczyk et al. 2011, Parkitna et al. 2011). Proper Fenton's reaction necessitates acidic reaction in the environment. The optimum level of pH ranges from 3 to 6. In the most of the cases, the reaction occurs especially at the value of pH = 3. Reaction time depends on other parameters of the process and on the type of the eliminated substrate.

Fenton's reaction rate improves with higher temperatures of the process. It is generally recommended that Fenton's reagent should be used within the range of temperatures of from 20 to 30°C. Fenton's reaction, which is one of the methods of advanced oxidation, considerably modifies the sewage sludge parameters which determine its environmental nuisance. Among other applications, it is used for decomposition of organic matter, degradation of toxic substances, colour removal, limitation of odour nuisance (Kowalczyk et al. 2011, Parkitna et al. 2011, Zaleska et al. 2008). Fenton's reaction can be used for preparation of raw sludge immediately from settling tanks. Conditioning can be also carried out by means of adding structure-forming substances to the sludge e.g. ashes, carbon or diatomaceous earth preparations. Addition of these substances forming a specific scaffold in the sludge significantly improves susceptibility of sludge to dewatering. Previous studies conducted in Poland and abroad have demonstrated that the use of polyelectrolytes, ashes or gypsum in sludge conditioning improves filtration properties, thus causing changes in sludge structure (Lee & Liu 2001, Wu et al. 1997, Zhao 2003).

A significant improvement in sludge dewatering was obtained when using a carbon suspension. It is used due to a higher density of carbon particles, which ensures higher resistance to shear strain and compression during mechanical dewatering in filtration presses. Other authors (Sander et al. 1989; Broeckel et al. 1996) have used carbon dust in conjunction with polyelectrolytes in order to improve the efficiency of the process of sewage sludge dewatering. Further, Smollen & Kafaar (1997) used powder carbon for the methods combined with polyelectrolytes in order to improve dewatering of sewage sludge. They found that adding powder carbon to lose sludge flocs that were formed during flocculation results in creation of porous, permeable and rigid structure of the net. Volatile ashes, cement dust and other materials have also been used as 'skeleton-forming' materials (Nelson & Brattlof 1979, Benitez et al. 1994, Lin et al. 2001).

These substances are claimed to form a permeable and rigid structure of a net, which is likely to remain porous during mechanical dewatering in filtration process. It intensifies the effect of sludge dewatering. In order to improve the effect of dewatering, Zhao (2002) used 'skeleton-forming' gypsum connected with polyelectrolytes. Addition of brown coal to sewage sludge (Durie 1991) causes an increase in caloric value of sewage sludge, which, after dewatering, can be combusted without supplying additional fuel. Sludge susceptibility to dewatering as a result of conditioning and the amount of conditioning agents to be used are determined in an experimental manner. Sludge dewatering is assessed after adding scaffold-building substances, using a test for specific filtration resistance. Appropriate effects of dewatering are typically observed when specific filtration resistance does not exceed the level of from 2 to $3 \cdot 10^{12}$ m/kg. Specific filtration resistance decreases with the higher doses of addition substance.

Sometimes sludge conditioning with these methods do not produce the expected effects or necessitates high amounts of substance. Combined methods of previous addition of conditioning chemicals are recommended in such cases (Zielewicz—Madej 2001).

2 MATERIAL AND METHODS

The study was performed on the sludge from the Warta Sewage Treatment Plant in Częstochowa.

The analysis focused on initial sludge mixed with excess thickened sewage sludge with dry matter content of 30 g/dm³, organic matter—77%, mineral matter—23%, capillary suction time—330 s, pH = 6.8. The sludge was conditioned with ultrasound field. Varied ultrasound disintegration times and varied disintegration wavelength were used. An ultrasonic disintegrator with power of 750 W and wavelength of 61 μm was used for conditioning, which produced the energy with density of 47.4 kJ/l (for the highest exposure time) and wavelength of 30.5 μm, which generated the energy with density of 10.1 kJ/l. Exposure times were 2, 4 and 6 minutes. Other chemical substances used for sludge preparation were gypsum and ashes and one of the methods of advanced oxidation i.e. Fenton's reaction.

The process of advanced oxidation using Fenton's reaction was carried out according to the following protocol:

• Sewage sludge was acidified through addition of a 50% solution of sulphuric (IV) acid (VI) (H_2SO_4) until it reached pH = 3.0.
• 1 g of iron (II) ions (Fe^{2+}) were dosed; the source of iron ions was iron (II) sulphate (VI); 1 g of iron ions = 5 g of iron (II) sulphate (VI).
• 4 g of hydrogen peroxide H_2O_2 in the form of Perhydrol was also added.
• Then, sewage sludge was mixed in a flocculator for 1 hour with the rate of 100 rpm.
• Next, the samples were neutralized to pH 7.0 using a 50% solution of NaOH.

In the case of cement and ashes a dose of 2 g was used. A varied order of preparation and different combinations of the methods were used during the experiment. The sludge where then centrifuged (5,000 rpm for 5 minutes) and dried at the temperature of 105°C. The obtained results and their comparison is aimed at finding the most suitable method of sewage sludge conditioning, which allows for reaching the highest degree of reduction in final hydration.

3 RESULTS

The aim of the study was to investigate whether using ultrasonic energy, advanced oxidation and chemical substances allow for changing the structure of sewage sludge, particle packing and whether it increases the amount of free water in sewage sludge.

As was initially assumed, the use of initial sonication as a physical method of modification of sewage sludge conditioned by one of the methods of advanced oxidation (Fenton's reaction) and chemical substances was a factor that intensified the process of sludge flocculation. The increase in efficiency of mechanical dewatering was obtained by means of using combined methods of sewage sludge preparation, Fenton's reaction and chemical substance dosing. Sewage sludge dewatering conditioned in ultrasound field in the process of centrifuging caused insignificant decrease in final dewatering compared to sewage sludge.

The differences that amounted to 1.5% of final hydration between the raw sludge and sonicated sludge were obtained for the highest parameters of ultrasound field (amplitude of 61 μm and time of 6 minutes) (Fig. 1). The use of only Fenton's reaction for sewage sludge conditioning caused a reduction in final hydration of 85.1—%. The increase in reduction of final hydration of sludge was obtained for using Fenton's reaction and ultrasound field with the following order of preparation: Fenton's

reaction, ultrasound disintegration for 6 minutes with amplitude of 61 μm. Final hydration of the sludge prepared in this manner was by 3.1—% lower compared to the level obtained for the raw sludge (Fig. 1).

The use of ultrasound field combined with the components (gypsum and ashes) for sewage sludge conditioning resulted in lower values of the parameter discussed in this study. The best effects were obtained in the case of sonication for 6 minutes and addition of 2 g gypsum to the sludge. Final hydration in this case amounted to 84.7% and was lower than the level obtained for the raw sludge (Fig. 2).

Conditioning of sewage sludge with Fenton's reaction and 2 g of gypsum allowed for obtaining final dewatering at the level of 82.1%, which was the

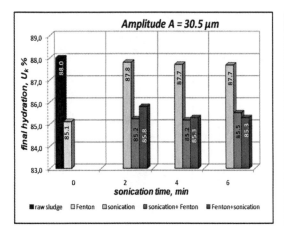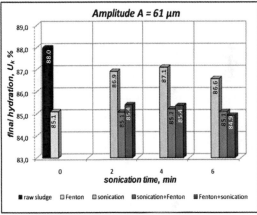

Figure 1.   Changes in final hydration of sewage sludge conditioned with ultrasound field with wavelength of 30.5 and 61 μm and Fenton's reaction obtained after centrifuging process.

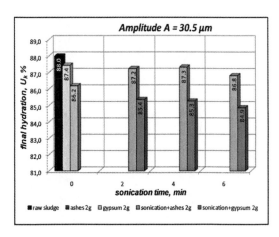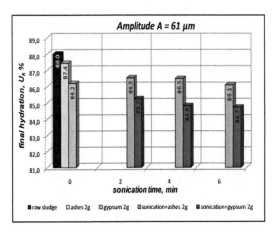

Figure 2.   Changes in final hydration of sewage sludge conditioned with ultrasound field with amplitude of 30.5 and 61 μm and 2 g of gypsum and ashes

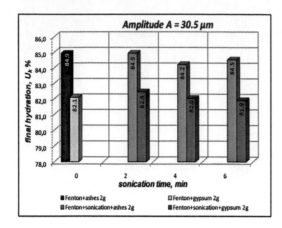
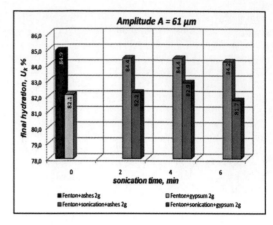

Figure 3. Changes in final hydration of sewage sludge conditioned with Fenton's reaction, ultrasound field with amplitude of 30.5 and 61 μm and gypsum and ashes in the process of centrifuging.

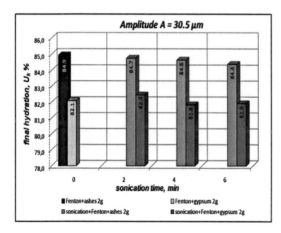
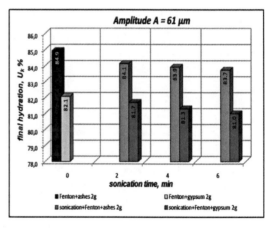

Figure 4. Changes in final hydration of sewage sludge conditioned with ultrasound field with amplitude of 30.5 and 61 μm, Fenton's reaction and gypsum and ashes in the process of centrifuging.

value by 5.9% lower compared to the level obtained for raw sludge (Fig. 3). The best effect of reduction in final hydration was obtained when conditioning sludge with Fenton's reaction with the component in the form of gypsum for any combinations with ultrasound field. The most efficient preparation was found for Fenton's reaction, exposure to ultrasound field for 6 minutes and the amplitude of 61 μm and 2 g of gypsum. The obtained final hydration was 81.7%, which was the level by 6.3—% lower compared to the value obtained for the raw sludge (Fig. 3). In the case of the use of a reversed combination, with the first conditioning factor being ultrasound field with maximum parameters followed by advanced oxidation and addition of gypsum, the lowest final hydration was obtained after the process of centrifuging (81.0%) (Fig. 4).

For each of the combination used for sewage sludge conditioning with ashes the higher final hydration was obtained compared to the use of gypsum.

4 CONCLUSIONS

The results obtained in the study confirmed the assumptions made at initial phase of the study, concerning the relationships between the methods of conditioning and the order of the conditioning factors and the effects obtained during centrifuging of sewage sludge expressed with the values of final hydration. Analysis of the obtained results allowed for formulation of the following conclusions:

• The use of ultrasound field for conditioning of sewage sludge affects the reduction in final hydration after the process of centrifuging. Regardless of the amplitude used (30.5 or 61 μm), the value of the parameter discussed decreased in proportion to the elongation of the time of exposure to ultrasound field.

- Using Fenton's reaction as the only procedure in sludge conditioning yields changes in final hydration after the process of centrifuging compared to raw sludge by 2.9%.
- The lowest final hydration in the sludge samples conditioned with ultrasound field combined with chemical substances in the form of ashes and gypsum was obtained as a results of application of ultrasound field with amplitude of 61 μm for 6 minutes and 2 g of gypsum. The final hydration was by 3.3% lower compared to raw sludge and it was 84.7%.
- The highest reduction in final hydration after the process of centrifuging was obtained after sewage sludge conditioning by means of combined methods that used ultrasound field with the amplitude of 61 μm for 6 minutes followed by Fenton's reaction and addition of 2 g of gypsum.

ACKNOWLEDGEMENTS

This work was supported by the Faculty of Environmental Protection and Engineering (Czestochowa University of Technology) BS/PB-401-301/11.

REFERENCES

Benitez J., Rodriguez A., Suarez A. (1994): Optimisation technique for sewage sludge conditioning with polymer and skeleton builders. *Water Res.*, 28 (10), 2067–2073.

Bień J., Szparkowska I. (2004); Wpływ dezintegracji ultradźwiękowej osadów ściekowych na przebieg procesu stabilizacji beztlenowej; *Inżynieria i Ochrona Środowiska*; 7(3–4); str. 341–352.

Bień J., Wolny L., Wolski P. (2001); Działanie ultradźwięków i polielektrolitów w procesie odwirowania osadów ściekowych; *Inżynieria i Ochrona Środowiska*; vol. 4 No. 1.

Broeckel U., Roemer R., Meyer J., Mittelstrass A., Witt R., Schubert E., Blei P., Fessler J. (1996): Method for filtering and dewatering of sewage sludge, EP, BASF A.-G., Germany, Patent No. 690.

Durie R.A. (1991): The Science of Victorial Brown Coal: Structure, Properties and Consequences for Utilisation. Butterworth-Heinemann Ltd.

Kowalczyk M., Parkitna K., Kamizela T., Milczarek M. (2011), Zastosowanie reakcji Fentona w technologii osadów ściekowych, Zeszyty Naukowe Uniwersytetu Zielonogórskiego nr 141, Inżynieria Środowiska-21, Zielona Góra, str. 98–112.

Kowalczyk M., Parkitna K., Kamizela T. (2011); Efektywność separacji i utleniania zanieczyszczeń w wyniku reakcji Fentona, Gospodarka Odpadami komunalnymi, Komitet Chemii Analitycznej PAN, tom VII, Koszalin, str. 265–276.

Lee C.H., Liu J.C. (2001): Sludge dewaterability and floc structure in dual polymer conditioning. *Adv. Environ. Res.*, 5, 129–136.

Lin Y.-F., Jing S.-R., Lee D.-Y. (2001): Recycling of wood chips and wheat dregs for sludge processing. *Biores. Technol.*, 76 (2), 161–163.

Nelson R.F., Brattlof B.D. (1979): Sludge pressure filtration with fly ash addition. J. Wat. Pol. Cont. Feder., 51 (5), 1024–1031.

Parkitna K., Kowalczyk M., Kamizela T. (2011), Stopień redukcji ładunku organicznego realizowany za pomocą reakcji Fentona, Materiały Konferencyjne Młodych Naukowców—Młodzi naukowcy dla polskiej nauki, Część 2—Nauki przyrodnicze, tom III, Kraków, str. 115–124.

Sander B., Lauer H., Neuwirth M. (1989): Process for producing combustible sewage sludge filter cakes in filter presses, US, BASF Aktiengesellschaft, Fed. Rep. Germany, Patent No. 4,840,736.

Smollen M., Kafaar A. (1997): Investigation into alternative sludge conditioning prior to dewatering. Water Science and Technology. Proceedings of the 1997 International Specialized Conference on Sludge Management 38 (11), 115–119.

Wu Ch.Ch., Huang Ch., Lee D.J. (1997): Effects of polymer dosage on alum sludge dewatering characteristics and physical properties. Coll. Surf. A. Physicochemical and Engineering Aspects, 122, 89–96.

Wolny L., Wolski P., Bień B., Zawieja I. (2006); Zmiany parametrów reologicznych w wybranych procesach kondycjonowania osadów ściekowych przed ich odwadnianiem; *Inżyieria i Ochrona Środowiska*; 9(1); str. 79–87.

Zaleska A., Grabowska E. (2008); Podstawy technologii chemicznej, Nowoczesne procesy utleniania—ozonowanie, utlenianie fotokatalityczne, reakcja Fentona. Politechnika Gdańska, Gdańsk.

Zawieja I., Bień J., Worwąg M. (2009); Przebieg procesu hydrolizy w stabilizacji beztlenowej osadów ściekowych. *Gaz, Woda i Technika Sanitarna*, nr 07–08.

Zhao Y.Q. (2003): Correlations between floc physical properties and optimum polymer dosage in alum sludge conditioning and dewatering. *Chem. Eng. J.*, 92, 227–235.

Zhao Y.Q. (2002): Enhancement of alum sludge dewatering capacity by using gypsum as skeleton builder. *Coll. Surf. A. Physicochemical and Engineering Aspects*, 211, 205–212.

Zielewicz-Madej E. (2001): Zastosowanie dezintegracji ultradźwiękowej do intensyfikacji produkcji lotnych kwasów tłuszczowych z osadu wtórnego. *Inż. i Ochr. Środ.*, 4, 2, 231–237.

Environmental Engineering IV – Pawłowski, Dudzińska & Pawłowski (eds)
© 2013 Taylor & Francis Group, London, ISBN 978-0-415-64338-2

Verification process for sewage sludge treatment

G. Borowski

Department of Fundamentals of Technology, Lublin University of Technology, Lublin, Poland

ABSTRACT: For disposal of waste sewage sludge they were agglomerated in a hydraulic press to provide cylindrical-shaped briquettes. We produced the samples of waste in the amount of about 42% by weight mixed with crushed glass (about 50% by weight) and cement (about 8% by weight). The briquettes were heated in a laboratory chamber furnace (at 1050°C, 1100°C and 1150°C) and cooled in water to get vitrified body. The results of mechanical tests (compressive strength—3.9 MPa, and resistance to gravitational drop—94,1%) as well the results of the performed tests of freeze resistance (1,2%), absorbability (14,4%) and leaching metals in aqueous extract were determined. Strength was sufficient in terms of assembly the requirements for materials to be used as a substitute aggregate for building industry (it was more than 2.5 MPa). There was no exceed the limits of hazardous substances contained in waste as well leaching of heavy metals in aqueous extract. So, we obtained environmentally safe products which can be used as a building material.

Keywords: sewage sludge, heavy metals, briquetting, vitrification

1 INTRODUCTION

Vitrification is an effective method of disposing of hazardous waste containing heavy metal compounds. Advanced vitrification waste may be toxic and may have very complex chemical composition, as well as unfavorable physical properties. It is usually difficult to be utilized by products of industrial processes such as radioactive waste, fly ash, slag, medical waste, asbestos (Bingham & Hand 2006, Connelly et al. 2011, Jackson et al. 2003, Pelino 2000). Sanitation of polluted soils is a big problem. Due to high contents of silicon, aluminum and other elements that the ground consists of, it is a good material for vitrification (Cocic et al. 2010). More and more frequently sewage sludge waste is subjected to vitrification (Bernardo & Dal Maschio 2011, Ingunza et al. 2011).

Vitrification process involves melting the material in high temperature heating furnace, and then quick cooling in cold water to convert the resulting liquid phase into a glassy structure. The energy expenditure depends on the type of material and furnace used. The resulting homogeneous glassy substance is characterized by high mechanical strength, low chemical reactivity and non-toxicity. Using waste vitrification of certain substances having the characteristic traits of the products enable them to be reused mainly in construction (Lin & Weng 2001, Żygadło & Latosińska 2008).

The process involves the formation of an impermeable and durable layer of glass.

Hazardous substances are disposed of in the following processes:

1. liquefaction of particles and incorporation into a crystalline structure of glass,
2. encapsulation of infusibility compounds.

Liquefaction can permanently bind such elements as phosphorus, boron and silicon. During heating to the liquid phase and cooling, the elements become an integral part of the crystal lattice of the glass. They encapsulate the waste components involved, such as cobalt, lead, sodium, magnesium, lithium and cesium. These ingredients penetrate into the crystal lattice of glass acting as intrusive and infusible accessories. Encapsulation can occur both during the heating of shredded waste sludge with glass, as well as during cooling (Xu et al. 2010).

The organic compounds found in the waste are disposed during heating. It requires special heating crucibles in the heating system where the material is melted by an inductive current flow at high intensity. The dried deposit in these crucibles is to be provided and it is heated to the temperature of 1300–1450°C (Zou et al. 2011). Partial liquefaction of waste occurs there, and then it continues to soak until the component migration process is completed. In this installation crushed waste glass must be added as a flux and it will facilitate bringing the mixture to the form of a semi-fluid, or, often, to the form of a paste. Liquid phase of the waste goes to the vats in which the cooling of the vitrified blocks takes place (Huang et al. 2007). To modify the process of vitrification, granulation of waste materials is performed in a compact body of uniform shape and size. Granules are heated in rotary kilns or grate furnaces. After the phase of surface transitions the bodies are cooled in water, which makes them produce glossy and tight coat. Thus, the original shape of nuggets is retained, in contrast to melting of blocks in the vats. Heating granules containing waste is usually carried out in a rotary kiln at the temperature range of 1100–1200°C, resulting in lower energy expenditure compared to the conventional process. This requires the preparation of mixtures for granulation and agglomeration process parameter selection (Żygadło & Latosińska 2008).

The paper presents the agglomeration of waste containing sludge by briquetting. The resulting bodies were directed to a ceramic kiln chamber and the vitrification of their surfaces followed. After cooling it was studied in terms of resistance to mechanical damage and chemical reactivity of the deposited environment. The most important effects of vitrification sludge process was waste volume reduction and the possibility of using it as a useful raw material in construction.

2 MATERIAL AND METHODS

The research material was sludge from the municipal wastewater treatment plant "Hajdów" in Lublin. These deposits contain large amounts of periodical heavy metal compounds and pathogenic organisms, hindering their agricultural use. Properties of the sediment are shown in Table 1.

The homogenized mixture contains dried sewage sludge in the amount of about 42% by weight, crushed waste glass (about 50% by weight) and

Table 1. Properties of sewage sludge from "Hajdów" plant.

Parameter	Unit	Content
Water content	%	82–86
Flammable substances	% d.m.	64–72
Heavy metals:		
Lead	mg/kg d.m.	200–500
Cadmium	mg/kg d.m.	10–20
Zinc	mg/kg d.m.	3000–4000
Copper	mg/kg d.m.	200–500
Nickel	mg/kg d.m.	200–300
Chromium	mg/kg d.m.	300–500
Calorific value	kcal/kg	3600–4000
Melting point	°C	1300–1320
Graining	mm	0,1–2,0

cement (about 8% by weight). Glass waste from used lamps and glass lighting businesses were added in the form of dust by up to 0.2 mm fraction. Their use is preferred due to lower thermal reaction temperature.

The homogenization process involved mixing the ingredients in a mixer paddle and adding portions of water until the humidity content reached 5–8%. After homogenization the mixture was compacted by the stamp hydraulic press to be given cylindrical shape, and its volume was about 14 cm³. These compacts were sent to a laboratory chamber furnace with a capacity of 1400 W. Heating was performed at 1050°C, 1100°C and 1150°C. Warm-up time was 30, 60 and 90 minutes. After the warm-up the products were cooled in a bathtub filled with water, and then they were tested. The studies identified:

- mechanical toughness,
- frost resistance and water absorption,
- leaching of hazardous substances.

Mechanical toughness of sinters acknowledged was based on the measurement of compact breaking load, as well as resistance to gravity drop. The value of the load force causing damage to the product was determined experimentally by placing it horizontally between flat surfaces of hydraulic testing machine and squeezing until the destruction of the structure. The device resistance to gravity discharge was assessed by the percentage of weight loss after three batches of samples dropped from the height of 2.0 m onto a 20 mm thick steel plate.

Frost was studied with the use of an indirect method specified by the Polish standard PN-88/B-06250. The study involved cyclic freezing and defrosting in the air and in the water. One cycle lasted 6 hours, and the sample was subjected to three such cycles. Then it was tested for compressive strength of sintering. The degree of frost resistance was based on the ratio of strength to weight loss.

Absorbability study, i.e., that of the ability to absorb water through the material at atmospheric pressure involved gradual immersion of the samples in water. Absorbability was defined as the ratio of mass of water absorbed to the mass of dry sample material.

Leaching of hazardous substances, such as chromium, cadmium, copper, lead, nickel, zinc and mercury was studied by plasma emission with the use of spectrometry method.

3 RESULTS AND DISCUSSION

First of all, the influence of temperature and heating time on the state of the surface vitrification of briquettes were investigated. Figure 1 shows the heating curves at selected temperatures. It was found that heating at 1050°C for 30 min does not ensure the creation of a consolidated glassy phase on the surface of the briquette. This surface was porous and cracked (Fig. 2a). Increasing the temperature while maintaining the warm-up time did not solve the problem of creating a homogeneous glassy phase due to the rapid release of gases from the interior of the briquette, so the inside pores are visible (Fig. 2b).

Extending the heating time to 90 min and using the temperature of 1150°C resulted in almost complete dissolution of a substantial change in the original shape of the briquette. Homogeneous glassy surface layer of the sample was obtained at 1100°C and in the heating time of 60 min. The visual assessment of the briquette surface was uniformly vitrified with little pores filled with glass (Fig. 3a) as well as inside the body (Fig. 3b). These samples were sent for further studies to determine the basic physical and mechanical properties. The results of these tests are presented in Table 2. Analyzing the results presented in the above Table, it can be stated that the vitrified products have high strength values of the load and resistance to discharge by gravity of the vitrification. The products meet the requirements for materials intended for

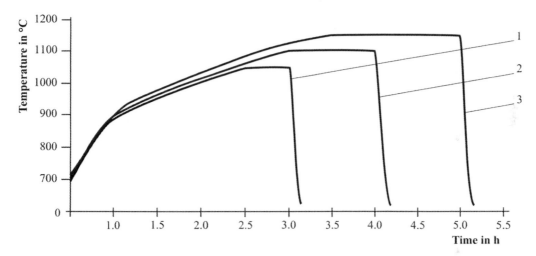

Figure 1.   Heating curves, depending on the temperature and time: 1—not glassy surface, 2—homogeneous glassy surface, 3—molten surface of the briquette.

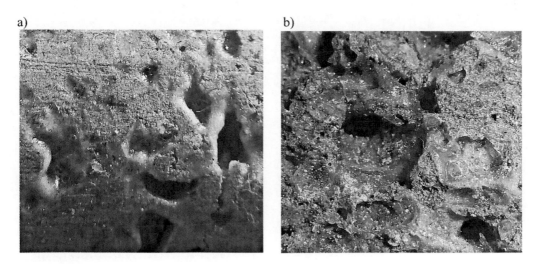

Figure 2.   Photo of abnormal product: a—surface, b—sectional view.

187

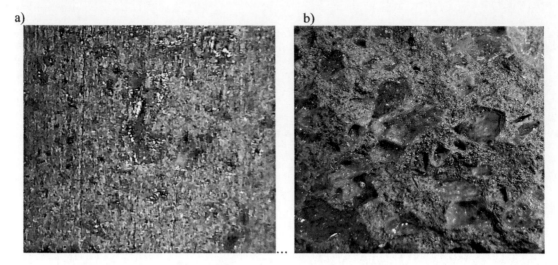

a)   b)

Figure 3.   Photo of the correct body: a—surface, b—sectional view.

Table 2.   Properties of vitrified products.

No.	Parameter	Unit	Measured values	The limit values for road construction
1	Compressive strength	MPa	3,9	>2,5
2	Resistance to gravitational drop	%	94,1	>90,0
3	Bulk density	kg/m^3	440,0	400,0–550,0
4	Freeze resistance	%	1,2	<2,0
5	Absorbability	%	14,4	<37,0
6	Leaching metals in aqueous extracts:			
	Cadmium	mg/dm^3	0,005	<0,1
	Mercury	mg/dm^3	0,050	<0,2
	Chromium	mg/dm^3	0,021	<0,5
	Lead	mg/dm^3	0,051	<0,5
	Copper	mg/dm^3	0,162	<2,0
	Zinc	mg/dm^3	0,320	<2,0
	Ferrum	mg/dm^3	0,350	<10,0

paved roads fabrication, where the load resistance is at least 2.5 MPa. Good mechanical properties of products correspond to the resistance to changing weather conditions. The results of frost resistance and water absorption measurements showed lower values than the limit.

Using vitrification of granulated sewage sludge, other researchers received LECA products applied in construction as aggregates (Żygadło & Latosińska 2008). In measuring, the leaching of heavy metal ions in aqueous extracts was also lower than the limit values. It means that the vitrified product is safe for the environment. It was confirmed that heavy metal compounds are built permanently into the structure of crystalline silicate and are not washed out. The presented results showed that the use of vitrification significantly reduced heavy metals leaching from waste. Similar results were obtained in the study of lightweight aggregates by vitrification of mining residues, heavy metal sludge, and incinerator fly ash (Huang et al. 2007). The only drawback was the necessity of incurring significant energy expenditure per unit mass of product. Vitrification waste in the body form is a process of less intense energy due to the use of approximately 200°C lower heating temperature compared to the vitrification of waste in powder form.

4 CONCLUSION

On the basis of the results of the vitrified briquettes with sewage sludge test the following conclusions were formulated:

1. No risk of dangerous substances migrating away from briquettes as well as heavy metals leaching under external climatic factors was noticed.
2. Products containing sewage sludge after vitrification process are environmentally safe.
3. The products are suitable for widespread use as replacement for aggregate based on course construction of paved roads, embankments and banks.

REFERENCES

Bernardo E. & Dal Maschio R. 2011. Glass-ceramics from vitrified sewage sludge pyrolysis residues and recycled glasses. *Waste Management* 31(11): 2245–2252.

Bingham P.A. & Hand R.J. 2006. Vitrification of toxic wastes: A brief review. *Advances in Applied Ceramics* 105(1): 21–31.

Cocic M., Logar M., Matovic B. & Poharc-Logar V. 2010. Glass-ceramics obtained by the crystallization of basalt. *Science of Sintering* 42(3): 383–388.

Connelly A.J., Hand R.J., Bingham P.A. & Hyatt N.C. 2011. Mechanical properties of nuclear waste glasses. *Journal of Nuclear Materials* 408(2): 188–193.

Huang S.-C., Chang F.-C., Lo S.-L., Lee M.-Y., Wang C.-F. & Lin J.-D. 2007. Production of lightweight aggregates from mining residues, heavy metal sludge, and incinerator fly ash. *Journal of Hazardous Materials* 144(1–2): 52–58.

Ingunza M.P.D., Duarte A.C.L. & Nascimento R.M. 2011. Use of sewage sludge as raw material in the manufacture of soft-mud bricks. *Journal of Materials in Civil Engineering* 23(6): 852–856.

Jackson M.J., Mills B. & Hitchiner M.P. 2003. Controlled wear of vitrified abrasive materials for precision grinding applications. *Sadhana* 28(5): 897–914.

Lin D.F. & Weng C.H. 2001. Use of sewage sludge ash as brick material. *Journal of Environmental Engineering* 127(10): 922–927.

Pelino M. 2000. Recycling of zinc-hydrometallurgy wastes in glass and glass ceramic materials. *Waste Management* 20: 561–568.

Xu G.R., Zou J.L. & Li G.B. 2010. Stabilization of heavy metals in sludge ceramsite. *Water Research* 44(9): 2930–2938.

Zou J.L., Dai Y., Yu X.J. & Xu G.R. 2011. Structures and metal leachability of sintered sludge-clay ceramsite affected by raw material basicity. *Journal of Environmental Engineering* 137(5): 398–405.

Żygadło M. & Latosińska J. 2008. Addition of sewage sludge—to raw material used in LECA production. In: Pawłowska M., Pawłowski L. (Eds.) Management of pollutant emission from landfills and sludge: 231–238, London: Taylor & Francis Group.

Environmental Engineering IV – Pawłowski, Dudzińska & Pawłowski (eds)
© *2013 Taylor & Francis Group, London, ISBN 978-0-415-64338-2*

The influence of sonication parameters on the disintegration effect of activated sludge

T. Kamizela & M. Kowalczyk
Faculty of Environmental Engineering and Biotechnology, Czestochowa University of Technology, Czestochowa, Poland

ABSTRACT: This paper presents preliminary experiments involving the characterization of organic substrates produced by sonication. In the experiments used an activated sludge from the biological wastewater treatment plants receiving municipal wastewater. In the experiments, the changes in concentration of carbon and nitrogen forms and volatile fatty acids in the after sonication liquids were monitored. Characteristics of the obtained disintegration product determine technological characteristics of wastewater treatment, which can be modified by installing an ultrasonic apparatus. The study was conducted to determine the optimal dose of ultrasound as a disintegrating agent at range acoustic energy < 200 kJ.

Keywords: activated sludge, disintegration, sonication

1 INTRODUCTION

Ultrasonic wave propagation called sonication plays role as a factor supporting the traditional methods of wastewater treatment and sludge treatment. However, the dominant application area of sonication is the sludge processing in anaerobic stabilization process. Ultrasonic cavitation causes excess sludge cell lysis, the release of organic substances into the liquid phase and the increase of the availability of substrate. An increase in biogas production and reduction of organic matter is obtained by the application of ultrasonic disintegration (Bougrier et al. 2006, Naddeo et al. 2009, Salsabil et al. 2009, Zawieja et al. 2008). Floc destruction and cell disruption by the sonication can be also useful in activated sludge process. Ultrasound application in the external or internal recirculation system of biological reactor creates the possibility for amount reduction of produced sludge. It is also justified to use ultrasound as a disintegration system in order to obtain easily assimilated carbon source for the dephosphatation and denitrification processes (Zhang et al. 2007, He et al. 2011).

In the primary way ultrasound potential can be expressed by the concentration measurement of dissolved organic substances in the solution. Biological sludge disintegration influence on increase in the concentrations of polysaccharides, proteins, nucleic acids in sludge liquids (Wang et al. 2006, Feng et al. 2009). The most common way for determination of strength and effects of sonification is the measurement of Soluble Chemical Oxygen Demand (SCOD) of substrates after ultrasonic disintegration.

The measurements of SCOD are helpful in determining the most effective parameters of sonication. Show et al. (2007) appointed the optimal ultrasonic density of 0.52 W/mL, sonication time about 1 minute at a concentration of sludge solids in the range from 2.3% to 3.2%. Zhang et al. (2008) prefer sonication times in a row 5–15 minutes using an ultrasonic wave intensity of 158–251 W/cm². In the case of experiments conducted by Pham et al. (2009) the most favorable conditions of sonication (obtained SCOD increment of 56%) is a 60 min. exposure time of the secondary sludge with a mass concentration 23 g/L with intensity 0.75 Wcm². El-Hadj et al. (2007) considered the best value of the specific energy of sonication ES = 11000 kJ/kg TS. According to Feng et al. (2009) optimal value is above the level of ES = 26000 kJ/kg TS. Simultaneously increasing the concentration of dissolved carbon, nitrogen and phosphorus compounds. At ultrasound dose E_S = 45000 kJ/kg TS concentrations of these indicators were 1600 mg/L, 900 mg/L and 400 mg/L, respectively (Kim et al. 2010). The objectives of this study were to optimize the dose of ultrasonic energy to improve the disintegration and solubilization of waste activated sludge. The study was focused on the impact of the sonication energy levels on the amount of carbon, nitrogen and phosphorus compounds released to dissolved phase of sludge.

2 MATERIALS AND METHODS

2.1 Waste activated sludge (WAS)

The research was carried out on activated sludge from Wastewater Treatment Plant (WWTP) in Czestochowa (Poland). The WWTP treats municipal wastewater and the maximum wastewater inflow is 2000 cubic meters per day. The collected sludge from oxic zone of activated sludge tank was immediately transferred to the laboratory and tested. The Total Solids (TS) concentration of activated sludge varied between 3,51 g/L and 6,15 g/L and the Volatile Solids (VS) were about 71%.

2.2 Sonication

Sonication was carried out with a low-frequency (20 kHz) ultrasonic processor VC750 (Sonics, USA), with a 19 mm titanium tip. Sludge samples with a volume (V_S) of one liter was sonicated at ambient temperature (20°C). Batch experiments were carried out in a 1,1 L glass beaker without temperature regulation. Cross-sectional area for the sonication vessel was $A' = 78,5$ cm^2. The acoustic energy delivered to the sample was adjusted by varying the input amplitude (A) and sonication time (t_S). The applied ultrasonic probe (horn) allowed the formation of ultrasonic waves with a maximum amplitude of 61.0 μm. The quantity of the amplitude is set as a percentage of maximum amplitude and kept constant by generator. One level of the amplitude A = 61.0 μm (100%) were tested in the study. Sludge sonication times were 240, 480, 720, 960 and 1200s. In order to analyze the results was used the additional operating variable of sonication process. It was specific energy (E_S) calculated by equation 1.

$$E = \frac{E}{V_s * TS'} \frac{kJ}{kg} TS \tag{1}$$

where: E: amount of acoustic energy in Joules (watts * seconds) that is being delivered to the probe (energy monitor of ultrasonic processor), J; V_S: sample volume, L; TS: total solids concentration in treated WAS samples, g/L.

2.3 Analysis of methods

Total Carbon (TC), Total Organic Carbon (TOC) and Total Nitrogen (TN) were measured by multi N/C analyzer (AnalytikJena UK). All measurements were done on soluble fractions of sludge. The soluble fraction was defined as the fraction resulting from the centrifugation and filtration. The samples were centrifuged at 10.000 rcf for 15 min. The supernatant was filtrated through a cellulose nitrate membrane (0.45 μm pore size). Concentration of Total Phosphorus (TP), Volatile Fatty Acids (VFA), Nitrite (N-NO$_2$), Nitrate (N-NO$_3$) and Ammonium Nitrogen (N-NH$_4$), were measured using a spectrophotometer tests and device HACH DR 5000. Total Solids (TS) and Volatile Solids (VS) were performed according to the Standard Methods.

2.4 Regression analysis

The obtained results were provided graphically in the form of central tendency which was arithmetic means (AVG). The statistical dispersion of data around the arithmetic mean was characterized by the Standard Deviation (STDV). In order to determine the nature of the relationship between a dependent variable (concentrations of TC, TOC, TN, TP, N-NH$_4$, N-NO$_3$, N-NO$_2$ and VFA) and explanatory variable (specific energy) regression analysis was used. Two most popular parametric models were used, namely the linear and polynomial regressions. The article does not provide regression equations that performed a task of prediction and forecasting, but only the values of coefficient of determination for linear regression R^2 (L) and polynomial regression R^2 (P).

3 RESULTS AND DISCUSSION

Specific energy (E_S) is an operational parameter linking the characteristics of treated waste activated sludge to the acoustic energy (E) of propagated ultrasonic wave. The results of WAS sonication were presented in graphs as a function of specific energy in intervals of up 10000 kJ/kg TS (Fig. 1).

The cyclic increase in energy sonication caused a clear increase in concentration of carbon in liquid of the sonicated sludge. When the specific

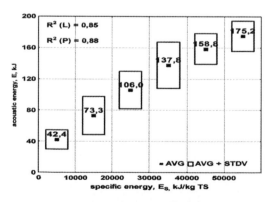

Figure 1. Acoustic energy of the ultrasonic wave propagated in the WAS samples at intervals of specific energy.

energy was greater than $E_S > 50000$ kJ/kg TS the average concentration was TC = 560.4 mg/L (Fig. 2). For these conditions of the ultrasonic treatment the total organic carbon concentration was TOC = 520.6 mg/L (Fig. 3). These values may indicate that disintegration process could increase the biodegradability of treated sludge. Initially, the results suggested the potential application of WAS sonication as a method for supporting the biological processes of denitrification or oxidation of accruing excess sludge.

The difference between the concentration of TC and TOC represents part of Inorganic Carbon (IC). Over the entire applied specific energy $E_S = 0 \div 6000$ kJ/kg TS, the IC concentration was rather stable and may vary between 40 mg/L and 50 mg/L but it is worth to analyze the ratio of TOC/IC. At ultrasound dose ES < 10000 kJ/kg TS the concentration of organic carbon was TOC≈118 mg/L, thus, the ratio was of TOC/IC≈3:1. At specific energy in range $E_S = 40000–50000$ kJ/kg TS this ratio

was 12:1. From a technological point of view, the concentration of organic carbon are the most desirable effect of sonication. For these reasons, ultrasonic disintegration of activated sludge at low energies (less than 10000 kJ/kg TS) may not be recommended. Nitrogen as well as carbon is the main chemical constituent of activated sludge. Therefore, the lysis of microorganisms under the influence of sonication resulted in the release to the solution of significant amounts of nitrogen (nitrogen compounds) (Fig. 4). The intensity of nitrogen emission to the solution (increased concentrations of total nitrogen) was directly dependent on the specific energy. The increase in specific energy from 10000 kJ/kg TS to 60000 kJ/kg TS caused about 3.5-fold increase in concentration of total nitrogen (from 61.2 to 214.8 mg/L). In comparison, the concentration of total carbon in the liquid increased only 3 fold (from 187.3 to 560.4).

An extremely important indicator is the ratio of TC/TN. It is a measure of the quality of the resulting product of sonication as a possible nutrient solution for microorganisms of activated sludge. When acoustic energy was less than 10000 kJ/kg TS the ratio was almost TC/TN = 3. For the acoustic energy more than 500000 kJ/kg TS decreased to a value of 2.6. These values were unfavorable because the resulting product of ultrasonic disintegration of WAS is characterized by deficiency of carbon as a source for cell synthesis and metabolism energy. Therefore, sonication of activated sludge with an energy less of $E_S < 60000$ kJ/kg TS can be considered as a method of enrichment of the solution (dissolved phase) in nitrogen compounds.

Nitrogen compounds dissolved in the liquid were mainly in the form of various macromolecular substances. During the test, the concentration of proteins was not determined but low concentrations of mineral forms of nitrogen were found. The effectiveness of biological method of nitrogen

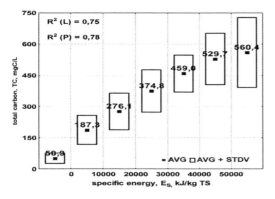

Figure 2. Influence of specific energy on Total Carbon (TC) concentrations.

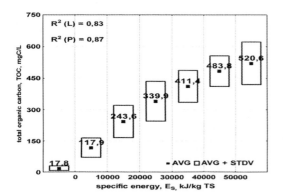

Figure 3. Influence of specific energy on Total Organic Carbon (TOC) concentrations.

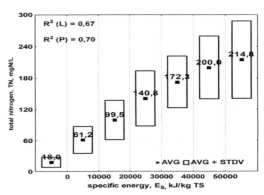

Figure 4. Total nitrogen release profiles to the liquid phase as a result of the WAS disintegration.

removal depends not only on their form but also their concentration. Hence, nitrogen compounds released into solution by sonification of waste activated sludge will require intensive mineralization. In such situation, the impact of the concentration of ammonium nitrogen produced by ammonification remains unclear and can inhibit the nitrification process. The second important aspect arising from the carbon and nitrogen concentrations is interdependence in the *growth* of heterotrophic and nitrifying bacteria, which compete for common substrate (ammonium nitrogen and oxygen). However, these problems will not be explained in the article because they were not objective of the present research and it will require the selection of other analytical methods.

The sonication of activated sludge caused an increase in concentration of ammonia nitrogen with great input of energy (Fig. 5). The concentration of ammonia nitrogen, as the most easily available form of nitrogen by microorganisms does not exceed 10 mg/L (ES = 50000÷60000 kJ/kg TS). Contrary to that, the propagation of ultrasonic wave caused reduction in concentrations of nitrate nitrogen. In such case, it was found that exceeding dose of energy $E_S = 10000$ kJ/kg TS resulted in the average concentrations of nitrate nitrogen in the range 3.8÷4.5 mg/L (Fig. 6). The increase in nitrite nitrogen concentration in the solution was also determined after WAS sonication. It was observed that $N-NO_2$ concentrations also remained constant, regardless of the value of specific energy. The range of observed concentrations amounted to 0.69÷1.01 mg/L (Fig. 7). The presence of nitrate nitrogen in the liquid may adversely affect the biological process of dephosphatation.

The cyclic increase in sonication and thus specific energy caused an increase in concentrations of Total Phosphorus (TP) released to the liquid phase of treated sludge (Fig. 8). The conditions of

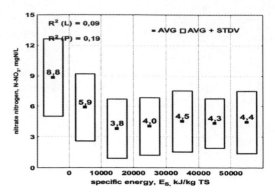

Figure 6. Nitrate nitrogen release profiles to the liquid phase as a result of the WAS disintegration.

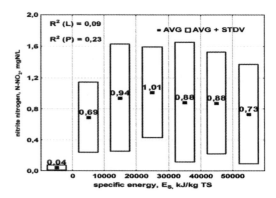

Figure 7. Nitrite nitrogen release profiles to the liquid phase as a result of the WAS disintegration.

WAS sonication corresponding to specific energy ES = 50000 kJ/kg TS generated a significant level of TP, which was TP≈50 mg/L. As a result, a significant concentration of liquid of biogenic compounds was determined. The increase in specific energy also changed the value of the ratio TC/TP, and it was a decreasing trend from 12.7 to 11.4. As with ratio TC/TN, there were also observed that the after-sonication product was characterized by an insufficient amount of available carbon per unit nitrogen and phosphorus.

The biodegradability of the product of the WAS disintegration can be highly evaluated through the concentration of volatile fatty acids (Fig. 9). VFA are regarded as an easily available carbon source, especially important in dephosphatation and denitrification processes. At an ultrasound dose about $E_S = 50000$ kJ/kg, TS it is possible to obtain VFA concentrations above 100 mg/L. However, concentration may be too low relatively to the amount of phosphorus released, hence the biological process dephosphatation may be overloaded. The

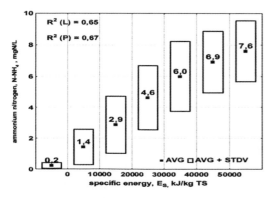

Figure 5. Total nitrogen release profiles to the liquid phase as a result of the WAS disintegration.

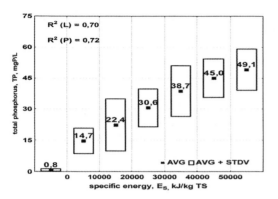

Figure 8. Effect of specific energy of ultrasonic treatment on total phosphorus solubilisation.

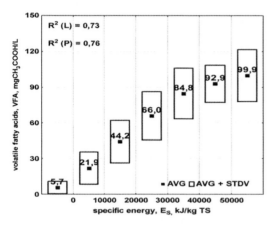

Figure 9. Effect of specific energy of ultrasonic treatment on volatile fatty acids solubilisation.

concentrated solution of the substances arising from the ultrasonic disintegration of activated sludge was the effect of sonication. Generally, with increasing sonication energy (independent variable) the increase in the concentration of the examined indicators (dependent variables) were obtained. The relationship between these variables had the most character of the polynomial regression. The highest coefficient of determination $R^2 = 0.87$ were obtained for the data set of TOC concentration as function of sonication energy. For dependent variables such as concentration of TC, TP and VFA the coefficient of determination ranged $R^2 >$ 0.7÷0.8. Polynomial regression of changes in TC evaluated the real data in 60–70%. Linear regression was also a good statistical tool to model the relationship between a variables. The correlation coefficients for linear regression were about 0.03 lower than for the polynomial regression. It seems that, by excluding the data obtained at sonication

dose 50000–60000 kJ/kg TS form the analysis, the linear regression would be a more accurate method to predict the values of dependent variables. Energy sonication row 50000 kJ/kg TS may be a threshold value above which the intensity of ultrasonic disintegration will be reduced. Thus, it can be a value at which the linear regression will change the value of the line slope. This was particularly evident, for example in changes of the VFA concentration. The increase in energy sonication of ES = 10000 kJ/kg TS induced an increase in VFA of approximately 7 mg/L (from 92.9 to 99.9 mg/L, ES = 50000 kJ/kg TS).

4 CONCLUSIONS

During the study the effect of solublization of organic and inorganic matter under the influence of sludge disintegration was observed. The increase of specific energy contributed to the increase of the concentration of dissolved form of organic carbon and volatile fatty acids in the liquid sludge. This shows that the disintegration of activated sludge can be a method that generates an easily assimilable carbon source for microorganisms. The question remains whether it will be satisfactory and economically justified effect of activated sludge disintegration. For the value of specific energy $E_S = 60000$ kJ/kg TS a significant energy dose was used. The second application of the ultrasonic disintegration of activated sludge (excess sludge) as a method of their quantity reduction also seems to be limited. Biological reactor and the biochemical processes of oxidation and reduction will be exposed to additional pollutant load, which arises from the disintegration of activated sludge. High concentrations of nitrogen and phosphorus can cause overload of bioreactor and loss treatment efficiency by dosing the disintegrated sludge. On the basis of the results, it cannot be determined whether the ultrasonic disintegration of activated sludge could create easy biodegradable substrate. The measurement of concentrations of released compounds into the liquid or the determination of the disintegration or solublisation degree are the first steps of the optimization of the ultrasonic disintegration. Proper evaluation of the effects of sludge disintegration and the characteristics of the resulting product will definitely require the use of biochemical tests. Functional groups in the active microbial biomass: include heterotrophic organisms, nitrifying and denitrifying bacteria and organisms capable of storing the excess phosphorus. They are able to perform biochemical oxidation and to reduce the appropriate type of substances contained in the disintegrated sludge. Therefore, the most suitable for the qualitative evaluation of ultrasonic disintegration of wastewater sludge seems to be the application of biotechnological testing, such as Oxygen Uptake Rate (OUR), Ammonia Utilization Rate (AUR), Nitrate Utilization Rate (NUR) and Phosphorus Release and Uptake Rate (PRUR). This test will be the most suitable to characterize the activity of the biomass and express changes in technology that will occur as a result of dosing of sonicated sludge to the biological reactor. The use of ultrasonic disintegration in the system, based on a laboratory flow or sequencing reactors, will be the best method of assessing the effectiveness and the suitability of the sonication process.

ACKNOWLEDGEMENTS

This work was supported by the Faculty of Environmental Protection and Engineering (Czestochowa University of Technology) BS/PB-401-301/11.

REFERENCES

Bougrier C., Albasi C., Delgenes J.P., Carrere H. 2006. Effect of ultrasonic, thermal and ozone pre-treatments on waste activated sludge solubilisation and anaerobic biodegradability. *Chemical Engineering and Processing* 45: 711–718.

El-Hadj T.B., Dosta J., Marquez-Serrano R., Mata-Alvarez J. 2007. Effect of ultrasound pretreatment in mesophilic and thermophilic anaerobic digestion with emphasis on naphthalene and pyrene removal. *Water Research* 41: 87–94.

Feng X., Lei H., Deng J., Yu Q., Li H. 2009. Physical and chemical characteristics of waste activated sludge treated ultrasonically. *Chemical Engineering and Processing* 48: 187–194.

He J., Wan T., Zhang G., Yang J. 2011. Ultrasonic reduction of excess sludge from activated sludge system: Energy efficiency improvement via operation optimization. *Ultrasonics Sonochemistry* 18: 99–103.

Kim D.H., Jeong E., Oh S.E., Shin H.S. 2010. Combined (alkaline + ultrasonic) pretreatment effect on sewage sludge disintegration. *Water Research* 44: 3093–3100.

Naddeo V., Belgiorno V., Landi M., Zarra T., Napoli R.M.A. 2009. Effect of sonolysis on waste activated sludge solubilisation and anaerobic biodegradability. *Desalination* 249: 762–767.

Pham T.T.H. Brar S.K., Tyagi R.D., Surampalli R.Y. 2009. Ultrasonication of wastewater sludge—Consequences on biodegradability and flowability. *Journal of Hazardous Materials* 163: 891–898.

Salsabil M.R., Prorot A., Casellas M., Dagot C. 2009. Pre-treatment of activated sludge: Effect of sonication on aerobic and anaerobic digestibility. *Chemical Engineering Journal* 148: 327–335.

Show K., Mao T., Lee D. 2007. Optimization of sludge disruption by sonication. *Water Research* 41: 4741–4747.

Standard Methods for the Examination of Water and Wastewater, 20th ed. 1998. American Public Health Association, Washington DC, USA.

Wang F., Lu S., Ji M. 2006. Components of released liquid from ultrasonic waste sludge disintegration. *Ultrasonics Sonochemistry* 13: 334–338.

Zawieja, I., Wolny, L., Wolski, P., 2008. Influence of excessive sludge conditioning on the efficiency of anaerobic stabilization process and biogas generation. *Desalination* 222: 374–381.

Zhang G., Zhang P., Yang J., Chena Y. 2007. Ultrasonic reduction of excess sludge from the activated sludge system. *Journal of Hazardous Materials* 145: 515–519.

Zhang G., Zhang P., Yang J., Liu H. 2008. Energy-efficient sludge sonication: Power and sludge characteristics. *Bioresource Technology* 99: 9029–9031.

Environmental Engineering IV – Pawłowski, Dudzińska & Pawłowski (eds)
© *2013 Taylor & Francis Group, London, ISBN 978-0-415-64338-2*

Mobility of heavy metals in sewage sludge from diversified wastewater treatment plant

J. Gawdzik & J. Latosińska
Faculty of Civil and Environmental Engineering, Kielce University of Technology, Kielce, Poland

ABSTRACT: The aim of this study was to evaluate the mobility of heavy metals in sewage sludge from different wastewater treatment plants in Świętokrzyskie Voivodeship. Stabilized SS from the wastewater treatment plant was analyzed in accordance with the extraction method proposed by the Community Bureau of Reference. It should be strongly emphasized that organometallics and aluminosilicates constitute the most prevalent forms of heavy metals in sewage sludge. The results for stabilized sewage sludge confirmed a trend observed in HMs concentration in the immobile fractions. It should be noted that HMs immobilized in the fraction F-III, may pose a potential hazard to soil.

Keywords: sewage sludge (SS), heavy metals (HM), sequential extraction, metals mobility

1   INTRODUCTION

The upgrading of municipal wastewater treatment plants has led to increased sludge production. The content of organic substances, nutrients (N, P, K) and microelements predisposes the sewage sludge to its environmental use (Rogeres 1996, Wang 1997). Landfill and incineration of sewage sludge deprives soils, which are low in organic matter content, of a potential source of organic material. Except for its desirable constituents, essential for agricultural use, sewage sludge consists of toxic substances, including heavy metals. The heavy metals in sewage sludge can be dissolved, precipitated with metal oxides, and adsorbed or associated on the particles in biological debris. Heavy metals are found in the form of oxides, hydroxides, sulphides, sulphates, phosphates, silicates, organic connections forming complexes with humic compounds and complex sugars (Alvarez et al. 2002). Polish regulations (Journal of Laws of the Polish Republic No. 137, item 924) specifying the maximum levels of heavy metals in municipal sewage sludge used for agricultural purposes refer to the total content of lead, cadmium, mercury, nickel, zinc, copper and chromium.

1.1   *Heavy metal sources*

Heavy metals cycle and migration in the natural environment is primarily connected with processes such as, e.g. rock weathering, volcanic eruptions, evaporation of oceans, forest fires, and soil-forming processes. Man's activities contribute to the increased exposure of the environment to toxic trace elements. The anthropogenic sources of the environmental contamination with heavy metals include industries, power engineering, transport, public utilities, waste disposal sites, fertilisers and wastes used in soil manuring (Table 1) (Boruszko et al. 1995). HMs that originate in those sources are dissipated in the environment, where they contaminate soils, waters, air, and get into humans' and animals' organisms either directly or through plant consumption (Bień et al. 2008).

Environmental contamination with heavy metals can result, among others, from:

– discharge of sewage from tanning, galvanizing, aviation and automotive industries,
– flow from farmed land areas, especially those where fertilisers and crop protection chemicals have been used intensively,
– precipitation,
– household sludge, industrial sludge and compost from municipal wastes from which metals are lixiviated,
– compost containing high amounts of municipal wastes used as fertiliser in agriculture is a serious source of heavy metal emissions.

1.2   *Toxicity of heavy metals*

The toxicity of HMs results not only from the level of the environmental contamination, but also from the biochemical role they play in metabolic processes and the degree to which living organism absorb and excrete them (Boruszko et al. 1995).

Plants, the main recipients of mineral nutrients from soils and waters, including toxic metals, are at the same time a major source of those metals

Table 1. Selected heavy metals and industries being sources of emission to the environment (Boruszko et al. 1995).

Metal	Industries
Cd	Galvanizing plants, production of dyes, batteries, paints, plastics, polymer stabilisers, chemical industry, crop protection chemicals industry, printing works
Pb	Production of dyes, batteries, chemical fertilisers, motorisation, power engineering industry, crop protection chemicals industry, electrochemical industry
Cr	Galvanizing, tanning, wood preservation, textile industries, production of dyes and plastics,
Cu	Metallurgy, dyes, textile industries, production of crop protection chemicals and chemical fertilisers
Hg	Production of batteries, phosphoric acid, caustic soda, cellulose manufacturing facilities, production of crop protection chemicals, metallic mercury production
Ni	Galvanizing, paper industries, refineries, steel works, chemical fertiliser factories
Zn	Production of batteries, paints, textile, plastics, polymer stabilisers industries, printing works

in people's and animals' food. The hazard posed by heavy metals involves mainly their entering the food chain. The migration of heavy metals into higher links of the food chain depends on natural biological barriers (The Act on Wastes of 27th April 2001).

On the basis of potential hazard posed by heavy metals for living organisms, four groups of those metals were differentiated (Sadecka et al. 2007):

- elements posing a very high risk: Cd, Hg, Pb, Cu, Sn, Cr, Ag, Sb,
- elements posing a high risk: Bi, U, No, Ba, Mn, Ti, Se, Te,
- elements posing a medium risk: Rb, As, W,
- elements posing a low risk: Sr, Nb, Zr.

The toxicity of heavy metals also depends on their biochemical and biological properties, in particular on:

- susceptibility to bioaccumulation from soil (Cd, Sn, Zn) or from water (Pb, Cd, Hg, Zn, Cu, Sr),
- possibility of concentration in biolits due to geotechnical processes (Cr, Pb, Ba),
- easy absorption from the digestive track (Hg, Cd),
- possibility of permeation through placenta (Cd, Hg, Pb, Zn),
- possibility of permeation to blood and brain (Hg, Pb).

Even the minimal concentration of toxic metals in the organism results in metabolic disorders, reduced organism physiological competence, deficient immune logical response and disturbed enzymatic activity processes, which in turn can lead to many diseases, and in some cases may even cause death.

The aim of this study was to evaluate the mobility of heavy metals in sewage sludge from different wastewater treatment plants in Świętokrzyskie Voivodeship. Stabilized sewage sludge from the wastewater treatment plant was analyzed in

accordance with the extraction method proposed by the Community Bureau of Reference (BCR). Zinc, cadmium, lead and nickel were determined by means of the standard addition with the use of the Perkin-Elmer 3100-BG FAAS atomic absorption spectrophotometer (with the background correction function turned on). Chromium and copper were tested using the FAAS technique. In order to determine mercury the CVAAS method was employed.

2 MATERIALS AND METHODS

2.1 Sample collection and pre-treatment

The tests were conducted on municipal sewage sludge collected (in accordance with PN-EN ISO 5667-13:2004) from nine municipal sewage treatment plants located in central Poland (Table 2).

2.2 The sequence extraction

The tests were conducted in accordance with the four-step BCR sequential extraction procedure (Alvarez et al. 2002, Pitt 1999) introducing a change in the method of residual fraction mineralisation, i.e. aqua regia was used in the process of mineralisation (PN-EN ISO 15587:2002).

Step one: acid soluble/exchangeable fraction (F-I) A 2 g sample of sewage sludge was placed in a 100 cm^3 test-tube for centrifuging. Then, 40 cm^3 of 0.11-molar acetic acid solution was added. The sample was shaken for 16 hours at room temperature. The extract was separated from the sewage sludge by centrifuge (4000 rpm). The content of the soluble metals in the water was marked in the liquid.

Step two: reducible fraction (F-II). Sewage sludge was washed in 20 cm^3 of distilled water (shaken and centrifuged). Subsequently, 40 cm^3 of 0.1-molar hydroxylamine hydrochloride solution,

Table 2. Sewage sludge from S1÷S6 different throughput water treatment plants.

Sample notation	Location	Type of water treatment plant	EP	Sludge stabilization method	Sludge use
S1	Gnojno	M-B with increased removal of nutrient	850	Aerobic sludge stabilization	Soiless land reclamation
S2	Pacanów	Mechanical-biological EvU-Perl	3140	Aerobic sludge stabilization	Land reclamation
S3	Barcza	Mechanical-biological EvU-Perl	3833	Aerobic sludge stabilization	Isolating layers on the disposal ground
S4	Opatów	Mechanical-biological	15250	Anaerobic mezophilic sludge stabilization	Soiless land reclamation
S5	Ostrowiec Świętokrzyski	Mechanical-biological	50150	Anaerobic mezophilic sludge stabilization	Isolating layers on the disposal ground
S6	Starachowice	Mechanical-biological	99000	Anaerobic mezophilic sludge stabilization	Isolating layers on the disposal ground

of pH = 2, was added to the sewage sludge. Nitric acid was used for the correction of the pH value. The procedure was the same as in step one, the mixture was shaken and centrifuged. Fraction II metals were marked in the liquid.

Step three: oxidation fraction (F-III). The sewage sludge was carried over quantitatively to a quartz evaporating dish and 10 cm^3 of 30% hydrogen peroxide was added. The contents of evaporating dish were heated in a water bath at 85° C for one hour.

The process was repeated with the addition of 10 cm^3 of 8.8-molar hydrogen peroxide solution to the sewage sludge. After drying, the sewage sludge sample was transferred to test-tubes to be centrifuged and then 50 cm^3 of ammonium acetate solution (1 mol/dm^3, pH = 2; nitric acid was used to correct the pH value) was added. The sample was shaken for 16 hours and afterwards the sewage sludge was separated from the extract. Fraction III metals were marked in the solution.

Step four: residual fraction (F-IV). The sludge was washed and dried to a solid state. The mineralization of the residual fraction was conducted with aqua regia; 30 cm^3 of concentrated hydrochloric acid and 10 cm^3 of concentrated nitric acid were added carefully to a 300 cm^3 conical flask together with 0.5 g of sludge. The conical flask was heated for 30 min and subsequently evaporated to dryness. After cooling, 25 cm^3 of 5% hydrochloric acid were added. The sewage sludge was dissolved, carried over to a metal measuring flask and topped up with 50 cm^3 of distilled water. Then the sample was mixed and strained to a dry dish. In the filtrate, the metal forms, Fraction IV, were marked.

The heavy metals in the extracts obtained were determined in accordance with ISO 9001:2000 using a Perkin-Elmer 3100 FAAS-BG atomic absorption spectrophotometer (impact bead). Each determination was repeated four times.

3 RESULTS AND DISCUSSION

The conducted sequential analysis proved that different forms of heavy metals are present in the sewage sludge. The obtained results of heavy metal mobility in the sewage sludge are presented in Figures 1–6.

The highest mobility reported for zinc in the tested sludge S2. The sum of the fraction F-I and F-II exceeded 87.3% for Zn (Fig. 2). High mobility of zinc (66%) has been reported for sludge S1 also.

The average mobility reported for zinc in the whole tested sludge S1 ÷ S6 exceeded 37.5%. The lowest HMs mobility has been noted for chromium in S4 (1%). Slightly higher mobility of copper equal to 1.1% was detected in a sample of sediment S6. All kinds of the tested sewage sludge had a very small amount of mobile copper fractions (fractions I-II) in comparison with immobile ones (F-III-IV). As far as fraction (F-III) is concerned, copper is temporarily immobile because its maintenance can vary according to the level of mineralization in the ground. The mobile fraction (F-I + F-II) of copper equalled from 1.1% up to 21.6% of the overall amount of copper (Table 3). The analysed samples had different concentration of nickel: 20.0 ÷ 83.5 mg/kg d.m. (Table 4). The preponderant fractions of nickel were both the temporarily immobile

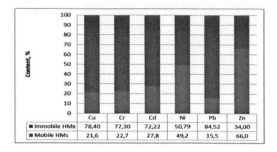

Figure 1. Analysis of heavy metal mobility based on metal speciation in sewage sludge from Gnojno wastewater treatment plant (S1).

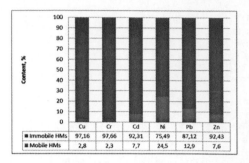

Figure 4. Analysis of heavy metal mobility based on metal speciation in sewage sludge from Opatów wastewater treatment plant (S4).

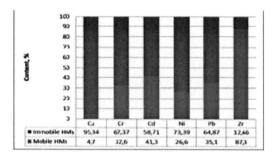

Figure 2. Analysis of heavy metal mobility based on metal speciation in sewage sludge from Pacanów wastewater treatment plant (S2).

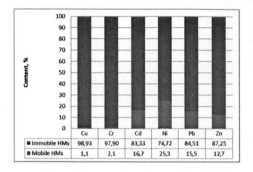

Figure 5. Analysis of heavy metal mobility based on metal speciation in sewage sludge from Ostrowiec Sw. wastewater treatment plant (S5).

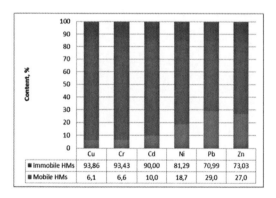

Figure 3. Analysis of heavy metal mobility based on metal speciation in sewage sludge from Barcza wastewater treatment plant (S3).

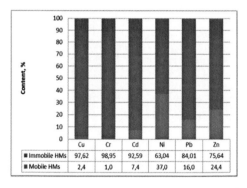

Figure 6. Analysis of heavy metal mobility based on metal speciation in sewage sludge from Starachowice. wastewater treatment plant (S6).

fraction (F-III) and the immobile fraction (F-IV). Despite this, the mobility of Ni was significantly higher than the mobility of Pb, Cd, Cu and Cr in samples S1÷S6. The average mobility of chromium in the tested sludge was very diversified (1.0 ÷ 32.6%) (Table 3). F-III fraction was predominant for S1, S2 and S3 sewage sludge. F-IV fraction was predominant for S4, S5 and S6 sludge. The possible to grasp the differences may result from the method of the sludge stabilisation. For S2 and S3 sludge, lead was present exclusively in an mobile fraction F-II and, respectively F-I. For S1, S2 and

Table 3. Mobility of heavy metals in municipal sewage sludge S1 ÷ S6, %.

Metal	Average mobility	Standard deviation	Minimum mobility	Maximum mobility
Cu	6.6	5.1	1.1	21.6
Cr	11.2	11.0	1.0	32.6
Cd	18.5	10.7	7.4	41.3
Ni	30.2	8.6	18.7	49.2
Pb	20.7	7.6	12.9	35.1
Zn	37.5	26.1	7.6	87.3

S6 sludge, fraction (F-III) of chromium also had a high contribution (from 54.8 up to 69.1%). The high immobility of chromium in the sewage sludge results from the presence of chromium in a form of indissoluble salts. Furthermore, the soil organic substance is a limiting factor of chromium bioactivity in the sewage sludge.

The total concentration of cadmium in the sewage sludge S3÷S6 was at a similar level. The immobile fraction (F-IV) was predominant for S4, S5 and S6 sludge. S1, S2 and S3 sludge contained the highest number of temporarily immobile cadmium

Table 4. Statistical results* of each fraction of heavy metals in sewage sludge S1 ÷ S6.

Speciation	Heavy metals [mg/kg d.m.]					
	Cu	Cr	Cd	Ni	Pb	Zn
Wastewater treatment plant S1						
Fraction I	3,1 ± 0,1	0,0 ± 0,1	2,4 ± 0,2	21,0 ± 0,3	6,1 ± 0,7	41,3 ± 0,7
Fraction II	29,4 ± 0,5	7,5 ± 0,2	0,0 ± 0,1	13,2 ± 0,2	6,8 ± 0,8	123,7 ± 0,9
Fraction III	116,3 ± 0,9	23,0 ± 0,5	5,1 ± 0,3	33,2 ± 0,4	46,0 ± 5	83,4 ± 0,8
Fraction IV	1,6 ± 0,1	2,7 ± 0,1	1,1 ± 0,1	2,1 ± 0,1	24,5 ± 3	1,6 ± 0,1
ΣFI ... IV	150,4	33,2 ± 0,6	8,6 ± 0,3	69,4	83,3 ± 9	249,9
Wastewater treatment plant S2						
Fraction I	1,5 ± 0,1	0,0 ± 0,1	0,0 ± 0,1	2,6 ± 0,1	3,7 ± 0,4	328,9 ± 0,9
Fraction II	25,6 ± 0,2	24,1 ± 0,3	4,2 ± 0,1	19,6 ± 0,3	14,0 ± 2	743,2 ± 2,3
Fraction III	551,4 ± 0,9	45,1 ± 0,4	5,1 ± 0,1	57,0 ± 0,6	6,0 ± 0,7	152,3 ± 0,9
Fraction IV	4,7 ± 0,1	4,7 ± 0,1	0,8 ± 0,1	4,3 ± 0,2	26,7 ± 3	3,1 ± 0,1
ΣFI ... IV	583,3	74,0	10,1	83,5	50,3	1228
Wastewater treatment plant S3						
Fraction I	4,0 ± 0,1	0,9 ± 0,1	0,1 ± 0,1	3,4 ± 0,2	4,5 ± 0,6	64,0 ± 0,5
Fraction II	2,4 ± 0,1	1,0 ± 0,1	0,1 ± 0,1	1,8 ± 0,1	0,2 ± 0,1	124,0 ± 1,2
Fraction III	82,5 ± 0,3	10,6 ± 0,2	1,1 ± 0,2	3,7 ± 0,3	3,2 ± 0,5	348,0 ± 2,3
Fraction IV	15,3 ± 0,2	16,4 ± 0,3	0,7 ± 0,2	18,9 ± 0,5	8,3 ± 0,8	161,0 ± 1,1
ΣFI ... IV	104,2	29,0	2,0	27,8	16,2	697,0
Wastewater treatment plant S4						
Fraction I	3,3 ± 0,2	2,0 ± 0,3	0,3 ± 0,1	3,5 ± 0,1	5,2 ± 0,1	79,4 ± 0,7
Fraction II	1,8 ± 0,1	1,1 ± 0,1	0,3 ± 0,1	1,4 ± 0,1	0,5 ± 0,2	122,8 ± 2,6
Fraction III	57,1 ± 1,1	16,1 ± 0,7	1,9 ± 0,1	5,9 ± 0,1	7,8 ± 0,1	323,8 ± 1,5
Fraction IV	22,8 ± 0,7	22,0 ± 0,7	1,1 ± 0,8	9,2 ± 0,3	54,7 ± 9,4	170,8 ± 1,3
ΣFI ... IV	85,0	41,2	3,6	20,0	68,2	696,8
Wastewater treatment plant S5						
Fraction I	3,1 ± 0,2	1,1 ± 0,2	0,0 ± 0,1	6,3 ± 0,3	5,7 ± 0,5	48,0 ± 0,5
Fraction II	2,2 ± 0,2	0,9 ± 0,2	0,2 ± 0,1	2,4 ± 0,2	2,3 ± 0,3	57,0 ± 0,6
Fraction III	123,2 ± 3,1	20,7 ± 0,7	0,5 ± 0,1	6,5 ± 0,3	7,1 ± 0,6	787,4 ± 5,5
Fraction IV	58,2 ± 0,8	62,9 ± 0,9	1,9 ± 0,2	20,3 ± 0,9	47,0 ± 4,4	495,0 ± 5,1
ΣFI ... IV	186,7	85,6	2,6	35,5	62,1	1387
Wastewater treatment plant S6						
Fraction I	0,9 ± 0,2	2,1 ± 0,2	0,1 ± 0,1	7,9 ± 0,3	7,8 ± 0,8	79 ± 1
Fraction II	1,2 ± 0,2	0,9 ± 0,1	0,3 ± 0,1	3,2 ± 0,2	1,0 ± 0,2	275 ± 3
Fraction III	124 ± 0,8	78,1 ± 0,8	0,9 ± 0,2	9,8 ± 0,5	4,5 ± 0,5	1491 ± 21
Fraction IV	69,9 ± 0,6	61,5 ± 0,5	1,1 ± 0,3	23 ± 0,9	43,5 ± 4,5	932 ± 82
ΣFI ... IV	196,0	142,6	2,4	43,9	56,8	2777

fractions (F-III). However, S2 sludge contain mobile cadmium of fractions F-II (41.3%). The content of zinc in the tested sewage sludge was the most diversification (249.9 for S1 to 2777 mg/kg d.m. for S6). In S3, S4, S5 and S6 sludge, the temporarily immobile fraction FIII had the highest contribution. The immobile fraction (F-IV) had the highest contribution percentage in the sludge S5 (35,7%). The levels of heavy metals in S1 ÷ S5 sewage sludge did not exceed the admissible limits (35) valid in Poland for sludge designed for environmental usage, including the agricultural usage. S6 sludge cannot be used in agriculture because of the exceeded admissible quantity of zinc. Due to the amount of zinc, S6 sewage sludge can be used in the process of land reclamation for other than agricultural purposes.

4 CONCLUSIONS

The sewage sludge is apparently abundant in nutrients. However, heavy metals found in it restrict or even make it impossible to use sludge for biological purposes.

Metals occur in various physical and chemical forms, which to a large extent determines their toxicity. In the sewage sludge, heavy metals are found in mobile forms which migrate to the soil, and in immobile forms that do not significantly affect the ground and the water environment.

Speciation analysis is based on the sequential extraction carried out in accordance with BCR methodology. The sequence analysis revealed the presence of heavy metals in all fractions (F-I, F-II, F-III, F-IV). It should be strongly emphasised that organometallics and aluminosilicates constitute the most prevalent forms of metals under consideration. Those, according to BCR, make fraction F-III and fraction F-IV, respectively. The capacities of the plants did not demonstrate any explicit influence on the forms of the heavy metals presence. It was hypothesised that copper forms in the sewage sludge are a feature characteristic of this chemical element. The mobility of lead, cadmium, nickel and chromium in the BCR fractions depend slightly on the method of sludge stabilization. The results for anaerobic stabilized sewage sludge confirmed the trend being observed in heavy metals concentration in the immobile fraction F-IV. The results for aerobic stabilized sewage sludge was

high HMs concentration in temporarily immobile fraction F-III. Nevertheless, it should be noted that heavy metals immobilized in the fraction F-III, may pose a potential risk to the soil in the aeration zone. The high mobility of zinc may be a source of concern. This problem occurred only for the aerobic system of the sewage sludge stabilization (S1, S2). Unfortunately, the high mobility of zinc in this case correlates well with a significant amount of zinc. With the reference to the sewage treatment plant in liquidation, there is largely uncontrolled discharge of wastewater into the water and the soil. It should be noted that the positive effects of treatment balances in full to the possible adverse effects on the environment.

REFERENCES

Alvarez, E.A. et al. 2002. Heavy metal extractable forms in sludge from wastewater treatment plants. *Chemosphere,* 47:765–775.

Bień, J. & Wystalska, K. 2008. *Thermal processes in sewage sludge neutralisation* (in Polish), Częstochowa, Publisher University of Czestochowa.

Boruszko, D., et al. 1995. Effect of some heavy metal ions in static fermentation processes of sewage sludge (in Polish), in: Bień B.J., *Problems of sludge management in sewage treatment plants*, Częstochowa, International Scientific-Technical Conference, UNi-Serwice Sp. Z o.o.

Pitt, R. et al. 1999. Groundwater contamination potential from storm water infiltration practices, *Urban Water*, 1:217–236.

PN-EN Iso 5667-13:2004. *The quality of water-collection of samples*-Part 13: Procedure guidelines concerning the collection of sludge samples from sewage treatment plants and water purification plants.

Rogers, H.R. 1996. Sources, behaviour and fate of organic contaminants during sewage treatment and in sewage sludge, *The Science of the Total Environment*, 185:3–26.

Sadecka, Z. & Myszograj, P. 2007. *Sewage purification and sewage sludge processing* (in Polish), Zielona Góra, Publisher University of Zielona Góra.

The Act on Wastes of 27th April 2001 (Journal of Laws of the Polish Republic No. 62, item 628).

The regulation of Poland's Minister of the Environment of 13th July 2010 on communal sewage sludge (Journal of Laws of the Polish Republic No. 137, item 924).

Wang, M.J. 1997. Land application of sewage sludge in China, *The Science of the Total Environment*, 197: 149–160.

Environmental Engineering IV – Pawłowski, Dudzińska & Pawłowski (eds)
© *2013 Taylor & Francis Group, London, ISBN 978-0-415-64338-2*

Double-agent method of sludge conditioning

L. Wolny
Faculty of Environmental Engineering and Biotechnology, Czestochowa University of Technology,
Czestochowa, Poland

ABSTRACT: In the paper the results of a study on sludge conditioning process with simultaneous application of two polyelectrolytes (low—and high cationic) were presented. The aim of the research was to determine whether the simultaneous application of two polyelectrolytes increases the sludge dewaterability. The combined action of two polyelectrolytes (Praestol 610 BC & Praestol 655 BC) was effective for certain dosage proportion. The obtained final hydration value of the tested sludge was about 18% lower in comparison with unprepared sludge. Moreover, it was determined that the course of conditioning process is significantly influenced by the sequence of polyelectrolyte dosage.

Keywords: sewage sludge, polyelectrolytes, co-conditioning, dewatering

1 INTRODUCTION

One of the main problem in sludge utilization is an effective separation of solid and liquid phase of the sludge. Effective dewatering decreases the sludge volume and also has a positive impact on the further stages of sludge treatment. Conditioning is one of the processes facilitating the sludge dewatering. In spite of different conditioning methods that are currently available, the chemical approach based on polyelectrolytes application tends to be used more frequently than other methods. However, for some kinds of sludge, additional agents, apart from the sole polyelectrolyte, are required in the conditioning process because this way of preparation is not always effective from technological and economical point of view (Bień et al. 2008, Stępniak & Wolny 2004, Stępniak et al. 2003).

Conditioning processes are intended to favorably alter the sludge properties. Overall, in the effort to convert 'sludge' into 'biosolids', environmental and economic issues must be taken into account. In many cases, conditioning is closely incorporated into treatment sequence, so it is not an independent technological stage, but is preferably situated between and within other processes. The conditioning process can also have significant impact on other sludge properties, not only structure. Considering the process strategies that should improve the sludge susceptibility to various disposal routs, the pathogen content and odour potential can constitute crucial factors, which have to be taken into consideration while choosing the conditioning processes (Dentel 2001, Glover et al. 2004, Zawieja et al. 2005).

Cationic flocculants represent the majority of the chemicals used in the sludge conditioning before dewatering. Flocculation of sludge is the step in the process during which destabilized particles are agglomerated in aggregates called the flocs. Flocculants, with their very high molecular weights (long chains of monomers) and their varied ionic charge, fix the destabilized particles on their chain. Therefore, the particle size in the aqueous phase increases through the course of the flocculation process and result in the formation of flocs. Those flocs facilitate a release of the water. This water will thus be easily eliminated during the dewatering step. The origin of destabilized particles varies significantly and essentially depends on the nature of the sludge. The charge that flocculant brings is usually selected according to the type of destabilized particles presented in the sludge to be treated. For example, medium cationic charge is favourable for mixed sludge and high cationic charge for biological sludge. Organic flocculants are characterized by five main parameters: the type of charge, the charge density, the molecular weight, the molecular structure and the type of monomer. Although the fraction of polyelectrolyte in the sludge is small, its role in the performance of subsequent treatment or disposal processes requires further examination (Bien & Wolny 1997, Wolny 2005). The aim of the research was to establish the influence of the dual method of sludge conditioning on the filtration process parameters, especially on sludge resistivity and final hydration and to determine the most effective combination of polyelectrolyte doses.

2 MATERIALS AND METHODS

The sludge used as a research material was taken from municipal wastewater treatment plant. It is mechanical and biological wastewater treatment plant based on the activated sludge method. Samples were taken after an aerobic stabilization process. Table 1 presents a physicochemical characteristic of the tested sludge.

The scope of investigation concerns the physicochemical analysis of the tested sludge, the selection of polyelectrolytes as well as their optimal doses and the determination of vacuum filtration parameters. The sludge was conditioned by 0.1% solution of polyelectrolytes. The samples were treated under the following conditions: sludge sample of 100 cm³ was prepared with an appropriate dose of the polyelectrolyte and mixed together for two minutes. The experiment was carried out at the constant temperature of 291 K. Two cationic polyelectrolytes were used in the experiment: Praestol 610BC and Praestol 655BC (Table 2). Flocculants 'Praestol' of cationic charge are the acrylamid copolimers with the cationic co-monomer predominant. The cationic groups contained in the polymer have positive charges in the water solution.

The low cationic polyelectrolyte Praestol 610BC is based on polyacrylamide and it is supplied in a micro-bead form. Praestol 655BC, as a second conditioner, is based on polyacrylamide and sodium acrylate. Table 2 presents the basic characteristic of these compounds. The most advantageous dose of polyelectrolytes was determined on the basis of the Capillary Suction Time (CST). Dewatering of sludge was carried out with the use of a laboratory vacuum filter. During this process the sludge parameters were established, mainly the

Table 1. Physico—chemical characteristic of the tested sludge.

Smell	–	Soily
Colour	–	Black
Initial hydration	%	98.21
Dry matter content	g/dm³	17.93
Mineral matter content	g/dm³	7.15
Organic matter content	g/dm³	10.78
Capillary suction time	s	2231.0
Filtration parameters		
Resistivity	m/kg · 10¹³	36.8
Final hydration	%	78.72
Velocity	cm³/s	0.66
Filtration output	kg/m² · h	1.75
pH	–	8.35

Table 2. Basic properties of polyelectrolytes applicated in the tests.

Parameter	Praestol 610 BC	Praestol 655 BC
Molecular weight, g/mol	High $6–12 \cdot 10^6$	High $6–12 \cdot 10^6$
Charge type	Low cationic	Very high cationic
pH range	6–13	1–14

sludge resistivity, final hydration, filtration velocity and productivity. All the measurements were done three times.

3 RESULTS AND DISCUSSION

At the beginning the sludge was conditioned by means of single polyelectrolyte. Even a small dose of a flocculant induced a significant decrease of the CST value regardless of the polyelectrolyte type used. For example, for a 1.0 mg/d.m. Praestol 610BC dose the capillary suction time was 363.7 seconds and respectively 974.6 seconds for Praestol 655BC. In comparison with untreated sludge (Table 1) the CST was reduced to the value of 84% for Praestol 610BC and to 56% for Praestol 655BC.

Table 3 presents parameters of the sludge obtained in the filtration process before and after conditioning. The final hydration of unconditioned sludge was 78.72% and resistivity was $36.8 \cdot 10^{13}$ m/kg. After polyelectrolyte Praestol 610BC application in the dose of 2.5 mg/g d.m. the final hydration was 64.20% and resistivity $6.7 \cdot 10^{13}$ m/kg.

In case of the second polyelectrolyte Praestol 655BC, the value of final hydration was 67.9% (dose 2.5 mg/g d.m.) and resistivity was equal $4.5 \cdot 10^{13}$ m/kg.

Further experiments were conducted with the simultaneous application of two polyelectrolytes. The research was performed for selected doses of polyelectrolytes: 1.5 mg/d.m. and 2.5 mg/d.m, according to the following combinations: dosage proportion 1:1 (combination A), proportion 1:3 (combination B) and proportion 3:1 (combination C) of Praestol 610BC and Praestol 655BC, respectively. The selected combinations of polyelectrolytes doses were verified in the vacuum filtration process. The obtained results were compared to the characteristic of unprepared sludge (Table 4). Moderate decrease of the sludge resistivity was observed. The best resistivity was achieved for the flocculant dose of 2.5 mg/d.m. (combination B and C) i.e. $4.8 \cdot 10^{13}$ m/kg, which was about

Table 3. Comparison of conditioned and filtrated sludge parameters.

Polyelectrolyte dose, mg/g d.m.	CST[a], s	Final hydration, %	Resistivity, m/kg · 10^13	Filtration output, kg/m² · h	Velocity, cm³/s
0.0	2231.0	78.72	36.8	1.75	0.0158
Single conditioning with polyelectrolyte Praestol 610BC					
1.5	522.3	69.13	15.0	1.15	0.019
2.5	212.5	64.20	6.7	1.06	0.02
Single conditioning with polyelectrolyte Praestol 655BC					
1.5	686.7	66.21	8.6	1.09	0.019
2.5	280.6	67.94	4.5	1.18	0.02

[a]—capillary suction time.

Table 4. Characteristic of the sludge conditioned by the dual method after filtration process.

Polyelectrolyte dose, mg/g d.m.	CST, s	Final hydration, %	Resistivity, m/kg · 10^13	Filtration output, kg/m² · h	Velocity, cm³/s
Unconditioned sludge					
0.0	2231.0	78.72	36.8	1.75	0.016
Single conditioning method					
1.5	522.3	69.13	15.0	1.15	0.019
2.5	212.5	64.20	6.7	1.06	0.02
Dual conditioning method					
Combination A					
1.5	382.4	64.70	7.2	1.11	0.019
2.5	261.5	63.70	5.6	1.15	0.019
Combination B					
1.5	315.6	63.17	5.3	1.07	0.02
2.5	205.8	64.85	4.8	1.06	0.02
Combination C					
1.5	188.6	60.59	5.6	1.09	0.02
2.5	216.9	65.67	4.7	1.06	0.02

87% lower in comparison with the unconditioned sludge. The best dewatering result was observed in combination C (dose 1.5 mg/d.m.), for which the final hydration of the sludge was 60.59% that was about 18% lower value in comparison to the unprepared sludge.

Sludge pre-conditioned with cationic polyelectrolyte could form more compact aggregates and that impact on the better dewaterability result (Dentel 2001). Cationic polyelectrolyte, through bridging effect, could increase rigidity of the flocs, which resulted in increased dimensions of the flocs and tails (Langer et al. 1994). In addition, the more compact primary flocs were formed due to the electrostatic attraction between the cationic polyelectrolyte and the negatively charged sludge surfaces.

Although the fraction of polyelectrolyte in the sludge is small, its role on the performance of subsequent treatment processes is very important.

4 CONCLUSIONS

The aim of the investigation was to assess the possibility to apply the dual-polymer methods to sewage sludge conditioning. In this study two cationic polyelectrolytes (low and very high cationic) were used simultaneously. The influence of the selected polyelectrolytes combinations on sludge dewatering was observed on the basis of capillary suction time and filtration process. Results obtained in dual methods were compared with those for the single polymer method.

Cationic polyelectrolyte has positive impact on the conditioned sludge properties, both in single as well as in dual conditioning method.

The best dewatering result was obtained for the dosage proportion of Praestol 610BC and Praestol 655BC 3:1. Better result was obtained in spite of maintaining the same total flocculant dose as in the single polyelectrolyte method. The results of the experiments were also influenced by the sequence of the processes.

ACKNOWLEDGEMENTS

Scientific work founded by BS-PB-401-303/12 resource.

REFERENCES

Bień, J.B & Kowalczyk, M. & Kamizela, T. 2008. Dewatering efficiency of sludge conditioned by chemicals and ultrasound. *Engineering and Protection of Environment*. Wydawnictwo Politechniki Częstochowskiej, Częstochowa, 11(1) 65–72.

Bień, J.B. & Wolny, L. 1997. Changes of some sewage sludge parameters prepared with ultrasonic field. *Wat. Sci. Technol.* 36(11), 101–106.

Dentel, S.K. 2001. Conditioning, thickening and dewatering: research update/research need; in: Conference proceeding. Sludge Management Entering the 3rd Millennium. Taipei, Taiwan; 1–8.

Glover, S.M. & Van, Y. & Jameson, G.J. & Biggs, S. 2004. Dewatering properties of dual polymer—flocculated systems. *International Journal of Mineral Processing*, 73; 145–160.

Langer, S. & Klute, R. & Hahn H.H. 1994. Mechanism of floc formation in sludge conditioning with cationic polyelectrolytes. *Wat. Sci. Technol.* 30; 129–138.

Stępniak, L. & Wolny, L. 2004. Ultrasonic energy application to the intensification of coagulants and polyelectrolytes action. *Ecological Chemistry and Engineering A*, 11; 1215–1224.

Stepniak L. & Wolny, L. & Kowalczyk, M. 2003. Ultrasound—aided processes of water treatment and sludge dewatering. IcheaP—6, *Chemical Engineering Transactions*, 2, Pisa, Italy; 647–652.

Wolny, L. 2005. Ultrasound—aided process of sewage sludge preparation prior to dewatering. Wyd. Polit. Czestochowskiej, Czestochowa (in Polish).

Zawieja, I. & Wolny, L. & Wolski P. 2008. Influence of excessive sludge conditioning on the efficiency of anaerobic stabilization process and biogas generation. *Desalination*, 222, 374–381.

Environmental Engineering IV – Pawłowski, Dudzińska & Pawłowski (eds)
© *2013 Taylor & Francis Group, London, ISBN 978-0-415-64338-2*

An experimental research on the influence of the main properties of sewage sludge on syngas composition

S. Werle & R.K. Wilk
Institute of Thermal Technology, Silesian University of Technology at Gliwice, Gliwice, Poland

ABSTRACT: The latest trends in the field of sludge management, i.e. combustion, pyrolysis, gasification and co-combustion, have generated significant scientists' interest. Gasification has attracted considerable interest from water utilities as an alternative technology with the advantage of destruction of pathogenic bacteria and volume reduction, and the additional benefits of energy recovery and lower-cost atmospheric emission control. Gasification takes place in an environment with low levels of oxidizers to prevent the formation of diox-ins and large quantities of sulphur and nitrogen oxides. The paper presents the experimental investigation of sewage sludge gasification process. Installation with downdraft gasifier was used. An analysis of the influence of the main properties of different sewage sludge samples (composition, volatile matter content, water content and heavy metal content) on the composition of the gas obtained in the autothermal gasification process was conducted. Results presented as a function of amount of gasification agent.

Keywords: sewage sludge, gasification, gas composition, downdraft gasifier

1 INTRODUCTION

Nowadays, there is a rising interest in many countries in biomass utilization (e.g. combustion, co-combustion, gasification and pyrolysis) (Lindzen 2010). This is a result of the limited reserves of fossil fuels (and because of security of energy supplies in the world) and environmental and climate regulations on CO_2 emissions (Pienkowski 2012, Dasgupta 2011, Hoedl 2011). Table 1 presents annual global energy consumption (Williams et al. 2012).

Resources vary from country to country and depend on geographic location, the climate, the population density and the degree of the industrialization of the country. The sludge production range (Table 2) is quite large (16–94 g/(person·day)),

thus indicating the different approach to wastewater treatment and sludge management in different countries. In countries where technology is less developed, direct agricultural application or landfilling are the typical ways for secure sludge outlet from the wastewater treatment plants. In countries where the stakeholders and policy makers practically have forbidden land application (e.g. the European Union) and where the landfill directive implementation into the national legislation was

Table 1. Annual global energy consumption Gtoe (1 toe = 41.9 GJ) (Williams et al. 2012).

Fossil fuels	10.45
Oil	4.03
Coal	3.56
Natural gas	2.86
Nuclear	0.63
Renewables	0.94
Hydro	0.78
Wind, biomass, solar	0.16
Total global energy consumption	12.02

Table 2. Per capita sludge production (g/(person·day) in different countries (Cao & Pawlowski 2012, Werle 2012a, Werle 2012b).

Country	Sludge production g/(person·day)
China	16
Slovenia	20
Brazil	33
Italy	38
Poland	42
Hungary	48
Austria	55
Portugal	60
Turkey	60
Canada	76
Finland	94
Medium value	49

stringent, only the high temperature destruction methods are available.

Biomass is the term used to describe renewable organic-rich material. Sewage sludge, originating from the waste water treatment process, is the residue generated during the primary (physical and/or chemical), the secondary (biological) and the tertiary (additional to secondary, often nutrient removal) treatment. It is rich in organic matter (cellulose, hemicelluloses and lignin). Therefore, sewage sludge is often considered as a biomass (Cao & Pawlowski 2012). Recently, interest in energy recovery form sewage sludge is increasing. The *6th Environment Action Programme 2002–2012* of the Euro-pean Commission has been described as a major factor in reducing sewage sludge disposal by 50% from 2000 by 2050. In the future, thermal methods of sewage sludge will be the predominant method of utilization. The latest trends in the field of sludge management, (i.e. combustion, pyrolysis, gasification and co-combustion) have generated significant scientific interest. Gasification has attracted considerable interest from water utilities as an alternative thermal treatment technology. It has several advantages over traditional combustion. In gasification, most of the energy of the sewage sludge is not yielded as heat but is captured as chemical energy in the form of a combustible gas (syngas), a liquid and solid char. The energy amount of the produced syn-gas depends on many factors, such as the reactor type and feed type. Gasification gives the possibility to destroy the pathogenic bacteria and to reduce volume, and the additional benefits of energy recovery and lower-cost atmospheric emission control. It takes place in an environment with low levels of oxidizers to prevent the formation of chlorinated dibenzodioxins and dibenzofurans and large quantities of sulphur, nitrogen oxides, heavy metals and fly ash. Therefore, gasification technology can be applied to convert the sewage sludge into a useable energy and to reduce the environmental problems (Fytili & Zabanitou 2008). Few studies have been conducted on sewage sludge gasification (Dogru et al. 2002, Werle 2012a, Werle 2012b, Werle 2012c, Werle 2012d, Werle & Wilk 2010). Therefore, fundamental research regarding the effects of sludge on gasification is important in attempts to obtain a syn-gas for production of useable form of energy.

The paper presents the experimental investigation of sewage sludge gasification process. Installation with downdraft gasifier was used. An analysis of the influence of the main properties of different sewage sludge samples (composition, volatile matter content, water content and heavy metal content) on the composition of the gas obtained in the autothermal gasification process was conducted.

2 MATERIALS AND METHODS

2.1 Materials

A commercial predried sludge form two Polish wastewater treatment plants (granulated sewage sludge 1—GS1 and 2—GS2), the properties of which are reported in Table 3, was investigated.

Fuel particles were produced by drying and granulating the raw sludge. The fuel particles were sieved in the 4.75–6.50 mm size range.

2.2 Apparatus and procedure

For the purpose of experimental investigations, a laboratory system was designed and developed by Werle shown in Figure 1 (Werle & Wilk 2011, Werle & Wilk 2012). The main part of the installation is a stainless downdraft Gasifier (G) 150 mm internal diameter and the total height of 250 mm. The maximum capacity is 5 kg of granular Sewage Sludge (SS).

For this study, granular sewage sludge was fed into the reactor from the top, while air was supplied by a Blower (B) from the bottom. The sewage sludge feedstock moved in a countercurrent direction to the flow gas and passed through the drying, pyrolysis, reduction and combustion zones. The moisture was evaporated in the drying zone. In the pyrolysis zone, the sewage sludge was thermally decomposed to volatiles and solid char. In the reduction zone, carbon was converted, and CO and H_2 were produced

Table 3. Properties of the fuels tested.

Fuel	Granulated sludge 1 (GS1)	Granulated sludge 2 (GS2)
Proximate analysis, % (as received)		
Moisture	9.00	25.00
Volatile matter	52.50	44.50
Ash	32.35	31.50
Ultimate analysis, % (dry basis)		
C (dry)	31.83	33.78
H (dry)	5.30	4.92
O (dry)	23.76	22.89
N (dry)	4.50	4.25
S (dry)	0.35	0.85
P (dry)	1.79	1.81
LHV, MJ/kg (on dry basis)	13.43	10.92
Heavy metals content, ppm		
C_u	170.5	490.1
N_i	18.2	102.9
C_r	489.7	192.0
P_b	67.7	119.2
Z_n	1031.9	912.3
C_d	4.2	3.9

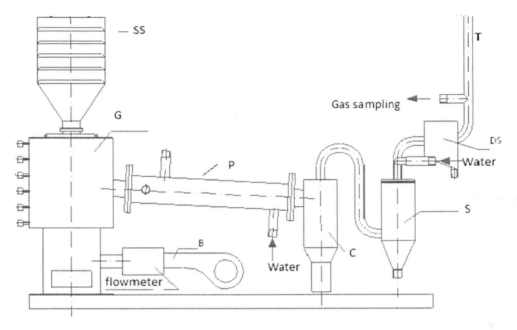

Figure 1.   Schematic diagram of the experimental system.

as the main components of the syngas. In the combustion zone, the remaining char was combusted, providing heat for endothermic reactions in the upper zones.

The temperature of the gasifier interior was measured by six N-type thermocouples located along the vertical axis of the gasifier at different heights. Additionally, the temperature of the syngas at the outlet of the gasifier was measured. The air flow rate supplied into the gasifier was measured by a flow meter. Syngas was transported from the gasifier by the Pipe (P).

At the outlet of the installation, there was a syngas sampling point where the syngas sample was collected and then supplied to CO and H_2 analyzers. The syngas was cleaned by a Cyclone (C), Scrubber (S) and Drop Separator (DS). Molar fractions of the main combustible species were measured online at the experimental stand and, for one specified experimental sample; the composition of syngas was investigated by chromatographic analysis.

The experimental procedure started by turning on the gasifier. The blower was switched on and sewage sludge was placed into the gasifier. After ap proximately 2 hours, the gasifier was heated and the experimental measurements started. The air flow rate to the gasifiers was adjusted to ensure a specified air excess ratio. Once syngas production began, measurements of key variables were taken. During the experiments, the samples were taken

twice form the outlet of the gasifier and the outlet of the water scrubber in order to analyze the produced wet gas.

3   RESULTS AND DISCUSSION

3.1   Influence of the air flow rate on the composition of syngas

Figure 2 presents the results obtained during the gasification of sewage sludge 1 (GS1) using various air flow rate. The results achieved for closely related types of sewage sludge were similar. On the basis of the data in the diagram, the amount of methane decreases as the air flow rate increases. For the majority of the range of ratios evaluated, the methane content was low. The percentage of hydrogen in the obtained gas was variable but the changes were not very drastic (z_{H2} ranged from 0.07–0.12). The volumetric fraction of hydrogen peaked when the air flow rate equaled to 1.85 kg/h (air excess ratio was $\lambda = 0.42$). The molar fraction of carbon monoxide slightly decreased from approximately 27% to 19%.

The molar fraction of carbon dioxide increased with increases in air flow rate. The amount of nitrogen also increases continuously with the increase of the air flow rate, as has been confirmed by theoretical calculations presented earlier (Werle 2012c).

3.2 Influence of sewage sludge on the composition of syngas

Figure 3 presents a comparison of the gasification results for two different granular sewage sludge samples (GS1 and GS2). Case 1 (GS1) is marked by the solid line, and case 2 (GS2) is marked by the dotted line. In case of GS1, the molar fraction of CO was within the range of 16–28%, while in the case of GS2, the molar fraction of CO ranged between 19–27%. All of the parameters of the gasification process were the same in both cases and the difference in the molar fractions of CO was likely due to the reactivity of the fuel. The molar fractions of hydrogen in the syngas were in both cases, ranging from 5–15%. The molar fraction of hydrogen was higher in the case of GS2, which was characterized by increased moisture in the sample. The moisture content of the fuel has a greater impact on both the operation of the gasifier and the quality of the produced gas.

3.3 Influence of air flow rate on the caloric value of the obtained gas

Figure 4 presents the dependence of the heating (caloric) value of the obtained gas on the air flow rate

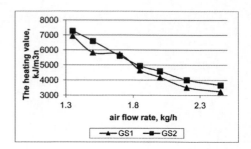

Figure 4. The heating value of syngas against the air excess ratio.

for the sewage sludge samples that were investigated. As expected, an increase in the air flow rate caused a decrease in the heating value. A greater amount of oxidizer increases the amounts of non-combustible species and the volumetric fraction of nitrogen, thus decreasing the heating value of the obtained gas.

4 CONCLUSIONS

New, original experimental results on sewage sludge gasification are presented in this study. Air sewage sludge gasification was investigated in a small scale gasification system under specified experimental conditions. An analysis of the influence of sewage sludge composition, volatile matter content, and water content on the composition of the gas obtained in the autothermal gasification process was conducted. The results, presented as a function of the amount of gasification agent, show that greater oxygen content in sewage sludge causes a reduction in the reaction temperature. Paradoxically, this effect causes an increase in the quantity of combustible components in the gas. As expected, increasing the air flow rate caused a decrease in the heating value of the produced gas. A higher amount of oxidizer increases the amounts of noncombustible species and volumetric fraction of nitrogen, thus reducing the heating value of the obtained gas.

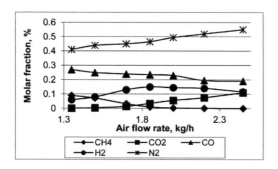

Figure 2. Molar fraction of hydrogen and carbon monoxide against the air flow rate.

ACKNOWLEDGMENTS

The paper has been prepared within the framework of the National Science Center, project no. N N523 737540.

REFERENCES

Cao Y., Pawlowski A., 2012, Sewage sludge-to-energy approaches based on anaerobic digestion and pyrolysis: brief overview and energy efficiency assessment, *Renewable and Sustainable Energy Reviews*, 16,: 1657–1665.

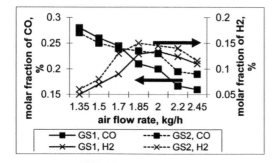

Figure 3. Molar fraction of hydrogen and carbon monoxide against the air excess ratio.

Dasgupta P., Taneja N., 2011, Low Carbon Growth: An In-dian Perspective on Sustainability and Technology Transfer, *Problemy Ekorozwoju—Problems of Sustainable Development*, 6(1): 65–74.

Dogru M., Midilli A., Howarth C.R., 2002, Gasification of sewage sludge using a throated downdraft gasifier, *Fuel Processing Technology*, 75,: 55–82.

Fytili D., Zabanitou A., 2008, Utilization of sewage sludge in EU application of old and new methods— A review, *Renewable and Sustainable Energy Reviews*, 12,: 116–140.

Hoedl E., 2011. Europe 2020 Strategy and European Recovery, *Problemy Ekorozwoju/Problems of Sustainable Development* 6(2): 11–18.

Pienkowski D., 2012, The Jevons Effect and the Consumption of Energy in the European Union, *Problemy Ekorozwoju/Problems of Sustainable Development* 7(1): 105–116.

Werle S., 2012, A reburning process using sewage sludge-derived syngas, *Chemical Papers*, 2,: 99–107.

Werle S., 2012, Modeling of the reburning process using sewage sludge-derived syngas, *Waste Management*, 4,: 753–758.

Werle S., 2012, Possibility of NOx emission reduction from combustion process using sewage sludge gasification gas as an additional fuel, *Archives of Environmental Protection*, 3,: 81–89.

Werle S., 2012, Analysis of the possibility of the sewage sludge thermal treatment, *Ecological Chemistry and Engineering A*, 19,: 137–144.

Werle S., Wilk R.K., 2010, Review of methods for the thermal utilization of sewage sludge: The Polish perspective, *Renewable Energy*, 35,: 1914–1919.

Werle S., Wilk R.K., 2011, Analysis of use a sewage sludge derived syngas in the gas industry, *Rynek energii*, 5,: 23–27.

Werle S., Wilk R.K., 2012, Experimental investigation of the sewage sludge gasification process in the fixed bed gasifier, *Chemical Engineering Transactions*, 29,: 715–720.

Williams A., Jones J.M., Ma L., Pourkashanian M., 2012, Pollutant from the combustion of solid biomass fuels, *Progress in Energy and Combustion Science*, 38,: 113–137.

Environmental Engineering IV – Pawłowski, Dudzińska & Pawłowski (eds)
© 2013 Taylor & Francis Group, London, ISBN 978-0-415-64338-2

Use of sewage sludge-compost in remediation of soil contaminated with Cu, Cd and Zn

Z.M. Gusiatin

Department of Environmental Biotechnology, University of Warmia and Mazury, Olsztyn, Poland

ABSTRACT: The effect of compost at 5 and 10% rates on stabilization of Cu, Cd and Zn in artificially contaminated loamy sand was investigated. Metal immobilization was assessed on the basis of their redistribution in soil, Risk Assessment Code (RAC) and binding Intensity (I_R) parameters. With the compost application rate of 5 and 10%, total metal concentration and soil pH slightly decreased, whereas organic matter increased. In non-amended soil all metals prevailed in exchangeable and acid soluble fraction: Cu (62.2%), Cd (89.2%) and Zn (80%). During 2-year stabilization, redistribution was the most visible for Cu at 10% compost rate, resulting in mobility decrease from high (RAC = 38.4%) to medium category (RAC = 25.6%). For Cd and Zn, it did not change significantly and remained at very high level. The I_R values for Cu increased initially over 6 months and gradually decreased over the time probably as a result of organic matter decomposition.

Keywords: compost, metal, redistribution, RAC, I_R

1 INTRODUCTION

Soil contamination with heavy metals is a worldwide problem. In the United States, about 70% of sites from the U.S. National Priority List are affected by heavy metals, therein Pb, Cr, Cu, Cd and Zn (Dermont et al. 2008). According to recent estimates, in Europe 37.3% of nearly 250000 sites are contaminated with heavy metals and need to be remediated (EEA 2007). In Poland, heavy metal contaminated areas occur locally and concern mainly metallurgical industry that has become one of the largest in Europe. Most metals in polluted sites characterized by light soils, poorly rich in organic matter and clay minerals can prevail in mobile chemical forms. To prevent uncontrolled metal migration, suitable activities should be undertaken.

For metals occurring in mobile forms, solidification/stabilization (S/S) can be recommended as treatment technology. In the United States, *ex-situ* and *in-situ* S/S technology represents 80% and 21%, respectively of all remedial actions performed at contaminated sites (Dermont et al. 2008). Stabilization refers to conversion of heavy metals into a less soluble, immobile and toxic form as a result of their adsorption to mineral surfaces, formation of stable complexes with organic ligands, surface precipitation and ion exchange (Kumpiene et al. 2008). On the contrary, solidification is based on encapsulation of contaminated soil into a monolithic solid of high structural integrity using cement and asphalt as main amendments (USEPA

2001, Khan et al. 2004). Stabilization is more advantageous than solidification as reduction of metal solubility proceeds without strong alteration of soil properties and enables site revegetation (Dermont et al. 2008).

Until now, for stabilization of heavy metals mainly mineral amendments were tested, including iron compounds, aluminum and manganese oxides, clay minerals, lime and fly ashes (Kumpiene et al. 2008). In Poland, for restoration of degraded lands usually limestone and organic fertilizers (i.e. peat and lignite) are used. Recently, also composts are potentially considered as the appropriate materials for that purpose. The production from sewage sludge make composts low-cost and readily available product. They are commonly used as soil fertilizer improving soil vitality, aeration, water and nutrient holding capacity, the pH and soil revegetation (Cl:AIRE 2008). Due to increasing volume of sewage sludge, in near future utilization of produced composts in other than agriculture sectors will be expected. According to The Waste Resource Application Program, between 2006 and 2010 more than 100.000 tones of compost were going to be used in the regeneration and reclamation of brown field and derelict sites in the UK (Cl:AIRE 2008). Composts are recognized as a potentially important source of humic substances having a high affinity for binding heavy metals (Smith 2009). The properties follow from the presence of negative charges coming from dissociated functional groups, i.e. carboxyl and hydroxyl on the surface of humic substances (Stevenson 1994). Formation of insoluble organometallic complexes with metals can decrease their mobility and crop uptake (Paré et al. 1999; Udom et al. 2004), whereas, the presence of dissolved organic ligands in form of low to medium molecular weight carboxylic acids, amino acids and fulvic acids may lead to the increase of metal solubility. Therefore, not all kinds of compost are able to redistribute and immobilize metals in soils. For example, van Herwijnen et al. (2007) revealed that compost from sewage sludge increased leaching of Cd and Zn from a highly polluted soil, whereas compost from garden green waste immobilized these metals. In particular, the effect of different kinds of Municipal Solid Waste (MSW) composts has been studied for their capacity of increasing metal availability in soil and their uptake by plants (Castaldi et al. 2005, Liu et al. 2009, Farrel, Jones 2010). The lack of consistent data on compost effect on metal mobility, especially in terms of their redistribution in soil during long term stabilization, imposes the need for more research in this area.

Therefore, the aim of present study was to evaluate the effect of compost produced from sewage sludge, in which lignocellulosic materials were straw rape, wooden chips, and grass, on Cu, Cd and Zn redistribution in artificially contaminated soil. Metal stabilization within 2 years time was performed at two compost rates: 5 and 10% (w/w). The efficiency of metal immobilization was established on the basis of changes in their chemical forms over time, Risk Assessment Code (RAC) and binding Intensity (I_R) parameters.

2 MATERIALS AND METHODS

2.1 Soil characteristic

The uncontaminated surface soil samples (depth 0–30 cm) were collected from agricultural area in Baranowo, Warmia and Mazury Province, north-eastern Poland. The soil was air-dried and ground to pass through a 1-mm sieve, then homogenized and stored until the analysis. Physicochemical characteristic of the soil is shown in Table 1.

The uncontaminated soil was classified by its granulometric composition as loamy sand. It showed neutral pH, low Cation Exchange Capacity (CEC) and Organic Matter (OM) content. Total metal concentrations (Cu, Cd and Zn) were well below the permissible limit values (OME 2002).

2.2 Soil contamination

The soil spiking procedure was adapted from Chaiyaraksa and Sriwiriyanuphap (2004). The soil was contaminated by metal mixture. For that purpose, to 500 g of air-dried soil prepared according to p. 2.1. 500 mL of salt solution containing $CuSO_4 \cdot 5H_2O$, $3CdSO_4 \cdot 8H_2O$ and $ZnSO_4 \cdot 7H_2O$ at suitable concentration was added. The samples were mixed for homogeneity by shaking overnight (100 rpm) on mechanical, horizontal shaker, left at room temperature for one month with frequent thorough mixing and rinsing with distilled water. Finally, the contaminated soil was dried at 103–105°C to a constant mass, and finally ground manually and passed through a 1-mm sieve. The metal concentrations in soil were expected to be: 500 mg Cu/kg, 50 mg Cd/kg and 1000 mg Zn/kg.

2.3 Compost characteristic

Compost consisted of dewatered municipal sewage sludge (60%), inoculation (3%), lignocellulosic materials: straw rape (7%), wood chips (15%) and grass (15%) was used as soil amendment. It was produced in two-stage system with maturation phase below 12 months. The compost samples collected directly from a windrow were dried at 105°C and ground in a RETSCH SM-100

Table 1. Physicochemical characteristic of uncontaminated soil (standard deviation from the mean value in parentheses, $n = 3$).

Sand (%)	Silt (%)	Clay (%)	OM (%)	CEC (cmol(+)/kg)	pH_{H2O} (–)	Cu (mg/kg)	Cd (mg/kg)	Zn (mg/kg)
64	16	20	1.6	12.3	7.46	36 (±1.0)	0.3 (±0.07)	57 (±2.0)

Table 2. Physicochemical characteristic of compost (standard deviation from the mean value in parentheses, $n = 3$).

OM (%)	CEC (cmol(+)/kg)	pH_{H2O} (–)	Cu (mg/kg)	Cd (mg/kg)	Zn (mg/kg)
32.1	14.5	6.33	38.5 (±1.1)	0.0 (±0.0)	451.9 (±0.12)

cutting mill (0.5 mm) in order to be homogenous. Physicochemical characteristic of compost is given in Table 2.

The used compost was rich in organic matter and showed slightly acidic reaction. The content of heavy metals varied. The highest concentration was stated for Zn, whereas Cd was not detected.

2.4 Pot experiments

The soil prepared according to p. 2.2 was amended with two compost rates (as w/w): 5% (25 g of compost + 475 g of soil) and 10% (50 g of compost + 450 g of soil). All the mixtures were placed in plastic pots and mixed thoroughly on mechanical shaker. Soil without compost was a control (non-amended) sample. The experimental setup of metal stabilization in soil is shown in Figure 1. During the whole process, soil moisture was regularly controlled. At suitable time intervals soil samples were collected for the analysis of metals in different chemical forms. Soil samples collected at time t = 2 days referred to the beginning of the experiment.

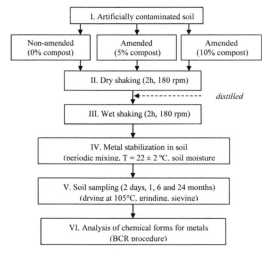

Figure 1. Experimental setup of metal stabilization in soil.

2.5 Analytical methods

Soil particles size was determined using the Mastersizer 2000 analyzer. The pH of soil and compost was measured at 1:2.5 (w/v) soil/water ratio using Hi 221 pH-meter (HANNA Instruments). Organic matter content was measured by loss on ignition at 550°C.

The Cation Exchange Capacity (CEC) was calculated as a sum of hydrolytic acidity (in 1 N $Ca(CH_3COO)_2$) and exchangeable bases (in 0.1 M HCl). Total Cu, Cd and Zn in soil and compost was determined by FAAS (Varian AA280FS, Australia) with earlier microwave digestion of samples in aqua regia (3:1 $HCl:HNO_3$) using MarsXPRESS oven (CEM Corporation, USA).

The chemical fractions of metals were determined according to BCR sequential extraction procedure (Table 3). Sequential extractions were carried out in triplicates. Metal recoveries calculated as the sum of metal in individual fractions to the total metal concentration ranged from 85.1 to 108% in non-amended soil and from 82.4 to 106.4% in amended soil.

2.6 Statistical analysis

Data were statistically evaluated using STATISTICA 9.0 (StatSoft, Inc.). Significantly different means ($p < 0.05$) were assessed by a post hoc Tukey's Least Significant Difference (LSD) test.

3 RESULTS AND DISSCUSION

3.1 Physicochemical properties of non-amended and amended soil

The selected physicochemical properties of soil without and with compost addition are given in Table 4. The data obtained indicate that compared to non-amended soil, addition of compost to soil affected the pH and Organic Matter content (OM). With the increasing compost rate, soil pH decreased from neutral (control soil) to slight acidic (amended soil). The soil pH value strongly affects metal mobility and availability. For effective

Table 3. Sequential extraction procedure (Rauert et al. 2000).

Step	Fraction	Extractant	Extraction conditions
1	Exchangeable and acid soluble (F1)	0.11M CH_3COOH	m/V = 1/40, 16h, 22–24°C
2	Reducible (F2)	0.5M $NH_2OH \cdot HCl$	m/V = 1/40, 16h, 22–24°C
3	Oxidisable (F3)	8.8M H_2O_2 1M CH_3COONH_4	m/V = 1/10, 2h, 85°C m/V = 1/50, 16h, 22–24°C
4	Residual (F4)	$HCl:HNO_3$	Microwave digestion, 0.5h, 170°C

Table 4. Physicochemical characteristic of non-amended and amended soil (standard deviation from the mean value in parentheses, $n = 3$).

Sample type	pH_{H2O} (–)	OM (%)	Cu (mg/kg)	Cd (mg/kg)	Zn (mg/kg)
Control soil	7.46	1.6	570.3 (±13.1)	52.9 (±2.24)	1156.4 (±222.8)
Soil with 5% compost	6.90	4.3	498.0 (±4.7)	50.6 (±1.1)	1017.3 (±5.3)
Soil with 10% compost	6.48	7.2	489.9 (±3.4)	49.0 (±1.6)	977.5 (±8.1)

stabilization, the high pH is the most desirable as it enables formation of precipitates, increasing the number of adsorption sites and metal stability with humic substances (Petruzzelli 1989, Pigozzo et al. 2006). Mendoza et al. (2006) announced that significant increase of pH in soil amended with organic amendments occur when the original pH of the soil is relatively low. In present study the original soil pH was neutral, whereas for compost it was slight acidic. Therefore, pH of amended soil slightly decreased compared to non-amended soil.

In comparison with non-amended soil, the OM content in soil with 5% compost increased by 2.7-fold, whereas with 10% compost rate by 4.5-fold. Similar results were reported by other authors. Castaldi et al. (2005) showed that total organic carbon increased from 3.2% (control sample) to 4.7% in sandy loam containing compost from olive husks, sewage sludge and vegetal waste at rate of 10%. According to Liu et al. (2009) with the change of chicken manure compost rate from 3 to 12% in soil artificially contaminated with Cd at concentration of 50 mg/kg, the increase of organic matter in range from 3.9 to 7.6 g/kg was observed. Achiba et al. (2009) using different rates of municipal solid waste compost stated that the increase of organic matter content in soil was proportional to the application rate. Compost application had also a slight influence on decreasing of total Cu, Cd and Zn compared to non-amended soil (Table 4). With the increase of compost rate, metal concentration tend to decrease. It was attributed to dilution of metal content in soil as compost did not contain Cd, whereas Cu and Zn concentration was visibly lower than in control sample. Castaldi et al. (2005) observed that addition of compost to soil did not statistically modify the total concentrations of Cd and Zn, while the concentration of Pb was lower.

3.2 Metal distribution in non-amended and amended soil

The results on metal fraction distribution in non-amended soil and amended with compost at the beginning of stabilization (t = 2 days) are shown in Figure 2. The investigations have revealed that in control soil, all metals were primarily occurred

in the exchangeable and acid soluble fraction (F1), especially Cd and Zn (Fig. 2b and c). The second, most abundant, was reducible form (F2). The content of organic (F3) and residual (F4) form was the lowest. Cadmium did not occurred in both fractions (Fig. 2b). Similar distribution of Cu, Cd and Zn was observed by Gusiatin & Klimiuk (2012) in loamy sand, despite total metal concentration it was 2-fold higher than in present study. Generally, metals in the exchangeable fractions are considered readily and potentially mobile, while the reducible and oxidizable fractions are relatively stable under normal soil conditions. Metals in the residual fraction are entrapped within the crystal structure of the minerals and represent the least mobile form (Miretzky et al. 2011).

Addition of compost had the highest influence only on changes in Cu distribution in soil. With the increase of compost rate from 5 to 10%, the content of F1 fraction gradually decreased, whereas for the other forms increased. The compost slightly increased relative amount of Cu bound to the organic fraction from 1.1% (control) to 2.2% (5% compost) and 6.3% (10% compost). The results differed from those reported by Perez et al. (2007). According to the authors, Cu concentration associated with organic matter increased significantly with the application rate of composted municipal waste from 12.5 to 100 t/ha.

The differences with the results obtained in present study probably resulted from the fact that the authors performed the experiment at shorter time (1 year) and higher compost pH (7.7–7.8). In the case of Cd and Zn, regardless of compost rate, the fractional pattern for both metals remained similar as in non-amended soil. For Zn, the content of F1 fraction in soil with compost did not change significantly and even slightly increased at compost rate of 5%. It was probably resulted from relatively high content of F1 fraction for Zn in compost, amounting to 36% of its total concentration (data not shown).

3.3 Redistribution of Cu, Cd and Zn in soil within time

Cu, Cd and Zn concentration changes in individual fractions (F1, F2, F3 and F4) in non-amended and

a)

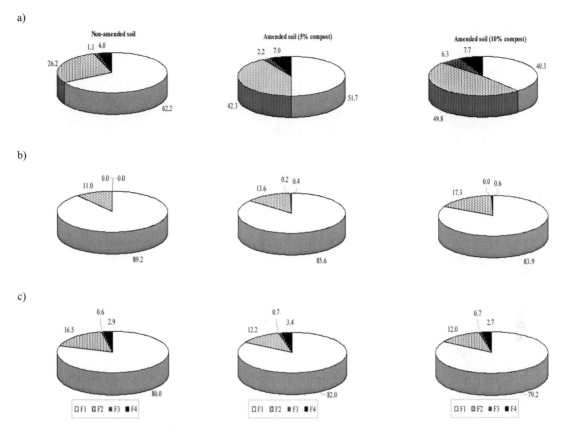

b)

c)

Figure 2. Metal distribution in individual fractions (in %) in non-amended and amended with compost soil at the beginning of stabilization (t = 2 days): a) Cu, b) Cd, c) Zn.

amended soil during 2-year stabilization are shown in Figures. 3, 4 and 5, respectively.

3.3.1 Copper

The investigations have revealed that compost addition influenced Cu concentration in individual fractions during stabilization. Within time, the F1 content decreased in all samples, but the changes were more visible in amended soil, especially during 6 months. 2-year stabilization led to decrease of the content of F1 fraction at 5% compost rate by 31.1%, whereas at 10% rate by 33.5%. In control soil, it was reduced only by 16.4%. Similarly, Paradelo et al. (2011) obtained higher reduction of Cu soluble forms in agricultural soil using municipal solid waste compost at rate of 6% than 3%. However, after 2-year experiment, at 10% compost rate a slight increase of Cu in F1 fraction (Fig. 3a) was observed (p < 0.05) that, from environmental point of view, it is unfavorable phenomenon. Achiba et al. (2009) observed increase of Cu abundance in the exchangeable fraction in soil from 0.9 to 1.26 mg/kg at compost rate of 40 and 120 t/ha, respectively

after 5-year stabilization. Higher compost rate and long stabilization time favored more substantial transformations of Cu in reducible (F2) fraction. Since the 1st month of stabilization, the concentration of F2 visibly increased and the differences were statistically significant (p < 0.05). At lower compost rate, the content of F2 increased only to the 6th month of stabilization (p < 0.05). In control soil, reducible fraction gradually increased (p < 0.05), but the level of concentrations was lower than in soil containing compost (Fig. 3b).

Changes of Cu concentration in organic and residual fractions in amended soil proceeded in similar way within time (Fig. 3c and d). During the first 6 months of stabilization, the content of F3 form increased significantly (p < 0.05), regardless of compost rate. However, at the end of stabilization, decrease of F3 fraction was observed probably as a result of organic matter decomposition as throughout the experiment compost was added only once. From literature review it follows that soil type also influences Cu binding in organic fraction (Smith 2009). Vaca-Paulín et al. (2006) showed

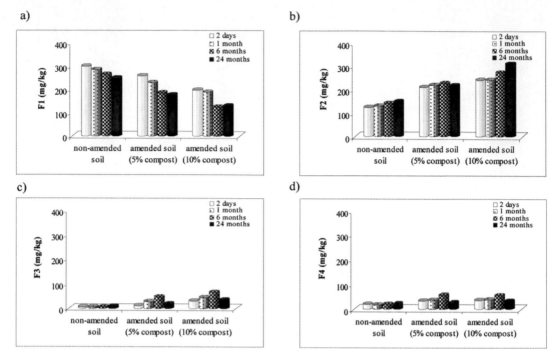

Figure 3. Cu concentration in individual fractions in soil during stabilization: a) F1, b) F2, c) F3, d) F4.

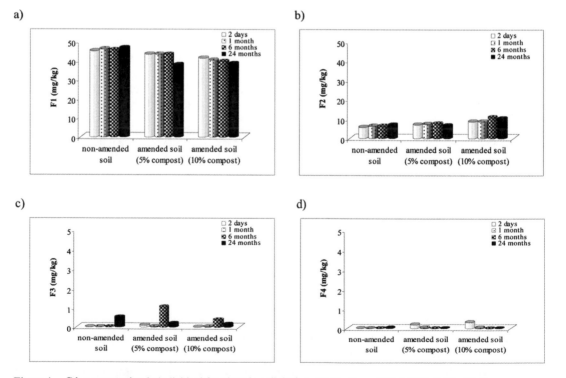

Figure 4. Cd concentration in individual fractions in soil during stabilization: a) F1, b) F2, c) F3, d) F4.

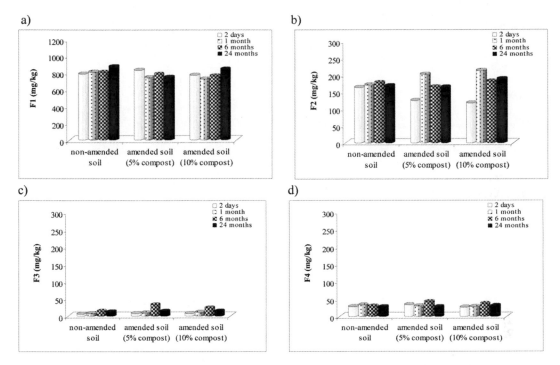

Figure 5.   Zn concentration in individual fractions in soil during stabilization: a) F1, b) F2, c) F3, d) F4.

that in clay loam at pH 6.0 amended with sludge and composted sludge at rate of 200 t/ha, Cu were bound in stable fractions as a result of metal sorption to mineral lattice and clays. In present study, loamy sand was used containing 3-fold more sand than clay that potentially could limit Cu redistribution in F3 and F4 fractions within time.

3.3.2 Cadmium

Redistribution of Cd depended on compost rate. Only at higher compost rate (10%), gradual decrease of F1 and increase of F2 fraction over time were observed (Fig. 4a and b), however, the changes were statistically insignificant (p > 0.05). No significant differences in F1 and F2 fractions within 6 months were noticed at compost rate of 5%. Decline of their concentration after 24 months was much more likely to be caused by low metal recovery (88.1%) rather than the effect of stabilization time. In non-amended soil, Cd concentration in both fractions remained on nearly constant level during the whole stabilization time: 44.8–46.6 mg/kg (F1) and 5.5–6.8 mg/kg (F2) (Fig. 4a and b). Cadmium was not found in organic and residual forms after compost application. Some fluctuations of concentration during stabilization were an evidence that sequential extraction may not be sensitive enough at particularly low metal concentration. The investigations performed by Liu et al. (2009)

revealed that at comparable as in present study total Cd concentration, application of compost at rates of 30, 60 and 120 g/kg to clay-loam soil containing 60.4% silt and only 8.5% sand decreased the amount of soluble/exchangeable Cd by 71.8–95.7% during 4 months. Additionally, increases by 0.9–7.8 times were found for organic fraction. In this study, the composition of sandy soil (16% silt, 64% sand) may limit the effective Cd redistribution into more stable fractions. Another reason might be an insufficient content of organic matter for the effective stabilization, especially in case when metal prevails in F1 fraction. Hanc et al. (2009) noticed that using of triple amount of compost from sewage sludge and wood chips significantly decreased the available Cd in soil. Therefore, the frequency of compost application in sandy soil to provide enough organic matter content might be also an important issue.

3.3.3 Zinc

In contrast to other metals, the increase of compost rate to 10% resulted in rising of F1 content within time from 729.4 mg/kg (t = 1 month) to 843.7 mg/kg (t = 24 months) (p < 0.05). Similar tendency was observed in non-amended soil (Fig. 5a). On the contrary, the lower compost rate did not influence significantly F1 content within time (p > 0.05) and the changes could be affected by incomplete metal recovery (below 100%).

Zinc is frequently the element where the largest increase in labile forms is reported when composted residuals or sewage sludge are applied to soil, particularly under acidic soil conditions (Planquart et al. 1999). One of the reasons of increased leachability of Zn from soil amended with composted sewage sludge is the possibility of complexes formation between metal and dissolved organic matter (van Herwijen et al. 2007). On the other hand, Illera et al. (2000) explained that Zn leachability is connected with pH effect. A pH increase, indeed, results in a decrease of the Zn bioavailability in soils (Jordao et al. 2006), whereas its decrease in the soil solution by only 1 unit increases the solubility of Zn by hundred times (Alloway 1993). The increase of F1 content within time in soil at compost rate of 10% could be affected by changes of soil pH as a result of organic matter decomposition.

Compost derived from sewage sludge had a favorable effect on Zn concentration in reducible fraction, but only for a short time (1 month) (Fig. 5b). Zheliakow and Warman (2004) confirmed that Zn in soil amended with MSW-compost had a high affinity for sorption on the surface of Fe and Mn oxides, particularly as the soil pH is raised. In present study, since the 6th month, the F2 content decreased, but remained nearly constant to the end of the experiment. Summarizing, among all metals, Cu revealed the highest tendency to be redistributed in amended soil, especially at compost rate of 10%. Long stabilization time facilitated its decrease and increase in F1 and F2 fractions, respectively. The changes among F3 and F4 forms were observed for a shorter time. The results obtained indicate that compost from sewage sludge is able to decrease Cu availability in soil and can be used in practice, especially in areas degraded by the metallurgical industry, i.e. Legnicko-Glogowski district (LGOM) in Poland, where Cu occurs at extremely high concentration with significant content of F1 fraction.

3.4 Assessment of metal stabilization in soil based on RAC and I_R

3.4.1 Risk assessment code

The RAC is defined as the percentage content of metal associated with the soil in exchangeable and carbonate fractions (in the BCR procedure it refers to the F1 form) in relation to the total metal concentration. It measures the tendency of metals to be released into the soil and become more rapidly bioavailable. Depending on the F1 fraction content for a given metal, the RAC can change in range: <1% (no risk), 1% to 10% (low risk), 11% to 30% (medium risk), 31% to 50% (high risk), and >50% (very high risk) (Rodríguez et al. 2009).

The changes of metal mobility during 2-year stabilization process in non-amended and amended soil are shown in Figure 6. Before the stabilization, all metals in soil without compost were characterized by very high mobility (RAC >> 50%) caused by significant content of F1 fraction.

Compared to non-amended soil, application of increasing compost rate had differential influence on metal mobility already at the beginning of stabilization (t = 2 days): visible decrease for Cu (Fig. 6a), slight decrease for Cd (Fig. 6b) and slight increase for Zn (Fig. 6c). Based on RAC, 2-year stabilization was the most efficient for Cu, especially at compost rate of 10%, at which metal mobility decreased from "high" category to "medium". For Cd and Zn, both in non-amended and amended soil their mobility was still on "very high" level owing to poor redistribution of F1 fraction into more stable forms. In general, the addition of organic amendments to soil can contribute to metal immobilization through formation of stable complex with OH and COOH groups of organic polymer. However, the increase of metal mobility may be also a sign of insufficient compost maturity (Madrid et al. 2007).

3.4.2 Reduced partition index

Second parameter, used in assessing metal stabilization, is I_R index, developed by Han et al. (2003) to describe the relative binding intensity of metals based on sequential extraction. The I_R index is defined as:

$$I_R = \frac{\sum_{i=1}^{k} i^2 F_i}{k^2} \tag{1}$$

where i is the index number of the extraction step, progressing from 1 (for the weakest) to the most aggressive extractant (in the BCR procedure, $k = 4$), and F is percentage content of the considered metal in fraction i. The range of I_R values is between 0–1: high values indicate metal stability in soils resulting from their occurrence in residual fractions; low values represent distributions with a high proportion of soluble forms; whereas intermediate represent patterns involving metal partitioning among all solid-phase components.

In Figure 7, there are presented the changes of I_R values. From data obtained, it follows that in non-amended soil no significant changes in I_R were observed. In soil containing compost, the binding intensity for Cu increased by the first 6 months of stabilization more stronger than for Cd and Zn. With duration of the experiment it tend to decrease. Cu and Cd binding intensity changes were more visible at higher compost rate (Fig. 7a and b), whereas for Zn they did not depend on compost rate and were comparable with non-amended soil (Fig. 7c). It means that for the effective stabilization of Cu and even Cd

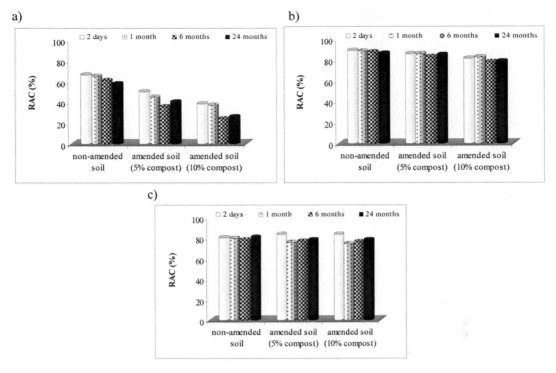

Figure 6. Changes of metal mobility (as RAC) in soil during stabilization: a) Cu, b) Cd, c) Zn.

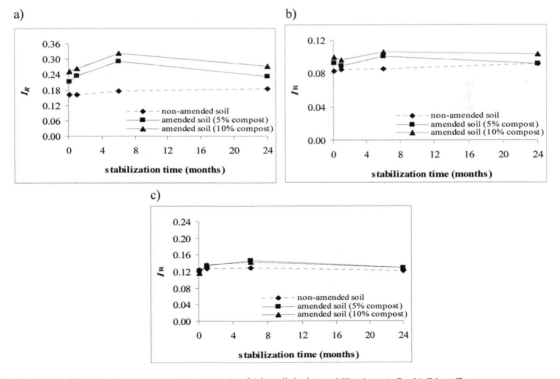

Figure 7. Changes of metal binding intensity (as I_R) in soil during stabilization: a) Cu, b) Cd, c) Zn.

in sandy soil compost should be used at higher rate and added repeatedly during stabilization to overcome losses caused by organic matter decomposition. In case of Zn other factors than organic matter can affect its redistribution, i.e. suitable pH.

Han et al. (2003) in metal salt-spiked and sludge amended soils obtained the lowest values of I_R for Cd, indicating its high bioavailability, whereas intermediate for Zn and Cu. Moreover, the I_R values for metals were higher in loessial than sandy soil. In present study, use of loamy sand may explain the lowest I_R values for Zn and Cd.

To sum up, metal stabilization in soil was assessed on the basis of RAC and I_R. However, there are some differences between these two parameters. The RAC includes metal changes only in F1 fraction and is the most appropriate when metal recovery is complete ($\approx$100%). Owing to the specificity of sequential extraction, the acceptable recoveries range for BCR procedure is 80–120% (Chang et al. 2009). So, when metal prevails in F1 fraction, as in present study, such wide range of recovery, can significantly influence RAC value and therefore metal mobility. In contrast, the I_R parameter refers to transformations among all fractions, with the emphasis on the most stable forms (F3 and F4). As during redistribution, metal are transformed from labile into more stable forms over time, therefore, the I_R may be considered as better tool for metal stabilization assessment in soil.

4 CONCLUSIONS

Redistribution of Cu, Cd an Zn in sandy soil during 2-year stabilization proceeded more efficiently in amended than non-amended soil. The process strongly depended on compost rate and metal species. The most susceptible to transformation between different fractions was Cu at compost rate of 10%, which significantly reduced its mobility in soil. Regardless of compost rate, content of exchangeable and acid soluble fraction for Cd and Zn was still considerable. The binding intensity (as I_R) increased apparently for Cu only during the first 6 months of stabilization suggesting short-term role of organic matter in controlling metal behavior in soil. However, for Zn it did not depend on compost rate and was comparable with non-amended soil. An effective metal immobilization in sandy soil within long time may need high compost rates applied several times during stabilization.

ACKNOWLEDGEMENTS

This study was partially supported by the Ministry of Science and Higher Education of Poland in the frame of the Project No. N523 740640.

REFERENCES

Achiba, W.B., Gabteni, N., Lakhdar, A., Laing, G.D., Verloo, M., Jedidi, N., Gallali, T. 2009. Effects of 5-year application of municipal solid waste compost on the distribution and mobility of heavy metals in a Tunisian calcareous soil. *Agriculture, Ecosystems and Environment* 130: 156–163.

Alloway, B.J. 1993. *Heavy Metals in Soils*. New York: Black Academic.

Castaldi, P., Santona, L., Melis, P. 2005. Heavy metal immobilization by chemical amendments in a polluted soil and influence on white lupin growth. *Chemosphere* 60: 365–371.

Chaiyaraksa, C. & Sriwiriyanuphap, N. 2004. Batch washing of cadmium from soil and sludge by a mixture of $Na_2S_2O_5$ and Na_2EDTA. *Chemosphere* 56: 1129–1135.

Chang, C.-Y., Wang, C.-F., Mui, D.T., Cheng, M.-T., Chiang, H.-L. 2009. Characteristics of elements in waste ashes from a solid waste incinerator in Taiwan. *Journal of Hazardous Materials* 165: 766–773.

CL:AIRE. 2008. The use of compost in the regeneration of brownfield land. *Sustainable Urban Brownfield Management: Integrated Management Bulletins, SUB* 10: 1–4.

Dermont, G., Bergeron, M., Mercier G., Richer-Laflèche, M. 2008. Metal-contaminated soils: remediation practices and treatment technologies. *Practice Periodical of Hazardous, Toxic, and Radioactive Waste Management* © ASCE 12: 188–209.

European Environment Agency (EPA). 2007. Progress in management of contaminated sites (CSI 015), http://www.eea.europa.eu/data-and-maps/indicators/progress-in-management-of-contaminated-sites/progress-in-management-of-contaminated-1/(available on line).

Farrell, M. & Jones, D.L. 2010. Use of composts in the remediation of heavy metal contaminated soil. *Journal of Hazardous Materials* 175: 575–582.

Gusiatin, Z.M., Klimiuk, E. 2012. Metal (Cu, Cd and Zn) removal and stabilization during multiple soil washing by saponin. *Chemosphere* 86: 383–391.

Han, F.X., Banin, A., Kingery, W.L., Triplett, G.B., Zhou, L.X., Zheng, S.J. 2003. New approach to studies of heavy metal redistribution in soil. *Advances in Environmental Research* 8: 113–120.

Hanc, A., Tlustos, P., Szakova, J., Habart, J. 2009. Changes in cadmium mobility during composting and after soil application. *Waste Management* 29: 2282–2288.

Illera, V., Walker, I., Souza, P., Cala, V. 2000. Short-term effects of biosolid and municipal solid waste application on heavy metals distribution in a degraded soil under a semi-arid environment. *Science of the Total Environment* 255: 29–44.

Jordao, C.P., Nascentes, C.C., Cecon, P.R., Fontes, R.L., Pereira, J.L. 2006. Heavy metals availability in soil amended with composted urban solid wastes. *Environmental Monitoring and Assessment* 112: 309–326.

Khan F.I., Husain T., Hejazi R. 2004. An overview and analysis of site remediation technologies. *Journal of Environmental Management* 71: 95–122.

Kumpiene, J., Lagerkvist, A., Maurice, C. 2009. Stabilization of As, Cr, Cu, Pb and Zn in soil using amendments—a review. *Waste Management* 28: 215–225.

Liu, L., Chena, H., Caia, P., Lianga, W., Huanga, Q. 2009. Immobilization and phytotoxicity of Cd in contaminated soil amended with chicken manure compost. *Journal of Hazardous Materials* 163: 563–567.

Madrid, F., López, R., Cabrera, F. 2007. Metal accumulation in soil after application of municipal solid waste compost under intensive farming conditions. *Agriculture, Ecosystems & Environment* 199: 249–256.

Mendoza, J., Garrido, T., Castillo, G., Martin, N.S. 2006. Metal availability and uptake by sorghum plants grown in soils amended with sewage from different treatments. *Chemosphere* 60: 1033–1042.

Miretzky, P., Rodriguez Avendaño, M., Muñoz, C., Carrillo-Chavez, A. 2011. Use of partition and redistribution indexes for heavy metal soil distribution after contamination with a multi-element solution. *Journal of Soils and Sediments* 11: 619–627.

OME—Ordinance of the Minister of Environment on soil and ground quality standards. 2002. *Journal of Law* 165 (1359): 10561–10564 (in Polish).

Paradelo, R., Villada, A., Barral, M.T. 2011. Reduction of the short-term availability of copper, lead and zinc in a contaminated soil amended with municipal solid waste compost. *Journal of Hazardous Materials* 188: 98–104.

Paré, T., Dinel, H., Schnitzer, M. 1999. Extractability of trace metals during cocomposting of biosolids and municipal solid wastes. *Biology and Fertility of Soils* 29: 31–37.

Pérez, D.V., Alcantra, S., Ribeiro, C.C., Pereira, R.E., Fontes, G.C., Wasserman, M.A., Venezuela, T.C., Meneguelli, N.A., Parradas, C.A.A. 2007. Composted municipal waste effects on chemical properties of Brazilian soil. *Bioresource Technology* 98: 525–533.

Petruzzelli, G. 1989. Recycling wastes in agriculture: heavy metal bioavailability. *Agriculture, Ecosystems and Environment* 27: 493–503.

Pigozzo, A.T.G., Lenzi, E., Junior, J.L., Scapin, C., Da Costa, A.C.S. 2006. Transition metal rates in latosol twice treated with sewage sludge. *Brazilian Archives of Biology and Technology* 49: 515–526.

Planquart, P., Bonin, G., Prone, A., Massiani, C. 1999. Distribution, movement and plant availability of trace metals in soils amended with sewage sludge composts: application to low metal loadings. *Science of the Total Environment* 241: 161–179.

Rauert, G., López-Sánchez, J.F., Sahuquillo, A., Barahona, E., Lachica, M., Ure, A.M, Davidson, C.M., Gomez, A., Lück, D., Bacon, J., Yli-Halla, M., Muntau, H., Quevauviller, Ph. 2000. Application of a modified BCR sequential extraction (three-step) procedure for the determination of extractable trace metal contents in a sewage sludge amended soil reference material (CRM 483), complemented by three-year stability study of acetic acid and EDTA extractable metal content. *Journal of Environmental Monitoring* 2: 228–233.

Rodríguez, L., Ruiz, E., Alonso-Azcárate, J., Rincón, J. 2009. Heavy metal distribution and chemical speciation in tailings and soils around a Pb–Zn mine in Spain. *Journal of Environmental Management* 90: 106–116.

Smith, S.R. 2009. A critical review of the bioavailability and impacts of heavy metals in municipal solid waste composts compared to sewage sludge. *Environment International* 35: 142–156.

Stevenson, F.J. 1994. *Humus Chemistry, genesis, composition, reactions.* Canada, John Wiley & Sons Inc.

Udom, B.E., Mbagwu, J.S.C., Adesodun, J.K., Agbim, N.N. 2004. Distributions of zinc, copper, cadmium and lead in tropical ultisol after long-term disposal of sewage sludge. *Environment International* 30: 467–470.

United States Environmental Protection Agency (USEPA). 2001. A Citizen's guide to Solidification/Stabilization (EPA 542-F-01-024), http://www.epa.gov/superfund/community/pdfs/suppmaterials/treatmenttech/solidification_stabilization.pdf (available on line).

Vaca-Paulín, R., Esteller-Alberich, M.V., Lugo-de la Fuente, J., Zavaleta-Mancera, H.A. 2006. Effect of sewage sludge or compost on the sorption and distribution of copper and cadmium in soil. *Waste Management* 26: 71–81.

van Herwijnen, R., Hutchings, T.R., Al-Tabbaa, A., Moffat, A.J., Johns, M.L., Ouki, S.K. 2007. Remediation of metal contaminated soil with mineral-amended composts. *Environmental Pollution* 150: 347–354.

Zheljazkov, V.D. & Warman, P.R. 2004. Phytoavailability and fractionation of copper, manganese, and zinc in soil following application of two composts to four crops. *Environmental Pollution* 131: 187–95.

Environmental Engineering IV – Pawłowski, Dudzińska & Pawłowski (eds)
© 2013 Taylor & Francis Group, London, ISBN 978-0-415-64338-2

Variability of nutrients in co-digestion of sewage sludge and old landfill leachate

A. Montusiewicz, M. Lebiocka & A. Szaja
Faculty of Environmental Engineering, Lublin University of Technology, Lublin, Poland

ABSTRACT: The study examined the variability of N and P concentrations in bioaugmented and non-bioaugmented anaerobic co-digestion of sewage sludge and old landfill leachate. On the basis of different nutrient forms (TN, N-NH$^+$, TP, P-PO$_4^{3-}$) a verification was made whether adding leachate from the matured landfill, as well as bioaugmentation with the use of commercial product Arkea®, could lead to the changes of nutrient content in the digest and its supernatant. The results indicated that bioaugmented and non-bioaugmented co-digestion of sewage sludge and old landfill leachate had no essential effect on the digester supernatant quality. However, a significant influence was found considering digest composition. The much lower concentration of total nitrogen and total phosphorus was achieved applying bioaugmented co-digestion.

Keywords: co-digestion, bioaugmentation, sewage sludge, landfill leachate, concentration of N and P compounds

1 INTRODUCTION

In municipal Wastewater Treatment Plants (WWTPs), anaerobic digestion (also defined as biomethanization) has commonly been used for both primary and waste sludge treatment. Biomethanization of such substrates (usually mixed) results in a high release of the nitrogen and phosphorus compounds into the supernatant, recycled further to the suspended-growth bioreactors carrying out advanced removal of C, N and P (Guo et al. 2010, Montusiewicz et al. 2010). Although the flow of the rejected supernatant is relatively small, it could contribute from 10 to 30% of the nitrogen load and from 10 to 80% of the phosphorus load in the bioreactor (van Loosdrecht & Salem 2006). A high N and P content in the supernatant stream can result in its overloading, leading to diminishing the nutrient removal efficiency and an increase of the cost of the total treatment. In view of this, investigation of the variability of N and P concentrations in anaerobic digestion using different substrates seems to be of great scientific and practical significance.

Among many technologies that improve the effectiveness of anaerobic treatment, co-digestion is currently the most widely studied since it provides increased biogas yields as well as a stable performance of the process compared to one-substrate biomethanization. Co-digestion refers to the anaerobic digestion of selected biodegradable substrate/substrates (preferably organic waste) with a base

substrate that system was designed to typically be fed. In WWTP this is sewage sludge. Many different types of co-substrates have been proposed in the literature. The most common are: organic fraction of municipal solid waste (Edelmann et al. 2000, Schmit & Ellis 2001, Sosnowski et al. 2003) and its selected group known as fruit and vegetable waste (Dinsdale et al. 2000, Lahdheb et al. 2009), fats, oils and grease (Davidsson et al. 2008, Kabouris et al. 2009, Luostarien et al. 2009, Silvestre et al. 2011), and waste from agricultural and food industry (Murto et al. 2004, Pecharaply et al. 2007, Romano & Zhang 2008). Co-digestion of sewage sludge with landfill leachate from controlled sanitary landfills of different decomposition stage (and thus different age of landfill) has also been investigated (Hombach et al. 2003, Montusiewicz & Lebiocka 2010, Lebiocka et al. 2010). However, studies regarding the problem of nitrogen and phosphorus release during co-digestion have not been found, except for our previous paper (Montusiewicz & Lebiocka 2010).

Currently, an advanced approach incorporates a new technique known as bioaugmentation to improve the co-digestion effects. A bioaugmented system is especially worth studying in the case of involving the low-biodegradable substrate which contains a high concentration of refractory compounds (e.g. old landfill leachate). Bioaugmentation is defined as an application of a specific strain or a consortium of microorganisms to enhance a required biological activity in the system under investigation (Rittmann & Whiteman 1994). This method has been involved in many different environmental areas, including bioremediation of contaminated soils and groundwater (D'Annibale et al. 2005, Hamdi et al. 2007), aerobic and anaerobic treatment systems as well as odor control. In wastewater treatment plants bioaugmentation has most commonly been used for increasing the population of nitrifying bacteria (Head & Oleszkiewicz 2005). Besides, it has been applied to the start-up of bioreactors (Guo et al. 2010, Schauer-Gimenez et al. 2010) and to recovering their stable performance after shock loads, fluctuations in pH, toxic agents or temperature changes. This technique has also been successfully involved in anaerobic systems to accelerate a digester's start-up, reduce an odor (Duran et al. 2006, Tepe et al. 2008) and improve both the process stability and biogas yields (Saravanane et al. 2001). Moreover, it has been used to decrease the recovery period of anaerobic digesters exposed to toxic events (Schauer-Gimenez et al. 2010). Taking into account the advantages mentioned above, it is worth investigating the effect of bioaugmentation on co-digestion of sewage sludge and old landfill leachate, known so far to be ineffective (Lebiocka et al. 2010).

In the present study, the variability of N and P concentrations in bioaugmented and non-bioaugmented anaerobic co-digestion of sewage sludge and old landfill leachate has been examined to verify whether adding leachate originating from the matured landfill, as well as bioaugmentation using commercial product Arkea®, could lead to the changes of nutrient content in the digest and its supernatant.

2 MATERIALS AND METHODS

Sewage sludge (primary and waste sludge after thickening) was obtained from the Puławy municipal wastewater treatment plant (Poland). Thickened sludge was sampled once a week in the WWTP, then transported and mixed at a volume ratio of 60:40 (primary:waste sludge) under laboratory conditions. The sludge samples were homogenized, screened through a 3 mm screen and stored at 4°C in a laboratory fridge for a week at the longest. Sludge prepared in this manner fed the digester as mixed Sewage Sludge (SS). The main characteristics of SS during experiments are presented in Table 1.

Leachate was sampled once as a collected sample achieved from a storage tank of old leachate in the Rokitno municipal solid waste landfill (the age of landfill exceeded 10 years). Under laboratory conditions it was homogenized and partitioned, then frozen and stored at −25°C in a laboratory

Table 1. Sewage sludge (SS) composition (the mean value and standard deviation are given).

Parameter	Unit	Experiment 1	Experiment 2
COD	g m^{-3}	41703 ± 4623	44895 ± 7958
Total Solids (TS)	g kg^{-1}	33.2 ± 4.31	36.7 ± 6.02
Volatile Solids (VS)	g kg^{-1}	25.3 ± 3.41	26.3 ± 5.05
pH	–	6.89 ± 0.96	6.55 ± 0.33
Total Nitrogen (TN)	g m^{-3}	2228 ± 468	3324 ± 327
Ammonia Nitrogen (N-NH$_4^+$)	g m^{-3}	64 ± 43	49 ± 29
Total Phosphorus (TP)	g m^{-3}	464 ± 131	582 ± 221
Phosphate Phosphorus (P-PO$_4^{3-}$)	g m^{-3}	122 ± 29	118 ± 26

Table 2. Leachate composition (the mean value and standard deviation are shown).

Parameter	Unit	Mean value
COD	g m^{-3}	5605 ± 156
BOD$_5$/COD	–	0.05 ± 0.01
Total Solids (TS)	g kg^{-1}	25.4 ± 0.95
Volatile Solids (VS)	g kg^{-1}	14.3 ± 0.62
pH	–	7.95 ± 0.06
Total Nitrogen (TN)	g m^{-3}	7200 ± 123
Ammonia Nitrogen (N-NH$_4^+$)	g m^{-3}	6915 ± 114
Total Phosphorus (TP)	g m^{-3}	71 ± 15.1
Phosphate Phosphorus (P-PO$_4^{3-}$)	g m^{-3}	39 ± 8.4

freezer. The leachate samples were thawed daily for 6 hours at 20°C in indoor air. The Old Leachate (OL) composition is presented in Table 2.

The commercial product Arkea® was taken for bioaugmentation and prepared as a liquor in continuous mode using the procedure given by ArchaeaSolutions Inc. The average TS and VS of the Arkea® liquor were 0.43 and 0.07 g kg^{-1}, respectively, during Experiment 1, whereas 0.47 and 0.19 g kg^{-1} in Experiment 2.

The study was carried out in reactors operating at the temperature of 35°C in semi-flow mode. The laboratory installation consisted of three completely mixed anaerobic digesters (with an active volume of 40 dm^3) working in parallel, equipped with a gaseous installation, an influent peristaltic pump and storage vessels. The gas system consisted of pipelines linked with a pressure equalization unit and a mass flow meter. The laboratory installation is shown in Figure 1.

An inoculum for the laboratory reactors was obtained from the Puławy wastewater treatment plant as a digest collected from an anaerobic digester operating at mesophilic temperature. The adaptation of the digester biomass was achieved after 30 days.

The study was divided into two experiments taking into consideration a statistically significant difference between the average concentration of TN in a base substrate (sewage sludge) during the entire investigation time. Each experiment consisted of three runs lasting 90 days (30 days for a complete exchange of reactor volume after a switch in the feed content and 60 days for measurements) and was carried out simultaneously in three parallel

Figure 1. Laboratory installation of anaerobic digestion of waste in a wet system.

systems. The digesters were supplied regularly once a day with an applied volume of the SS or mixture of SS and OL with or without an addition of Arkea® in the designed dose. The feed composition and operational parameters are shown in Table 3.

In the sewage sludge and its supernatant the following parameters were analyzed once a week: total nitrogen, ammonia nitrogen (N-NH$_4^+$), nitrite and nitrate nitrogen (N-NO$_x^-$), total phosphorus, ortho-phosphate phosphorus (P-PO$_4^{3-}$), pH and alkalinity. The supernatant samples were obtained by centrifuging the sludge at 4000 r min^{-1} for 30 min.

The leachate composition was determined once after it arrived at the laboratory. The value of the parameters were examined using the SS schedule.

In the digest and its supernatant, the specified parameters were determined twice a week, in accordance with the timetable accepted.

Most analyses were carried out in accordance with Polish Standard Methods. The analyses of N-NH$_4^+$, N-NO$_x^-$ and P-PO$_4^{3-}$ were performed with FIASTAR 5000 using FOSS analytical methods. Ammonium was determined according to ISO 11732, nitrite and nitrate according to ISO 13395, and orthophosphate in accordance with ISO/FDIS 15681-1.

3 RESULTS AND DISSCUSSION

The results of the study are shown in Tables 4 and 5 (average data are reported) and in Figures 2 and 3. It should be noticed that feed conditions varied throughout the experiments, which was attributed both to the changes in sludge characteristics and the leachate and Arkea® liquor addition. Moreover, nitrite and nitrate nitrogen was not detected.

Total nitrogen concentration (TN) significantly increased in the digester feed as an effect of

Table 3. Feed composition and operational regime during experiments.

Run	Feed composition	Feed volume (dm^3)	HRT (d)	OLR (kg VS m^{-3}d^{-1})
Experiment 1				
R 1.1 (control)	SS (sewage sludge)	2.0	20	1.27
R 1.2	SS + OL 95:5 v/v	2.0 + 0.1	19.1	1.30
R 1.3	SS + OL + Arkea® 90:5:5 v/v	2.0 + 0.1 + 0.1	18.2	1.30
Experiment 2				
R 2.1 (control)	SS	2.0	20	1.32
R 2.2	SS + OL 95:5 v/v	2.0 + 0.1	19.1	1.35
R 2.3	SS + OL + Arkea® 87:4.3:8.7 v/v	2.0 + 0.1 + 0.2	17.4	1.33

* HRT—hydraulic retention time, OLR—organic loading rate.

Table 4. Concentration of TN and TP, as well as nutrient removal efficiency η, during experiments (concentrations given in g m⁻³).

	Experiment 1						Experiment 2					
	Runs						Runs					
	R 1.1		R 1.2		R 1.3		R 2.1		R 2.2		R 2.3	
Parameter	Feed	Digest	Feed	Digest	Feed	Digest	Feed	Digest	Feed	Digest	Feed	Digest
TN	2228	2197	2465	2026	2356	1887	3324	3260	3509	3270	3210	2920
η_{TN}	1.2		17.8		19.9		1.9		8.1		8.9	
TP	464	412	446	398	425	332	582	557	558	463	509	443
η_{TP}	11.2		10.8		17.7		11.7		11.5		12.9	

Table 5. Alkalinity and pH value of digest (concentrations given in g $CaCO_3$ m⁻³).

	Experiment 1			Experiment 2		
	Runs			Runs		
	R 1.1	R 1.2	R 1.3	R 2.1	R 2.2	R 2.3
Parameter	SS	SS + OL	SS + OL + Arkea®	SS	SS + OL	SS + OL + Arkea®
Alkalinity	2916	3333	3359	2650	3034	3020
pH	7,56	7,62	7,49	7,67	7,77	7,61

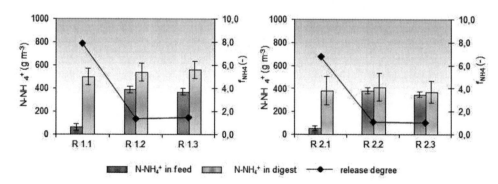

Figure 2. Concentration of ammonia nitrogen and release degree f_{NH4} during experiments (error bars represent confidence levels, $\alpha = 0,05$).

leachate addition (runs R 1.2 and R 2.2), however, its content in digest was much lower (in Experiment 1) or similar (in Experiment 2) to those achieved for sewage sludge. This indicated a higher removal efficiency confirmed by η_{TN} value (Table 4). Using Arkea® apart from leachate, the analogous tendency was observed, while TN removal was improved.

In contrast, a decrease of total phosphorus concentration (TP) was found in the feed during runs with leachate addition. In such cases digest contained less phosphorus as compared to sewage sludge, but removal efficiency η_{TP} sustained at the same level. Introducing Arkea® deeply decreased TP feed concentration, but simultaneously enhanced its removal efficiency. One of the reasons of improved nutrient removal throughout co-digestion and bioaugmented co-digestion could be the fact that such systems operating at shorter Hydraulic Retention Time (HRT) (Table 3), thus nitrogen and phosphorus incorporated in microorganism

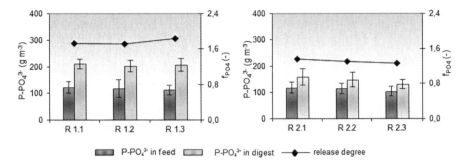

Figure 3. Concentration of orthophosphorus phosphate and release degree f_{PO4} during experiments (error bars represent confidence levels, $\alpha = 0{,}05$).

cells were washout faster from digesters. It should also be noticed that variable data obtained during Experiment 1 and 2 could be resulted from statistically different TN concentration in sewage sludge.

Despite the significant variability of ammonia nitrogen concentration in the digester feed caused by an old landfill leachate addition (such a tendency did not appear with regard to the orthophosphate phosphorus), quite a consistent quality of the digest was achieved in all investigated runs (Figs. 2, 3). To evaluate the difference between concentrations of the inorganic forms of N and P both in the feed and digest, the release degree was calculated considering $N-NH_4^+$ and $P-PO_4^{3-}$. This was defined as a ratio of the reactor effluent load to influent load and indicated as f_{NH4} and f_{PO4}, respectively.

On determining the f_{NH4}, a clearly visible release of $N-NH_4^+$ was confirmed. During control runs (R 1.1, R 2.1) the high values of 7.9 and 6.8 were achieved, whereas in the runs with leachate and optionally with Arkea® liquor addition, the release degree was much lower and equaled, respectively, 1.4 and 1.5 in Experiment 1 (runs R 1.2 and R 1.3) and 1.1 in Experiment 2 (runs R 2.2 and R 2.3). This observation was consistent with the authors' previous research (Montusiewicz & Lebiocka 2011) regarding co-digestion of sewage sludge and intermediate landfill leachate. Such releases appeared also for the orthophosphate phosphorus. In this case the values of f_{PO4} were comparable throughout experiments, confirming the analogous release efficiency in different systems. In Experiment 1 the release degree was 1.74 (R 1.1), 1.71 (R 1.2) and 1.84 (R 1.3), respectively, whereas somewhat lower values were obtained during Experiment 2: 1.35 (R 2.1), 1.30 (R2.2) and 1.27 (R 2.3).

The analysis of the nitrogen loads indicated that ammonification released the load of 827 mg d^{-1} in R 1.1 (control), 317 mg d^{-1} in R 1.2 and 416 mg d^{-1} in R 1.3. During Experiment 2 the loads released by ammonification decreased to 652 mg d^{-1} in R 2.1 (control) and even to 65 and 53 mg d^{-1},

respectively, in R 2.2 and R 2.3. Considering the above mentioned, introducing the old landfill leachate as a co-substrate in anaerobic digestion caused a minor release of $N-NH_4^+$ into the supernatant as compared to sewage sludge. Bioaugmentation using commercial product Arkea® (apart from OL) did not influence the effect of the co-digestion process since the loads released were quite similar to those calculated for non-bioaugmented runs with OL addition. One reason of this state of affairs could be the fact that the released ammonium nitrogen during co-digestion and bioaugmented co-digestion could have been precipitated as struvit ($MgNH_4PO_4$). The latter explanation complies with a lower $P-PO_4^{3-}$ concentration (Fig. 3) and a higher pH value, as well as alkalinity, in the digest supernatant after co-digestion of SS and OL (Table 5). Although in bioaugmented systems (runs R 1.3 and R 2.3) a lower pH value was found, the alkalinity increased significantly as compared to sewage sludge (runs R 1.1 and R 2.1) that had the beneficial effect due to the specific buffering conditions. The observed $N-NH_4^+$ values in the digest supernatants under investigation were similar to those reported by Song et al. (2004) during mesophilic biomethanization of sewage sludge, however much higher $P-PO_4^{3-}$ concentrations were achieved in our study.

4 CONCLUSIONS

The results indicated that bioaugmented and non-bioaugmented co-digestion of sewage sludge and old landfill leachate had no essential effect on the digester supernatant quality. However, a significant influence was found considering digest composition. The much lower concentration of total nitrogen and total phosphorus was achieved applying bioaugmented co-digestion with commercial product Arkea®. In the control runs and systems without bioaugmentation comparable results were obtained.

ACKNOWLEDGEMENTS

The authors thank for the financial support from the National Centre of Science (Poland), No. 7405/B/T02/2011/40.

REFERENCES

D'Annibale, A., Ricci, M., Leonardi, V., Quaratino, D., Mincione, E., Petruccioli, M. 2005. Degradation of aromatic hydrocarbons by white-rot fungi in a historically contaminated soil. *Biotechnol. Bioeng.* 90 (6): 723–731.

Davidsson, Å, Lövstedt, C., la Cour Jansen, J., Gruvberger, C., Aspegren, H. 2008. Co-digestion of grease trap sludge and sewage sludge. *Waste Manage* 28: 986–992.

Dinsdale, R.M., Premier, G.C., Hawkes, F.R., Hawkes, D.L. 2000. Two-stage co-digestion of waste activated sludge and fruit/vegetable waste using inclined tubular digesters. *Bioresour. Technol.* 72: 159–168.

Duran, M., Tepe, N., Yurtsever, D., Punzi, V.L., Bruno, C., Mehta, R.J. 2006. Bioaugmenting anaerobic digestion of biosolids with selected strains of Bacillus, Pseudomonas, and Actinomycetes species for increased methanogenesis and odor control. *Appl. Microbiol. Biotechnol.* 73: 960–966.

Edelmann, W., Engeli, H., Gradenecker, M. 2000. Co-digestion of organic solid waste and sludge from sewage treatment. *Water Sci. Technol.* 41(3): 213–221.

Guo, H.C., Stabnikov, V., Ivanov, V. 2010. The removal of nitrogen and phosphorus from reject water of municipal wastewater treatment plant using ferric and nitrate bioreductions. *Bioresour. Technol.* 101 (11): 3992–3999.

Hamdi, H., Benzarti, S., Manusadžianas, L., Aoyama, I., Jedidi, N. 2007. Bioaugmentation and biostimulation effects on PAH dissipation and soil ecotoxicity under controlled conditions. *Soil Biol. Biochem.* 39 (8): 1926–1935.

Head, M.A. & Oleszkiewicz, J.A. 2005. Bioaugmentation with nitrifying bacteria acclimated to different temperatures. *J. Environ. Eng.* 131(7): 1046–1051.

Hombach, S.T., Oleszkiewicz, J.A., Lagasse, P., Amy, L.B., Zaleski, A.A., Smyrski, K. 2003. Impact of landfill leachate on anaerobic digestion of sewage sludge. *Environ. Technol.* 24: 553–560.

Kabouris, J.C, Tezel, U., Pavlostathis, S.G., Engelmann, M., Dulaney, J.A., Todd, A.C., Gillette, R.A. 2009. Mesophilic and thermophilic anaerobic digestion of municipal sludge and fat, oil, and grease. *Water Environ. Res.* 81(5): 476–485.

Lahdheb, H., Bouallagui, H., Hamdi, M. 2009. Improvement of activated sludge stabilization and filterability during anaerobic digestion by fruit and vegetable waste addition. *Bioresour. Technol.* 100: 1555–1560.

Lebiocka, M., Montusiewicz, A., Zdeb, M. 2010. Anaerobic co-digestion of sewage sludge and old landfill leachate. *Monogr. Pol. J. Environ. Stud.* 2: 141–145.

Luostarinen, S., Luste, S., Sillanpää, M. 2009. Increased biogas production at wastewater treatment plants through co-digestion of sewage sludge with grease trap sludge from a meat processing plant. *Bioresour. Technol.* 100: 79–85.

Montusiewicz, A., Lebiocka, M., Rożej, A., Zacharska, E., Pawłowski, L. 2010. Freezing/thawing effects on anaerobic digestion of mixed sewage sludge. *Bioresour. Technol.* 101: 3466–3473.

Montusiewicz, A. & Lebiocka, M. 2011. Co-digestion of intermediate landfill leachate and sewage sludge as a method of leachate utilization. *Bioresour. Technol.* 102: 2563–2571.

Murto, M., Björnsson, L., Mattiasson, B. 2004. Impact of food industrial waste on anaerobic co-digestion of sewage sludge and pig manure. *J. Environ. Manage* 70: 101–107.

Pecharaply, A., Parkpian, P., Annachhatre, A.P., Jugsujinda, A. 2007. Influence of anaerobic co-digestion of sewage and brewery sludges on biogas production and sludge quality. *J. Environ. Sci. Health Part A* 42: 911–923.

Pitman, A.R. 1999. Management of biological nutrient removal plant sludges—change the paradigms. *Water Res.* 33: 1141–1146.

Rittmann, B.E. & Whiteman, R. 1994. Bioaugmentation: a coming of age. *Water Qual. Int.* 1: 12–16.

Romano, R.T. & Zhang, R. 2008. Co-digestion of onion juice and wastewater sludge using an anaerobic mixed biofilm reactor. *Bioresour. Technol.* 99: 631–637.

Saravanane, R., Murthy, D.V.S., Krishnaiah, K. 2001. Bioaugmentation and treatment of cephalexin drug-based pharmaceutical effluent in an upflow anaerobic fluidized bed system. *Bioresour. Technol.* 76: 279–281.

Schauer-Gimenez, A., Zitomer, D., Maki, J., Struble, C. 2010. Bioaugmentation for improved recovery of anaerobic digesters after toxicant exposure. *Water Res.* 44: 3555–3564.

Schmit, K.H. & Ellis, T.G. 2001. Comparison of temperature phased and two phase anaerobic co-digestion of primary sludge and municipal solid waste. *Water Environ. Res.* 73(3): 314–321.

Silvestre, G., Rodriguez-Abalde, A., Fernández, B., Flotats, X., Bonmatì, A. 2011. Biomass adaptation over anaerobic co-digestion of sewage sludge and trapped grease waste. *Bioresour. Technol.* 102: 6830–6836.

Song, Y.-C., Kwon, S.-J., Woo, J.-H. 2004. Mesophilic and thermophilic temperature co-phase anaerobic digestion compared with single-stage mesophilic and thermophilic digestion of sewage sludge. *Water Res.* 38: 1653–1662.

Sosnowski, P., Wieczorek, A., Ledakowicz, S. 2003. Anaerobic co-digestion of sewage sludge and organic fraction of municipal solid waste. *Adv. Environ. Res.* 7: 609–616.

Tepe, N., Yurtsever, D., Mehta, R.J., Bruno, C., Punzi, V.L., Duran, M. 2008. Odor control during post-digestion processing of biosolids through bioaugmentation of anaerobic digestion. *Water Sci. Technol.* 57 (4): 589–594.

van Loosdrecht, M.C.M. & Salem, S. 2006. Biological treatment of sludge digester liquids. *Water Sci. Technol.* 53: 11–20.

Environmental Engineering IV – Pawłowski, Dudzińska & Pawłowski (eds)
© *2013 Taylor & Francis Group, London, ISBN 978-0-415-64338-2*

Co-digestion of Organic Fraction of Municipal Solid Waste with different organic wastes: A review

A. Grosser, M. Worwąg, E. Neczaj & T. Kamizela

Institute of Environmental Engineering, Czestochowa University of Technology, Czestochowa, Poland

ABSTRACT: The paper summarizes research results of Organic Fraction of Municipal Solid Waste (OFMSW) co-digestion with various wastes. The possible use of sewage sludge, livestock waste and other organic waste (for instance olive mill effluents, organic solid poultry slaughterhouse waste) as a co-substrate in co-digestion process has been evaluated. The presented data shows huge potential and numerous benefits of co-digestion of organic fraction of municipal solid waste with different wastes. In most cases, the addition of co-substrate to the OFMSW digester resulted in an improved the biogas yield as well as volatile solid removal.

Keywords: organic fraction of municipal solid waste, co-digestion, organic waste, manure, sewage sludge

1   INTRODUCTION

For a long time storing Municipal Solid Waste (MSW) was the main way of its disposal all over the world. For example, in 2006 approximately 61%, 63%, 67%, 69%, and 75% of the total MSW generated was landfilled, in America, Asia, Europe, Africa, and Oceania respectively (Pognani et al. 2009). At the present, landfilling is no longer the advisable solid waste management method in European countries due to stricter waste storage regulations (e.g. Landfill Directive 1999/31/EC and Working Document on Biological Treatment of Biowaste 2nd Draf) as well as the scarcity of land (Dereli et al. 2010, Pognani et al. 2009, Bień et al. 2007). All these factors increased the interest in biological methods of biodegradable fraction of the landfilled MSW (Organic Fraction of Municipal Solid Waste—OFMSW) stabilization (Ağdağ & Sponza 2005, Pognani et al. 2009, Mata-Alvarez 1996. Liu et al. 2007). Thus, composting was one of the main options for processing the organic fraction of municipal solid waste (Bień et al. 2010, Bień et al. 2011, Nguyen et al. 2007, Pognani et al. 2009). However, new EU policy supports the increase of the renewable energy share in the general energy balance, which caused that anaerobic digestion of OFMSW has gone rapidly growing (Pognani et al. 2009, Esposito et al. 2012). The use of AD for treating organics waste offers the following benefits: the reduction of sludge volume and organic content of the sludge, improvement of sludge dewaterability, destruction of most pathogens and production of a renewable and inexpensive energy in the form of methane as well as generation of a potentially valuable byproduct that can be used for agricultural purposes (Esposito et al. 2012, Pognani et al. 2009, Bień et al. 2004, Zawieja et al. 2008, Griffin et al. 1998, Liu et al. 2007, Yu and Schanbacher 2010, Nguyen et al. 2007).

An interesting option for improving yields of AD of OFMSW is co-digestion, namely the simultaneous anaerobic digestion of a homogeneous mixture of at least two components (Del Borghi et al. 1999, Mata-Alvarez 2000, Bień et al. 2010, Neczaj et al. 2012, Liu et al. 2009, Cuetos et al. 2011). The article presents the recent results of research into organic fraction of municipal waste anaerobic co-digestion and show current knowledge of how co-substrates affect processes.

2   CHARACTERISTIC OF ORGANIC FRACTION OF MUNICIPAL SOLID WASTE AND WASTE GENERATION RATES

Municipal solid waste is the nonliquid, nonhazardous waste generated in a community. Thus, MSW includes waste from households, institutions (e.g. schools, hospitals, universities etc.), and commercial (e.g. from markets, hotels, shops etc.) (Bień et al. 2011). Nevertheless, municipal solid waste is typically composed of the following fractions (Zaher et al. 2007):

- Inert fraction, e.g. metal, sand, glass or stones;
- Low or non-digestible organic fraction, e.g. wood, paper, cardboard;

Table 1. Characteristic of different selected organic fraction of MSW (Dong et al. 2010).

Parameters	Unit	WS-OFMSW	SS-OFMSW	MS-OFMSW
Particle size	mm	≤10	–	≤30
Density	g l^{-1}	933	–	295
TS	g kg^{-1}	184	170–370	172
VS	% of TS	61.6	81–92	43
Ash	% of TS	38.4	8–19	57
Proteins	% of TS	14.2	10–18	–
Lipids	% of TS	9.6	10–18	–
C:N ratio	–	11.4	15.5–20.5	11.9
Carbohyd-rates	% of TS	37.8	–	–
Heat value	MJ kg^{-1} TS	21	19–22	–

WS-OFMSW—water sorted organic fraction of municipal solid waste; SS-OFMSW—source sorter organic household; MS-OFMSW—mechanically selected organic fraction of MSW.

Table 2. Waste generation rates and physical composition of MSW in selected countries (Jayarama Reddy 2011, den Boer E. et al. 2010).

Country	MSW generation rate (kg/capita/day)	Physical composition of MSW					
		Organic	Paper	Plastic	Metal	Glass	Other
Canada	1.80	34	28	11	8	7	13
USA	2.0	23	38	9	8	7	16
Japan	1.12	26	46	9	8	7	12
Norway	1.40	18	31	6	5	4	36
Turkey	1.09	64	6	3	1	2	24
France	1.29	25	30	10	6	12	17
Poland	0.93	37.8	34.5	6.8	7.1	6.1	9.9
Australia	1.89	50	22	7	5	9	8
Spain	0.99	44	21	11	4	7	13
Switzerland	1.10	27	28	15	3	3	24

- Digestible organic fraction—readily biodegradable organic matter, e.g. grass cuttings, food residue, kitchen scraps, food processing wastes, etc.

One of the major factors affecting the characteristics of Municipal Solid Waste (MSW) is collection system. Separately collected fractions of MSW contain a higher content of organic fraction, with low participation of non-biodegradable contaminants (plastic, metal and glass), while the selected organic fraction of MSW is more contaminated (Table 1) (Castillo et al. 2006, Hartmann & Ahring 2006).

The generation rate and composition of municipal waste can be affected by various factors, including the economic situation, climate, industrial structure, collection frequency, "degree of urbanization", season, wastes management regulations as well as life style (Jayarama Reddy 2011, Forster-Carneiro et al. 2007, Li et al. 2011, Pichtel 2005, Hanc et al. 2011). Table 2 provides waste generation rates and compositions in selected countries.

3 METHODS OF WASTE DISPOSAL

Figure 1 shows methods of waste disposal in Poland in 2009. Thermal disposal of OFMSW (e.g. direct combustion) is economically unviable due to high moisture content and low calorific value of this waste. For this reason, a much better alternative to the use of OFMSW is biological treatment, such as composting or anaerobic digestion. The main advantages of AD as compared to composting are the following (Hartmann & Ahring 2006, Baldasano & Soriano 2000, Edelmann et al. 2000, Khalid et al. 2011, Bidlingmaier et al. 2004):

- positive energy balance—in the case of composting, energy consumption is approximately 30–35 kWh Mg^{-1} of waste output, while by AD process can have a net energy yield equivalent to100–150 kWh Mg^{-1} of input waste.
- lower area requirements.
- minimized emission of odors and greenhouse gases into the environment due to the fact that the whole process is conducted in the reactors.

Landfilled; 78%

Composted; 7%

Recycled; 14%

Incinerated; 1%

Figure 1. Disposal of municipal solid waste in Poland in 2009 (according to Main Statistical Office).

Table 3. Biogas potential of different organic waste.

Substrate	Methane yield (m³ Mg⁻¹ VS$_{added}$)	Reference
Sewage sludge	260–460	Bień et al. 2010
Sewage sludge	220	Rintala and Jarvinen 1996
OFMSW	310	Rintala and Jarvinen 1996
OFMSW	382	Ponsá et al. 2011
Grease trap sludge	845–928	Bień et al. 2010
Household waste	350	Ferrer et al. 2011
MSW	200	Walker et al. 2009
MSW	360	Vogt et al. 2002
Fruit and vegetable wastes	420	Bouallagui et al. 2005
Swine manure	337	Ahn et al. 2009
Organic fraction of source sorted MSW	387.5	Del Borghi et al. 1999
Sewage sludge	186	Del Borghi et al. 1999
Source sorted OFMSW	275–410	Davidsson et al. 2007
Source sorted-OFMSW	400	Li et al. 2011
Pig manure	200–300	Pesta 2007
Raw glycerol (biodiesel)	690–720	Pesta 2007
Fat (separators)	700 (1000)	Pesta 2007

Moreover, if the digested material will be stored, the application of fermentation offers the following advantages: reduces the weight and volume of waste, inactivates the biological and biochemical processes, reduces gas and odors emissions from landfills, reduces sedimentation and immobilizes of pollution (Fricke et al. 2005, Esposito et al. 2012, Khalid et al. 2011, Jayarama Reddy 2011).

Generally, the organic fraction of municipal waste is considered a difficult waste to fermenta-tion due to high content of dry matter (fluctuates between 30 and 50%), high carbon-nitrogen ratio, low content of macro-and micronutrients and content of toxic compounds (e.g. heavy metals, phthalates) (Zaher et al. 2007, Hartmann et al. 2002). In addition, the organic fraction of municipal waste is not a homogeneous mixture. Therefore, production of biogas in the fermentation process depends not only on the operating parameters of the reactor but also on the composition of the waste (Zaher et al. 2007). For this reason, co-digestion of OFMSW with another organic waste produced at large amounts seems to be attractive alternative for sustainable management (Ersahin et al. 2011). There is abundant literature about co-digestion of Organic Fraction of Municipal Solid Waste (OFMSW) with other organic wastes, such as: sewage sludge (Sosnowski et al. 2003, Edelmann et al. 2000, Zhang et al. 2008, Hartmann et al. 2002, Mhamadi et al. 2012, Hamzawi et al. 1998, Pahl et al. 2008, Oleszkiewicz & Poggi-Varaldo 1997), animal manure (Zaher et al. 2007, Hartmann & Ahring 2005, Yu & Schanbacher 2010), agricultural residues (Kübler et al. 2000, Converti et al. 1997, Nordberg & Edström 2005), energetic crop (Pognani et al. 2009), solid Slaughterhouse Waste (SHW) (Cuetos et al. 2008), crude glycerol (Fountoulakis & Manios 2009). Table 3 compares biogas potential of different organic waste. The most important challenge in the selection of co-substrates is balancing parameters used substrates. The main issue for co-fermentation lies in C:N ratio but also should take into account: pH, the content of macro—and micronutrients, content of biodegradable organic fractions, dry weight and the presence of inhibitors of the fermentation process (Hartmann H. et al. 2002).

4 CO-DIGESTION OF ORGANIC FRACTION OF MUNICIPAL SOLID WASTES AND SEWAGE SLUDGE

Sewage sludge are reported to be the most commonly used co-substrate for digestion of OFMSW, probably due to the complementary characteristics of wastes (Table 4) as well as possibility effectively implement in existing wastewater

Table 4. Comparison of properties of OFMSF, sewage sludge and manure (Hartmann et al. 2002).

	OFMSW	Sewage sludge
Content of macro-and micronutrients	Low	High
C:N ratio	High	Low
Content of biodegradable organic matter	High	Low

Table 5. Results of organic fraction of municipal waste co-digestion with sewage sludge.

Kind and ratio of substrates	Type of reactor	Operational conditions	VS reduction (%)	Y biogas/methane ($m^3 kg^{-1} VS_{add}$)	Methane content (%)	Reference
SS:OFMSW 75:25[1]	CSTR, UASB	Two-stage quasi-continuous, 56°C (CSTR), 36°C (UASB), OLR = 0,669, 2.76 and 3.084 g VS l⁻¹ d⁻¹ HRT = 17,33, 11.1 8.9 d (CSTR) and 44,2 17.3 20.9 d (UASB)	NR	0,532/NR (27%)[2, K1]	60	Sosnowski et al. 2003
OFMSW:SS 50:50[3]	CSTR	37°C, HRT = 14.5 d, OLR = 0.56 to 5.0 g COD l⁻¹ d⁻¹	57	0.6/NR	60	Del Borghi et al. 1999
		55°C, HRT = 12 d OLR = 0.56 to 5.0 g COD l⁻¹ d⁻¹	64	0.36/NR	50	
MSW:SS NR	Batch-type reactor operated in semi-continuous mode	Room temperature varying from 26 to 36°C, HRT = 25 d, OLR = 0.5, 1.0, 2.3, 2.9, 3.5 and 4.3 g VS l⁻¹ d⁻¹	88.1	0.13–0.36/NR	68–72	Elango et al. 2007
SS:KS 1:4[4]	CSTR	36°C, HRT = 15, 20, 25, 30 and 35 d, OLR = 2,17, 2.53, 3.03, 3.79 and 5.06 g VS l⁻¹ d⁻¹	51.1–68.3	0.4–1.84/NR	60–67	Zajda et al. 2011
SS:HS NR	Full scale in two mesophilic anaerobic digesters	Average quantity of domestic organic waste to municipal sludge digester in the range 1.0–2.08 m³ d⁻¹, HRT = 20 d, average OLR = 0.8 g VSS l⁻¹ d⁻¹	81 (+14%)[2]	0,6 (1,5x)[2]/NR	NR	Zupančič et al. 2008
SSKS:SS 20:80, 40:60, 50:50, 60:40, 70:30[3]	CSTR	36°C	45.7–63.6 (form 6.8 to 48.6%)[2]	Max 0.527 (max 46%)[2]/NR	NR	Kuglarz and Mrowiec 2007
AS:FVW 65:35, 35:65, 30:70, 20:80, 15:85[3]	ASBRs	35°C, HRT = 10, 20 d OLR = 0.3, 0.43, 1.03, 1.55, 1.87 8 g VSS l⁻¹ d⁻¹	65.1–88 (form 17.5 to 58.8%)[2]	0.52–0.72 (from 0 to 38.5%)[2]/NR	59–60 (from 1.7 to 3.4%)[2]	Habiba et al. 2009
MSW:SS 100:0, 90:10, 80:20, 70:30 and 60:40[1]	Simulated landfill reactor	All reactors were operated at room temperature of 32–38°C	75.44–83.43[5] (from 373 to 423%)[6]	0.169–0.394[7] (from 96.5 to 358%)[6, K2]	47–59 (from 194 to 269%)[6]	Sanphoti et al. 2007

234

			COD degradation	Methane production rate / SGP	TVS removal	Reference
WAS:OFMSW 25:75[3]	CSTR	37°C, OLR = 3.2 to 17 kg COD m^{-3}day^{-1}	58–88[8]	0.25–0.378[9]	43–63	Flor et al. 2004
AS:WC 90:10, 80:20[10]	CSTR	35°C, HRT = 30 d	63.2, 68.2 (+20.6 and 30.15)[2]	0.072, 0.0045 (+63,6 and 2.3%)[2]	NR	Saev et al. 2009
WAS:KS 20:80, 40:60, 50:50, 50:40 and 70:30[3]	Batch	36°C, 35 d	45.8–63.6 (from 7 to 48.6%)[6]	Max 0.336 (+121%)[2]	NR	Bohdzie-wicz et al. 2010
MSW:STS About 3:1[11]	Bioreactor landfill simulators	36°C	59.13 (+30%)[6]	0.6[12] (+87.5%)[6]	NR	Valencia et al. 2009
SOW:PS NR	CSTR	35°C, HRT = 20 d, OLR = 2.5 to 6.0–6.5 g VS l^{-1} d^{-1}	62.9 (+70,5%)[2]	0.538 (+70.8%)[2]/ 0.310 (+52%)[2]	60–65	Purcell and Stentiford 2000
MSW:SS 1:1, 1:2[3]	Anaerobic simulated landfilling bioreactors	35–40°C	87 and 89 (+4.8 and 7.2)[5,6]	NR/6.89 and 8.5 (+37.8 and 70%)[6,12]	70 and 72 (+18.6 and 22%)[6]	Ağdağ and Sponza 2007
SS:OFMSW 75:25[1]	Large scale laboratory reactor	Batch experiments, 23–40 days	NR	NR/0.439[1] (+38%[2] 87.6%)[K3]	NR	Sosnowski et al. 2008
PS:FVW[K4]	Laboratory reactor	Static conditions and with different mixing conditions, OLR = 2.5–9.0 g VS fed l^{-1} d^{-1}, HRT = 37, 40 and 47 d, mesophilic	51–58.5	0.6–0.8[K5]/NR	NR	Gómez et al. 2006
FVW:WAS 30:70[1]	Anaerobic sequencing batch reactors	35°C, HRT = 10 d, OLR = 2.51 g VS l^{-1} d^{-1}	85.4 (+12%)[6]	0.49 (+58%)[6]/NR	NR	Bouallagui et al. 2009
PW:SS[K6]	Laboratory—scal jars	36°C, 55°C, OLR = 5.0 g COD l^{-1} d^{-1}	NR	NR/0.21 and 0.35 (+35 and 45.2%)[2]	60–64	Grasmung et al. 2003

1—v/v, 2—improvement as compared to anaerobic stabilization of sewage sludge alone, 3—based on total solid, 4—at the weight ratio, 5—COD degradation degree (%), 6—improvement as compared compared to anaerobic stabilization of OFMSW alone, 7—methane production rate (l CH_4 d^{-1}), 8—TVS removal (%), 9—specific methane yield (l g^{-1} TVS_{added}), 10—ratio in percentage, 11—based on wet weight, 12—g CH_4 $COD_{removed}$ kg^{-1} VS_{added}, 13—l g^{-1} $VSS_{removed}$; OLR—organic loading rate, SGP—specific gas production, SS—sewage sludge, NR not reported, KS—kitchen biowaste, HS—households waste, SSKS—source separated kitchen waste, AS—activated sludge, FVW—fruit and vegetable waste, WAS—Waste Activated Sludge, WC—wasted vegetables, STS—septic tank sludge, SOW—supermarket organic wastes, PS—primary sludge, PW—process water separated from OFMSW, K1—SGP in two stage system (0,5321 g^{-1} VSS_{added}) was much higher than in the batch experiment (0,427 g^{-1} VSS_{added}), K2—Authors noted shortened lag phase time (from 52 to 32 d), K3—Observed improved the buffering capacity of fermentation broth and acceleration of biogas production, K4—The mixture for co-digestion was prepared with an average total solid (TS) concentration of 6%, with the PS supplying 22% of the 6%of dry solid content, K5—at low mixing conditions, K6—20% (w/v) of PW in sewage sludge mixture.

Table 6. The parameters of existing MSW and sewage sludge co-digestion installations in world.

Location	Process	Reactor volume (m³)	Capacity of plant (Mg/a)	Daily gas production (m³)	Methane content (%)	Energy generation	Biogas yield (m³ kg⁻¹VS)	Reference
Zgorzelec (Poland)	Single stages 35°C, HRT = 21 d	2 000	10 000	2300	70	1400–2800[1]	0,40–0,63	Bień et al. 2010
Puławy (Poland)	Single stages 39°C, HRT = 21 d	2 × 3 130	22 000	4200	67	3200[1]	0,73–0,805	
Kayseri WWTP (Turkey)	Single stages HRT = 15.4	6 750	70 000	16 300[2]	NR	47 500[3]	NR	Dereli et al. 2011
	Single stages HRT = 17.3		68 600	19 500[2]	NR	56 000[3]	NR	

1—annual, MWh, 2—methane, NR—not reported.

treatment plants (WWTPs) (Bień et al. 2010,). The main benefits of co-digestion of these wastes were: improved biogas/methane yields as well as increased degradation degree of treated substances and methane content (Table 5). Many research has suggested that the process efficiency can be increased through pretreatment of sludge prior to anaerobic digestion such as: separation of acidogenic and methanogenic stages during co-digestion or alkaline or acid pretreatment (Del Borghi et al. 1999, Cavinato et al. 2007, Sosnowski et al. 2003, Hartmann et al. 2002).

The huge potential of co-digestion of municipal solid wastes with sewage sludge can be confirmed by the fact that most of full-scale installations process these waste (Bolzonella et al. 2006 a, b, Krupp et al. 2005). Table 6 shows the parameters of selected co-digestion plants.

5 CO-DIGESTION OF ORGANIC FRACTION OF MUNICIPAL SOLID WASTES AND MANURE

Manure may be used for anaerobic co-digestion with OFMSW due to its availability and suitable physicochemical characteristics such as high buffering capacity as well as high content of water and macro—and micronutrients (Cuetos et al. 2011, Angelidak & Ellegaard 2003). Hartmann and Ahring (2005) performed a laboratory scale OFMSW and manure co-digestion in two thermophilic (55°C) wet lab scale reactor systems at a hydraulic retention time (HRT) of 14–18 d and an organic loading rate (OLR) of 3.3–4.0 g VS l⁻¹d⁻¹. In both reactors the addition of OFMSW in the feedstock was gradually increased up to 50% (on VS basic). However, in the first reactor from day 84 to 139 the ratio of OFMSW to manure was slowly increased to 100%. Then, recirculation of process liquid started in the reactor (supernatant of the effluent for dilution of OFMSW). These experiments showed that co-digestion as well as separate digestion of OFMSW with process liquid recirculation can reach stable operation despite fluctuations in the feed volume. The biogas yield for both configurations (0.63–0.71 l g⁻¹VS) exceeded the separate anaerobic digestion of OFMSW (0.18–0.22 l g⁻¹VS).

Macias-Corral et al. (2008) also observed increase of digestion efficiency during anaerobic stabilization of dairy Cow Manure (CM) and OFMSW. In comparison with the control reactors (digestion of single waste—CM or OFMSW) an increase of methane production from 0.62 (for CM) or 0.37 (for OFMSW) to 1.72 (co-digestion mixture) l g⁻¹ of dry waste) has also been noted. Moreover, they found that the addition of a co-substrate to the feedstock promotes synergistic effects resulting in higher mass conversion and lower weight and volume of digested residual. Alvarez and Liden (2008) presented research results in which compared semi-continuous co-digestion of solid slaughterhouse waste, manure and fruit and vegetable waste with single substrate digestion of mentioned substrates. Authors noted for all analyzed co-digestion mixtures a higher methane production than for digestion of pure substrates alone. Methane yields for the mixtures were in the range 0.27–0.35 l g^{-1} VS_{added}, with methane content between 50–56% and a 56.6–67.4% reduction of VS with an OLR in the range 0.3–1.3 g VS $l^{-1}d^{-1}$.

What is more, studies carried out by Bohdziewicz et al. (2011) on co-digestion of SS-OFMSW and manure at mesophilic temperatures resulted in higher efficiency of anaerobic stabilization process. In comparison with separate anaerobic digestion of manure the authors obtained an increased biogas production rate by approximately of 35% for an optimum ratio of wastes (30% municipal biowaste and 70% pig manure based on TS). According of the authors, addition of biowaste in the amount of 20–50% did not lead to deterioration significantly susceptibility of fermented biomass to dewatering.

However, Callaghan et al. (2002) conducted co-digestion of vegetable wastes with chicken manure an HRT of 21 days and OLR maintained in the range 3.62–5.22 kg VS $m^{-3}d^{-1}$. They observed that the presence of organic waste residues improved volatile solid removal (from about 30% to 50%) as well as the methane yield from 0.23 to 0.45 m^3 kg^{-1} VS_{added}. Also, results showed that higher proportion of chicken manure in the feedstock caused deterioration in digester performance (probably due to ammonia inhibition).

6 CO-DIGESTION OF ORGANIC FRACTION OF MUNICIPAL SOLID WASTES WITH ORGANIC WASTES FROM INDUSTRY

Lipid-rich wastes seem an attractive substrate for biogas production (Table 3). Nevertheless, treatment of this waste by anaerobic digestion is often hampered due to inhibitory effect of intermediate compounds (LCFAs—long chain fatty acids) and operational problems, such as: sedimentation clogging, scum formation, hindrance and flotation of biomass (Worwąg et al. 2011). The application of the anaerobic co-digestion process to lipid-rich wastes gives the possibility of treating this waste. For instance, Fernández et al. (2008) investigated anaerobic co-digestion of a simulated organic fraction of municipal solid wastes and increasing

amounts of fats of different origin. Process was conducted in semi-continuous regime in the mesophilic range (37°C) and the Hydraulic Retention Time (HRT) equaled 17 days. After an adaptation period, the authors noted that the yields of biogas and methane for co-digestion mixtures were similar to that observed during single substrate digestion of OFMSW alone. Whereas, methane content in biogas was marginally higher in the presence of fats (increased from 58 to 61%). Results showed also a high level of total fat removal throughout the experiment (above 88%).

In another study, Martín-González et al. (2010) reported an increased biogas production for a mixture of OFMSW and grease, trap, sludge (FOG). The digestion was examined in semi-continuous mode at hydraulic retention time of 14.5 days and OLR of 4.5 kg VS m^{-3} d^{-1}. The methane yield, methane content in biogas, VS reduction and total FOG reduction observed during the process were 0.35 ± 0.03 l g^{-1} VS feed, 63 ± 1% and 65 ± 3%, 56 ± 3, respectively. Solid slaughterhouse waste (SHW) has also been evaluated as a digester feedstock by Cuetos et al. (2008). In the study, fermentation was carried out in pilot plant operating semi-continuously in the mesophilic temperature and HRT from 52 to 50 days. The result showed that the application of OFMSW to the co-digestion system improved fat and VS removal (about 4.6% and 35%, respectively) and doubled daily biogas production. The authors also noted that stable methane production could be reached only when the SHW was progressively acclimated to high fats and LCFA concentrations—that was achieved by gradually decreased the HRT from 50 to 25 days. Edstrom et al. (2003) also noted an increased biogas production for a mixture of animal byproducts from slaughterhouses and food waste. Processes conducted in continuous-flow stirred tank reactors at laboratory and pilot scale under mesophilic conditions. The authors achieved stable processes at HRT less than 40 days and organic loading rates OLRs exceeding 2.5 g of VS $l^{-1} \cdot d^{-1}$. The specific yield of biogas for co-digestion mixtures were in the in the range of 0.75 to 0.86 l g^{-1} VS. However, Fountoulakis & Manios (2009) evaluated the possibility of the use of glycerol as a co-substrate. In the experiment, biogas production of the digester treating OFMSW was raised by 49.6% to 2094 ml d^{-1} by the addition of glycerol.

Sometimes the organic fraction of municipal waste is used as a co-substrate that allows for increase of the effectiveness of stabilization hardly decomposable substance. For instance, Ağdağ (2011) examined the effects of Organic Fraction of Municipal Solid Waste (OFMSW) addition on AD of the olive-mill pomace. Results showed that addition of OFMSW promoted biodegradability

Table 7. Results of organic fraction of municipal waste co-digestion with other organic waste.

Co-substrates	Operational conditions	Y biogas/methane ($m^3 kg^{-1}VS_{add}$)	VS reduction (%)	Reference
Vegetable oil	37°C, in sealed aluminium bottles (1 l)	NR/0.699 (+83%)[2]	NR	Ponsá et al. 2011
Animal fat		NR/0.508 (+33%)[2]	NR	
Cellulose		NR/0.254	NR	
Protein		NR/0.288	NR	
Algal sludge	OLR = 2–6 g VS l^{-1} d^{-1}, 35 °C HRT = 10 d (4 l)	NR/max1601[4] (+179%)[5]	NR	Yen and Brune 2007
Industrial sludge and cattle manure	35 °C, Batch, 65 d	NR/15–225[5]	32–59[5]	Capela et al. 2008
Food waste	35 °C, [1]CSTR (4 l) OLR = 3 g VS $g^{-1} \cdot d^-$	NR/0.49[7]	74.9[7]	Lin et al. 2011
Energy crops and cow manure slurry and agro-industrial waste	Full-scale, wet, 55°C, HRT = 40 d	0.567/NR	NR	Pognani et al. 2009
Meat residues	HRT = 15–20 d, OLR = 2.4 and 2.7 g COD $l^{-1}d^{-1}$ (30l)	0.12–0.64/NR	NR	Garcia-Peña et al. 2011

1—on dry weight, 2—improvement as compared to anaerobic stabilization of OFMSW alone, 3—based on volatile solid, 4—methane production rate ml l^{-1} d^{-1}, 5—increase in compared to algal sludge digestion alone, 5—specific methane yield l g^{-1} TVS_{added}, 6—TVS removal, 7—results for the optimum mixture ratio, NR—not reported.

of olive-mill pomace. While, Angelidaki & Ahring (1997) investigated the potential of anaerobic digestion for the treatment of olive oil mill waste-waters (OME) through co-digestion with house-hold waste. The OME-utilization degrees observed during the process was about 55% with 73% lipid reduction obtained during the process.

Nevertheless, results showed that process can be conducted with high dilution of the waste (in experiment—OME dilution with water—1:2 to 1:5 OME: total-volume). Overview of the anaerobic digestion processes of organic fraction of MSW with other organic waste are presented in Table 7.

7 CONCLUSION

Anaerobic stabilization of co-digestion mixture has a positive effects both on process efficiency (e.g. increase biogas production and degradation degree of treated substances, it improves the process stability) as well as on the treatment economy (e.g. disposal of various waste streams in centralized large-scale facilities) and environmental aspects (Mata-Alvarez 2000, Ersahin et al. 2011, Hartmann H. et al. 2002, Bień et al. 2010, Esposito et al. 2012, Khalid et al. 2011, Edelman et al. 2005). Furthermore, the addition of substrate to digester for solid wastes could be beneficial due to: dilution of potential toxic compounds, increased load of biodegradable fraction, improved nutrient balance and content of macro—and micronutrients, help in establishing the required moisture content of the digester feed and the buffering capacity of fermentation broth, adjustment of pH; supply of the necessary buffer capacity to the mixture as well as adjustment of C:N ratio feedstock to optimal range for methane production (20:1–30:1) (Bień et al. 2010, Esposito et al. 2012, Fernández et al. 2005, Buendía et al. 2009). Nevertheless, some limitations of process also exist. The main drawbacks of co-digestion technology are: transport costs of co-substrate, the problems arising from the harmonization of the waste generators and additional pre-treatment facilities, decrease in quality of the digester effluent (Nayono 2009, Bień et al. 2010). Improvements in collection system reduce some of these barriers.

ACKNOWLEDGEMENTS

The present work was supported by the Faculty of Environmental Protection and Engineering (Czestochowa University of Technology) BS/MN-401-316/11 and BS/PB-401-303/11.

REFERENCES

Ağdağ, O.N. 2011. Biodegradation of olive-mill pomace mixed with organic fraction of municipal solid waste. Biodegradation. 22: 931–938.

Ağdağ, O.N., Sponza, D.T. 2005. Co-digestion of industrial sludge with municipal solid wastes in anaerobic simulated landfilling reactors. Process Biochemistry. 40: 1871–1879.

Ağdağ, O.N., Sponza, D.T. 2007. Co-digestion of mixed industrial sludge with municipal solid wastes in anaerobic simulated landfilling bioreactors. Journal of Hazardous Materials. 140: 75–85.

Ahn, H.K., Smith, M.C., Kondrad, S.L., White, J.W. 2009. Evaluation of biogas production potential by dry anaerobic digestion of switchgrass–animal manure mixtures. Appl. Biochem. Biotechnol. 160: 965–975.

Alvarez, R., Liden, G. 2008. Semi-continuous co-digestion of solid slaughter house waste, manure, and fruit and vegetable waste. Renew Energy. 33: 726–734.

Angelidaki, I., Ahring, B.K. 1997. Codigestion Of Olive Oil Mill Wastewaters With Manure, Household Waste Or Sewage Sludge. Biodegradation. 8: 221–226.

Angelidaki, I., Ellegaard, L. 2003. Codigestion of Manure and Organic Wastes in Centralized Biogas Plants Status and Future Trends. Applied Biochemistry and Biotechnology. 109: 95–105.

Baldasano, J.M. Soriano, C. 2000. Emission of greenhouse gases from anaerobic digestion processes: comparison with other municipal solid waste treatments. Water Science and Technology. 41 (3): 275–282.

Bidlingmaier, W., Sidaine, J.M., Papadimitriou, E.K. 2010. Separate collection and biological waste treatment in the European Community. Reviews in Environmental Science & Bio/Technology. 3: 307–320.

Bień, J., Grosser, A., Neczaj, E., Worwąg, M., Celary, P. 2010. Co-digestion of sewage sludge with different organic wastes: a review. Polish Journal of Environmental Studies. 2: 24–30.

Bień, J.B., Kacprzak, M., Neczaj, E., Wystalska, K. 2007. Amendment of the UE legislation on biowaste management. Environment Protection Engineering. 33 (2): 71–78.

Bień, J.B., Malina, G., Bień, J.D., Wolny, L. 2004. Enhancing anaerobic fermentation of sewage sludge for increasing biogas generation. Journal of Environmental Science and Health—Part A Toxic/Hazardous Substances and Environmental Engineering. 39 (4): 939–949.

Bień, J.B., Milczarek, M., Neczaj, E., Grosser, A. 2011. Kowalczyk, M., Worwag, M. Co-composting of municipal solid waste and sewage sludge. In proceeding of International Conference Environmental (Bio) Technologies. 5–8.09.2011. Gdańsk:Poland.

Bień, J.B., Milczarek, M., Neczaj, E., Gałwa-Widera, M. 2010. Co—composting of industrial sewage sludge from mineral water production plant "Jurajska" in Postep and municipal sewage sludge. Polish Journal of Environmental Studies. 2: 7–11.

Bień, J.B., Milczarek, M., Sobik-Szołtysek, J., Okwiet, T. 2011. Optimization of the thermophilic phase in the process of co—composting sewage sludge and municipal waste. Nauka Przyr. Technol. 5 (4): 33.

Bohdziewicz, J., Kuglarz, M., Mrowiec, B. 2010. Assessment of kitchen biowaste and sewage sludge susceptibility to methanogenic co-digestion in batch tests. Ecological Chemistry and Engineering A. 17 (11): 1405–1413.

Bohdziewicz, J., Kuglarz, M., Mrowiec, B. 2011. Intensification of pig manure digestion by co-substrate addition in the form of municipal biowast. Nauka Przyr. Technol. 5 (4): 53 [in Polish].

Bolzonella, D., Battistoni, P., Susini, C., Ceccchi, F. 2006a. Anaerobic codigestion of waste actived sludge and OFMSW: the experiences of Viareggio and Treviso plants (Italy). Water Sci. Technol. 53 (8): 203–211.

Bolzonella, D., Pavan, P., Mace, S., Cecchi, F. 2006b. Dry anaerobic digestion of differently sorted organic municipal solid waste: a full scale experience. Water Science and Technology. 53 (8): 23–32.

Bouallagui, H., Touhami, Y., Ben Cheikh, R., Hamdi, M. 2005. Bioreactor performance in anaerobic digestion of fruit and vegetable wastes. Process Biochem. 40: 989–995.

Bouallagui, H., Lahdheb, H., Romdan, E.B., Rachdi, B., Hamdi, M. 2009. Improvement of fruit and vegetable waste anaerobic digestion performance and stability with co-substrates addition. Journal of Environmental Management. 90: 1844–1849.

Buendía, I.M., Fernández, F.J, Villaseñor, J., Rodríguez, L. 2009. Feasibility of anaerobic co-digestion as a treatment option of meat industry wastes. Bioresource Technology. 100: 1903–1909.

Callaghan, F.J., Wase, D.A.J., Thayanithy, K., Foster, C.F. 2002. Continuous co-digestion of cattle slurry with fruit and vegetable wastes and chicken manure. Biomass and Bioenergy. 22 (1): 71–77.

Capela, I., Rodrigues, A., Silva, F., Nadais, H., Arroja, L. 2008. Impact of industrial sludge and cattle manure on anaerobic digestion of the OFMSW under mesophilic conditions. Biomass and Bioenerg. 32 (3): 245–251.

Cavinato, C., Pavan, P., Fatone, F, Cecchi, F. 2007. Bioenergy from waste activated sludge and marketwaste: single and two phase thermophilic codigestion. In 8th international conference on chemical and process engineering. NAPLES, vol. 2, pp. 869–874, Convegno: ICheaP-8, Naples, 24–27 June.

Converti, A., Drago, F., Ghiazza, G., Borghi, M., Macchiavello, A. 1997. Co-digestion of municipal sewage sludges and pre-hydrolysed woody agricultural wastes. Journal of Chemical Technology & Biotechnology. 69: 231–239.

Cuetos, M.J., Fernández, C., Gómez, X., Morán, A. 2011. Anaerobic co-digestion of swine manure with energy crop residues. Biotechnology and Bioprocess Engineering. 16 (5): 1044–1052.

Cuetos, M.J., Gomez, X., Otero, M., Moran, A. 2008. Anaerobic digestion of solid slaughterhouse waste (SHW) at laboratory scale: Influence of co-digestion with the organic fraction of municipal solid waste (OFMSW). Biochemical Engineering Journal. 40: 99–106.

Davidsson, A., Gruvberger, C., Christensen, T.H., Hansen, T.L., Jansen, J.C. 2007. Methane yield in source-sorted organic fraction of municipal solid waste. Waste Management. 27: 406–414.

Del Borghi, A., Converti, A., Palazzi, E., Del Borghi, M. 1999. Hydrolysis and thermophilic anaerobic digestion of sewage sludge and organic fraction of municipal solid waste. Bioprocess Engineering. 20: 553–560.

den Boer, E., Jędrczak, A., Kowalski, Z., Kulczyck, J., Szpadt, R. 2010. A Review of Municipal Solid Waste Composition and Quantities in Poland. Waste Management. 30 (3): 369–377.

Dereli, R.K., Ersahin, M.E.,.Gomec, C.Y., Ozturk, I., Ozdemir, O. 2010. Co-digestion of the organic fraction of municipal solid waste with primary sludge at a municipal wastewater treatment plant in Turkey. Waste Management and Research. 28 (5): 404–410.

Dong, L., Zhenhong, Y., Yongming, S. 2010. Semi-dry mesophilic anaerobic digestion of water sorted organic fraction of municipal solid waste (WS-OFMSW). Bioresource Technology. 101: 2722–2728.

Edelmann, W., Engeli, H., Graddenecker, M. 2000. Co-digestion of organic solid waste and sludge from sewage treatment. Water Sci. Technol. 41 (3): 213–221.

Edelmann, W., Baier, U., Ehgeli, H. 2005. Environmental aspects of anaerobic digestion of organic fraction of municipal solid wastes and of agricultural wastes. Water Sci. Technol. 52: 203–208.

Edström, M., Nordberg, A., Thyselius, L. 2003. Anaerobic Treatment of Animal Byproducts from Slaughterhouses at Laboratory and Pilot Scale. Applied Biochemistry and Biotechnology. 109: 127–138.

Elango, D., Pulikesi, M., Baskaralingam, P., Ramamurthi, V., Sivanesan, S. 2007. Production of biogas from municipal solid waste with domestic sewage. J Hazard Mater. 141: 301–304.

Esposito, G., Frunzo, L., Giordano, A., Liotta, F., Panico, A., Pirozzi, F. 2012. Anaerobic co-digestion of organic wastes. Rev Environ Sci Biotechnol. 11 (4): 325–341.

Ersahin, M.E., Gomec, C.Y., Dereli, R.K., Arikan, O., Ozturk, I. 2011. Biomethane production as an alternative: Bioenergy source from codigesters treating municipal sludge and organic fraction of municipal solid wastes. Journal of Biomedicine and Biotechnology. 8: 1–8.

Fernández, A., Sánchez, A., Font, X. 2005. Anaerobic co-digestion of a simulated organic fraction of municipalsolid wastes and fats of animal and vegetable origin. Biochemical Engineering Journal. 26: 22–28.

Ferrer, I., Garfí, M., Uggetti, E., Ferrer-Marti, L., Calderon, A., Velo, E. 2011. Biogas production in low-cost household digesters at the Peruvian Andes. Biomass and Bioenergy. 35 (5): 1668–1674.

Flor, A., Coelho, N., Arroja, L., Capela, I. 2004 Co-digestion of organic fraction of municipal solid waste and waste activated sludge in a continuous reactor, In proceeding of 10th World Anaerobic Digestion Conference. 29.08–02.09.2002. Montrèal:Canada.

Forster-Carneiro, T., Pèrez, M., Romero, L.I., Sales, D. 2007. Dry-thermophilic anaerobic digestion of organic fraction of the municipal solid waste: Focusing on the inoculum sources. Bioresource Technology. 98: 3195–3203.

Fountoulakis, M., Manios, T. 2009. Enhanced methane and hydrogen production from municipal solid waste and agro-industrial by-products co-digested with crude glycerol. Bioresour. Technol. 100: 3043–3047.

Fricke, K., Santen, H. Wallmann, R. 2005. Comparison of selected aerobic and anaerobic procedures for MSW treatment. Waste management. 25: 799–810.

Garcia-Peña, E.I, Parameswaran, P., Kang, D.W., Canul-Chan, M., Krajmalnik-Brow, R. 2011.

Anaerobic digestion and co-digestion processes of vegetable and fruit residues: Process and microbial ecology. Bioresource Technology. 102: 9447–9455.

Gómez, X., Cuetos, M. J., Cara, J., Morán, A., Garcia, A.I. 2006. Anaerobic co-digestion of primary sludge and the fruit and vegetable fraction of the municipal solid wastes. Conditions for mixing and evaluation of the organic loading rate. Renew. Energ. 31: 2017–2024.

Grasmung, M., Roch, A., Braun, R., Wellacher, M. 2003. Anaerobic co-digestion of pre-treated organic fraction of municipal solid waste with municipal sewage sludge under mesophilic and thermophilic conditions. Engineering and protection of environment. 6 (3–4): 267–273.

Griffin, M.E., McMahon, K.D., Mackie, R.I., Raskin, L. 1998. Methanogenic population dynamics during start-up of anaerobic digesters treating municipal solid waste and biosolids. Biotechnol Bioeng. 57 (3): 342–355.

Habiba, L., Hassib, B., Moktar, H. 2009. Improvement of activated sludge stabilisation and filterability during anaerobic digestion by fruit and vegetable waste addition. Bioresource Technol. 100 (4): 1555–1560.

Hamzawi, N., Kennedy, K.J., Mclean, D.D. 1998. Anaerobic digestion of co-mingled municipal solid-waste and sewage sludge. Water Sci. Technol. 38 (2): 127–132.

Hanc, A., Novak, P., Dvorak, M., Habart, J., Svehla, P. 2011. Composition and parameters of household biowaste in four seasons. Waste Manage. 31: 1450–1460.

Hartmann, H., Angelidaki, I., Ahring, B.K. 2002. Co-digestion of the organic fraction of municipal waste with other waste types. In Mata-Alvarez, J. (ed) Biomethanization of the Organic Fraction of Municipal Solid Wastes, IWA Publishing.

Hartmann, H., Ahring, B.K. 2005. Anaerobic digestion of the organic fraction of municipal solid waste: influence of co-digestion with manure. Water Research. 39 (8): 1543–1552.

Hartmann, H., Ahring, B.K. 2006. Strategies for the anaerobic digestion of the organic fraction of municipal solid waste: an overview. Water Sci. Technol. 53 (8): 7–22.

Jayarama Reddy, P. 2011. Municipal Solid Waste Management: Processing—Energy Recovery—Global Examples. CRC Press.

Khalid, A., Arshad, M., Anjum, M., Mahmood, T., Dawson, L. 2011. The anaerobic digestion of solid organic waste. Waste Manag. 31 (8): 1737–1744.

Krupp, M., Schubert, J., Widmann, R. 2005. Feasibility study for co-digestion of sewage sludge with OFMSW on two wastewater treatment plants in Germany. Waste Management. 25 (4): 393–399.

Kübler, H., Hoppenheidt, K., Hirsch, P., Kottmair, A., Nimmrichter, R., Nordsieck, H., Mücke, W., Swerev, M. 2000. Full-scale co-digestion of organic waste. Water Science and Technology. 41 (3): 195–202.

Kuglarz, M., Mrowiec, B. 2009. Co-digestion of municipal biowaste and sewage sludge for biogas production, In Płaza, E., Levlin, E. (eds). Research and application of new technologies in wastewater treatment and municipal solid waste disposal in Ukraine, Sweden and Poland: proceedings of Polish-Ukrainian-Swedish seminar: 177–184, Stockholm, Sweden.

Li, Y., Park, S.Y., Zhu, J. 2011. Solid-state anaerobic digestion for methane production from organic waste. Renewable and Sustainable Energy Reviews. 15: 821–826.

Lin, J., Zuo, J., Gan, L., Li, P., Liu, F., Wang, K., Chen, L., Gan, H. 2011. Effects of mixture ratio on anaerobic co-digestion with fruit and vegetable waste and food waste of China. Journal of Environmental Sciences. 23: 1403–1408.

Liu, G. Zhang, R., Sun, Z., Li, X., Dong, R. 2007. Research Progress in Anaerobic Digestion of High Moisture Organic Solid Waste. Agricultural Engineering International: the CIGR EJournal. 13 (9).

Liu, K., Tang, Y.Q., Matsui, T., Morimura, S., Wu, X.L., Kida, K. 2009. Thermophilic anaerobic co-digestion of garbage, screened swine and dairy cattle manure. Journal of Bioscience and Bioengineering. 107 (1): 54–60.

Macias-Corral, M., Smani, Z., Hanson, A., Smith, G., Funk, P., Yu, H., Longworth, J. 2008. Anaerobic digestion of municipal solid waste and agricultural waste and the effect of co-digestion with dairy cow manure. Bioresour., Technol. 99: 8288–8293.

Martín-González, L., Colturato, L.F., Font, X., Vicent, T., 2010. Anaerobic codigestion of the organic fraction of municipal solid waste with FOG waste from a sewage treatment plant: Recovering a wasted methane potential and enhancing the biogas yield. Waste Management. 30 (10): 1854–1859.

Mata-Alvarez, J. 1996. Biological household waste treatment in Europe: second Aalborg international conference. Resour Conserv Recycl. 17: 67–73.

Mata-Alvarez, J., Macé, S., Llabrés, P. 2000. Anaerobic digestion of organic solid wastes. An overview of research achievements and perspective. Bioresource Technology. 74: 3–16.

Mhamadi, A.T., He, Q., Ntakirutimana, T., Li, J. 2012. Start-up performance of a pilot—scale Integrated Reactor for treating domestic garbage and Sewage Sludge from Treatment Plant. Journal of American Science. 8 (6): 132–138.

Muhammad, N.I., Mohd Ghazi, T.I., Omar, R. 2012. Production of biogas from solid organic wastes through anaerobic digestion: a review. Appl Microbiol Biotechnol. 95: 321–329.

Neczaj, E., Bień, J., Grosser, A., Worwąg, M., Kacprzak, M. 2012. Anaerobic treatment of sewage sludge and grease traps sludge in continuous co-digestion. Global NEST Journal. 14 (2): 141–148.

Nguyen, P.H.L., Kuruparan, P., Visvanathan, C. 2007. Anaerobic digestion of municipal solid waste as a treatment prior to landfill. Bioresource Technology. 98 (2): 380–387.

Oleszkiewicz, J.A., Poggi-Varaldo, H.M. 1997. High-solids anaerobic digestion of mixed municipal and industrial waste. J. Environ. Eng. 123 (11): 1087–1092.

Pahl, O., Firth, A., Macleod, I., Baird, J. 2008. Anaerobic co-digestion of mechanically biologically treated municipal waste with primary sewage sludge—A feasibility study. Bioresource Technology. 99: 3354–3364.

Pesta, G., 2007. Anaerobic Digestion of Organic Residues and Wastes. In V. Oreopoulou, W. Russ (Eds.), Utilization of by-products and treatment of waste in the food industry. 53–72, Springer.

Pichtel, J. 2005. Waste Management Practices Municipal, Hazardous, and Industrial. Boca Raton, FL. CRC Press.

Pognani, M., D'Imporzano, G., Scaglia, B., Adani, F. 2009. Substituting energy crops with organic fraction of municipal solid waste for biogas production at farm level: a full-scale plant study. Process Biochem. 44: 817–821.

Ponsá, S., Gea, T., Sánchez, A. 2011. Anaerobic co-digestion of the organic fraction of municipal solid waste with several pure organic co-substrates. Biosystems Engineering. 108: 352–360.

Purcell, B., Stentiford, E.I. 2000. Co-digestion—enhancing recovery of organic waste, ORBIT Journal, htpp://www.orbit-online.net/downloads/articles/01_01_06.

Rintala, J.A., Järvinen, K.T. 1996. Full-scale mesophilic anaerobic co-digestion of municipal solid waste and sewage sludge: methane production characteristics. Waste Management Res. 14 (2): 163–170.

Saev, M., Koumanova, B., Simeonov, I. 2009. Anaerobic co-digestion of wasted vegetables and activated sludge. Biotechnology & Biotechnological Equipment. 23 (2): 832–835.

Sanphoti, N., Towprayoon, S., Chaiprasert, P., Nopharatana, A. 2007. Anaerobic co-digestion of organic municipal solid waste and sewage sludge, In proceedings of International Conference on Engineering and Environment (ICEE 2007), 10–11 May 2007, Phuket:Thailand.

Satoto Endar Nayono. 2009. Anaerobic digestion of organic solid waste for energy production, Karlsruhe KIT Scientific Publ.

Schmit, K.H., Ellis, T.G. Comparison of the temperature phased and other state of the art processes for anaerobic digestion of municipal solid waste. http://public.iastate.edu/~tge/schmit.pdf.

Sosnowski, P., Klepacz-Smolka, A., Kaczorek, K., Ledakowicz, S. 2008. Kinetic investigations of methane cofermentation of sewage sludge and organic fraction of municipal solid wastes. Bioresource Technology. 99: 731–5737.

Sosnowski, P., Wieczorek, A., Ledakowicz, S. 2003. Anaerobic co-digestion of sewage sludge and organic fraction of municipal solid wastes. Advances in Environmental Research. 7 (3): 609–616.

Valencia, R., den Hamer, D., Komboi, J., Lubberding, H.J., Gijzen, H.J. 2009. Alternative treatment for septic tank sludge: Co-digestion with municipal solid waste in bioreactor landfill simulators. Bioresource Technology. 100 (5): 1754–176.

Vogt, G.M., Liu, H.W., Kennedy, K.J., Vogt, H.S., Holbein, B.E. 2002. Super blue box recycling (SUBBOR) enhanced two-stage anaerobic digestion process for recycling municipal solid waste: laboratory pilot studies. Bioresour. Technol. 85: 291–299.

Walker, M., Banks, C.J., Heaven, S. 2009. Two-stage anaerobic digestion of biodegradable municipal solid waste using a rotating drum mesh filter bioreactor and anaerobic filter. Bioresour. Technol. 100: 4121–4126.

Worwąg, M., Neczaj, E., Grosser, A., Krzemińska, D. 2011. Methane production from fat-rich materials. Civil and Environmental Engineering Reports. 6: 147–162.

Yen, H.W., Brune, D.E. 2007. Anaerobic co-digestion of algal sludge and waste paper to produce methane. Bioresour. Technol. 98: 130–134.

Yu, Z. Schanbacher, F.L. 2010. Production of methane biogas through anaerobic digestion. In O.V. Singh and S.P. Harvey (eds.), Sustainable Biotechnology: renewable resources and new perspectives: 105–127. Springer, The Netherland.

Zajda, A., Kuglarz, M., Mrowiec, B. 2011. Methanogenesis efficiency in the conditions of sewage sludge and kitchen biowaste co-digestion. Nauka Przyr. Technol. 5 (4): 62.

Zawieja, I., Wolny, L., Wolski, P. 2008. Influence of excessive sludge conditioning on the efficiency of anaerobic stabilization process and biogas generation. Desalination. 222: 374–381.

Zaher, U., Cheong, D., Wu, B., Chen, S. 2007. Producing Energy and Fertilizer From Organic Municipal Solid Waste. Department of Biological Systems Engineering, Washington State University. Ecology Publication No. 07-07-024.

Zhang, P., Zeng, G., Zhang, G., Li, Y., Zhang, B., Fan, M. 2008. Anaerobic co-digestion of biosolids and organic fraction of municipal solid waste by sequencing batch process. Fuel Processing Technology. 89: 485–489.

Zupančič, G.D., Uranjek-Ževart, N., Roš, M. 2008. Full-scale anaerobic co-digestion of organic waste and municipal sludge. Biomass and Bioenergy. 32: 162–167.

Environmental Engineering IV – Pawłowski, Dudzińska & Pawłowski (eds)
© *2013 Taylor & Francis Group, London, ISBN 978-0-415-64338-2*

Indicators influencing the course of the thermophilic phase of composting process

M. Milczarek, E. Neczaj, K. Parkitna & M. Worwąg
Institute of Environmental Engineering, Czestochowa University of Technology, Czestochowa, Poland

ABSTRACT: One of the many factors determining the use of compost in nature is an aspect hygienization. In the composting process the degree of hygienization determines obtained temperature and duration of thermophilic phase. Temperature dynamics in composted mass is the result of (indicator) biochemical (energy) transformations of organic matter. The appropriate degree of aeration will start the thermophilic phase, characterized by a rapid increase in temperature (average of about 60–70°C), evaporation of moisture contained in the material, increasing the pH to 7.5–9.0 and the change in the properties of organic waste, which take the appearance of brown, with the scent of the crushed mass of forest litter. Thermophilic phase usually can last from 4 to 7 weeks. Improving the process of composting involves in particular the selection of the composition (use of different types of organic additives—in order to obtain the corresponding C/N ratio) and to adjust aeration piles.

Keywords: composting, thermophilic phase, organic wastes

1 INTRODUCTION

Composting is a natural process by which microorganisms decompose organic matter into simpler nutrients. Aerobic composting is the process where decomposition takes place in the presence of oxygen. As the quickest way to produce high quality compost, aerobic composting is a widely accepted way of stabilizing organic wastes and converting them to a usable, and value added compost product. It is widely accepted that temperature is an important environmental variable in composting efficiency (Namkoong & Hwang 1997, Joshua et al. 1998). For proper composting process it is necessary to ensure appropriate process parameters such as moisture content of 45–60%, pH in the range 6.5–7.5, the concentration of oxygen in the air leek > 15% (Bien 2011). Temperature increase within composting materials is a function of initial temperature, metabolic heat evolution and heat conservation (Miller 1992). The achievement of minimum temperature levels is essential to an effective composting process (Finstein et al. 1986) and contributes substantially to the high rates of decomposition achieved during processing. Composting occurs most effectively in the temperature range 45–60°C. Temperature level, its duration and the amount of air supplied determine the availability of quickly degradable organic compounds (Jedrczak 2007, Liang et al. 2003, Sidelko 2005). Organisms that have an optimum temperature (Top t) of 45–60°C are

called thermophiles. Thermophiles have not only an elevated Tmax, but also an elevated minimum temperature of growth (Tmin). However, the temperature range in which these organisms grow is similar to that of the mesophiles which grow at 'normal' temperatures. It means that the overall temperature range at which any organism can grow is rather narrow, approximately 30°C (Maheshwari 2005).

1.1 *pH range*

Centrally collected household waste is often acidic, with pH normally ranging between 4.5 and 6 (Eklind et al. 1997). The acidity is due to the presence of shortchain organic acids, mainly lactic and acetic acid (Beck-Friis et al. 2001). These acids are found in the raw material, and their concentrations increase during the initial phase of composting (Beck-Friis et al. 2003). The presence of short-chain fatty acids under acidic conditions and their absence during alkaline conditions indicate that they are a key factor regulating the pH in composts (Choi & Park 1998, Beck-Friis et al. 2003). A number of authors have noted stagnation or decline in microbial activity in the transition from mesophilic to thermophilic conditions in laboratory-scale compost reactors (Day et al., 1998; Schloss & Walker 2000; Beck-Friis et al. 2001, Weppen 2001). The stagnation in the microbial activity has in some cases been observed to coincide with low pH in the material (Day et al.

1998; Beck-Friis et al. 2001). Beck-Friis et al. (2001) noted that the change from mesophilic to thermophilic conditions during the initial stage of composting coincided with a change in pH from acidic (pH 4:5–5:5) to alkaline (pH 8–9). Different methods have been used to increase the rate of degradation when acidic materials are composted. Nakasaki et al. (1993) composted organic household waste at 60°C and observed an increased degradation rate when pH was prevented from decreasing below 7 through liming. Nakasaki et al. (1996) demonstrated that the degradation rate at the initial stage of composting can be significantly increased by inoculation with acid-tolerant thermophilic bacteria. Choi and Park (1998) observed that the growth of thermophilic bacteria in food waste compost at 50°C was stimulated by an addition of thermophilic yeast that breaks down organic acids. In recent experiments in a composting reactor, Smars et al. (2002) showed that the time of the initial acidic phase could be reduced if the process temperature was kept below 40°C until the pH value in the condensate is above 5. The reason for this was that the microbial respiration in the well-controlled composting reactor was seriously inhibited if the temperature increased above 40°C, while the substrate was still acidic.

1.2 Structure of initial material and the composition of mixture

The term structure of composted matter concerns the size of individual particles of the material forming the charge. This affects directly the amount of contained water and air in the pores of the material. Increased as a consequence of moisture inhibits the growth and development of microorganisms and may lead to anaerobic decomposition of organic matter which is proved by the decrease in demand for oxygen. Indicator characterizing the structure of compost mass is part of the space between the clumps of waste without water. The share of the free space of the material to the total pore volume during composting should be 25–35% (Sidelko 2005). As shown Sidelko (2005), Bien et al. (2011) compacting the compost feedstock expressed decreasing porosity from 40 to 10% more air flow resistance and five times thus limits the availability of oxygen for thermophilic bacteria, reducing the warm phase of composting. Taking into account the air flow resistance and the availability of oxygen for the microorganisms, fragmentation of composted material should be 2.0–6.0 cm. Das et al. (2003) have shown that adding to the compost pile as a bulking agents 5 cm rings of PP (polypropylene) at 50% d.m., there were decreases in temperature gradients, and oxygen content. The satisfactory operation of composting was introduced in the amount of PP rings 7% d.m. Research of Bien et al (2011) confirmed that the mixture can have a significant effect on the degree hygienization in the compost. Factor determining the extending the duration of the thermophilic phase and obtained, the maximum temperature level was the addition of green waste. The study shows that the optimum composition of the mixtures of compost is to use proportions by weight: 20%—sewage sludge, 40%—green waste, 30%—organic fraction of municipal waste, 10%—bulking agent. This follows from the fact that the thermophilic phase parameters for this composition were similar to those for a larger share of green waste, so it is justified to use higher doses of these wastes, because their availability is seasonal. Studies have shown that at higher doses of sewage sludge generated in the thermophilic phase parameters were insufficient for full hygienization, although there was a substantial reduction in the number of parasital eggs, which disqualifies such natural compost for use (Bien et al 2011, 2010).

1.3 Moisture content

Moisture content of the composting blend is an important environmental variable as it provides a medium for the transport of dissolved nutrients required for the metabolic and physiological activities of microorganisms (McCartney & Tingley, 1998). Very low moisture content values would cause early dehydration during composting, which will stop the biological process, thus giving physically stable but biologically unstable composts (Bertoldi et al., 1983). On the other hand, high moisture may produce anaerobic conditions from water logging, which will prevent and halt the ongoing composting activities. Many investigators have conducted experiments and identify that 50–60% moisture content is suitable for efficient composting (Tiquia et al. 1998). Liang et all (2003) suggests that there is a minimal moisture content requirement for active biosolids composting, and 50% seems to be the lower limit, moisture above 60% seems to be optimal for biosolids composting and this experimental design could not detect an upper limit. In large scale composting systems, high moisture contents impact the system in two ways, (1) it limits oxygen diffusion into the composting matrix and (2) increases pliability of the materials leading to a potential for the compaction of the composting matrix (Das & Keener, 1997). Microbial activities measured in biosolids blends at controlled temperature and moisture settings show that moisture content has a greater influence on activity than temperature. The enhancement of microbial activities induced by temperature increment can also be realized by increasing moisture content alone. Liang et al. (2003) shows that temperature proves to be an important factor impacting microbial activity, but its effect is less influential than moisture content. At lower moisture contents of 30% and 40%, 43°C appears to be the optimal temperature setting for the 10-day cumulative O_2 uptake. However, at higher moisture contents (50%, 60%, and 70%), this is not always the case. For example at 50% moisture, 29°C has significantly higher cumulative O_2 uptake than 43°C, while at 60% moisture, there is no significant difference in the 10-day cumulative O_2 uptake between 43 and 57°C, both of which have the highest in comparison with other temperature settings. Moisture content appears to have a compensating or offsetting effect for temperature. In most cases, higher temperatures induced an earlier initiation of increased microbial activity as well as a higher average O_2 uptake rate over time.

1.4 Aeration intensity

Quantity of air is the parameter indicating the intensity of biochemical processes. Moreover, the corresponding amount of air inhibits anaerobic processes and is a factor in regulating the heat balance during the composting process. The amount of air flowing through the mass of compost involves elevating the problem of nitrogen and water vapor. With the air fed to the composted material, is supplied oxygen guarantees proper course of composting, while the stream is discharged thermal energy generated from the transformation of a biochemical exothermic reactions. Assessed the studies performed it was found that the amount of energy transferred to the system in the form of heat is about 12 MJ/kg composted mass (Sidelko et al. 1994). Bech-Friis et al. (2001) said that the quantity of air will affect the value achieved during the composting temperature and is a polynomial. Two characteristic points for the system temperature/oxygen flow are described in coordinates: 40°C for 4.0 mg $O_2 \cdot g^{-1}$ c.l $\cdot h^{-1}$ and 60°C for 5 mg $O_2 \cdot g^{-1}$ c.l $\cdot h^{-1}$. It follows that by increasing the oxygen flow was achieved by one mg $O_2 \cdot g^{-1}$ c.l $\cdot h^{-1}$ increase in temperature of 20°C. The amount of the applied oxygen at 6.0 mg $O_2 \cdot g^{-1}$ c.l $\cdot h^{-1}$ caused a rise in temperature only to 66°C. It was shown that there is a limit beyond which increasing the air flow is not justified in terms of technological and economic (Bech-Friis et al. 2001, Schultz 1960).

1.5 The content of nutrients

Type of organic compounds resulting from their chemical structure is a key factor in determining susceptibility to biodegradation. Nitrogen and phosphorus, in addition to carbon atoms, is an essential component of microbial protoplasm (Tang et al. 2007). Both elements are present in the organic matter of municipal waste, even though their participation may vary. As a result of biochemical changes occurring in the compost mass, nitrogen content decreases systematically with respect to baseline. Loss of nitrogen mainly in the form of ammonium ions, is related to the intensity of water evaporation due to temperature increase inside the composted mass (Ros et al. 2006). It is particularly undesirable phenomenon, causing a change ratio C:N, which has a substantial effect on the kinetics of biochemical processes and consequently to increase or decrease of thermophilic phase. Nutrient balance, expressing the ratio between organic carbon and nitrogen and phosphorus, is the primary criterion for assessing the suitability of the material for biodegradation under aerobic conditions. Particularly important is the ratio C:N because of its low value which indicates a high concentration of nitrogen, the excess may pass in the form of N-NH$_3$ inhibiting microbial growth process, in turn, too large value of C:N indicates low nitrogen content (Sidelko 2012). Li et al. (2011) shows that co-composting of poultry manure with municipal solid waste and corn straw at low C/N ratio of 15, 20, 25 resulted in the compost reaching maturity after 32 days, 24 days and 12 days, respectively. However, treatment at a low initial C/N ratio of 15 affected the behaviors of a number of important parameters significantly during co-composting. As authors shown in two samples, temperature reached 50°C and entered the thermophilic phase on the 7th and 3rd day of composting, indicating quick establishment of microbial activities in the composting pile. However, the other sample required about 11 days, a comparatively longer time, to reach a temperature of 50°C. Thermophilic phase of samples continued for 11 days but for the other samples it lasted for only 7 and 6 days. The reason why the temperature rose slowly was that the raw material cannot supply enough carbon resource for microorganisms to metabolism for low C/N ratio (Sidelko 2012).

2 CONCLUSION

Municipal waste and sewage sludge create risk to the environment due to possible contamination of air, ground and surface water with pathogenic microorganisms, which are medium. Most municipal wastes contain 40–50% organic matter content, in the sludge there can be 51–81% of dry matter (Bien et al. 2010). For proper composting process it is necessary to provide respective process parameters as moisture content of 45–60%, pH in the range 6.5–7.5, the concentration of oxygen in the air leek > 15%. Composting takes place most effectively in the temperature ranged between 45–60°C. Temperature level, its duration and the amount of air supplied, determine the availability of organic compounds. The resulting compost process is a valuable fertilizer, which can be used in nature. One of the many factors determining the use of compost, according to the Minister of Agriculture and Rural Development, dated from the 6th of August 2008, is an aspect of hygienization. The main factor determining the extent of hygienization is temperature achieved and the duration of thermophilic phase. It is important if the process introduces sewage sludge and municipal waste, posing a health risk due to the possibility of occurrence of these pathogenic bacteria and parasite eggs. Optimization of the composting process focuses on the proper selection of both the technological parameters and the composition of the mixtures in order to intensify the thermophilic phase of the maintenance of chemical and biological quality of obtained compost. Research of Bien et al. (2011) confirmed that the mixture can have a significant effect on the hygienization degree in the compost. Research is needed to further understanding of correlation between the factors causing the increase in temperature and the nature of the course of thermophilic phase.

ACKNOWLEDGEMENTS

The present work was financially supported by the statutory research found of the Institute of Environmental Engineering, Czestochowa University of Technology No. BS/MN 401-313/11 and BS/PB 401-303/11.

REFERENCES

Beck-Friis B., Smars S., Johnsson H., Kirchmann H., 2001, Gaseus emission of CO2, N-NH3 and N-NO2 from organic household waste in a compost reactor under different temperature regiments, Journal of Agriculture Engineering Research, Vol. 78, 423–430.

Beck-Friis, B., Sm_ars, S., Jonsson, H., Kirchmann, H., 2001. Gaseous emissions of carbon dioxide, ammonia and nitrous oxide from organic household waste in a compost reactor under different temperature regimes. Journal of Agricultural Engineering Research 78 (4), 423–430.

Beck-Friis, B., Sm_ars, S., Jonsson, H., Eklind, Y., Kirchmann, H., 2003. Composting of source-separated household organics at different oxygen levels: Gaining an understanding of the emission dynamics. Compost Science and Utilization 11 (1), 41–50.

Bertoldi, M.D., Vallini, G., Pera, A., 1983. The biology of composting. Waste Manage. Res. 1, 157–176.

Bień J.B., Milczarek M., Sobik—Szołtysek J., Okwiet T., 2011, Optymalizacja fazy termofilowej w procesie współkompostowania osadów ściekowych i odpadów komunalnych, Nauka Przyr. Technol., t.5, nr 4.

Bień J.B., Grosser A., Neczaj E., Worwąg M., Celary P., 2010, Co-digestion of sewage sludge with different organic wastes: a review, Polish journal of environmental studies, 2, 24–30.

Bień J.B., Worwąg M., Neczaj E., Grosser A., 2010, The possibilities of receiving biogas from industrial sewage sludge, Polish journal of environmental studies, 2, 12–16.

Choi, M.H., Park, Y.H., 1998. The influence of yeast on thermophilic composting of food waste. Letters in Applied Microbiology 26, 175–178.

Day, M., Krzymien, M., Shaw, K., Zaremba, L., Wilson, W.R., Botden, C., Thomas, B., 1998. An investigation of the chemical and physical changes occurring during commercial composting. Compost Science and Utilization 6 (2), 44.

Das, K., Keener, H.M., 1997. Moisture effect on compaction and permeability in composts. J. Environ. Eng. 123 (3), 275–281.

Das K.C., Tollner E., Eiteman M., Comparison of synthetic and natural bulking agents in food waste composting. Compost Science and Utilization. Vol 11, No. 1 (27–35).

Eklind, Y., Beck-Friis, B., Bengtsson, S., Ejlertsson, J., Kirchmann, H., Mathisen, B., Nordkvist, E., Sonesson, U., Svensson, B.H., Torstensson, L., 1997. Chemical characterization of source-separated organic household waste. Swedish Journal of Agricultural Research 27, 167–178.

Finstein, M.S., Morris, M.L., 1975. Microbiology of municipal solid waste composting. In: Perlman, D. (Ed.), Advances in Applied Microbiology, Vol. 19. Academic Press Inc., New York, pp. 113–151.

Finstein, M.S., Miller, F.C., Strom, P.F., 1986. Waste treatment composting as a controlled system. Biotechnology 8, 396–443.

Joshua, R.S., Macauley, B.J., Mitchell, H.J., 1998. Characterization of temperature and oxygen profiles in windrow processing systems. Compost Sci. Util. 6, 15–28.

Jędrczak A., 2007. Biologiczne przetwarzanie odpadów, PWN, Warszawa.

Liang C., Das K.C., Mclendon R.W., 2003. The influence of temperature and moisture contents regimes on the aerobic microbial activity of a biosolids composting blend, Bioresource Technology 86: 131–137.

Li Q., Wang Y., Lv B., 2011, Impact of C/N Ratio on Nitrogen Changing during Composting with Poultry Manure, Energy Procedia, 11, 3175–3179.

McCartney, D., Tingley, J., 1998. Development of a rapid moisture content method for compost materials. Compost Sci. Util. 6, 14–25.

Miller, F.C., 1992. Composting as a process based on the control of ecologically selective factors. In: Blaine-Metting, F. (Ed.), Soil Microbial Ecology: Applications in Agriculture Environment Management. Marcel Dekker Inc., New York, p. 646.

Maheshwari R., 2005, Life at high temperatures, Resonance, 23–36.

Neczaj E., Bień J.B., Grosser A., Worwąg M., Kacprzak M., 2012, Anaerobic treatment of sewage sludge and grease traps sludge in continuous co-digestion, Global NEST Journal, 14, 2, 141–148.

Namkoong, W., Hwang, E.Y., 1997. Operational parameters for composting night. Compost Sci. 5 (4), 46–51.

Nakasaki, K., Uehara, N., Kataoka, M., Kubota, H., 1996. The use of Bacillus licheniformis HA1 to accelerate composting of organic wastes. Compost Science and Utilization 4 (4), 47–51.

Nakasaki, K., Yaguchi, H., Sasaki, Y., Kubota, H., 1993. Effects of pH control on composting of garbage. Waste Management and Research 11 (2), 117–125.

Ros M., Garcia C., Hernandez T., 2006, A full-scale study of treatment of pig slurry by composting: Kinetic changes in chemical and microbial properties, Waste Manag. 26, 1108–1118.

Schloss, P.D., Walker, L.P., 2000. Measurement of process performance and variability in inoculated composting reactors using ANOVA and power analysis. Process Biochemistry 35 (9), 931–942.

Sundberg C., Smars S., Jonson H., 2004, Low pH as an inhibiting factor in the transition from mesophilic to thermophilic phase in composting, Bioresource technology 95, 145–150.

Schultz K.L., 1960, Rate of oxygen consumption and respiratory quotients Turing aerobic decomposition of a synthetic garbage, Compost Science and utilization, 1, 36–40.

Sidełko R., 2005. Kompostowanie—optymalizacja procesu i prognoza jakości produktu, Wydawnictwo Uczelniane Politechniki Koszalińskiej, Koszalin.

Sidelko R., Ewartowska Z., Szymanski K., 1994, Charakterystyka odpadów komunalnych z terenu miasta Kołobrzegu. II Konferencja Nauk.-Tech. Nt. Gospodarka odpadami Komunalnymi.

Sidelko R., 2012, Kompostowanie frakcji organicznej odpadów komunalnych, V konf. Szk. Mech. Biol. Przetw. Odp., 155–170.

Tiquia, S.M., Tam, N.F.Y., Hodgkiss, I.J., 1998. Changes in chemical properties during composting of spent pig litter at different moisture contents. Agric. Ecosys. Environ. 67, 79–89.

Tang J-C., Shibata A., Zhou Q., Katayama A., 2007, Effect of temperature on reaction and microbial community in composting of cattle manure with rice straw, Journal of Bioscience and Bioengineering, 104,(4) 321–328.

Venglovsky J., Sasakova N., Vargova M., Pacajova Z., Placha I., Petrovsky M., Harichova D., 2005, Evolution of temperature and chemical parameters during composting of the pig slurry solid fraction amended with natural zeolite, Bioresource Technology, 96, 181–189.

Weppen, P., 2001. Process calorimetry on composting of municipal organic wastes. Biomass and Bioenergy 21 (4), 289–299.

Environmental Engineering IV – Pawłowski, Dudzińska & Pawłowski (eds)
© 2013 Taylor & Francis Group, London, ISBN 978-0-415-64338-2

Humic acids complexes with heavy metals in municipal solid wastes and compost

B. Janowska

Faculty of Civil Engineering, Environmental and Geodetic Sciences, Koszalin University of Technology, Koszalin, Poland

ABSTRACT: This study examined the heavy metal content occurring in municipal waste and composts at various stages of degradation. A sequential extraction protocol was used to measure the distribution of metals: Copper (Cu), Zinc (Zn), Lead (Pb) and Nickel (Ni) in water-soluble, exchangeable, organically complex, organically bound, solid particulate and residual fractions. During composting, major portions of Cu, Pb, Zn and Ni appeared in the organically-bound, solid particulate, and residual fractions, respectively. The complexing ability humic and fulvic acids results mainly from their content of oxygen functional groups (phenolic—OH and carboxylic—COOH), which show high affinity to those metals. The raw material for compost production and fresh compost contained Cu, Zn, Pb and Ni bound to fulvic acid fraction. There was an increased % of metals bound to humic acid bound metals present in mature composts.

Keywords: heavy metals, sequential extraction, composting, humic acids, fulvic acids

1 INTRODUCTION

Composting of organic wastes generates carbon dioxide, water and organic matter comprising of humic substances and ash. Composting reduces original material volume and mass and the end-product may be used as a fertilizer improving soil properties (Ben Achiba et al. 2010, Flyhammar 1997, Dasgupta 2011, Pawłowski 2011, Shan 2012).

Excessive concentration of heavy metals in the municipal waste composting products determines usability of such material (compost) in agricultural practices. Usually, the total metal amount in compost should be in accordance with Polish legislature but in reality such assessment is only approximate and may be misleading. Data on heavy metal fractioning in municipal waste-derived compost are rare. Therefore, there is a need to examine fractionation processes as all heavy metal forms do not penetrate plants and thus may not all exert adverse effects. (Han et al. 2003, Szymanski et al. 2005). The current study was limited to the identification of several compost fractions in which these metals occur. The current method for the determination of fractionation of heavy metals proposed by Tessiere et al. (1979) was developed for the soil environment. Therefore, this method fails to comply with all the requirements needed specifically for heterogeneous material such as municipal waste-derived compost. This methodology of Tessiere et al (1979) (Kowalkowski & Buszewski, 2009) as well as other methodologies do not take into account the biochemical processes that occur during composting and compost maturing (Turek et al. 2005).

The composting of the organic waste such as municipal waste, sewage sludge or others leads to organic material decomposition which forms organic compounds including humus precursors (Koivula & Hänninen 2001). The soil humus is composed mainly of humic compounds (50%) and polysaccharides (up to 30%). The Humic Substances (HS) are high-molecular weight organic compounds comprising numerous amino acids and heterocyclic forms of nitrogen, other hydrocarbons, alcohol groups, methoxy, phenol, carboxylic, carbonyl, quinone and amino acid origin alkaline groups (Jansen et al. 1995, Pandey et al. 2000). The Humic Acids (HA) are weakly dissociating polybasic organic acids with molecular weight varying from 2 000 Da (easily soluble forms) to 50 000 Da (sparingly soluble forms). HA show cation exchange properties that originate from the presence of carboxyl or hydroxyl (phenol) functional groups (Kinniburgh et al. 1999, Zbytniewski & Buszewski 2005). HA are darker in color than Fulvic Acids (FA) and have a higher carbon and oxygen content. Binding of metal with organic matter compounds exerts a significant influence on the total metal content in the surface/ground waters and soil. Organic matter affects the bioavailability and toxicity of heavy metals through formation of organometallic complexes as once complex the metals are not active (Ingelmo et al. 2012; Kinniburgh et al. 1999; Veeken et al. 2000). Therefore, the bioavailability of heavy metals for plants and microorganisms as well as migration of these metals in water and soil environment is correlated with metal affinity to HA (Gondar et al. 2006, Hernandez et al. 2006). The aim of this study was to determine the bioavailability of Copper (Cu), Zinc (Zn), Lead (Pb) and Nickel (Ni) forms existing in the municipal waste-derived compost and measure binding of metals with HA formed during composting processes.

2 MATERIALS AND METHODS

Samples of municipal waste and composts were taken from the Municipal Waste Management Facility in Kołobrzeg. The composting facility operates in a two-stage system based on the DANO process. Two fractions of municipal waste were selected for tests. The sample fractions were received from the screen analysis where the medium (Fm) comprised of waste organic parts

$(1 \times 1$ cm $< x < 10 \times 10$ cm$)$ and Fine fraction (Ff) $(<1 \times 1$ cm$)$. The fresh Compost samples (Cf) were taken directly at the outlet from the biostabilizer, whereas samples of varied maturity compost were obtained from compost piles. Compost samples after 4-month (C4) and 36-month (C36) storage period in compost piles were taken for tests. The samples were dried at 105°C and than analyzed. To perform comparable tests *Argo*MAT Compost CP-1 (MC) Certified Reference Material (SCP SCIENCE Canada) was used.

The total organic carbon content in the tested samples was determined with the use of the Tiurin technique in accordance with the Polish standard PN-91/Z-15005. The total content of tested metals Cu, Pb, Zn and Ni was determined with the use of the flame atomic absorption spectrometry method. The samples of compost and waste were mineralized with a mixture of concentrated HNO_3, $HClO_4$ acids and H_2O_2 using microwave energy.

Fractioning of the tested material was carried out using the sequential extraction scheme (Fig. 1) developed by Hsu and Lo (2001) who categorized 6

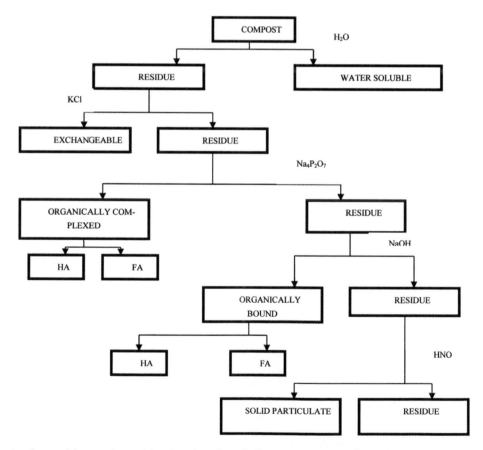

Figure 1.   Sequential extraction and fractionation of metals and organic carbon scheme (Hsu & Lo 2001).

Table 1. Analytical procedure scheme (Hsu & Lo 2001).

Fraction	Extractant	Extraction conditions	
		Shaking time	Temperature
F1—water soluble	20 cm³ H_2O	24 h	Room
F2—exchangeable	20 cm³ KCl 1 M	24 h	Room
F3—organically complexed	20 cm³ $Na_4P_2O_7$ 0,1 M + 250 mg KCl (flocculant)	24 h	Room
F4—organically bound	20 cm³ NaOH 0,1 M	24 h	Room
F5—soild particulate	20 cm³ HNO_3 4 M	24 h	Room
F6—residual	5 cm³ 65% HNO_3 + 1 cm³ 30% H_2O_2 + 1 cm³ 75% $HClO_4$	Microwave mineralisation	

fractions with which the heavy metals were combined: water soluble, exchangeable, organically complex, organically bound, soild particulate and the residual. One gram of a material samples were subjected to sequential extraction. The extraction conditions and extractants applied are shown in Table 1. The samples were centrifuged at 5000 rpm for 10 min. after each extraction stage and Cu, Pb, Zn and Ni content were determined in the obtained supernatant by FAAS method. Furthermore, in each eluate organic carbon content was determined with the use of the Tiurin method (Klejment & Rosiński 2008). Sulfuric acid 3M (pH = 1) was added into two supernatants received after the third ($Na_4P_2O_7$ extractant) and fourth (NaOH extractant) extraction stages and left for 24 hour at room temperature (Hsu & Lo, 2001). After 10 min. (5000 rpm) of centrifugation, the metal fraction bound to FA was separated and content was determined using the FAAS method. The concentration of metals bound to HA was presented as the difference between metal concentration in given fraction and content in the metal-Fulvic Acid (FA) compartment.

All results were presented as the mean of two composting piles, and data for each pile was the mean of three replicates. Statistical analysis was conducted with the STATISTCA software.

3 RESULTS AND DISCUSSION

3.1 *Total metal content*

Table 2 shows total content of metals in the samples analyzed. As it appears from Table 2 the Ni and Pb concentrations in two-year old compost samples as well as Ni levels in the organic and fine fractions municipal waste samples exceeded the admissible concentrations in the compost defined in the Ordinance by the Polish Minister of Agriculture and Rural Development of June 18, 2008.

Table 2. Total content of selected heavy metals in municipal waste and composts [mg kg⁻¹ d.m].

Sample	Copper	Lead	Zinc	Nickel
Fm	41.23	63.51	1128.6	144.79
Ff	45.44	199.35	1721.9	135.52
Cf	51.33	67.39	387.7	49.25
C4	108.21	85.40	590.48	19.81
C36	276.94	266.22	708.39	112.58
MC	180.22	29.50	191.73	30.39
Admissible values*	–	140	–	60

*The Ordinance by the Minister of Agriculture and Rural Development of 18 June 2008.

The total organic carbon and selected metals (Cu, Pb, Zn and, Ni) concentrations were determined in the reference material samples (*Agro-MAT Compost CP-1*) which, according to the supplier originated from organic material of plant origin (Table 3).

3.2 *Organic carbon*

The total organic carbon content reached the maximal levels characteristic for organic waste and fresh compost fractions (above 20%, presented in Table 4). With aging of compost the organic carbon content decreased.

3.3 *Distribution of metals*

3.3.1 *Copper*

Soluble Cu compound content in both fractions of municipal waste (Fm and Ff) amounted to approximately 7% of the total content. The percentage of Cu in the fresh compost fraction was approximately 16% (Fig. 2). In aged composts Cu fell to approximately 5%. The ion exchangeable Cu forms in the tested samples did not exceed 4% of

Table 3. Comparison of the parameters determined in the reference material with the certified values.

Parameter	Unit of measure	Experimental value	Certified value	Tolerance interval
Organic carbon	%	15.09	28	14–42
Copper	mg kg^{-1}	180.22	227	153–301
Lead	mg kg^{-1}	29.50	33	24–42
Zinc	mg kg^{-1}	191.73	240	193–287
Nickel	mg kg^{-1}	30.39	30	16–44

Table 4. Percentage of organic carbon content in particular municipal waste and compost fractions.

	C [% of total content]						
Sample	F1	F2	F3	F4	F5	F6	Total content
Fm	4.11	1.05	0.97	1.55	3.78	9.98	21.44
Ff	1.79	0.74	0.22	0.58	1.84	8.15	13.32
Cf	3.16	0.13	0.47	0.45	1.18	18.18	23.57
C4	0.93	0.67	0.63	0.47	1.03	12.1	15.83
C36	0.84	0.67	1.19	0.69	1.34	10.13	14.86
MC	0.84	0.60	0.56	0.37	1.28	11.44	15.09

Copper

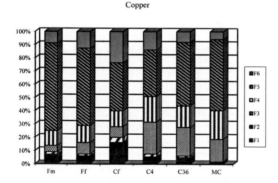

Figure 2. Cu percentage in particular fractions of the tested samples.

Lead

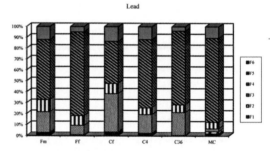

Figure 3. Pb percentage in particular fractions of the tested samples.

the total metal content. In all tested samples the highest % Cu forms occurring in fraction F5 (solid particulate) were noted.

3.3.2 Lead

Lead is an element showing fairly high stability of organometallic compounds, therefore, the percentage of Pb% in mobile fractions is low. The highest % of this metallic form was found in fresh compost (Fig. 3) and did not exceed 4% of the total Pb content. The soluble Pb% decreased with compost age. In the remaining samples Pb did not exceed 2%. The lowest Pb concentrations were present in the ion exchangeable fraction in all tested samples. Pb was detected in highest % in

fraction F3 (organically complexed). Complexed Pb compound % in fractions F3 and F4 was higher in fresh compared to aged compost.

3.4 Distribution of metals and organic carbon

The highest organic carbon content was found in the water-soluble fraction of organic waste and fresh compost fractions. This fraction is comprised of low molecular weight organic compounds such as polysaccharides, amino acids, hemicellulose, organic acids and proteins, which quickly decompose at the early composting stage due to intense biochemical and microorganism activities. As compost ages the water-soluble organic compound quantity decreases (Hsu & Lo 2001, Koivula & Hänninen 2001).

As the compost maturing process proceeds forward, the organic carbon share in fractions F3 and F4 associated with humic substances increases. Organic carbon content in fraction F6 (Table 4) increases insignificantly. It is quite likely that, following a resynthesis process, high-molecular weight organic compounds, which do not decompose at the extraction stage 5, are formed.

Metals in the water-soluble compounds fraction make the most mobile and bioavailable fraction in the environment. Potential hazard for the environment is caused also by the ion exchangeable and mobile organic complex forms of metals. The organo-metallic compounds and solid particulate bound metals (adsorbed on Fe/Mn oxides surface, bound with carbonates or in the form of sparingly soluble inorganic compounds) are not very mobile, therefore, hard to access for live organisms (Flyhammar 1997, Han et al. 2003). Metals occurring in the remaining fraction cannot be released into the soil environment conditions.

There are considerable differences in distribution between fractions among heavy metals that arise from the differences in properties. They originate from various degree of affinity to functional groups occurring in the soil compounds and their tendency to coprecipitation or sorption on the surface of active particles. (Flyhammar 1997, Hsu & Lo 2001). As it is clearly seen, the test results indicate that specific element forms depend on the age of the compost.

Copper compounds content in fraction 5 (solid particulate) was lower for composts than for the municipal waste. Copper content occurring in the form of organic complexes and organometallic compounds was increasing with compost maturity. It is associated with the formation of humic substances. Copper shows high affinity to carboxylic groups that occur in the humic substances (Kinniburgh et al. 1999, Powell & Fenton 1996, Wei et al. 2004). The percentage of copper occurring in the form of organic complexes increases with the degree of compost maturity and it is higher than the percentage of copper forming the organo-metallic combinations. During the composting process the copper content reduction, in the remaining fraction, (Fig. 2) also occurs.

The statistical analysis has shown that the percentage of copper forms in fraction F5 (Table 5) strongly correlated with the organic carbon content. The percentage of copper in fractions F2 and F6 does not depend on the organic carbon content, therefore, one can expect that it occurs mainly in the form of inorganic compounds.

Lead occurred in all the tested samples but first and foremost in fraction F5 and the higher value occurred in the fine waste parts (Fig. 3). It may have originated from the fact that in this

Table 5. Percentage share correlation factors for particular metals with relation to organic carbon C_org percentage.

Metal	F1	F2	F3	F4	F5	F6
Cu	0.50	−0.81	0.54	−0.50	0.93*	−0.68
Pb	0.24	0.41	0.03	0.62	0.26	−0.89*
Zn	0.31	−0.40	0.14	−0.65	0.70	0.47
Ni	−0.02	0.03	0.81	−0.25	−0.51	−0.07

fraction inorganic particles such as argillaceous materials, where lead can be sorbed, are present (Szymanski et al. 2005). Fraction F6 showed strong negative correlation with organic carbon content, therefore, it can be assumed that, in this specific fraction, mainly sparingly soluble inorganic lead compounds occur (Table 5). It has been noted that particular lead forms existing in fraction 6 are not clearly correlated with organic carbon content.

Zinc is mainly associated with organic complexes but no significantly high correlation between zinc concentration in particular fractions and organic carbon concentration has been established (Table 5).

A high correlation between the concentration of nickel in the form of organic complex compounds and organic carbon content has been noted (Table 5).

3.5 Combination of metals with fulvic and humic acids

The humic acids structure varies considerably and depends on the analysed material (compost, peat, soil). Therefore, qualitative and quantitative differences in the formation of organo-metallic combinations may occur (Flyhammar 1997, Jansen et al. 1995, Soler-Rovira et al. 2010). The humic substance fractions (HA and FA) may differ in affinity to particular metal ions. Generally Humic Acids (HA) make more stable combinations with metals than Fulvic Acids (FA) (Nantsis & Carter, 1997, Veeken et al. 2000). The percentage of metals tested in fractions after $Na_4P_2O_7$ and NaOH extractions have been presented in Table 6.

Copper in municipal waste (organic and fine parts) in fraction F3 occurred mostly in connection with humic acids, whereas in compost samples, with fulvic acids. The percentage of copper associated with humic acids increased with compost maturity. In the organically bound fraction (F4) Cu was associated with FA acids in all tested samples. Copper is often sorbed by carboxylic and ketone groups that occur in humic substance molecules. The ability to bind copper by humic substances is determined by the carbon skeleton of

Table 6. Percentage of metals bound with Fulvic (FA) and Humic (HA) acids in metal fractions making organic complexes (Na$_4$P$_2$O$_7$) and in organo-metallic fraction (NaOH).

Sample	Cu [%]				Pb [%]			
	Na$_4$P$_2$O$_7$		NaOH		Na$_4$P$_2$O$_7$		NaOH	
	FA	HA	FA	HA	FA	HA	FA	HA
Fm	36.0	64	67.7	32.3	94.3	5.7	100	0.0
Ff	49.8	50.2	71.7	28.3	94.9	5.1	99.3	0.7
Cf	79.6	20.4	67.4	32.6	95.2	4.8	100	0.0
C4	36.2	63.8	53.7	46.3	77.9	22.1	100	0.0
C36	54.6	45.4	62.6	37.4	50.4	49.6	86	14
MC	100	0.0	59.8	40.2	100	0.0	100	0.0
	Zn [%]				Ni [%]			
Fm	88.2	11.8	87.3	12.7	100	0.0	51.4	48.6
Ff	82.9	17.1	79.9	20.1	59.4	40.6	93.5	6.5
Cf	100	0.0	83.6	16.4	42.3	57.7	41.3	58.7
C4	80	20	86.0	14	68.6	31.4	7.50	92.5
C36	91	9.0	85.6	14.4	89.5	10.5	83.4	16.6
MC	100	0.0	86.1	13.9	66	34	83.6	16.4

given acid. Copper more readily forms complex compounds with humic acids than fulvic because the amount of carboxylic and phenole groups in humic acids is higher (Kang et al. 2011, Powell & Fenton 1996, Pandey et al. 2000).

Lead in fraction F3 occurred mostly in form of complex compounds with fulvic acids (Table 6). Complex compounds with FA made in the waste and fresh compost samples over 90% of that metal content. The increase of humic acids bound lead has been observed with compost maturity degree. In two-year compost 50% of lead was bound with HA. In fraction F4 lead formed exclusively organo-metallic compounds with fulvic acids. It appears from the tests performed that lead shows higher affinity to the fulvic acids. It may appear from the fact that this metal shows higher affinity to the phenol groups than to carboxyl ones (Nantsis & Carter, 1997).

Zinc in compost samples occurred mostly as complex compounds with Fulvic Acids (FA) in fraction F3 (Table 6). The percentage of zinc forming complexes with FA decreased with compost age. Zinc complex compounds with FA made in the municipal waste over 80% of the total content. In extracts containing zinc organo-metallic compounds approx. 15% were the compounds with humic acids (for all tested samples). Zinc and fulvic acids form mostly compounds of organo-metallic character. The organo-metallic compounds with HA acids make approx. 15% of total Zn content in fraction 4. Zinc shows high affinity to carboxyl groups making the fulvic acids structure (Nantsis & Carter 1997, Dabkowska-Naskret 2003).

4 CONCLUSIONS

The total heavy metal content in a compost, as a criterion of its quality assessment, defines in an insignificant degree the level of hazard associated with the introduction of this material into the soil environment. The fractionation of waste and compost samples allows for the identification of mobility and bioavailability in the metal forms. The occurrence of various heavy metal forms depends on physical and chemical parameters of compost and material from which it was produced. The metal bioavailability in the tested municipal waste and compost samples decreases in Cu > Zn > Pb > Ni series. The highest percentage of water-soluble compounds has been found for copper. Zinc is bound mostly with fraction F3, where plant easily assimilable organic complexes can be found. Lead and nickel occurred in fractions, which were hardly accessible for the environment.

The applied analytic procedure scheme allows for tracing the destination of heavy metals bound with organic matter during biochemical processes leading to decomposition of the organic matter in the course of composting. Two organic fractions have been distinguished here, namely, metal organic complexes and the fraction containing organo-metallic compounds. From those fractions metals bound with fulvic and humic acids can be extracted. In the sequential extraction scheme proposed by Tessier et al. (1979) only one organic fraction in which metal organic and sulphide compounds are extracted, is obtained. Therefore, the concentrations of metal bound with humic compounds cannot be clearly defined.

In the municipal waste samples being the raw material for compost production and in fresh compost the tested metals were bound with the fulvic acids fraction. In this fraction there can also occur some compounds that do not show the humic substance character such as low-molecular weight polysaccharides, amino acids and phenol acids (Kang et al. 2011, Koivula & Hänninen 2001). Increased percentage of humic acids bound metals has been found in mature composts.

ACKNOWLEDGEMENTS

The present research is part of the project no 504.01.15.

REFERENCES

Ben Achiba, W. Lakhdar, A. Gabteni, N. Du Laing, G. Vreloo, M. Boeckx, P. Van Cleemput, O. Jedidi, N. & Gallali T. 2010. Accumulation and fractionation of trace metals in Tunisian calcareous soil amended with farmyard manure and municipal solid waste compost. *Journal of Hazardous Material* 176: 99–108.

Dasgupta, P. & Taneja, N. 2011. Low Carbon Growth: An Indian Perspective on Sustainability and Technology Transfer. *Problemy Ekorozwoju/Problems of Sustainable Development* 6(1): 65–74.

Dabkowska-Naskret, H. 2003. The role of organic matter in association with zinc in selected arable soils from Kujawy Region, Poland. *Organic Geochemistry* 34: 645–649.

Gondar, D. Iglesias, A. Lopez, R. Fiol, S. Antelo, J.M. & Arce, F. 2006. Copper binding by peat fulvic and humic acids extracted from two horizons of an ombrotrophic peat bog. *Chemosphere* 63: 82–88.

Flyhammar, P. 1997. Estimation of heavy metal transformations in municipal solid waste. *Science Total Environmental* 198: 123–133.

Han, F.X. Banin, A. Kingery, W.L. Triplett, G.B. Zhou, L.X. Zheng, S.J. & Ding, W.X. 2003. New approach to studies of heavy metal redistribution in soil. *Advance Environmental Research* 8: 113–120.

Hernandez, D. Plaza, C. Senesi, N. & Polo, A. 2006. Detection of copper (II) and zinc (II) binding to humic acids from pig slurry and amended soils by fluorescence spectroscopy. *Environmental Pollution* 143: 212–220.

Hsu, J.H. & Lo, S.L. 2001. Effect of composting on characterization and leaching of copper, manganese, and zinc from swine manure. *Environmental Pollution* 114: 119–127.

Ingelmo, F. Molina, M.J. Soriano, M.D. Gallardo, A. & Lapena, L. 2012. Influence of organic matter transformations on the bioavailability of heavy metals in a sludge based compost. *Journal of Environmental Management* 95: 104–109.

Jansen, S. Paciolla, M. Ghabbour, E. Davies, G. & Varunum, J.M. 1995. The role of metal complexation in the solubility and stability of humic acid. *Materials Science Engineering C* 4: 181–187.

Kang, J. Zhang, Z. & Wang, J.J. 2011. Influence of humic substances on bioavailability of Cu and Zn during sewage sludge composting. *Bioresource Technology* 102: 8022–8026.

Kinniburgh, D.G. Van Riemsdijk, W.H. Koopal, L.K. Borkovec, M. Benedetti M. F. & Avena, M.J. 1999. Ion binding to natural organic matter: competition, heterogeneity, stoichiometry and thermodynamic consistency. *Colloids and Surfaces A* 151: 147–166.

Klejment, E. & Rosiński, M. 2008. Testing of thermal properties of compost from municipal waste with a view to using it as a renewable, low temperature heat source. *Bioresource Technology* 99: 8850–8855.

Koivula, N. & Hänninen, K. 2001. Concentrations of monosaccharides in humic substances in the early stages of humification. *Chemosphere* 44: 271–279.

Kowalkowski, T. & Buszewski, B. 2009. Soli reclamation by municipal sewage compost: Heavy metals migration study. *Journal of Environmental Science and Health Part A* 44: 522–527.

Nantsis, E.A. & Carter, W.R. 1997. Molecular structure of divalent metal ion-fulvic acid complexes. *Theochem* 423: 203–212.

Pandey, A.K. Pandey, S.D. & Misra, V.M. 2000. Stability constants of metal-humic acid complexes and its role in environmental detoxification. *Ecotoxicology and Environmental Safety* 47: 195–200.

Pawłowski L., 2011. Sustainability and Global Role of Heavy Metals. *Problemy Ekorozwoju/Problems of Sustainable Development* 6(1): 59–64.

Powell, H.K. & Fenton, E. 1996. Size fractionation of humic substances: Effect on protonation and metal binding properties. *Analytica Chimica Acta* 334: 27–38.

Shan, S. & Bi, X. 2012. Low Carbom Development of China's Yangtze River Delta Region. *Problemy Ekorozwoju/Problems of Sustainable Development* 7(2): 33–41.

Soler-Rovira, P. Madejon, E. Madejon, P. & Plaza, C. 2010. In situ remediation of metal-contaminated soils with organic amendments: Role of humic acids in copper bioavailability. *Chemosphere* 79: 844–849.

Szymański, K. Janowska, B. Sidelko, R. 2005. Estimation of bioavailability of copper, lead and zinc in municipal solid waste and compost. *Asian Journal of Chemistry* 17: 1646–1660.

Tessier, A. Cambell, D.G. & Bisson, M. 1979. Sequential Extraction Procedure for the speciation of Particulate Trace Metals. *Analytical Chemistry* 51: 844–851.

Turek, M. Korolewicz, T. & Ciba, J. 2005. Removal of heavy metals from sewage sludge used as soil fertilizer. *Soil & Sediment Contamination* 14: 143–154.

Veeken, A. Nierop, K. Wilde, V. & Hamelers, B. 2000. Characterisation of NaOH-extracted humic acids during composting of biowaste. *Bioresource Technology* 72: 33–41.

Wei, Y.L. Lee, Y. CH. Yang, Y.W. & Lee, J.F. 2004. Molecular study of concentrated copper pollutant with compost. *Chemosphere* 57: 1201–1205.

Zbtyniewski, R. & Buszewski, B. 2005. Characterization of natural organic matter (NOM) derived from sewage sludge compost. Part 1: chemical and spectroscopic properties. *Bioresource Technology* 96: 471–478.

Zhou, P. Yan, H. & Gu, B. 2005. Competitive complexation of metal ions with humic substances. *Chemosphere* 58: 1327–1337.

Environmental Engineering IV – Pawłowski, Dudzińska & Pawłowski (eds)
© 2013 Taylor & Francis Group, London, ISBN 978-0-415-64338-2

Influence of EM BIO™ on changes in contents of macroelements in the compost produced from municipal sewage sludge mixed with different components

M. Gibczyńska, M. Romanowski & A. Zwierko
General and Ecologic Chemistry Department, The West-Pomeranian University of Technology in Szczecin, Szczecin, Poland

ABSTRACT: Assessment of the influence the influence of EM BIO™ on changes in content of calcium, magnesium, potassium, acidity and dry mass, during composting of sewage sludge mixed with coniferous tree sawdust or wastepaper has been presented in the paper. In the composted material the main component was sewage sludge from a municipal wastewater in Nowogard. The two-factor experiment covered the following experimental factors: 1—type of composted material, 2—a dose of preparation—EM BIO™. Adding to the composted sewage sludge, sawdust of coniferous, wastepaper and EM BIO™ microbiological preparation resulted in a statistically significant increase in dry matter. On the basis of the changes of pH, calcium, magnesium, potassium and dry matter mineralization processes of organic compounds with the greatest intensity occurred in the sludge in conjunction with the composted wastepaper and a higher dose of microbial preparation EM BIO™.

Keywords: macroelements, sewage sludge, sawdust, wastepaper, microbiologic preparation EM BIO™

1 INTRODUCTION

In 2010 year the municipal sewage treatment plants have been produced 526.7 thousand Mg dry weight of sludge (Rocznik Statystyczny 2011). Biochemical processing in the composting process to the methods that are becoming more widely used in the art of organic waste recycling. However, this method is important for the quality of waste criteria and requirements for the management of produced compost. It is expected that the composting of sewage sludge in the next few years will provide the main processing technology with the aim: the destruction of pathogens, production of high-quality organic fertilizer and giving the fertilizer a practical and convenient to use.

Sewage sludge can be composted with addition of various structure-creating components, as e.g. sawdust or wastepaper, and the process itself can be enhanced with microbiologic preparations containing the Effective Microorganisms. Based on controlled experiments, when determining changes taking place during composting sewage sludge mixed with sawdust, straw, silvers and waste of hemp, Czekała (2009) showed that rate and directions of quantitative changes and solubility of components were determined not only by the composting time, but by type and participation of the composted organic materials.

The Effective Microorganisms Technology was invented in 1982 by a Japanese scientist, Teruo Higa, of the Agricultural University of Ryukyus, Okinawa (Mau 2003). Microorganisms used in support, are able to reproduce and to protect various ecosystems. They support the basic life processes, including: generation of humus, germination, assimilation of food by plants and animals, decomposition of organic matter. EM preparations contain over 80 species existing in the natural environment, as e.g. photosynthesizing bacteria, milk acid bacteria, yeasts, fungi and Actinomycetes (www.emgreen.pl). The basic groups of microorganisms included in the preparations with effective microorganisms include: milk acid bacteria, photosynthesizing bacteria, yeasts, Actinomycetes and fermentation bacteria (Baranowski 2004). All these organisms occur in the natural environment worldwide. Upon selection, the Effective Microorganisms are evaluated and will never be detrimental to humans, animals or plants (Onyszyk 2003).

Milk acid bacteria—are a strong sterilizing agent, thanks to which they hamper the development of detrimental microorganisms and fungi, or *Fusarium,* causing many dangerous diseases in crops. Photosynthesizing bacteria—are independent, self-maintaining organisms. They use atmospheric CO_2, sunlight and the soil heat to live, and they also synthesize necessary substances from plant root secretions, organic matter or hazardous gases (e.g. H_2S). Yeasts—synthesize antibiotic and beneficial substances from amino-acids and sugars. *Actinomycetes*—their structure suggests their placement among bacteria and fungi. They produce substances from amino-acids secreted by photosynthesizing bacteria and from organic matter. Thanks to their actions, nitrogen is bound better by bacteria *Azotobakter*. Fermenting fungi—here we include *Aspergillus* and *Penicillinum* that bring about quick decomposition of organic matter, producing alcohol, esters and substances fighting detrimental microbes (Mau 2003).

Plants absorb calcium from the soil solution in the calcium ion form, hence availability of the exchangeable forms in the soil is more important than the total content of it. Content of CaO makes for 2% to 5% d.m. of the sewage sludge and depends on type and quality of sewage or the technology of its treatment. When composting some sewage sludge from wastewater treatment in Sierpc, mixed with grain straw, Siuta et al. (2007) obtained calcium content in the subsequent years within the range of 1.4–4.4% d.m. In sewage sludge, magnesium content may fluctuate between a couple of and a dozen or so grams per a kilogram of dry mass (3–12 g $Mg \cdot kg^{-1}$ d.m.) and this is a value comparable to magnesium content in soils (Nowak et al. 1998). Generally, potassium content in the K_2O form in sewage sludge is 0.2–0.4% d.m. When composting some sludge from wastewater treatment in Sierpc, mixed with grain straw, Siuta et al. (2007) obtained potassium content in the subsequent years within the range of 0.06 do 1.2% d.m. In order to improve the quality of wastewater from the microbiologic point of view, one has to apply methods contributing to extinction of such pathogenic organisms as bacteria, virus, yeasts, fungi, protozoa cysts or worm eggs. To maximize elimination of pathogenic organisms, sewage sludge is made a subject to numerous stabilizing processes. One of them is composting with different components (Marcinkowski 2003).

Assessment of the influence the influence of EM BIO™ on changes in content of calcium, magnesium, potassium, acidity and dry mass, during composting of sewage sludge mixed with coniferous tree sawdust or wastepaper has been presented in the paper.

2 MATERIAL AND METHODS

2.1 Conditions for the implementation experience

The experiment with composting sewage sludge started on 17 January 2011 in the vegetation hall at the Faculty of Environmental Management and Agriculture of the West-Pomeranian University of Technology in Szczecin. In the composted material the main component was sewage sludge from a municipal wastewater in Nowogard. The two-factor experiment covered the following experimental factors:

Factor 1—type of composted material:

1. sewage sludge
2. sewage sludge + coniferous tree sawdust
3. sewage sludge + white wastepaper

Factor 2—a dose of microbiologic preparation—EM BIO™

1. 0 $dm^3 \cdot m^{-3}$ of the compost mass
2. 1 $dm^3 \cdot m^{-3}$ of the compost mass
3. 2 $dm^3 \cdot m^{-3}$ of the compost mass.

The tests covered nine objects (Table 1).

The volume pots used in the experiment was 10 dm^3. The structure-making material, i.e. coniferous tree sawdust or white wastepaper, accounted for 30% of the volume weight of compost. The mixture of compost was mixed with a mechanic agitator made by Dedra. Calculated according to the vase capacity, preparation doses were applied with a manual sprinkler made by Kwazar (www.kwazar.com.pl). During composting, the content of the pots was mixed every two weeks. In the vegetation hall, where the composted material containers were stored, the average 24-h temperature was 15°C.

EM BIO™ has been distributed in Poland since 2009 by one distributor, Greenland Spółka z o.o. in Trzcianka near Puławy. EM BIO™ is a fully natural, biotechnological product for municipal purposes. In sewage sludge EM BIO™ impedes the process of microbiologic decomposition of organic substances, blocking at the same time the emergence of odors accompanying the composting process. In Polish climatic conditions this process takes from three to five months. After the transformation process, the sewage sludge is already stabilized and the matter, with applied EM BIO™,

Table 1. Object number and applied components.

Object	Components
1	Sewage sludge, control
2	Sewage sludge + EM BIO 1 dm^3
3	Sewage sludge + EM BIO 2 dm^3
4	Sewage sludge + sawdust
5	Sewage sludge + sawdust + EM-BIO 1 dm^3
6	Sewage sludge + sawdust + EM-BIO 2 dm^3
7	Sewage sludge + wastepaper
8	Sewage sludge + wastepaper + EM-BIO 1 dm^3
9	Sewage sludge + wastepaper + EM-BIO 2 dm^3

neither undergoes further rotting processes nor emits odors. At correct application of EM BIO™ it is possible to resign from hygienization of the sewage sludge with lime. With EM BIO™ we also achieve reduction of the sludge volume and mass (www.emgreen.pl).

Samples of the compost mass, to be used for chemical analysis, were collected on: 17 February, 17 March, 18 April 2011. From each pot, at the above mentioned terms, using the Enger stick, 5 samples of the total weight of 250 g were collected. The samples were determined: pH, dry mass and total content of calcium, magnesium and potassium.

2.2 Methodology of chemical analysis

The percentage of dry mass in the sewage sludge was calculated as per standard PN-EN 12880:2004. Content of calcium and magnesium in the composted sewage sludge was determined under standard PN-EN ISO 7980:2002, and potassium content—according to standard PN-ISO 9964-2/Ak:1997. pH was determined potentiometrically (pH_{KCl}), as per PN-EN 12176:2004. All mean values of pH, as logarithmic functions, were calculated by going back to the value of concentration of Hydrogen ions $[H^+]$, and, having established the mean values, $-\log [H^+]$, or pH, was calculated again. To process the obtained results, the method of variance and half-interval values calculated with the application of Tukey test was used, and the relevance level $\alpha = 0.05$, using FR-ANALWAR software prepared by prof. Franciszek Rudnicki.

3 RESULTS AND DISCUSSION

3.1 Changes in moisture of the composted sewage sludge

The dry matter content in the sewage sludge is a significant characterizing parameter and it presents a high diversification. When analyzing the fertilizing value of sewage sludge, Czekała (1999) specifies the dry mass content within the range 16.2–48.7%. Undoubtedly, the differences are a result of the wastewater treatment technology and handling the product after its treatment. The sewage sludge that was used for the experiment, presented a dry mass content within the range between 24.43–28.79%. For comparison, sewage sludge collected in waste treatment plants in Siedlce and Łuków, being the subject of studies conducted by Wysokiński & Kalembasa (2011), showed a lower content of dry mass—on average 150 and 145 g of dry mass in 1 kg of fresh mass. Addition of coniferous sawdust, wastepaper and the microbiologic preparation EM BIO™ to the sludge brought about a statistically significant increase of dry mass. An increased dose of the preparation is more effective, increasing the quantity of the dry mass in the composted sewage sludge (Table. 2). Dry mass content in the composted sewage sludge reached maximally 37.68%, which was observed for the sewage sludge with wastewater and a double dose of the microbiologic preparation EM BIO™.

When analyzing the influence of the added components to the composted sewage sewage on dry mass changes, it was assumed that 100% represented a mean value for three terms, for the same sludge. The graph clearly reflects the fact that the

Table 2. Changed in moisture of the composted sewage sludge at subsequent terms.

Factor I—type of material	Factor II	I term	II term	III term	Mean
		% d.m.			
Sewage sludge	0	24,48	25,45	28,79	26,24
Sewage sludge	1	25,65	27,29	32,85	28,60
Sewage sludge	2	27,50	31,61	33,70	30,94
Sewage sludge + sawdust	0	26,83	27,53	29,40	27,92
Sewage sludge + sawdust	1	27,73	29,35	31,27	29,45
Sewage sludge + sawdust	2	28,37	31,70	32,59	30,89
Sewage sludge + wastepaper	0	31,48	32,66	34,23	32,79
Sewage sludge + wastepaper	1	32,74	33,60	35,49	33,94
Sewage sludge + wastepaper	2	33,20	34,49	37,68	35,12
LSD$_{0,05}$	Factor I	0,525	0,302	0,270	n.c.**
	Factor II	0,910	0,523	0,468	n.c.
	IxII	i.d.*	i.d.	i.d.	n.c.
	IIxI	i.d.	i.d.	i.d.	n.c.

i.d.*—insignificant difference.
n.c.**—not counted.

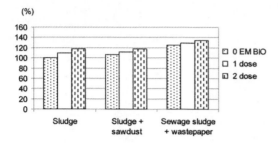

(%)

160
140
120
100
80
60
40
20
0

Sludge   Sludge +   Sewage sludge
          sawdust   + wastepaper

□ 0 EM BIO
□ 1 dose
□ 2 dose

Figure 1. Comparison of changes in the quantities of dry mass as a result of addition of sawdust, wastewater and the microbiologic preparation EM BIO™ to the sludge, quantity of dry mass in the sludge itself equals 100%.

highest dry mass content was observed in the sewage sludge mixed with wastepaper and the larger dose of microbiologic preparation EM BIO™, which can indicate that in this case the mineralization of organic compounds was more intensive. Obtained effects are coherent with the statement made by Wysokiński & Kalembasa (2011) that mixes of sewage sludge with all the additives (lignite coal, straw and sawdust, calcium oxide) contained more dry mass than sludge with no additives. In turn, while composting sewage sludge and plant remnants, mixed with sawdust from beech wood, Zaha et al. (2011) perceive the sawdust as a factor hampering activity of microorganisms in decomposition of complex compounds in that material.

3.2   Changes in pH of composted sewage sludge

pH is the main factor determining for solubility of metal compounds in the environment. Inter alia, pH value decides on shifting the balance coefficient between sorption and desorption of metal and Hydrogen ions. The correct pH of soils and the hydrolytic acidity, which condition sorptive and buffering qualities, have also grave importance for maintenance of soil abundance with nutritious components and, consequently, for creating fertility of the soils (Jakubus 2006). Literature of the subject provides diversified ranges of pH for tested sewage sludge; most often, however, it is a slight acidic or neutral value (Czekała 2002). In tests on their impact on changes in physico-chemical and chemical qualities of soil, sewage sludge used by Jakubus (2006) presented a pH range from 6.5 to 7.2. Sewage sludge tested in the experiment by Czekała (1999) showed a wider pH from 5.9 to 7.8. According to Sweeten and Auvermann (2008), with initial pH 6.5–7.2 in the sewage sludge, composting effects are the best. Then, due to release of free organic acids, pH may go down to 6.0. After completion of composting, values of pH of the

final compost should remain within the range from 7.5 to 8.5. Composted sewage sludge presented pH within the range of 6.23–6.86, which is slightly acidic, but does not disqualify it as fertilizing material (Table 3). Combination of the composted sewage sludge with coniferous sawdust brought acidification of the tested material. As a result of adding wastepaper to the composted sewage sludge, the material became more alkaline, which must be attributed to presence of calcium in the additive. Comparing the pH values for different dates, one has to notice a gradual increase in pH values, which results from the fact that with time, due to decomposition of organic compounds, participation of the alkaline metals (Ca, Mg and K) in the entire quantity of the tested material grows. On the other hand, when describing the sewage sludge for their use for pro-nature purposes, Kaniuczak et al. (2009) state that the sewage sludge in question featured slightly acidic or neutral reaction, and the pH values of the composted sludge were inversely proportional to the time of their composting. After 3 months of composting, the sludge reached pH 8.1, and after 15 months—pH 7.3.

Addition of microbiologic preparation EM BIO™ to the sludge caused alkalization of the material and an increased dose of the preparation acted more effectively (Table 3).

3.3   Total calcium content in the composted sewage sludge

In the arable layer of soils in Poland, the total calcium level varies between 0.07 and 3.6% Ca (Krzywy 2007). The content of CaO basically makes for 2–5% of the dry mass of the sludge and depends on type of the sludge and technology of its treatment. When composting some sludge from wastewater treatment in Sierpc, mixed with corn straw, Siuta et al. (2007) obtained calcium content of 1.4–4.4% d.m. in subsequent years.

The sewage sludge used in the experiment presented low content of calcium—0.5%, which is close to low abundance of the soil (Table 4). According to Krutul (1998), pine wood contains highest levels of calcium, potassium and magnesium, and far lower values of manganese, iron, zinc and cupper. Wood of conifers is characterized by the abundance of calcium within the range from 0.5 to 1,5 g in one kilogram of dry mass (Krutul 1998). Combination of the composted sewage sludge with sawdust had no effect in the total calcium content changes in the concerned material, which should be attributed to the similar concentration of this element in the materials used for the experiment.

Addition of wastepaper to the sewage sludge resulted in a significant increase of the total calcium in the composted material and the observed increase was about 50% (Table 4).

Table 3. Changes in pH of composted sewage sludge on subsequent terms.

Factor I—type of material	Factor II	I term	II term	III term	Mean
		pH			
Sewage sludge	0	6,34	6,43	6,48	6,41
Sewage sludge	1	6,39	6,50	6,54	6,48
Sewage sludge	2	6,44	6,53	6,61	6,53
Sewage sludge + sawdust	0	6,23	6,26	6,33	6,27
Sewage sludge + sawdust	1	6,28	6,36	6,42	6,35
Sewage sludge + sawdust	2	6,29	6,42	6,47	6,39
Sewage sludge + wastepaper	0	6,58	6,64	6,74	6,65
Sewage sludge + wastepaper	1	6,61	6,73	6,83	6,72
Sewage sludge + wastepaper	2	6,73	6,79	6,86	6,79
$LSD_{0,05}$	Factor I	0,036	0,030	0,035	n.c.**
	Factor II	0,046	0,051	0,065	n.c.
	IxII	i.d.*	i.d.	i.d.	n.c.
	IIxI	i.d.	i.d.	i.d.	n.c.

i.d.*—insignificant difference.
n.c.**—not counted.

Table 4. Changes in subsequent terms, the total calcium content in composted sewage sludge.

Factor I—type of material	Factor II	I term	II term	III term	Mean
		g Ca·kg⁻¹ d.m.			
Sewage sludge	0	5,28	5,63	5,43	5,45
Sewage sludge	1	5,56	5,82	5,53	5,64
Sewage sludge	2	5,84	6,12	5,97	5,98
Sewage sludge + sawdust	0	4,94	4,73	6,46	5,38
Sewage sludge + sawdust	1	5,20	4,84	6,72	5,59
Sewage sludge + sawdust	2	5,43	4,93	6,98	5,78
Sewage sludge + wastepaper	0	8,52	8,14	6,86	7,84
Sewage sludge + wastepaper	1	8,77	8,30	7,11	8,06
Sewage sludge + wastepaper	2	8,99	8,49	7,28	8,25
$LSD_{0,05}$	Factor I	0,051	0,222	0,145	n.c.**
	Factor II	i.d.*	i.d.	i.d.	n.c.
	IxII	i.d.*	i.d.	i.d.	n.c.
	IixI	i.d.	i.d.	i.d.	n.c.

i.d.*—insignificant difference.
n.c.**—not counted.

Comparing abundance of the material in question on subsequent terms, a gradual increase of the total calcium in the composted sewage sludge can be observed. Similar dependences were noticed by other authors in their studies (Drozd et al. 1996; Czekała 1999; Patorczyk-Pytlik et al. 1999). The above relationship can be explained with the fact that in the course of time, due to decomposition of organic compounds, rapid reduction of carbon content takes place, which is documented in many publications (Ciećko et al. 2001; Wysokiński and Kalembasa 2011), and at the same time participation of calcium in the general mass of composted sewage sludge grows.

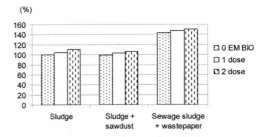

Figure 2. Comparison of changes in calcium content as a result of adding sawdust, wastepaper and preparation EM BIO to the sewage sludge; quantity of calcium in the sludge itself equals 100%.

Table 5. Changes in subsequent terms, the total magnesium content in composted sewage sludge.

Factor I—type of material	Factor II	I term	II term	III term	Mean
		g Mg·kg⁻¹ d.m.			
Sewage sludge	0	3,11	3,58	3,22	3,31
Sewage sludge	1	3,21	3,82	3,52	3,52
Sewage sludge	2	3,33	3,73	3,49	3,52
Sewage sludge + sawdust	0	2,41	2,70	3,04	2,72
Sewage sludge + sawdust	1	2,80	2,82	3,49	3,04
Sewage sludge + sawdust	2	2,86	2,56	3,02	2,81
Sewage sludge + wastepaper	0	3,18	2,86	2,89	2,98
Sewage sludge + wastepaper	1	3,49	3,09	3,41	3,33
Sewage sludge + wastepaper	2	3,52	2,86	3,11	3,17
$LSD_{0,05}$	Factor I	0,136	0,380	i.d.	n.c.**
	Factor II	i.d.*	i.d.	i.d.	n.c.
	IxII	i.d.*	i.d.	i.d.	n.c.
	IIxI	i.d.	i.d.	i.d.	n.c.

i.d.*—insignificant difference.
n.c.**—not counted.

Addition of microbiologic preparation EM BIO™ to the composted sewage sludge caused an increase in content of total calcium in the material and increasing the dose of the preparation brought better effects. This dependence, however, was not confirmed by statistical calculations (Table 4).

Analyzing the effects of the components added to the composted sewage sludge in changes of total calcium, the mean value for three dates was assumed as 100% in the very sludge. The figure shows clearly the fact that the highest content of total calcium was present in the sewage sludge composted together with wastepaper and an increased dose of the microbiologic preparation. In this case, the processes of mineralization of organic compounds with the greatest intensity occurred.

3.4 Content of total magnesium in the composted sewage sludge

In the arable layer of soils in Poland, the total magnesium level varies between 0.06 and 1.2% (Krzywy 2007). Magnesium content in the sewage sludge may vary between a couple of to over a dozen grams per a kilogram of dry mass (Nowak et al.1996). Studying changes in content of magnesium and zinc available to plants in soils after the application of sewage sludge and mixes of sewage sludge with peat, Chowaniak & Gondek (2009) used sludge containing mgnesium at 4.86 g Mg·kg⁻¹. Coniferous tree wood has a lower level of mgnesium than of calcium, i.e. between 0.08 and 0.20 g in one kilogram of dry mass (Krutul 1998). When composting sewage sludge containing 0.26% of total magnesium with sawdust,

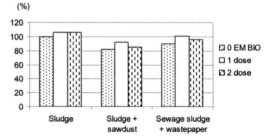

Figure 3. Comparison of changes in magnesium content as a result of addition of sawdust, wastepaper and the microbiologic preparation EM BIO™ to the sewage sludge, quantity of magnesium in the sludge itself equals 100%.

Ciećko et al. (2003) stated that the concentration of magnesium in the sawdust was lower than the one in the sewage sludge. The sewage sludge used in the experiment was characterized by a low magnesium content of 0.3%—which is close to the low soil abundance (Table 5). As a result of adding sawdust and wastepaper to the composted sewage sludge, a significant reduction of the level of general magnesium in the material in study was observed on the second and third date after addition of the wastepaper, and on all the three terms after addition of the sawdust (Table 5). Similarly to the changes in calcium content, one has to notice a regular increase in the general magnesium level in the composted sewage sludge. This regularity finds its confirmation in information provided in the literature of the subject (Ciećko et al. 2003; Drozd et al. 1996;

Table 6. Changes of total potassium content in the composted sewage sludge on subsequent terms.

Factor I—type of material	Factor II	I term	II term	III term	Mean
		g K·kg⁻¹ d.m.			
Sewage sludge	0	5,93	7,06	6,29	6,43
Sewage sludge	1	6,12	7,26	6,31	6,56
Sewage sludge	2	6,46	7,34	7,18	6,99
Sewage sludge + sawdust	0	5,43	4,89	6,09	6,47
Sewage sludge + sawdust	1	5,80	5,50	6,24	5,85
Sewage sludge + sawdust	2	5,66	5,82	6,62	6,03
Sewage sludge + wastepaper	0	6,22	5,18	7,65	6,35
Sewage sludge + wastepaper	1	5,79	5,73	7,70	6,41
Sewage sludge + wastepaper	2	5,94	5,92	7,98	6,61
$LSD_{0,05}$	Factor I	0,256	0,428	0,365	n.c.**—n.c.**
	Factor II	i.d.*	i.d.	i.d.	n.c.
	IxII	i.d.*	i.d.	i.d.	n.c.
	IIxI	i.d.	i.d.	i.d.	n.c.

i.d.*—insignificant difference.
n.c.**—not counted.

Czekała 1999; Licznar et al. 1999; Patorczyk-Pytlik et al. 1999; Stuczyński 1992).

Addition of a single dose of microbiologic preparation EM BIO™ to the composted sludge brought about an increase in the level of general magnesium in the material. Increased doses of the preparation did not find any reflection in the form of clearly observed changes in magnesium content in the composted sludge (Table 5, Fig. 3).

3.5 Content of total potassium in the composted sewage sludge

In the soils in Poland, the general Magnesium level varies between 0.8 and 2.5% K (Krzywy 2007). As potassium is the most deficit component in the sludge, additional amounts of it have to be added while using the sludge as a fertilizer (Ignatowicz et al. 2011). The relatively low level of potassium in sludge is a result of good solubility of most of potassium compounds and of the fact that they do not settle as sediments (Gorlach & Gambuś 1998; Oleszczuk 2008). Participation of potassium fluctuates below the content of this element in the other organic fertilizers, being generally between 0.1 and 0.56% s.m. (Bień 2002).

In pine wood, potassium is several times less than calcium and its quantity remains within the scope of 0.05–0.45 g in one kilogram of dry mass (Krutul 1998).

The sewage sludge used in the experiment contained 0.5% of total potassium, which is below the lower limit of soil abundance (Table 6). Combination of the composted sewage sludge

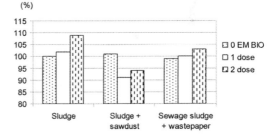

Figure 4. Comparison of changes in potassium content as a result of addition of sawdust, wastepaper and the preparation EM BIO™ to the sludge, quantity of potassium in the sludge itself equals 100%.

with sawdust brought about reduction of the total potassium level in the material and it amounted to 4.89–6.62 g K·kg⁻¹ d.m. This can be attributed to the fact that softwood sawdust contains ca. 0.4 g of potassium in one kilogram of dry mass (Ciećko et al. 2005). The addition of wastepaper to the sewage sludge did not have an unambiguous effect on levels of total potassium in the composted material (Table 6).

Comparing abundance of the material in the experiment on subsequent terms, one has to note the steady increase of total potassium levels in the composted sewage sludge—similarly to the other macroelements. This regularity finds its confirmation in the results presented by Ciećko et al. (2005) that after completion of the composting process, the potassium content in compost mixed with 40% sawdust and in compost mixed with some manure was higher by 100% and 82%, comparing

Table 7. The values of the equivalent relations for the averages of the three terms.

Factor I—type of material	Factor II	Equivalent relations	
		Ca/Mg	K/(Ca+Mg)
Sewage sludge	0	0,99	1,20
Sewage sludge	1	0,96	1,17
Sewage sludge	2	1,02	1,21
Sewage sludge + sawdust	0	1,19	1,20
Sewage sludge + sawdust	1	1,10	1,19
Sewage sludge + sawdust	2	1,23	1,25
Sewage sludge + wastepaper	0	1,58	1,02
Sewage sludge + wastepaper	1	1,45	0,97
Sewage sludge + wastepaper	2	1,56	1,00

to the initial material. In the compost without any organic additives the increase was lower by 50%. The increase of total potassium during the composting of sewage sludge was observed also by a number of other authors (Bauman-Kaszubska & Buraczewski 1999; Drozd et al. 1996; Licznar 1999; Stuczyński 1992).

Addition of microbiologic preparation EM BIO™ to the composted sludge resulted in the increase of total potassium content in the material, while an increased dose of the preparation was more effective. This regularity, however, was not confirmed by statistical calculations (Table 6). To analyze the influence of the components added to the composted sewage sludge on the levels of total potassium, the mean value for three terms for the same sludge was assumed as 100%. The figure presents clearly the fact that the higher abundance, as far as total potassium is concerned, characterizes the sewage sludge composted without any additional components, but with a larger dose of the microbiologic preparation.

3.6 Analysis of equivalent relationships Ca/Mg and K/(Ca+Mg)

Because of the obvious degree of oxidization of the macroelements discussed herein (calcium and magnesium +2 and potassium +1) it is possible to calculate the equivalent relationships Ca/Mg and K/(Ca+Mg). To calculate the equivalent relationships, we divide content of the given element by its atom mass and oxidization degree.

Analysis of the equivalent relationships Ca/Mg in the material in study confirmed the effect of wastepaper added to the sewage sludge, expressed in an increase in the abundance of the composted material with calcium, compared to magnesium (Table 7). The higher calcium content in the compost with wastepaper, compared to its level in the

sewage sludge, resulted in a slight reduction of relationship K/(Ca+Mg). This result makes for an indicative to additional application of potassium to soils upon using composted sludge as fertilizing material.

4 CONCLUSIONS

1. Adding to the composted sewage sludge, softwood sawdust of coniferous, wastepaper and EM BIO™ microbiological preparation resulted in a statistically significant increase in dry matter.
2. Addition of wastepaper to the sewage sludge resulted in a significant increase of the total calcium in the composted material and the observed increase was about 50%.
3. Combination of sewage sludge with sawdust resulted in the reduction of total magnesium and potassium contents in the composted material.
4. Comparing od the different dates contents of the composted sewage sludge, a regular increase in calcium, magnesium and potassium levels and their alkalization were observed, which can be attributed to the increased participation of mineral compounds in general amount of component in the compost under the study.
5. The higher level of calcium in the compost mixes with wastepaper, comparing to its levels in the sewage sludge itself, reduced the relationship K/(Ca+Mg).
6. Based on the changes, reaction and the calcium, magnesium, potassium and dry mass levels, processes of mineralization of organic compounds were more intense in the composted sewage sludge mixed with wastepaper and a higher admixture of microbiologic preparation EM BIO™.

REFERENCES

Baranowski, A. 2004. Preparat "Efektywne mikroorganizmy (EM)"—próby zastosowania w rolnictwie, [The product "Effective microorganisms (EM)"—try to apply in agriculture], Przegląd Hodowlany 72(4): 26–27.

Bauman-Kaszubska, H. & Buraczewski, G. 1999. Kompostowanie i użytkowanie kompostu, [Composting and use of compost] Mat. I Konf. Nauk.-Tech., Puławy-Warszawa 155.

Bień, J.B. 2007. Osady ściekowe teoria i praktyka, [Sludge theory and practice], 1–309, Częstochowa, Wyd. PCz.

Chowaniak, M. & Gondek, K. 2009. Changes in the available magnesium and zinc contents of soils after the application of sewage sludges and sewage sludge—peat mixtures, J. Cent. Eur. Agric. 10(1): 79–88.

Ciećko, Z. & Harnisz, M. & Najmowicz, T. 2001. Dynamika zawartości węgla i azotu w osadach ściekowych podczas ich kompostowania, [Dynamics of carbon and nitrogen in sewage sludge during their composting], Zesz. Probl. Post. Nauk Rol., 475: 253–262.

Ciećko, Z. & Harnisz, M. & Wyszkowski, M. & Najmowicz, T. 2003. Zmiany zawartości magnezu w osadach ściekowych podczas ich kompostowania z dodatkiem różnych substancji, [Changes in magnesium content in sewage sludge during their composting with the addition of various substances], Inżynieria Ekologiczna, 25: 95–101.

Ciećko, Z. & Harnisz, M. & Najmowicz, T. & Wyszkowski, M. 2005. Dynamics of potassium content in sewage sludge during its composting with different additives, Polish Journal of Soil Science. XXXVIII(1): 31–40.

Czekała, J. 1999. Osady ściekowe źródłem materii organicznej i składników pokarmowych, [Sludge source of organic matter and nutrients], Fol. Univ. Agric. Stetin. 200, Agricultura, 77: 33–38.

Czekała, J. 2002. Wybrane właściwości osadów ściekowych z oczyszczalni regionu Wielkopolski. Cz. I. Odczyn, sucha masa, materia i węgiel organiczny oraz makroskładniki, Acta Agrophys., 70: 75–82.

Czekała, J. 2009. Ocena składu chemicznego i wartości nawozowej kompostów wyprodukowanych z osadów ściekowych z dodatkiem odpadów organicznych, [Evaluation of the chemical composition and fertilisation value of composts produced from sewage sludges supplemented with organic wastes], J. Research Applic. Agric. Enginn. 54(3): 43–48.

Drozd, J. & Licznar, M. & Patorczyk-Pytlik, B. & Rabikowska, B. & Jamroz, E. 1996. Zmiany zawartości węgla i azotu w procesie dojrzewania kompostów z odpadów miejskich, [Changes in the content of carbon and nitrogen in the compost maturation of urban waste], Zesz. Probl. Post. Nauk Rol., 437: 123–130.

Gorlach, E. & Gambuś, F. 1998. Evaluation of sewage sludge as fertilizer in experiment, Acta Agr. Silv., Ser. Agr., 36: 9–21.

Ignatowicz, K. & Garlicka, K. & Breńko, T. 2011. Wpływ kompostowania osadów ściekowych na zawartość wybranych metali i ich frakcji, [The inf luence of sewage sludge composting for content of chosen metals and their fractions], Inżynieria Ekologiczna, 25: 231–241.

Jakubus, M. 2006. Wpływ wieloletniego stosowania osadu ściekowego na zmiany wybranych własności chemicznych gleby, [Effect of long-term applications of sewage sludge on changes in selected soil chemical properties], Zesz. Prob. Post. Nauk Rol., 512: 209–219.

Kaniuczak, J. & Hajduk, E. & Zamorska, J. & Ilek, M. 2009. Charakterystyka osadów ściekowych pod względem przydatności do przyrodniczego wykorzystania,. [Characteristics of sewage sludge in terms of suitability for use in nature], Zesz. Nauk., 11: 89–94.

Krutul D. 1998. Zawartość substancji ekstrakcyjnych i mineralnych na przekroju poprzecznym oraz podłużnym pni sosnowych (Pinus sylvestris L.), [Content of the extractiwe and mineral substsnces in pine wood (Pinus sylvestris L.)]. Folia Forestalia Polonica, 29: 5–17.

Krzywy E. 2007. Żywienie roślin, [Plant nutrition], Szczecin Wyd. Nauk. AR.

Licznar, M. & Drozd, L. & Jamroz, E. & Licznar, S.E. & Weber, L. 1999. Przemiany makro—i mikroskładników w procesie dojrzewania kompostów produkowanych z odpadów miejskich w warunkach kontrolowanych, [Transformation of macro-and micronutrients in the process of maturation of compost produced from municipal waste under controlled conditions], Zesz. Nauk. AR Szczecin, 77: 201–205.

Mau, P. 2003. Fantastische Erfahrungen mit EM. Ed. Emiko.

Marcinkowski, T. 2003. Wpływ stabilizacji osadów wtórnych wodorotlenkiem wapnia na ich skład biologiczny, [Effects of secondary sludge stabilization with calcium hydroxide to the composition of a biological], Ochrona Środowiska, 2: 49–55.

Nowak, G. & Wierzbowska, J. & Klasa, A. 1996. Zawartość magnezu, wapnia, strontu i baru w osadach z oczyszczalni ścieków na terenie Pojezierza Mazurskiego oraz w recyrkulowanej nim gleby, [The content of magnesium, calcium, strontium and barium in sediments, sewage treatment plant in the Mazury Lake District and the recycled soil], Biul. Magnezol., 2: 12–16.

Nowak, D. & Wójcik-Szwedzińska, M. & Bień, J.B. 1998. Charakterystyka osadów w aspekcie mikrobiologicznym. Osady ściekowe w praktyce, Characteristics in terms of microbiological deposits. Sludge in practice], VII Konferencja Naukowo-Techniczna, Częstochowa: 16–18.

Oleszczuk, P. 2008. Phytotoxicity of municipal sewage sludge composts related to physico-chemical properties, PAHs and heavy metals, Ecotoxicology and Environmental Safety, 69: 496–505.

Onyszk, B. 2003. Zastosowanie EM w produkcji roślinnej i zwierzęcej, [The use of EM in crop and livestock production], Wieś Mazowiecka, 11: 18, 31.

Patorczyk-Pytlik, B. & Spiak, Z. & Gediga, K. 1999. Ocena możliwości rolniczego wykorzystania osadów ściekowych z zakładów przetwórstwa drobiowego. Cz. I. Wpływ procesu kompostowania na zmiany składu chemicznego osadów przemysłu drobiowego, [Evaluation of the possibility of the land application of sewage sludge from poultry processing plants. Part. I. The impact of the composting process changes the chemical composition of sediments poultry industry], Fol. Univ. Agric. Stetin., Agricultura, 77: 11–316.

PN-ISO 9964-2/Ak:1997 Oznaczanie sodu i potasu— Oznaczanie potasu w ściekach metoda absorpcyjnej spektrometrii atomowej, [Determination of sodium and potassium—atomic absorption spectrometry method].

PN-EN ISO 7980:2002—Oznaczanie wapnia i magnezu— Metoda atomowej spektrometrii absorpcyjnej, [Determination of calcium and magnesium—atomic absorption spectrometry method].

PN-EN 12880:2004—Charakterystyka osadów ściekowych. Oznaczanie suchej pozostałości i zawartości wody, [Characteristics of sewage sludge. Determination of dry residue and water content].

PN-EN 12176:2004—Charakterystyka osadów ściekowych—Oznaczanie wartości pH, [Characteristics of sewage sludge. Determination of the pH].

Rocznik Statystyczny 2011 Warszawa, GUS [Statistical Yearbook 2011 Warsaw GUS].

Sweeten, J.M. & Auvermann, B.W. 2008. Composting Manure and Sludge http://AgriLifebooksstore.org. E-479.

Siuta, J. & Dusik, L. & Lis, W. 2007. Kompostowanie osadu ściekowego w Sierpcu, [Composting of sewage sludge in Sierpc], Inżynieria Ekologiczna, 19: 97–105.

Stuczyński, T. 1992. Wpływ stosowania różnego rodzaju inoculum na przebieg procesu kompostowania i jakość uzyskanego produktu, [The effect of different types of inoculum on the composting process and the quality of the resulting product], Pamiętnik Puławski, 100: 217–225.

Wysokiński, A. & Kalembasa S. 2011. Wpływ dodatków mineralnych i organicznych do osadów ściekowych oraz kompostowania uzyskanych mieszanin na ich wybrane właściwości, [The influence of mineral and organic additives to the sewage sludge and composting process of those mixtures on their selected properties], Inżynieria Ekologiczna, 27: 240–249.

Zaha, C. & Manciulea, I. & Sauciuc, A. 2011. Reducing the volume of waste by composting vegetable waste, sewage sludge and sawdust, Environmental Engineering and Management Journal, 10(9): 1415–1423.

Internet sources

www.emgreen.pl (4.03.2011r.)

www.kwazar.pl (9.01.2012r.)

Environmental Engineering IV – Pawłowski, Dudzińska & Pawłowski (eds)
© *2013 Taylor & Francis Group, London, ISBN 978-0-415-64338-2*

The influence of the compost mixture composition on the reduction of PAHs concentration in the process of composting

I. Siebielska

Faculty of Civil Engineering, Environmental and Geodetic Sciences, Technical University of Koszalin, Koszalin, Poland

ABSTRACT: For many years a continuous growth in the amount of the sewage sludge, as well as the municipal waste and its organic fraction, has been observed. Their direct use for fertilization is limited by the presence of many pollutants, including the Polycyclic Aromatic Hydrocarbons (PAHs). Therefore, it is vital to utilize the biodegradable waste to obtain a product that is environmentally safe. The composting process is one of the most commonly used methods for that purpose. Due to the presence of the organic pollutants, including PAHs, in the substrates for that process, the monitoring of concentration changes of these compounds in the composting process is very important, as well as their concentration in the final product. The aim of this study was to examine the influence of varied percentage composition of the compost mixture on the reduction of the concentration of 16 selected PAHs.

Keywords: polycyclic aromatic hydrocarbons, composting process, gas chromatography, sewage sludge, organic fraction of municipal waste

1 INTRODUCTION

Composting is one of the most popular methods for the utilization of the biodegradable waste including mainly the sewage sludge and also the organic fraction of the municipal waste. That process can be carried out both in the nonreactor and reactor systems. The bioreactors can be static or dynamic, most often with forced oxygenation (Jędrczak 2007). The aim of composting is to obtain a fertilizer from the biodegradable waste, with properties similar to humus containing up to 50% of the organic substance, nutrients for plants and microorganisms which improve the microflora and the microfauna in the soil (Epstein 1997).

The decomposition and transformation of the organic matter takes place in the process of mineralization, hygienization, and finally maturing of the compost, which is displayed in the decreasing C:N ratio or in COD_{Cr} (Epstein 1997). At the same time, in these conditions, the biodegradation of the organic pollutants such as the polycyclic aromatic hydrocarbons occurs. These compounds are highly toxic both for the people and for the environment (Boström et al. 2002). For their lyophobic structure, the polycyclic aromatic hydrocarbons show a high tendency for sorption on solids. They are very sensitive to changes in the external factors like the temperature and pH, and to the presence of many chemical compounds, surface-active substances and pesticides among others. PAHs oxidize easily, both under the influence of UV and the oxidants like oxygen or ozone, in such a way they form the quinones (Dutkiewicz et al. 1988). In the case of compounds with a higher number of condensed rings, the peroxides can be the final product (Namieśnik 1995). The biodegradation process of the polycyclic aromatic hydrocarbons plays probably the most important role in the reduction of these compounds concentration in the natural environment. When it comes to the hydrocarbons with three condensed aromatic rings, the cis-dihydrodiol is the intermediate product and the catechol, which oxidizes further to the carboxylic acids and the aldehydes, is the final one. These compounds are used by the microorganisms for the energy production, for example, by oxidizing to the carbon dioxide. Some bacteria, responsible for the decomposition of the hydrocarbons with a fewer number of aromatic rings, also participate in the biodegradation process of the hydrocarbons with four and more condensed aromatic rings. In the suggested way of the biological decomposition, the ring scission takes place in the first stage, first of all in the position 9, 10. Then the diols are the product of the oxidation, which are afterwards oxidized to the carboxylic acids (Antizar-Ladislao et al. 2006).

Due to the industrial development, the concentration of PAHs has been increasing gradually, both in the sewage sludge and in the municipal waste. The highest concentration of PAHs in Europe was found in Germany, where the sum of the concentrations ranged from 2 to 80 mg/kg DM. In Switzerland it ranged from 1 to 10 mg/kg DM, in France (the waste treatment plant for Paris) from 15 to 30 mg/kg DM and in Spain from 2 to 6 mg/kg DM (Blanchard et al. 2004, Perez S. et al. 2001). Similarly to other European countries, according to the studies taken in various research centres, the sum of the PAHs concentrations in Poland ranged from 1 to 6 mg/kg DM (Janosz-Rajczyk et al. 2006). The direct use of sewage sludge, polluted in this way in agriculture, could lead to additional contamination of the natural environment.

The studies taken in many research centres show that, in the process of composting, a significant reduction in the sum of the PAHs concentrations is observed (Amir S. at al. 2005, Antizar-Ladislao et al. 2004, Lazzari et al. 2005, Oleszczuk 2005, Siebielska 2009). The presence of humus-like substances probably accelerates the desorption of PAHs, which results in the intensification of these compounds biodegradation. The humus substance has a greater impact on the hydrocarbons degradation process than the temperature (Sayara et al. 2010). The composting process enables the use of the resulting product for cropland's manuring, due to the fact that the sum of the concentrations of 11 PAHs is usually lower than the number allowed by the European Union—6 mg/kg DM for the sewage sludge used in the agriculture (CEC, Working Document on Sludge, 3rd Draft, 2000).

The aim of this study was to analyse the changes in the concentration of the selected polycyclic aromatic hydrocarbons in the composting process of the organic fraction mixture of the municipal waste and the sewage sludge from the municipal waste treatment plant, which take place in a static bioreactor. The succeeding cycles differed only in the percentage composition of the mixture.

2 METHODS

2.1 *Experimental details*

6 composting cycles were conducted in a static bioreactor with forced aeration. The bioreactor's volume was 60 l. In the first stage of the temperature rise (30 days), there was a 2,5 dm³/min. of air supply. The maximum temperature was 60°C. During the stage of compost maturing (154 days), the air supply was reduced to 0,5 dm³/min because the intensity of treatment was decreased. In all the cycles the process lasted 182 days. The ratio of the composted mass surface to its volume was 15,7:1 m²/m³, which falls in the range characteristic for the bioreactors working on a laboratory scale. This indicator is much lower in the case of the bioreactors working on a technological scale. Due to those measurements, the difference in the

course of composting on the laboratory scale and the technological scale lies in the length of the thermophilous phase. That difference can be slightly reduced by the use of a bioreactor, in which the rise in the temperature during the thermophilous phase results from the biochemical processes. However, it requires good insulation from external conditions which, in turn, limits chilling and delaying the composting process. A system prepared in this way enables controlling and even limiting the influence of variable factors on the composting process. Whereas the conduction of several cycles of the process in repeatable conditions on the technological scale causes numerous problems and is very often impossible (Mason et al. 2005).

The substance allocated for composting was a mixture of the sewage sludge and the organic fraction of the municipal waste. In the first two cycles the batch consisted of 30% of the organic fraction of the municipal waste and 70% of the sewage sludge. In the following two cycles, 50% of both the organic fraction of the municipal waste and the sewage sludge were processed. The remaining two cycles consisted of 70% of the organic fraction of waste and 30% of the sewage sludge. The volume of the batch was 50 l. The sewage sludge was collected from the municipal waste treatment plant operating for a city of 100.000 inhabitants. The sludge humidity, after dehydration, levelled at 70–75%. The organic fraction of the municipal waste came from the municipal landfill also operating for a city of 100.000 inhabitants. Non-separated waste was first run through a sieve with 4 cm mesh openings, and then through a sieve with mesh of 2 cm.. The fraction collected was the organic fraction of the municipal waste. Its humidity was about 50%. Unfortunately, apart from the biodegradable waste, glass and plastic could also be found there, which influenced the structure-building substance but with a negative impact on the quality of the compost produced.

The samples of the composted mass were collected in accordance with the standard PN-Z-15011-1 called: "Compost from the municipal waste. The intake of samples". The frequency of the samples collection depended on the intensity of the process. For the first 4 weeks the samples were collected after every 7 days, for the following 6 weeks after every 14 days, for the next 12 weeks after every 21 days, and the last sample was collected after 28 days.

2.2 *The methodology of physical and chemical determinations*

2.2.1 *Total nitrogen*
The Total Nitrogen in the fresh sample was determined with the use of the elemental analysis

technique. In the process of the catalytic oxygen combustion in the helium atmosphere and in the temperature of 900°C the nitrogen compounds decompose. In such conditions not only the nitrogen but also nitric oxides are produced, which are then reduced to the nitrogen in the reduction range in the temperature of 830°C. The macro analyzer VARIOMAX CN, which was used for the analysis, was equipped with a thermal conductivity detector. The lowest content of nitrogen determined in a sample was 0,02 mg (AN-A-220205-E-01).

2.2.2 *Total organic carbon*
A dried sample was the subject for the elementary analysis. In helium atmosphere carbon compounds decompose in the process of the oxygen combustion in the temperature of 900°C. In such conditions, not only the carbon dioxide but also carbon monoxide and methane are produced, which are then oxidized to the carbon dioxide in the oxidizing range in the temperature of 900°C. The macro analyzer VARIOMAX CN, which was used for the analysis, was equipped with a thermal conductivity detector. The lowest content of carbon determined in a sample was 0,02 mg (AN-A-220205-E-01).

2.2.3 *Humidity*
The content of water in a sample was analysed with the use of gravimetric method. The composted mass was dried in the temperature of 105°C. (Hermanowicz W. et al. 1999).

2.2.4 *Chemical oxygen demand (COD)— with the use of chromatometry*
The content of the organic compounds oxidizing in the acidic environment (dipping acid VI) was estimated in a dry sample, in the presence of a catalytic agent and in elevated temperature. The potassium dichromate was a strong oxidant which was reduced to the chromium cation (III). The silver (I) sulfate (VI) was the catalyst. In order to standardize the sample, the granules diameter was not bigger than 0,1 mm (Hermanowicz W. et al. 1999).

2.3 *PAHs determination methodology*
A sample of approx. 10 g of dried compost was extracted with acetone for one hour using ultrasound. 40 cm^3 of distilled water and then 5 cm^3 of isopropanol was added to the extract. The obtained solution was passed through the pre-conditioned sorption columns in accordance with the Solid Phase Extraction (SPE) technique. The sorption columns were filled with the octyl phase. The absorbed aromatic hydrocarbons were eluted with the Tetrahydrofuran (THF). The elution obtained was condensed in the nitrogen atmosphere up to the volume of 0,5 cm^3 (Bakerbond Application

Note 385). The final stage was to determine the quantity and quality of a gas chromatograph (AT 7890) equipped with a jet separation injector and a mass detector (AT 5975C VL MSD). The analysis was performed on the HP-5MS chromatographic column with the use of helium as the carrier gas. The initial 60°C temperature of the GC-MS oven was maintained for 1,5 min. Later, it increased to 160°C at the rate of 30°C per minute. Then, with the rate of 5°C per minute, it rose to 195°C. Finally, it reached 280°C at the rate of 3°C per minute. The final temperature was kept for 18 minutes (Amir et al. 2005). The MS transfer line was 280°C. The MS source was 230°C. The injector temperature was kept at 300°C. The data obtained was analyzed with the use of Selected Ion Monitoring (SIM) mode. The quality and quantity indication were made on the basis of the external standard of the 16 PAHs (acenaphthene, acenaphthylene, anthracene, benz(a)anthracene, benzo(b)fluoranthene, benzo(k)fluoranthene, benzo(ghi)perylene, benzo(a)pyrene, chrysene, dibenz(a,h)anthracene, fluoranthene, fluorene, indeno(1,2,3-cd)pyrene, naphthalene, phenanthrene, pyrene). The method's sensitivity was 0,001 mg/kg DM.

3 RESULTS AND DISCUSSION

3.1 *Physical and chemical parameters*
The optimum humidity for composting should range between 40 and 60% (Epstein E. 1997). In the six cycles carried out, the highest humidity (70%) was on the first day for the batch consisting of 30% of the organic fraction in cycle II. In other cases it was approx. 60%. During the process, the composted mass humidity stayed at a level of 50%. The lowest humidity, at approx. 40%., was observed during the composting of a batch with 30% of the organic fraction in cycle I and of a batch with 50% of the organic fraction in cycle II, which took place on the last, 182nd, day. In all the cycles the humidity during the process did not exceed 70% and did not fall below 40%. In the six cycles carried out, pH of the composted mass ranged between 6,64 and 7,04.

The Chemical Oxygen Demand (COD_{Cr}) is the indicator of the compost stability level. A stable compost will have COD_{Cr} reduced to 80% of the original value (Epstein E. 1997). Figure 1 shows the results of the determinations for all six cycles.

The highest value of this parameter was observed for the mixture of 30% of the organic fraction and 70% of the sewage sludge. 4800 mg O_2/g DM in cycle I and 5133 mg O_2/g DM in cycle II respectively. The lowest value of the COD_{Cr} was characteristic for the batch of 70% of the organic

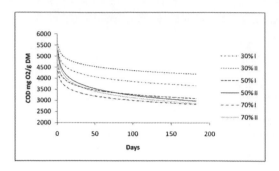

Figure 1.   Changes of the COD index.

fraction and 30% of the sewage sludge. Respectively, 4062 mg O_2/g DM in cycle I and 4518 mg O_2/g DM in cycle II. The COD_{Cr} for the mixture of 50% of each component was 4600 mg O_2/g DM in both cycles. Those values indicate the sewage sludge as the basic source of the easily degradable organic carbon. For the mixture of 30% of the organic fraction, the batch from the 90th day of the process in cycle I and from the 112th day in cycle II, can be considered as a stable compost. In the case of the mixture of 50% of both components, the compost may be considered stable after 70 days of the process in both cycles. The 56th day is the moment from which the compost is stable in cycles I and II of composting for the batch consisting of 70% of the organic fraction and 30% of the sewage sludge. The content of the sewage sludge in the mixture, and thus the amount of the organic carbon, determines the length of the mineralization phase in the process. The lower the organic carbon content in the composted batch is, the sooner the compost can be considered stable. It is likely that the organic compounds in the sewage sludge undergo mineralization much easier than the organic substance in the organic fraction of the municipal waste.

Another parameter describing the compost stability is the C:N ratio. The compost is considered stable when this parameter is 10:1, which is a characteristic feature for humus (Kulikowska et al. 2009). For all the cycles, on the first day of composting the C:N ratio was lower than 15:1. Such a low value of that parameter probably resulted from a high total nitrogen content, which is typical for the sewage sludge. Composting with such a low initial value of this parameter causes high emissions of an odor-generating ammonia in the first days of the process (Epstein 1997, Jędrczak 2007).

In cycle I of composting of a mixture with 30% of the organic fraction of waste and 70% of the sewage sludge, the C:N ratio went down from 13:1 on the first day to 10:1 on the last day. In cycle II of composting of the same kind of mixture, the

value went down from 10:1 to 7:1. In cycle I, the analysed parameter went down from 11:1 to 8:1 and in cycle II from 10:1 to 8:1 for a batch containing 50% of both components. During composting of a mixture with 70% of the organic fraction and 30% of the sewage sludge, the initial ratio was 11:1 and the final one 9:1 in cycle I. In cycle II, at the beginning, the C:N ratio was 12:1 and at the end 9:1. In the contrary to the COD_{Cr} parameter, the C:N parameter includes not only the easily oxidizing organic compounds but also all the organic substances, together with the hardly decomposing ones. In all the cycles of the process only a slight decrease in the C:N ratio was observed. The concentration of the carbon compounds decreased more than the concentration of the nitrogen compounds. It resulted most probably from a high nitrogen content in hardly decomposing organic compounds present in the organic fraction of the municipal waste. Whereas the higher reduction of carbon might have resulted from the degradation of easily decomposing organic compounds which are characteristic for the sewage sludge. However, this relativeness is not as evident as for the COD_{Cr} parameter.

3.2   The aromatic hydrocarbons

The 16 analysed PAHs were present in the composted mixture on the first day of the process in all the cycles.

At the beginning of the process, the concentration of naphthalene ranged between 0,093 mg/kg DM and 0,256 mg/kd DM. The concentration of the hydrocarbon did not depend on the percentage composition of the composted mixture. The highest concentration of naphthalene was observed in cycle I for the batch with 50% of both components. Whereas the lowest concentration was observed for the same kind of mixture, but in cycle II. The results are shown in Figure 2. The naphthalene's concentration during composting was reduced by

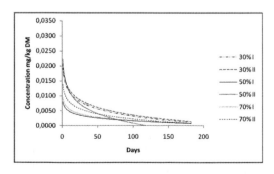

Figure 2.   Naphthalene concentration changes in the compost.

approx. 70% in all the cycles. For the mixture of 70% of the organic fraction of waste that value was 68,5%, and for the mixture of 50% of the organic fraction—75%.

The correlation between the composition of the batch and the degree of the reduction of naphthalene concentration was not observed. The identical results were obtained by Cai et al. During composting of the sewage sludge, the concentration of naphthalene was reduced by 71% (Cai et al. 2007). Under the microaerobic conditions, in the presence of fungi, the degree of the reduction of naphthalene concentration was 46,5% (Silva et al. 2009).

The highest reduction was observed during the first 50 days of the process, i.e. during the stage of mineralization. The biodegradation is, most probably, the main phenomenon responsible for the naphthalene degradation. The naphthalene, with only two condensed aromatic rings, is a PAH which is most easily decomposed by bacteria due to its lowest hydrophobicity. In addition, the weakest sorption for that group of hydrocarbons favours that process as well (Hafidi et al. 2008).

Acenaphthylene, acenaphthene, fluorene, phenanthrene and anthracene represent the PAHs with three condensed rings. In the samples examined, the last two hydrocarbons showed the highest concentration on the first day of the process. In the mixture of 30% of the organic fraction, in cycle I, the concentration of the phenanthrene was 1,272 mg/kg DM and the anthracene 2,057 mg/kg DM. In the samples examined, the lowest concentration of the acenaphthene was observed (0,010 mg/kg DM) in cycle I for the mixture with 70% of the organic fraction of waste. In the remaining cycles, among all the other PAHs with three condensed rings, the concentration of the acenaphthene was also the lowest. Analogically to the case of the naphthalene, the degradation of the above mentioned PAH took place during the mineralization stage in all the cycles of the process. The reduction in the concentration of the five compounds, depending on the cycle, ranged between 60% and 75%. In the research carried out by Wong et al., the reduction in the concentration of the PAHs with three rings during manure composting was 95% in the first three weeks (Wong et al. 2002). Hafidi et al. observed the highest reduction in the concentration of the PAHs with three condensed rings during the mineralization stage (Hafidi et al. 2008). In comparison to other PAHs, the presence of only three condensed rings effected in a relatively low hydrofobicity and an inconsiderable tendency to sorption. (Pignatello et al. 1996). Such a structure enables higher biodegradation intensity, hence probably such a high reduction in the concentration of these substances during the composting process. Figure 3 presents diagrams describing the

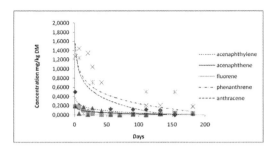

Figure 3. Changes of PAH with 3 rings concentration in the compost.

reduction of acenaphthylene, acenphthene, fluorene, phenenthrene and anthracene.

Fluoranthene, pyrene, chrysene and benz(a) anthracene are the PAHs with four condensed aromatic rings. The presence of these compounds was observed in all the cycles of the composting process. The highest concentration of the fluoranthene (3,183 mg/kg DM) was observed in cycle I of composting of a mixture with 30% of the organic fraction of waste. In the remaining cycles, it was also the fluoranthene that showed the highest concentration of numbers. Whereas the lowest concentration was noticed in cycle II for the batch with 30% of the organic fraction of waste in the case of the benz(a)anthracene and it was 0,068 mg/kg DM. In general, the lowest concentration of the chrysene and the benzo(a)anthracene was observed in all the cycles. The reduction in the concentration of the hydrocarbons with four aromatic rings did not exceed 70% after 182 days in all the cycles,. It ranged between 47% and 69%. The composted mixture composition did not affect the degree of the hydrocarbons' degradation. The compounds structure was of greater importance. The presence of four aromatic rings increases the hydrophobicity and the sorption ability (Caristrom et al. 2003). These features reduce the biodegradation which plays the most important role in the hydrocarbons degradation. The smaller reduction of the concentration, when compared to the PAHs with maximum three condensed rings, confirms the fact. The research led by Lashermes et al. presented that the composted mass ability for sorption decreased in time, which caused the intensification of the biodegradation process in the maturing stage. Therefore, the fluoranthene's reduction depends more on the presence of specific microbial biomass than on the bioavailability (Lashermes et al. 2010).

The last group of the hydrocarbons consists of compounds with at least five condensed rings. That includes: benzo(b)fluoranthene, benzo(k)fluoranthene, benzo(a)pyrene and dibenzo(a,h)anthracene with five rings and also benzo(ghi)perylene and indeno(1,2,3-cd)pyrene with six condensed rings. When it comes to a mixture with 70% of the organic fraction of waste, the lowest concentration of the PAHs with five aromatic rings was observed in both cycles. For other batches, the concentration ranged from 0,080 mg/kg DM to 0,669 mg/kg DM on the first day of the process. The quantity of the sewage sludge did not affect the content of analysed substances. Similarly, the lowest concentration of the hydrocarbons with 6 aromatic rings was observed for composting of a mixture with 70% of the organic fraction of waste. It was 0,022 and 0,024 mg/kg DM in cycle I, 0,017 and 0,015 mg/kg DM in cycle II. The higher the percentage of the sewage sludge in the composted mixture, the higher the concentration of the benzo(ghi)perylene and the indeno(1,2,3-cd) pyrene on the first day of the process.

The structure of the PAHs with at least five rings significantly reduces the bioavailability (Pignatello et al. 1996). These compounds are the most hydrophobic among the hydrocarbons analysed, and thus they are the least water-soluble and show the highest tendency for sorption (Pignatello et al. 1996). The higher reduction of the high weight molecular PAHs is an effect of a strong bond with the compost rather than of the biodegradation (Oleszczuk 2007). After approx. 100 days of composting it results in the increase in the concentration. A similar phenomenon was observed by Lazzari et al. (Lazzari et al. 2000). The lowest reduction of individual substances additionally confirms the fact. For the PAHs with five rings the numbers ranged from 37% to 63%, whereas the values over 60% were observed only for the benzo(a)pyrene during cycle II for a mixture with 30% of the organic fraction, and for the dibenzo(a,h)anthracene for the same kind of a mixture, but in cycle I. In the case of the benzo(ghi)perylene and the indeno(1,2,3-cd) pyrene there was even smaller reduction observed. The reduction in the concentration of these compounds did not exceed 56%. In most tests it was approx. 40%.

3.3 The comparison of changes in the reduction of the PAHs in all the cycles of composting

The changes in the concentration of the PAHs during composting of three different kinds of mixtures were compared with the use of the principal component analysis. The first stage of the analysis determined the eigenvalues. The highest eigenvalue for the first factor was 116,056 and it accounted for 49,6% of the total variance. The second factor's eigenvalue was 56,885 and it accounted for 23,4% of the total variance. The third factor's eigenvalue was 28,885 and it stood for 12,3% of the total variance. On the whole, the three factors described stood for 86,25% of the total variance. Therefore, a three-dimensional space was assumed with the loss of less than 14% of data.

The first factor was the most correlated with the concentration of the majority of the PAHs. The concentration of the analyzed compounds from the 56th till the 91st day had the reverse influence on that factor, which was indicated by the positively charged agents. The PAHs with six condensed rings were exceptions because during the whole process of composting they had the same impact on the first factor. Therefore, the transformations of the PAHs during the process of composting for a mixture containing the sewage sludge and the organic fraction of the municipal waste can be assumed to be the first factor. The second factor was the most correlated with the values of the C:N ratio. During the whole process of composting, the impact of the ratio on the second factor remained at the same level. The identical signs for the charge agents confirmed the fact. It is most probable that this factor describes the degree of changes in all the compounds in the stage of compost stabilization and maturing.

The third and the last factor illustrated the highest correlation with the COD_{Cr} parameter. Similarly like for the C:N ratio, the COD_{Cr} parameter had the same impact on the third factor during the process, which was indicated by the positively charged agents. The changes of the easily degradable organic substance can be assumed to be the third factor.

The graph in Figure 4 presents three evident groups of different composting cycles. The first group includes the cycles with 70% of the organic fraction of waste, cycle I for a mixture containing 50% of both components and cycle II for a mixture of 30% of the organic fraction of waste. The second group is cycle II for a mixture with 50% of both components and the third group is cycle I for a mixture with 30% of the organic fraction. The first and the third group are characteristic for similar changes in the PAHs concentrations and analogical decrease in the C:N ratio, in other words for the degradation of the organic substances including the nitrogen compounds. In cycle II for a mixture with 50% of the organic fraction of waste, the reduction of the PAHs was similar to the cycles described above. The decrease in the C:N ratio was different, however, it differed in the degradation of the organic substances, including the nitrogen compounds. The ultimate cycle I for a mixture with 30% of the organic fraction of the municipal waste was characteristic for a different reduction of the PAHs than in the other cycles. However, the decrease in the C:N ratio was analogical to the results of the first group. The graph in Figure 5 additionally shows the third factor, correlated with the changes in the easily degradable organic compounds. The degradation of these substances was different only

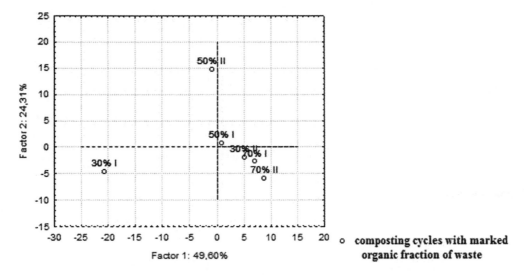

Figure 4. The graph of the coordinates for factors 1 and 2 in all the composting cycles.

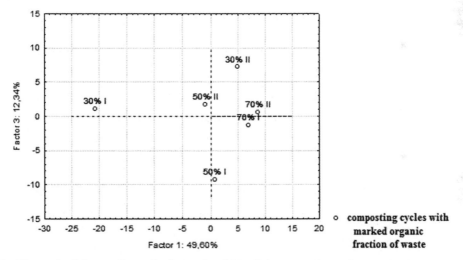

Figure 5. The graph of the coordinates for factors 1 and 3 in all the composting cycles.

in cycle I for a mixture with 50% of the organic fraction of waste. In the remaining cycles, analogical changes in the value of the COD_{Cr} were observed.

4 CONCLUSIONS

- Differences in the percentage composition did not influence the reduction of the PAHs.
- Two cycles of composting of a mixture with 50% of both components differ from the others in changes of the parameters connected to the degradation of the organic substance.

- The highest reduction in the hydrocarbons was observed for the compounds with the maximum of four condensed aromatic rings.
- The reduction of the PAHs with at least five condensed aromatic rings was much smaller and the increase in their concentration at the end of the process could be a result of condensing.

ACKNOWLEDGMENTS

This study was supported by The Committee for Scientific Research within the framework of the project No. 4483/B/T02/2009/36.

REFERENCES

Amir S., Hafidi M., Merlina G., Hamdi H., Revel J.C. 2005, Fate of polycyclic aromatic hydrocarbons during composting of lagooning sewage sludge, *Chemosphere*, 58,: 449–458. AN-A-220205-E-01.

Antizar-Ladislao B., Lopez-Real J.M., Beck A.J. 2004, Bioremediation of polycyclic aromatic hydrocarbons (PAH)—contaminated waste using composting approaches, *Crit. Rev. Envirom. Sci. Technol.* 34,: 249–289.

Antizar-Ladislao B., Lopez-Real J., Beck A.J., 2006, Degradation of polycyclic aromatic hydrocarbons (PAHs) in an aged coal tar contaminated soil under in-vessel composting conditions, *Environmental pollution* 141,: 459–468.

Bakerbond Application Note 385.

Blanchard M., Teil M.J., Ollivon D., Legenti L., Chevreuil M. 2004, Polycyclic aromatic hydrocarbons and polychlorobiphenyls in wastewaters and sewage sludges from the Paris area (France), *Environmental Research*, 95,: 184–197.

Boström C.E., Gerde P., Hanberg A., Jernström B., Johansson Ch., Kyrklund T., Rannug A., Törnqvist M., Victorin K., Westrholm R., 2002, Cancer risk assessment, indicators and guidelines for polycyclic aromatic hydrocarbons in the ambient air, *Environ. Health Perspect.* 110,: 451–488.

Cai Q.Y., mo C.H., Wu Q.T., Zeng Q.Y., Katsoyiannis A., Férard J.F., 2007, Bioremediation of polycyclic aromatic hydrocarbons (PAHs)-contaminated sewage sludge by different composting processes, *Journal of Hazardous Materials* 142,: 535–542.

Caristrom C.J., Tuovinen O.H. 2003, Mineralization of phenanthrene and fluoranthene in yard waste compost, *Environ. Pollut.* 124,: 81–91.

CEC, Council of the European Community, Working Document on Sludge, 3rd Draft, 2000 Brussels, 27 April, 20p.

Dutkiewicz T., Lebek G., Masłowski J., Mielżyński D., Ryborz S., 1988, *Wielopierścieniowe węglowodory aromatyczne w środowisku przyrodniczym*, Warszawa, PWN.

Epstein E., 1997, *The science of composting*, Lancaster, Technomic Publishing Company Inc.

Hafidi M., Amir S., Jouraiphy A., Winterton P., Gharous M.El., Merlina G., Revel J.-C., 2008, Fate of polycyclic aromatic hydrocarbons during composting of activated sewage sludge with green waste, *Bioresource Technology*, 99,: 8819–8823.

Hermanowicz W., Dojlido J., Dożańska W., Koziorowski B., Zerbe J., 1999, *Fizyczno-chemiczne badanie wody i ścieków* Warszawa, Arkady.

Janosz-Rajczyk M., Dąbrowska L., Rosińska A., Płoszaj J., Zakrzewska E., 2006, *Zmiany ilościowo—jakościowe PCB, WWA i metali ciężkich w kondycjonowanych osadach ściekowych stabilizowanych biochemicznie*, Częstochowa, Monografia nr120 Politechniki Częstochowskiej.

Jędrczak A., *Biologiczne przetwarzanie odpadów*, 2007, Warszawa, PWN.

Kulikowska D., Bilicka K., 2009, Analiza przemian materii organicznej i związków azotu podczas kompostowania osadów ściekowych *Czasopismo Techniczne*, 11,: 101–110.

Lashermes G., Houot S., Barriuso E., 2010, Sorption and mineralization of organic pollutants during different stages, *Chemosphere* 79,: 455–462.

Lazzari L., Sperni L., Bertin P., Pavoni P., 2000, Correlation between inorganic (heavy metal) and organic (PCBs and PAHs) micropolutant concentrations during sewage sludge composting processes, *Chemosphere* 41,: 427–435.

Mason I.G., Milke M.W., 2005, Physical modelling of the composting environment: A review. Part 1: Reactor systems, *Waste Management* 25,: 481–500.

Namieśnik J., Jaśkowski J., 1995, *Zarys ekotoksykologii*, Gdańsk.

Oleszczuk P., 2007, Changes of polycyclic aromatic hydrocarbons during composting with chosen physicochemical properties and PAHs content, *Chemosphere*, 67,: 582–591.

Perez S., Guillamon M., Barcelo D., 2001, Quantitative analysis of polycyclic aromatic hydrocarbons in sewage sludge from wastewater treatment plants, *Journal of chromatography A*, 938,: 57–65.

Pignatello J.J., Xing B., 1996, Mechanisms of slow sorption of organic chemicals to natural particles, *Environ. Sci. Technol.*, 30,: 1–11.

Sayara T., Sarrà M., Sánchez A., 2010, Effects of compost stability and contaminant concentration on the bioremediation of PAHs-contaminated soil through composting, *Journal of Hazardous Materials*, 179,: 999–1006.

Siebielska I., 2009, Kinetics of polyaromatic hydrocarbons transformation process during composting, *Polish Journal of Environmental Studies Series of Monographs*, 6,: 102–106.

Silva I.S., Grossman M., Durrant L.R., 2009, Degradation of polycyclic aromatic hydrocarbons (2–7 rings) under microaerobic and very-low-oxygen conditions by soil fungi, *International Biodeterioration & Biodegradation*, 63,: 224–229.

Smith E.C. M.J., Lethbridge G., Burns R.G., 1997, Bioavailability and biodegradation of polycyclic aromatic hydrocarbons in soils, *FEMS microbial. Lett.*, 152,: 141–147.

Wong J.W.C., Wan C.K. Fang M., 2002, Pig manure as a co-composting material for biodegradation of PAH-contaminated soil, *Environ. Technol.*, 23,: 15–26.

Environmental Engineering IV – Pawłowski, Dudzińska & Pawłowski (eds)
© *2013 Taylor & Francis Group, London, ISBN 978-0-415-64338-2*

Mitigation of BTEXs and p-cymene emission from municipal solid waste landfills by biofiltration—preliminary results of field experiment

M. Zdeb & M. Pawłowska
Lublin University of Technology, Faculty of Environmental Engineering, Lublin, Poland

ABSTRACT: The preliminary results of field-scale experiment on biofiltration of landfill gas, aimed to remove selected volatile organic compounds (benzene, toluene, ethylbenzene, xylenes and p-cymene), are presented in the paper. Examinations were carried out at the closed municipal solid waste landfill in Kraśnik. The mixture of municipal solid waste compost with expanded clay pellets (1:1 volumetric ratio) was used as filter bed material. The planned duration time of entire experiment is one year but the presented results refer only to the first two months. The aim of the experiment is to evaluate the biofiltration efficiency of toxic and odorous substances removal from landfill gas in various conditions. In late spring season (May, June), in which the research was carried out, when the average concentrations of benzene, toulene, ethylobenzene, o-xylene, m-xylene and p-cymene were: 3.23; 1.79; 3.52; 2.11; 4.15 and 12.64 $\mu g\ m^{-3}$ respectively, any emission of examined compounds from biofilter surface was observed. The gas released from the biofilter did not generate odour nuisance. The rate of trace gas removal determined in batch tests decreased with gas molecular mass growth, reaching the highest values for benzene and toluene and the lowest for xylenes and p-cymene.

Keywords: landfill, BTEXs, p-cymene, biofilter, microbial oxidation, removal rate

1 INTRODUCTION

The compounds belonging to the group of aromatic hydrocarbons called BTEXs (benzene, toluene, ethylobenzene and xylenes) are the starting materials in many industrial processes and the major components of gasoline (Rene et al. 2012). Therefore, they are released from various industrial plants and processes. They are also emitted from solid waste landfills as a trace component of Landfill Gas (LFG). Trace compounds are released from hazardous materials deposited in the landfill due to volatilization process or biodegradation of materials deposited in the landfill (Scheutz & Kjeldsen 2003). Atmospheric emissions of BTEXs from industry, motor vehicle exhaust and landfills cause odour nuisance. What is more important, BTEXs constitute a health hazard. There is sufficient evidence that benzene is carcinogenic to man (IARC, Durmusoglua et al. 2010). It is classified in the 1st group of carcinogens (IARC). Toluene and xylenes are suspected to be carcinogenic. They are belonging to 3rd group. This category is used for agents, mixtures and exposure circumstances for which the evidence of carcinogenicity is inadequate in humans and inadequate or limited in experimental animals. BTEXs are classified as priority environmental pollutants by the Environmental Protection Agency in the USA (Yeom & Daugulis 2001).

Technologies tending to remove BTEXs from air are needed to maintain good air quality.

BTEXs elimination from biogas can be conducted in many ways, which depends on several parameters, i.e. the composition of the gas, its pressure and temperature, sort of pollutants and their concentration. BTEXs removal can be lead via physical, chemical and biological methods (Corbit 1990, Muneron de Mello et al. 2010). The disadvantages of physical and chemical methods are high operating costs, high chemicals prices and problems with the chemical wastes (products of reaction) disposal (Abumaizar et al. 1998, Hassan & Sorial 2009, Yeom & Daugulis 2001). Thus, the biological processes seem to be the most attractive methods for BTEXs removal from contaminated gases, as they appear to have a potential to overcome some of the drawbacks of the physicochemical methods. Microorganisms play a fundamental role in biological methods: they convert contaminants into harmless compounds like CO_2, H_2O and biomass (Rene et al. 2012). Biological methods are useful because of their low cost and absence of negative effects on atmosphere (Delhoménie et al. 2002, Abumaizar et al. 1998, Devinny et al. 1999, Delhoménie & Heitz 2005, Shareefdeen & Singh 2005). Moreover, microbiological processes can be conducted in ambient temperatures by the atmospheric pressure, which eliminates the necessity to heat up the system and decreases the energy costs to a minimum; biological methods also do not generate secondary waste streams (Delhoménie et al. 2002, Abumaizar et al. 1998, Mudliar et al. 2010). The most common biological methods of gases treatment are biofiltration, biotrickling filtration and bioscrubbing (Zilli et al. 1993, 1996). All these systems contain three phases: a solid, usually organic phase, a liquid phase and a gas phase, which are in contact with each other (McNevin & Barford 2000). Biofiltration is used to treat large amounts of gases with low levels of pollutants (Abumaizar et al. 1998, Crocker & Schnelle 1998, Devinny et al. 1999, Delhoménie & Heitz 2005).

Several studies have been conducted on BTEXs removal from model gas mixture on laboratory scale. Little research, however, has been carried out on BTEXs removal from mixture simulating landfill gas (containing methane) or from landfill gas in a field—scale. The aim of the present work was to evaluate the efficiency of biodegradation of BTEXs emitted from municipal solid waste landfill, in field-scale biofilter ($2 \times 2 \times 1.5$ m). The filter bed consists of a mixture of municipal waste compost and expanded clay pellets—keramsite (mix ratio 1:1 by volume).

2 METHODS

2.1 Field-scale experiment description

The experiment was performed on the closed municipal solid waste landfill "Wilcze Doły" in Kraśnik, in the eastern part of Poland. Poland is a country, which is located in the temperate climate zone, with relatively cold winters and warm summers. The average annual air temperatures are about 6–8.5°C; the annual rainfall is 500–700 mm and snow constitutes 5–20% of it (Weather Online).

The municipal solid waste landfill "Wilcze Doły" was closed in 2006. The area of the landfill is about 3.08 hectare. There is no sealing in the bottom of the landfill, which causes percolating of waste disposal leachate into the ground. The gas is collected by gas extraction system connected to flare. In 2006, the height of deposited wastes was 8–14 m. The capacity of landfill is about 350 000 m³ and the mass of deposited wastes is about 251 500 Mg. The wastes were covered by the layer of sand and soil overgrown by the grass and wild herbaceous plants.

The experimental biofilter was situated in the vicinity of the landfill. It was connected to the gas extraction system, which collects landfill gas from all gas extraction wells. The biofilter ($2 \times 2 \times 1.5$ m) was built as a double walled chamber (box in box system). There was an insulating layer of mineral

wool (12 cm) placed between the walls of boxes. Inside the biofilter, on the height of 40 cm from the bottom, there was a perforated PVC plate partition, with a density of 1600 holes per m², and hole diameter of 20 mm, covered with a plastic mesh. It constitutes a frontier between the filter bed material part (upper) and gas distribution part (lower).

The gas was supplied to the biofilter by PVC tube (ø 50–200 mm), tightly connected to the landfill gas extraction system. The gas flow regulator (GCR-B9 KS-BS30 Vögtlin) was installed on tube, connecting the biofilter with extraction system. Gas flow rate ranged from 2520 to 3200 cm³/min. Continuous gas supply to biofilter was provided by turbine (Systemair, type EX 140-2C), which was sucking the gas from the wells. Before entering the biofilter, the tube was divided into two parallel pipes (ø 150 mm of each) placed at a distance of 50 cm from each other. There were three parallel rows of holes (ø 6 mm) drilled in each pipe, in the inside section of the biofilter. The biofilter filling bed consists of two layers: the lower (drainage) part, which constitutes the layer of expanded clay pellets (ø 5 mm) with a thickness about 15 cm, and the upper part—actual filter bed material. The filter layer consists of mixture of the expanded clay pellets and municipal waste compost, mixed in equal volumetric proportions.

The properties of municipal waste compost are shown in Table 1.

The total height of biofilter filling at the beginning of the experiment was 0.65 m. Vertical and horizontal sections of biofilter are shown in Figure 1.

Biofilter was working for 2 months (May, June) at various conditions of gas flow rate and at different weather conditions (temperature, rainfall, atmospheric pressure).

2.2 Batch tests

The capacity of filter bed material for analyzed trace gases removal was evaluated on laboratory scale. After 2 months of the biofilter activity, the 0.5 kg sample of bed material was collected, from

Table 1. Properties of municipal waste compost.

Parameter	Unit	Value
Average moisture	%	30.2
Dry mass content	%	69.8
Organic dry mass content	%	16.24
Mineral dry mass content	%	53.56
Total carbon content	% of dry mass	11.4
Total nitrogen content	mg/kg	2.17
Total phosphorus content	mg/kg	0.85

which subsamples (2 g) were separated. Subsamples were placed in glass vials of 40 cm³. The examined trace gases were injected separately into closed vials and decrease of their concentration in time was analyzed.

Based on obtained results, the removal rates of particular trace gases in relation to bed material mass were calculated. Examinations were carried out at temperature of $22 \pm 1°C$, in three repetitions for each gas.

2.3 Bacterial identification

There was a preliminary microbial analysis of the filter bed material conducted, before introducing

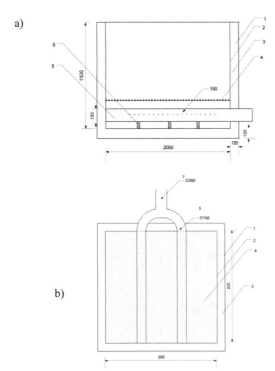

Figure 1. The biofilter scheme: a) vertical section, b) horizontal section. Signatures: (1) outside wall, (2) inside wall, (3) insulation of mineral wool, (4) horizontal perforated partition, (5) inner perforated pipe ϕ150, (6) strengthening elements, (7) outer pipe ϕ200.

it into the biofilter. Initial suspensions were prepared by an addition of mixture of clay pellets and municipal waste compost samples to sterile saline in Erlenmeyer flasks.

The suspensions were shaken at 200 rmp for one hour. Serial dilutions of initial suspensions were made. Aliquots of each dilution were inoculated on specific growth media by pour plate technique. Petri plates were incubated in thermostats in appropriate temperatures for required periods of time (Table 2).

After incubation, several colonies were picked randomly and streaked onto fresh medium. The initial bacteria identification was made using API tests (API NE, API E, API CH). The API system consists of a plastic strip with miniaturized tests, each containing a specific reagent intended for determinations of metabolic capabilities of different bacteria.

2.4 Analytical procedures

The concentration of main gases in the biogas before and after the biofilter was measured using portable biogas analyzers Gas Data 430 and GA-2000. The concentration of BTEXs in the biogas before and after the biofilter was measured using gas chromatography (GC/MS Trace Ultra-PolarisQ chromatograph with a Rtx Dioxin column to BTEXs concentrations determination, initial temperature 45°C and accretion of 8°C/min, Xcalibur ver. 2.0 programme to peaks integration).

The SPME method was used for gas samples taking (15 minutes SPME fibres exposition to the landfill gas). The measurements were done once every two weeks. The SPME fiber was introduced to the tube entering the biofilter, through the rubber stopper placed in the tube wall. After the biofilter, the BTEXs was absorbed from the gas accumulated in a plexiglas chamber box placed on the bed surface. In order to prevent pressure increase inside the chamber, a plastic pipe (inner diameter of 0.6 mm) was placed at upper surface of the chamber.

The BTEXs biofiltration efficiency was calculated on the basis of the particular BTEXs concentrations in the gas stream supplied to and leaving the biofilter. Weather condition data

Table 2. Conditions of microorganisms incubation.

Group of microorganisms	Growth basis	Temperature of incubation	Time of incubation
Mesophilic bacteria	Nutritious agar	37°C	24h/48h
Pseudomonas	Agar for *Pseudomonas*	37°C	48h
Total number of microorganisms	PCA	30°C	48h

(air temperature, atmospheric pressure, precipitation level) were obtained from the administration of landfill.

3 RESULTS AND DISCUSSION

3.1 Concentration of main components of LFG entering and leaving the biofilter

The measurements of landfill gas entering and leaving the biofilter took into consideration main LFG components concentrations. It was found that LFG composition was unstable during the experiment, which depended probably on weather conditions. There were observed very low concentrations of CH_4 and CO_2 entering the biofilter, which could be explained by diluting of LFG in air, during the gas suction through the turbine. Concentrations of main components of LFG entering and leaving the biofilter are shown in Table 3.

The average inlet concentration of methane was 1.8 vol.%; the average concentration of carbon dioxide was 0.8 vol.%. As it concerns the LFG leaving the biofilter, average concentrations of CH_4 and CO_2 were 1.0 and 1.2 vol.%, respectively.

3.2 Potential of trace gas removal in biofilter

The concentration of particular BTEXs and p-cymene in landfill gas entering and leaving the biofilter were measured five times. It was found that biofilter had an ability to remove trace gases from LFG. The concentrations of analyzed volatile organic compounds in landfill gas entering the biofilter were in lower range of the values measured in landfill gas (Table 4).

The high content of air in gas stream entering the biofilter influenced the BTEXs removal capacity. It promoted oxidation process, providing oxygen supply for bacteria responsible for the process. Due to that fact, BTEXs concentrations were practically not detected in LFG leaving the biofilter. Therefore, it was stated that LFG dilution in air resulted in almost 100% efficiency of BTEXs removal. The various weather conditions during the measurements were observed (Table 5). The temperature ranged from 15.6 to 30.3°C, atmospheric pressure: from 889.9 to 987.3 hPa, air humidity: from 41 to 82%, and the daily rainfall: from 0 to 9.3 mm. Because of 100% efficiency of trace gas removal, it is not possible to determine the removal capacity, related to the volume (or surface area) of biofilter filling bed. What is more, the influences of meteorological parameters on the biofilter activity could not be determined. It could be only stated that at trace gases concentrations in gas input, the variability of meteorological parameters (in examined range) did not affect the biofilter.

3.3 Rate of BTEXs and p-cymene removal

Rates of particular gases removal in material taken from surface part of field biofilter, examined in batch tests were lower than the values presented in relevant literature. The filter material in upper part of the biofilter was exposed on extremely small concentration of analyzed gases which can explain the disparities. The highest rates were observed in the cases of benzene (9.91 mg kg^{-1}d^{-1}) and toluene (6.68 mg kg^{-1}d^{-1}). Other gases showed significantly lower degradability. For comparison, the removal rates of toluene in studies conducted by Scheutz et al. (2003, 2004) were in range 33.6 to 39 mg kg^{-1}d^{-1}. In literature, there are inaccurate pieces of information information about BTEXs susceptibility to microbial decomposition. Generally, toluene is considered to be the most susceptible to biodegrability. According to Gülensoy & Alvarez (1999), toluene, in a biodegradation hierarchy, is the most easily degraded, followed by

Table 3. Concentrations of main components of LFG entering and leaving the biofilter.

| LFG component | Measurement number | | | | | | | |
| | I | | II | | III | | IV | |
	Before biofilter	After biofilter	Before biofilter	After biofilter	Before biofilter	After biofilter	Before biofilter	After biofilter
Air [vol.%]	97.736	97.788	97.558	98.158	97.390	98.700	96.707	96.650
Methane (CH_4) [vol.%]	1.520	1.455	1.169	0.0	2.155	0.200	2.413	2.443
Carbon dioxide (CO_2) [vol.%]	0.744	0.757	1.273	1.847	0.455	1.100	0.881	0.907

Table 4. Concentrations of BTEXs and p-cymene in LFG entering and leaving the biofilter.

Gas concentrations [µg m⁻³]	I Before biofilter	I After biofilter	II Before biofilter	II After biofilter	III Before biofilter	III After biofilter	IV Before biofilter	IV After biofilter	V Before biofilter	V After biofilter
Benzene	12.98	n.d.	2.97	Traces amount	0.21	0	0	Traces amount	0	0
Toluene	7.83	n.d.	0.14	Traces amount	0.60	Traces amount	0.37	Traces amount	Traces amount	0
Ethylobenzene	0.83	n.d.	2.13	Traces amount	4.86	0	5.99	0	3.79	0
o-xylene	0.64	n.d.	1.68	Traces amount	2.86	0	3.16	0	2.18	0
m-xylene	1.84	n.d.	0.74	Traces amount	6.57	0	7.12	0	4.32	0
p-cymene	5.19	n.d.	3.89	Traces amount	15.50	Traces amount	24.78	Traces amount	13.83	Traces amount

n.d.—not detected.

Table 5. Weather conditions during the measurements.

Parameter [unit]	Measurement number I	II	III	IV	V
Atmospheric pressure [hPa]	889.9	987.3	986.2	975.6	983.3
Temperature [°C]	30.3	21.6	21.8	15.6	29.8
Air humidity [%]	41	43	52	82	55
Rainfall [mm]	0	1.5	0	9.3	2.1

p-xylene, m-xylene, benzene, ethylobenzene, and o-xylene. Biodegrability of compounds examined in this paper decreased with the molecular mass and the complexity of molecular structure of the compounds (Table 6).

Obtained removal rates might be affected by various initial concentrations of trace gases, that is why the next series of examination will be conducted by similar initial concentrations, which will enable the more reliable evaluation of removal efficiency of trace gases in biofilter filling material.

The rates of removal were calculated for the interval of 16–20 hours after the start of the tests. In that time depletion of all gases concentrations in the headspaces had near linear trend (Fig. 2). Significant decline of gas removal rate was observed after that time, probably due to a decrease in the concentration of reactants: both analyzed gas, and oxygen.

3.4 Bacterial identification

The bacterial strains found in the filter bed material before introducing it into biofilter (after initial qualitative analysis) are the following: *Pseudomonas fluorescens, Pseudomonas putida Flavimonas oryzihabitans, Sphingomonas paucimobilis, Aeromonas salmonicida, Comamonas testosteroni, Brevundimonas vesicularis, Stenotrophomonas maltophila, Achromobater denitrificans, Chryseobacterium indologenes, Bacillus sp.*

These are partial results of bacterial identification, and the analyses (after prior DNA isolation) are underway. All these bacteria are aerobic. Most of them are commonly found in different natural and anthropogenic environments. Some of them are pathogenic for humans (Table 7) and some (e.g. *P. putida*) are capable to grow on atypical substrates (like aromatic compounds), which are used as a sole source of carbon and energy. Therefore, they are exploited in biotechnology, for example in bioremediation of soils contaminated with hydrocarbons.

279

Table 6.   Average removal rates of particular BTEXs and p-cymene.

Compound	Removal rate [mg kg⁻¹d⁻¹]	Molecular mass [g/mol]	Molecular formula	Structure
Benzene	9.91	78.11	C_6H_6	
Toluene	6.68	92.14	$C_6H_5–CH_3$	
Ethylobenzene	0.70	106.17	$C_6H_5–C_2H_5$	
o-xylene	0.48	106.16	$C_6H_4–(CH_3)_2$	
m-xylene	0.68			
p-cymene	0.14	134.22	$CH_3–C_6H_4–CH(CH_3)_2$	

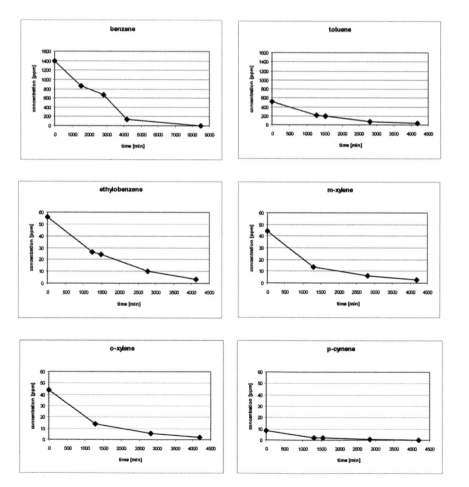

Figure 2.   Time-dependent decrease of gas concentration in the headspaces observed during batch tests.

280

Table 7. Characteristics of bacteria isolated from the material used as filling bed of field biofilter (sources: EDL base; Abraham & Simon 2007, Sgrelli et al. 2012).

Bacteria	Occurence	Patogenic influence for human
Pseudomonas fluorescens	Soil, water, plants, and contaminated foodstuffs	Infections only in immuno-compromised patients
Pseudomonas putida	Soil, water, plants, and foodstuff	Commonly regarded as nonpathogenic microorganism
Flavimonas oryzihabitans (*Pseudomonas oryzihabitans*)	Water, soil, rice paddies, sink drains, hospitals apertures	Rarely encountered in human infections
Sphingomonas paucimobilis		Catheter infections in humans
Aeromonas salmonicida	Freshwater, chlorinated water, polluted water, brackish water, estuarine water, and sewage	Causing gastroenteritis and wound infections in humans
Comamonas testosteroni	Water, soil, and on plants, hospital oxygen humidifier reservoirs	Commonly regarded as nonpathogenic microorganism
Brevundimonas vesicularis	Water and clinical specimens, as blood	Rarely encountered in human infections
Stenotrophomonas maltophila	Water, soil and plants	Nosocomial infection (hospital-acquired infection) difficult to treat
Achromobater denitrificans	Water and soil	Renal infections
Chryseobacterium indologenes	Raw meat, poultry, vegetables, dairy products, soil and water	Nosocomial (hospital-acquired) infections
Bacillus sp.	Soil, dust, water, plants, humans and animals	The majority of species are nonpathogenic or opportunistic pathogens for humans. Exceptions: *B. anthracis*, causes anthrax, and *B. cereus*—an agent of food poisoning.

4 CONCLUSIONS

The field study carried out in the biofilter working at the municipal solid waste landfill "Wilcze Doły" in Kraśnik near Lublin (Poland) showed that the biofiltration could be considered as an effective method for trace gases removal from the landfill gas. The BTEXs removal efficiency measured in the biofilter was near 100%. However, the concentrations of BTEXs in gas stream entering biofilter were very small, and the maximum rate of its removal could not be determined. In the case of preliminary bacterial identification, few bacterial strains were detected in the municipal waste compost before introducing it into biofilter. The identification of other microorganisms (after prior DNA isolation) is still lasting. Moreover, identification of the bacteria consortium developing in the filter bed during the experiment is planned. It can be predicted that mainly strains growing on LFG components will be found. The rate of trace gas removal determined in batch tests decreased with molecular mass growth, reaching the higher value for benzene and toluene and the lowest for xylenes and p-cymene.

ACKNOWLEDGEMENTS

We would like to thank mgr Ewelina Staszewska and mgr Paweł Golianek for their substantial assistance in microbiological assays and trace gases chromatographic analysis.

This research was financed from a research project Nr 7413/T02/2011/40 entitled: "Removal of selected volatile organic compounds from landfill gas by biofiltration".

REFERENCES

Abraham, J.M. & Simon, G.L. 2007. Comamonas testosteroni Bacteremia: A Case Report and Review of the Literature. *Infectious Diseases in Clinical Practice* 15 (4): 272–273.

Abumaizar, R.J., Kocher, W. & Smith, E.H. 1998. Biofiltration of BTEX contaminated air streams using compost-activated carbon filter media. *Journal of Hazardous Materials* 60: 111–126.

Corbit, R.A. 1990. Standard handbook of environmental engineering. McGraw-Hill, New York: 4.1–4.110.

Crocker, B. & Schnelle, K. 1998. Air pollution control for stationary sources. In Meyers, R.A. (Ed.) Encyclopedia of Environmental Analysis and Remediation: 151–213, John Wiley and Sons, Inc., New York.

Delhoménie, M.C., Bibeau, L., Bredin, N., Roy, S., Broussau, S., Brzezinski, R., Kugelmass, J.L. & Heitz, M. 2002. Biofiltration of air contaminated with toluene on a compost-based bed. *Advances in Environmental Research* 6: 239–254.

Delhoménie, M.C. & Heitz, M. 2005. Biofiltration of air: a review. *Critical Reviews in Biotechnology* 25: 53–72.

Devinny, J., Deshusses, M. & Webster, T. 1999. Biofiltration for air pollution control CRC Press, Boca Raton, FL: 51–81.

Durmusoglua, E., Taspinarb, F. & Karademira, A. 2010. Health risk assessment of BTEX emissions in the landfill environment. *Journal of Hazardous Materials* 176: 870–877.

EDL Environmental Diagnostic Laboratory http://www.edlab.org/glossary.html, access data: 12.2012.

Gülensoy, N. & Alvarez, P.J.J. 1999. Diversity and correlation of specific aromatic hydrocarbon biodegradation capabilities. *Biodegradation* 10: 331–340.

Hassan, A.A. & Sorial, G. 2009. Biological treatment of benzene in a controlled trickle bed air biofilter. *Chemosphere* 75: 1315–1321.

IARC, Monographs on the evaluation of carcinogenic risks to humans, Benzene, toluene, ethylbenzene and xylenes, vol. 29, 71 and 77.

McNevin, D. & Barford, J. 2000. Biofiltration as an odour abatement strategy. *Biochemical Engineering Journal* 5: 231–242.

Mudliar, S., Giri, B., Padoley, K., Satpute, D., Dixit, R., Bhatt, P., Pandey, R., Juwarkar, A. & Vaidya, A. 2010. Bioreactors for treatment of VOCs and odours—A review. *Journal of Environmental Management* 91: 1039–1054.

Muneron de Mello, J.M., de Lima Brandão, H., Ulson de Souza, A.A., da Silva, A. & de Arruda Guelli Ulson de Souza, S.M. 2010. Biodegradation of BTEX compounds in a biofilm reactor—Modeling and simulation. *Journal of Petroleum Science and Engineering* 70: 131–139.

Rene, E.R., Mohammad, B.T., Veiga, M.C. & Kennes, C. 2012. Biodegradation of BTEX in a fungal biofilter: Influence of operational parameters, effect of shock-loads and substrate stratification. *Bioresource Technology* 116: 204–213.

Scheutz, C., Bogner, J., Chanton, J., Blake, D., Morcet, M. & Kjeldsen, P. 2003. Comparative oxidation and net emissions of methane and selected non-methane organic compounds in landfill cover soils. *Environmental Science and Technology* 37 (22): 5150–5158.

Scheutz, C. & Kjeldsen, P. 2003. Capacity for biodegradation of CFCs and HCFCs in a methane oxidative counter-gradient laboratory system simulating landfill soil covers. *Environmental Science & Technology* 37: 5143–5149.

Scheutz, C., Mosbæk, H. & Kjeldsen, P. 2004. Attenuation of methane and volatile organic compounds in landfill soil covers. *Journal of Environmental Quality* 33: 61–71.

Sgrelli, A., Mencacci, A., Fiorio, M., Orlandi, C., Baldelli, F. & De Socio, G.V.L. 2012. Achromobacter denitrificans renal abscess. *The New Microbiologica* 35 (2): 245–247.

Shareefdeen, Z. & Singh, A. 2005. Biotechnology for odor and air pollution control. Springer Berlin, Heidelberg, New York.

Weather Online. Available: (www.weatheronline.pl/reports/climate/Poland.html, access data: 03.2012.

Yeom, S.-H. & Daugulis, A.J. 2001. Benzene degradation in a two-phase partitioning bioreactor by Alcaligenes xylosoxidans Y234. *Process Biochemistry* 36: 765–772.

Zilli, M., Converti, A., Lodi, A., Borghi, M.D. & Ferraiolo, G. 1993. Phenol removal from waste gases with a biological filter by Pseudomonas putida. *Biotechnology and Bioengineering* 41: 693–699.

Zilli, M., Fabiano, B., Ferraiolo, A. & Converti, A. 1996. Macro-kinetics investigation on phenol uptake from air by biofiltration: influence of superficial gas flow rate and inlet pollutant concentration. *Biotechnology and Bioengineering* 49: 391–398.

Zou, S.C., Lee, S.C., Chan, C.Y., Ho, K.F., Wang, X.M., Chan, L.Y. & Zhang, Z.X. 2003. Characterization of ambient volatile organic compounds at a landfill site in Guangzhou, South China. *Chemosphere* 51(9): 1015–1022.

Air protection and quality

Environmental Engineering IV – Pawłowski, Dudzińska & Pawłowski (eds)
© 2013 Taylor & Francis Group, London, ISBN 978-0-415-64338-2

Associative rules for daily air pollutant concentration profiles

R. Jasiński

Faculty of Environmental Engineering and Biotechnology, Czestochowa University of Technology, Czestochowa, Poland

ABSTRACT: The purpose of the study was to indicate the associative rules for the simultaneous occurrence of specific types of daily air pollutant concentration profiles in successive days of the year. These profiles were determined using the cluster analysis by the k-means method. Single objects, in the form of daily air pollutant concentrations, were grouped in such a way that the centres of the obtained clusters in the 24-dimensional Euclidean space each formed 5, possibly different, daily air pollutant concentration variations. Then, the associative rules for the concentration variations of selected air pollutants in the successive days of the year were found. The support and confidence of the discovered rules were calculated. The obtained results have indicated the existence of distinct relationships for the co-occurrence of specific types of daily air pollutant concentrations. The employed methodology enables the estimation of the missing data.

Keywords: cycles, daily profile, air pollutant, cluster analysis, associative rules, missing data

1 INTRODUCTION

Measurements of the concentrations of selected air pollutants and meteorological parameters, conducted in automatic air monitoring stations, are conducted in a continuous mode. Obtained data is averaged for mean one-hour periods and then stored in extensive monitoring databases. In this type of data, a daily cyclicity is observed (Seinfeld & Pandis 1998, Mayer 1999, Jasinski 2006). This phenomenon is associated mainly with cyclically recurring meteorological conditions (such as solar radiation intensity, relative humidity) and the pollutant emission intensity variation related to the daily human activity (Jasinski 2011). It has been noticed that the variability of air pollutant concentration levels, as observed in daily cycles, is not random but is well ordered in nature and carries the information about the occurrence of factors that determine it in the location of taking the measurements (Jasinski 2012). This variability can be defined using the profiles of daily air pollutant concentration variations. The shape of daily air pollutant concentration profiles depends on numerous factors including meteorological parameters, seasons, the measuring station's location, the proximity of emission sources, etc. It has been found that in some specific, recurring conditions the daily variations of air pollutant concentrations have comparable profiles (shapes) (Mazzeo et al. 2005, Jasinski 2009).

In the present study, several typical averaged daily profiles were determined for each of the selected air pollutants. Then, an analysis was made in order to discover the rules for the simultaneous occurrence of specific types of daily air pollutant concentration profiles in successive days of the year. The discovery of the frequent cases of co-occurrence of some types of daily profiles (with high values of support and confidence) will significantly complement the knowledge of the interrelations between individual parameters measured on air monitoring stations. This approach does not entail the necessity of calculating the parameters of either a linear or non-linear relationship between the parameters but the analysis results may assume the form of event patterns (Kwasnicka & Switalski 2005). The occurrence, on a given day, of specific daily profile types of selected air pollutants, called predecessors, may, with a certain probability, indicate the occurrence of a specific type of daily variation of another pollutant, so called successor. So far, daily air pollutant concentration profiles have been represented as time series (in the two-dimensional space). In the present study, these profiles are treated as single objects in the 4-dimensional Euclidean space. For the determination of the averaged daily profile types the cluster analysis made by the k-means method was used (Mao et al. 2004), while for determining the associative rules for the co-occurrence of individual daily profile types the Market Basket Analysis method was employed (Kwasnicka & Switalski 2005, Winarko & Roddick 2007, Adhikari & Rao 2008).

2 MATERIAL AND METHODS

The data from the period of 2006–2010 in the form of one-hour mean values of the concentrations of CO, NO_2, PM_{10}, SO_2 and O_3 derived from the Automatic Air Monitoring Station in Katowice were used for the study. The measuring station is part of the Silesian Air Monitoring System and is supervised by the Silesian Provincial Environmental Protection Directorate in Katowice. The missing values of a measurement gap, whose length did not exceed 6, were made up for by the one-dimensional linear interpolation method (Junninena et al. 2004, Plaia & Bondi 2006, Hoffman & Jasiński 2008, 2009). In the case where the measurement gap magnitude exceeded 6 values, the whole 24-hour period was not considered in the analysis.

Then the data were segregated to the form of daily variation matrices. Five separate matrices of one-hour mean concentration values of CO, NO_2, PM_{10}, SO_2 and O_3 were created, in which respective columns represented successive hours of the day, from 1 to 24, while 1825 rows (objects, cases) in the matrices represented successive days in the 5-year period. Data prepared in such a way were subjected to cluster analysis by the k-means method, where the cases (rows) were grouped. $k = 5$ clusters were taken and a maximum of 10 iterations were assumed, for all pollutants, with the preliminary cluster centres being searched for by sorting the distances and taking observations with a constant interval.

From the data clustering process, 5 groups of object clusters were obtained for each of the air pollutants considered. The centres of those clusters in the 24-dimensional Euclidean space are represented by possibly different types of daily variation of a given pollutant. It was identified to which group each of the individual variable objects belonged. Therefore, only one out of the 5 daily concentration variation profile types of individual air pollutants was assigned to each day. The following symbols were assigned to the obtained groups (clusters) for NO_2: N1, N2, N3, N4 and N5; for CO: C1, C2, C3, C4 and C5; for SO_2: S1, S2, S3, S4 and S5; for PM_{10}: P1, P2, P3, P4 and P5; and for O_3: O1, O2, O3, O4 and O5 respectively. The results of assigning the daily variation types to individual pollutants in successive days of the period of 2006–2010 were summarized in a tabulated form (Table 1).

Monitoring data processed in such a way were subjected to the associative rule discovering process. In the above analysis, the 24-hour period was regarded as a single event, while the elements of this event were profile types for individual pollutant occurring during this period. The total number of events in the five-year period amounted 1826.

Table 1. A fragment of the tabulated summary of the obtained daily profile types in successive days in the years 2006–2010.

| Object no. | Date | Profile type | | | | |
		NO_2	CO	SO_2	PM_{10}	O_3
1	2006-01-01	N1	C3	S1	P2	O5
2	2006-01-02	N1	C2	S1	P1	O5
3	2006-01-03	N1	C3	S1	P3	O5
...	...	...	...	...	...	...
...	...	...	...	...	...	...
1824	2010-12-29	N1	C2	S1	P5	O5
1825	2010-12-30	N1	C3	S2	P2	O5
1826	2010-12-31	N1	C4	S3	P3	O5

The quality of the obtained associative rules was determined using two parameters: support (s) and confidence (c), which define, respectively, the frequency and the accuracy of occurrence of a given rule. For one-element associative rules, i.e. the ones with one predecessor and successor, $A \Rightarrow B$ (if A then B), the support s was defined as the number of events including particular elements, e.g. A and B, occurring in the set of all one-element events (1).

$$s = P(A \cap B) = \frac{Number\ of\ events\ including\ A\ and\ B}{Total\ number\ of\ events}$$

(1)

The confidence for a given associative rule $A \Rightarrow B$ was defined as the percent of events including elements A that included also elements B (2).

$$c = P(B|A)$$
$$= \frac{P(A \cap B)}{P(A)} = \frac{Number\ of\ events\ including\ A\ and\ B}{Number\ of\ events\ including\ A}$$

(2)

For a greater number of elements in an event, the support and the confidence were defined in a similar manner. In the first place, rules with a one-element predecessor and successor were investigated. Then, two-, three- and four-element sets of predecessors with single successors were searched for. The ARMADA algorithm operating in the MATLAB environment was used for discovering associative rules (Winarko & Roddick 2007).

3 RESULTS AND DISCUSSION

The daily variations of air pollutant concentrations in the multi-dimensional analysis approach can be interpreted as set of single elements in the 24-dimensional Euclidean space (Jasinski 2011). By using the clustering analysis by the k-means method, 5 clusters for each of the NO_2, CO, SO_2, PM_{10} and O_3 concentrations were created with the highest possible level of mutual differentiation and the greatest intra-group similarity. The geometrical centres of these clusters in the 24-dimensional Euclidean space were obtained by determining their mean value. As a result, 5 possibly different types of daily variation profiles were generated for each of the pollutants under consideration (Fig. 1).

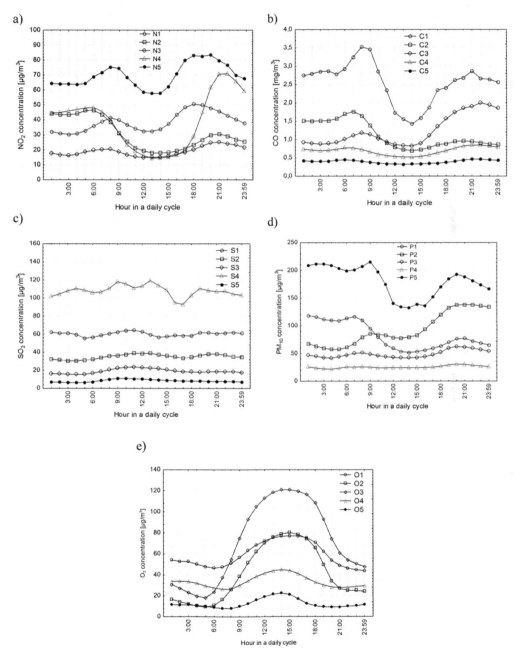

Figure 1. Graph of the means of each of the clusters, corresponding to 5 types of daily concentration profiles (the Katowice Monitoring Station, 2006–2010): a) NO_2, b) CO, c) SO_2, d) PM_{10}, e) O_3.

Table 2 presents the number of cases (days) assigned to each cluster. For NO_2 concentrations, the type corresponding to the profile N3 (37.9% cases) constitutes the most numerous group, while the type N5 (4.7% cases) the least numerous one. In the case of CO concentrations, the type corresponding to C5 (52.9% cases) makes the most numerous group, with the C1 group (1.8% cases) being the least numerous one. Also for the remaining pollutants, the most numerous groups and the least numerous could be distinguished, namely: for SO_2 concentrations—S5 (50.4%) and S4 (1.2%); for PM_{10} concentrations—P4 (40.5%) and P5 (2.0%); and for ozone concentrations—O5 (29.0%) and O3 (11.3%). For all of the pollutants considered, the most numerous groups corresponded to the profiles with the lowest average concentration levels, whereas the least numerous groups corresponded

Table 2. The number of cases assigned to each cluster of daily courses of NO_2, CO, SO_2, PM_{10} and O_3 concentrations.

Air pollutant	The cluster number	The number of cases in a cluster	The number of cases in a cluster [%]
NO_2	N1	386	21.1
	N2	295	16.2
	N3	692	37.9
	N4	203	11.1
	N5	86	4.7
	Rejected	164	9.0
CO	C1	33	1.8
	C2	139	7.6
	C3	99	5.4
	C4	534	29.2
	C5	965	52.9
	Rejected	56	3.1
SO_2	S1	74	4.0
	S2	212	11.6
	S3	528	28.9
	S4	21	1.2
	S5	920	50.4
	Rejected	71	3.9
PM_{10}	P1	128	7.0
	P2	110	6.0
	P3	515	28.2
	P4	740	40.5
	P5	36	2.0
	Rejected	297	16.3
O_3	O1	271	14.8
	O2	310	17.0
	O3	206	11.3
	O4	398	21.8
	O5	530	29.0
	Rejected	111	6.1

to the profiles with the highest concentration levels. For example, for CO concentrations the most numerous group was created by the C5 type profile (52.9% cases), for which the maximum value did not exceed the level of mg/m³, while the least numerous group was made up by the profile of the C1 type (1.8% cases), for which the maximum value reached 3.5 mg/m³ at 8:00 hours. This distribution of individual groups indicates that the high concentration values of individual air pollutants could have been much rarely observed.

For each NO_2 profile type, there were increases in concentration levels in the morning and in the evening, and decreases in concentration levels in the afternoon (12:00–14:00). The most characteristic profile is the N4 type (11.1% cases), in which a dramatic increase in NO_2 concentrations by over 55 µg/m³ occurred during the hours 17:00–21:00. All types of daily CO concentration profiles noted a daily minimum falling in the afternoon hours (13:00–14:00) and an increase in values in the morning and in the evening, which can be shown. This is particularly visible in the C1 type profile, where the difference between the daily minimum and maximum exceeds 2 mg/m³. For SO_2 concentrations, individual profiles are characterized by distinctly differentiated mean values with low variability over the successive hours of the day. Only for the S4 type profile a reduction in SO_2 concentration levels by more than 25 µg/m³ occurred from 13:00 to 17:00, followed by an increase by approx. 20 µg/m³ up to 19:00 o'clock. For PM_{10} concentrations, daily profile types can be distinguished, which have morning and evening maxima characteristic of NO_2 and CO concentrations, with the P2 type profile showing a clearly increasing trend; while the P1 type profile—a decreasing trend in the daily cycle. Unlike the profiles obtained for NO_2, CO and PM_{10}, individual profile types obtained for ozone reached their daily maxima in the afternoon (14:00–15:00 hrs.), while the daily minima in the morning and in the evening.

Table 3 presents discovered 1-, 2-, 3- and 4-element associative elements which were graded by the highest support and confidence values. Due to the limitation on the table size, only the first 49 rules with the highest support values are shown. Rules of relatively low support levels have been obtained, which do not exceed 40% for one-element rules and 6.1% in the case of four-element rules. Rules with the highest support relate primarily to the pollutant profile types that occur in the largest numbers in individual groups. This is consistent with the "A-priori" algorithm, according to which multi-element rules are generated on the basis on the most frequently occurring one-element sets. The most numerous clusters forming individual profile types are made up of groups composed of variations with the lowest concentration levels in

Table 3. Discovered 1-, 2-, 3- and 4-element associative elements graded by the highest support and confidence values, higher than 50%.

One-element rules	Sup. [%]	Conf. [%]	Two-element rules	Sup. [%]	Conf. [%]	Three-element rules	Sup. [%]	Conf. [%]	Four-element rules	Sup. [%]	Conf. [%]
S5 → C5	38,8	77,0	S5 P4 → C5	26,4	90,9	N3 S5 P4 → C5	18,3	92,3	N3 S5 P4 O4 → C5	6,1	94,1
C5 → S5	38,8	73,4	C5 P4 → S5	26,4	76,9	N3 C5 P4 → S5	18,3	78,4	N3 C5 S5 O4 → P4	6,1	84,7
P4 → C5	34,3	84,7	C5 S5 → P4	26,4	68,1	N3 C5 S5 → P4	18,3	77,9	C5 S5 P4 O4 → N3	6,1	76,0
C5 → P4	34,3	65,0	N3 S5 → C5	23,5	88,1	C5 S5 P4 → N3	18,3	69,3	N3 C5 P4 O4 → S5	6,1	71,6
N3 → C5	30,5	80,5	N3 C5 → S5	23,5	77,0	N3 P4 O4 → C5	8,5	91,2	N3 S5 P4 O1 → C5	6,0	94,8
C5 → N3	30,5	57,7	C5 S5 → N3	23,5	60,6	N3 C5 O4 → P4	8,5	84,7	C5 S5 P4 O1 → N3	6,0	92,4
P4 → S5	29,0	71,6	N3 P4 → C5	23,3	88,9	C5 P4 O4 → N3	8,5	78,3	N3 C5 P4 O1 → S5	6,0	82,7
S5 → P4	29,0	57,6	N3 C5 → P4	23,3	76,5	S5 P4 O4 → C5	8,0	93,6	N3 C5 S5 O1 → P4	6,0	82,1
N3 → S5	26,7	70,4	C5 P4 → N3	23,3	67,9	C5 S5 O4 → P4	8,0	82,0	N3 C5 S5 O5 → P4	2,3	89,4
S5 → N3	26,7	52,9	N3 P4 → S5	19,8	75,6	C5 P4 O4 → S5	8,0	73,7	N3 S5 P4 O5 → C5	2,3	77,8
N3 → P4	26,2	69,2	N3 S5 → P4	19,8	74,3	N3 S5 O1 → C5	7,3	95,0	C5 S5 P4 O5 → N3	2,3	73,7
P4 → N3	26,2	64,7	S5 P4 → N3	19,8	68,3	C5 S5 O1 → N3	7,3	85,9	N3 C5 P4 O5 → S5	2,3	70,0
P2 → C4	14,7	52,2	P4 O4 → C5	10,8	88,0	N3 C5 O1 → S5	7,3	77,0	N3 S3 P4 O4 → C5	2,0	87,8
C4 → P3	14,7	50,4	C5 O4 → P4	10,8	78,9	N3 P4 O1 → C5	7,3	88,7	C5 S3 P4 O4 → N3	2,0	87,8
N1 → O5	14,3	67,9	N3 O4 → C5	10,0	85,9	C5 P4 O1 → N3	7,3	83,6	N3 C5 S3 O4 → P4	2,0	83,7
O4 → C5	13,7	63,1	C5 O4 → N3	10,0	72,9	N3 C5 O1 → P4	7,3	76,4	N3 S5 P4 O3 → C5	1,9	100,0
O4 → P4	12,3	56,5	S5 O4 → C5	9,7	82,0	N3 S5 O4 → C5	7,2	90,3	N3 C5 P4 O3 → S5	1,9	92,1
O → C5	12,3	82,7	C5 O4 → S5	9,7	70,9	C5 S5 O4 → N3	7,2	73,6	N3 C5 S5 O3 → P4	1,9	68,6
O4 → S5	11,9	54,5	N3 O1 → C5	9,5	88,8	N3 C5 O4 → S5	7,2	71,6	C5 S5 P4 O3 → N3	1,9	55,6
O4 → N3	11,7	53,5	C5 O1 → N3	9,5	77,7	S5 P4 O1 → C5	6,5	95,2	N1 C4 S3 P3 → O5	1,8	68,8
O2 → S5	10,8	63,9	N3 O4 → P4	9,3	79,8	C5 S5 O1 → P4	6,5	76,3	N1 S3 P3 O5 → C4	1,8	68,8
N1 → C4	10,8	51,3	P4 O4 → N3	9,3	75,6	C5 P4 O1 → S5	6,5	74,8	C4 S3 P3 O5 → N1	1,8	60,0
O1 → N3	10,7	72,3	P4 O1 → C5	8,7	88,3	N3 S5 O4 → P4	6,5	81,4	N1 C4 S3 O5 → P3	1,8	56,9
N2 → C5	9,9	61,4	C5 O1 → P4	8,7	71,0	S5 P4 O4 → N3	6,5	75,6	N2 S5 P4 O2 → C5	1,8	100,0
O1 → P4	9,9	66,4	S5 O1 → C5	8,5	94,5	N3 P4 O4 → S5	6,5	69,4	N2 C5 P4 O2 → S5	1,8	94,1
O2 → C5	9,7	57,4	S5 O4 → P4	8,5	71,9	S5 P4 O1 → N3	6,4	92,8	N2 C5 S5 O2 → P4	1,8	62,7
N2 → S5	9,5	59,0	C5 O1 → S5	8,5	69,6	N3 S5 O1 → P4	6,4	82,3	N3 S5 P4 O2 → C5	1,6	93,5
O1 → S5	9,0	60,9	P4 O4 → S5	8,5	69,3	N3 P4 O1 → S5	6,4	77,3	N3 C5 P4 O2 → S5	1,6	90,6
O3 → C5	8,9	78,6	P4 O1 → N3	8,2	83,3	S5 P4 O2 → C5	4,5	93,3	N3 C5 S5 O2 → P4	1,6	59,2
O3 → S5	7,3	64,6	N3 O1 → P4	8,2	76,5	C5 P4 O2 → S5	4,5	85,6	N2 S5 P3 O3 → C5	1,4	96,3
N4 → S5	6,7	60,1	C5 O2 → S5	8,2	83,7	C5 S5 O2 → P4	4,5	55,7	N1 S5 P3 O5 → C4	1,4	89,7
S2 → O5	6,3	54,2	S5 O2 → C5	8,2	75,3	N2 S5 P4 → C5	4,3	98,7	N2 C5 P3 O3 → S5	1,4	78,8
N4 → O2	5,7	51,2	N3 O4 → S5	7,9	68,1	N2 C5 P4 → S5	4,3	85,7	N1 C4 S5 P3 → O5	1,4	76,5

(Continued)

Table 3. (Continued).

One-element rules	Sup. [%]	Conf. [%]	Two-element rules	Sup. [%]	Conf. [%]	Three-element rules	Sup. [%]	Conf. [%]	Four-element rules	Sup. [%]	Conf. [%]
C2 → O5	5,0	66,2	S5 O4 → N3	7,9	66,8	N1 C4 P3 → O5	4,3	72,2	N1 C4 S5 O5 → P3	1,4	68,4
P2 → O5	4,6	76,4	N2 S5 → C5	7,9	82,8	N1 P3 O5 → C4	4,3	70,3	C4 S5 P3 O5 → N1	1,4	65,0
C3 → O5	4,4	80,8	N2 C5 → S5	7,9	79,6	C4 P3 O5 → N1	4,3	60,5	C5 S5 P3 O3 → N2	1,4	61,9
P1 → C2	4,1	57,8	S5 O1 → N3	7,7	85,5	N1 C4 O5 → P3	4,3	58,6	N2 C5 S5 O3 → P3	1,4	50,0
C2 → P1	4,1	53,2	N3 O1 → S5	7,7	71,9	N2 C5 S5 → P4	4,3	54,2	N2 S5 P4 O3 → C5	1,2	100,0
N5 → O5	3,7	79,1	N1 C4 → O5	7,3	67,2	N3 S3 P4 → C5	4,2	79,4	N2 C5 P4 O3 → S5	1,2	77,8
P1 → S3	3,5	50,0	C4 O5 → N1	7,3	60,5	N3 C5 S3 → P4	4,2	73,3	N3 S3 P4 O1 → C5	1,1	66,7
P2 → N1	3,5	57,3	N1 O5 → C4	7,3	50,8	C5 S3 P4 → N3	4,2	66,4	N3 C5 S3 O1 → P4	1,1	58,8
C3 → N1	3,3	61,6	C5 P3 → S5	7,1	69,7	S5 P4 O3 → C5	3,5	100,0	C5 S3 P4 O1 → N3	1,1	57,1
C3 → P2	3,3	60,6	P3 O5 → C4	7,1	69,4	C5 P4 O3 → S5	3,5	86,3	N4 C5 P4 O2 → S5	1,0	90,5
P2 → C3	3,3	54,5	C4 O5 → P3	7,1	58,6	C5 S5 O3 → P4	3,5	51,2	N4 S5 P4 O2 → C5	1,0	86,4
S1 → O5	2,6	63,5	S5 P3 → C5	7,1	56,3	N3 C5 O5 → P4	3,3	88,2	N4 C5 S5 P4 → O2	1,0	61,3
P5 → O5	1,8	91,7	S5 O1 → P4	6,8	75,8	N3 P4 O5 → C5	3,3	76,9	N4 C5 P3 O2 → S5	0,9	81,0
P5 → N5	1,6	83,3	P4 O1 → S5	6,8	69,4	C5 P4 O5 → N3	3,3	74,1	N1 C4 S2 P3 → O5	0,9	77,3
C1 → O5	1,6	87,9	S5 O3 → C5	6,7	92,5	N1 C4 S3 → O5	3,2	65,9	N1 C4 S2 O5 → P3	0,9	58,6
C1 → N5	1,5	81,8	C5 O3 → S5	6,7	75,9	C4 S3 O5 → N1	3,2	61,1	N1 S2 P3 O5 → C4	0,9	56,7

the daily cycle. So, the rules with the highest support relate to those 24-hour periods in which the lowest levels of individual air pollutant concentration occur. These are rules that include profile types designated with the symbols N3, C5, S5, P4 and O5 in a varying configuration. This does not mean that these rules are not significant but they could be regarded as obvious, or even so called trivial rules.

The most interesting, from the point of view of the study of co-occurrence of daily air pollutant patterns, are the rules including profile types with a lower incidence but with a high support. It means that a given situation does not occur too frequently in the measurement period but it gives the answer to the question about what profile type of a specific pollutant is the most probable with a given configuration of the remaining pollutant profiles. It should be noted that some of the daily profile types occurred only a few dozen times out of 1826 cases, which means that the incidence was of several percent. Therefore, associative rules including daily concentration variation types that do not occur very often in the measurement period will also have low support levels. In the case where low support values exist for a given rule, the selection of the best associative rule is determined by its confidence value.

The most interesting, from the point of view of the study of co-occurrence of daily air pollutant patterns, are the rules including profile types with a lower incidence but with a high support. It means that a given situation does not occur too frequently in the measurement period but it gives the answer to the question about what profile type of a specific pollutant is the most probable with a given configuration of the remaining pollutant profiles. It should be noted that some of the daily profile types occurred only a few dozen times out of 1826 cases, which means that the incidence was of several percent. Therefore, associative rules including daily concentration variation types that do not occur very often in the measurement period will also have low support levels. In the case where low support values exist for a given rule, the selection of the best associative rule is determined by its confidence value.

4 CONCLUSIONS

In accordance with the performed computational experiment, one may claim that the proposed method of clustering by the k-means approach is suitable to transforming monitoring data from the form of time series to the form of quantified daily profile types. The division into individual cluster groups was done in accordance with the average levels of individual pollutant concentrations. This transformation enabled the continuation of the investigation aimed at discovering the associative rules for the patterns of occurrence of individual daily profile types.

The obtained associative rules provide a very simple, though useful, form of rule patterns for the exploration of monitoring data. These rules do not define the causative relationship but reveal potential relationships between individual air pollutants in the daily cycle. It should be emphasized, however, that the set of associative rules does not provide a consistent model that would enable conclusions to be drawn in a consistent manner for all days in the measurement period. Very extensive rules have been obtained, the majority of which are rules related to low levels of air pollutant concentrations. It means that rules with a high support are not necessarily interesting. Moreover, some of the rules may predict different outcomes for identical initial conditions. In such a case, an important decisive factor is the confidence of the rule.

The applied procedure enables the tracing of specific regularities (behaviour patterns) in pollutant concentration variations, which can be used for searching for interesting analogies of historic data with data from ongoing air monitoring. The obtained associative rules, along with their support and confidence values, could be used in expert systems designed for the prediction and completion of missing monitoring data in daily cycles. In the expert systems the expert knowledge would be in that case selected associative rules discovered in the archival database. The discovery of the associative rules for the co-occurrence of specific types of daily air pollutant concentration profiles on the selected automatic air monitoring stations will enhance modelling of one of the parameters, which is temporarily not measured on the measuring station concerned. These rules may also be applied in the event where some data are missing in the measuring series.

ACKNOWLEDGEMENTS

This work was supported by Czestochowa University of Technology project no BS-PB-402-301/11.

REFERENCES

Adhikari, A. & Rao, P.R. 2008. Synthesizing heavy association rules from different real data sources. *Pattern Recognition Letters* 29: 59–71.

Hoffman, S. & Jasinski, R. 2008. Completing missing data in air monitoring stations using diurnal courses of regional pollution concentrations. *Archives of Environmental Protection* 34(3): 133–142.

Hoffman, S. & Jasinski, R. 2009. Classification of Air Monitoring Data Gaps. *Polish J. of Environ. Stud.* 18(2B): 177–181.

Jasinski, R. 2006. The Types of Seasonal Changes in Daily Concentration of Some Air Pollutants in the Region of Upper Silesia Agglomeration. *Environment Protection Engineering* 32(4): 85–90.

Jasinski, R. 2009. Patterns of Air Pollution Concentrations Diurnal Courses in Different Regions of Poland. *Polish J. of Environ. Stud.* 18(2B), 170–176.

Jasinski, R. 2011. Multidimensional analysis of daily variations in air pollutants and meteorological parameters derived from the upper silesian urban area. *Pol. J. of Environ. Stud.* 20(4A): 104–109.

Jasinski, R. 2012. Directions of air pollution inflows as a method for evaluation of representativeness of automatic air monitoring stations area. *Environment Protection Engineering* 38(2): 99–108.

Junninena, H.; Niskaa, H.; Tuppurainenc, K.; Ruuskanena, J.; Kolehmainena, M. 2004. Methods for imputation of missing values in air quality data sets. *Atmospheric Environment* 38: 2895–2907.

Kwasnicka, H. & Switalski, K. 2005. Discovery of association rules from medical data—classical and evolutionary approaches. XXI Autumn Meeting of Polish Information Processing Society, Conference Proceedings: 163–177.

Mao, R.; Yin, Y.; Pei, P. 2004. Data Mining and Knowledge Discovery. Kluwer Academic Publishers.

Mayer, H. 1999. Air pollution in cities. *Atmospheric Environment* 33: 4029–4037.

Mazzeo, N.A.; Venegas, L.A.; Choren, H. 2005. Analysis of NO, NO2, O3 and NOx concentrations measured at a green area of Buenos Aires City during wintertime. *Atmospheric Environment* 39: 3055–3068.

Plaia, A. & Bondi, A.L. 2006. Single imputation method of missing values in environmental pollution data sets. *Atmospheric Environment* 40: 7316–7330.

Seinfeld, J.H. & Pandis, S.N. 1998. Atmospheric *Chemistry and Physics*. John Wiley & Sons.

Winarko, E. & Roddick, J.F. 2007. ARMADA—An algorithm for discovering richer relative temporal association rules from interval-based data. *Data & Knowledge Engineering* 63: 76–90.

Environmental Engineering IV – Pawłowski, Dudzińska & Pawłowski (eds)
© 2013 Taylor & Francis Group, London, ISBN 978-0-415-64338-2

Concentrations and chemical composition of ambient dust at a traffic site in southern Poland: A one-year study

W. Rogula-Kozłowska, K. Klejnowski, P. Rogula-Kopiec & J. Błaszczyk
Institute of Environmental Engineering, Polish Academy of Science, Zabrze, Poland

ABSTRACT: Diurnal samples of PM_{10} were collected in Katowice (southern Poland) near a highway. Three PM_{10} combined (monthly) samples were prepared for each calendar month of 2010. Each monthly sample was analyzed for one of three groups of chemicals: carbon (organic and elemental, Behr C50 IRF carbon analyzer), water soluble ions (Na^+, NH_4^+, K^+, Mg^{2+}, Ca^{2+}, Cl^-, NO_3^-, SO_4^{2-}, Herisau Metrohm AG ion chromatograph), three elements (Ti, Al, Fe, Varian Liberty 220 spectrometer). To perform the PM_{10} mass closure calculations, the PM_{10} chemical components were categorized into Organic Matter (OM), Elemental Carbon (EC), Secondary Inorganic Aerosol (SIA), Crustal Matter (CM), Marine Components (MC) and Unidentified Matter (UM). The PM_{10} concentrations in the vicinity of the highway were higher than at the urban background station in Katowice by 8 $\mu g/m^3$. Both the concentrations and the chemical composition suggest combustion of fossil fuels for industrial and domestic energy production as the main source of PM_{10}.

Keywords: PM_{10}, organic matter, elemental carbon, SIA, crustal matter, chemical mass closure, traffic site, highway

1 INTRODUCTION

The strongest relations between health and air pollution may be observed when studying PM_{10} and $PM_{2.5}$ health effects. The health effects of PM in humans depend on the dust chemical composition and the respiratory system part they reach (Massolo et al. 2002, Englert 2004, Dockery et al. 2006, de Kok et al. 2005, 2006, Zhang et al. 2011, López-Villarrubia et al. 2012).

The last quarter-century is a period of significant reduction in dust emission, especially in the highly developed countries. In Europe, the PM_{10} primary emission decreased by about 18% in average between 1990 and 2000 (http://ec.europa.eu/environment/archives/air/cafe/activities/). On the other hand, the traffic PM_{10} emission grows all over the world (Slezakowa et al. 2007). In Europe and in the USA, emission from vehicles becomes the greatest source of air pollutants (Chow et al. 1996, Paoletti et al. 2003). In Europe, the transport and the energy production shares in the PM_{10} emission are in the average proportion of 0.8 (http://reports.eea.europa.eu/technical_report_2008_7/en). Since the 1990s, the PM industrial emission decrease has also been observed in Poland, especially in Upper Silesia. However, it brought only slight lowering of the PM_{10} concentrations, still maintained high by low municipal sources and growing road traffic

(Pastuszka et al. 2003, 2010, Klejnowski et al. 2009). In Poland, the emissions of PM from transport and energy production are in proportion of only 0.17 (http://reports.eea.europa.eu/technical_report_2008_7/en), but the number of cars *per capita* approaches continuously the European average. Moreover, the imported obsolete cars (e.g. without diesel particulate filter) from the EU countries add to the emission of the finest particles, in the urbanized regions more.

The road traffic emits mainly $PM_{2.5}$ (Spurny 1996, Zhu et al. 2002, Vallius 2005), but it is also a serious source of $PM_{2.5-10}$ (Hueglin et al. 2005). The greater part of the ambient finest particles, including PM_1, come from diesel engines (Morawska et al. 1999a, b, Morcelli et al. 2005). The share of road traffic in the $PM_{2.5}$ concentrations is estimated as 32% (some authors claim that it is more than 50% (Harrison et al. 2004)), in the $PM_{2.5-10}$ concentrations—as 13%. The mass contributions of elemental and organic carbon to PM in areas affected by heavy traffic are greater than in other areas (Hueglin et al. 2005). For example, in street canyons in Bern and Zurich-Wiedikon (Switzerland) elemental carbon was 14–18% of the PM mass; in areas beyond the traffic effect it was 5–10% (Hueglin et al. 2005). The ambient concentrations of the soil-derived and trace elements are greater in vicinities of roads than elsewhere. The greatest differences are between Si, Fe and Cu concentrations (Janssen et al. 1997).

Investigations carried out at the beginning of the 21st century in the Katowice centre neither $PM_{2.5}$ nor PM_{10} evidenced significant differences between their concentrations and properties near and far from a busy road (Pastuszka et al. 2003). In 2005–2007, it was proved that the emission from busy roads and crossroads not only increases the PM concentrations but also changes the PM granular composition, PM-related trace element concentrations (in the air and dust), profile and concentration of PM adsorbed polycyclic aromatic hydrocarbons, morphology and optical properties of PM particles (Rogula-Kozłowska et al. 2009, 2011). However, there is a shortage of knowledge about the chemical composition of PM near traffic-related sources, especially on elemental carbon content of PM in Poland. The goal of the present work was to perform a whole-year analysis of the concentrations and chemical composition of PM_{10} at a site directly affected by traffic.

2 MATERIAL AND METHODS

2.1 PM_{10} sampling

During 2010, 358 diurnal samples of PM_{10} were taken at the traffic sampling point in Katowice (southern Poland)—Figure 1. The point was located almost on a curbside of a very busy highway (30000 vehicles per day).

North of the station, the Katowice centre begins, where big public buildings prevail over dwelling houses. Southward, an open plane area stretches from the highway to the high buildings of the clinic located 300 m from the station. The terrain gradually declines east from the sampling point to the vast buildings of the Regional Police Headquarters and the high blocks of flats behind. Also eastward, behind the highway, high public buildings and shopping centers are. The terrain elevates west to

Figure 1.   Location of the measuring point.

a)　　　　　　　　　　　　　　　　　　b)

Figure 2.   View towards the measuring point from west (a); PM_{10} sampler (Atmoservice PNS) at the measuring point (b).

the very high, 22 floor, building and other public buildings (Figs. 1 and 2a, b).

PM_{10} was collected onto 47 mm-diameter quartz filters with the use of an Atmoservice PNS sampler. Before and after exposure, all the filters were conditioned in a weighing room (48 hours, relative air humidity 45 ± 5%, air temperature 20 ± 2°C) and weighed twice, with a 24 h period between, on a Mettler Toledo microbalance (resolution 2 μg).

2.1.1   Chemical analyses

Three equal parts were cut out from each exposed substrate filter (diurnal dust sample). Taken one per day in a calendar month and combined together, the sample parts made a monthly sample. For each month, three analyses were done: for Organic (OC) and Elemental Carbon (EC), for water soluble ions (Na^+, NH_4^+, K^+, Mg^{2+}, Ca^{2+}, Cl^-, NO_3^-, SO_4^{2-}) and for metals (Ti, Al, Fe), each applied to one of the three monthly samples.

The OC and EC contents were determined using a Behr C50 IRF. The CO_2 content in the gas stream was determined by means of Non-Dispersive Infrared spectrometry (NDIR, $\lambda = 4.26$ μm). The gas mixture with certified CO_2 content was used to calibrate the apparatus. The detection limit for the method was 60.03 μg of CO_2. The standard recovery (RM 8785 NIST and RM 8786 NIST) for this calibration was 119% for EC and 79% for OC.

The ion content in the extracts was determined using a Metrohm ion chromatograph (Herisau Metrohm AG). The method was validated against the CRM Fluka products nos. 89316 and 89886, the standard recovery was 92–109% of the certified value, the detection limits were: 0.02 mg/l for NH_4^+, 0.05 mg/l for Cl^-, SO_4^{2-} and K^+, 0.07 mg/l for NO_3^- and Na^+, 0.12 mg/l for Ca^{2+} and Mg^{2+}. The water extracts of PM_{10} were made by ultrasonizing

the filters with the samples in 50 cm³ of de-ionized water for 60 min at temperature 15°C, then shaking for about 12 hours (18°C, 60 r/min). Before the extraction, the surface of the filters was moistened with 0.1 cm³ of ethanol (96%, pure).

To analyze metal content, a monthly sample was put into a Teflon container and mineralized gradually by successively adding concentrated acids: 12 mL of HNO_3, 5 mL of HF and 2 mL of $HClO_4$, 2 mL of $HClO_4$, about 5 mL of H_2O and 2 mL of HCl. Fe, Al and Ti, were determined with the use of Inductively Coupled Plasma-Atomic Emission Spectroscopy (ICP-AES) on a Varian ICP Liberty 220 spectrometer. The limits of detection of the method, found by analyzing blanks (clean filter substrates) according to PN-EN 14902 2006, were: 3.15 ng/m³ for Ti, 82.6 ng/m³ for Al, 3.24 ng/m³ for Fe (the flow rate 55 m³/24h).

Analyzes of the NIST SRM 1648a standard were done to check on the methods. The recoveries of the standards were from 91% (Al, Fe) to 99% (Ti).

The mass closure analysis was done for the PM_{10}-related substances categorized into six classes. Secondary Inorganic Aerosol (SIA) is the sum of PM_{10}-related SO_4^{2-}, NO_3^-, NH_4^+. The Marine Component (MC) is the sum of Na^+ and Cl^-. The mass [OM] of organic matter OM is equal to 1.8 of the mass [OC] of OC (Grosjean & Friedlander 1975, Turpin & Lim 2001), EC is a category in itself. Crustal Matter (CM) is the sum of CO_3^{2-}, SiO_2, Al_2O_3, Mg^{2+}, Ca^{2+}, K_2O, FeO, Fe_2O_3 and TiO_2 ([FeO] and [Fe_2O_3] calculated stechiometrically assuming equal Fe mass distribution between FeO and Fe_2O_3, [SiO_2] = 2[Al_2O_3] (Querol et al. 2001), and [CO_3^{2-}] = 1.5[Ca^{2+}]+2.5[Mg^{2+}] assuming CO_3^{2-} Earth crust occurrence mainly in calcite and magnesite). The Unidentified Matter (UM) is

PM$_{10}$-SIA-MC-OM-EC-CM. The mass closure for PM$_{10}$ was examined separately for each month of 2010.

3 RESULTS AND DISCUSSION

3.1 Concentrations of PM$_{10}$

The yearly PM$_{10}$ concentration at the traffic measuring point in Katowice in 2010 was 60.66 µg/m^3, 1.5 times the yearly permissible PM$_{10}$ concentration (40 µg/m^3). It did not diverge much from values at traffic sites within the Silesian agglomeration in previous years (Rogula-Kozłowska et al. 2008, 2009).

The very high concentrations in the heating season (especially in winter), when the monthly concentrations were 2–4 times greater than in summer (Table 1), decided the yearly PM$_{10}$ concentration. The highest monthly PM$_{10}$ concentrations, exceeding 100 µg/m^3, occurred in February and December 2010 (the highest daily concentration, 318.72 µg/m^3, occurred in January). In Upper Silesia, such bad air quality conditions occur almost regularly at the beginning of a year (Ośródka et al. 2006, Juda-Rezler et al. 2011); yet higher PM$_{10}$ concentrations were noted in January 2006 at crossroads in Zabrze, 30 km west of Katowice (Pastuszka et al. 2010).

The concentrations of air pollutants in a particular area are an effect of several independent factors, including emission, weather conditions, and geographic location (Lu 2002, Ravindra et al. 2008). Figure 3 illustrates the dependencies of daily PM$_{10}$ concentrations in Katowice in 2010 on daily air temperature and precipitation. The linear correlation coefficient for monthly PM$_{10}$ concentrations and air temperature is high (R^2 = 0.91). The monthly PM$_{10}$ concentrations are not significantly correlated with precipitation, direction and velocity of wind, or insolation (R^2 within 0.33–0.61)—it may indicate the domination of local pollution sources (Ragosta et al. 2008) and, consequently, that the strong correlation between monthly PM$_{10}$ concentrations and air temperature may be a result of elevated emission of PM and its precursors at low air temperatures. In Poland, the energy production bases on combustion of hard and brown coal, more intense in winter. While flue gases from power plants, containing precursors of PM$_{10}$ (SO$_x$, NO$_x$, organic compounds), affect PM$_{10}$ concentrations during a whole year (more in winter), the main factor affecting the PM$_{10}$ concentrations in winter, and a whole heating season, in big cities (like Katowice) is local municipal sources.

In the heating season of 2010 (January–March and October–December), more than 50% of the daily PM$_{10}$ concentrations in each month (from 14 to 27) exceeded 50 µg/m^3 (Table 1). In this period, 75% of 172 valid daily concentrations (valid—measurement covered at least 75% of a day) exceeded 50 µg/m^3; during the rest of the year it was no more than 16%.

In the warm season of 2010 (May–September) the daily PM$_{10}$ concentrations were between 14.87 and 58.82 µg/m^3. Consequently, the monthly PM$_{10}$ concentrations were much lower than in winter. In the period April–September the daily PM$_{10}$ concentrations varied less than in the rest of the year—the standard deviation was 15 µg/m^3.

Despite the seasonal differences between the concentrations, the monthly medians did not differ much from the monthly averages because of relative stability of the daily PM$_{10}$ concentrations in each month.

The PM$_{10}$ concentrations at the investigated traffic point were compared with the PM$_{10}$ concentrations measured at the same time at the urban background station of the Regional Inspectorate of Environmental Protection in Katowice (http://stacje.katowice.pios.gov.pl/monitoring/). In general, the diurnal PM$_{10}$ concentrations at the traffic site were greater. The null hypothesis, stating that the diurnal PM$_{10}$ concentrations at the two sites have the same distributions was rejected by the sign test and the Wilcoxon signed-rank test for the two dependent samples at the level of significance p = 0.05 (Statistica 7.1).

The monthly PM$_{10}$ concentrations at both sites were strongly linearly correlated (R^2 = 0.91, Fig. 4). Their differences were between −7.46 (April) and 29.43 µg/m^3 (January). In May-September, the period of lower effect of municipal emission, the concentrations were higher at the traffic site by the values from 0.77 (June) to 4.58 (September) µg/m^3. In 2010, the average PM$_{10}$ concentration at the traffic site was greater than at the urban background site by 8 µg/m^3 (about 14%).

The proportion of the difference between the monthly PM$_{10}$ concentrations at the traffic and backgrounds sites to the concentrations at the background site, and 0 if the difference is less than 0, indicates how much higher the monthly PM$_{10}$ concentrations may be at sites directly affected by traffic. In Katowice this value varies between 0 (April, June, July) and 36% (January). Probably, the 0 s in June and July are caused by lower holiday traffic. The condition of being representative of such an urbanized area as Katowice imposes on an urban background sampling point the property of being affected by traffic-related PM$_{10}$ sources by definition. It may mean that big traffic-related PM sources have weaker effect on PM$_{10}$ concentrations in Polish than in other European cities. It is confirmed by the data from 2005–2007, when in the Silesian Agglomeration the daily PM$_{10}$ concentrations near heavily trafficked roads

Table 1. Monthly statistics of daily PM_{10} concentrations at the traffic site (highway) in Katowice in 2010.

Parameter	January	February	March	April	May	June	July	August	September	October	November	December
Number of daily concentrations[*]	31	28	28	30	31	30	27	29	27	27	29	29
Valid data [%]	100.00	100.00	90.32	100.00	100.00	100.00	87.10	93.55	90.00	87.10	96.67	93.55
Number of concentrations greater than 50 µg/m³[**]	25	27	20	14	3	1	3	2	5	20	14	22
Minimum [µg/m³]	23.46	43.60	26.24	18.25	14.87	16.19	18.67	14.87	16.96	31.28	17.01	21.82
Day of minimum occurrence	09.01	28.02	01.03	05.04	17.05	27.06	24.07	28.08	29.09	05.10	13.11	10.12
Meteorological conditions:												
Air temperature [°C]	0.0	5.6	6.4	6.6	7.1	16.8	18.0	14.6	8.0	10.3	13.1	−3.8
Wind speed [m/s]	4.1	4.5	6.8	1.5	3.8	1.5	4.0	2.9	3.0	3.8	6.6	4.5
Wind direction	NE	S	SW	NW	W	NE	W	W	W	E	SW	W
Precipitation [mm]	10.3	2.5	0.0	9.9	68.5	0.0	4.8	9.6	7.4	2.2	0.5	0.8
Maximum [µg/m³]	318.72	168.33	148.73	86.42	55.82	50.15	52.14	58.82	63.21	141.36	114.22	219.03
Date of maximum occurrence	27.01	16.02	08.03	09.04	26.05	11.06	15.07	02.08	25.09	11.10	26.11	04.12
Meteorological conditions:												
Air temperature [°C]	−14.0	−3.6	−6.1	8.4	10.9	24.7	24.8	22.0	14.4	4.6	−1.0	−10.7
Wind speed [m/s]	2.7	1.0	2.6	2.0	1.3	2.7	1.7	1.1	1.3	0.9	1.1	1.3
Wind direction	S	S	E	W	SW	SW	S	S	E	E	SE	SW
Precipitation [mm]	0.0	0.2	0.0	0.0	0.1	0.0	0.0	0.0	0.0	0.0	0.0	0.0
Average [µg/m³]	110.43	106.49	73.25	51.54	35.26	28.23	34.98	31.81	36.58	65.34	48.84	105.18
Standard deviation [µg/m³]	69.91	39.29	29.34	14.33	10.32	9.02	10.65	11.17	13.47	26.32	23.09	58.87
Percentile 25 [µg/m³]	59.01	75.98	49.81	42.43	30.45	21.36	25.24	24.85	27.51	48.40	30.25	53.42
Median [µg/m³]	97.63	103.52	66.67	48.29	34.59	25.13	34.07	32.28	33.88	63.10	43.60	108.26
Percentile 75 [µg/m³]	131.35	137.69	90.52	58.91	42.66	33.44	43.61	37.10	46.61	71.93	66.72	135.32

[*] measuring time at least 75% of a day.

[**] 50 µg/m³—permissible daily PM_{10} concentration; it may be exceeded no more than 35 times in a year.

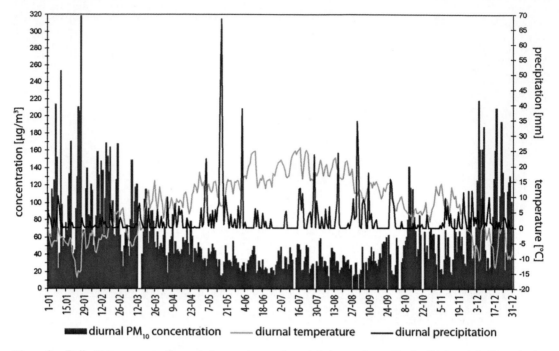

Figure 3. Daily PM$_{10}$ concentrations, air temperature and precipitaion at the traffic site (highway) in Katowice in 2010.

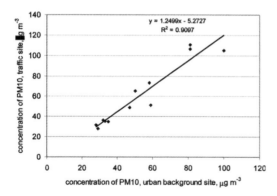

Figure 4. Linear correlation between monthly PM$_{10}$ concentrations at the urban background and the traffic sites in Katowice in 2010.

(30000–40000 vehicles per day) were by about 20% higher than at urban background sites. It was different at crossroads (35000 vehicles per day), where the concentrations were higher by 40%, and in rush hours by 55% (Rogula-Kozłowska et al. 2011). However, the effect of stationary sources may be higher at the urban background than at the traffic site and, consequently, the estimated traffic effect on the PM$_{10}$ concentrations at the traffic site may be underestimated.

But still does the municipal emission (stationary combustion) seem to be responsible for high level of air pollution in Katowice, and the 14% monthly PM$_{10}$ concentration increase is specific only of the areas directly affected by big traffic PM$_{10}$ sources. High PM$_{10}$ concentrations, especially in winter, indicate hazard to the inhabitants and the risk from exposure to PM$_{10}$ is greater in the neighborhood of traffic-related objects and to traffic participants.

3.2 Chemical composition of PM$_{10}$

At the sampling point in Katowice, like in other urbanized regions, mineral matter (mainly Al and Fe), organic carbon (hundreds of compounds), elemental carbon, sulfates, nitrates and ammonia account for almost all PM$_{10}$ mass (Sillanpää et al. 2006, Zhang et al. 2007, Yin and Harrison 2008, Perez et al. 2008, Carbone et al. 2010, Putaud et al. 2010, Kong et al. 2012). Enormously high, hardly to be found beyond seaside areas, ambient concentrations of Na and Cl ions occurred in Katowice (Table 2). Because of the highest Na and Cl concentrations occurred in January and the lowest in spring and summer, and their monthly concentrations were strongly linearly correlated (R^2 = 0.83), their great part may be assumed to be secondary pollutants coming from salt application to road deicing.

Table 2. Statistics of the monthly concentrations of selected PM_{10}-related substances at the traffic site in Katowice in 2010, ng/m³.

	Number of monthly concentrations	Percent of valid data	Minimum	Month of minimum occurrence	Maximum	Month of maximum occurrence	Average	Standard deviation	Median
Titanium (Ti)	12	100.00	13.32	Dec	62.28	Feb	30.35	14.76	26.02
Aluminum (Al)			241.34	May	1611.56	Jan	592.78	440.44	400.07
Iron (Fe)			561.27	Dec	1055.68	Oct	772.55	149.90	733.97
Sodium (Na⁺)	12	100.00	418.61	Aug	4066.50	Jan	1553.89	1266.94	966.72
Potassium (K⁺)			23.08	June	705.68	Jan	271.13	227.69	208.29
Calcium (Ca²⁺)			53.81	June	621.71	May	387.71	150.80	404.54
Magnesium (Mg²⁺)			41.54	June	565.46	March	171.02	165.53	117.89
Ammonium ion (NH₄⁺)			593.19	June	5664.89	Jan	2784.12	1897.44	2651.45
Sulfates (SO₄²⁻)			440.20	May	10450.19	Jan	5335.73	2721.17	5000.48
Nitrates(NO₃⁻)			224.23	May	7426.57	Feb	3538.37	2477.10	3652.05
Chlorides (Cl⁻)			121.09	May	8857.71	Jan	3285.98	3172.10	2027.81
Elemental carbon (EC), µg/m³	12	100.00	1.42	Oct	36.21	Jan	9.75	11.93	3.92
Organic carbon (OC), µg/m³			1.56	May	21.11	Jan	7.43	6.20	4.86

The monthly concentrations of ions of K, a component of anticacking agent mixed with salt, confirm it by assuming the maximum, 10 times greater than the minimum, in January (Table 2). In Poland, Magnesium Chloride ($MgCl_2$) and Calcium Chloride ($CaCl_2$) are also used as deicing agents, but the maxima of monthly ambient Mg and Cl concentrations occurred in spring (Table 2). The linear correlation between monthly ambient Mg and Cl concentrations confirm at least partial Mg and Cl origin from the same source ($R^2 = 0.58$).

Thus, their ambient concentrations are more related with soil emission than deicing. Instead, Al in PM_{10} was related with the road salting in winter. Al, together with silica (SiO_2), Ca, and Fe, is a basic component of sand. The monthly ambient Al concentrations were higher in January than in spring and summer. Moreover, the linear correlation coefficient R^2 was 0.6 for the Al and Na concentrations, and 0.55 for the Al and Cl concentrations. Therefore, sprinkling of the highway with salt, sand or mixture of both seems to have significantly affected the chemical composition of PM_{10}, yet more the Na and Cl than the mineral matter content of PM_{10}.

The mineral components of PM_{10} such as Al, SiO_2 (estimated from Al content of PM_{10}), Ti, Fe, K, Ca and Mg are in CM. In winter, some part of them may come from road deicing. Both the present study and Rogula-Kozłowska et al. (2012) suggest that fuel combustion (low-quality coal in obsolete domestic furnaces and stoves and wood) elevates concentrations of PM_{10}-related Na, Cl, K, Ca, and Mg (in winter, at some locations in Upper Silesia they occur mainly in $PM_{2.5}$, Rogula-Kozłowska et al. 2012). From the air pollution point of view, all these sources are hardly distinguishable in an area with such a dense arrangement of emission sources as Katowice, and the classification of the PM_{10} components may appear arbitrary and must be done cautiously in such an area.

The linear correlations between the monthly ambient concentrations of K, Ca, Mg and the concentrations of EC, Na or Cl in Katowice in 2010 are very weak (all R^2 less than 0.2) what precludes combustion from being the main source of CM. In winter, the monthly contribution of CM to the PM_{10} mass was about 9% in average. In spring and summer the monthly contributions of CM to the PM_{10} mass were between 12 and 19%, the greatest was in May. Thus, in Katowice, the high ambient concentrations of mineral components of PM_{10} in winter result rather from the high concentrations of PM itself, and the monthly CM shares in PM_{10} suggest the soil origin of CM.

The coefficients of the linear correlation between the monthly ambient concentrations of Cl and Na and EC were 0.59 and 0.55, respectively. MC, containing the Na and Cl ions, had the highest share in the PM_{10} mass in January–March (Fig. 5). Partly, MC came from combustion of fossil fuels, but in October–December (in heating season) the average share of MC in the PM_{10} mass did not exceed 6.5%. Therefore, at least some part of PM_{10} might have come from road sprinkling (snowfalls and low air temperature at the beginning of 2010, Fig. 3).

OC and EC were the key components of PM_{10} at the highway in Katowice in 2010. The maximum monthly concentrations of OC and EC were 21.11 and 36.21 µg/m³, respectively (Table 2). The minimum concentrations did not exceed 1.6 µg/m³. Among all the averages (over the measuring period) of the ambient PM_{10}-related EC concentrations at traffic points listed in Table 3, only in France, Taiwan, and Egypt they are greater than in Katowice. The average OC concentration in Katowice is in the middle of this list and resembles the concentrations in the European countries. The proportions of the monthly maxima and average OC concentrations to the maxima and average EC concentrations (OC/EC), respectively, are lower than 1. Analogously computed OC/EC for minima and medians are slightly greater than 1. It may mean higher shares of EC than of OC in PM_{10}-related total carbon (TC = OC+EC), particularly in the periods of high PM concentrations. In general, OC/EC at traffic sites listed in Table 3 are lower than at urban sites, let alone suburbs or rural sites (Castro et al. 1999, Decesari et al. 2001, Aymoz et al. 2004, Duan et al. 2005, Hueglin et al. 2005, Aymoz et al. 2007, Yttri et al. 2007, Schwarz et al. 2008). Proximity of a great traffic-related source (highway in Katowice) may explain this phenomenon. The car exhaust gas contains particles of elemental carbon, aromatic hydrocarbons from incomplete combustion of fuel, and fuel residuals. In the mountain tunnels Allegheny and Tuscarora, USA, about 80% of the PM mass was carbon compounds (Pierson & Brachaczek 1983). In road tunnels, carbon (elemental and organic) may have the greatest share in the PM mass, EC may be 50–70% of diesel exhaust gas and 30–40% of gasoline exhaust gas (Weingartner et al. 1997, Kirchstetter et al. 1999, Allen et al. 2001, Geller et al. 2005). So, the traffic may add to EC from other sources and

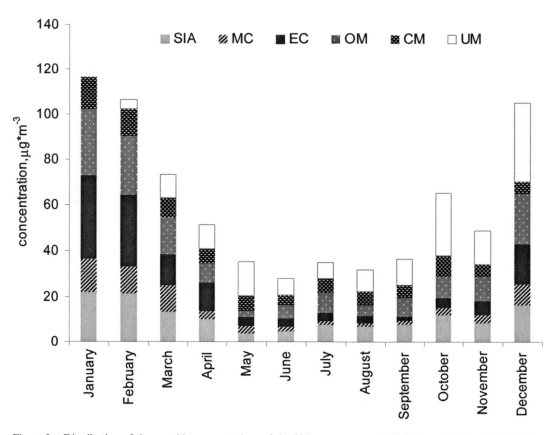

Figure 5. Distribution of the monthly concentrations of the PM_{10} components at the highway in Katowice in 2010.

Table 3. Average concentrations of PM_{10}-related EC and OC and average OC/EC at various traffic sites in the world.

Site	Investigation period	EC, $\mu g/m^3$	OC, $\mu g/m^3$	OC/EC	Method	Source
Paris, France	Summer-autumn 1997	13.6	34.6	2.5[A)	Thermal/ colo-rimetric	Ruellan et al. 2001
Brno, Swiss	01.04.1998– 31.03.1999	5.6	6.4	1.1[A)	Thermal/ colo-rimetric	Hueglin et al. 2005
Zurich, Swiss	07/08.1998, 01/02.1999	7.7	6.4	0.8[A)		
Basel, Swiss	1.04.1998– 31.03.1999	5.4	5.5	1.02[A)	Thermo-graphy/ IR	Röösli et al. 2001
Helsinki, Finland	07.2000 07.2001	1.3	4.2	3.2[A)	Thermal-optical/FID	Viidanoja et al. 2002
London, Birmingham, England	04.2000–01.2002	9.19	6.36	0.69[A)	Thermal/IR	Harrison et al. 2004
Chamois, Alp, France (1038 m a.s.l)	21.02.2001– 3.07.2003	1.89	8.97	4.97	Thermal-optical	Aymoz et al. 2007
London, England	01.2002– 06.2004	3.38	4.51	1.33[A)	Automatic carbon analyzer	Jones et al. 2005
Birmingham, England	16.05– 20.11.2005	3.8	4.6	1.21[A)	Thermal-optical/FID	Yin & Harrison 2008
Birmingham, England	Summer 2005 Autumn 2005	3.0 4.6	4.0 5.3	1.3[A) 1.1[A)		
Mediolan, Italy	2007	6,7	16	2,39[A)	Thermal-optical	Larsen et al. 2012
Taipei, Taiwan	10.1999 06.2000	12.9 17.3	16.7 19.4	1.46 1.14	Thermal/NDIR	Li & Lin 2002
Hong Kong, China	11.2000–02.2001 06–08.2001	7.0 6.3	12.1 11.1	1.7 1.8	Catalytic thermal (MnO_2) Thermal-optical	Ho et al. 2006
Hong Kong, China	01–02.2002	7.0	12.7	1.8	Thermal-optical	Cao et al. 2003
Hong Kong, China	06–07.2002	4.7	7.4	1.6	Thermal-optical	Cao et al. 2004
Cairo, Egypt	21.02.– 3.03. 1999 29.10.– 27.11. 1999	20.3	48.5	2.4[A)	Thermal-optical	Abu-Allaban et al. 2002
Dar Es Salaam, Tanzania	Rainy season 2005	4.3[B)	11.4[B)	–	Thermal-optical	Mkoma et al. 2009
Thessalonica, Greece	12.2006–02.2007 06–09.2007	2.93 2.91	8.73 6.40	2.98[A) 2.20[A)	Thermal-optical	Terzi et al. 2010
Dar Es Salaam, Tanzania	08–09.2005 (dry season)	4.7	17	3.64[A)	Thermal-optical	Mkoma et al. 2010
Katowice, Poland	01–12. 2010	9.7	7.4	0.76[A)	Thermal/NDIR	This study

[A) proportion of average concentrations.
[B) median.
[C) computed from TC = OC+EC, TC and OC/EC known.

affect the EC ambient concentrations and shares in TC in Katowice. But in Katowice, the highest EC share in the PM_{10} mass occurred in January and February (29–33%), in May–October it was not greater than 10%. The monthly shares of OM in PM_{10} (computed as 180% of OC) showed slightly different behavior (despite $R^2 = 0.87$ for monthly

OC and EC concentrations). Except for April, May, August, and October, in all months of 2010 the share of OM in the PM_{10} mass exceeded 20%, being the greatest in January, February, March, June, July, September. Clearly, the traffic intensity on the highway did not decide the behavior of the monthly concentrations and shares in the PM_{10}

mass of both OM and EC in Katowice. They were affected by traffic, but in general, determined by combustion of fuels for heating (i.e. indirectly by the air temperature), and the OM ambient concentrations and shares in the PM_{10} mass were affected by the factors such as emission of volatile organic compounds and insolation, which control the transformations of gaseous precursors of secondary organic aerosol.

The results, and therefore the reasoning, about EC and OC are likely to have a methodological flaw. Some amount of organic compounds carbonized during the OC determination is included into EC by the NDIR (Behr) software. It may be seen from the results presented in Table 3. Thermal analyzers (like NDIR) yield low, often lower than 1, OC/EC; thermal-optical analyzers (that include carbonization product into OC) almost always give OC/EC close to or even higher than 2. Nevertheless, the assessment of the TC share in PM_{10} and of the behavior of the monthly concentrations of PM_{10}-related TC are correct. No matter if the EC share in PM_{10} is overestimated and the OC share in PM_{10} is underestimated, the behavior of the monthly ambient concentrations of EC and OC at the traffic site in Katowice suggests local fossil fuel combustion, not traffic, as their principal source. The ambient PM_{10}-related EC concentrations in summer were between 1 and 2 μg/m³ and the traffic contribution to the EC concentrations cannot be greater in this area. Neither might the OC concentrations have been significantly affected by the emission from the highway.

Sulfate, nitrate and ammonium ions together (SIA) were the second contributor to the PM_{10} mass in Katowice in 2010. (Fig. 5). Their lowest monthly concentrations occurred in May–June, and the highest—in January and February.

Regardless of the monthly SIA ambient concentrations, almost all monthly shares of SIA in PM_{10} mass were close to 20% (only in May it was 10%; in July and August they were greater than 22%, Fig. 5).

SIA is the part of PM_{10} that arises from transformations of gaseous precursors such as SO_2 and NO_x. In the air, SO_2 is oxidized to gaseous SO_3 or liquid H_2SO_4, then neutralized to ammonium sulfate (($NH_4)_2SO_4$) or ammonium bisulfate (NH_4HSO_4). NO_x oxidize photochemically to HNO_3 then they are neutralized to ammonium nitrate (NH_4NO_3). In Katowice, the concentrations of SO_2 and NO_2 were strongly correlated with SIA ($R^2 = 0.85$ and 0.83 respectively).

The meteorological conditions in a hot season are not favorable for ammonium nitrate formation in the air because ammonium ions (NH_4^+) tend to neutralize sulfates first. When it is cold, ammonium ions are more probable to occur in the amount

sufficient to neutralize sulfuric and nitric acids, and then ammonium nitrate arises. It may be seen in Figures 6 and 7. The monthly ambient concentrations of SO_2, SO_4^{2-} and SIA are very close and probably the majority of SO_2 was first oxidized to sulfuric acid and then neutralized to ammonium sulfate or ammonium bisulfate. The concentrations of NO_2, NO_3^- and SIA were closest in April and May (in May closer) what may indicate ammonium nitrate formed in the air as the main source of NO_3^-. This period was very rainy (Fig. 3). Washing out of sulfuric and/or nitric acids from the atmosphere might enable the ammonium ions to neutralize the compounds arising from NO_x and SO_2 in the air.

The neutralization ratio (NR, $[NH_4^+]/([SO_4^{2-}]+[NO_3^-])$, where $[NH_4^+]$, $[SO_4^{2-}]$, $[NO_3^-]$ are in normal-equivalents/m³) was 7.9. Therefore, ammonium ions were in amount sufficient for neutralizing sulfuric and nitric acids totally and PM_{10} was alkaline in this period. NR were close to or exceeded 1 also in January, February, March, October, November, and December. In June, July, August and September, NR were about 0.4.

It confirms the effect of traffic on the sampling site. In winter, early spring and late autumn, when the emission from local furnaces and stoves

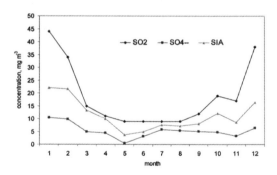

Figure 6. Monthly ambient SO_2, SO_4^{2-} and SIA concentrations in Katowice in 2010.

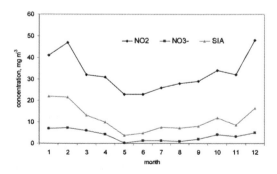

Figure 7. Monthly ambient NO_2, NO_3^- and SIA concentrations in Katowice in 2010.

overshadowed the traffic emission and exerted the strongest effect on the PM composition and ambient nitrate ion concentrations, the secondary inorganic part of PM at the sampling point almost entirely consisted of ammonium sulfate and ammonium nitrate, and PM_{10} was alkaline. In the non-heating season, when the highway emitted approximately the same amounts of SO_2 and NO_x as in the heating season, the anthropogenic ammonium emission ceased and PM_{10} was acidic. It was caused by nitric acid prevailing over nitrate ions in the air because sulfuric acid had been almost totally neutralized. The ratio SO_4^{2-}/NH_4^+ for $(NH_4)_2SO_4$ is 2.67, and in Katowice, in June–September, its monthly values for PM_{10} varied between 2 and 2.7. Therefore, in summer, in the vicinity of the highway in Katowice, the shortage of ammonium ions occurred and nitric acid could not be neutralized. However, during the whole year, some, or even total, nitric acid might be neutralized by cations, Na^+ or Ca^{2+}, through the formation of $NaNO_3$ and $Ca(NO_3)_2$, specially in summer.

4 CONCLUSIONS

Big traffic-related sources, such as highways, increase PM_{10} concentrations in their vicinity. In Katowice, in 2010, the PM_{10} concentrations at the highway were greater than at the urban background measuring point by about 8 µg/m³ (14%) in average. The relations between monthly concentrations of the PM_{10} components in this traffic-affected area suggest origin of at least several percent of the PM_{10} mass in winter from salt and sand sprinkled to the highway for deicing.

In the non-heating season, a period of lower than in winter ammonia emission, PM_{10} was acidic. The acidity of PM_{10} derived from the excess of nitric acid over ammonium ions in the air.

The ambient concentrations of the main PM_{10} component, carbon compounds, are probably determined by municipal emission (local combustion of fossil fuels). In general, the municipal emission is the principal source of PM and it is responsible for poor air quality in Katowice, especially in winter.

The high PM_{10} concentrations, especially in winter, cause health hazard to the Katowice inhabitants, and the hazard is greater in the neighborhood of traffic-related objects, like highways, and to road traffic participants.

ACKNOWLEDGEMENTS

All the above considerations are based on the results of the investigations carried out within the frame of the project "Analysis of the air pollution by PM_{10} and $PM_{2.5}$; chemical composition of PM and effects of natural sources" realized jointly by the Institute of Environmental Engineering of the Polish Academy of Sciences, Institute for Ecology of Industrial Areas, Institute of Meteorology and Water Management—National Research Institute, Institute of Environmental Protection—National Research Institute at the request of the Chief Inspectorate of Environmental Protection and financed by the National Fund for Environmental Protection and Water Management.

The work was partially supported by the grant No. N N523 564038 from the Polish Ministry of Science and Higher Education.

REFERENCES

Abu-Allaban, M., Gertler, A.W. & Lowenthal, D.H. 2002, A preliminary apportionment of the sources of ambient PM10, PM2.5, and VOCs in Cairo. *Atmospheric Environment* 36: 5549–5557.

Allen, J.O., Mayo, P.R., Hughes, L.S., Salmon, L.G & Cass, G.R. 2001. Emission of size-segregated aerosols from on-road vehicles in the Caldecott Tunnel. *Environmental Science and Technology* 35: 4189–4197.

Aymoz, G., Jaffrezo, J.L., Jacob, V., Colomb, A. & George C. 2004. Evolution of organic and inorganic components of aerosol during Saharan dust episode observed in the French Alps. *Atmospheric Chemistry and Physics* 4: 2499–2512.

Aymoz, G., Jaffrezo, J.L., Chapuis, D., Cozic, J. & Maenhaut W. 2007. Seasonal variation of PM10 main constituents in two valleys of the French Alps. I: EC/OC fractions. *Atmospheric Chemistry and Physics* 7: 661–675.

Cao, J.J., Lee, S.C., Ho, K.F., Zhang, X.Y., Zou, S.C., Fung, K., Chow, J.C. & Watson, J.G. 2003. Characteristics of carbonaceous aerosol in Pearl Delta Region, China during 2001 winter period. *Atmospheric Environment* 37: 1451–1460.

Cao, J.J., Lee, S.C., Ho, K.F., Zou, S.C., Fung, K., Li, Y., Watson, J.G. & Chow, J.C. 2004. Spatial and seasonal variations of atmospheric organic carbon and elemental carbon in Pearl Delta Region, China. *Atmospheric Environment* 38: 4447–4456.

Carbone, C., Decesari, S., Mircea, M., Giulianelli, L., Finessi, E., Rinaldi, M., Fuzzi, S., Marinoni, A., Duchi, R., Perrino, C., Sargolini, T., Vardè, M., Sprovieri, F., Gobbi, G.P., Angelini, F. & Facchini, M.C. 2010. Size-resolved aerosol chemical composition over the Italian Peninsula during typical summer and winter conditions. *Atmospheric Environment* 44: 5269–5278.

Castro, L.M., Pio C.A., Harrison R.M. & Smith D.J.T.; 1999, Carbonaceous aerosol in urban and rural European atmospheres: estimation of secondary organic carbon concentrations. *Atmospheric Environment* 33: 2771–2781.

Chow, J.C., Watson, J.G., Lowenthal, D.H. & Countess R.J. 1996. Sources and chemistry of PM-10 aerosol in Santa Barbara county, CA. *Atmospheric Environment* 30: 1489–1499.

de Kok, T.M.C.M., Hogervorst, J.G.F., Briedé, J.J., van Herwijnen, M.H., Maas, L.M., Moonen, E.J., Driece, H.A.L. & Kleinjans, J.C. 2005. Genotoxicity and physicochemical characteristics of traffic-related ambient particulate matter. *Environmental and Molecular Mutagenesis* 46: 71–80.

de Kok, T.M.C.M., Driece, H.A.L., Hogervorst, J.G.F. & Briedé, J.J. 2006. Toxicological assessment of ambient and traffic-related particulate matter. A Review of Recent Studies. *Mutation Research* 613: 103–122.

Decesari, S., Facchini, M.C., Matta, E., Lettini, F., Mircea, M., Fuzzi, S., Tagliavini, E. & Putaud J.P. 2001. Chemical features and seasonal variation of fine aerosol water-soluble organic compounds in the Po Valley, Italy. *Atmospheric Environment* 35: 3691–3699.

Dockery, D.W., Cunningham, J., Damokosh, A.I., Neas, L.M., Spengler, J.D., Koutrakis, P., Ware, J.H., Raizenne, M. & Speizer, F.E. 1996. Health effects of acid aerosols on North American children: respiratory symptoms. *Environmental Health Perspective* 104: 500–505.

Duan, F., He, K., Ma, Y., Jia, Y., Yang, F., Lei, Y., Tanaka, S. & Okuta, T. 2005. Characteristics of carbonaceous aerosols in Beijing, China. *Chemosphere* 60: 355–364.

Englert, N. 2004. Fine particles and human health—a review of epidemiological studies. *Toxicology Letters* 149: 235–242.

Geller, M.D., Sardar, S.B., Phuleria, H., Fine, P.M. & Sioutas, C. 2005. Measurements of particle number and mass concentrations and size distributions in tunnel environment. *Environmental Science and Technology* 39: 8653–8663.

Grosjean, D. & Friedlander, S.K. 1975. Gas-particle distribution factors for organic and other pollutants in the Los Angeles atmosphere. *Journal of the Air Pollution Control Association* 25: 1038–1044.

Harrison, R.M., Jones, A.M. & Barrowcliffe, R. 2004. Field study of the influence of meteorological factors and traffic volumes upon suspended particle mass at urban roadside sites of differing geometries. *Atmospheric Environment* 38: 6361–6369.

Harrison, R.M., Jones, A.M. & Lawrence R.G. 2004. Major component of PM10 and PM2.5 from roadside and urban background sites. *Atmospheric Environment* 38: 4531–4538.

Ho, K.F., Lee, S.C., Cao, J.J., Li, Y.S., Chow, J.C., Watson, J.G. & Fung, K. 2006. Variability of organic and elemental carbon, water soluble organic carbon, and isotopes in Hong Kong. *Atmospheric Chemistry and Physics* 6: 4569–4576.

http://ec.europa.eu/environment/archives/air/cafe/activities/.

http://reports.eea.europa.eu/technical_report_2008_7/en.

http://stacje.katowice.pios.gov.pl/monitoring.

Hueglin, C., Gehrig, R., Baltensperger, U., Gysel, M., Monn, C. & Vonmont, H. 2005. Chemical characterization of PM2,5, PM10 and coarse particles at urban, near-city and rural sites in Switzerland. *Atmospheric Environment* 39: 637–651.

Janssen, N.A.H., Van Mansom, D.F.M., Van Der Jagt, K., Harssema, H. & Hoek, G. 1997. Mass concentration and elemental composition of airborne particulate matter at street and background locations. *Atmospheric Environment* 31: 1185–1193.

Jones, A.M. & Harrison, R.M. 2005. Interpretation of particulate elemental and organic carbon concentrations at rural, urban and kerbside sites. *Atmospheric Environment* 39: 7114–7126.

Juda-Rezler, K., Reizer, M. & Oudinet, J.P. 2011. Determination and analysis of PM10 source apportionment during episodes of air pollution in Central Eastern European urban areas: The case of wintertime 2006. *Atmospheric Environment* 45: 6557–6566.

Kirchstetter, T.W., Harley, R.A., Kreisberg, N.M., Stolzenburg, M.R. & Herring, S.V. 1999. On-road measurement of fine particles and nitrogen oxide emission from light- and heavy-duty motor vehicles. *Atmospheric Environment* 33: 2955–2968.

Klejnowski, K., Rogula-Kozłowska, W. & Krasa, A. 2009. Structure of atmospheric aerosol in Upper Silesia (Poland)—contribution of PM2.5 to PM10 in Zabrze, Katowice and Częstochowa in 2005–2007. *Archives of Environmental Protection* 35: 3–13.

Kong, S., Ji, Y., Lu, B., Bai, Z., Chen, L., Han, B. & Li, Z. 2012. Chemical compositions and sources of atmospheric PM10 in heating, non-heating and sand periods at a coal-based city in northeastern China. *Journal of Environmental Monitoring* 14: 852–865.

Larsen, B.R., Gilardoni, S., Stenström, K., Niedzialek, J., Jimenez, J. & Belis, C.A. 2012. Sources for PM air pollution in the Po Plain, Italy: II. Probabilistic uncertainty characterization and sensitivity analysis of secondary and primary sources. *Atmospheric Environment* 50: 203–213.

Li, C.-S. & Lin, C.-H. 2002. PM1/PM2.5/PM10 characteristics in the urban atmosphere of Taipei. *Aerosol Science and Technology* 36: 469–473.

López-Villarrubia, E., Iñiguez, C., Peral, N., García, M.D. & Ballester, F. 2012. Characterizing mortality effects of particulate matter size fractions in the two capital cities of the Canary Islands. *Environmental Research* 112: 129–138.

Lu, H.C. 2002. The statistical characters of PM10 concentration in Taiwan area. *Atmospheric Environment* 36: 491–502.

Massolo, L., Muller, A., Tueros, M., Rehwagen, M., Franck, U., Ronco, A. & Herbath, O. 2002. Assessment of mutagenicity and toxicity of different-size fractions of air particulates from La Plata, Argentina, and Lepzig, Germany. *Environmental Toxicology* 17: 219–231.

Mkoma, S.L., Maenhaut, W., Chi, X., Wang, W. & Raes, N. 2009. Characterization of PM10 atmospheric aerosols for the wet season 2005 at two sites in East Africa. *Atmospheric Environment* 43: 631–639.

Mkoma, S.L., Chi, X. & Maenhaut, W. 2010. Characteristics of carbonaceous aerosols in ambient PM10 and PM2.5 particles in Dar es Salaam, Tanzania. *Science of the Total Environment* 408: 1308–1314.

Morawska, L., Thomas, S., Jamriska, M. & Johnson, G. 1999. The modality of particle size distribution of environmental aerosols. *Atmospheric Environment* 33: 4401–4411.

Morawska, L., Thomas, S., Gilbert, D., Greenaway, Ch. & Rijnders, E. 1999. A study of the horizontal and vertical profile of submicrometer particles in relation to a busy road. *Atmospheric Environment* 33: 1261–1274.

Morcelli, C.P., Figueiredo, A.M., Sarkis, J.E., Enzweiler, J., Kakazu, M. & Sigolo, J.B. 2005. PGEs and other traffic-related elements In roadside soils from Sao Paulo, Brazil. *Science of the Total Environment* 345: 81–91.

Ośródka, L., Klejnowski, K., Wojtylak, M. & Krajny, E. 2006. Smog episodes analysis in winter season in Upper Silesia Region. Konieczyński J. (eds), Air protection in theory and in practice: 197–207. Zabrze, Poland: Polish Academy of Sciences (in Polish).

Paoletti, L., De Berardis, B., Arrizza, L., Passacantando, M., Inglessis, M. & Mosca, M. 2003. Seasonal effects on the physico-chemical characteristics of PM2,1 in Rome: A study by SEM and XPS. *Atmospheric Environment* 37: 4869–4879.

Pastuszka, J.S., Wawroś, A., Talik, E. & Paw, U. 2003. Optical and chemical characteristics of the atmospheric aerosol in four towns in southern Poland. *The Science of the Total Environment* 309: 237–251.

Pastuszka, J.S., Rogula-Kozłowska, W. & Zajusz-Zubek, E. 2010. Characterization of PM10 and PM2,5 and associated heavy metals at the crossroads and urban background site in Zabrze, Upper Silesia, Poland, during the smog episodes. *Environmental Monitoring and Assessment* 168: 613–627.

Perez, N., Pey, J., Querol, X., Alastuey, A., Lopez, J.M. & Viana, M. 2008. Partitioning of major and trace components in PM10-PM2,5-PM1 at an urban site in Southern Europe. *Atmospheric Environment* 42: 1677–1691.

Pierson, W.R. & Brachaczek, W.W. 1983. Particulate matter associated with vehicles on the road. II. *Aerosol Science and Technology* 2: 1–40.

Putaud, J.P., Van Dingenen, R., Alastuey, A., Bauer, H., Birmili, W., Cyrys, J., Flentje, H., Fuzzi, S., Gehrig, R., Hansson, H.C., Harrison, R.M., Herrmann, H., Hitzenberger, R., Hüglin, C., Jones, A.M., Kasper-Giebl, A., Kiss, G., Kousa, A., Kuhlbush, T.A.J., Löschau, G., Maenhaut, W., Molnar, A., Moreno, T., Pekkanen, J., Perrino, C., Pitz, M., Puxbaum, H., Querol, X., Rodriguez, S., Salma, I., Schwarz, J., Smolik, J., Schneider, J., Spindler, G., ten Brink, H., Tursic, J., Viana, M., Wiedensohler, A. & Raes, F. 2010. A European aerosol phenomenology—3: Physical and chemical characteristics of particulate matter from 60 rural, urban, and kerbside sites across Europe. *Atmospheric Environment* 44: 1308–1320.

Querol, X., Alastuey, A., Rodriguez, S., Plana, F., Ruiz, C.R., Cots, N., Massague, G. & Puig, O. 2001. PM10 and PM2.5 source apportionment in the Barcelona Metropolitan area, Catalonia, Spain. *Atmospheric Environment* 35: 6407–6419.

Ragosta, M., Caggiano, R., Macchiato, M., Sabia, S. & Trippetta, S. 2008. Trace elements in daily collected aerosol: Level characterization and source identification in a four-year study. *Atmospheric Research* 89: 206–217.

Ravindra, K., Stranger, M. & Van Grieken, R. 2008. Chemical characterization and multivariate analysis of atmospheric PM2.5 particles. *Journal of Atmospheric Chemistry* 59: 199–218.

Rogula-Kozłowska, W., Pastuszka, J.S. & Talik, E. 2008. Influence of vehicular traffic on concentration and particle surface composition of PM10 and PM2.5 in Zabrze, Poland. *Polish Journal of Environmental Studies* 17: 539–548.

Rogula-Kozłowska, W. 2009. Ph.D. thesis. Silesian University of Technology. Gliwice, Faculty of Energy and Environmental Engineering (in Polish).

Rogula-Kozłowska, W., Pastuszka, J.S. & Talik, E. 2011. Properties of atmospheric aerosol from road traffic, Works and Studies No.80. Zabrze, Poland: Polish Academy of Sciences (in Polish).

Rogula-Kozłowska, W., Klejnowski, K., Rogula-Kopiec, P., Mathews, B. & Szopa, S. 2012. A study on the seasonal mass closure of ambient fine and coarse dusts in Zabrze, Poland. *Bulletin of Environmental Contamination and Toxicology* 88: 722–729.

Röösli, M., Theis, G., Künzli, N., Staehelin, J., Mathys, P., Oglesby, L., Camenzind, M. & Braun-Fahrländer, C. 2001. Temporal and spatial variation of the chemical compositions PM10 at urban and rural sites in the Basel area, Switzerland. *Atmospheric Environment* 35: 3701–3713.

Ruellan, S. & Cachier, H. 2001. Characterization of fresh particulate vehicular exhausts near a Paris high flow road. *Atmospheric Environment* 35: 453–468.

Schwarz, J., Chi, X., Maenhaut, W., Civis, M., Hovorka, J. & Smolik, J. 2008. Elemental and organic carbon in atmospheric aerosols at downtown and suburban sites in Prague,. *Atmospheric Research* 90: 287–302.

Sillanpää, M., Hillamo, R., Saarikoski, S., Frey, A., Pennanen, A., Makkonen, U., Spolnik, Z., Van Grieken, R., Braniš, M., Brunekreef, B., Chalbot, MC., Kuhlbusch, T., Sunyer, J., Kerminen, V.M., Kulmala, M. & Salonen, R.O. 2006. Chemical composition and mass closure of particulate matter at six urban sites in Europe. *Atmospheric Environment* 40: 212–223.

Slezakowa, K., Pereira, M.C., Reis, M.A. & Alvim-Ferraz, M.C. 2007. Influence of traffic on the composition of atmospheric particles of different sizes—Part 1: Concentrations and elemental characterization. *Journal of Atmospheric Chemistry* 58: 55–68.

Spurny, K.R. 1996. Chemical mixtures in atmospheric aerosols and their correlation to lung diseases and lung cancer occurrence in the general population. *Toxicology Letters* 88: 271–277.

Terzi, E., Argyropoulos, G., Bougatioti, A., Mihalopoulos, N., Nikolaou, K. & Samara, C. 2010. Chemical composition and mass closure of ambient PM10 at urban sites. *Atmospheric Environment* 44: 2231–2239.

Turpin, B.J. & Lim, H.-J. 2001. Species contributions to PM2.5 mass concentrations: Revisiting common assumptions for estimating organic mass. *Aerosol Science and Technology* 35: 602–610.

Vallius, M. 2005. Characteristics and sources of fine particulate matter in urban air. Kuopio, Finland: National Public Health Institute.

Viidanoja, J., Sillanpää, M., Laakia, J., Kerminen, V.-M., Hillamo, R., Aarnio, P. & Koskentalo, T. 2002. Organic and black carbon in PM2.5 and PM10: 1 year of data from an urban site in Helsinki, Finland. *Atmospheric Environment* 36: 3183–3193.

Weingartner, E., Keller, C., Stahel, W.A., Burtscher, H. & Batenspelger, U. 1997. Aerosol emission in a Road Tunnel. *Atmospheric Environment* 31: 451–462.

Wichmann, H.-E. 2004 Health effects of particles in ambient air. *International Journal of Hygiene and Environmental Health* 207: 399–407.

Yin, J. & Harrison, R.M. 2008. Pragmatic mass closure study for PM1.0, PM2.5 and PM10 at roadside, urban background and rural sites. *Atmospheric Environment* 42: 980–988.

Yttri, K.E., Aas, W., Bjerke, A., Cape, J.N., Cavalli, F., Ceburnis, D., Dye, C., Emblico, L., Facchini, M.C., Forster, C., Hanssen, J.E., Hansson, H.C., Jennings, S.G., Maenhaut, W., Putaud, J.P. & Torseth, K. 2007. Elemental and organic carbon in PM10: a one year measurement campaign within the European Monitoring and Evaluation Programme (EMEP). *Atmospheric Chemistry and Physics* 7: 5711–5725.

Zhang, W., Lei, T., Lin, Z.Q., Zhang, H.S., Yang, D.F., Xi, Z.G., Chen, J.H. & Wang, W. 2011. Pulmonary toxicity study in rats with PM10 and PM2.5: differential responses related to scale and composition. *Atmospheric Environment* 45: 1034–1041.

Zhang, X., Zhuang, G., Guo, J., Yin, K. & Zhang, P. 2007. Characterization of aerosol over the Northern South China Sea during two cruises in 2003. *Atmospheric Environment* 41: 7821–7836.

Zhu, Y., Hinds, W.C., Seongheon, K., Shen, S. & Sioutas, C. 2002. Study of ultrafine particles near a major highway with heavy-duty diesel traffic. *Atmospheric Environment* 36: 4323–4335.

Environmental Engineering IV – Pawłowski, Dudzińska & Pawłowski (eds)
© 2013 Taylor & Francis Group, London, ISBN 978-0-415-64338-2

Odour impact range assessment of a chosen distillery on the basis of field measurements in the plume

I. Sówka, M. Skrętowicz & J. Zwoździak
Institute of Environmental Protection Engineering, Wroclaw University of Technology, Wroclaw, Poland

ABSTRACT: The results of field measurements in the plume made in July 2011 in selected areas around of the distillery are presented. The research intensity of the odors were conducted based on guidelines developed by the Association of German Engineers VDI 3940. The measurements results analysis show that the smells from the plant under study reach out to the surrounding neighborhoods inhabited by people (up to 2000 m from the plant), It should be noted also that the odors present in the area inhabited by people generally did not occur during the 10-minute measurement period at all times of measurements.

Keywords: odour, intensity, field inspections

1 INTRODUCTION

Odour nuisance derived from industrial sources may cause a significant reduction of a life comfort of people living in the vicinity of the odour emission source. This problem refers especially to residents of large, industrialized cities where residential areas may be affected by several industrial plants that are odour emitters. Prolonged exposure to odours, not only reduces the quality of life but can also cause headaches, nausea and vomiting in extreme cases as well as anxiety or depression (Zarra et al. 2010) (Sówka et al. 2011). The main sources of odours are: agriculture, municipal facilities and industrial plants (Wasąg et al. 2009) (Kulig et al. 2010) (Szklarczyk et al. 2010). The group of industrial facilities that may emit odours include i.e. chemical and petrochemical plants, textile and cellulose plants, paint and foundry plants as well as food processing plants, including distilleries. The impact of industrial plants in terms of odour emission can be assessed by measurements 'at the source' (emission) or in the form of field research (odour concentration at reception point). In the first case, odour concentration measurement is carried out, then with the knowledge of the gas volume flow, odour emission is determined. To determine the impact of odours on a given area, computational tools, such as gaussian plume model (using Pasquills' formula)—a reference model in Poland, should be used. These calculations however, may be subjected to large calculation error, resulting from various factors i.e. insertion of a long-term wind rose to a model. Thus, measurements at the source can be replaced or supplemented by field research (Ribeiro et al, 2010)

In accordance with German guidelines VDI3940, field measurements can be conducted through research in the grid and in the plume. Research in the grid is based on measurements in constant, predetermined computational grid, which allows to assess the air quality at the chosen point or on the given area (measuring square). Tests should be performed periodically for some time and as a result, information about, so-called 'odour hours', are obtained. Through research in the plume, particularly following the wind vector, a range of odour plume emitted by chosen industrial facility can be determined. At the same time, other odours that have an impact on examined area and contribute to odour nuisance can be also identified.

2 MATERIAL AND METHODS

The examined area is a typical industrial-service center on the outskirts of a big city with a population exceeding 500000. Except for few industrial plants (i.e. examined distillery or food processing plants) there are also shopping centers, snack bars and residential estates on the tested area; there are also the freeway and communication hub which run through it. Residential buildings are mostly low-detached, occasionally blocks of flats occur (not exceeding height of three storeys). The studied area is of approximately 3 km^2 and analyzed industrial plant is located on its north-western part. The measurement area was selected on the basis of a wind rose (vast predominance of north-western winds) and current meteorological situation at the time of measurement.

In order to identify main types of odours derived from selected distillery and to determine their sources before measurements, a visit in the Plant and discussions with its representatives were carried out. The following odour emission sources were identified: gluten dryer, fodder dryer, fermenters and sewage sludge (odour derived from the adjacent sewage treatment facility).

Field inspection of odour nuisance was performed on the 7th–9th of July 2011, in accordance with the methodology included in German guidelines VDI3940. Due to the objective of measurements, which was to determine the range of the odour impact, measurements in the plume were carried out. In this type of measurements the decision on the area of research is taken ad-hoc depending on wind direction after the arrival of the team in place of measurements. During the study, after the arrival of the team on the examined area, 5 measurements were carried out. Places of measurements are presented in Figure 1.

The measurement was attended by seven properly qualified people—field inspectors and 2 operators. Field inspectors were selected according to the methodology included in PN-EN 13725. Additionally, they were trained before the measurement and acquainted to different types of smells coming from the Plant. Operators' task was to indicate the area of research and measurement points and provide the coordinates and meteorological data at each point of measurement.

Wind direction and speed and air temperature and humidity constituted the recorded meteorological. The evaluator left at the designated point, set in the direction of the wind, measured the time with an electronic stopwatch every 10 seconds for 10 minutes, recorded the intensity of sensed odour and its type on a specially prepared protocol (in 7-scale, according to VDI3882), in accordance with the indications given in Tables 1 and 2. Additionally, recorded meteorological parameters mentioned before were wrote down on a protocol. Off-site distance between measurements points was 100 m, on-site 50 m.

Odours qualified as deriving from the Plant:

'L'—smell of baked bread—smell deriving from gluten drying process.
'M'—gluten (pasta)—smell deriving from fodder drying process.
'N'—alcohol (yeast)—smell deriving from fermentation process.
'O'—sewage sludge—odour deriving from adjacent sewage treatment facility.

In the first series of measurements on the area of the Plant, additional type of odour was identified and marked as 'R'—the odour from the refinery (Refinery Plant cooling towers). The odour

Table 1. Odour intensity scale (by VDI3940).

Odour intensity	Description
0	No odour
1	Very weak odour
2	Weak odour
3	Distinct odour
4	Strong odour
5	Very strong odour
6	Extremely strong odour

Table 2. The character of sensed odour with odours identified from analyzed distillery (by VDI3940).

Indication of odour	Character of odour
A	No odour
B	Agricultural odour (difficult to define)
C	Agricultural odour: Manure
D	Agricultural odour: Pig breed
E	Agricultural odour: Cattle breed
F	Agricultural odour: Horse breed
G	Agricultural odour: Hay, grain
H	Car exhaust odour
I	Smell of burned coal from domestic furnaces
J	Grass smell
K	Other: specify
L	Smell of baked bread
M	Gluten (pasta)
N	Alcohol (yeast)
O	Sewage sludge

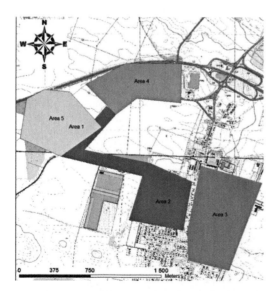

Figure 1. Field research area.

resembled sewage sludge odour. During the day when measurements were carried out, mentioned odour was intense, therefore it was justified to include it in the group of recognized odours related with the activities of the Plant. In five measurement series, the following number of measurements were obtained series 1: 46 measurements, series 2: 40 measurements, series 3: 48 measurements, series 4: 22 measurements, series 5: 36 measurements.

3 RESULTS AND DISCUSSION

On the basis of completed protocols, databases were established. Data were statistically analyzed, i.e. maximum intensity at the selected measuring point and the percentage of occurrence of the odor in the air at the time of measurements were calculated. Properly prepared databases were introduced into the programme ArcGIS v. 10.0 by ESRI. With geographical coordinates taken from the GPS device, it was possible to visualise all data points. Figure 2 shows the results of measurements performed on the sample area (area 2).

Measurements taken on the area of the Plant (area 1 on Fig. 1) enabled the evaluators to acquaint even better with selected odours related with the activities of the Plant and also allowed to recognize formerly unidentified odour marked as 'R'. Due to the fact that on the area of the Plant all odours were sensed with different intensity (depending on the distance from the measuring point to selected emission sources), measurement on the area 2 was conducted in the south-eastern direction from the Plant at a distance up to 1500 m, among single-family housing (Fig. 2). North-western winds dominated during the measurements.

Odours derived from the distillery were sensed at each measurement point with maximum intensity marked as 4 and frequency of odour occurrence in the measurement above 15% up to the distance of 800 m. Farther, odours were sensed occasionally, depending on temporary wind direction, frequency of odours occurrence was generally below 10%. Among odours emitted by the distillery, odour derived from the gluten drying process, marked as 'L' and 'M' (odours derived from drying processes) occurred most often, rarely odour 'N' occured (odour derived from fermentation process). Odour derived from adjacent sewage treatment facility was detectable only at the beginning of the measurement, up to the distance of 100 m from the Plant, while odours derived from refining process did not occur outside the distillery. Figure 4 shows measuring points at which occurrence of selected odours was recorded. Measurement on the area 3 was carried out at a distance up to 2500 m from the area of a distillery. Initially, southern and south-western winds occurred. During the measurement, although the wind direction changed to western and north-western odours from the Plant were later perceptible on a tested area at a distance up to 2000 m. Except for the 'baked bread' smell, occasionally 'gluten, pasta' odour occurred as well as 'alcohol, yeast'. Odours deriving from adjacent sewage treatment facility and refining processes ('R') did not occur.Measurements on the area 4 were carried out contrary to the wind direction. Area 4 is located north-east of the Plant and the wind blew from north-western direction. Despite the proximity to the distillery (the farthest point was located at a distance of 1000 m), with such wind direction any odours from the Plant were not identified, except for one measurement

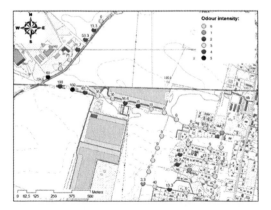

Figure 2. Maximum values of intensity of all odours derived from the Plant recorded during measurement on the area 2 and their incidence rates at selected measuring points

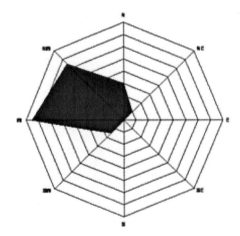

Figure 3. Wind rose based on observations of instanta neous wind directions at measuring points on the area 2.

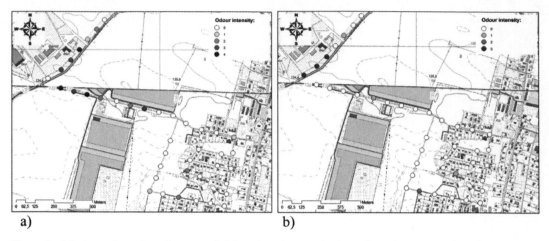

a)   b)

Figure 4.   Maximum intensity values recorded for selected types of odours from the Plant: a) from fodder drying process; b) from fermentation process.

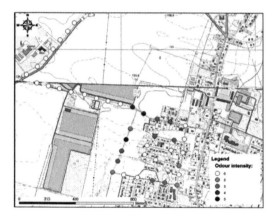

Figure 5.   Maximum intensity values of an odour deriving from different from analyzed industrial plant ('burnt popcorn' odour) identified during measurement performed on the area 2.

point located closest to the distillery were odour 'N' occurred, but only by 13% of the measurement time. That could be induced by local turbulence of air masses or instantaneous change of wind direction. Area 5 was the region around the Plant beyond its borders. Southern and south-eastern winds were dominant during this measurement. All of the selected odours related to the activities of the Plant were indentified. These odours were perceptible to the north, north-east and north-west from the Plant, mainly along the highway that runs close. During measurements on some areas (e.g. areas 2, 5), odours not related to the activities of analyzed distillery were identified but also those derived from industrial plants. Exemplary

are odours described as 'burnt popcorn' or 'sweet caramel'. Figure 5 shows an area on which one of the odours mentioned above was identified and maximum intensity values with which it was sensed at particular measuring points.

Various industrial plants may have an impact, in terms of odour emission, on the examined areas. The impact of different sources of odours can cause an intensification of discontent of people living on nearby residential areas and increase their aversion to plants emitting odours (Zarra 2010).

4   CONCLUSIONS

Field research conducted in the plume, in accordance with VDI3940, allowed to determine odour impact range of analyzed distillery. Performing field research, an impact of a chosen plant was assessed by taking into account the intensity of an odour as well as types of selected odours coming from odour emission sources on the given area.

Beyond the area of the Plant, odours reached surrounding residential areas (within up to 2000 m from the Plant). However, in any of the measurements taken on these areas, maximum intensity value (value 6) did not occur, the highest values recorded during measurements did not exceed the value of 4. Among odours derived from the Plant and occurring beyond its area, odours 'L' described as 'baked bread' predominated in the range up to about 2000 m, to a lesser extent 'M' described as 'gluten, pasta' in the range up to about 1800 m and 'N' described as 'alcohol, yeast' in the range of 1800 m.

In any measurement taken on residential areas, no odours deriving from adjacent sewage treatment facility as well as refining process were sensed.

It means that main cause of odour nuisance related to the activities of the Plant are the processes of gluten and fodder drying and alcohol fermentation. Additionally, on the tested area (mainly areas 2 and 5), odours derived from other industrial processes than examined are perceptible. Measurements were performed mainly in south-eastern direction from the Plant, it was related primarily to the wind rose for the tested area. A freeway that runs north of the distillery can be a barrier to the spread of odours from this direction. Subsequent tests need to be carried out on residential areas located north from the Plant, taking into account winds blowing from southern and south-western direction.

ACKNOWLEDGEMENTS

Work carried out under the project 'Enterprising graduate student—an investment in the innovative development of the region'. The project was performed with the financial support of European funds—Operational Programme "Human Capital, Priority VIII Regional human resources, the action 8.2. Knowledge Transfer, Sub-8.2.2. Regional Innovation Strategies".

UNIA EUROPEJSKA
EUROPEJSKI
FUNDUSZ SPOŁECZNY

KAPITAŁ LUDZKI
NARODOWA STRATEGIA SPÓJNOŚCI

REFERENCES

Kulig A., Lelicińska-Serafin K., Podedworna J., Sinicyn G., Heidrich Z., Czyżkowski B. 2010. Identyfikacja, inwentaryzacja i charakterystyka źródeł odorantów w gospodarce komunalnej w Polsce. In Wydawnictwo Naukowo-Techniczne (ed.), *Współczesna problematyka odoró:* 14–25. Warszawa.

PN-EN 13725. 2007. Jakość powietrza—określanie stężenia substancji zapachowych metodą olfaktometrii dynamicznej. PKN (ed.).Warszawa.

Ribeiro C., Varela H., Coutinho M., Borrego C. 2010. Dispersion modeling and field inspections approach to evaluate the odor impact of a composting plant in Lisbon. In *13th Conference on Harmonisation within Atmospheric Dispersion Modelling for Regulatory Purpose, Proc. intern. symp., Paris, 1–4 June. 2010.* 882–886.

Sówka I., Skrętowicz M., Szklarczyk M., Zwoździak J. 2011. Evaluation of nuisance of odour from food industry. *Environment Protection Engineering* 37. 5–12.

Szklarczyk M., Zwoździak J., Sówka I. 2010. Przemysłowe źródła emisji zapachów. In Wydawnictwo Naukowo-Techniczne (ed.), *Współczesna problematyka odorów*: 54–84. Warszawa.

VDI 3940 Part 2. 2006. *Determination of odorants in ambient air by field inspections.* VDI Guidelines Department (ed.). Düsseldorf.

VDI 3882 Part 2. 1994. *Olfactometry—Determination of hedonic odour tone.* VDI Guidelines Department (ed.). Düsseldorf.

Wasąg H., Czerwiński J., Guz Ł. 2009. Ograniczenie uciążliwości zapachowej zakładów utylizacji odpadów zwierzęcych za pomocą filtrów z jonitami włóknistymi. *Chemik* 11. 435–439.

Zarra T., Naddeo V., Giuliani S., Belgiorno V. 2010 Optimalization of field inspection method for odour impact assessment, *Chemical Engineering Transactions* 23. 93–98.

Environmental Engineering IV – Pawłowski, Dudzińska & Pawłowski (eds)
© *2013 Taylor & Francis Group, London, ISBN 978-0-415-64338-2*

Opportunities of assessing the atmospheric air pollution by using the object-oriented programing and statistics

A. Duda, S. Korga & Z. Lenik

Department of Fundamentals of Technology, Faculty of Fundamentals of Technology, Lublin University of Technology, Lublin, Poland

ABSTRACT: Industrial emission from transport and domestic furnaces affects the condition of atmospheric air in urban agglomeration. The main pollutants of atmospheric air are: sulfur, nitrogen and carbon oxides, hydrocarbons and dust containing toxic metals: lead, zinc, arsenic, selenium, manganese and others. The object-modeling system concerns two interweaving processes, namely the air quality assessment as well as the preparation and implementation of repair programs to protect the air. Tools and sources of information in these processes will be the measurements of pollution emissions to the atmospheric air and the object-oriented modeling of pollutants' spread in the atmosphere. In the literature, there are lists and divisions of the models of pollutants dispersion in atmospheric air. Basic models which describe the pollutants dispersion are: Euler's differential, Langrange's integral and Gaussian statistical models. The object-modeling has emerged in the 1970s through the use of new object-oriented programming languages (Simula, Smalltalk, and Ada).

Keywords: atmospheric air pollution, object-oriented programming, statistics, computer modeling

1 INTRODUCTION

Fumes, liquids and solids present in the air, together with substances present in increased amounts in comparison with the natural composition of the air, are air pollutants. However, they are not the natural ingredients of the air. In general, air pollutants can be divided into dusty and gaseous. The main source of air pollution is anthropogenic emission (point, industrial and linear), which consists of emission from industrial activity, communal and household sector, and also from the traffic emission (Badyta et al. 2006). The Framework Directive's provisions 96/62/WE on the assessment and management of the air quality, and derivative directives concerning the limit values for the atmospheric pollutants is the fundamental document defining the requirements for the assessment and management of air quality are (Environmental Monitoring Library 2011).

In Poland, the Ordinance of the Minister of Environment from the 8th of March 2008 concerning the levels of certain substances in the air is in force (Journal of Law 2008 No 47, item 281). Whereas, the Law on Environmental Protection Inspectorate legally defines routine environmental monitoring and testing methods of the air pollutants (Dz. U. 2008 Nr 47, poz. 281).

The national surveillance system managed by the State Inspectorate for Environmental Protection,

the direct executors, namely the Voivodship Inspectorates for Environmental Protection, and the Departments of Sanitation, possesses the largest share in control of the air pollution. Carrying out such works enables to undertake the research concerning the analysis of the air pollution changes. They allow to verify whether in the air there are no physical, chemical, and biological changes, which are harmful to life and health of living organisms (Lesiak & Świsulski 2002).

The object-oriented systems are used as integrated environments of designing and programming the virtual measuring devices. Programming, based on the objects that use graphical language "G", allows to build complex and specialized applications that support the research processes (Council of Ministers 2002). The flexibility of the object-oriented systems enables to build metrological applications and tools that support the measuring and decision-making processes. The designed models of the object-oriented systems are being developed in order to facilitate research and to use the environmental assessment.

The acquisition of the data from measuring equipment is often characterized by its multiple occurrence, which hinders, in a great way, the analysis of the phenomenon. Therefore, some actions were undertaken to build a computer application to evaluate the changeability of pollutants concentrations with the use of the data from the environmental monitoring analysis. One of its objectives should to be answer to the question whether the content of pollutants in the atmospheric air in a given area is the dependence changing in time (Stadler 2002, Świsulski 2004).

2 MATERIALS AND METHODS

In the experimental part of this work, the following analyses and measurements were performed:

- traffic intensity in selected places with a characteristic automobile traffic intensity,
- the analysis of ambient concentration measurements: Nitric Oxides (NO_x), Sulfur Dioxide (SO_2), Carbon Monoxide (CO),
- the investigation of the results of ambient concentration measurements in the continuous monitoring station,
- statistical evaluation of the received analysis and measurements' results.

In order to assess the variability of individual pollutants concentration contents changing in the air, in particular time and space; and a detailed analysis of individual substances concentrations should be conducted at appropriate seasons and in a different area. For the purpose of this work, basic atmospheric air pollutants, such as: NO, NO_2, NO_x, CO_2, SO_2, were qualitatively assessed. The data acquisition system records the results of the atmospheric air pollution concentrations in a tabular form.

The statistical analysis of the main atmospheric air pollutants level could be carried out at various time-intervals. The analyzed set of specimens relates to 5-minute pollutants concentrations. To confirm the observed regularities in the appearance of individual pollutants concentrations, the analysis and statistical evaluation of measurements was carried out. Moreover, sampling frequency of the data acquisition process has been set at equal time-intervals. In order to develop the evaluation of changes in pollutants concentrations appearance in atmospheric air, there was a program generated, which allows the visual and statistical analysis of the distribution of pollutants concentration. Figure 1 presents a statistical analysis of air pollutants samples.

This program provides the possibility to obtain information needed to assess the state of air pollution and its changes o. For instance, the analysis of air pollution distribution shown in Figure 1 revealed that in depicted pollutants\ distribution there were no significant changes observed (air pollution in a given area and in a given time remained stable).

This program provides the possibility to obtain information needed to assess the state of air pollution and its changes o. For instance, the

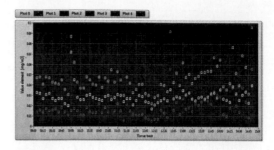

Figure 1. The view of program created to perform the statistical samples analysis.

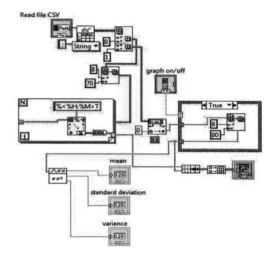

Figure 2. Block program scheme to statistical samples analysis.

analysis of air pollution distribution shown in Figure 1 revealed that in depicted pollutants\distribution there were no significant changes observed (air pollution in a given area and in a given time remained stable).

The program is based on five algorithms downloading the stored data in CSV file, which allow to conduct a comparative analysis of the examined elements concentrations which are changeable in particular time (Tłaczała, 2002).

Algorithms, developed with the help of the object-oriented programming, enable to download the data in the form of date and the content of pollutants concentrations is automatically scaled at the axes of the system. It allows to perform the process of continuous monitoring of pollutants concentrations for various values of the data and time intervals.

The view of block program scheme is shown in Figure 2.

The establishment of average content of pollutants concentration, the calculation of standard deviation and variance enable to find the periodic, seasonal changeability of the atmospheric air pollutants concentrations (Winiecki et al. 2001).

Statistical analysis of the atmospheric air pollution emission on the continuous monitoring stations presents significant differences in measured substances conducted at the individual stages of the research.

3 CONCLUSION

The monitoring process which use the data processing algorithms allows to reveal changes in the presence of analyzed atmospheric air pollutants in a given time together with the identification of their cyclicity and changeability. The authors suggested the solution in the form of applications presented, which enable the realization of this task in a simple, fast and effective way. This program, if needed, can be upgraded with mathematical models based on the method of objective estimation considering also the size of point, areal and linear emission.

REFERENCES

Badyta A., Majewski G. (2006): Analiza stężeń zanieczyszczeń komunikacyjnych na tle natężenia ruchu pojazdów i podstawowych elementów meteorologicznych, Ochrona powietrza w teorii i praktyce, Instytut Podstaw Inżynierii Środowiska PAN, Zabrze; 1–11.

Biblioteka Monitoringu Środowiska (2011): Raport o stanie środowiska województwa lubelskiego roku 2010, Biblioteka Monitoringu Środowiska, Lublin.

Dz. U. 2008 Nr 47, poz. 281.

Lesiak P., Świsulski D. (2002): Komputerowa technika pomiarowa w przykładach, Agenda Wydawnicza PAK, Warszawa.

Rada Ministrów (2002): Polityka ekologiczna państwa na lata 2003–2006 z uwzględnieniem perspektywy na lata 2007–2010, Warszawa.

Stadler A.W. (2002): Systemy akwizycji i przesyłania danych, Oficyna Wydawnicza Politechniki Rzeszowskiej, Rzeszów.

Świsulski D. (2004): Systemy pomiarowe. Laboratorium, Wydawnictwo Politechniki Gdańskiej, Gdańsk.

Tłaczała W. (2002): Środowisko LabVIEW w eksperymencie wspomaganym komputerowo, WNT, Warszawa.

Winiecki W., Nowak J., Stanik S. (2001): Graficzne zintegrowane środowiska programowe do projektowania komputerowych systemów pomiarowo-kontrolnych, MIKOM, Warszawa.

Environmental Engineering IV – Pawłowski, Dudzińska & Pawłowski (eds)
© 2013 Taylor & Francis Group, London, ISBN 978-0-415-64338-2

Magnetometry application to the analysis of air pollutants accumulated on tree leaves

S.K. Dytłow & B. Górka-Kostrubiec

Institute of Geophysics, Polish Academy of Science, Warsaw, Poland

ABSTRACT: The paper presents results of a study of air pollutants accumulated on the surface of the tree leaves in the urban environment of the city (Warsaw). The study used magnetic method to identify magnetic particles of traffic and industrial pollution. The samples were taken from the chestnut trees along the main roads of varied traffic density and the low level of pollution. High magnetic susceptibility values ($19 \cdot 10^{-8}$ m^3 kg^{-1}) were obtained for leaves along roads with heavy traffic and the low values ($7 \cdot 10^{-8}$ m^3 kg^{-1}) for gardens and parks. Increased susceptibility was observed for autumn samples compared to the spring ones. We find the presence of maghemit as the main mineral in air pollution. The magnetic parameters indicated two different sources of pollution for autumn and spring. The greater concentration of SP grains was observed for spring dust.

Keywords: environmental magnetism, leaves, biomonitoring, susceptibility

1 INTRODUCTION

Several studies reported a negative impact of atmospheric pollution on the health of residents in urban cities, which can penetrate deep into the human lungs and cause respiratory illness due to their fine and ultra-fine size. At present, vehicle emission is an important source of particle pollution in urban dust. Vehicles can generate non-spherical magnetic particles via exhaust emission, and as a product of abrasion and corrosion processes of engine, vehicle body, tires, brake linings and asphalt. Pollution produced by road traffic can be a complex of different components hazardous to human health, among them heavy metals, which can be inhaled and become absorbed into specific organs. From this point of view, it is very important to find new method for monitoring the level of pollution in urbanized areas.

In several studies (Szönyi et al. 2007, 2008) the authors have applied magnetic technique to identify atmospheric pollution and to demonstrate the possibility of monitoring the distribution of anthropogenic and natural Particulate Matter (PM) in time and space. Magnetic method identifies all minerals heaving magnetic behaviour, therefore, magnetic particles of pollution can be discriminated within the samples. Analysis of magnetic mineral of pollution is a reliable rapid, non-destructive and inexpensive alternative method to conventional atmospheric pollution monitoring sensitive to the low detection levels. Different plant materials: leaves of tree (Maher et al. 2008,

Mitchell et al. 2010, Matzka et al. 1999, Szönyi et al. 2007, 2008, Moreno et al. 2003, Kardel et al. 2011, Hansard et al. 2011), needles life-span (Lehndorff et al. 2006, Urbat et al. 2004), tree trunk core (Zhang et al. 2008) can be used for biomonitoring of pollution by means of magnetic method. For our research we used tree leaves as many earlier works (Kardel et al. 2011, Moreno et al. 2003, Sagnotti et al. 2009, Szönyi et al. 2007, 2008) well documented correlation between ambient PM and the magnetic properties of the leaves. These studies pointed out that the application of magnetic measurement of tree leaves is a powerful tool to determine the level of PM level in town and to identify their sources.

The aim of our study was to use biomonitoring to determine the level of air pollution deposited during warm period of year in highly urbanized city as Warsaw, and to compare the level of pollution and composition of magnetic minerals gathered in different periods of year. This work reported preliminary results devoted to recognition of possibilities of method.

2 MATERIAL AND METHODS

2.1 Sampling methods

The set of leaf samples was collected in the centre of Warsaw, in places with different traffic intensity and in urban green areas (parks and gardens). Samples were collected at the beginning (May) and the end (October) of growing season. Additionally, soil samples have been taken under the trees from which leaves have been collected. In Table 1 the sample names were listed together with a brief description of sampling details.

Leaves were collected from the outer canopy of a tree at a height of 1.5–2.0 m. For this research, the leaves of horse chestnut tree were selected, as the most abundant in the urban environment of Warsaw (e.g. gardens, parks, road-side). Moreover, the leaves of horse chestnut due to big and hairy surface efficiently retain particles of pollution falling from atmosphere. Thirteen leaves of similar size and age were taken for one sample. A sample consists of discs of diameter d = 35 mm cut from the centre of each leaf. Then, the discs were dried, crushed and packed to cubic plastic boxes of volume of 10 cm³.

2.2 Measurements

Magnetic measurements were carried out in the Paleomagnetic Laboratory of the Institute of Geophysics, Polish Academy of Science in Warsaw. For all samples of leaves and soils, magnetic susceptibility was measured at three different frequencies of magnetic field (200 A/m): 976 Hz, 3900 Hz and 15600 Hz; using the bridge MFK1-FA Multi-Function Kappabridge (AGICO). In order to determine the grain size, domain structure and mineralogy of magnetic particles of pollution accumulated on the leaves, hysteresis loops and thermomagnetic curves were measured. The measurements of hysteresis loops were conducted with the use of the Vibrating Sample Magnetometer (Molspin). The parameters of hysteresis (saturation isothermal remanence (Mrs), saturation magnetization (Ms), coercivity (Bc) and coercivity of remanence (B_{cr})) were determined from this measurement after paramagnetic slope correction. Thermomagnetic curves of saturation isothermal remanent magnetization SIRM were performed

with the use of a device built by the TUS (Poland) at temperature range from 20 to 700°C. A sample was magnetized in a field of 9T, high enough to saturate hard magnetic minerals. For each sample, two consecutive curves of SIRM (T) were made. The temperature at which magnetic remanence was unblocked was used to identify the kind of magnetic minerals. The high-temperature measurement of magnetic susceptibility was performed using the MFKA1-FA Kappabridge working with the CS-3 high-temperature furnace. Susceptibility of powdered sample was measured during heating to 700°C and cooling to room temperature.

3 RESULTS AND DISCUSSION

Figure 1 shows the magnetic susceptibility normalized by mass (χ) for samples collected from 6 locations in the city centre at the beginnings (May) and at the end (October) of the growing season. The magnetic susceptibility is higher for samples collected in October than in May at about 80%. It is seen in Figure 1 that the susceptibility exhibits the highest value of $19 \cdot 10^{-8}$ m³ kg⁻¹ near roads with heavy traffic. Samples taken from the park and the old city centre, where traffic is limited have the lowest values of χ. Initial values of χ for dust accumulated on leaves at the beginning of growing season do not show relation to traffic intensity. The highest value was measured for local street with medium traffic level (locality P4-K). The increase of χ during growing season is almost twice bigger for leaves collected at places with heavy traffic than in places with low traffic. The smallest increase was observed for park locality.

These results are consistent with the work of Power et al. (2009), who studied the mean concen-

Table 1. Samples names and a brief description of places of their collection.

Sample name	Description of sampling location
P3-K	Local street in Warsaw centre (low traffic volume)
P4-K	Local street in Warsaw centre (medium traffic volume)
P5-K	Major road in Warsaw centre (high traffic volume)
P7-K	Park in Warsaw centre (low pollution area)
P8-K	Major road in Warsaw centre (high traffic volume)
P11-K	Plaza in Old Town (low pollution area)

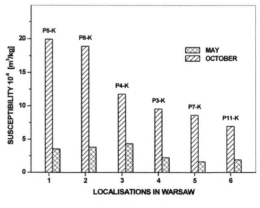

Figure 1. Changes of magnetic susceptibility of pollution deposited on horse chestnut leaves in spring and autumn in Warsaw.

tration of magnetic particles accumulated on leaves in relation to traffic intensities in Wolverhampton, UK. They found that sites having lower magnetic concentration value are located far from the road. Moreover, the sites displaying high magnetic concentration were situated close to the roads with the highest traffic volume. This study suggests that the deposition of PM on the road-side tree leaves varies due to the different intensity of road traffic.

Soil accumulates contaminants over a longer period than the leaves so it can give information about the total level of contamination. On the contrary, leaves due to short growing season give information about the short-time contamination which make it possible to follow the trend of changes during warm part of year. In Figure 2 one can see that the long-term level of soil contamination does not coincide with the level of contamination measured on the leaves during the growing season and do not correlate with traffic level. It can be explained by the presence of additional sources of pollution during the winter and autumn when there are no leaves on the trees. In Warsaw, it is heating season, when extra pollution is emitted to the atmosphere.

In order to examine the structure of the magnetic particles present in pollution accumulated on the leaves, the dependence of magnetization remanence on the magnetic field in the range of 0–1 T was determined. Curves of M_r (H) are an indicator of hardness of magnetic phases present in the samples. As it is seen in Figure 3 the M_r is saturated between 200 mT and 300 mT. It means that soft magnetic minerals like magnetite or maghemite contribute significantly to creation of dust.

Figure 4 shows example of curve of thermal demagnetization of SIRM. The first heating curve

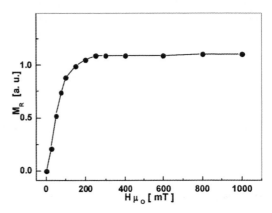

Figure 3. Examples of acquisition of isothermal remanent magnetization curves.

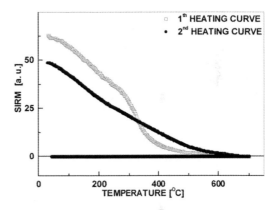

Figure 4. Curves of continuous thermal demagnetization of saturation isothermal remanence—SIRM (T).

shows total thermal demagnetization of SIRM at unblocking temperature >600°C corresponding to temperature range characteristic for maghemite. The second heating unblocks SIRM at temperature 680°C characteristic for hematite. It is visible that initial intensity of SIRM of second curve is lower than that for the first heating. This behaviour indicates the formation of hematite during heating. The results of thermomagnetic analysis show that the dust contains low-coercivity magnetic grains of maghemite. This is consistent with the results obtained for the magnetic particles in PM collected in Munich (Germany) by Muxworthy et al. 2002, who identified maghemite and metallic iron (coming from tram lines) as responsible for the observed magnetic properties. Other magnetic studies of tree leaves (Moreno et al. 2003, Szönyi et al. 2007) mention magnetite as the main carrier of magnetic signal in street dust but a small contribution of harder magnetic phase like hematite was also observed by

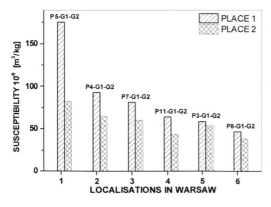

Figure 2. Susceptibility of soils taken from two opposite places (dashed and sparse boxes) under the trees leaves of which were collected in autumn.

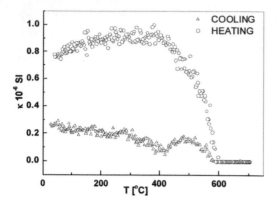

Figure 5. High-temperature curve of magnetic susceptibility during heating and cooling.

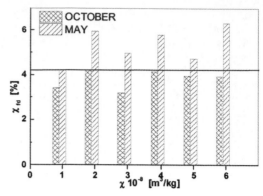

Figure 6. Distribution of χ_{fd} for spring and autumn samples.

other authors (Gautam et al. 2005, Hansard et al. 2011).

Figure 5 presents high-temperature curves of susceptibility. The heating curve shows decrease of susceptibility between room temperature and 430°C. Then it is visible broad peak at 500°C and drastic drop at Curie temperature about 580°C, after which samples show paramagnetic behaviour. During cooling susceptibility rapidly increases after passing Curie temperature until about 450°C and then starts to decrease slowly. The inflection point at 430°C on heating curve may be connected with structure transformation of maghemite to hematite which has lower susceptibility. The same course of the heating curve was observed by Sagnotti et al. 2009 for street dust accumulated on leaves in Rome. The presence of maghemite for leaves samples was confirmed by SIRM(T) (see Fig. 4) curves. The broad maximum at 500°C probably resulted from production of magnetite from maghemite/hematite during heating in reducing atmosphere characteristic for $\chi(T)$ experiments made in MFKA1-FA Kappabridge working with the CS-3 high-temperature furnace (Jeleńska et al. 2010). Decrease of susceptibility on cooling curve below 400°C can point to the presence of fine grains on the threshold between SP and SD, which while cooling become magnetically ordered SD grains.

Frequency dependence of the susceptibility (χ_{fd}) is indicative of the presence of SP grains in PM. χ_{fd} has higher values for spring dust than for autumn dust (Fig. 6). For the spring dust, χ_{fd} is greater than 4.2% reaching 6.3% for the park locality (P11-K), whereas for autumn dust χ_{fd} is lower than 4.2%. The smallest value 3.2% belongs to dust from heavy traffic location (P5-K). This is evidence for greater contribution of SP grains in spring dust and indicates different sources of pollution in spring and in autumn. Figure 7 shows relationship for both

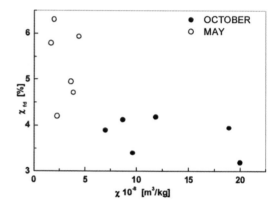

Figure 7. Relationship between χ and χ_{fd}.

seasons between χ and χ_{fd}. The date for autumn and spring occupy different area on the plot. It is additional evidence for different magnetic composition of both dusts.

Figure 8 shows no correlation between concentration dependent parameters χ, SIRM and M_s. The data grouped in two separated areas. It means that dust from strongly polluted area has different magnetic composition than dust from low polluted location and park. Gain in χ during growing season is twice larger for heavy polluted places (P5-K, P8-K) than for low polluted parks or local street.

In order to describe the domain state of magnetic particle, hysteresis loops (Fig. 9a) were measured. Hysteresis loops for all samples are characterized by the magnetic mineral saturated in magnetic field between 300 mT and 400 mT. A similar result for the street dust was reported in earlier works by Muxworthy et al. (2002) and Szönyi et al. (2008). On the basis of the hysteresis loop (Fig. 9a), the set of parameters M_s, M_{rs}, B_c i B_{cr} was determined. Values of B_c and B_{cr} range

320

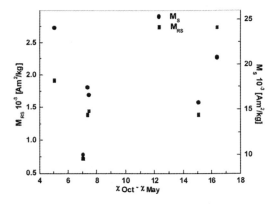

Figure 8. Relationship between SIRM and M_S versus difference of χ_{Oct} and χ_{May}.

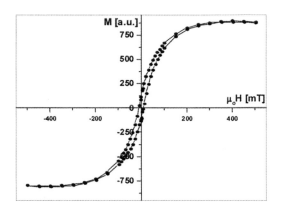

Figure 9(a). Example of hysteresis loop measured for polluted dust deposited on leaves of horse chestnut.

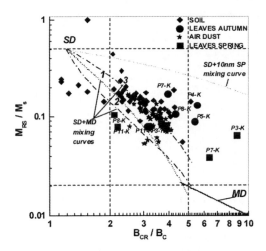

Figure 9(b). Distribution of hysteresis parameters on the Day plot modified by Dunlop (Day et al. 1977, Dunlop et al. 2002a, 2002b). Data for leaves (■, ●) were shown together with data for soil (◆) (Górka-Kostrubiec et al. 2012) and air filter (★) (Górka-Kostrubiec et al. 2011) samples collected in Warsaw (Day et al. 1977).

from 5 to 19 mT and from 5 to 70 mT, respectively. The corresponding ratios of hysteresis parameters Mrs/Ms and B_{CR}/B_C placed on the Day diagram (Day et al. 1977) modified by Dunlop (Dunlop et al. 2002a, 2002b) allow to determine the type of domain structure of grains. The diagram (Fig. 9b) shows that in the case of leaves collected in autumn data for four samples are located in the area between MD+SD mixing curve and SD+SP mixing curve. Two samples lie along the mixing curve of SD+MD grains of magnetite. Four samples collected in spring occupied two distinct areas. The data for spring leaves have very low susceptibility values and hysteresis parameters were not possible to measure for all samples. The data for autumn leaves are in agreement with studies conducted in Rome by Szönyi et al. (2007). They found that data for leaves collected along high traffic roads lies in the same region of the Day plot between the theoretical curves for mixture of SD and MD grains and of SD and SP grains of magnetite. Sagnotti et al. (2009) identified this region as characteristic of the particles of pollution coming from the abrasion of the brake linings. It is emphasized that three samples of the leaves taken from low polluted areas lie within the data obtained for air filters in Warsaw (Górka-Kostrubiec et al., 2011).

4 CONCLUSION

Magnetic susceptibility of dust deposited on the leaves of trees is a good proxy of level and source of air pollution evaluation in great cities. χ is low for green area ($7 \cdot 10^{-8}$ m^3 kg^{-1}), much higher for the low traffic areas ($9 \cdot 10^{-8}$ m^3 kg^{-1}) and highest for major roadside trees ($19 \cdot 10^{-8}$ m^3 kg^{-1}).

Relationship between susceptibility and frequency dependence of the susceptibility of dust accumulated on the leaves indicated different sources of pollution for autumn and spring. χ_{fd} values for spring dust indicate greater concentration of SP than for autumn dust.

The lack of correlation between concentration dependent parameters usually observed for one source pollutants and different relationship between these parameters for green areas and highly polluted roads indicate different magnetic composition of pollution in these two areas.

The presence of maghemite was suggested by total thermal demagnetisation of SIRM at unblocking temperature >600°C.

REFERENCES

Day, R., Fuller, M., Schmidt, V.A., 1977. Hysteresis properties of titanomagnetites: grain size and compositional dependence. Physics of the Earth nad Planetary Interiors, 13 pp. 260–267.

Dunlop D.J., 2002a. Theory and application of the Day plot (Mrs/Ms versus Hcr/Hc): 1. Theoretical curves and tests using titanomagnetite data. *J. Geophys. Res.*, 107(B3), 2056, DOI: 10.10292001JB000486.

Dunlop, D.J., 2002b. Theory and application of the Day plot (Mrs/Ms versus Hcr/Hc): 2. Application to data for rocks, sediments, and soils. *J. Geophys. Res.*, 107(B3), 2057 DOI: 10.1029/2001JB000487.

Gautam, P., Blaha, U., Appel, E., 2005. Magnetic susceptibility of dust-loaded leaves as a proxy of traffic-related heavy metal pollution in Kathmandu city, Nepal. *Atmospheric Environment* 39, 2201–2211.

Górka-Kostrubiec, B., Król, E., Jeleńska, M., 2011. Dependence of air pollution on meteorological conditions based on magnetic susceptibility measurements: a case study from Warsaw. *Studia geophysica and geodaetica*, vol. 56, 861–877.

Górka-Kostrubiec, B., Król, E., Teisseyre-Jeleńska, M., 2012. Magnetic susceptibility as an indicator of traffic pollution in some Warsaw localities. Współczesne problemy inżynierii i ochrony środowiska, 54, 67–82.

Hansard, R., Maher, B.A., Kinnersley, R., 2011. Biomagnetic monitoring of industry-derived particulate pollution. *Environmantal Pollution,* 159, 1673–1681.

Jeleńska, M., Hasso-Agopsowicz, A., Kopcewicz, B., 2010. Thermally induced transformation of magnetic minerals in soils based on rock magnetic study and Mössbauer analysis. Physics of the Earth and Planetary Interiors, 179, 164–177.

Jordanova, D., Petrovsky, E.P., Hoffman, V., Gocht, T., Panaiotu, C., Tsacheva, T., Jordanowa, N. Magnetic signature of different vegetation species in polluted environment. Stud. Geophys. Geod., 54, 417–442.

Kardel, F., Wuyts, K., Maher, B.A., Hansard, R., Samson, R., 2011. Leaf saturation isothermal remanent magnetization (SIRM) as a proxy for particulate matter monitoring: inter-species differences and in-season variation. *Atmospheric Environment* 45, 5164–5171.

Lehndorff, E., Urbat, M., Schwark, L., 2006. Acumulation histories of magnetic particles on pine needles as function of air quality. *Atmospheric Environment* 40, 7082–7096.

Maher, B.A., Moore, C., Matzka, J., 2008. Spatial variation in vehicle-derived metal pollution identified by magnetic and elemental analysis of roadside tree leaves. *Atmospheric Environment* 42, 364–373.

Matzka, J., Maher, B.A., 1999. magnetic biomonitoring of roadside tree leaves: identification of spatial and temporal variations in vehicle-derived particulates. *Atmospheric Environment* 33, 4565–4569.

Mitchell, R., Maher, B.A., Kinnersley, R., 2010. Rates of particulate pollution deposition onto leaf surfaces: temporal and inter-species magnetic analyses. *Environmental Pollution* 158, 1472–1478.

Moreno, E., Sagnotti, L., Dinares-Turell, M., Winkler, A., Cascella, A., 2003. Biomonitoring of traffic air pollution in Rome using magnetic properties of tree leaves. *Atmospheric Environment*, 37, 267–2977.

Muxworthy, A., Schmidbauer, E., Petersen, N., 2002. Magnetic properties and Mössbauer spectra of urban atmospheric particulate matter, a case study from Munich, Germany, *Geophysical. Journal International* 150, 558–570.

Power, A.L., Worsley, A.T., Booth, C. Magneto-biomonitoring of inra-urban spatial variations of particulate matter using tree leaves, 2008. *Environ Geochem Health* 31, 315–325.

Sagnotti, L., Taddeucci, J., Winkler, A., Cavallo, A., 2009. Compositional, morphological, and hysteresis characterization of magnetic airbone particulate matter in rome, Italy, 2009. G3 Geochemistry, Geophysics, Geosystems, *An Electronic Journal Of The Earth Sciences,* 10 (8).

Szönyi, M., Sagnotti, L., Hirt, A.M., 2007. On leaf magnetic homogeneity in particulate matter biomonitoring studies. Geophysical Research Letters 34, L06306.

Szönyi, M., Sagnotti, L., Hirt, A.M., 2008. A refined biomonitoring study of airbone partculate matter pollution in Rome, with magnetic measurements on Quercus Ilex tree leaves. GJI, Geomagnetism, Rock Magnetism and Paleomagnetism, 173, 127–141.

Urbat, M., Lehndorff, E., Schwark, L., 2004. Biomonitoring of air quality in the Cologne conurbation using pine needles as a passive sampler-part I: magnetic properties. *Atmospheric Environment* 38, 3781–3792.

Zhang, Ch., Huang, B., Piper, J.D.A., Luo, R., 2008. Science Of The Total Environment, 393, 177–190.

Environmental Engineering IV – Pawłowski, Dudzińska & Pawłowski (eds)
© 2013 Taylor & Francis Group, London, ISBN 978-0-415-64338-2

The magnetometric study of indoor air pollution inside flats located in Warsaw and its suburbs

E. Król, B. Górka-Kostrubiec & M. Jeleńska
Institute of Geophysics, Polish Academy of Science, Warsaw, Poland

ABSTRACT: The magnetometric monitoring of indoor air pollution in Warsaw flats is presented in this paper. The mass magnetic susceptibility (χ) of home dust is a good proxy for evaluation of indoor air pollution by magnetic particles and its spatial distribution. The dust samples from 143 private flats of different Warsaw areas have been studied. It was proved that the higher values of magnetic susceptibility of home dust appear in flats situated near pollution sources such as roads with intensive traffic or places with emission of individual combustion products. Correlations between the magnetic parameters and heavy metals content were observed.

Keywords: indoor air pollution, dust, magnetic susceptibility

1 INTRODUCTION

In recent years, pollution of atmosphere in urban agglomerations has been studied intensively. The negative influence of Particulate Matter (PM) contents in air on human health have directed the attention to its regular monitoring and to study its mineral composition, ways of distribution and natural and anthropogenic sources. Besides traditional (mainly chemical) methods of determination of such pollutants, the new approach—magnetometry appeared some years ago, an application of magnetic parameters for measuring pollutions containing iron-reach magnetic compounds. A short description of magnetometric method applied for the evaluation of air and soils pollution (with the literature review of important studies in this matter) can be found in our preliminary study devoted to indoor air pollution in Warsaw (Jeleńska et al. 2011). Up to now the majority of papers have been devoted to study the outdoor air pollution in big agglomerations with application of magnetometric methods.

Pollution of indoor air in relation to outdoor and indoor sources was studied by Fisher et al. (2000), Kingham et al. (2000) and by He et al. (2004). They found a significant correlation between outdoor and indoor concentrations of pollutants examining the content of particular matter (PM_{10}, $PM_{2.5}$) together with chemical, organic pollutants (polycyclic aromatic hydrocarbons—PAH and volatile organic compounds—VOC) but they did not obtain reliable correlations of these factors with compounds—VOC), but they did not obtain reliable correlations of these factors

with the traffic-related air pollutants. Halsall et al. (2008) have combined magnetic method with determination of PAH presence to indicate the relation between indoor and outdoor pollution. Jordanova et al. (2006) have studied composition of magnetic fraction in cigarette ashes. They found that the ashes contain large amount of magnetite grains of fine and coarse dimensions. The same authors examined indoor and outdoor dusts from Bulgaria by magnetic and aerobiological methods and have determined that mineralogy of an anthropogenic magnetic fraction in dust was built from magnetite, maghemite and hematite (Jordanova et al. 2010). Heavy metals content in house dusts from Warsaw area have been determined by chemical methods by Tatur et al. (2006), but authors did not examine the spatial distribution of metals contamination in flats in the city.

The preliminary part of this research started in 2010 and has been conducted on small amount of dust samples collected in 12 Warsaw flats (Jeleńska et al, 2011). In spite of insufficient number of samples, it was stated that mass magnetic susceptibility is a good proxy for monitoring indoor air pollution. The magnetic susceptibility χ changed for studied samples of dust in the range: $40–200 \cdot 10^{-8}$ m^3 kg^{-1}. Magnetic parameters of hysteresis loop, which gave information about grain size and domain state of magnetic particles, show that the dust samples contain a mixture of Single Domain (SD) and Multi Domain (MD) grains with variable contribution of Superparamagnetic (SP) grains. The preliminary study determined mineralogy of magnetic fraction of indoor dust as build mainly from magnetite of the unblocking temperature $T_{ub} = 575°$ C with the small contribution of fine meta-stable Single Domain grains (SD), close to Superparamagneic (SP) state (with T_{ub} around 150° C) and some maghemite ($T_{ub} = 380°$ C) presence. The dust magnetic mineralogy has been compared with typical magnetic mineralogy of unpolluted and polluted soils in Warsaw area. As magnetic method appeared to be a good proxy for monitoring indoor air pollution, the new, bigger collection of indoor dust samples has been gathered during 2011/2012 years in Warsaw and in some surrounding suburbs. The results

obtained for this new collection are presented in this paper.

2 MATERIALS AND METHODS

2.1 Collection

Samples of house dust have been taken from vacuum cleaner bags from flats or houses located in all Warsaw districts and in some city suburbs. The main collection consists of 193 samples taken in 143 places (Fig. 1). Some samples have been taken twice

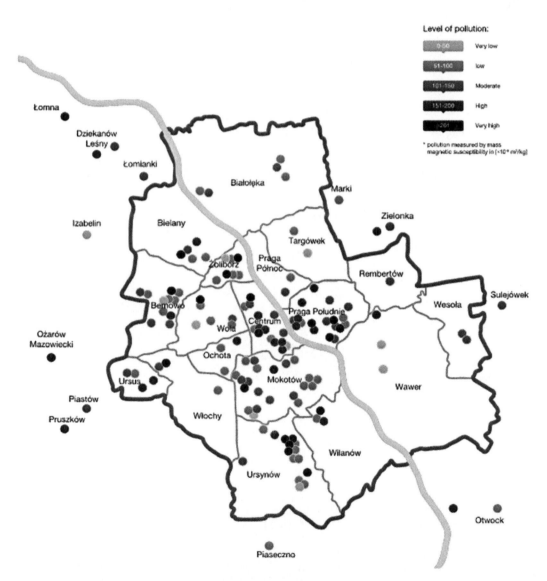

Figure 1.   Distribution of localities in which samples of home dust have been collected in Warsaw agglomeration. The gradation of gray colors defines the level of indoor air pollution in the five-range scale.

or occasionally more times from the same locations. The distribution of samples in the Warsaw area is not uniform: we had obtained 134 samples from 106 localities situated in the area on the left bank of the Vistula river and 58 samples of dust from 37 localities on the right bank of the river. The number of flats, where samples were taken, situated on particular floors varied (Table 1).

The majority of new samples have been collected during the spring-early summer season of 2011, samples repeated for part of flats has been obtained during the late autumn and wintertime of 2011/2012. On an average, the time span of dust collection in vacuum cleaners changed between 2 weeks and one month, according to individual habits of flat owners.

Samples have been collected using mostly modern, hermetic (compact) vacuum cleaners with the multi-layers bags and with the HEPA system of filters on the exit of a cleaner.

2.2 Measurements of magnetic parameters

The measurements of magnetic parameters of dust samples have been conducted in paleomagnetic laboratory of the Institute of Geophysics of Polish Ac. Sc. in Warsaw. The mass magnetic susceptibility (χ^- in the unit 10^{-8} m^3 kg^{-1}) was measured using the KLY-2 kappa bridge (made by Geofizyka-Brno, Czech Republic) and by the modern multi function kappa-bridge MFK1-FA (produced by the AGICO company in Brno). Dust samples were sewed, put into small plastic boxes (V = 10 cm^3) and weighted before measurements. To evaluate the quality of data, the standard deviation of χ measurements for two data sets was calculated: one set was obtained for dust samples collected 9 times during more than 1.5 year in one house, situated in the suburb, and second set was obtained for 6 samples collected (during one year) in the flat located close to the city canter. The average values of χ for the two flats, was similar: $\chi_1 = 122.1 \cdot 10^{-8}$ m^3 kg^{-1}.(for the house situated in the suburb) and $\chi_2 = 128.4 \cdot 10^{-8}$ m^3 kg^{-1} (for the flat by the street

with a moderate car motion, in the city canter). The standard deviations of both series of measurements differ significantly (s.d.$_1 = \pm 44.5 \cdot 10^{-8}$ m^3 kg^{-1} and s.d.$_2 = \pm 17.2 \cdot 10^{-8}$ m^3 kg^{-1}). It was supposed that big standard deviation in the suburb house was caused by seasonal changes related to local heating system, whereas no such effect was observed for flats situated in the city center, where dominates the central heating system. Nevertheless, these two series of magnetic susceptibility measurements confirmed that the results are representative with acceptable errors.

2.3 Chemical analysis of the chosen dust samples

The content of 20 elements have been determined by chemical analysis made for 10 dust samples. The samples has been chosen according to their mass magnetic susceptibility and covered the range from the lowest value $24 \cdot 10^{-8}$ m^3 kg^{-1} from the suburb Wawer to 100 times greater value of χ, obtained for exceptional sample from the flat situated on Mokotów. The extraction of those metals has been done using the supra pure nitric acid in a proportion of 5 ml to 0.5 g of dust specimen. Total amount of trace elements: Ag, Al, As, Ba, Be, Cd, Co, Cr, Fe, Mg, Mn, Mo, Ni, P, Pb, Sr, Ti, U, V, Zn has been measured using the emission spectroscopy method ICP MS. Measurements have been performed at the Chemical Laboratory of Department of Geography at the Warsaw University. The results were obtained in p.p.m. of dried mass (i.e. mg/kg of dry mass).

3 RESULTS AND DISCUSSION

3.1 Magnetic susceptibility of indoor dust specimens

The mass magnetic susceptibility of indoor dusts from flats and houses in the Warsaw agglomeration changes in very wide range between $18.6 \cdot 10^{-8}$ m^3 kg^{-1} and $1540.5 \cdot 10^{-8}$ m^3 kg^{-1}. All data have been divided into five groups, according to the values of magnetic susceptibility: very low, low, medium, high and very high. The estimation of indoor air pollution range has been done by the comparison of obtained results of χ with the typical mass magnetic susceptibilities for unpolluted and polluted soils. We also compared the obtained χ values for indoor dusts with the values of annual magnetic susceptibility of outdoor air pollution deposited on air-filters during four years 1977, 1980, 1981 and 1982 in Warsaw (Górka-Kostrubiec et al. 2012). The χ showed decreasing trend from $250 \cdot 10^{-8}$ m^3 kg^{-1} in 1977 to $115 \cdot 10^{-8}$ m^3 kg^{-1} in 1985. The value of χ in 1985 is close to

Table 1. The number of flats situated on the different floors, from which the dust samples have been collected.

Floor level	Number of flats
First floor	35
Second floor	26
Third floor	22
Four floor	9
5, 6 and 7 floors	14
8–15 floors	12

nowadays-obtained average value for indoor air in Warsaw. The values cited allow us to accept *per analogiam* that the mass magnetic susceptibility of home dust up to $50 \cdot 10^{-8}$ m³ kg⁻¹ indicates the background level of indoor air pollution (or: the very low pollution), up to $100 \cdot 10^{-8}$ m³ kg⁻¹ determines the low level of pollution, up to $150 \cdot 10^{-8}$ m³ kg⁻¹ can be related to the medium pollution and up to $200 \cdot 10^{-8}$ m³ kg⁻¹ and higher to the high and the very high level of air pollution. The ranges for this five point scale and amount of localities in the each group of pollution level is shown in Figure 2. This distribution shows that almost in the half of flats (~60) $\chi 100 \cdot 10^{-8}$ m³ kg⁻¹ was observed, more than half of flats show the medium and the high pollution level.

$(\chi \geq 150 \cdot 10^{-8}$ m³ kg⁻¹$)$, and only 10 samples have $\chi \leq 50 \cdot 10^{-8}$ m³ kg⁻¹.

The extremely high magnetic susceptibilities ($>500 \cdot 10^{-8}$ m³ kg⁻¹) have been found for 8 samples collected in the 6 flats randomly localized in the city. These exceptionally high magnetic susceptibilities probably were caused by the reasons not connected with outdoor conditions but with the unique situation inside these flats. For example, the very high content of ferromagnetic compounds found inside one of flats is connected with its renovation shortly before collecting the dust sample. In other place, the special sort of job conducted by flat owner caused also very high value of dust magnetic susceptibility. For this reason, it was decided to exclude these extremely high susceptibilities from the farther consideration. The average value of mass magnetic susceptibility for dust calculated in that way is $122 \cdot 10^{-8}$ m³ kg⁻¹. All results of χ have been placed in the Warsaw administrative map (Fig. 1).

The city area was divided into 10 areas according to the space distribution of collection places,

Table 2. The 10 areas of Warsaw agglomeration, where studied dust samples in flats have been collected.

Areas of localities	Number of flats	Number of dust samples
1. Bemowo (Be)	11	12
2. Sub-urbs of the right bank of Vistula river (S-u r)	19	32
3. Bielany (Bi) and Żoliborz (Z)	13	14
4. Centre (C)	15	20
5. Mokotów (M) and Wilanów (Wl)	25	39
6. Praga-Centre (P-C)	19	27
7. Wola (Wo) and Ochota (O)	7	8
8. Ursus (U) And Włochy (W)	10	11
9. Ursynów (Urs)	17	20
10. Suburbs of the left bank of Vistula river	7	10
Together	143	193

Samples from Białołęka (B), Targówek (T), Remberów (R), Wawer (Wa), Wesoła (We), Marki, Sulejówek, Otwock and Zielonka are grouped as from the area 2, named Sub-urbs of the right bank of Vistula river. The samples from the area 10 named Sub-urbs of the left bank of Vistula river have been collected in Dziekanów Leśny, Izabelin, Łomna, Łomianki and Piaseczno. Some dust samples from Ożarów Maz., Piastów and Pruszków are joined to the area 8.

which only roughly cover the district division of the city (Table 2). This sub-division of results in the urban space assured more logical distribution of samples according to their proximity in space, not connected with the exact administrative districts. The average values of magnetic susceptibility for the 10 areas described above are presented in the Figure 3 in the growing order, together with minimal and maximal χ values for each area. The average values of χ calculated for each of the ten mentioned areas are in the narrow range between 101.5 and $154.7 \cdot 10^{-8}$ m³ kg⁻¹. If the value of $\chi \leq 100 \cdot 10^{-8}$ m³ kg⁻¹ was accepted as the limit for unpolluted place, all areas exceeded this value. The relatively low magnetic susceptibility are for peripheral area of city on the right side of the Vistula River (Białołęka, Rembertów, Targówek, Wawer, Wesoła, Marki, Zielonka, Sulejówek and Otwock) and for Bemowo, Mokotów with Wilanów and Ursynów—on the left bank of the river.

The Centre and the areas close to it (Wola with Ochota), Żoliborz, Bielany and the canter of Praga belong to the highly polluted areas of Warsaw agglomeration. The same situation is observed for suburbs on the left side of Vistula River and for

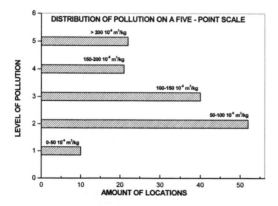

Figure 2. Distribution of magnetic susceptibility values for studied dust samples in Warsaw.

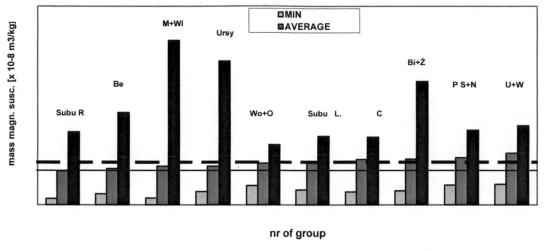

Figure 3.   The minimal, average and maximal mass magnetic susceptibily in the ten area of Warsaw.

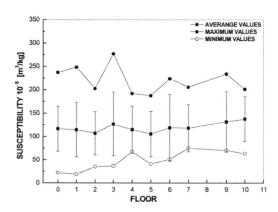

Figure 4.   Magnetic susceptibility of dust samples in function of flat location over ground level.

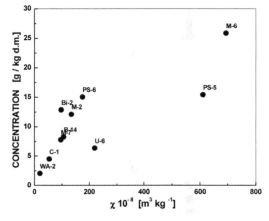

Figure 5.   The correlation between magnetic suscepti-bility of indoor dust and: concentration of total sum of concentration of 20 elements in the 10 samples.

Ursus with Włochy together with the two suburbs close to the two quarters (where the maximal aver-age value of χ was revealed).

The obtained distribution of the level of indoor air pollution in the Warsaw area should be treated as the result of many factors, among which the very big concentration of road traffic in central districts looks as very important. For areas with-out the city central heating, local emission of com-bustion products causes the seasonal increase of pollution.

The topography and the ventilation channels of the city are the other important factors, which modulate the distribution of pollution of outdoor air, which penetrates to indoor spaces. As vertical distribution of pollution (Fig. 4) shows no correla-tion between floor level of sampling flats and χ we

found that there is no influence of this condition on the pollution extent. However, it is visible that dispersion of obtained results for the lower floors is greater than for the higher floors. It means that at high level above ground the space distribution of pollution is more homogeneous than near to the ground.

The results of chemical analysis of the chosen 10 samples of home dust revealed that the content of 20 elements varies significantly from flat to flat. The total concentration of all 20 elements corre-lates very well with the mass magnetic susceptibil-ity χ with r = 0.92 (Fig. 5).

Letters indicate the names of sample groups for different city area. Solely the AVERAGE values of m.m.s. are in the increasing order. The two straight lines mark:

a. dashed line—the average level of indoor air pollution for the whole collection of dust samples,
b. full line—the low level of indoor air pollution ($=100 \cdot 10^{-8}$ m^3 kg^{-1}) in the five range scale

These results confirm the conclusion that mass magnetic susceptibility can serve as easy measured magnetic parameter for the estimation of the level of pollution by Fe compounds and other trace metals commonly present in indoor dust.

4 CONCLUSIONS

The presented study confirmed that the mass magnetic susceptibility of home dust is a good proxy for evaluation of indoor air pollution by in-organic particular matter (mainly Fe compounds and heavy metals) of anthropogenic origin. This conclusion corresponds to the high correlation of total amount of Fe compounds and heavy metals found in home dust with χ. Moreover the five point scale of the level of indoor air pollution has been proposed, which is based on comparison of the data with mass magnetic susceptibility of unpolluted and polluted soils and air filters depositions.

The χ values distribution shows that in each area in Warsaw agglomeration there are flats with low and high level of pollution depending on such factors as the local intensity and the concentration of road traffic and on the presence of sources of combustion products. None of the areas of Warsaw has the level of pollution less than $100 \cdot 10^{-8}$ m^3 kg^{-1} although 10 flats have χ smaller than $50 \cdot 10^{-8}$ m^3 kg^{-1}.

Vertical distribution of χ does not show any correlation with the floor level on which sampled flat was situated.

ACKNOWLEDGEMENTS

This study was financially supported by the Miele Co. Ltd.—Polish Section in Warsaw, thanks the agreement with the Institute of Geophysics of Polish Academy of Sciences regarding the common project named "Undusted Warsaw".

The authors thank to P. Szwarczewski Ph. D. of the Dept. of Geography at the Warsaw University for making the chemical analysis of dust samples.

REFERENCES

Fisher, P.H., Hock, G., Van Reeuvijk, H., Briggs, D.J., Leber, E., Van Vijnen, J.H., Kingham, S. & Elliott, P.E. 2000. Traffic related differences in outdoor and indoor concentrations of particles and volatile organic compounds in Amsterdam. *Atmospheric Environment* 34; 3713–3722.

Górka-Kostrubiec, B., Król, E. & Teisseyre-Jeleńska M. 2012a. Podatność magnetyczna jako wskaźnik zanieczyszczeń komunikacyjnych w wybranych lokalizacjach w Warszawie. Prace Naukowe Politechniki Warszawskiej (Inżynieria Środowiska);z. 59; 67–82.

Górka-Kostrubiec, B., Król, E. & Jeleńska, M. 2012b. Dependence of air pollution on meteorological conditions based on magnetic susceptibility measurements, a case study from Warsaw. Stud. Geophys. Geod.; 56(3); 861–877.

Halsall, C.J., Maher, B.A., Karloukovski, V.V., Shah, P. & Watkins S.J. 2008. A novel approach to investigating indoor-outdoor pollution links. Combined magnetic and PAH measurements. *Atmospheric Environment* 42; 8902–8909.

He, C., Morawska, L., Hitchins, J. & Gilbert, D. 2004. Contribution from indoor sources to particle number and mass concentration in residential houses. *Atmospheric Environment* 38; 3405–3415.

Jeleńska, M., Górka-Kostrubiec, B. & Król E. 2011. Magnetic properties of dust as indicators of indoor air pollution: preliminary results. in: Management of Indoor Air Quality—Dudzińska (ed) ©Taylor & Francis Group, London, 129–136.

Jordanova, N., Jordanova, D., Yankova, R., Petrov, P., Popov, T. & Tsacheva, T. 2006. Magnetism of cigarette ashes. *J. Magn. Mater.* 301; 50–66.

Jordanova, N., Jordanova, D., Yankova, R., Petrov, P., Popov, T. & Tsacheva, T. 2010. Magnetic and aerobioloical studies o indoor and outdoor dust from Bulgaria. List of abstracts. 12 Castle Meeting on Paleo-, *Rock and Environmental Magnetism.*

Kingham, S., Briggs, D.J., Elliott, P., Fisher, P. & Lebret, E. 2000. Spatial variations in the concentrations of traffic—related pollutants in indoor and outdoor air in Huddesfield, England. *Atmospheric Environment* 34; 905–916.

Magiera, T., Lis, J., Nawrocki, J., Strzyszcz, Z.,(Editors) 2002. Podatność magnetyczna gleb Polski.(Atlas) Wyd. Państwowego Instytutu Geologicznego i Instytutu. Podstaw Inżynierii Środowiska, Warszawa.

Tatur, A., Gromadka, P. & Wasilewska, A. 2006. Heavy metals in house dust from Warsaw. *Ecological Chemistry and Engineering* 13(7); 695–702.

Environmental Engineering IV – Pawłowski, Dudzińska & Pawłowski (eds)
© *2013 Taylor & Francis Group, London, ISBN 978-0-415-64338-2*

Assessment of fuel suitability for low-emission biomass combustion

R. Kobyłecki, M. Ścisłowska, M. Wichliński, A. Kacprzak & Z. Bis
Department of Energy Engineering, Czestochowa University of Technology, Czestochowa, Poland

ABSTRACT: The article presents the research results and the evaluation of physicochemical parameters of various fuels, mainly belonging to the group of biomass and agromass (*i.e.* agricultural biomass). All fuels were investigated with respect to the ultimate and proximate analysis as well as their HHV (high heating value) and potential to form agglomerates during combustion. As a result of laboratory tests, it was found that both biomass and agromass, as well as their ashes, contain quite significant amount of alkali elements and chlorine. The studies of ash properties indicated that the combustion of selected types of agromass at temperatures typical for fluidized beds may cause operational difficulties and even boiler shutdown due to the formation of sinters and agglomerates in the bed. The results indicated that in each case comprehensive analysis is required before the combustion of any new fuel in fluidized bed boilers.

Keywords: biomass, agromass, combustion, fuel, fluidized bed boilers

1 INTRODUCTION

The biomass is the largest source of renewable energy. It includes biodegradable substances derived from waste products, production forestry, agriculture, and industry for the recycling of its products and waste from industry biodegradable or urban origin (Wils 2012). Nowadays, the conversion of forestry biomass, agricultural biomass (agromass) and waste biomass into heat and power encompasses a wide range of different conversion options, end-use applications and infrastructure requirements (Arbon 2002, Ghani 2009). Although the environmental and other benefits of using biomass to displace fossil fuels are well known, it seems that they still cannot be completely effective in the market without tax credits, subsidies and other artificial measures. A drawback to the on increment of biomass use for power generation is the fact that it is actually converted into power with thermal efficiency much lower than in the case of coal. Overall efficiencies of biomass-to-power tend to be rather low, at typically 15% for small plants up to 30% for larger plants.

These reference values are very far from the typical efficiencies of the most efficient energy conversion plant: the natural gas combined cycle plant (Beer 2000). Moreover, the combustion of biomass releases various different chemical pollutants so that the environmental effects of burning biomass are generally considered less harmful than those associated with coal but more harmful than those associated with natural gas (Evrendilek 2003, Atimtay 2010). Despite these drawbacks,

from an energy conversion point of view, the use of biomass presents a lot of favorable aspects (Franco 2005, Biagini 2009). The biomass is particularly suitable for cofiring with coal in fluidized bed combustors. However, due to its composition the biomass may cause several operational problems to the boiler, such as defluidization, fouling, etc. Accordingly, the properties of biomass must be carefully investigated to avoid any unexpected operational troubles (Bis 2010).

The current paper presents the research results and the evaluation of physicochemical parameters of various biomass and agromass samples in order to cofire them at low-emission combustion facilities. The fuel samples were also investigated to determine their potential to agglomerate during the fluidized bed combustion.

2 MATERIALS AND METHODS

Twelve fuel types of biomass and agromass origin were chosen for the present investigation.

The ultimate and proximate analysis was conducted according to the commonly used standards and procedures for biomass and agromass fuels. In order to avoid any 'evaporation' of volatile ash components the ash content was determined for samples burnt at 575°C. In order to determine the characteristic ash softening temperatures, the fuel ash samples were specially formed and placed in an electrically-heated furnace. The samples were then heated up with an assumed heating rate. During sample heating simultaneous observation

Table 1. The fuel samples investigated in the present study.

Sample No.	Fuel type	Sample	Sample No.	Fuel type	Sample
1	mix of pellets (1)		7	mix of pellets (2)	
2	sunflower husk		8	oat hulls pellets	
3	sunflower husk		9	sunflower husk pellets (2)	
4	sawdust		10	sunflower husk pellets (3)	
5	sunflower husk pellets (1)		11	sunflower husk pellets (4)	
6	pomace oil		12	sunflower husk pellets (5)	

of the changes in the shape of the ash sample were conducted. The appropriate characteristic temperatures (softening, melting and liquid) were then determined from the change of the sample shape.

3 RESULTS AND DISCUSSION

The results of proximate analysis of the investigated fuel samples are shown in Table 2. The values are in most cases quite typical, however, it is worth noting that the values of external moisture of some samples were negative, thus indicating that they remained in very dry environment. The ash content in majority of the samples was quite low (<7%) but it was significant in samples No. 1, 6 and 7. Those samples seem thus to be more suitable for large-scale CFB combustion since the addition of fuel ash and/or other inert material to the bed is crucial for the control of bed temperature and the emission of pollutants.

The results of the elemental analysis of the samples are shown in Tables 3 and 4. As indicated by the results in Table 3 the samples No. 7, 1 and 2 contain quite significant amount of sulfur and their combustion in large-scale CFB boilers may thus require the addition of limestone to meet the sulfur emission standards. Some main elements chosen in the fuel samples are shown in Table 4. The values are in the typical range for biomass and agromass. However, samples No. 1 and No. 6 are characterized by quite high chlorine content and may bring about an increased rate of chlorine corrosion and

damages to the heat transfer surfaces, as well as the formation on dioxins and other organohallogen compounds during 'uncontrolled' combustion in large-scale CFBC. The results of the ash analysis are shown in Table 5. The main ash-forming elements are Si (up to 24%), Ca (<33%) and K (<32%) compounds. Since high content of potassium in the fuel ash is usually responsible for the formation of low-temperature eutectics, the combustion of the investigated fuels may thus bring about the formation of agglomerates in the furnace. In order to avoid any agglomeration, the properties of the ashes must be investigated more in detail, especially with respect to the ash behavior and softening.

The corresponding tests were conducted and the results are summarized in Table 6 and Figure 1. The results in Table 6 indicate that ashes from the burning of samples No. 1, 5, 8 and 9 are slightly molten after the combustion at 900°C i.e. at temperature typical for fluidized bed combustion.

Therefore, the combustion of those fuels may bring about operational troubles in real CFB boiler. In order to determine more precisely the ash softening temperature, the fuel ash samples were heated up to the temperature where symptoms of ash softening were visually determined. The results are shown in Figure 1. The samples No. 2, 3, 4, 6, 7, 10, 11 and 12 did not show any softening in the studied temperature range (<1200°C). However, the softening temperatures of some fuel samples (No. 1, 5, 8 and 9) were relatively low, roughly between 800–900°C. The data confirm the results shown in Table 6 and indicate that the combustion of those

Table 2. Proximate analysis of the fuel samples (air-dry, only the external and inherent moisture are given for the 'as-received' state).

No.	Fuel type	External moisture [%]	Inherent moisture [%]	Total moisture [%]	Volatiles [%]	Ash [%]	Fixed carbon [%]	HHV [MJ/kg]
1	Mix of pellets (1)	1.3	6.8	8.0	63.2	14.6	15.4	16.03
2	Sunflower husk	−0.1	6	5.9	66.9	6.9	20.1	17.94
3	Sunflower husk	−1.3	2	0.7	76.2	2.6	19.2	18.95
4	Sawdust	49.1	2.5	50.4	80.1	3.1	14.3	16.34
5	Sunflower husk	−0.1	1.8	1.7	76.1	2.7	19.4	19.04
6	Pomace oil	4.3	6.2	10.2	66.1	11.3	16.4	17.66
7	Mix of pellets (2)	6.1	7.2	12.9	63.8	17.5	11.5	16.00
8	Pellets oat hulls	−1.8	3.2	1.5	81.0	3.7	12.1	16.60
9	Sunflower husk pellets (2)	−0.8	1.8	1.0	77.0	2.8	18.4	19.19
10	Sunflower husk pellets (3)	13.7	4.9	17.9	71.9	5.5	17.7	17.95
11	Sunflower husk pellets (4)	−2.5	2.4	0	75.5	2.8	19.3	18.95
12	Sunflower husk pellets (5)	−0.8	1.9	1.1	80.0	2.6	15.5	18.58

Table 3. Ultimate analysis of the fuel samples (dry).

No.	Fuel type	C [%]	H [%]	N [%]	S [%]	O [%]	Ash$_{575}$ [%]
1	Mix of pellets (1)	42.30	5.29	2.45	0.31	35.05	14.6
2	Sunflower husk	48.88	5.73	0.83	0.22	37.45	6.9
3	Sunflower husk	49.77	6.19	0.54	0.11	40.79	2.6
4	Sawdust	48.25	5.92	0.31	0.01	43.03	3.1
5	Sunflower husk	49.77	6.25	0.56	0.11	40.61	2.7
6	Pomace oil	46.62	5.77	1.47	0.09	34.75	11.3
7	Mix of pellets (2)	41.01	5.72	3.62	0.70	31.45	17.5
8	Pellets oat hulls	45.83	6.44	0.99	0.09	42.95	3.7
9	Sunflower husk pellets (2)	50.43	6.28	0.35	0.10	40.05	2.8
10	Sunflower husk pellets (3)	47.93	5.97	0.52	0.10	39.98	5.5
11	Sunflower husk pellets (4)	49.68	6.18	0.69	0.11	40.55	2.8
12	Sunflower husk pellets (5)	48.81	6.17	0.31	0.09	42.03	2.6

Table 4. Main elements in the fuel samples [wt%].

Sample no.	1	2	3	4	5	6	7	8	9	10	11	12
Na	0.148	0.057	0.015	0.023	0.014	0.042	0.135	0.02	0.003	0.027	0.003	0.01
K	3.177	1.544	1.505	0.237	1.281	4.875	0.841	1.123	1.273	2.152	1.952	1.609
Cl	0.527	0.132	0.127	0.045	0.098	0.5	0.138	0.129	0.11	0.172	0.125	0.108
Al	0.6	0.315	0.023	0.195	0.017	0.309	0.78	0.023	0.023	0.346	0.015	0.016
Si	4.096	1.017	0.092	1.667	0.06	1.193	3.425	1.549	0.083	1.221	0.048	0.062
Ca	3.052	2.361	0.78	1.42	0.596	3.266	6.258	0.262	0.714	1.664	0.729	0.758
Mg	0.577	0.353	0.391	0.102	0.314	0.525	0.422	0.203	0.34	0.684	0.4	0.365
Fe	1.506	0.431	0.13	1.052	0.101	0.623	4.37	0.053	0.065	0.373	0.051	0.048

Table 5. Summary results of the analysis of ash samples (the values in wt%, ashing @ 575⁰C).

	Sample no.											
	1	2	3	4	5	6	7	8	9	10	11	12
Na	1.074	0.662	0.488	0.309	0.317	0.438	0.428	0.508	0.753	0.848	0.535	0.698
K	8.418	7.12	32.51	3.296	27.01	11.23	3.184	13.42	18.46	18.69	29.03	14.705
P	3.551	1.452	1.536	<0.001	2.182	0.594	4.775	5.166	2.967	2.266	3.053	3.754
Cl	0.279	0.049	1.229	0.051	0.764	0.098	0.202	<0.001	0.428	0.693	0.679	0.058
Al	2.716	3.08	0.237	2.763	0.214	1.516	2.473	0.133	0.304	2.469	0.113	0.238
Si	24.53	12.12	0.984	20.59	0.934	5.714	12.73	28.9	1.669	9.971	0.739	1.361
Ca	11.43	30.93	21.01	21.49	24	41	26.22	4.28	29.58	19.93	21.42	33.361
Mg	2.072	3.759	5.842	1.154	7.151	5.002	1.478	1.382	9.086	5.867	7.544	9.355
Fe	5.585	5.986	0.941	5.661	2.701	3.309	15.45	2.444	2.838	4.93	3.209	2.560

Table 6. View of the ash samples.

Number of fuel samples	1	2	3	4	5	6
Ash after combustion @ 575°C						
Ash after combustion @ 900°C						
Number of fuel samples	7	8	9	10	11	12
Ash after combustion @ 575°C						
Ash after combustion @ 900°C						

molten

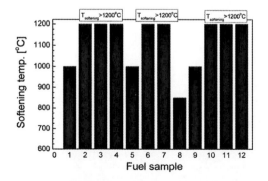

Figure 1. The range of the softening temperature of the investigated fuel samples.

fuels may cause the operational problems in the boiler, particularly in the zones where the temperature is increased e.g. due to poor fluidization. The cofiring of those fuels should by thus done with particular care and continuous monitoring of the bed hydrodynamics.

4 CONCLUSIONS

The analysis of the research results described in the present paper allows to formulate the following conclusions:

1. Twelve biomass and agromass samples were selected for laboratory testing of the fuel

properties in order to determine the possibility to cofire those fuels in CFB boilers. The total moisture content in the samples was mainly low (<20%)—only the sawdust sample contained roughly 50% of water. The chlorine content in the samples did not exceed 0.6%. Most chlorine was contained in the samples No. 1 (pellets from the mixture) and No. 6 (pomace oil). The HHV of the samples was in the range of roughly 16–19 MJ/kg.

2. The ash content in the fuel samples was between 2.6% (sunflower husk) and 17.5% (mix of pellets 2). The highest ash content was determined for the pomace oil pellets. The combustion of fuels with high ash content is recommended for CFB boilers since in such case no addition of any external solids is required to maintain the bed and control its temperature.

3. The results of the determination of ash softening temperature indicated that some fuel ashes (sample No. 1, 5, 8 and 9) may soften at temperatures below 900°C. Since such temperature level is typical for CFB combustion the cofiring of those fuels may bring about increased formation of agglomerates in the fluidized bed. In order to control the process continuous monitoring of the bed hydrodynamics is recommended.

The results clearly indicate that detailed analysis of the fuel properties is very important and the parameters of any new fuel should be determined before its application for large scale CFB combustion. Otherwise, operational troubles may occur leading even to an emergency shutdown of the boiler.

REFERENCES

Arbon I.M., 2002. Worldwide use of biomass in power generation and combined heat and power schemes, *J. Power Energy* 216: 41–57.

Atimtay, Aysel T. 2010. Combustion of agro-waste with coal in a fluidized Bed, *Clean Technologies and Environmental Policy,* 12(1): 43–52.

Beer J.M., 2000. Combustion technology developments in power generation in response to environmental challenges, *Progr. Energy Combust. Sci.* 26: 301–327.

Biagini E., Simone M., Tognotti L., 2009. Characterization of high heating rate chars of biomass fuels, Proceedings of the Combustion Institute, 32(2): 2043–2050.

Bis Z., 2010. Fluidized bed boilers, theory and practice (in Polish): 365–375, Czestochowa University of Technology, Czestochowa, Publishing House of the Czestochowa University of Technology.

Evrendilek F., Ertekin C., 2003. Assessing the potential of renewable energysources in Turkey, *Renewable Energy* 28: 2303–2315.

Franco A., Giannini N., 2005. Perspectives for the use of biomass as fuel in combined cycle power plants, International *Journal of Thermal Sciences* 44: 163–177.

Ghani, W. Alias, A.B., Savory, R.M., Cliffe K., 2009. Co-combustion of agricultural residues with coal in a fluidised bed combustor, *Waste Management* 29(2): 767–773.

Wils, A. Calmano, W. Dettmann P. Kaltschmit, M. Ecke H., 2012. Reduction of fuel side costs due to biomass co-combustion, *Journal of Hazardous Materials*, 207–208: 147–151.

Environmental Engineering IV – Pawłowski, Dudzińska & Pawłowski (eds)
© 2013 Taylor & Francis Group, London, ISBN 978-0-415-64338-2

Low emission high temperature air combustion technology (HTAC)—numerical analysis of methane ignition phenomenon

S. Werle & R.K. Wilk

Institute of Thermal Technology, Silesian University of Technology at Gliwice, Gliwice, Poland

ABSTRACT: Ignition delay is an important parameter in the most modern and advanced combustor concepts designed for low-NO_x emission. This work presents the results of a modeling study of the ignition process of methane under High Temperature Air Combustion (HTAC) conditions. This technology is a promising solution for energy saving, flame stability enhancement and decreasing of harmful substances emission (NO_x, CO, C_nH_n). A mathematical model was formulated to predict the dependence of the equivalence ratio and temperature of oxidizer on the temperature increment. The mathematical model incorporates the basic principles of the energy and mass balance. The theoretical predictions were correlated with the experimental data. The increment of temperature ($\Delta T = T_{max} - T$) was assumed as parameter characterizing the process of ignition. The influence of initial temperature of oxidizer (t = 687–961°C) and equivalence ratio is analyzed and discussed. The results of calculation were compared with the experimental ones which have been done on Constant Volume Bomb (CVB) reactor. It is shown that in order to achieve the effective reaction of ignition (taking into account the maximal value of increment of temperature ΔT), it is not necessary to maximize the oxidizer temperature. There are optimal values of temperature oxidizer (t ≈ 830°C) in which parameters mentioned above reaches its extreme values.

Keywords: methane, HTAC Technology, modeling, kinetic mechanisms

1 INTRODUCTION

Ignition, which is one of the most important parts of combustion and directly affects combustion efficiency and emission, has been the subject of many studies that try to elucidate the ignition characteristic. Ignition can depend on physical, chemical, mixing and transport features of a problem, and in some cases on heterogeneous phenomena. Excellent description of ignition phenomenon can be found in (Lefebvre 1999, Richards et al. 2001), however, ignition in general is an enormous subject, and the present work cannot provide a through treatment. Many previous investigators have focused their attention on analytical studies and modeling of the methane ignition. Hydrocarbons are certainly the best studied class of compounds for which reliable and detailed kinetics models for combustion exist. It is not surprising that the bulk of automotive fuels are comprised almost exclusively of hydrocarbons. Recent advances in experimental and kinetics modeling capabilities have provided new insights into ignition, offering new possibilities to develop new combustors. Most of the previous modeling and experimental results led to development of GRI-Mech 1.2, Gri-Mech 2.1 (Petersen 1999), Gri-Mech 2.11 and later

GRI-Mech 3.0 (Smith et al.). The last version, Gri-Mech 3.0 (Wu et al. 1999, Safta et al. 2006, Matynia et al. 2009), consist of 325 elementary chemical reactions for the 53 species optimized to perform over the ranges 1000–2500 K, 1–1000 kPa, and equivalence ratio from 0.1 to 5.0 (Zhou et al. 2010). The ignition of methane-oxygen mixtures of equivalence ratio 0.4–1.0, was determined by (Jee et al. 1999) at 1520–1940 K and in reflected shock waves with initial pressures of 2.7 kPa. The results were modeled with GRI-Mech 1.2 with which they were in satisfactory agreement. A 38 species, 190-reaction kinetics model (RAMEC), based on the Gas Research Institute's GRI-Mech 1.2 Mechanism was developed by (Petersen et al. 1999) using additional reactions that are important in methane oxidation.

(Mertens et al. 1999) has studied the reaction kinetics in shock heated methane/oxygen/argon mixtures, at 2880–3030 K and 100–152 kPa by comparing emission traces against simulated results of radicals concentrations based on a GRI-Mech 2.11. (Stamatov et al. 2005) have studied ignition behavior of methane/air mixtures. A mathematical model has been formulated to predict the dependence of the explosion-delay time on the radiative power flux and gas composition. The elementary steps of methane combustion were taken from GRI-Mech 2.11. In (Brett et al. 2001) investigating the chemical process of prediction ignition of methane in rapid compression machine was presented. Two complete mechanism for methane oxidation were used to run the model: GRI-Mech 2.11, which has 279 reactions of 48 reactive species and (Leeds mechanism) ver. 1.3 considers 35 reactive species and 190 reactions. Another example of the Leeds mechanism is its version 1.4 (Simmie 2003) which consist of 351 reactions of 37 species, built with an overall philosophy akin to that of GRI-Mech and using much the same set of experiments of laminar flame speeds, ignition delay measurements and species profiles in laminar flames. Another purely electronic detailed reaction mechanism form methane and natural gas combustion due to (Konnov, Coppens et al. 2007, Chaumiex et al. 2007) also deals with C_2, C_3 hydrocarbons and their derivates. The mechanism comprises some 1200 reactions and 127 species and is extensively validated against a large dataset of experiments including species profiles and ignition delay times in shock waves, laminar flame species profiles and laminar flame speeds. (Davidenko et al. 2002) analyzed a number of methane mechanisms: GRI-Mech 1.2, GRI-Mech 3.0, Princeton (Sang 1998), Leeds 1.5 (Hughes 2001) and LLNL (Chaumiex et al. 2007) on their way to producing skeletal models which they could employ in multi-dimensional simulations of complex reacting flows.

In this paper, we examine the ignition process of methane under High Temperature Air Combustion technology (HTAC) conditions. Combustion in high temperature air can achieve significant energy saving. The HTAC has been applied in many industrial furnaces (especially fired with natural gas). Furthermore, it was also proved that this technology may be used for combustion of light liquid fuels. Also, the first trials have been made to burn solid fuels under HTAC technology. However, HTAC is still a new promising combustion technology. So, it is still a subject of scientist's research in many countries. Previously, many of the characteristic of high temperature air combustion have been investigated in laboratory-scale systems but the knowledge of both experimental and numerical studies of the gaseous fuel ignition phenomenon under high temperature is an example of the gap in the research of HTAC.

(The Leeds methane oxidation mechanism) was used to run the model. This mechanism considered 35 reactive species and 190 reactions. Because the Chemkin is the most widely used software for combustion simulations with detailed chemistry, we have chosen it in our computations.

2 MATERIALS AND METHODS

It is convenient to assume that during the ignition process either the density or the pressure remains constant. Here, the authors want to consider the constant density case. Taking into consideration a closed vessel at constant volume V containing a given mass m, the density $\rho = m/V$ remains constant during the ignition process. It was assumed that the process takes place in the model of a constant volume cylindrical vessel depicted in Figure 1. The cold fuel was injected to the vessel filled initially with hot air at atmospheric pressure.

In the present work, time-dependent computations were done assuming laminar flow. Thus, the turbulent mixing and associated effects on chemical reactions were not considered. Presented computations model were limited to the onset of the ignition. The system is analyzed "after loaded" ($\dot{m}_{in} \, d\tau = 0$). Heat transfer to the chamber walls is analyzed (combustion chamber was not assumed to be adiabatic). Gases are treated as ideal gases, the governing equations concerns the gaseous combustible mixture. The transient state is taken into consideration. Gravitational acceleration and viscous dissipation were neglected. The governing equations consisted of the mixture mass continuity equation, energy equation and Gibbs function equation. These equations are given below.

The total mass balance can be written in the following way (Szlęk 2004)

$$\dot{m}_{in} \, d\tau = \dot{m}_{out} \, d\tau + dm_s \qquad (1)$$

Taking into consideration that this system is analysed "after loaded" ($\dot{m}_{in} \, d\tau = 0$), it can be expressed as:

$$dm_s = -\dot{m}_{out} \, d\tau \qquad (2)$$

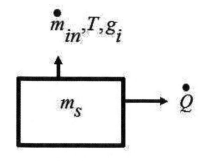

Figure 1. Scheme of the analyzed system.

Appending to the right side an element illustrating the yield of the source reaction following $V \dot{\omega}_i M d\tau$ and using the dependence $1/\rho = V/m_s$, for the i-th substance we can be written as:

$$\frac{dg_i}{d\tau} = \frac{1}{\rho} \dot{\omega} M_i \tag{3}$$

In compliance with the first law of thermodynamics, the energy balance is expressed in its general form suggested in (Szlęk 2004):

$$\dot{E}_{in} d\tau = \dot{E}_{out} d\tau + dE_s \tag{4}$$

Taking into consideration that this system is analysed "after loaded" ($\dot{E}_{in} d\tau = 0$) it may be written as:

$$dE_s = -\dot{E}_{out} d\tau \tag{5}$$

Introducing to $dE_s = dU$ and $-\dot{E}_{out} d\tau = \dot{I}_{out} + \dot{Q}$ we get:

$$dU = -\dot{I}_{out} d\tau - \dot{Q} d\tau \tag{6}$$

Using Gibbs function quoted by (Szargut 2000)

$$u = i - pv \tag{7}$$

and using equation (2) and assuming that $vm = V$ and $d(pV) = 0$ we have:

$$\frac{di}{d\tau} = \frac{-\dot{Q}}{m} \tag{8}$$

Taking into consideration that $i = \sum g_i i_i$ and $di/d\tau = c_{p_i}(dT/d\tau)$, after a few simplifications and using eq. (2), we get the following function:

$$\frac{dT}{d\tau} = \frac{\dfrac{-\dot{Q}}{m} - \sum i_i \dfrac{1}{\rho} \dot{\omega}_i M_i}{\sum g_i c_{p_i}} \tag{9}$$

where $-\dot{Q}/m$ is given by [21]

$$-\frac{\dot{Q}}{m} = \frac{A(T_{max}^4 - T^4)}{V\rho} \tag{10}$$

The species and temperature changes and heat radiation parameter A were computed with the use of Chemkin code.

3 RESULT AND DISCUSSION

3.1 Experimental results

An example of the determination of the increment of temperature ΔT is presented in Figure 2 (Werle & Wilk 2007). Point "0" is interpreted as the moment of the gas injection and the start of ignition.

The temperature distance between point "0" and the point in which the temperature reaches its maximum is determined as the increment of temperature ΔT. The maximal value of standard deviation of increment of temperature ΔT equals $\sigma_{\Delta T} = 6.22$ K. The signals from thermocouples were collected every 0.001 s. The oxidizer temperature t was measured with an enlarged uncertainty equal $\delta t = 2$ K (Agilent 1999). The temperature inside the reaction chamber (after gas injection) T_{max} was measured with uncertainty equal $\delta T_{max} = 1.2$ K (Agilent 1999) Error of calculations of the equivalence ratio $\delta\varphi$ equaled $\delta\varphi = 0.02\%$.

Experimental tests were performed on the special CVB stand (Fig. 3), which has been described before (e.g. Werle & Wilk 2007, Werle & Wilk 2010, Werle & Wilk 2010a, Werle 2011, Szlęk et al. 2009, Werle 2010).

Next Figures illustrate the dependence of the increment of temperature ΔT as a function of the equivalence ratio at the oxidizer temperature equal 716, 831 and 961°C (Fig. 4) and as a function of oxidizer temperature for equivalence ratio equal 0.50, 0.91 and 1.43 (Fig. 5).

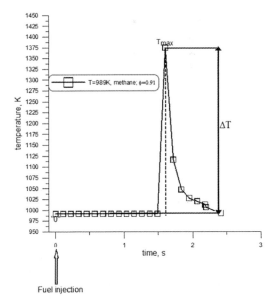

Figure 2. Example of the increment of the temperature ΔT as a function of the duration of the test.

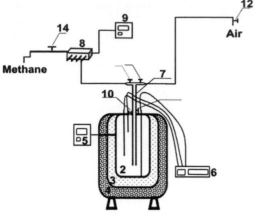

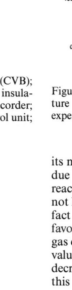

Figure 3. Scheme of constant volume bomb (CVB); 1, reaction chamber; 2, heating coil; 3, thermal insulation; 4, microprocessor control unit; 5, digital recorder; 6, conduit supplying gas; 7, electric valve; 8, control unit; 9, safety vent; 10–13, valves.

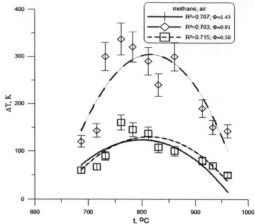

Figure 5. Dependence of the increment of temperature as a function of temperature oxidizer; results of experiment.

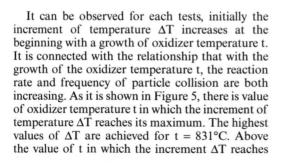

Figure 4. Dependence of the increment of temperature as a function of equivalence ratio; results of experiment.

It can be observed for each tests, initially the increment of temperature ΔT increases at the beginning with a growth of oxidizer temperature t. It is connected with the relationship that with the growth of the oxidizer temperature t, the reaction rate and frequency of particle collision are both increasing. As it is shown in Figure 5, there is value of oxidizer temperature t in which the increment of temperature ΔT reaches its maximum. The highest values of ΔT are achieved for t = 831°C. Above the value of t in which the increment ΔT reaches

its maximum (ΔT_{max}), this parameter is decreasing due to smaller level of density (concentration) of reacting gases at constant pressure (chamber is not hermetic). It can be concluded that despite the fact that increment of oxidizer temperature ΔT favours growth of reaction rate, the decrease of gas density is much stronger. As a result, above the value of t, in which $\Delta T = \Delta T_{max}$, this parameter is decreasing. It seems that preheating oxidizer above this value of temperature oxidizer, which equals approximately t ≈ 830°C, is unsubstantiated. Similar diagrams have been plotted for all analyzed values of equivalence ratio but it is well known that most combustion properties have simple maxima or minima in the neighborhood of φ = 1. It is worthwhile to emphasize one important feature. Methane as majority of alkanes inhibits its own ignition (Imbert et al. 2008) when rate of branching of chain branching (as essential reaction for the high temperature combustion) is higher than rate of tearing off of chain branching. Hydrogen atoms and alkyl radicals (the most important products of chain branching) in different conditions are characterized by different properties. It can be noticed in Figure 4 that for rich mixtures (φ > 1) alkanes are inhibitors of ignition reaction due to important rerouting of hydrogen atoms from chain branching reaction. This feature is predominating over ignition so increment temperature decreases with the growth of φ. For lean mixtures (φ < 1), not very often collisions between oxidizing and reducing species (small presence of fuel) are limiting the rerouting of hydrogen atoms (favorable factor). Here, alkanes behave like promoters of their own ignition. In this case, increment temperature increases with the growth of φ. For values φ in which ΔT reaches its maximum, there is a balance between the amount of reactive radicals and not very well reactive hydrogen atoms.

3.2 Comparison of experimental results with numerical simulations

Figures 6 and 7 show the measured and calculated increment temperatures using Leeds Methane

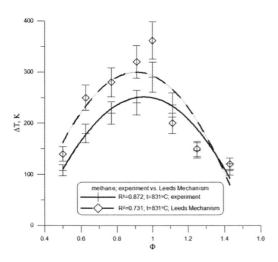

Figure 6. Dependence of the increment of temperature as a function of equivalence ratio; comparison of the experiment results and calculation using Leeds Mechanism.

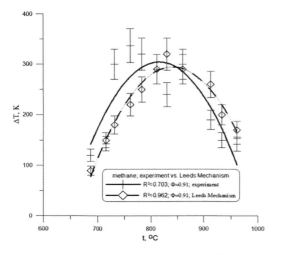

Figure 7. Dependence of the increment of temperature as a function of oxidizer temperature; comparison of the experiment results and calculation using Leeds Mechanism.

oxidation Mechanism. As can be seen, the calculation using the Leeds Mechanism show good agreement with the measurements over the range of oxidizer temperature and equivalence ratio studied.

4 CONCLUSIONS

In this work the ignition of methane was investigated with the use of different high temperature oxidizers. Moreover, Constant Volume Bomb (CVB) was applied in the investigation. Taking into consideration analyzed parameter (ΔT), it seems that preheating the oxidizer without limits seems to be unsubstained. There is the value of temperature oxidizer (t ≈ 830°C) in which analyzed parameter reaches its extreme values. Very good agreement between the measurements and calculations using Leeds Mechanism of methane oxidation is observed. Over the entire operational range of temperatures of oxidizer and equivalence ratio used in present study, the differences between the measured and calculated values of the increment temperature are less than 25%.

ACKNOWLEDGMENTS

The paper has been prepared within the frame of the statute research of the Institute of Thermal Technology, Silesian University of Technology.

NOMENCLATURE

A	radiation constant,
c_p	specific heat at constant pressure [J/kgK],
$\dot{E}$	flux of energy [W],
g	mass fraction,
i	specific enthalpy [J/kg],
$\dot{I}$	flux of enthalpy [W],
$\dot{m}$	flux of mass [kg/s],
m	mass [kg],
M	molar mass [kg/kmol],
p	pressure [Pa],
Q	flux of heat [W],
u	specific internal energy [J/kg],
U	internal energy [J],
v	specific volume [m³/kg],
V	volume [m³],
t	temperature [°C],
T	temperature [K],

Greek letters:

ρ	density [kg/m³],
τ	time [s],
ω	yield of source reaction [kmol/m³ s],

Subscripts:
i   for "i-th" substance,
in   inlet,
out   outlet
max   maximal,
s   system,
w   wall.

REFERENCES

Agilent 34970A Data Acquisition/Switch Unit, Heweltt-Packard, 1999.

Brett L., Macnamara J., Musch P., Simmie J.M., 2001, Simulation of methane autoignition in a rapid compression machine with creviced pistons, Combustion and Flame, 124,: 326–329.

Chaumiex N., Pichon S., Lafosse F., Paillard C.-E., 2007, Role of chemical reaction on the detonation properties of hydrogen/natural gas/air mixtures, International Journal of Hydrogen Energy, 32,: 2216–2226.

Coppens F.H.V., De Ruyck J., Konnov A.A., 2007, The Effects of Composition on the Burning Velocity and Nitric Oxide Formation in Laminar Premixed Flames of CH4 + H2 + O2 + N2, Combustion and Flame, 149,: 409–417.

Davidenko D., Gökalp I., Dufour E., Gaffiè D., 2002, Kinetic mechanism validation and numerical simulation of supersonic combustion of methane-hydrogen fuel, AIAA 2002 (paper number 5207).

Hughes K.J., Turănyi T., Clague A.R., Pilling M.J., 2001, Development and testing of a comprehensive chemical mechanism for the oxidation of methane, International Journal of Chemical Kinetic, 33,: 513–538.

Imbert B., Lafosse F., Catoire L., Paillard C-E., Khasainov B., 2008, Formulation reproducing the ignition delays simulated by a detailed mechanism: Application to n-heptane combustion, Combustion and Flame, 155,: 380–408.

Jee S.B., Kim W.K., Shin K.S., 1999, Shock-tube and modeling study of ignition in methane, Journal of the Korean Chemistry Society, 43,: 156–160.

Konnov A., Version 0.5. Available http://homepages.vub.ac.be/~akonnov.

Lefebvre A.H., 1999, Gas Turbine Combustion, Second ed., Philadelphia, PA, Taylor & Francis.

Matynia A., Delfau J.-L., Pillier L., Vovelle C., 2009, Comparative study of the influence of CO2 and H2O on the chemical structure of lean and rich methane-air flames at atmospheric pressure, Combustion, Explosion and Shock Waves, 45,: 635–645.

Methane oxidation mechanism. Available at http:www.chem.leeds.ac.uk/combustion/combustion.html.

Mertens J.D., A shock tube study of CH reaction in CH4 and C2H2 oxidation, 22nd International Symposium on Shock Waves, 1999 (paper 4510).

Petersen E.L., Davidson D.F., Hanson R.K., 1999, Kinetics Modeling of shock-induced ignition in low-dilution CH4/O2 mixtures at high pressures and intermediate temperatures, Combustion and Flame, 117,: 272–290.

Richards G.A., McMillian M.M., Gemmen R.S., Rogers W.A., Cully S.R., 2001, Issues for low-emission, fuel-flexible power systems, Progress in Energy and Combustion Science, 27,: 141–169.

Safta C., Madnia C.K., 2006, Autoignition and structure of nonpremixed CH4/H2 flames: detailed and reduced kinetic model, Combustion and Flame, 144,: 64–73.

Sang C.J., Li B., Law C.K., 1998, Structure and sooting limits in counterflow methane/air and propane/air diffusion flames from 1 to 5 atmospheres, Proceedings of the Combustion Institute 27,: 1523–1530.

Simmie J.M., 2003, Detailed chemical kinetic models for the combustion of hydrocarbon fuels, Progress in Energy and Combustion Science, 29,: 599–634.

Smith G.P., Golden D.M., Frenklach M., Moriarty N.W., Eiteneer B., Goldenberg M, Bowman C.T., Hanson R.K., Soonho Song, Gardiner W.C., Lissianski V.V. Jr., Qin Z., Avialable at http://www.me.berkeley.edu/gri_mech/.

Stamatov V.A., King K.D., Zhang D.K., 2005, Explosions of methane/air mixtures induced by radiation-heated large inert particles, Fuel 84,: 2086–2092.

Szargut J., 2000, Technical thermodynamics, Gliwice, Silesian University of Technology (in Polish).

Szlęk A., 2004, Mathematical modeling of gas combustion kinetics, Gliwice, Silesian University of Technology (in Polish).

Szlęk A., Wilk R.K., Werle S., Schaffel N., 2009, Czyste technologie pozyskiwania energii z węgla oraz perspektywy bezpłomieniowego spalania, Rynek Energii 4,: 39–45.

Werle S., 2010, Study on ignition of gaseous fuels using different types of reactors, Architecture, Civil Engineering, Environment, 4,: 115–124.

Werle S., 2011, Evolution of hydrocarbons ignition delay time over HTAC (high temperature air combustion) conditions, High Temperature Materials and Processes, 1–2,: 145–149.

Werle S., Wilk R.K., 2007, Self-ignition of methane in high temperature air, Chemical and Process Engineering, 28,: 399–412.

Werle S., Wilk R.K., 2010, Ignition of methane and propane in high-temperature oxidizers with various oxygen concentrations, Fuel, 89,: 1833–1839.

Werle S., Wilk R.K., 2010, A constant-volume bomb and co-flow reactor investigation of ignition phenomenon of hydrocarbon fuels in high temperature oxidizer, International Journal of Thermodynamics, 2,: 43–49.

Wu C.-Y., Chao Y.-C., Cheng T.S., Chen C.-P., Ho C.-T., 2009, Effects of CO addition on the characteristics of laminar premixed CH4/air opposed-jet flames, Combustion and Flame, 156,: 362–373.

Zhou X., Chen C., Wang F., 2010, Modeling of non-catalytic partial oxidation of natural gas under conditions found in industrial reformes, Chemical Engineering and Processing: Process Intensification, 49,: 59–64.

Environmental Engineering IV – Pawłowski, Dudzińska & Pawłowski (eds)
© 2013 Taylor & Francis Group, London, ISBN 978-0-415-64338-2

Testing and evaluation of novel CO_2 adsorbents

I. Majchrzak-Kucęba
Czestochowa University of Technology, Faculty of Environmental Engineering and Biotechnology,
Częstochowa, Poland

ABSTRACT: The increasing attractiveness of the sorption methods of CO_2 capturing from coal power plant flue gas, which can be observed in recent years, is linked directly with the appearance and development of new, efficient CO_2 adsorbents. The preparation and physicochemical properties of CO_2 adsorbents derived from fly ash are described in the paper. The CO_2 adsorbents were prepared by modifying the fly ash-based porous materials with imine. The materials (before and after being loaded with imine) were examined by the X-ray powder diffraction, nitrogen adsorption/desorption, FT-IR and SEM methods to assess their physico-chemical properties. The CO_2 adsorption/desorption tests of the adsorbents were carried out by thermogravi-metric methods. The CO_2 adsorption experiments on modified fly ash-based materials, carried out within this work, demonstrated that these materials exhibited high CO_2 adsorption capacity.

Keywords: CO_2, adsorbents, MCM-41, PEI, fly ash

1 INTRODUCTION

Due to the global obligations to switch over to low-emission technologies, the development of clean technologies must be maintained in parallel to the technologies of enhancing efficiencies and reducing CO_2 emissions by CO_2 capture and storage, among other things. Although CO_2 separation technologies are known from various sectors of industry, the separation of CO_2 is a new problem for the power industry, and therefore there is no industrial practice in this matter. In addition, new CO_2 separation technologies are not cost-effective when considered in the contexts of their application in large power plants. Hence, a need arises for studies toward the development of the existing technologies with the aim of reducing the costs and energy consumed in the CO_2 capturing process. Tests to verify the suitability of the separations methods on a commercial scale (demonstration installations) are also necessary. The activities being undertaken currently towards the development of CO_2 separation methods include the development of CO_2 solvents (absorbents) and adsorbents to enhance their stability, sorption capacity and regeneration ability (Metz 2005, Kather 2008).

The success of an adsorbent depends on the development of the material that, under flue gas temperature conditions, will have high sorption capacity and selectivity for CO_2. At the same time, the ease of regeneration and the usable lifetime of the adsorbent is of key importance. The more easily an adsorbent is regenerated, the more effective the overall CO_2 capture process becomes. Similarly, the lifetime of an adsorbent is important when determining the number of cycles in which the adsorbent will be used, which obviously influences the economy of the process. The increasing attractiveness of the sorption methods of CO_2 capturing from coal power plant flue gas (compared to the preferred currently absorption methods), which can be observed in recent years, is linked directly with the appearance and development of new, efficient CO_2 sorbents (Xu 2003 & 2005, Son 2008, Khalil 2012, Olivares-Marin 2011, Belmabkhout, Shafeeyan 2011). These sorbents include amine-modified mesoporous silica, in which the porous substrate (mesoporous silica) is modified with amine. The conducted tests cover both various types of mesoporous materials (MCM-41, MCM-48, SBA-15, SBA-12, SBA-16, KIT-6), as well as different types of amines (PEI, TEPA, APS, DEA). The processes of mesoporous material modification can be conducted in different ways, e.g. by impregnation, grafting amine functional groups, or surface coating. By saturation of solid adsorbents with amines, the regeneration costs are reduced, eliminate the corrosion problems and ensure a stronger reaction of amines with CO_2 (chemical adsorption). In addition, chemical adsorption allows operation at higher temperatures and higher separation selectivity, after all it is essential that the sorbent operate at the flue gas temperature.

Particularly often, the MCM-41 mesoporous material (Xu, 2003 and 2005, Son, 2008) is proposed as the solid support and PEI as the amine, together with a simple impregnation method. The reason are probably the properties of this material (the large volume of pores), which allow a large quantity of PEI to be introduced to it for capturing CO_2.

The attractiveness of polymer-modified mesoporous materials as adsorbents may continue to increase owing to the fact that there is a possibility of synthesizing mesoporous materials, such as MCM-41 using fly ash as the source of Si (Kumar 2001, Misram 2007, Jang 2009, Majchrzak-Kucęba 2009). Adsorbents produced from fly ash could become more attractive CO_2 adsorbents than those synthesized classically, owing to, for instance, their production cost. The utilization of fly ash as the source of Si in the synthesis of mesoporous molecular sieves creates a situation, where a combustion by-product becomes in part an efficient CO_2 adsorbent. This is confirmed by first studies carried out in this direction (Chandrasekar 2008, Jang 2009). Chandrasekar and co-workers studied CO_2 adsorption on the SBA-15 mesoporous material derived from fly ash. Produced based on ash extract, the SBA-15 was impregnated with PEI in the amount of 50 wt%, and showed a sorption capacity of 110 mgCO$_2$/g of adsorbent at a temperature of 75°C (with pure CO_2). For the commercial SBA-15, a slightly lower sorption capacity value of 120 mgCO$_2$/g of adsorbent was obtained. Jang and co-workers examined CO_2 adsorption on the MCM-48 mesoporous material produced from rice husk ash. The obtained MCM-48 they modified with APTS (3-aminopropyltriethoxysilane). They examined amine-grafted MCM-48 for CO_2 adsorption at temperatures of 25, 50, and 75°C in 1.04% CO_2 in He, obtaining sorption capacity at a level of 0.639, 0.596, 0,479 mmolCO$_2$/g sorbent, respectively. The TGA method is used by many (Xu, 2003 & 2005, Son 2008) as a straightforward, fast laboratory method enabling the evaluation of the sorption capacity of materials synthesized and the determination of their suitability for CO_2 separation. The thermogravimetric methods enables also the assessment of sorption/desorption profiles for test samples. In addition, the common use of this method allows sorption testing results obtained by different research centres for different sorbents to be compared.

The aim of the work was to synthesize the MCM-41 mesoporous material from fly ash, modify it with PEI to increase its sorption capacity, and to assess its sorption properties with respect to CO_2, particularly in a flow of a mixture of gases simulating the flue gas and at temperatures typical of flue gas.

2 MATERIAL AND METHODS

2.1 Materials and adsorbent preparation

Cetyltrimethylammonium bromide (CTAB:C$_{16}$H$_{33}$ (CH$_3$)$_3$ NBr), PEI (polyethyleneimine) sodium hydroxide, methanol and deionized water were used as raw materials. All chemicals were purchased from Aldrich. For comparison, sample commercial MCM-41 from Aldrich was used. Coal fly ash was obtained from an utility power station in Poland. The fly ash was chemically analyzed and used to prepare the mesoporous materials MCM-41.

In order to extract silicon (in the form of sodium silicate) from fly ash, a method consisting in the fusion of fly ash with sodium hydroxide was used. According to this methodology (Shigemoto 1994), fly ash F (25 g) and sodium hydroxide (30 g) were used for the sintering of fly ash with sodium hydroxide so as to obtain an NaOH:fly ash weigh ratio of 1:2. After being accurately weighted, sodium hydroxide was ground to obtain powder. Then, the weighted amount of ash was added and ground with NaOH until a homogeneous mixture was obtained. So obtained powder was heated according to the following programme: increasing temperature from 298 K to 773 K—1 K/min.; holding at 773 K for 1 hour. Then, the mixture was carefully comminuted by grinding and, after weighing, it was transferred to a large beaker, to which an appropriate amount of water was poured so as to obtain a water to product weight ratio of 4:1. The whole was stirred at room temperature for 24 hours, and then vacuum filtered. Thus obtained solution containing sodium silicate was subjected to ICP analysis (using a Varian ICP-OES VISTA-MPX emission spectrometer) for the following elements: Si, Al.

The obtained sodium silicate, after having been weighed, was transferred to a large polyethylene beaker and was stirred using a mechanized stirrer. After a 20 minutes, a surfactant (Cetyltrime-thylammonium bromide—CTAB—Aldrich) was started to be slowly added by drops in such an amount that the Si:CTAB weigh ratio was 20:5. After the surfactant had been added, the pH of the solution was adjusted so as to its value was 11. Sulphuric acid (VI) was used for this purpose. After obtaining the desired pH value, the whole was stirred for another 30 minutes, and then the pH value was checked again and adjusted accurately to 11. Next, the mixture was poured to a polyethylene bottle and put in an oven for 24 hours at a temperature of 373 K. After this time, the pH value was adjusted again to 11, and then the mixture was placed in the oven for another 24 hours at 373 K. Subsequently, the obtained solid substance was separated from the solution and flushed with approx. 200 cm^3 of distilled water. The obtained material was first dried at 333 K for 24 hours, and then calcined at 823 K (at a temperature increase rate of 5 K/min), with dried helium being passed through the oven at a rate of 60 cm^3/min during the whole period of temperature increase and during the first hour of calcination. Thus obtained material was designated as F-MCM-41.

The method proposed by Xu et al (2003) was applied to the obtained PEI-modified F-MCM-41 from fly ash and PEI-modified commercial MCM-41 (for comparision). Following this method, the appropriate amount of PEI was dissolved in 4 g of methanol under stirring for about 15 minutes, and finally 1 g of calcined F-MCM-41 (MCM-41) was added to the PEI/methanol solution.

The obtained mixture was stirred for about 30 minutes, and then dried at a temperature of 70°C for 16 hrs under a pressure of 700 mm Hg. Thus obtained adsorbent was designated as F-MCM-41-PEI-30 (MCM-41-PEI-30) and F-MCM-41-PEI-50, (MCM-41-PEI-50) where 30 and 50 denotes the percentage share of PEI in the test samples.

2.2 Characterisation of the adsorbents

The fly ash-based mesoporous materials F-MCM-41 (before and after being loaded with PEI) was characterized by powder X-ray diffraction, nitrogen adsorption/desorption, TGA, FT-IR and SEM analysis. XRD patterns were recorded on a Bruker AXS D8 Advance diffractometer using CuKα radiation of wavelength 0.15405 nm). Diffraction data were recorded between 1.4 to 8° 2Θ at an interval of 0.02° 2Θ and between 6 to 60° 2Θ at an interval of 0.05° 2Θ. The porous properties of the adsorbents were investigated by determining their N$_2$ gas adsorption and desorption isotherms at 77 K, using an ASAP 2010 Instrument (Micromeritics).

The calcined mesoporous samples were degassed at 300°C over night and the PEI loaded samples were out-gassed at 75°C for 48h on a high vacuum line prior to adsorption. The specific surface area was calculated by the BET method from the linear part of BET plot according to IUPAC recommendations using the adsorption isotherm (relative pressure (p/p$_o$) = 0.05–0.23). The pore size distribution was calculated by the BJH method and the pore volume was obtained from the maximum amount of adsorption at p/p$_o$ of 0.99. The FT-IR spectra of the samples were recorded at room temperature on a Nicolet 6700 spectrometer using KBr pellet technique. The samples were analyzed in the wavelength region 1800–400 cm^{-1}. The microstructures of the adsorbents were observed using a Electron Microscopes (EVO-40 Series, Carl Zeiss SMT).

2.3 CO$_2$ adsorption measurements

The examination of the adsorption and desorption of CO$_2$ on fly ash-synthesized F-MCM-41 before and after being loaded with PEI was carried out using a Mettler TGA/SDTA 851e thermobalance. For the examination of the adsorption/desorption process, temperature-programmed adsorption test were used.

In the temperature-programmed adsorption test, a sample of approx. 10 mg was used, which was placed in a platinum crucible and heated up to a temperature of 100°C under a nitrogen atmosphere (50 ml/min), and then heated isothermally at that temperature for about 15 minutes until a constant sample mass had been reached. Next, the temperature was changed to 25°C and the CO$_2$ adsorbate (the simulated flue gas composed of CO$_2$ (vol.10%), O$_2$ (10 vol.%), N$_2$ (80 vol.%) was passed for a period of 2 hours at a flow rate of 50 ml/min. At the next step, temperature was increased at a rate of 0.25°C min^{-1} from 25°C to 100°C (in order to determine the effect of temperature on the sorption capacity of the samples tested). After completion of the adsorption process the gas flow was switched from CO$_2$ adsorbate to N$_2$ at a flow rate of 50 ml/min to measure the CO$_2$ desorption performance.

The change in weight during the carrying out of a temperature programme applied enabled the determination of the sorption and desorption capacities of the adsorbents examined.

The adsorption capacity for the adsorbents tested was expressed in mg CO$_2$/g adsorbent, and the desorption capacity—in percents, based on the change of sample mass during the sorption/desorption processes. The desorption capacity in % was defined as the ratio of the quantity of desorbed gas to the quantity of adsorbed gas.

The CO$_2$ adsorption/desorption capacity of PEI modified fly ash-based F-MCM-41 were compared with the PEI modified commercial MCM-41.

3 RESULTS AND DISCUSSION

3.1 Characterization of the adsorbents

For the assessment of MCM-41 formation from fly ash extract, X-ray diffraction analysis was employed. The obtained diffraction pattern was compared with the diffraction pattern for the commercial MCM-41. Diffraction pattern for nanoporous phase shows reflections in a low-angle range. In Figure 1, well developed XRD reflections can be observed in the low-angle range—a very strong peak at approx. 2.1° 2Θ, resulting from the (100) index, which suggests that this fly ash-derived material is a good-quality MCM-41 preparation.

As shown in Figure 1, two weak peaks were also distinguished as peaks characteristic of the MCM-41 family, at 4.10° (110) and 4.8° (200), suggesting a hexagonal symmetry. For the fly ash-derived F-MCM-41 sample, reflections characteristic of zeolite materials were noticed in the high-angle range, which suggests that the sample was doped with zeolites; the details can be found in work (Arenillas, 2005). Figure 1 shows also XRD reflections of fly ash-derived F-MCM-41 after being PEI loaded −30 wt.% and 50 wt%. As can be seen from the figure, the impregnation with PEI has caused a reduction of the intensity of peaks characteristic of the MCM-41 material (which is greater, the larger amount of PEI has been introduced to the modified material sample). The obtained results confirm that, as a result of the impregnation process, the PEI polymer has been located in the pores of F-MCM-41. Similar results for the commercial MCM-41 were obtained by Xu et al. (2003, 2005).

Figure 2 shows adsorption isotherms for the MCM-41 material produced from fly ash and its modification with PEI (30 wt.% and 50 wt.%).

The adsorption isotherm for the F-MCM-41 is an isotherm of type II. Its shape is similar to the isotherm for the commercial MCM-41 (in the Fig. 2), which confirms that a good quality MCM-41 material has been obtained from the fly ash extract.

After loading 30 wt.% PEI into the pores of the F-MCM-41 sample, the nitrogen adsorption iso-

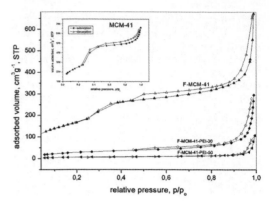

Figure 2. Nitrogen adsorption-desorption isotherms for fly ash-based F-MCM-41 before and after being PEI loaded (30 wt.% and 50 wt.%) and for commercial MCM-41 (for comparison).

therm becomes more flat, and the adsorbed volume rapidly decreases, which is obviously due to the fact that the pores of the PEI-modified material are filled with amine, and therefore blocked. The adsorption isotherm becomes even more flat after loading 50 wt.% PEI into the pores of F-MCM-41, which confirms that in this case the pores are filled with a still larger amount of the amine.

The chemical impregnation process results in a decrease of BET areas and pore volume of the fly ash F-MCM-41, indicating a blocking of some of the micro- and mesopore volume. The BET area for F-MCM-41 (610 m^2/g) has decreased to 109.2 m^2/g (30 wt.% PEI) and 26.5 m^2/g (50 wt.% PEI). In turn, the pore volume for F-MCM-41 (1.03 cm^3 g^{-1}) has decreased to 0.43 cm^3/g (30 wt.% PEI) and 0.13 cm^3/g (50 wt.% PEI). So, the degree of reduction, in terms of both the area and pore volume, increases with the increase in the amount of introduced amine. A similar reduction in the area and pore volume after impregnation, though for pure (not fly ash-derived) MCM-41, was found by Xu et al in their studies (2003, 2005). It has also been confirmed by the authors of the present work, who, for comparison, modified and tested a commercial MCM-41 sample.

Figure 3 shows (a) pore size distributions of F-MCM-41 before and after PEI loaded (30 wt% and 50 wt%) and (b) pore size distributions of commercial MCM-41.

Figure 3 indicates that the introduction of amine to the pores of the MCM-41 material results in a dramatic reduction of the volume of pores (both F-MCM-41 material mesopores, and micropores, as the product is doped with zeolite), and macropores present in the sample. The introduced amine causes their blocking.

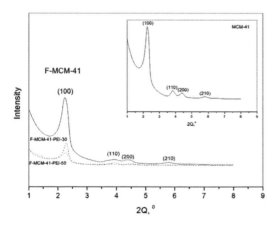

Figure 1. X-ray diffraction pattern of fly ash-based F-MCM-41 (before and after being PEI loaded −30 wt.% and 50 wt.%) and commercial MCM-41 (for comparison).

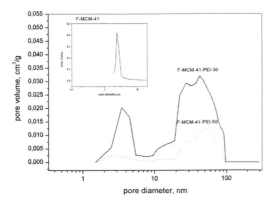

Figure 3. Pore size distributions for fly ash-based F-MCM-41 before and after being PEI loaded (30 wt.% and 50 wt.%).

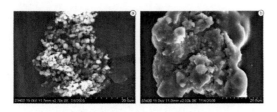

Figure 4. SEM photographs of commercial MCM-41: a) before PEI modification b) after PEI modification.

Figure 5. SEM photographs of fly ash-based F-MCM-41: a) before PEI modification b) after PEI modification.

Figures 4 and 5 show SEM images for commercial and fly ash based MCM-41 before and after being PEI loaded (50%). Figures 4a and 5a show SEM photograph of fly ash-derived MCM-41 and commercial MCM-41 (for comparison) before modification of PEI. The SEM image for sample shows a regular arrangement of channels. The particle size of the fly ash derived F-MCM-41 was 5–10 μm.

The photographs confirm the presence of the MCM-41 phase in the sample produced from fly ash. Figures 4b and 5b show a SEM photograph of fly ash MCM-41 and commercial MCM-41 after modification with PEI. Both in the case of

the commercial MCM-41 and in the case of the fly ash MCM-41, a considerable agglomeration of MCM-41 particles due to PEI impregnation and the location of the polymer on the outer surface of the modified materials are visible. Figure 6 shows the FT-IR profiles for the F-MCM-41 before and after being PEI loaded (30 and 50 wt.%).

The FT-IR spectra of the F-MCM-41 (before and after being PEI loaded) shows a characteristic absorption band at 1080 cm^{-1} and 800 cm^{-1} from asymmetric and symmetric stretching of silicon dioxide vibrations, which are characteristics of the mesoporous materials. Weak bands for NH_2 group at 1600 cm^{-1} and for CH_2 at 1460 cm^{-1} are present from PEI and confirm the presence of PEI in the modified samples (F-MCM-41-30PEI, F-MCM-41-50PEI). The XRD, N_2 adsorption/desorption, SEM, and IR examinations of the MCM-41 material (before and after being PEI loaded) confirmed that as a result of the impregnation of the F-MCM-41 material, the PEI used was located within the material pores.

3.2 Carbon dioxide sorption

Figure 7 grammed CO_2 adsorption/desorption test for the F-MCM-41 before and after being PEI modified (F-MCM-41-PEI-30 and F-MCM-41-PEI-50) (10 vol.% CO_2). As can be seen, the F-MCM-41 material exhibits a low sorption capacity with respect to CO_2 (5.9 mg CO_2/g adsorbent at 25°C). The CO_2 up-take on the fly ash-derived mesoporous materials decreased with increasing adsorption temperature. The starting mesoporous F-MCM-41 material exhibited a behaviour typical of physical adsorbents, that is its sorption capacity was the highest at a temperature of 25°C (though it was much lower than the sorption capacity of

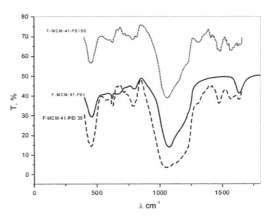

Figure 6. IR profiles for the fly ash-based F-MCM-41 before and after being PEI loaded (30 wt.% and 50 wt.%)

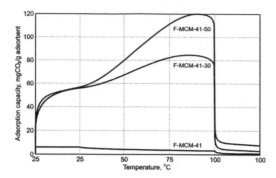

Figure 7. Temperature-programmed CO_2 adsorption/desorption test for the F-MCM-41and the modified F-MCM-41-PEI-30 and F-MCM-41-PEI-50 (for a mixture of gasses with 10 vol. % CO_2).

PEI-impregnated materials), and then rapidly decreased as the temperature increased. This trend is due to the fact that the physical CO_2 adsorption, which took place in this case, is strongly dependent on temperature, considerably decreasing with its increase. The low sorption capacity of the MCM-41, either synthesized classically or fly ash-derived, results from weak interaction with CO_2, particularly at higher temperatures.

The pure silica surfaces do not interact very strongly with CO_2 because the residual hydroxyl groups are not able to induce enough strong interactions and further specific adsorption sites are missing (Macario 2005). For PEI-impregnated sample, we can notice a different behaviour, namely an increase in sorption capacity with the increase in temperature. In this case, is presumably likely to be caused by the combination of physical adsorption inherent from the parent sample and chemical adsorption of the loaded amine groups. As indicated by Figure 7, the 50 wt.% PEI impregnated (F-MCM-41-PEI-50) sample exhibited the greatest increase in sorption capacity with increasing temperature, reaching a maximum (120 mg CO_2/g of adsorbent) at 90°C. At this temperature, the starting F-MCM-41 material had a sorption capacity of approx. 1 mg CO_2/g of adsorbent, so almost 120 times smaller. The 30 wt.% PEI impregnated (F-MCM-41-PEI-30) sample showed the smallest sorption capacity increase with the increase in temperature, reaching a maximum at a lower temperature of 80°C.

According to Xu et al. (2003, 2005), as the temperature increases, the PEI polymer becomes more flexible and more CO_2-affinity sites will be exposed to the CO_2 and thus CO_2 adsorption capacity increases. PEI-modified mesoporous materials are sorbents that can be classified as physicochemical sorbents. This is so because the solid physical adsorbent (which is mesoporous material) adsorbs physically, and the chemical substance, that it is impregnated with, adsorbs chemically. Desorption carried out for the test samples at 100°C was complete in each case, which confirms the reversible character of the process and the possibility of using the sorbent in the next cycle.

4 CONCLUSION

The CO_2 sorbents were produced by the impregnation of fly ash-derived mesoporous materials (F-MCM-41) with PEI. The good quality of the F-MCM-41 and the change of its properties after impregnation has been confirmed by numerous analyses. The performed tests of CO_2 adsorption on PEI-modified fly ash-derived mesoporous materials (F-MCM-41) have shown that these materials are characterized by high sorption capacity with respect to CO_2 at a CO_2 concentration typical of flue gas (10%) and at temperatures typical of flue gas (75–100°C).

The PEI-modified fly ash-derived F-MCM-41 material exhibits similar sorption capacity values to those of its commercial counterpart.

As follows from the presented data compared with conventional adsorbents, modified mesoporous materials (both those classically synthesized and those obtained from fly ash) are characterized by higher sorption capacity at relatively high temperatures (at temperature typical of flue gas). The 50 wt.% PEI impregnated (F-MCM-41-PEI-50) sample exhibited the greatest increase in sorption capacity with increasing temperature, reaching a maximum (120 mg CO_2/g of adsorbent) at 90°C. At this temperature, the starting F-MCM-41 material had a sorption capacity of approx. 1 mg CO_2/g of adsorbent, so almost 120 times smaller. The 30 wt.% PEI impregnated (F-MCM-41-PEI-30) sample showed the smallest sorption capacity increase with the increase in temperature, reaching a maximum at a lower temperature of 80°C.

In addition, these materials are even more attractive for CO_2 removal from flue gas, insomuch as the amine groups attached to the solid surface facilitate the regeneration stage. Therefore, based on the presented data it can be stated that the modified fly ash MCM-41 is suitable to be used for CO_2 separation of from flue gas. The utilization of fly ash as a source of Si in the synthesis of mesoporous molecular sieves creates a favourable situation, where a combustion by-product becomes in part an efficient CO_2 adsorbent.

REFERENCES

Arenillas A., K.M. Smith, T.C. Drage, C.E. Snape, 2005, CO_2 capture using some fly ash-derived carbon materials, *Fuel* 84: 2204–2210.

Belmabkhout Y., Serna-Guerrero R., Sayari A., 2011, Adsorption of CO_2-containing gas mixtures over amine-bearing pore-expanded MCM-41 silica: application for CO_2 separation, Adsorption, 17: 395–401.

Chandrasekar G., W.J. Son, W.S. Ahn, 2008, Synthesis of mesoporous materials SBA-15 and CMK-3 from fly ash and their application for CO_2 adsorption, *Journal of porous materials,* 16: 545–551.

Jang H.T., Park YK, Ko YS, Lee JY, Margandan B, 2009, Highly siliceous MCM-48 from rice husk ash for CO_2 adsorption, *International Journal of Greenhouse Gas Control*, 3: 545–549.

Kather A., et al. 2008, Research and development needs for clean coal deployment. IEA Clean Coal Centre, Report CCC/130.

Khalil S.H., Aroua M.K., Daud W.M.A.W. 2012, Study on the improvement of the capacity of amine-impregnated commercial activated carbon beds for CO_2 adsorbing, *Chemical Engineering Journal* 183:15–20.

Kumar P., N. Mal, Y. Oumi, K. Yamana, T. Sano, 2001, Mesoporous materials prepared using coal fly ash as the silicon and aluminium source, *J. Mater. Chem* 11:3285–3290.

Macario A, A. Katovic, A. Giordano, F. Iucolano, 2005, Synthesis of mesoporous materials for carbon dioxide sequestration, *Micropor. Mesopor. Mater.* 81:139–147.

Majchrzak-Kucęba I, W. Nowak, 2009, Studies on the properties of mesoporous materials derived from polish fly ashes, The 26th Annual International Pittsburgh Coal Conference, Proceedings CD.

Metz B., O. Davidson, 2005, Carbon Dioxide Capture and Storage. IPCC Special Report, Cambridge University Press, New York.

Misram H., R. Sing, S. Begum, M.A. Yarmo, 2007, Processing of mesoporous silica materials (MCM-41) from coal fly ash, *Journal of Materials Processing Technology*, 186: 8–13.

Misram H., S. Ramesh, M.A. Yarmo, R.A. Kamarudin, 2007, Non-hydrothermal synthesis of mesoporous materials using sodium silicate from coal fly ash, *Materials Chemistry and Physics*, 101: 344–351.

Olivares-Marin M., Garcia S., Pevida C., Wong M.S., Maroto-Valer M., 2011, The influence of the precursor and synthesis method on the CO_2 capture capacity of carpet waste-based sorbents, *Journal of Environmental Management* 92: 2810–2817.

Olivares-Marin M., Maroto-Valer M., 2011, Preparation of a highly microporous carbon from a carpet material and its application as CO_2 sorbent, *Fuel Processing Technology* 92: 322–329.

Shafeeyan M.S., Daud W.M.A.W., Houshmand A., Arami-Niya A., 2011, Ammonia modification of activated carbon to enhance carbon dioxide adsorption: Effect of pre-oxidation, *Applied Surface Science* 257: 3936–3942.

Shigemoto N.H. Hayashi, K. Miyaura, J. Mater. Sci. 28 (1993) 4781.

Son W.J et al., 2008, Adsorptive removal of carbon dioxide using polyethyleneimine-loaded mesoporous silica materials, *Micropor. Mesopor. Mater.* 113: 31–40.

Xu X., Ch. Song, J.M. Andresen, B.G. Miller, A.W. Scaroni, 2003, Prepared and characterization of novel "molecular basket" adsorbent based on polymer-modified mesoporous molecular sieve MCM-41. *Micropor. Mesopor. Mater.* 62: 29–45.

Xu X., Ch. Song, J.M. Andresen, B.G Miller, A.W. Scaroni, 2005, Adsorption of carbon dioxide from flue gas of natural-fired boiler by a novel nanoporous "molecular basket" adsorbent, *Fuel Processing Technology*, 86: 1457–1472.

Environmental Engineering IV – Pawłowski, Dudzińska & Pawłowski (eds)
© *2013 Taylor & Francis Group, London, ISBN 978-0-415-64338-2*

Removal of hydrogen sulphide from air by means of Fe-EDTA/Fiban catalyst

H. Wasąg
Faculty of Environmental Engineering, Lublin University of Technology, Lublin, Poland

E. Kosandrovich
Institute of Physical Organic Chemistry, National Academy of Sciences of Belarus, Minsk, Belarus

ABSTRACT: In the present paper removal of low concentration of hydrogen sulphide by its selective catalytic oxidation with Fe(III)-EDTA catalyst carried on fibrous ion exchangers has been described. The complex of trivalent iron converts hydrogen sulphide to elemental sulphur. Bivalent iron formed in the reaction is oxidized by the atmospheric oxygen, so complex of trivalent iron is continuously regenerated and the overall process can be accounted as pseudo-catalytic. The role of fibrous package is improving the mass-transfer between the gas and liquid, buffering the pH of absorbing solution and, probably, catalyzing the reactions of oxidation of sulphur and iron ions. It was proved that the filtering layers with anion exchange package are much more active in the catalytic processes of hydrogen sulphide removal than cation exchanger and inert materials. In the addition to the nature of the fibre–solution carrier the process of catalytic oxidation depends on concentration of hydrogen sulphide in the air, relative air humidity, the process time and the content of Fe-EDTA complex in the fibres. It has been established that application of the Fe(III)-EDTA/Fiban catalytic system, under appropriate conditions, led to nearly complete conversion of H_2S to elemental sulphur.

Keywords: hydrogen sulphide, catalytic oxidation, fibrous ion exchangers, air deodorization

1 INTRODUCTION

Hydrogen sulphide is a colourless compound which can be found in natural gases as well as in volcanic gases and hot springs. It is also produced in human and animal wastes and by different human industrial activities, such as rayon textiles manufacture, pulp and paper mills, oil refinery and natural gases treatment, meat rendering plant, etc. (Busca & Pistarino 2003). It may be present in waste gas streams at concentrations between 2 and 1000 ppm. Hydrogen sulphide has a very typical and irritating smell of rotten eggs and can be smelled at concentrations as low as 0.5 ppb (Meeyoo et al. 1998).

The removal of hydrogen sulphide is required for health reasons, odour problems, safety and corrosivity problems. The means of hydrogen sulphide removing mainly depend on its concentration and kind of medium to be purified. Nowadays, in order to remove sulphide hydrogen from wastewater streams, a number of physicochemical processes are in common use (Busca & Pistarino 2003). Many of the methods are based on direct air stripping, chemical precipitation and oxidation. Oxidation processes used for sulphide removal are aeration (catalysed and uncatalysed), chlorination,

ozonation, potassium permanganate treatment and hydrogen peroxide treatment. In all these processes, apart from elemental sulphur, thiosulphate and sulphate may also be formed as end products (Deng et al. 2009). For the removal of hydrogen sulphide from sour gases various well-established techniques are available. On the industrial scale, at high concentration, removal is effected by the Claus reaction or by the modification of the process itself (Rakmak et al. 2010). At lower concentrations, adsorption or oxidation to sulphur oxides is preferred (Piche et al. 2005). Molecular sieves, silica gel and activated carbons have been used to adsorb hydrogen sulphide from waste gas streams (Panza & Belgiorno 2010). Molecular sieves have the highest capacity for adsorption from dry gas streams (Soriano et al. 2009). Activated carbons are more efficient from wet gas streams, such as these generated by sewage (Wu et al. 2005).

In some cases, especially for large gas streams, the volume of the gas to be treated is reduced by concentrating the hydrogen sulphide. Concentration occurs by a chemical reaction or physical absorption under high pressure in an amine or glycol solution. The hydrogen sulphide rich gas (40–90 vol.%), then it is treated in a Claus plant (Jansen et al. 1999).

Since the conventional physicochemical methods for removing hydrogen sulphide from wastewaters and sour gases require large investment and operational costs (e.g. high pressures, high temperatures or special chemicals), the continuing search for more economical and more efficient methods has led to the investigation of new technologies for purifying hydrogen sulphide containing gases. There are numerous current studies devoted to application of catalytic processes in air deodorization technologies (Kawalczyk & Szynkowska 2012). It is known that complex Fe (III)—ethylenediaminetetraacetate (EDTA) is an efficient and selective catalyst of hydrogen sulphide oxidation (Piche & Larachi 2006). Advantages of these systems include the ability to treat both aerobic and non-aerobic gas streams, high H_2S removal efficiencies, great flexibility, essentially 100% turndown on H_2S concentration in feedstock (Karimi et al. 2010). The main disadvantage of these processes is the fact that the catalyst is active in strong alkali environment which causes serious corrosivity problems (Demmink et al. 2002). It seems that the problem can be solved with the help of fibrous ion exchangers as carriers of the catalyst. The base of these materials is fibrous ion exchanger with Fe(III)-EDTA complex immobilized on their functional groups. Having a strong alkali medium they absorb H_2S from the air and convert it to elemental sulphur due to the following redox process:

$$2Fe^{+3}EDTA^- + H_2S = S^0 + 2Fe^{+2} EDTA^- + 2H^+$$

The complex of bivalent iron formed in the reaction is oxidized by the atmospheric oxygen so the overall process can be accounted as pseudo-catalytic (Soldatov & Kosandrovich 2011).

2 MATERIAL AND METHODS

Fibrous ion exchangers Fiban used in the experiments were synthesized at the experimental production plant of the Institute of Physical Organic Chemistry National Academy of Sciences of Belarus (Minsk, the Republic of Belarus). The ion exchangers have been prepared by chemical modification of industrial polyacrylonitrile (PAN) fibre Nitron production of Polotsk petrochemical plant (the Republic of Belarus). The modification was performed with the staple fibres with the fibre length 65 mm and effective diameter 20–22 μm (Wasag et al. 2008). Chemically modified staple was processed to nonwoven needle punctured canvas with surface density 500 g/m² and 5 mm thickness.

Fiban K-5 fibre was obtained by reaction of cross-linking of polyacrylonitrile fibre with hydrazine and following hydrolysis in strong alkali at 90°C. Hydrolyzation causes formation of different nitrogen containing groups, most probable $-NH=NH_2$ and $-NH_2=N$ and predominant carboxylic functional groups formed in the second step of hydrolyzation. The cationic and anionic exchange capacities of the sample were 4.20 and 1.22 meq/g respectively (Soldatov et al. 2004).

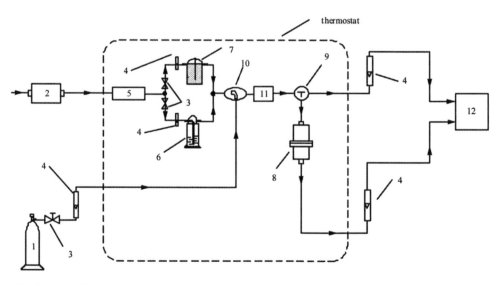

Figure 1. Scheme of experimental laboratory set-up (1—source of hydrogen sulphide, 2—air compressor, 3—valve, 4—flow meter, 5—heat exchanger, 6—humidifier, 7—dryer, 8—reaction chamber, 9—T-valve, 10—mixer, 11—thermo hygrometer, 12—H_2S analyser).

Fiban A-6 is a product of aminolysis of poly-acrylonitrile fibre by dimethylaminopropylamine followed by quaternization of the tertiary amino group with epichlorohydrine. The ion exchanger contains strong and weak basic groups with the capacities 1.80 meq/g and 0.65 meq/g respectively (Soldatov et al., 1996).

Fiban AK-22 is a polyampholyte containing primary and secondary amino groups and carboxylic acid groups (Shunkevich et al., 2001). It is obtained by chemical modification of polyacrylonitrile (PAN) fibre. The predominant functional groups are $RCOOH$ and $R-CO-NH_2-CH_2CH_2NH-CH_2CH_2NH_2$. The cationic exchange capacity is 1.5 meq/g, the anionic exchange capacity is 4.0 meq/g.

Preparation of the H_2S fibrous sorbent-catalysts was performed by soaking of the canvasses or by casting of the impregnation solution onto the surface of the canvas. The impregnation solution was prepared by mixing 0.3 M solution of sodium salt of ethylenediaminetetraacetic acid (EDTA) and $FeCl3 \cdot 6H_2O$ in molar ratio 1:2. Analytical grade reagents were purchased from POCH (Gliwice, Poland).

Removal of hydrogen sulphide from air was studied on the continuous flow experimental laboratory set presented in Figure 1.

Hydrogen sulphide (1000 ppm $H_2S + N_2$, pure, Linde AG) was supplied from a gas cylinder (1) and it was diluted to required concentrations by means of air from a compressor (2). The air before mixing with hydrogen sulphide was dried or humidified depending on required relative humidity of the prepared mixture. The relative humidity and the air temperature were measured by means of thermo hygrometer (AZ8721, Merazet, Poland). Concentrations of hydrogen sulphide in the gas stream before and after the sorption process were measured by GFM101 (Gas Data Ltd., U.K.) gas analyser (12). All experiments were performed at the constant temperature of 20°C. In the conducted laboratory experiments the influence of the following parameters on removal of hydrogen sulphide from the air was investigated: kind of fibrous carrier of the catalyst, relative humidity of the purified air, concentration of hydrogen sulphide and time of the process. The obtained results of laboratory experiments are shown in Tables 1 and 2 and are illustrated in Figures 2–7.

Table 2. Influence of the thickness of filtering layer on the efficiency of H_2S removal from the air by—Fe-EDTA/Fiban AK-22 material (RH = 80%, filtration time 1 hour, air flow velocity (r) 0.14 m/s, initial H_2S concentration 10 mg/m³).

Thickness of the filtering layer, mm (s)	Retention time of the air in the filtering layer, s (t)	C/C_o experiment	C/C_o calculated from equation (1)
2	0.014	0.60	0.59
4	0.029	0.34	0.34
6	0.043	0.20	0.20
8	0.057	0.10	0.12
10	0.072	0.05	0.07

Table 1. Removal of hydrogen sulphide on the 5 mm filtering layer of different fibrous materials after one hour of work (initial concentration of H_2S—60 mg/m³, filtration rate 0.05 m/s).

Carrier of Fe-EDTA (kind of fibrous material)	RH (%)	H_2S removal degree (%)
Fiban K-5 (Na⁺ form)	60	31
	70	45
	80	53
	90	68
Fiban AK-22 (OH⁻ form)	60	41
	70	70
	80	81
	90	92
Fiban A-6 (OH⁻ form)	60	74
	70	89
	80	94
	90	98

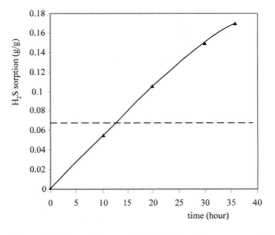

Figure 2. Dependence of the H_2S amount removed from the air per gram of fibre Fe-EDTA/ Fiban AK-22 as a function of the process time (initial concentration of H_2S—60 mg/m³, relative humidity 80%, filtration rate 0.05 m/s).

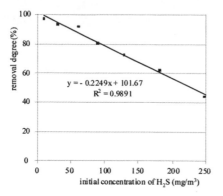

Figure 3. Removal degree of hydrogen sulphide from the air by Fe-EDTA/Fiban A-6 catalytic system as a function of initial concentration of H_2S (filtering layer thickness 5 mm, relative humidity 70%, filtration rate 0.05 m/s).

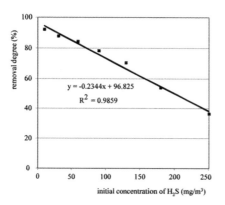

Figure 4. Removal degree of hydrogen sulphide from the air by Fe-EDTA/Fiban AK-22 catalytic system as a function of initial concentration of H_2S (filtering layer thickness 5 mm, relative humidity 80%, filtration rate 0.05 m/s).

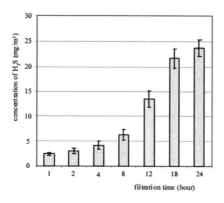

Figure 5. Influence of the process time on H_2S concentration in the purified air after filtration through 5 mm layer of fibre Fe-EDTA/Fiban A-6 (initial concentration of H_2S 30 mg/m³, relative humidity 70%, filtration rate 0.05 m/s).

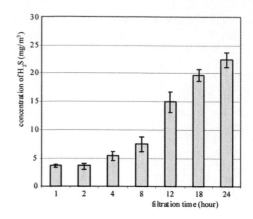

Figure 6. Influence of the process time on H_2S concentration in the purified air after filtration through 5 mm layer of fibre Fe-EDTA/Fiban AK-22 (initial concentration of H_2S 30 mg/m³, relative humidity 80%, filtration rate 0.05 m/s).

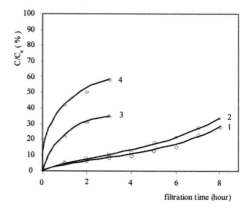

Figure 7. Influence of Fe(III) content in the fibre on the air purification from H_2S by the catalytic system on the base of Fiban AK-22 (initial concentration of H_2S—60 mg/m³, RH = 80%, filtration rate 0.05 m/s, content of Fe(III): 1—0.62; 2—0.35; 3—0.26; 4—0.18 mmol/g).

3 RESULTS AND DISCUSSION

There are many parameters affecting removal of H_2S from air by its catalytic conversion to elemental sulphur with the use of fibrous carrier of Fe-EDTA catalyst. The data presented show that the filtering layers with anion exchange package are much more active in the catalytic processes of H_2S removal than cation exchanger (see Table 1) and inert materials (Soldatov & Kosandrovich 2011). The latter worked only at RH 80–90% and higher. A higher catalytic activity of anion exchanges than that of cation exchanger and inert fibres proves that the Fe complex fixed on the cationic group of the fibre has a higher catalytic activity than that of the free complex. In addition to the nature of the fibre (carrier of the catalyst) relative air humidity is a critical factor determining efficiency of the material in the air purification from H_2S. The effect of RH on the percent of H_2S removal depends on the nature of ion exchanger (Table 1). From the data presented it could be seen that ion exchanger Fiban A-6 is only the one carrier efficient at relative humidity of the purified air below 80%. It could be used even for the RH values ranging 60% with very high degree of hydrogen sulphide conversion exceeding 70%. In case of fibrousion exchanger Fiban AK-22 similar results were obtained for the RH value 80%. Cation exchanger Fiban K-5 requires higher humidity but even at the RH value of 90% removal degree of H_2S is below 70%. Figure 2 shows that the removal of H_2S from the air proceeds mainly because of its catalytic conversion into elemental sulfur. The dotted line relates to the maximal possible absorption of H_2S by anionic groups of the fiber by reaction:

$$R\text{-}NH_2 + H_2S = R\text{-}NH_3^+HS^-.$$

The catalytic conversion of H_2S depends on iron concentration in the fiber. As could be observed from Figure 3, increasing concentration of iron in the fiber improves the catalyst performance at its low values and reaches nearly constant efficiency for iron contents higher than 0.35 mmole/g. The efficiency of air purification, which increases with the increase of thickness in the agreement with empirical formula (1) simply depends on the thickness of filtering layer as it is shown in Table 2.

$$C = C_o \cdot e^{-37,2\,t} \qquad (1)$$

Where t is the residence time of air in the filtering layer (s), it is function of layer thickness and flow velocity:

$$t = s/r \qquad (2)$$

As could be seen from Figures 3 and 4, the performance of the catalytic system Fe-EDTA/Fiban fibre depends on initial concentration of hydrogen sulphide in the air. At lower concentration of hydrogen sulphide the catalytic process allow to remove more than 90% of H_2S. However, the performance of the process decreases according to simple linear regression with the increase of H_2S concentration. The catalyst activity decreases with the increase of the process time (see Figs. 4 and 5). It could be assumed that the reason for deactivation of fibrous catalyst in the conversion process $2H_2S + O_2 \rightarrow 2H_2O + 2S^0$ is blockage of the catalyst surface with accumulating colloidal particles of elemental sulphur.

This negative effect could be minimized by adding surfactants and keeping in this way colloidal sulphur in the liquid phase. Investigations to study such a possibility are being conducted at the moment as a part of research on hydrogen sulphide removal by means of Fe-EDTA/Fiban catalytic system.

4 CONCLUSIONS

The data presented proved the possibility of obtaining fibrous filtering materials able to remove small concentrations of H_2S from the air. The filtering layers with anion exchange package are much more active in the catalytic processes of hydrogen sulphide removal than cation exchanger and inert materials. The role of fibrous package is improving the masstransfer between the gas and liquid, buffering the pH of absorbing solution and, probably, catalyzing the reactions of oxidation of sulphur and iron ions. The process of H_2S removal from the air includes several stages including its dissolution in the solution of Fe(III)-EDTA complex, oxidation of the S^{2-} to S^0 and oxidation of Fe(II) to Fe(III) by the atmospheric oxygen. All of them can proceed in the solution. The catalyst activity decreases with both the increase of the process time and the increase of concentration of hydrogen sulphide in the air. The observed deactivation of fibrous catalyst is probably caused by blockage of its surface with accumulating colloidal particles of elemental sulphur.

ACKNOWLEDGEMENTS

The research presented here was financed by the Polish Ministry of Science and Higher Education: project 7549/B/T02/2011/40 "Application of fibrous ion exchangers as a carrier of catalyst for oxidation of hydrogen sulphide in the air deodorization processes". Partially, the work was conducted within a cooperative research program between the Lublin University of Technology, Faculty of Environmental Engineering and the Institute of Physical Organic Chemistry National Academy of Sciences of Belarus.

REFERENCES

Busca G., Pistarino C., 2003, Technologies for the abatement of sulphide compounds from gaseous streams: a comparative overview, Journal of Loss Prevention in the Process Industry, 16, 363–371. DOI:10.1016/S0950-4230(03)00071-8.

Demmink J.F., Mehra A., Beenackers A.A.C.M., 2002, Absorption of hydrogen sulfide into aqueous solutions of ferric nitrilotriacetic acid: local auto-catalytic effects, Chemical Engineering Science, 57, 1723–1734. DOI:10.1016/S0009-2509(02)00012-6.

Deng L., Chen H., Chen Z., Liu X., Pu X., Song L., 2009, Process of simultaneous hydrogen sulfide removal from biogas and nitrogen removal from swine wastewater, Bioresource Technology, 100, 5600–5608. DOI:10.1016/j.biortech.2009.06.012.

Janssen A.J.H., Lettinga G., Keizer A., 1999, Removal of hydrogen sulphide from wastewater and waste gases by biological conversion to element sulphur. Colloidal and interfacial aspects of biologically produced sulphur particles, Colloids and Surfaces A: Physicochemical and Engineering Aspects, 151, 389–397. DOI:10.1016/S0927-7757(98)00507-X.

Karimi A., Tavassoli A., Nassernejad B., 2010, Kinetic studies and reactor modeling of single step H_2S removal using chelated iron solution, Chemical Engineering Research and Design, 88, 748–756. DOI:10.1016/j.cherd.2009.11.014.

Kowalczyk L., Szynkowska M.I. (2012). Oxidation of ammonia using modified TiO2 catalyst and UV-VIS irradiation. Chemical Papers, 66, 607–611. DOI:10.2478/s11696-012-0159-x.

Meeyoo V., Lee J.H., Trimm D.L., Cant N.W., 1998, Hydrogen sulphide emission control by combined adsorption and catalytic combustion, Catalysis Today, 44, 67–72. DOI:10.1016/S0920–5861(98)00174–6.

Panza D., Belgiorno V., 2010, Hydrogen sulphide removal from landfill gas, Process Safety and Environmental Protection, 88, 420–424. DOI: 10.1016/j.psep.2010.07.003.

Piche S., Larachi F., 2006, Dynamics of pH on the oxidation of H_2S with iron(III) chelates in anoxic conditions, Chemical Engineering Science 61, 7673–7683. DOI:10.1016/j.ces.2006.09.004.

Piche S., Ribeiro N., Bacaoui A., Larachi F., 2005, Assessment of a redox alkaline/iron-chelate absorption process for the removal of dilute hydrogen sulfide in air emissions, Chemical Engineering Science, 60, 6452–6461. DOI:10.1016/j.ces.2005.04.065.

Rakmak N., Wiyaratn W., Bunyakan C., Chungsiriporn J., 2010, Synthesis of Fe/MgO nano-crystal catalyst by sol-gel method for hydrogen sulfide removal, Chemical Engineering Journal, 162, 84–90. DOI:10.1016/j.cej.2010.05.001

Shunkevich A.A., Med'jak G.V., Martsinkevich R.V., Grachek V.I., Soldatov V.S., 2001. Fibrous chelating carboxylic acid ion exchangers. Sorption & Chromatography Processes, 1, 741–748.

Soldatov V.S., Kosandrovich E.G., 2011, Ion exchangers for air purification, in Ion Exchange and Solvent Extraction, A Series of Advances, Volume 20, Edited by Arup K. Sengupta, CRC Press, pp. 45–117. DOI:10.1201/b10813-3.

Soldatov V.S., Pawlowski L., Wasag H., Elinson I., Shunkievich A., 1996, Air pollution control with fibrous ion exchangers. Chemistry for the Protection of the Environment, 2, Plenum Press, New York, 55–66. DOI:10.1007/978–1-4613–0405–0_7.

Soldatov V.S., Pawlowski L., Wasag H., Schunkevich A., 2004. New materials and technologies for environmental engineering. Part I—Synthesis and structure of ion exchange fibres, Lublin (Poland). Monographs of the Polish Academy of Sciences, 21, 1–127.

Soriano M.D., Jimenez-Jimenez J., Concepcion P., Jimenez-Lopez A., Rodriguez-Castellon E., Lopez Nieto J.M., 2009, Selective oxidation of H_2S to sulphur over vanadium supported on mesoporous zirconium phosphate heterostructure, Applied Catalysis B: Environmental, 92, 271–279. DOI:10.1016/j.apcatb.2009.08.002.

Wasag H., Soldatov V., Kosandrovich E., Sobczuk H., 2008. Odour control by fibrous ion exchangers, Chemical Engineering Transactions, 15, 387–394.

Wu X., Kerchief A.K., Schwartz V., Overbuy H.S., Armstrong R.T., 2005, Activated carbons for selective catalytic oxidation of hydrogen sulfide to sulfur, Carbon 43, 1084–1114. DOI:10.1016/j.carbon.2004.11.033.

Indoor microclimate

Environmental Engineering IV – Pawłowski, Dudzińska & Pawłowski (eds)
© *2013 Taylor & Francis Group, London, ISBN 978-0-415-64338-2*

A ground source heat pump with the heat exchanger regeneration—simulation of the system performance in a single family house

J. Danielewicz, N. Fidorów & M. Szulgowska-Zgrzywa
Faculty of Environmental Engineering, Wroclaw University of Technology, Wrocław, Poland

ABSTRACT: In this paper issues concerning the performance of a ground coupled heat pump with borehole heat exchanger serving as an energy source have been discussed. The main part of the presented analysis included the regeneration of the vertical earth loop with the surplus heat from solar collectors or using the waste heat form natural cooling system. In this article the analysis concerning systems in an exemplary single family house has been performed. The possibility of coverage of the energy taken during the heating season by the accumulation of the waste energy from solar collectors and natural cooling in the ground has been assessed. Performed analysis revealed that it is possible to cover 25 to 28% of the energy taken during heating season with the waste energy from natural cooling system and solar collectors respectively.

Keywords: renewable energy sources, solar collector, heat pump, natural cooling, vertical ground heat exchanger, heat source regeneration

1 INTRODUCTION

The ground coupled heat pumps with vertical heat exchanger have been known and used for many years. It is also known that if the energy is only extracted or supplied to the ground the heating or cooling performance will drop because of the changes of the lower energy source temperature (Li et al. 2006). Because of the phenomena described above, the idea of the vertical heat exchanger regeneration is examined all over the world. The Underground Thermal Energy Storages (UTES) have chance to be used as the storage of solar and waste energy (Sanner et al. 2003). The multiple tools for the ground temperature fields analysis are constantly developed (Katsura et al. 2008). The simulation research related to the temperature distribution around the vertical heat exchanger are often performed (Li et al. 2009), the models describing the heat exchange in the loop, taking into account the heat exchanger shape and the underground water flow are created (Nam et al. 2008), and the comparative simulations of the long term exploitation of the boreholes in number of configurations, with negligible underground water flow, are also conducted (Lazzari et al. 2010). The whole systems also become the subject of simulation, both the ones using the heat exchanger regeneration based on the all year round system performance in heating and active cooling mode (Desideri et al. 2011), and the ones

using the surplus energy from solar collectors for regeneration (Kjellsson et al. 2010). The simulations are performed using the available computer programmes like TRNSYS (Desideri et al. 2011, Kjellsson et al. 2010) or FLUENT (Li et al. 2009), but the researchers also create the original tools, in which after the initial introduction of data the user receives the amount of energy that has been extracted and supplied to the exchanger as well as the electric energy consumption (Michopoulos & Kyriakis 2009). The computer simulations need to be verified by the experimental research and such examinations are also conducted (Li et al. 2006, Trillat-Berdal et al. 2006). The models, simulations or the research mentioned above are conducted in many European (Lazzari et al. 2010, Desideri et al. 2011, Kjellsson et al. 2010, Michopoulos & Kyriakis 2009, Trillat-Berdal et al. 2006) or Asian countries (Li et al. 2006, Katsura et al. 2008, Li et al. 2009, Nam et al. 2008, Karaback et al. 2010, Hwang et al. 2009) and usually concern local climate conditions, which change according to the research location. Most of the simulations concern small to medium systems in residential buildings (Li et al. 2006, Desideri et al. 2011, Kjellsson et al. 2010, Trillat-Berdal et al. 2006). In countries with hot climate the ground overheating while using active cooling is the problem (Li et al. 2006, Li et al. 2009, Karaback et al. 2010, Hwang et al. 2009), and not the overcooling related to the heat pump exploitation mainly in low external temperatures.

2 MATERIALS AND METHODS

2.2 *Ground coupled heat pumps with vertical heat exchangers*

A heat pump is a device that transfers the energy from a source that has lower temperature to the receiver, which has higher temperature, using the physical work according to the second law of thermodynamics. Heat pumps are widely used in heating, cooling and heat recovery installations. Nowadays, when the renewable energetics becomes more and more significant, heat pumps are used as devices that may get energy from renewable sources that have low thermal potential (whose temperature prevents their direct use). The simulation research described in the article is related to the ground coupled heat pumps with vertical closed loops in which the water ethylene glycol solvation serves as intermediary agent. Ground serves as good heat source and receiver for a heat pump. The availability of the energy from the ground is in principle coherent to the needs because the ground is wormer than the external air in winter and cooler in summer. The disadvantage of this source is the high investment cost (Szulgowska-Zgrzywa & Fidorów 2011).

Designing the vertical ground coupled heat exchanger is a complicated process which demands plenty of information and wide knowledge. In order to calculate the required quantity of boreholes and their length, and to propose their configuration the thermal characteristics of the ground as well as the heat pump system labour characteristics need to be known. The information about the ground might be taken from the geological research, however, the source is usually not precise enough. The geological characteristics are usually created for the areas that are several times greater than the exchanger area. Usually the test boreholes are made in order to know the ground structure and to examine its thermal properties. It is also known that those properties may change during the time. Those changes happen due to the natural seasonal change of temperature or underground water flow but also the change of ground characteristics due to the thermal exploitation of the boreholes. If the heat extraction from the ground is not balanced, the ground temperature may change permanently, especially when there is no ground water flow, or the water flow is negligible. The phenomena of the overcooling or overheating the ground are intensified by the neighbouring boreholes. The boreholes in the ground are placed in certain distance one from another, but it is known that the area of the heat exchanger is always limited and the boreholes may not be located extremely distant one from another, what would exclude their reciprocal

influence (Szulgowska-Zgrzywa & Fidorów 2011). In Poland the heat pumps, using the vertical boreholes as heat source, are used mostly for heating. This means that the energy is taken from the ground, what causes the vertical boreholes cool down and would influence the heat pump output. As a consequence, greater amount of energy will be derived from electric grid (Szulgowska-Zgrzywa & Fidorów 2011).

2.3 *Solar collectors*

Solar thermal collectors become, also in Poland, the standard solution among installations. Collectors may be installed both in the single and multi-family houses as well as in public buildings. Typical solution for the single family house consists of thermal solar collectors that support the preparation of domestic hot water. Systems that let also the supply of the central heating installations with the heat coming from the sun become more and more popular. In public buildings thermal solar collectors may be used for the same purposes as in residential buildings. In both cases the problem with the excess thermal energy produced periodically by the collectors often occurs. While designing the solar collectors installation it is important to keep that issue in mind and try to solve it by looking for the way of use of this surplus heat. One of the solutions is to heat the pool water, but it is obvious that not all buildings may be equipped with such heat receiver. The other solution is to store the heat and use it some other time (Szulgowska-Zgrzywa & Fidorów 2011).

2.4 *Natural cooling system*

Natural cooling or free-cooling is a system that lets produce some cooled water without the involvement of refrigeration compressors. Nowadays, it is most often present in air-cooled air conditioning units. Water or other agent may be also cooled, without compressors, in the ground, for example in the vertical closed loops that serve as a lower energy source for heat pumps in winter. The cooling agent obtained in such a way is much cooler than external air and the only generated energy cost is a result of pumping the medium through the heat exchanger. In order to apply a natural cooling system in a heat pump system, which supplies the residential building in heat, it is necessary to equip the device with additional heat exchanger, mixing valve and circulating pump. Cooling may be accomplished by the same devices that heat the building in winter, in particular by floor heating system or fan coils. Natural cooling system is not an air conditioning system. It allows for the air temperature reduction by 2°C to 4°C in comparison with the external temperature. Energy that is received from the spaces in the building is delivered to the ground heat exchanger and may be used in heating season (Szulgowska-Zgrzywa & Fidorów 2011).

2.5 Heating and cooling systems for a single family house

Solutions described above may be combined in a complete heating and domestic hot water preparation system including solar collectors cooperating with the heat pump (SYSTEM I) or cooling and heating system including separate solar domestic hot water preparation system and ground coupled heat pump with free cooling installation (SYSTEM II). The performance of such two systems has been analysed in the present article.

The heating system based on the heat pump with vertical borehole regeneration is planned for a hypothetical single-family house with living space of 200 m² and the design heat load of 8440 W. For the purposes of floor heating system, with parameters 45/35°C, Vitocal 108 BW 300 heat pump has been chosen. The heat pump heating power equals to 9.5 kW for the brine temperature of 7°C and the flow temperature of 50°C, the power of the ground heat exchanger for these parameters reaches a value of 6.8 kW and the COP of 3.55. Two boreholes 83 meters long each (assuming the active length of exchanger equal to 68 m and an average heat input capacity of the soil probes for double U-shaped pipe for water-saturated sedimentary layers at 50 W/m) constitute the lower energy source. The boreholes are located 6 meters from each other. The impact of the borehole to the ground was assumed within 3 meters from the borehole. Therefore, the theoretical volume of Underground Thermal Energy Storage (UTES) will be 3843 m³. In both analysed cases the same central heating system has been implemented.

Installation of hot water was provided for four people. In SYSTEM I four solar collectors with total area of 8.8 m² have been applied. An additional heat source is an electric heater with a power of 3000 W installed in domestic hot water cylinder with a capacity of 300 litres. In SYSTEM I the solar collectors cooperate with heat pump, the excess heat from the solar collectors will be directed to the heat pump boreholes. In the SYSTEM II the heat source for domestic hot water is also solar collectors set with an additional electric heater. In such a case, the two collectors with a total area of 4.4 m², the domestic hot water cylinder with a capacity of 300 litres and electric heater with a power of 3000 W have been applied. The domestic hot water preparation system is separated from the central heating system in this case.

The natural cooling system uses an existing installation of floor heating. System components that allow for the cooling are Vitotrans 100 heat exchanger, two changeover valves, mixing valve, circulating pump and an additional set of sensors supplied by the manufacturer of the heat pump. The cooling capacity of the natural cooling system has been estimated on the basis of the data provided by manufacturer of the system and depends on the active surface, the floor material and distance among pipes. For the designed system, the cooling power that can be derived from the natural cooling system at flow temperature of 14°C equals to 5.25 kW.

2.6 Model description

In the article the heat pump performance in two systems, which allow for the ground regeneration during periods of high external temperatures with waste heat from solar collectors or natural cooling system, has been discussed. The scheme of both systems performance has been presented in Figures 1 and 2. It is noticeable that both systems have common elements. The final model requires the development of computational approaches to the various system components, as shown below.

2.6.1 The model of the solar collectors cooperation with heat pump system (SYSTEM I)

Figure 3 presents a simplified computational scheme of the flat solar collector installation feeding ground heat exchanger of the heat pump. Introduced symbols were used in the formulas depicting the steps of the calculation algorithm that allow for the determination of the characteristic installation parameters. The steps of the algorithm for SYSTEM I are described below.

2.6.2 Calculations concerning the simulation of the heat pump heat exchanger discharge

In the first step the instantaneous heat demand for spatial heating has been calculated.

$$Q_{CH} = \frac{t_i - t_e}{t_i - t_{ed}} \cdot \Phi - Q_{IG}, W \qquad (1)$$

where:
Q_{CH}, W—instantaneous heat load,
t_i, °C—internal temperature,
t_e, °C—instantaneous external temperature,
t_{ed}, °C—design external temperature,
ϕ, W—design heat load,
Q_{IG}, W—internal solar gains.

In the second step the instantaneous temperature in the buffer tank has been determined $\{t_B, °C\}$.

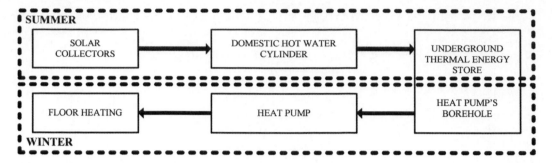

Figure 1. SYSTEM I—storage and utilization of waste solar energy.

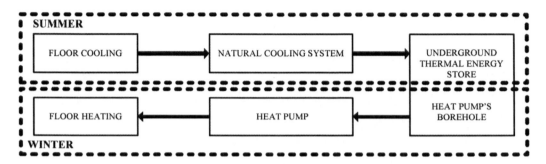

Figure 2. SYSTEM II—storage and utilisation of waste energy from natural cooling system.

In the third step, depending on the needs, the heat pump turning on and off was simulated in order to include the energy delivered to the system {Q_{HP}, W}. In the fourth step the amount of heat that has been taken from the lower heat pump energy source has been calculated.

In the third step, depending on the needs, the heat pump turning on and off was simulated in order to include the energy delivered to the system {Q_{HP}, W}. In the fourth step the amount of heat that has been taken from the lower heat pump energy source has been calculated.

$$Q_{G,Hp} = Q_{HP} \cdot \frac{COP-1}{COP} = Q'_0, W \qquad (2)$$

where:
$Q_{G,HP}$, W—lower energy source output,
Q_{HP}, W—heat pump output,
COP—heat pump coefficient of performance,
Q'_0, W—heat taken from the ground.

In the fifth step the resultant temperature of the buffer tank has been calculated.

$$t_B^{\tau+\Delta\tau} = t_B + \frac{\Delta\tau}{V_B \cdot \rho_w \cdot cp_w} \cdot (Q_{HP} - Q_{CH}), °C \qquad (3)$$

where:
$t_{B'}^{\tau+\Delta\tau}$, °C—resultant buffer tank temperature,
t_B, °C—buffer tank temperature,
$\Delta\tau$, s—time step,
V_B, m³—buffer tank capacity,
ρ_w, kg/m³—water density,
cp_w, kJ/(kgK)—specific heat of the water.

In the sixth step the ground temperature, showing the energy storage discharge, has been calculated.

$$t_{G'}^{\tau+\Delta\tau} = t_G - \frac{\Delta\tau}{V_G \cdot \rho_G \cdot cp_G} \cdot Qo', °C \qquad (4)$$

where:
$t_{G'}^{\tau+\Delta\tau}$, °C—resultant ground temperature,
t_G, °C—ground temperature,
V_G, m³—underground thermal energy storage capacity,
ρ_G, kg/m³—ground density,
cp_G, kJ/(KgK)—specific heat of the ground.

2.6.3 *Calculations concerning the simulation of the heat pumps heat exchanger charge (after discharge)*

In the first step the heat gained in the solar collectors have been calculated.

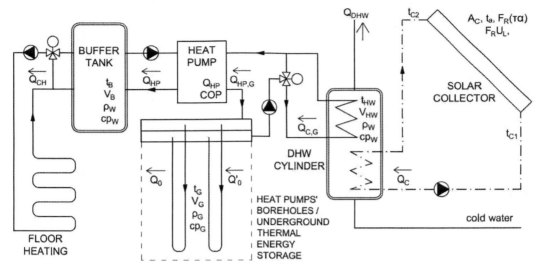

Figure 3.   Scheme of the solar collectors installation feeding the heat pump boreholes.

$$Q_C = A_C \cdot FR(\tau\alpha) \cdot G - A_C \cdot F_R U_L \cdot (t_{C1} - t_a), W \tag{5}$$

where:
Q_C, W—heat gained from solar collectors,
A_C, m²—solar collectors area,
FR $(\tau\alpha)$—solar collector optical efficiency,
G, W/m²—intensity of solar radiation,
$F_R U_L$, W/m²—solar collector heat loss coefficient,
t_{C1}, °C—solar collectots inlet temperature,
t_a, °C—external air temperature.

In the second step the instantaneous temperature in the domestic hot water cylinder has been determined {t_{HW}, °C}. In the third step the flow temperature of solar collectors have been calculated.

$$t_{c1} = t_{HW} + 5, °C \tag{6}$$

where:
t_{HW}, °C—domestic hot water temperature.

In the fourth step the heat distribution for domestic hot water preparation purposes have been assumed {Q_{DHW}, W}. In the fifth step, in case of excess energy occurrence in the cylinder the heat withdrawal from the tank in order to feed boreholes has been simulated {Q_0, W}. In the sixth step the resultant temperature of the hot water cylinder tank has been calculated.

$$t_{HW'}^{\tau+\Delta\tau} = t_{HW} + \frac{\Delta\tau}{V_{HW} \cdot \rho_w \cdot cp_w} \cdot (Q_C - Q_0 - Q_{DHW}), °C \tag{7}$$

where:
$t_{HW'}^{\tau+\Delta\tau}$,   °C—resultant domestic hot water temperature,
V_{HW}, m³—domestic hot water cylinder capacity,
Q_0, W—heat transferred to the ground,
Q_{DHW}, W—heat for domestic hot water preparation purposes.

In the seventh step the ground temperature, showing the energy storage charge, has been calculated.

$$t_{G'}^{\tau+\Delta\tau} = t_G + \frac{\Delta\tau}{V_G \cdot \rho_G \cdot cp_G} \cdot Q_0, °C \tag{8}$$

In the analysis the actual efficiency of the coil in the domestic hot water cylinder has been omitted, and the economic temperature difference of 5 K has been adopted as a simplification.

During the analysis of the domestic hot water cylinder performance the hot water distribution has been assumed as for the family of four. The minimum temperature in the domestic hot water cylinder has been set at the level of 40°C, below which the additional electric heater is turned on. If the water temperature rises above the 55°C the surplus energy is directed into the borehole. The borehole is fed only with the surplus energy if the temperature rise above the 55°C and the electric heater does not work.

The data necessary for the above equations solvation are the pieces of information of the main devices in the system: surface and parameters of solar collectors, size of the domestic hot water cylinder, weather data like the outdoor temperature and solar radiation, and information on energy requirement for the preparation of domestic hot water.

The calculations have been performed assuming the time step of one hour.

2.7 The model of the underground thermal energy store charge with the waste heat from natural cooling installation (SYSTEM II)

Figure 4 presents a simplified computational scheme of the ground source heat pump system with vertical heat exchanger serving as a heat source, and a natural cooling system. Natural cooling is also used for regeneration of ground heat exchanger. Symbols introduced in the drawing, were also used in the formulas depicting the steps of the calculation algorithm.

2.7.1 Calculations concerning the simulation of the heat pumps ground heat exchanger discharge

The phenomena of ground heat exchanger discharge will be exactly the same as in the system provided with solar collectors.

2.7.2 Calculations concerning the simulation of the heat pumps heat exchanger charge (after discharge)

For the purposes of the analysis, it has been assumed that the natural cooling system works with constant output equal to 5.25 kW. The installation is turned on when the outside temperature exceeds 22°C. Whole energy received by the cooling system is transferred directly into the borehole.

The ground temperature, showing the energy storage charge:

$$t_{G'}^{\tau+\Delta\tau} = t_G + \frac{\Delta\tau}{V_G \cdot \rho_G \cdot cp_G} \cdot Q_0, °C \qquad (9)$$

3 RESULTS AND DISCUSSION

The analysis has been performed for the heat pump system designed for a single family house with the living space of 200 m² and the design heat load of 8440 W. The analysis has been carried out in one hour step. Annual energy demand for heating and ventilation (including heat losses, internal gains and solar gains) is equal to 12 634 kWh.

The start date for simulation was assumed to be the 1st of October. An analysis of the heat consumption from the ground by the heat pump has been carried out. The heating season was assumed to end on 30th of April. The buffer tank capacity, between a heat pump and installation, was assumed to be 300 l. Size of underground thermal energy store has been set up as 3 m around the borehole. Therefore, the theoretical volume of underground heat storage tank will equal to 3843 m³. In addition, the analysis took into account the heat pump efficiency reduction caused by the soil cooling around the borehole. The parameters of the soil taken for calculation are listed in Table 1.

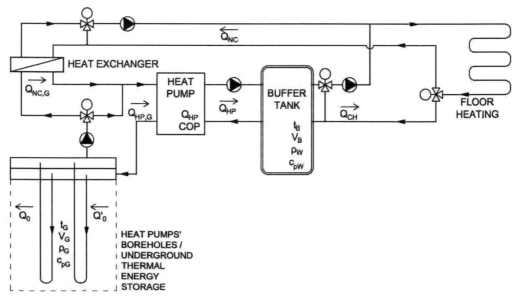

Figure 4.   Scheme of the natural cooling installation feeding the heat pump's boreholes.

The analysis carried out with the assumptions described above shows the temperature drop from 10.00°C to 6.36°C (Δt_G = 3.64 K) in theoretical underground thermal energy store. The obtained value is theoretical one, in fact, the temperature drop depends on the ground-water conditions in the area of heat pump boreholes and is very difficult to estimate. The heat pump will take 8 904 kWh energy from the ground during the heating season ($Q_{HP,G}$), which equals to 70% of energy demand for heating and ventilation. The remaining 30% constitute the electric energy supplied to the heat pump by the compressor.

The next step of the analysis objective was to check how much energy is possible to transfer to the heat pump's boreholes from the solar collectors installation. The start date for simulation was assumed to be 1st May. It has been carried out with one hour step for the calculations till the 30th September and during the heating season while the heat pump does not work. As mentioned before the solar collectors installation consists of four collectors (2.2 m² area each) mounted towards south

at an angle of 45°. The primary objective of the solar installation was to prepare the domestic hot water for the inhabitants. After the main objective is fulfilled the excess thermal energy is directed to the heat pump installation. For the purpose of the analysis, the family of four has been assumed with the domestic hot water consumption at 70 l per person per day. The monthly demand for the thermal energy for domestic hot water installation is about 235 kWh (2835 kWh per annum). It has been also assumed that the inhabitants leave the house for two weeks in July for the summer vacation.

The minimum temperature for the domestic hot water has been assumed at 40°C. The instant heat power directed to the borehole has been adapted to the heat pump installation design and has been set at 6800 W. The assumptions made for the installation let keep the water temperature in the domestic hot water cylinder, of 300 l capacity, at level below 85°C.

The analysis performed for such assumptions show that the ground that has been cooled down by 3.64 K during the heating season may be raised by regeneration by 1.03 K. In the figures below the monthly balances of the energy for the particular purposes and energy sources in single family house that has been the subject of analysis has been presented.

Solar collectors transfer 2523 kWh ($Q_{C,G}$) energy to the ground during one year of exploitation, which equals to 28% of the energy taken by the heat pump from the ground ($Q_{HP,G}$) (Fig. 5). Simultaneously solar collectors cover 58% ($Q_{C,DHW}$) of the thermal energy needs for the purpose of domestic hot

Table 1. Parameters of the soil around the heat pump borehole.

Parameters name	Value	Unit
Density of the soil (2/3 sand + 1/3 water)	1400	kg/m³
Soil thermal capacity (2/3 sand + 1/3 water)	1636	J/(kgK)

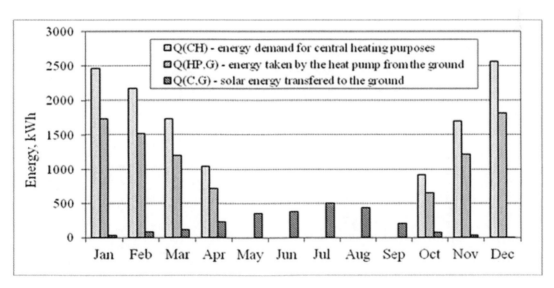

Figure 5. The energy balance: demand for central heating (Q_{CH}), gained from the ground ($Q_{HP,G}$) and transferred to the ground ($Q_{C,G}$).

water preparation (Q_{DHW}) (Fig. 6). In general 61% of energy transferred by solar collectors is directed to the ground ($Q_{C,G}$) and 39% is used by domestic hot water system ($Q_{C,DHW}$). Electric energy has 30% share in the energy demand for spatial heating ($Q_{EL,SH}$) and 42% share in the demand for energy for domestic hot water preparation ($Q_{EL,DHW}$).

The next analysis has been performed for the same single family house with ground coupled heat pump system with vertical heat exchanger and natural cooling system. Heat consumption from the ground during the heating season from the 1st of October till 30th April is the same as in the house with solar collectors cooperating with heat pump system. During the analysis the amount of energy that can be transferred from the natural cooling installation to the ground heat exchanger has been estimated. It has been assumed that existing installation of the floor heating will be used for cooling purposes in summer. As mentioned before

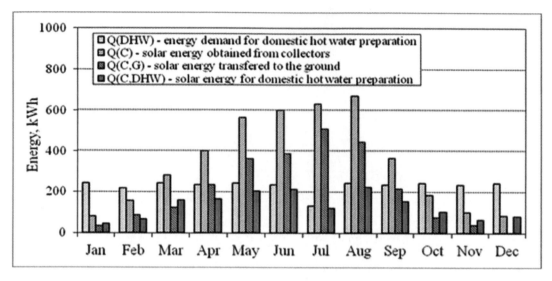

Figure 6. The energy balance: demand for domestic hot water preparation (Q_{DHW}), gained from the sun (Q_C), transferred to the ground ($Q_{C,G}$) and solar energy transferred to domestic hot water system ($Q_{C,DHW}$).

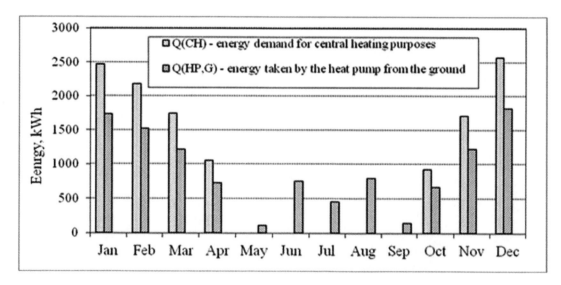

Figure 7. The energy balance: demand for spatial heating (Q_{SH}), gained from the ground ($Q_{HP,G}$) and transferred to the ground ($Q_{NC,G}$).

the natural cooling installation is turned on when the external temperature exceeds 22°C.

Analysis performed for the assumptions described above shows that the heat pump, exactly in the same way as in SYSTEM I, cools down the ground by 3.64 K during the heating season. The regeneration rises the ground temperature by 0.93 K during summer. Like in previous analysis the outcomes are theoretical and experimental research is required for their validation. Similarly

to the first analysis the monthly balances for the energy of particular purposes and sources have been presented in the Figures 7 and 8. Natural cooling system transfers 2263 kWh energy ($Q_{NC,G}$) to the underground thermal energy store, what equals to 25% of the energy that has been taken by the heat pump during the heating season ($Q_{HP,G}$) (Fig. 7). For the comparison purposes the monthly balances for domestic hot water system has also been presented. Solar collectors cover 51% ($Q_{C,DHW}$) of the energy

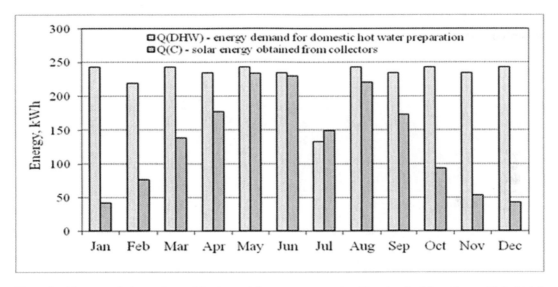

Figure 8. The energy balance: demand for domestic hot water preparation (Q_{DHW}), gained from the sun (Q_C) which is equal to the solar energy transferred to domestic hot water system ($Q_{C,DHW}$).

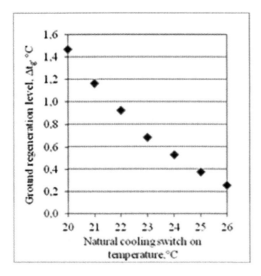 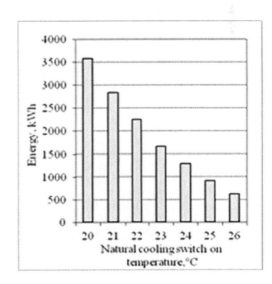

Figure 9. Ground regeneration level and the amount of the energy transferred to the ground in function of the switch on temperature of the natural cooling system

Table 2. The energy balance for the heating and domestic hot water purposes for the analysed single family house.

Energy source	SYSTEM I		SYSTEM II	
	kWh/a	%	kWh/a	%
SUM	15380	100	15380	100
Energy from the ground	8904	58	8904	58
Solar energy	1602	10	1629	11
Electric energy	4873	32	4846	31

demand for domestic hot water purposes (Q_{DHW}) (Fig. 8). Electric energy has 30% share in the energy demand for spatial heating ($Q_{EL,SH}$) and 42% share in the demand for energy for domestic hot water preparation ($Q_{EL,DHW}$). The amount of the energy that can be transferred to the ground depends on the switch on temperature of the natural cooling system. Figure 9 shows the results of the analysis concerning the amount of energy transferred to the ground and the level of ground regeneration in function of the switch on temperature of the natural cooling system. The range of the underground thermal energy store temperature rise is between 0.26 K to 1.47 K for switch on temperatures from 26°C to 20°C. The amount of transferred energy equals from 625 kWh to 3591 kWh for the same switch on temperatures range, which equals to the 7% to 40% of the energy that has been taken by the heat pump during the heating season ($Q_{HP,G}$).

4 CONCLUSION

In the Table 2 below, the structure of the shares of particular energy sources used in both analysed systems has been presented.

The results summarized in Table 2 show that the distribution of energy supplied to the analysed single family house is similar in both systems. Both regeneration systems, solar waste energy and natural cooling give similar results. In SYSTEM I additional solar collectors are required and in SYSTEM II some additional investment in natural cooling system and control system parts is needed. The ground regeneration with fewer solar collectors is, of course, possible but in case of above analysis those would give insignificant results because of the relatively large heat demand of single family house, and therefore a large heat pump and ground heat exchanger size. For installations similar to those analysed in the article, the reduction of the demand for heat will have a significant impact on the system efficiency. The regeneration

systems described in the article have advantages and disadvantages which the authors tried to list below.

The main advantage of both systems is the use of renewable sources to provide the house with energy. In SYSTEM I, using a fairly large area of solar collectors (8.8 m^2), a significant share of the solar energy in both, domestic hot water preparation and regeneration, has been obtained. In case of the house with smaller energy demand for heating the amount of collectors would not have to be so greatly increased. In SYSTEM II similar coverage of the energy demand for domestic hot water preparation purposes has been achieved with smaller collectors area (4.4 m^2) but they do not take part in the heat exchanger regeneration. The application of the regeneration to a large extent solves the problem of the lower energy source cooling down and decreasing of heat pump efficiency. Theoretical COP decrease equals to 9% during the heating season. After another year without regeneration it could reach, assuming extremely adverse conditions, 18%. This problem is particularly important in case of the boreholes located in the areas with no ground water flow (like rocks). Groundwater flow helps the ground to come back to natural temperature balance. The application of regeneration (SYSTEM I) allows also to solve the problem of domestic hot water cylinder overheat. In the analysed case the temperature in cylinder did not exceed 85°C. In the system without regeneration the cylinder temperature rises significantly above 100°C. The additional advantage is the possibility of smaller cylinder selection for the system because of the additional heat removal from solar collectors. Thanks to the regeneration (SYSTEM I) a solar energy store is established and the performance of heat pump is improved.

The system with natural cooling application (SYSTEM II) has similar advantages. It has no influence on the domestic hot water preparation system and does not use the solar energy but thanks to natural cooling system the ground heat exchanger temperature is getting balanced and the performance of the heat pump is improved, additionally the comfort of the building exploitation in the summer is also improved.

Disadvantage of the systems is that in both installations some rise in investment costs is unavoidable. It has to be noticed that the gains from the lower energy source regeneration are difficult to estimate. The performed analyses are general and more detailed research, which will introduce the problem well enough to allow predicting the earnings based on the type of soil, heat pump and other system parameters, is needed.

REFERENCES

Desideri, U., Sorbi, N., Arcioni, L., Leonardi, D., (2011), Feasibility study and numerical simulation of a ground source heat pump plant, applied to a residential building, *Applied Thermal Engineering*, 31: 3500–3511.

Hwang, Y., Lee, J.-K., Jeong, Y.-M., Koo, K.-M., Lee, D.-H., Kim, I.-K., Jin, S.-W., Kim, S.H., (2010), Cooling performance of a vertical ground-coupled heat pump system installed in a school building, *Renewable Energy*, 34: 578–582.

Karaback, R., Acar, S.G., Kumsar, H., Gökgöz, A., Kaya, M., Tülek, Z., (2010), Experimental investigation of the cooling performance of a ground source heat pump system in Denizli, Turkey, *International Journal of Refrigeration*, 34: 454–465.

Katsura, T., Nagano, K., Takeda, S., (2008), Method of calculation of the ground temperature for multiple ground heat exchangers, *Applied Thermal Engineering*, 28: 1995–2004.

Kjellsson, E., Hellström, G., Perers, B., (2010), Optimization of systems with the combination of ground-source heat pump and solar collectors in dwellings, *Energy*, 35: 2667–2673.

Lazzari, S., Priarone, A., Zanchini, E., (2010), Long-term performance of BHE (borehole heat exchanger) fields with negligible groundwater movement, *Energy*, 35: 4966–4974.

Li, S., Yang, W., Zhang, X., (2009), Soil temperature distribution around a U-tube heat exchanger in a multi-function ground source heat pump system, *Applied Thermal Engineering*, 29: 3679–3686.

Li, X., Chen, Z., Zhao, J., (2006), Simulation and experiment on the thermal performance of U-vertical ground coupled heat exchanger, *Applied Thermal Engineering*, 26: 1564–1571.

Michopoulos, A., Kyriakis, N., (2009), A new energy analysis tool for ground source heat pump systems, *Energy and Buildings*, 41: 937–941.

Nam, Y., Ooka, R., Hwang, S., (2008), Development of a numerical model to predict heat exchange rates for a ground-source heat pump system, *Energy and Buildings*, 40: 2133–2140.

Sanner, B., Karytsas, C., Mendrinos, D., Rybach, L., (2003), Current status of ground source heat pumps and underground thermal energy storage in Europe, *Geothermics*, 32: 579–588.

Szulgowska-Zgrzywa, M., Fidorów, N., (2011), Eksploatacja gruntowej pompy ciepła z regeneracją odwiertów pionowych w okresie lata, in: *"Współczesne metody i techniki w badaniach systemów inżynieryjnych pod red. Sergeya Anisimova"*: 133–139, Wrocław: Instytut Klimatyzacji i Ogrzewnictwa Wydział Inżynierii Środowiska. Politechnika Wrocławska.

Trillat-Bredal, V., Souyri, B., Fraisse, G., (2006), Experimental study of a ground-coupled heat pump combined with thermal solar collectors, *Energy and Buildings*, 38: 1477–1484.

Environmental Engineering IV – Pawłowski, Dudzińska & Pawłowski (eds)
© 2013 Taylor & Francis Group, London, ISBN 978-0-415-64338-2

Experimental study of thermal stratification in a storage tank of a solar domestic hot water system

A. Siuta-Olcha & T. Cholewa
Division of Indoor Environment Engineering, Faculty of Environmental Engineering, Lublin University of Technology, Lublin, Poland

ABSTRACT: An analysis of experimental research results of water temperature distribution at fifteen specified levels in the accumulation water tank of capacity 350 dm^3 with heating coil, cooperating with the vacuum tube solar collector of 3.9 m^2 area is presented.

For each day of July 2011 the quantity of thermal energy obtained from thermal conversion of solar radiation energy has been defined and the amount of heat accumulated in the tank has been estimated. The efficiency of energy accumulation in this solar system has been calculated. On the basis of the measurement data gathered, it was possible to determine the maximum water temperature in the tank on a particular day. The hourly mean degree of thermal stratification of water in the tank has been determined. Relation between the stratification number and the thermal gradient in the water tank has been considered. For selected operating cases the thermocline thickness in percentages has been estimated.

Keywords: efficiency of thermal energy accumulation, stratification number, thermal gradient, thermocline

1 INTRODUCTION

Systematic depleting of fossil fuels causes major risk for a development of a global energy management (Gurtowski 2011). Taking it into account, renewable source of energy utilization are recommended (Dasgupta & Taneja 2011, Shan & Bi 2012, Wall 2013). These activities can contribute to reduce greenhouse gas emissions. Review the science associated with global change was presented in Lindzen (2010).

Solar energy is one of the interesting directions of search for new energy sources. Photovoltaic systems can be applied to electrical energy generation (Holland et al. 2011, Moshnikov et al. 2012). Solar energy is also converted into heat energy in solar liquid or air collectors. The efficiency of energy accumulation in the solar system is one of the crucial parameters determining the efficiency of operation of the whole system and demonstrating correct selection of its elements. The thermal stratification of water in the storage tank affects the energy performance of the hot water solar heating system (Zachár et al. 2003, Shah & Furbo 2003, Shah et al. 2005, Furbo et al. 2005, Han et al. 2009, Rhee et al. 2010, Mawire & Taole 2011). Detailed description of the thermal stratification in solar water heating systems was presented by Ismail et al. 1997, Buzás et al. 1998, Shin et al. 2004, Duffie & Beckman 2006. Thermal stratification in the water tank means that the vertical gradient of the water is being stored, resulting from density differences (Castell et al. 2010). The thermal stratification

depends mainly on the convection movements and the heat conduction in the storage medium, on heat conduction along the tank wall, the heat losses via the tank surface to the surrounding environment as well as on forced and free convection in the storage tank caused by charging and discharging processes (Eames & Norton 1998, Hahne & Chen 1998, Alizadeh 1999). A transition layer of water, dividing the upper part of almost constant high temperature from the lower part of almost constant low temperature may occur in the heat storage tank. The thermocline thickness should be as low as possible in order to minimize the process of water mixing in the tank. It is defined on the basis of water temperature distribution inside the storage (Majid & Walujo 2010).

According to the experimental studies carried out, the influence of material and constructional parameters of the tank on thermal stratification has been established. In particular, it takes into consideration the height of diameter ratio z/D of the tank (the tank slenderness), difference of temperature at the inlet and outlet of the tank, different fluid mass flow rates in the charging circuit (Lavan & Thompson 1977, Bouhdjar & Harhad 2002, Spur et al. 2006). According to Jordan & Vajen 2000, Smolec 2000 and Cristofari et al. 2003, the installation equipped with the thermal stratification tank is 5–20% more efficient than a fully mixed water storage tank system, that is the system with uniform stored water temperature in the whole volume of the tank. In typical accumulation tanks thermal destratification, caused by water mixing during water intake, may result in over 23% decrease in heat efficiency per year (Knudsen 2002).

Computational and experimental researches on the phenomenon of thermal stratification have been performed mainly on storage water tanks directly incorporated into circulation of solar collectors or on mantle tanks. However, in most cases small hot water solar systems are equipped with vertical water storage tank, separated from the collector circulation through internal heat exchanger. Incorrectly chosen heating coil may contribute to the disturbance in the phenomenon of stratification. Moreover, placing spiral-tube heat exchanger in the lower part of the tank may cause the levelling of water temperature above the exchanger (Pluta 2000). To thoroughly examine the phenomenon of thermal stratification in water accumulation tank with internal heat exchanger, as well as energy performance of the system, operating studies of the solar heat system have been carried out in the climatic conditions of south-eastern Poland.

2 MATERIALS AND METHODS

2.1 Description of a hot water solar system

The object of the analysis is the experimental hot water solar system located in Lublin.

Figure 1 represents the schematic diagram of this solar installation (Siuta-Olcha & Cholewa 2011a).

When it comes to technical parameters of a vacuum solar collector and a water storage tank, they are presented in Table 1 (Siuta-Olcha & Cholewa 2010). The z/D ratio of the tank is circa 3.4. Inside the tank, in its lower section, there is a steel 26.9×2.3 mm spiral-tube heating coil with the length of 18 m. The surface area of the solar heating coil is 1.5 m^2 and its capacity is about 7 dm^3. The solar hot water system is filled with the water-propylene glycol mixture (50%). The solar installation is made of thermally insulated copper piping. The system is protected from excessive pressure with a diaphragm pressure expansion vessel of 18 dm^3, and a safety valve. The flow of working medium in the solar collector circuit is forced with operation of a pump with regulated performance.

An appropriate measurement system for experimental solar installation has been used. It allows to measure meteorological parameters (e.g. ambient temperature, solar radiation intensity, wind velocity), exploitation parameters in specific points of the solar heating system, and programmable control of the pump operation and the solenoid bleed valve. The hourly average values of measurement data recorded each 5 minutes were used in order to conduct detailed calculations. Elements of the

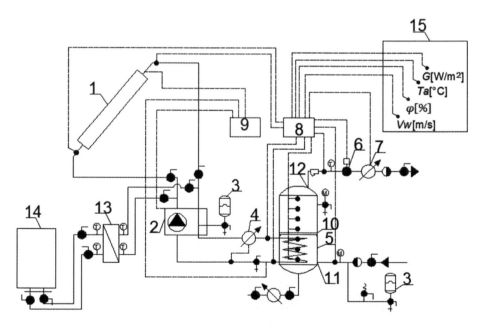

Figure 1. Schematic diagram of the experimental hot water solar system: 1—vacuum tube collector A_{col} = 3.9 m^2, 2—variable delivery pump, 3—diaphragm pressure expansion vessel, 4—heat meter, 5—water storage tank V = 350 dm^3, 6—electromagnetic release valve, 7—turbine flowmeter (600 impulse per dm^3), 8—regulating and measured module, 9—regulating device, 10—baffle inside of the tank, 11—cold water connection, 12—hot water connector pipe, 13—counter-flow heat exchanger, 14—ultrathermostat, 15—meteorological station.

Table 1. Technical parameters of the solar installation.

Parameter	Description
Vacuum tube collector type "heat-pipe" SG 1800/24	
Dimensions: length × width × depth (mm)	2040 × 1994 × 157
Gross area of collector (m²)	3.90
Glass thickness of a tube (mm)	1.6
Inside/outside diameter of a tube (mm)	47/58
Insulation thickness of a manifold (mm)	55
Insulation material	Mineral wool
Fluid conduits diameter (mm)	22
Flow channel material	Copper
Covering layers on absorbers	AL-N/AL
Absorber absorptance	0.95
Maximum efficiency of the collector related to absorber surface	0.623
Collector tilt (°)	38
Orientation	South-East
Water storage tank	
Tank material	Steel
Volume (dm³)	≈350
Height/diameter (m)	1.73/0.5
Insulation material	Mineral wool
Insulation thickness (mm)	100
External mantle material	Steel
Maximum working pressure (bar)	6
Maximum working temperature (water) (°C)	90

measurement system are given in Table 2 (Siuta-Olcha & Cholewa 2010). For continuous measurement of water temperature, 31 class-A resistance sensors Pt500 of fixed immersion depth and various length: 5, 15, 25 cm have been installed in the tank. In the solar circuit a heat meter has been installed. For measurement and temperature control, in specific points of the system, bimetallic thermometers with the measuring range of 0–120°C have been installed, as well as pressure gauges with the measuring range of 0–6 bar.

2.2 Method of studies

The amount of useful energy output of the solar collector depends on weather conditions, foremost on the irradiation conditions and temperature of the ambient outside air. Day Solar Irradiation for a Day (SID) has been determined daily according to the formula (1):

$$SID = \frac{\sum_{i=1}^{n}(G_i \cdot t_p)}{3600000} \quad (kWhm^{-2}d^{-1}) \quad (1)$$

The amount of thermal energy absorbed per 1 m² of collector area has been established according to the formula (2):

$$E_{sol} = 3.6 \cdot SID \cdot \eta_{col} \quad (MJm^{-2}d^{-1}) \quad (2)$$

Table 2. Data for sensors in the measurement system.

Parameter	Sensor
Interface AL154SAVDA5.6U.1L	
Meteorological probe:	Rotronic HygroClip
Humidity	Precision at 23°C: ± 1% rh / ±0.3 K Pt100 1/3 DIN
Temperature	Measurement interval: 0.7 s
Global solar radiation	Kipp&Zonnen Pyranometer CMP 6
	ISO Classification: First Class
	Directional error (at 80° with 1000 Wm⁻² beam): ±20 Wm⁻²
	Sensitivity: 15.40 μV per Wm⁻² at normal incidence on horizontal pyranometer
Wind velocity	Cup anemometer type 4531
	Precision: ±0.5 ms⁻¹, measuring range: 0...50 ms⁻¹
Air temperature in the room	AF11. Pt500. Cu
Interface AL154M1SAVDA5	
Fluid temperature in pipes	Alf-Sensor TOP-PKGKbm-21-1xPT500-A-6-G1/2-1.4571-50-3p-L2TS-6000-Z (unit: 6)
Water temperature in the tank	Alf-Sensor TOP-PKGKbm-21-1xPT500-A-6-G1/2-1.4571-50-3p-L2TS-6000-Z (unit: 15)
	Alf-Sensor TOP-PKGKbm-21-1xPT500-A-6-G1/2-1.4571-150-3p-L2TS-6000-Z (unit: 8)
	Alf-Sensor TOP-PKGKbm-21-1xPT500-A-6-G1/2-1.4571-250-3p-L2TS-6000-Z (unit: 8)
Water volume flow rate	Kobold Turbine Flowmeter
	Measuring range: 2...30 dm³min⁻¹, precision: ±1.5% of measuring range

The instantaneous thermal efficiency of a solar collector has been calculated according to the Equation (3):

$$\eta_{col} = \frac{\dot{m}_{col} \cdot c_f \cdot (T_{f2} - T_{f1})}{G \cdot A_{col}} \qquad (3)$$

Energy assessment of the system is done on the basis of the determination of the amount of energy stored in the water tank, heat losses to the surrounding environment, and energy performance.

Energy stored in the tank in each specified isothermal layer j from the start of a charging process has been calculated according to the formula (4):

$$E_{stj}(t) = (V \cdot \rho \cdot c_p)_j \cdot (T - T(t=0))_j \quad (\mathbf{MJ}) \qquad (4)$$

Water density as a function of temperature has been calculated according to the following formula (5):

$$\rho = 1000 - \frac{[0.00198582 \cdot (T+10) \cdot (T-277)^2]}{T - 205.8} \quad (\mathbf{kgm^{-3}}) \qquad (5)$$

Total thermal energy stored in the water tank has been established as the sum of thermal energy accumulated in each layer:

$$Q_{st}(t) = \sum_{j=1}^{J} (E_{stj}(t)) \quad (\mathbf{MJ}) \qquad (6)$$

Stratification number (Str) is a parameter that allows for the assessment of thermal stratification inside the accumulation water tank.

The stratification number is defined as the ratio of the mean of the temperature gradients at each time interval to that of the beginning ($t = 0$) (Equations 7–8) (Fernández-Seara et al. 2007, Mawire & Taole 2011):

$$Str = \frac{\left(\overline{\dfrac{\partial T}{\partial z}}\right)_t}{\left(\overline{\dfrac{\partial T}{\partial z}}\right)_{t=0}} \qquad (7)$$

where:

$$\frac{\overline{\partial T}}{\partial z} = \frac{1}{J-1} \cdot \left[\sum_{j=1}^{J-1} \left(\frac{T_{j+1} - T_j}{\Delta z} \right) \right] \qquad (8)$$

The phenomenon of thermocline has been used to determine the heat storage quality during tank charging. Thermocline thickness is defined as part of tank height, with diversified area attributed to it, which divides the area of almost homogenous lower temperature from the area of almost levelled higher temperature. In this layer, fast mixing of liquid layers of various temperatures takes place. To determine the thermocline thickness, one should define dimensionless temperature Θ according to the Equation (9) (Chung et al. 2008):

$$\Theta = \frac{T - T_c}{T_h - T_c} \qquad (9)$$

The thermocline thickness is the distance between two points where $0.9 \leq \Theta \leq 0.1$.

3 RESULTS AND DISCUSSION

3.1 Weather conditions

Variability of meteorological parameters considerably affects the thermal results of a solar collector operation. Figure 2 shows, for particular days of July 2011, mean solar radiation and mean ambient air temperatures, based on averaged values of these parameters from the period between 6 a.m. and 6 p.m. The lowest value of mean daily solar irradiation (42.34 Wm^{-2}) was recorded on July 30, and the highest 470.87 Wm^{-2} on July 18. Mean temperature of the ambient outside air was within the range 12.5°C–30.5°C.

Mean air relative humidity in July was 76.1%, and mean wind speed was 0.22 ms^{-1}.

Mean daily irradiation for July 2011 was 2.57 kWhm^{-2}, and for individual days it remained within the extent between 0.603 kWhm^{-2} (for July 30) and 5.41 kWhm^{-2} (for July 16). Daily solar irradiation rates for the entire month have been presented in Figure 3.

3.2 Daily evolution of the available solar energy and efficiency

In Figure 4 the collation of irradiation values per day at the solar collector and daily efficiency of the hot water solar system examined has been shown. Monthly energy gain in the system was 178 kWh. It is circa 55 kWh (23.6%) less than the respective figure for July 2010 (Siuta-Olcha & Cholewa 2011b).

Energy performance of the system depends not only on irradiation conditions but also on thermal efficiency of the solar collector. The mean thermal efficiency of the solar collector was established at 55.5%. Mean daily efficiency rates of the solar collector installed, determined for the hours 6 a.m.–5 p.m., have been shown in Figure 5.

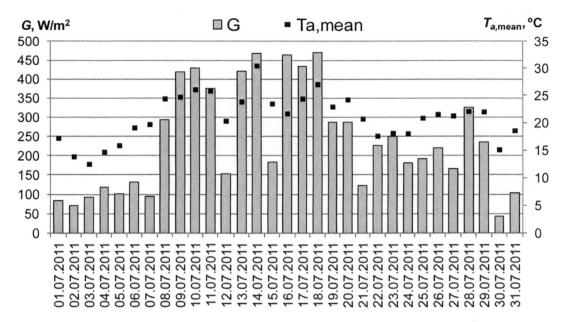

Figure 2.   Variability of weather parameters in July 2011.

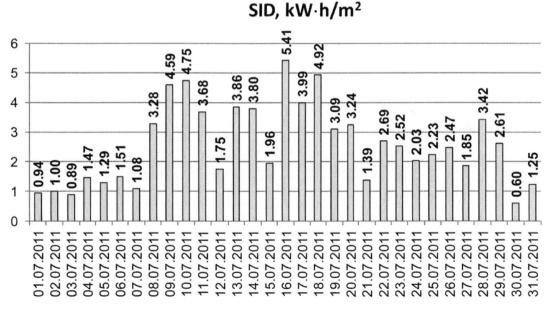

Figure 3.   Variability of a solar irradiation for a day.

Daily values of accumulation efficiency and amount of heat accumulated in the water storage tank in individual days of July 2011 have been shown in Figure 6.

Average efficiency of thermal energy accumulation for the month being analyzed was 27.2%.

The maximum accumulation efficiency in the solar heating system (44.2%) was obtained on July 27, and the minimum accumulation efficiency was 6.8% on July 2. The highest amount of thermal energy was stored on July 13 (19.2 MJd⁻¹), July 16 (19.9 MJd⁻¹) and July 28, 2011 (18.8 MJd⁻¹).

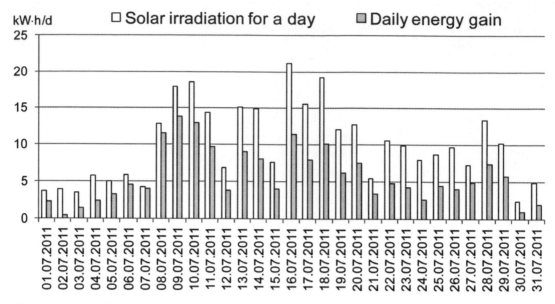

Figure 4. Energy efficiency of the experimental solar installation.

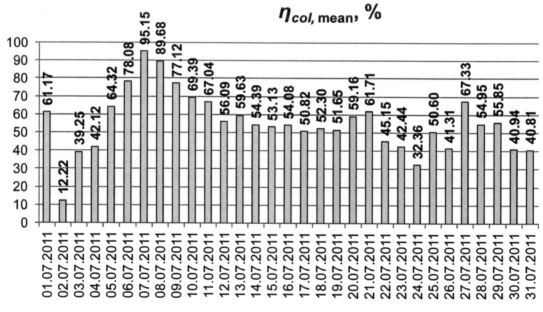

Figure 5. The collector efficiency variability per day.

All month long, the amount of 323.1 MJ (90 kWh) of thermal energy was stored in the tank. Figure 7 presents profiles of the daily hot water consumption. From 19.07. to 24.07.2011 no hot water usage was planned.

3.3 Daily evolution of water temperatures in the storage tank

Based upon the data gathered, it was possible to determine the maximum water temperature in the tank on a particular day (Fig. 8). The highest water

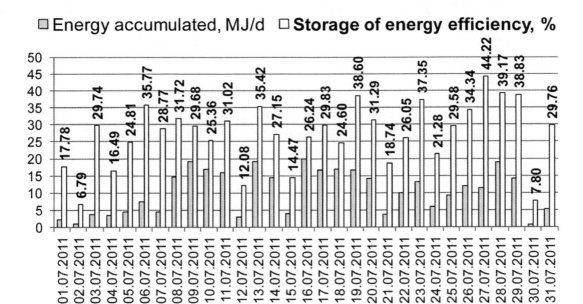

Figure 6. Energy accumulation efficiency in the solar system.

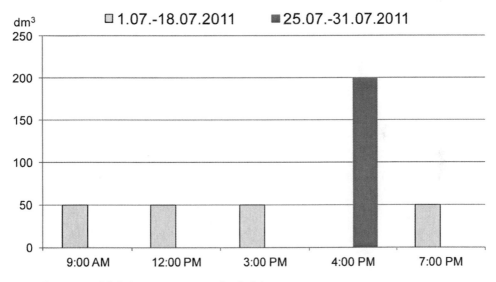

Figure 7. Histograms of daily hot-water consumption in July.

temperature in the tank, 53.3°C, was recorded on July 20. In the first week of July the maximum water temperature was low and amounted to 28.5°C on average. At the average in July, the maximum temperature in the tank amounted to 42°C. The hourly mean degree of thermal stratification of water in the tank has been determined for each day of the month. Stratification degree is defined as the difference between the highest water

temperature in the tank and its minimum temperature at a particular hour of a day. The highest degree of stratification, 12.6°C, was recorded on July 18. Monthly mean stratification degree was 6.7°C.

Changes of water temperature in the storage tank recorded by the sensors on 15 specified levels (T1–T7—the five-centimetres-long temperature sensors and T24–T31—the

375

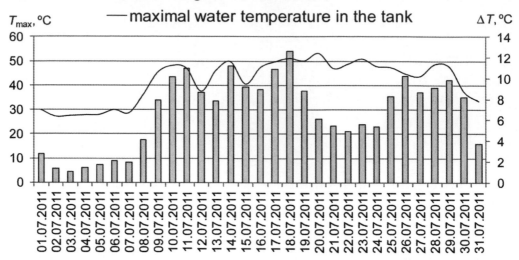

Figure 8. Mean degree of thermal stratification and evolution of maximal water temperature in the storage tank.

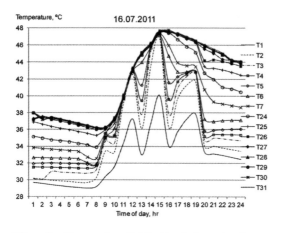

Figure 9.   Temperature distribution in the water storage tank on July 16.

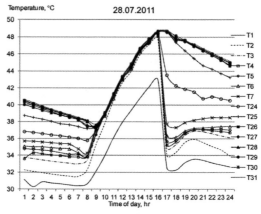

Figure 10.   Temperature distribution in the water storage tank on July 28.

twenty-five-centimetres-long temperature sensors) for three days are shown in Figures 9–11.

July 16 was a day with favourable weather conditions: $G_m = 464.6$ Wm^{-2}, $T_{a,m} = 21.7$°C. According to the schedule of hot-water consumption, water was being taken four times a day, 50 dm^3 per each time, which had a considerable effect on temperature distribution of water in the tank. Daily energy gain in the solar system was established as 11.4 kWh. On July 28 in similar weather conditions ($G_m = 326.5$ Wm^{-2}, $T_{a,m} = 22.1$°C), but during a single hot water intake of 200 dm^3 at 4 p.m. the water temperature in the tank was almost uniform during the tank charging. Considerably lower temperature of water occurred

only in the lower part of the tank constituting one section (T1). Daily energy gain in the solar system was established as 7.3 kWh. July 30 was a cold and cloudy day ($G_m = 42.3$ Wm^{-2}, $T_{a,m} = 15.1$°C). Energy gain was equal to circa 1 kWh.

Figures 12–14 present profiles of water temperature in the storage tank for the days studied. During tank charging the temperature in the whole tank was almost uniform. The thermal stratification occurs when the tank is not being charged, in general in the evening and night hours. In some operating cases, out of the tank charging phase, the thermocline phenomenon was noticed. The thermocline thickness varied between 0.5 m and 1.1 m.

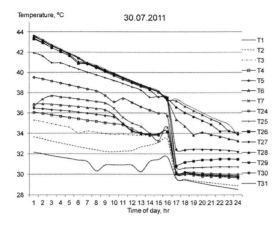

Figure 11.   Temperature distribution in the water storage tank on July 30.

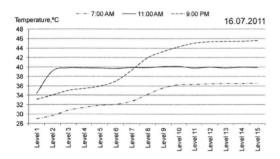

Figure 12.   Instantaneous changes of water temperature in the tank on July 16.

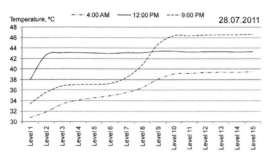

Figure 13.   Instantaneous changes of water temperature in the tank on July 28.

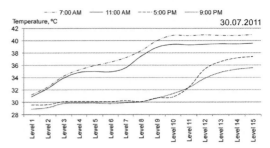

Figure 14.   Instantaneous changes of water temperature in the tank on July 30.

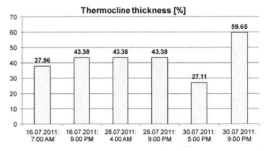

Figure 15.   Thermocline thickness in percentages.

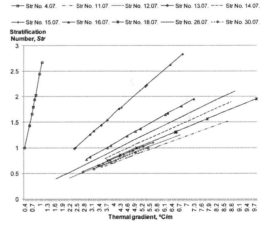

Figure 16.   Correlation between the stratification number and thermal gradient.

The percentage thickness of thermocline is shown in Figure 15.

The correlation between the stratification number and the thermal gradient along the height of the storage tank for several days in the month of July is presented in Figure 16. The thermal gradient represents the ratio of the water temperature difference between the top and the bottom of the water tank to the storage height. For all the days the stratification number is linearly dependent on the thermal gradient.

4   CONCLUSIONS

The results of experimental research have allowed to conduct a qualitative assessment of the hot water solar system operating in real conditions. The weather conditions in July 2011 were not far from standard ones. Daily mean solar energy for this month was 10 kWh, and month mean of daily energy gain was 5.74 kWh. Due to temperature sensors placed in the tank at 15th level it was possible to examine the water temperature distribution in the heat storage.

The research results have demonstrated that when charging the tank with internal heat exchanger the water temperature was virtually homogeneous in the whole volume. Differences in tank water temperature occurred after each hot water consumption from the storage tank and during its stagnation.

NOMENCLATURE

A_{col}	collector area (m²)
c_f	fluid specific heat (Jkg⁻¹K⁻¹)
c_p	water specific heat (Jkg⁻¹K⁻¹), $c_p = 4199$ Jkg⁻¹K⁻¹
D	diameter of the water storage tank (m)
E_{sol}	useful energy (MJm⁻²d⁻¹)
E_{st}	energy accumulated in the storage tank (MJ)
G	total solar irradiance (Wm⁻²)
J	number of water layers, $J = 15$
J	water layer
$\dot{m}_{col}$	circulated fluid mass-flow rate (kgs⁻¹)
Q_{st}	total energy accumulated in the storage tank (MJ)
T	water temperature in the tank (°C)
T_a	ambient temperature (°C)
T_{f1}	inlet fluid temperature (°C)
T_{f2}	outlet fluid temperature (°C)
t	time (s)
t_p	time period (s)
V	volume (m³)
V_w	wind velocity (ms⁻¹)
z	height of the storage tank (m), $z = 1.73$ m

Greek letters

ρ	water density (kgm⁻³)
η_{col}	solar collector efficiency
φ	air relative humidity (%)
Δz	distance between temperature sensors (m), $\Delta z = 0.10$ m

Subscripts

c	cold
h	hot
m	mean

REFERENCES

Alizadeh, Sh. 1999. An experimental and numerical study of thermal stratification in a horizontal cylindrical solar storage tank. *Solar Energy* 66: 409–410.

Bouhdjar, A., Harhad, A. 2002. Numerical analysis of transient mixed convection flow in storage tank: influence of fluid properties and aspect ratios on stratification. *Renewable Energy* 25: 555–567.

Buzás, J., Farkas, I., Biró, A., Németh, R. 1998. Modelling and simulation of a solar thermal system. *Mathematics and Computers in Simulation* 48: 33–46.

Castell, A., Medrano, M., Solé, C., Cabeza, L.F. 2000. Dimensionless numbers used to characterize stratification in water tanks for discharging at low flows rates. *Renewable Energy* 35: 2192–2199.

Chung, J.D., Cho, S.H., Tae, C.S., Yoo, H. 2008. The effect of diffuser configuration on thermal stratification in a rectangular storage tank. *Renewable Energy* 33: 2236–2245.

Cristofari, C., Notton, G., Poggi, P., Louche, A. 2003. Influence of the flow rate and the tank stratification degree on the performances of a solar flat-plate collector. *International Journal of Thermal Sciences* 42: 455–469.

Dasgupta, P., Taneja, N. 2011. Low Carbon Growth: An Indian Perspective on Sustainability and Technology Transfer. *Problems of Sustainable Development* 6(1): 65–74.

Duffie, J.A., Beckman, W.A. 2006. *Solar Engineering of Thermal Processes*. John Wiley & Sons, New York.

Eames, P.C., Norton, B. 1998. The effect of tank geometry on thermally stratified sensible heat storage subject to low Reynolds number flows. *Int. J. Heat Mass Transfer* 41(14): 2131–2142.

Fernández-Seara, J., Uhía, F.J., Sieres, J. 2007. Experimental analysis of a domestic electric hot water storage tank. Part I: Static mode of operation. *Applied Thermal Engineering* 27: 129–136.

Furbo, S., Andersen, E., Thür, A., Shah, L.J., Andersen, K.D. 2005. Performance improvement by discharge from different levels in solar storage tanks. *Solar Energy* 79: 431–439.

Gurtowski, S. 2011. Green Economy Idea—Limits, Perspectives, Implications. *Problems of Sustainable Development* 6(1): 75–82.

Hahne, E., Chen, Y. 1998. Numerical study of flow and heat transfer characteristics in hot water stores. *Solar Energy* 64(1–3): 9–18.

Han, Y.M., Wang, R.Z., Dai, Y.J. 2009. Thermal stratification within the water tank. *Renewable and Sustainable Energy Reviews* 13: 1014–1016.

Holland, E., Bedin, D., Rodrigo, A., Aguilera, J. Nofuentes, G., Terrados, J., Muñoz, V., Rassu, A.G., Demurtas, V., Lampadaris, K., Olchowik, J.M., Villa, A., Schrittwieser, W. 2011. *ADMINISTRATIVE HANDBOOK—The installation of ground PV plants: EU regulations, procedures and main country differences*. Venice, ISBN: 978-84-694-2317-2.

Ismail, K.A.R., Leal, J.F.B., Zanardi, M.A. 1997. Models of liquid storage tanks. *Energy* 22(8): 805–815.

Jordan, U., Vajen, K. 2000. Influence of the DHW load profile on the fractional energy savings: a case study of solar combi-system with TRNSYS simulations. *Solar Energy* 69: 197–208.

Knudsen, S. 2002. Consumers' influence on the thermal performance of small SDHW systems—theoretical investigations. *Solar Energy* 73(1): 33–42.

Lavan, Z., Thompson, J. 1977. Experimental study of thermally stratified hot water storage tanks. *Solar Energy* 19: 519–524.

Lindzen, R.S. 2010. Global Warming: The Origin and Nature of the Alleged Scientific Consensus. *Problems of Sustainable Development* 5(2): 13–28.

Majid, M.A.A., Waluyo, J. 2010. Thermocline thickness evaluation of stratified thermal energy storage tank of co-generated district cooling tank. *Journal of Energy and Power Engineering* 4(2): 28–33.

Mawire, A., Taole, S.H. 2011. A comparison of experimental thermal stratification parameters for an oil/pebble-bed thermal energy storage (TES) system during charging. *Applied Energy* 88: 4766–4778.

Moshnikov, V.A., Gracheva, I., Lenshin, A.S., Spivak, Y.M., Anchkov, M.G., Kuznetsov V.V., Olchowik, J.M. 2012. Porous silicon with embedded metal oxides for gas sensing applications. *Journal of Non-Crystalline Solids* 358: 590–595.

Pluta, Z. 2000. *Theoretical basis of solar energy thermal conversion* (in Polish). Warsaw: Warsaw University of Technology.

Rhee, J., Campbell, A., Mariadass, A., Morhous, B. 2010. Temperature stratification from thermal diodes in solar hot water storage tank. *Solar Energy* 84: 507–511.

Shah, L.J., Andersen, E., Furbo, S. 2005. Theoretical and experimental investigations of inlet stratifies for solar storage tanks. *Applied Thermal Engineering* 25: 2086–2099.

Shah, L.J., Furbo, S. 2003. Entrance effects in solar storage tanks. *Solar Energy* 75: 337–348.

Shan, S., Bi, X. 2012. Low Carbon Development of China's Yangtze River Delta Region. *Problems of Sustainable Development* 7(2): 33–41.

Shin, M.S., Kim, H.S., Jang, D.S., Lee, S.N., Lee, Y.S., Yoon, H.G. 2004. Numerical and experimental study on the design of a stratified thermal storage system. *Applied Thermal Engineering* 24: 17–27.

Siuta-Olcha, A., Cholewa, T. 2010. Research of thermal stratification processes in water accumulation tank in exploitive conditions of solar hot water installation. In *Proceedings of 41th International Congress on Heating, Refrigerating and Air-Conditioning*, Edited by Branislav Todorović: 391–400.

Siuta-Olcha, A., Cholewa, T. 2011a. Experimental studies of solar hot water installations for climatic conditions of Lublin. Part 1. Assessment of actual energy effects (in Polish). *District Heating, Heating, Ventilation* 9: 339–344.

Siuta-Olcha A., Cholewa, T. 2011b. Experimental studies of solar hot water installations for climatic conditions of Lublin. Part 2. Research of stratification phenomenon in water storage tank (in Polish). *District Heating, Heating, Ventilation* 10: 426–430.

Smolec, W. 2000. *Thermal conversion of solar energy* (in Polish). Warsaw: PWN.

Spur, R., Fiala, D., Nevrala, D., Probert, D. 2006. Influence of the domestic hot-water daily draw-off profile on the performance of a hot-water store. *Applied Energy* 83: 749–773.

Wall, G. 2013. Exergy, Life and Sustainable Development. *Problems of Sustainable Development* 8(1): 27–41.

Zachár, A., Farkas, I., Szlivka, F. 2003. Numerical analyses of the impact of plates for thermal stratification inside a storage tank with upper and lower inlet flows. *Solar Energy* 74: 287–302.

Environmental Engineering IV – Pawłowski, Dudzińska & Pawłowski (eds)
© 2013 Taylor & Francis Group, London, ISBN 978-0-415-64338-2

Critical flow factor of refrigerants

S. Rabczak & D. Proszak-Miąsik
Faculty of Civil and Environmental Engineering, Rzeszow University of Technology, Rzeszow, Poland

ABSTRACT: On the basis of the *Martin-Hou* EOS, an analytically predicted critical flow models has been verified. Measurements of the critical mass flow for R-410A were carried out on test-bench. An experimental verification of computing results referring to the critical flow function C^* value is a main goal of theoretical results examinations depending on the T_0 temperature and p_0 pressure at the assumed stagnation conditions. The ISO 9300 critical Venturi nozzles at the suitable cross section have been executed to assure conditions of the sonic flow in the closed loop system. The critical flow function of gas flow through the critical nozzle strongly depends on real gas effects. In this work, computational study using improved One Dimensional (1D) gas dynamic theory was employed. The model with boundary layer displacement thickness gives more sufficient prediction of critical flow function. An agreement between theory and experiment has been confirmed.

Keywords: refrigerant, critical flow, Venturi nozzle

1 INTRODUCTION

Growing energy consumption associated with local environmental and operating problems of electric, heating and cooling power systems have increased the interest in refrigerants not applied on the broader scale so far. Solutions utilizing low-temperature or waste energy sources in low boiling point applications are being introduced more and more commonly in the various systems of the energy conversion. Main group of above mentioned low boiling temperature agents are organic refrigerants as propane, butane and inorganic substances (carbon dioxide) or whole range of environmentally harmful replacements for chloro-fluorocarbons. Detailed data referring to the estimation of their critical flow characteristics are not available for many of them, particularly in the high pressures area (performing e.g. in transcritical cycles of CO_2). An implementation of ideal gas model to the calculation of properties and process in real compressible fluid leads to a large discrepancy between the measured and predicted values of parameters. It is a currently developed class of methods involving an application of well-known thermal and fundamental Equations of State (EOS) not only within thermodynamics of the process but in the fluid flow analysis. At the principal study and qualitative examination of physical processes the most popular is *van der Waals* (*vdW*) model. A main weakness of this approach is a poor quantitative level of obtained particular numerical results. It is caused by improper values of higher order partial derivatives predicted from *vdW* EOS, especially in the region of high-density gas, close to the thermodynamic critical state. Other cubic two parameter thermal EOS's, based on *vdW* model give the comparable results (*PR*, *RKS*). It should be mentioned that this class of EOS's is very useful for an elaboration of the hand-made calculation procedures (simple algebraic expressions for calculation of thermodynamic functions and their derivatives). More sophisticated methods involve an application of multi-parameter EOS's and special computer routines in calculation of necessary parameters. In an engineering practice the most popular and well known is *Martin-Hou* (*MH*) and modified by *Benedict-Webb-Rubin* (*MBWR*) EOS. These EOS's have been widely applied and actually used for new pure refrigerants and blends, for example R134a, R407C, R410A and R507 (Thompson & Sullivan 1977). Its correctness is confirmed by many authors and can be applied up to high-density region of one component fluid as well as the solutions and saturated state analysis. The main purpose of this work is to deliver some useful data and tools for a prediction of the critical flow conditions in a high-density region of refrigerant superheated vapour and other compressible fluids. Based on refrigerant R-410A data an experimental verification of the theoretical calculation were performed. In the compressible flow with a sufficiently high pressure ratio, the flow is choked at minimum cross sectional area of flow passage, in which the mass flow rate reaches a maximum value. The minimum pressure ratio for the choke flow is called the critical pressure ratio. If the pressure ratio is critical, flow will no longer depend on pressure change in downstream flow field. In this case, the mass flow is determined only by the stable conditions upstream of the flow passage. According to classical 1D gas dynamics, the mass flow rate is a function of the pressure and temperature at the upstream stagnation conditions, the diameter of nozzle throat and the ratio of specific heats of the gas (Hillbrath 1981). This phenomenon is one of the unique features of compressible internal flows.

The critical nozzle is defined as devices to measure the mass flow with only the nozzle supply conditions making use of the flow choking phenomenon at the nozzle throat. It is due to viscous and heat transfer effects of flows through the critical nozzle. Prediction of the mass flow rate and critical flow function is of practical importance since the mass flow rate is essentially associated with the limiting working gas consumption. Critical pressure ratio should be known to establish the operating conditions for safety valves and expansion units. For the high Reynolds numbers, the discharge coefficient approaches to unity, indicating that the 1D in viscid theory is valid for the prediction of mass flow. For the low Reynolds numbers the discharge coefficiently reduces considerably the unity mentioned below. It is due to the boundary layers effects in the flow through the sonic nozzle. Several experimental works have been made to investigate the discharge coefficient for low Reynolds numbers (Bignell 1996) in the flow of technical gases. For a small critical nozzle, the Reynolds number can be low, and in this case the prediction of the mass flow rate is not straightforward since it can be affected by downstream pressure change, even under the condition of the critical pressure ratio.

The objective of this work is to predict the critical flow function for R-410A and compare the computational data with experimental ones.

2 MATERIAL AND METHODS

Problem of calculation of the critical flow ("sonic flow") of dense gas or vapor was examined by many authors (Thompson & Sullivan 1977, Leung 1988, Bober & Chow 1977). For CFC refrigerants R12, R22 and R502, such analysis also was presented (Shumann 1990).

Gorski in his monograph (Gorski 1997) has proposed a unified approach for calculation of thermodynamic property and process in dense gases. The newly developed method of "Virial Compressibility Derivatives" (VCD), gave in this area more simple and directly related results to an ideal gas model.

At the case of dense gas 1D steady flow well-known relations presented in classical books on gas dynamics are generally considered as invalid. It is caused by strong variation of all physical constants such as the isentropic index, specific heats and all thermal and caloric properties involving the calculation of the process. By using the VCD formalism it is possible to find the basic relations between parameters in the isentropic flow at the similar form to an ideal gas (Rabczak 2007). The obtained approximate relations between critical flow parameters and the stagnation ones are to be shortly expressed:

$$\frac{T^*}{T^0} \cong \left(\frac{2}{\chi+1}\right)^{\frac{\Phi}{\chi-1}}, \frac{p^*}{p^0} \cong \left(\frac{2}{\chi+1}\right)^{\frac{k}{\chi-1}}, \frac{\rho^*}{\rho^0} \cong \left(\frac{2}{\chi+1}\right)^{\frac{1}{\chi-1}} \quad (1)$$

where:

$$\Phi = \frac{\rho}{T}\left(\frac{\partial T}{\partial \rho}\right)_s = \frac{z_T}{c_v}, k = \frac{\rho}{p}\left(\frac{\partial p}{\partial \rho}\right)_s = \gamma\frac{z_v}{z},$$

$$\chi = k + \frac{\rho}{k}\left(\frac{\partial k}{\partial \rho}\right)_s = 2\Gamma - 1 \quad (2)$$

and

$$z = \frac{pv}{RT}, z_v = z - v\left(\frac{\partial z}{\partial v}\right)_T = z + \rho\left(\frac{\partial z}{\partial \rho}\right)_T,$$

$$z_T = z + T\left(\frac{\partial z}{\partial T}\right)_v, \gamma = \frac{c_p}{c_v}, \bar{c}_v = \frac{c_v}{R} \quad (3)$$

In those three equations the *Grüneisen* parameter Φ and generalized isentropic index χ (or fundamental derivative Γ) as well as the *Poisson* ratio γ, isentropic exponent k and two virial compressibility derivatives z_T and z_v are introduced. These parameters should be found from an EOS which in the typical case is given as $p = p(T, v)$, as for example *MH* and *MBWR*. It is easy to prove that in an ideal gas limit all parameters appearing in the above presented relations are equal to its classical constants ($z = z_T = z_v = 1$, $\chi = k = \gamma$, $\Phi = R/c_v$). The important relation which allows to find limiting mass flow in a critical section of the conduit A^* for 1 D flow is given (Rabczak, 2007).

$$C^* = \frac{\dot{m}^*\sqrt{RT_0}}{A^*p_0} \cong \sqrt{\frac{k}{z_0}}\left(\frac{2}{\chi+1}\right)^{\frac{\chi+1}{2(\chi-1)}} \quad (4)$$

All parameters appearing in equations (1) to (4) are obviously functioning in the temperature and specific volume and vary along the isentrope between stagnation and critical state. Therefore, a calculation of a process needs an iterative procedure. It can be simplified when one initially assumes values for given stagnation state. After few steps calculation results converge to the mean values of process "constants". These approach confirmed PC procedures.

The complete analysis needs an iterative solution of the energy equation along an isentrope.

$$h_0 = h + \frac{w^2}{2} = h^* + \frac{a_*^2}{2} = \text{idem}. \tag{5}$$

where in sonic flow conditions $Ma = 1$, $w = a_*$.

The calculation procedures are based on actual data sheets and *Martin-Hou* EOS (Martin & Hou, 1955). Some example of calculation results in a critical flow function C* which is presented below confirm a great discrepancy between ideal gas values and more realistic data derived from thermal EOS's for selected refrigerants. It is evident that in practical calculation of critical function at the sonic flow of refrigerants and other fluids in dense gas region appears as great discrepancy between ideal gas model analysis and real gas simulation. The critical flow function in the refrigerants flow does change over 30% in a comparison to ideal gas flow where the sonic conditions depends only on *Poisson* constants for a particular gas (relating only to molecular mass of the fluid).

3 EXPERIMENTAL TESTS

Experimental verification of theoretical results and comparison of obtained data from test-bench to computational ones were the main goals of this work. The test-bench was made as a refrigeration circuit capable to measure the mass flow with the use of critical Venturi nozzles. Superheated vapour of R-410A was the medium which was allowed to prepare critical condition on the upstream side of nozzle.

The presented computational model was proceeded and tested for two diameter of critical

Venturi nozzles: 0,8 mm and 1,0 mm made as ISO 9300 standard (ISO 9300, 1990)—see Figure 1. The authors took 16 measurements points (see Fig. 4) to obtain in an experimental way the critical flow function and verify them to theoretical results. The equipment and gauges used to measurements have allowed to determine a maximum measurement error below 1,7%. While the test-bench was worked, the data were collect in a computer database with the time step of 1 second. Stabilised conditions in the circuit have been reached for about 4 to 5 hours, and then the right data were collected to obtain the critical mass flow and critical flow function. Schema of test-bench is presented in Figure 2.

The range of operating conditions of the test loop was limited up to $p_0 < 35$ bar and $p_0 < 110$ °C and behind of saturation line in order to avoid two-phase flow in the main line.

The critical nozzle of diameter 0,8 mm was used up to 15 bar stagnation pressure. Under 15 bar the nozzle of diameter 1,0 mm was needed to preserve technical heater limitation of calorimeter.

4 RESULTS AND DISCUSSION

Figure 4 presents a comparison of theoretical computation results of critical flow function and the experimental data. Obtained experimental data are with closely agreement with computational based on the *MH* equation of state. The average discrepancy between theoretical and experimental results is less then 1,0%. It confirms that the theoretical model is valid and formulated in accordance with the real physics. The detailed view of a main part of circuit loop was presented in Figure 3.

All experimental results shown in Figure 4 are taking into consideration the boundary layers effect. Boundary Layer displacement (*BL*) thickness was calculated by *Geropp* model (Gerop, 1971) as:

$$\frac{2\delta}{d} = 0,001309 + 1,72169 \frac{1}{\sqrt{\text{Re}}} \tag{6}$$

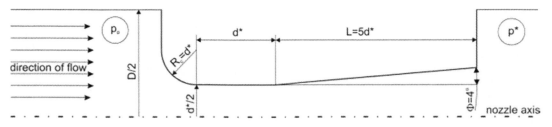

Figure 1.   Critical Venturi nozzle schema (ISO 9300, 1990).

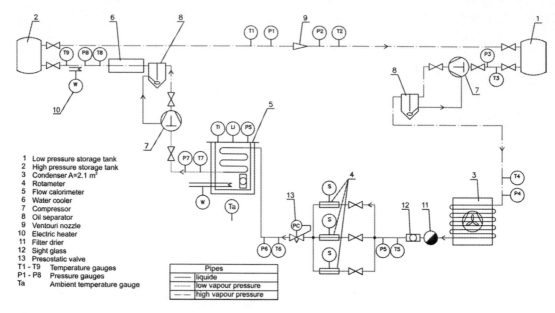

Figure 2. Schema of test—bench for measurement of critical flow function.

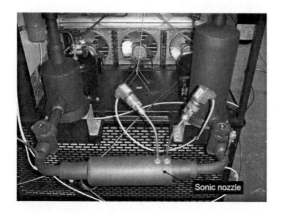

Figure 3. View on the sonic nozzle on test—bench.

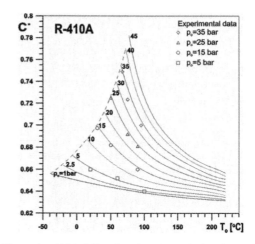

Figure 4. Critical flow function: theoretical and experimental data comparison.

Figure 5 shows the correlation of relative BL thickness as a function of the Reynolds number in a range $2 \cdot 10^5 \div 1{,}2 \cdot 10^6$. As it is visible on Figure 5 measured flow was strongly turbulent. Depending on nozzles used on test-bench the direct correlation between nozzle diameter an Reynolds number can be presented in Figure 6.

Boundary layer thickness is strongly increased due to Reynolds number below $6*10^5$. Under this value BL thickness for both nozzle diameters are changing in the narrow range. The nozzle throat diameter changing (0,8 to 1,0 mm) makes approximately 0,58 mm variation in BL thickness.

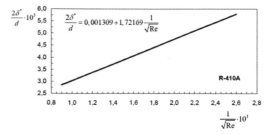

Figure 5. Displacement *BL* thickness vs. Reynolds No.

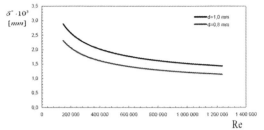

Figure 6. Correlation between nozzle diameter, *BL* thickness and Reynolds No.

5 CONCLUSION

The paper compares a computational model of dense gas flow through a critical nozzle with experimental results. The analytical results have been confirmed by experimental data. The final results are summarised in the following way:

1. The critical flow function for dense refrigerants strongly depends on the upstream stagnation conditions.
2. The computation results for *R-410A* and *R-507A* are in a satisfactory agreement with experimental data.
3. The proposed approximation of critical flow conditions for real dense gases is better than other comparable methods.
4. The Boundary Layer (*BL*) at the sonic nozzle throat is turbulent and affects the critical flow function.
5. Other theoretical results have been actually developed for *R-22*, *R-123*, *R-125*, *R-134a*, *R-143a*, *R-152a*, *R-227*, *R-404A*, *R-407C*, *R-410A*, *R-507A* and CO_2.

NOMENCLATURE

A	area [m²]
a	speed of sound [m/s]
C^*	critical flow function [–]
c_p, c_v	specific heats [J/kg K]
d	nozzle diameter [mm]
h	specific enthalpy [kJ/kg K]
k	isentropic exponent [–]
Ma	Mach Number [–]
$\dot{m}$	mass flow [kg/s]
p	pressure [Pa]
R	gas constants [J/kg K]
Re	Reynolds number [–]
T	temperature [K]
v	specific volume [m³/kg]
w	velocity of flow [m/s]
z	compressibility factor [–]
z_T, z_v	VCD's, eq. (3) [–]
Γ	fundamental derivative [–]
γ	Poisson constants [–]
δ^*	boundary layer displacement thickness [mm]
ρ	density [kg/m³]
χ	generalized isentropic exponent [–]
Φ	Grüneisen parameter [–]

Indices

s	isentropic
T	isothermal
v	isochoric
0	stagnation state
$*$	critical flow ($Ma = 1$)

REFERENCES

Bignell N., 1996. The use of small sonic nozzles at low Reynolds numbers. Flow Measurement Instrumentation. No 7, p. 109.

Bober W. & Chow W.L., 1977. Nonideal isentropic flow through converging-diverging nozzles. ASME Jour. Fluids Eng. No 111 (4), p. 455.

Geropp D., 1971. Laminare Crenzschichten in Ebenen und Rotation Ssymmetrischen Lavalduesen. Deutche Luft und Raumfahrt for Schungsbericht, p.71.

Gorski J., 1997. Modeling of real gas properties and its thermal-flow processes, (in Polish). Rzeszow: Oficyna Wyd. PRz.

Hillbrath H.E., 1981. The critical flow ventury—an update, flow: its measurement and control in science and industry. Instrum. Soc. Am., No 2, p. 407.

Leung J.C. & Epstein M., 1988. A generalized critical flow model for nonideal gases. AIChE Jour. No 34 (9), p. 1568.

Martin J.J. & Hou, Y.C., 1955. Development of an equation of state for gases. AIChE Jour., vol. 1, No. 2, p. 142–151.

Rabczak S., 2007. Thermal Equations of State in the Flow Analysis of New Refrigerants (in Polish), PhD Dissertation, Fac. of Environmental Engineering, Warsaw: Oficyna Wyd. PW.

Shumann S.P., 1990. Real gas critical flow factors for R12, R22 and R502. ASHRAE Trans. No 96 (2), p. 329.

Standard ISO 9300, 1990. Measurement of gas flow by means of critical flow Venturi nozzles.

Thompson P.A. & Sullivan D.A., 1977. Simple predictions for the sonic conditions in a real gas. ASME Jour. Fluids Eng. No 99 (1), p. 57.

Thompson P.A. & Sullivan D.A., 1977. Simple predictions for the sonic conditions in a real gas. ASME Jour. Fluids Eng. No 99 (1), p. 217.

Environmental Engineering IV – Pawłowski, Dudzińska & Pawłowski (eds)
© 2013 Taylor & Francis Group, London, ISBN 978-0-415-64338-2

Processes of heat transfer in variable operation of radiant floor heating system

T. Cholewa, M.R. Dudzińska & A. Siuta-Olcha
Faculty of Environmental Engineering, Lublin University of Technology, Lublin, Poland

Z. Spik & M. Rosiński
Faculty of Environmental Engineering, Warsaw University of Technology, Warsaw, Poland

ABSTRACT: So far, the most of research on radiant floor heating system was conducted in steady state conditions of the floor radiator. However, in real conditions, radiant floor heating systems are subjected to different kind of thermal or hydraulic forcing. Thereby, experimental research of the massive floor radiator in the semitechnical scale in dynamic conditions is performed. The investigated floor radiator was subjected to thermal and hydraulic forcing, which appear during regulation processes of this type of heating systems. Heat transfer coefficients in the radiation way $\alpha_R = 5.45$–5.81 Wm^{-2} K^{-1} and convection $\alpha_K = 2.77$–3.48 Wm^{-2} K^{-1} and also the total heat transfer coefficient $\alpha_{total} = 8.77$–9.95 Wm^{-2} K^{-1} from the surface of the massive floor radiator for the heating function of rooms were calculated. The difference (above 13%) between the value of total heat transfer coefficient from the surface of floor radiator, which was received in experimental way, and the value calculated by use of universally applied equation by design of radiant floor heating panel was noticed.

Keywords: floor heating system, heat transfer coefficient, dynamic conditions

1 INTRODUCTION

Nowadays, the radiant floor heating is more and more often applied in new buildings and in buildings after modernization, which gets out of many advantages characterizing this type of heat emitter. Analyzing methods of calculations of the heat flux density taken over from radiant floor heating surface in static and dynamic conditions (Kilkis et al. 1995, Kowalczyk & Strzeszewski 1999, Kowalczyk & Rosiński 2007), it can be noticed that one of the value which has a considerable influence on received results of above-mentioned physical value, is the heat transfer coefficient from the surface of the radiator. It may be the total heat transfer coefficient (α_{total}). However, for more detailed and exact thermal analyses convective heat transfer coefficient (α_K) values are necessary, as well as radiant heat transfer coefficient (α_R).

Materials for designing of the radiant floor heating systems base from the thermal side most often on the value of the total heat transfer coefficient from the radiant surface. This value is assumed in compliance with EN-1264-5 (2008) and is a constant value, equal to $\alpha_{total} = 10.8$ Wm^{-2} K^{-1}. Another way of α_{total} calculations for the radiant floor heating system, given in EN-1264-2 (2008),

is the use of Equation (1), which makes the value of α_{total} coefficient dependent on the temperature difference between the radiant surface and mean indoor air temperature.

$$\alpha_{total} = 8.92 \cdot \left(t_p - t_i\right)^{0.1} \tag{1}$$

However, in case of the radiant heat transfer coefficient value from the surface of the radiator, it is recommended to assume it as the constant value equal to $\alpha_R = 5.6$ Wm^{-2} K^{-1} (Karlsson & Hagentoft 2005, Karlsson 2006a, Karlsson 2006b, Causone et al. 2009), or else equal to $\alpha_R = 5.9$ Wm^{-2} K^{-1} (Caccavelli & Bedouani 1995).

Then, for α_K calculation, many authors formulated criterial dependencies, determined in the experimental way by use of the similarity theory of physical phenomena concerning heat transfer (Khalifa 2001a, Khalifa 2001b, Awbi & Hatton 1999, Awbi & Hatton 2000), which may be used for exergy analyzes of heating systems (Wall 2013). However, using these dependences, it can be ascertained that received results of α_K coefficient are divergent, which can cause the uncertainty by their use in engineer's practice or in detailed scientific analyses (Rosiński & Cholewa 2010,

Cholewa & Rosiński 2010). Recently, Rahim & Sabernaeemi (2011) investigated the participation of heat emitted in the radiation way and convection from the floor radiant surface. They ascertained that 75–80% of heat is emitted in the radiation way. However, authors did not appointed the value of heat transfer coefficients from radiant floor surface. The most of research on these kinds of heating systems was conducted in steady state conditions of radiant surface. However in real conditions, radiant floor heating systems are subjected to different kind of thermal and hydraulic forcing.

Therefore, very important problem becomes the recognition of characteristic parameters connected with the heat transfer during the dynamic operation of the system, and which have simultaneously the direct influence on the heat flux density taken over from the radiant surface. Taking the above into consideration, and the fact that floor radiators are used more and more often for heating rooms in designed and modernized buildings, we decided to do experimental research on the massive floor radiator in the semitechnical scale during dynamic operation of the system.

2 MATERIALS AND METHODS

2.1 Laboratory stand

Experimental research was done on the laboratory stand in the semitechnical scale. Analyzed floor radiator (1.56 × 1.56 m) is made in the wet technology and consists of the following layers (counting from indoor air):

- terracotta (1.0 cm thick, $\lambda = 1.050$ Wm^{-1} K^{-1}),
- layer of granolith (6.5 cm thick, $\lambda = 1.200$ Wm^{-1} K^{-1}),
- pipes 16 × 2.0 PE-RT/Al/PE-RT with the distance of 15 cm,
- styrofoam (6.0 cm thick, $\lambda = 0.045$ Wm^{-1} K^{-1}).

The floor radiator was placed in the climatic chamber with black walls, which emissivity is equal to $\varepsilon = 0.95$. Isolating the floor radiator from the unchecked influence of other parameters (e.g. the solar radiation) and maintenances of constant parameters during every experiments were realized by assurance of constant temperature conditions outside the chamber, situated in a heated room.

The experimental system consists of:

- 19 temperature sensors PT500 (accuracy equal to 0.1 K) made of stainless-steel and used for the temperature measurement in the layer of granolith (9 pieces), temperature of indoor air on different heights in the climatic chamber (7 pieces) and temperature of black globe (3 pieces);

- 19 temperature sensors PT500 (accuracy equal to 0.1 K) equipped into the copper plate and used for temperature measurement of radiant surface (9 pieces) and of walls of the climatic chamber (10 pieces). The sensors are stuck to the surface with the glue, which is characterized by the large value of heat conduction coefficient (the value is close to copper and equal to $\lambda = 400$ Wm^{-1} K^{-1});
- 3 black globes with diameter equal to 150 mm;
- 2 temperature sensors PT500 (accuracy equal to 0.1 K) used for measurement of the heating medium temperature on supply and on return from floor radiator;
- the turbine-flow meter for liquid with the impulse output (600 impulse per liter), which is used for the measurement of volume flow rate of heating medium;
- 2 sensors of heat flux density (accuracy equal to 5.0%) used for measurement of heat flux density emitted from the radiant surface;
- recorders adapted to register the measurement;
- ultrathermostat, which is a heat source for the investigated system.

The arrangement of sensors is shown in Figure 1.

2.2 Research methodology

Ten measuring series that allowed to calculate thermal parameters of the floor radiator during thermal and hydraulic forcing were performed, namely:

- series 1, 2, 3: warming up the floor radiator for $t_z = 35°C$ (series 1), for $t_z = 45°C$ (series 2), for $t_z = 55°C$ (series 3);
- series 4, 5, 6: cooling down the floor radiator, after cutoff of supplying it with the heating factor $t_z = 35°C$ (series 4), $t_z = 45°C$ (series 5), $t_z = 55°C$ (series 6);
- series 7: increase in the temperature of the heating medium supplying the radiant panel by 10 K from $t_z = 45°C$ to $t_z = 55°C$ by 3.3 dm^3 min^{-1};
- series 8: decrease in the temperature of the heating medium supplying radiant panel by 10 K from $t_z = 55°C$ to $t_z = 45°C$ by 3.3 dm^3 min^{-1};
- series 9: increase in the volume flow rate of the heating medium from 0.5 dm^3 min^{-1} to 1.0 dm^3 min^{-1} by $t_z = 55°C$;
- series 10: increase in the volume flow rate of the heating medium from 1.0 dm^3 min^{-1} to 3.3 dm^3 min^{-1} by $t_z = 55°C$.

Values measured with the use of calibrated sensors were archived with the time step equal to 5 minutes. All changes of the system characteristic parameters (series 7–10) and also the beginning/end of supplying the system (series 1–6) were realized in 1 hour of the measurement.

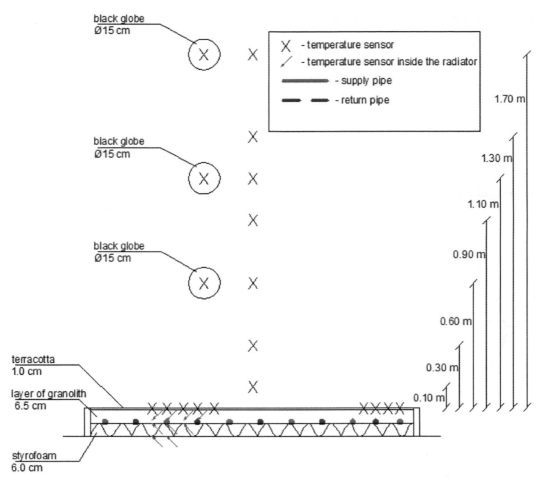

Figure 1.   The section of the centre of the radiant plate and climatic chamber.

Nevertheless, all changes of characteristic parameters were realized when system reached steady state after the earlier change of parameters. On the basis of archived measurement, it was possible to calculate the convective heat transfer coefficients (α_K) and radiant heat transfer coefficients (α_R) from the radiant surface in the way presented below.

2.2.1   *Radiant heat transfer coefficient (α_R)*
For the arrangement between the not concave surface and the surrounding surface the total heat flux density exchanged by radiation between the radiant floor surface and its surroundings can be counted on the basis of Equation (2).

$$q_R = \varepsilon_z \cdot \sigma \cdot \left(T_p^{\,4} - T_{ef}^{\,4}\right) \qquad (2)$$

The supplementary radiation emissivity is calculated by use of Equation (3). However, the average

unheated surface temperature (effective temperature) is appointed by means of Equations (4).

$$\varepsilon_z = \cfrac{1}{\cfrac{1}{\varepsilon_p} + \cfrac{A_p}{A_2}\left(\cfrac{1}{\varepsilon_2} - 1\right)} \qquad (3)$$

where:

$$T_{ef} = \sqrt[4]{\varphi_{1s,p} \cdot T_{1s}^{\,4} + \varphi_{2s,p} \cdot T_{2s}^{\,4} + \varphi_{3s,p} \cdot T_{3s}^{\,4} + \varphi_{4s,p} \cdot T_{4s}^{\,4} + \varphi_{sf,p} \cdot T_{sf}^{\,4}}$$

(4)

The view factor for two equal rectangular surfaces distant from each other by b (Fig. 2) is calculated on the basis of Equation (5).
However, the view factor for two perpendicular surfaces with common edge a (Fig. 2) is calculated on the basis of Equation (6).

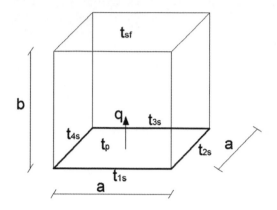

Figure 2. Schema of climatic chamber with the analyzed floor radiator.

$$\varepsilon_z \cdot \sigma \cdot \left(T_p^{\,4} - T_{ef}^{\,4}\right) = \alpha_R \cdot \sum_i \varphi_{i,p} \cdot \left(t_p - t_i\right)$$

$$\alpha_R = \frac{\varepsilon_z \cdot \sigma \cdot \left(T_p^{\,4} - T_{ef}^{\,4}\right)}{\sum_i \varphi_{i,p} \cdot \left(t_p - t_i\right)} \tag{9}$$

2.2.2 Convective heat transfer coefficient (α_K)

Taking into consideration the total heat flux density emitted from the radiant floor heating panel (q_{total}) and the heat flux density emitted in the radiation (q_R) way, the heat flux density emitted in the convection (q_K) way can be calculated on the basis of Equation (10).

$$q_K = q_{total} - q_R \tag{10}$$

$$\varphi_{i,p} = \frac{1}{\pi}\left[\frac{b^2}{a^2}\ln\frac{(a^2+b^2)(a^2+b^2)}{\left(2a^2+b^2\right)b^2} - \frac{2b}{a}arctg\frac{a}{b} - \frac{2b}{a}arctg\frac{a}{b} + \frac{4\sqrt{a^2+b^2}}{a}arctg\frac{a}{\sqrt{a^2+b^2}}\right] \tag{5}$$

$$\varphi_{i,p} = \frac{1}{\pi b}\left(b \cdot arctg\frac{a}{b} + a \cdot arctg\frac{a}{a} - \sqrt{b^2+a^2}\,arctg\frac{a}{\sqrt{b^2+a^2}}\right)$$
$$+ \frac{1}{4\pi ab}\ln\left\{\left[\frac{2a^2\cdot(a^2+b^2)}{(2a^2+b^2)a^2}\right]^{a^2} \times \left[\frac{(2a^2+b^2)b^2}{(a^2+b^2)(b^2+a^2)}\right]^{b^2}\left[\frac{(2a^2+b^2)a^2}{2a^2(b^2+a^2)}\right]^{a^2}\right\} \tag{6}$$

On the other hand, the amount of heat exchanged by radiation between grey surfaces may be calculated by the use of radiant heat transfer coefficient from Equation (7).

$$q_{Ri-p} = \alpha_R \cdot \varphi_{i,p} \cdot \left(t_p - t_i\right) \tag{7}$$

After location of the floor radiator in climatic chamber constructed from the uniform material with well-known emissivity ($\varepsilon = 0.95$), as shown in Figure 2, the total amount of heat exchanged in the radiation way can be calculated with the use of Equation (8).

$$q_{Ri-p} = \alpha_R \cdot \varphi_{1s,p} \cdot \left(t_p - t_{1s}\right) + \alpha_R \cdot \varphi_{2s,p} \cdot \left(t_p - t_{2s}\right)$$
$$+ \alpha_R \cdot \varphi_{3s,p} \cdot \left(t_p - t_{3s}\right) + \alpha_R \cdot \varphi_{4s,p} \cdot \left(t_p - t_{4s}\right)$$
$$+ \alpha_R \cdot \varphi_{sf,p} \cdot \left(t_p - t_{sf}\right) = \alpha_R \cdot \sum_i \varphi_{i,p} \cdot \left(t_p - t_i\right) \tag{8}$$

Comparing dependences (2) and (8) Equation (9) is received.

Therefore, the value of convective heat transfer coefficient from floor radiant surface could be appointed with the use of Equation (11).

$$\alpha_K = \frac{q_K}{t_p - t_i} \tag{11}$$

For closure of the balance, α_K may be defined directly in the experimental way with the use of the resemblances theory concerning taking over heat by convection, which will be an object of future research.

2.2.3 Total heat transfer coefficient (α_{total})

The value of total heat transfer coefficient from the radiant floor plate can be calculated with the use of Equation (12).

$$\alpha_{total} = \frac{q_{total}}{t_p - t_{op}} \tag{12}$$

In this case, it is recommended to use the operative temperature as the reference temperature, which may be calculated into the simplified manner (Equation 13) under the following conditions (ISO 7730: 2005, ISO 7726: 1998, ASHRAE Standard 55:2004):

- air velocity $v_a < 0.2$ ms⁻¹,
- $t_{mr} - t_i < 4$ K.

$$t_{op} = \frac{t_i + t_{mr}}{2} \tag{13}$$

In turn, the mean radiant temperature is calculated with the use of Equation 14 (ISO 7726:1998).

$$t_{mr} = \left[\left(t_g + 273 \right)^4 + 0.4 \times 10^8 \left| t_g - t_i \right|^{1/4} \times \left(t_g - t_i \right) \right]^{1/4} - 273 \tag{14}$$

3 RESULTS AND DISCUSSION

During the process of warming up the floor radiator (Fig. 3), it can be noticed that in the initial phase of this process the most of heat (above 50%) is transferred to the heated room in the convection way. However, in steady state conditions of the radiator, the participation of the convection in the total amount of heat transferred to the room falls and is within the range between 32–35%. The participation of this heat transfer mechanism depends, *inter alia*, on the temperature of heating factor which supplies the system. Because by its greater values, higher temperature of heating surface is received, which causes the increase of convective air flows.

Moreover, it was noticed that during the warming up of the floor radiator, the value of α_R increased (within the range 11–13%) stable in the time interval between power-up of floor radiator to the achievement the steady state by the system. For steady state conditions, values of heat transfer coefficient depending on the temperature of heating medium supplied the system, were listed in Table 1.

Analyzing data in Table 1, it can be noticed that values of heat transfer coefficients rise together with the increase of temperature of heating medium, which supplies the heating system. What is more, the difference (13–18%) between the value of total heat transfer coefficient from

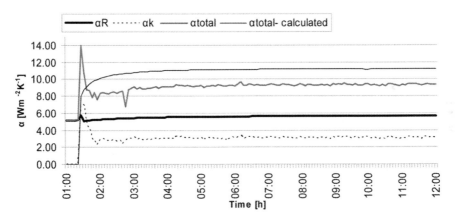

Figure 3. Values of heat transfer coefficients from the floor radiant surface by warming up the radiator for $t_z = 45°C$ (series 2).

Table 1. Values of heat transfer coefficient depending on the temperature of heating medium supplying the system in steady state conditions.

t_z/Series [°C]/[–]	α_R	$\alpha_{K\,0.1}$	$\alpha_{K\,1.1}$	$\alpha_{K\,1.7}$	α_{total}	$\alpha_{total—calculated}$
	[Wm⁻² K⁻¹]					
35/Series 1	5.45	3.28	2.77	2.76	8.77	10.69
45/Series 2	5.63	3.66	3.11	3.08	9.34	11.19
55/Series 3	5.81	4.01	3.48	3.44	9.95	11.51

the radiant surface received in experimental way and the value counted by use of Equation (1) was ascertained.

This difference can lead to design a system characterized by the smaller calorific effect in real conditions in comparison with computational conditions, what may contribute consequently to the decrease of thermal comfort in heated rooms.

Values of $\alpha_{K1.1}$ and $\alpha_{K1.7}$ coefficients, calculated with the use of the indoor air temperature at height 1.1 m and 1.7 m respectively, show small variability. Therefore, it can be assumed that above 1.1 m height, the value of the α_K coefficient and the indoor temperature are constant.

Analyzing the process of heat exchange arising after turning off the floor radiator (Fig. 4), it can be noticed that the values of heat transfer coefficients from radiant surface decrease and the participation of convection in the total amount of heat transferred to the room increase.

Analyzing courses of value of heat transfer coefficients during elementary thermal (Figs. 5–6) and hydraulic forcing (Figs. 7–8), it may be noticed that thermal forcing had a greater influence on changes of heat transfer coefficients from radiant surface than hydraulic forcing. During the thermal forcing, consisting in the change of temperature of heating medium from $t_z = 45°C$ to $t_z = 55°C$, the increase of

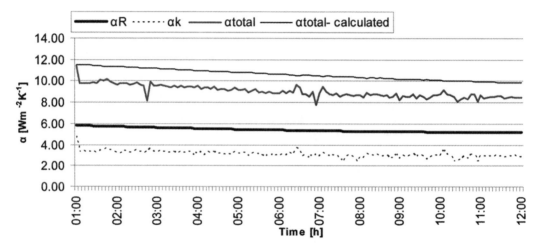

Figure 4. Values of heat transfer coefficients from the radiant surface by cooling down the floor radiator, after cutoff of supplying it with the heating factor for $t_z = 55°C$ (series 6).

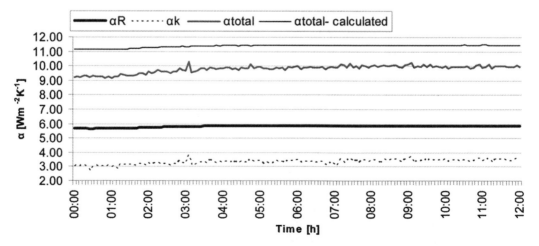

Figure 5. Values of heat transfer coefficients from radiant surface after change of heating medium temperature from $t_z = 45°C$ to $t_z = 55°C$ (series 7).

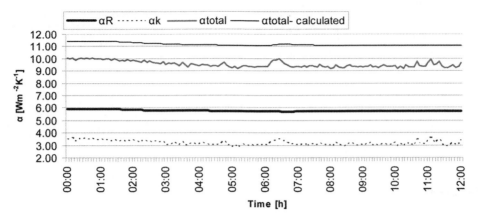

Figure 6. Values of heat transfer coefficients from radiant surface after change of heating medium temperature from $t_z = 55°C$ to $t_z = 45°C$ (series 8).

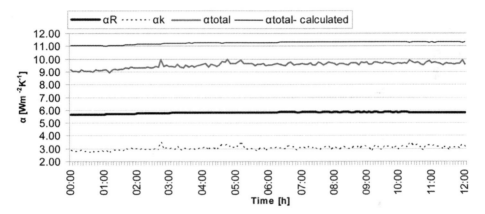

Figure 7. Values of heat transfer coefficients from radiant surface after change of volume flow rate of heating medium from 0.5 to 1.0 dm³ min⁻¹ by $t_z = 55°C$ (series 9).

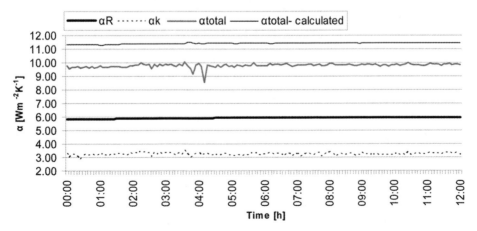

Figure 8. Values of heat transfer coefficients from radiant surface after change of volume flow rate of heating medium from 1.0 to 3.3 dm³ min⁻¹ by $t_z = 55°C$ (series 10).

α_R and α_K was equal to 4% and 17%, respectively. This increase was close to the results received during the hydraulic forcing in series 9 because then values α_R and α_K increased properly of about 3% and 12%. However, during hydraulic forcing in series 10 (Fig. 8), values α_R increase about 1.5% and the value α_K about 1.8%.

4 CONCLUSIONS

Experimental research of the massive floor radiator were performed in the semitechnical scale in dynamic conditions during different kind of thermal and hydraulic forcing, what let to determine the heat transfer coefficients courses from radiant floor heating surface. The participation of the convection in the total heat transferred to the room in steady state conditions is within the range between 32% and 35%. However, during the initial phase of warming up the floor radiator, the participation of the convection increases even to the level exceeding 50%.

On the basis of experimental research, it can be stated, that values of heat transfer coefficients from radiant surface for analyzed cases in steady state conditions were within the range $\alpha_R = 5.45$–5.81 Wm^{-2} K^{-1}, $\alpha_K = 2.77$–3.48 Wm^{-2} K^{-1} and $\alpha_{total} = 8.77$–9.95 Wm^{-2} K^{-1}. Higher values of these coefficients appeared by supplying the system with higher temperatures of the heating medium. The difference (13–18%) between the value of total heat transfer coefficient from the radiant surface received in experimental way and the value calculated with the use of the universally applied equation by design of this kind of heating system (Equation 1) was ascertained.

Nevertheless, it was noticed that thermal forcing has greater influence on the changes of heat transfer coefficients from the radiant surface than hydraulic forcing, which can lead to energy savings by correct supply of the floor radiator with heating medium.

ACKNOWLEDGEMENTS

This study was supported by project No. N N523 7452 40, financed by the Polish Ministry of Science and Higher Education.

NOMENCLATURE

A_p	area of radiant surface, [m²],
A_2	area of surfaces surrounding the radiant surface, [m²],
t_g	black globe temperature, [°C],
t_i	mean indoor air temperature outside the near wall layer of the fluid (h = 1.1 m), [°C],
t_{mr}	mean radiant temperature, [°C],
t_{op}	operative temperature, [°C],
t_p	radiant surface temperature, [°C],
T	absolute temperature, [K],
T_p	absolute temperature of radiant surface, [K],
T_{ef}	average unheated surface temperature (effective temperature), [K],
q_{total}	total heat flux density, [Wm⁻²],
q_K	convective heat flux density, [Wm⁻²],
q_R	radiant heat flux density, [Wm⁻²],
v_a	air velocity, [ms⁻¹],

Greek symbols

α_{total}	total heat transfer coefficient, [Wm⁻² K⁻¹],
α_K	convective heat transfer coefficient, [Wm⁻² K⁻¹],
α_R	radiant heat transfer coefficient, [Wm⁻² K⁻¹],
ε_p	emissivity of the floor heating surface,
ε_z	substitutive radiation emissivity,
ε_2	emissivity of surfaces surrounding the radiant surface,
λ	coefficient of thermal conductivity of the fluid, [Wm⁻¹ K⁻¹],
σ	Stefan–Boltzmann constant, [Wm⁻² K⁻⁴],
$\varphi_{i,p}$	view factor between radiant surface and i-surface

Subscripts

1s, 2s, 3s, 4s	walls adjoining with radiant surface
sf	ceiling above radiant surface
i	i-function ($i = 1, 2$)

REFERENCES

ASHRAE Standard 55:2004, Thermal Environmental Conditions for Human Occupancy.

Awbi H.B., Hatton A. 1999. Natural convection from heated room surfaces. *Energy and Buildings* 30: 233–244.

Awbi H.B., Hatton A. 2000. Mixed convention from heated room surfaces, *Energy and Buildings* 32: 153–166.

Caccavelli D., Bedouani B. 1995. Modelling and dimensioning a hot water floor heating system. IBPSA, 571–579.

Causone F., Corgnati S.P., Filippi M., Olesen B.W. 2009. Experimental evaluation of heat transfer coefficients between radiant ceiling and room. *Energy and Buildings* 41: 622–628.

Cholewa T., Rosiński M. 2010. Heat transfer in rooms with panel heating systems. In *Proceedings of 41th International Congress on Heating, Refrigerating and Air-Conditioning*, Edited by Branislav Todorović, Belgrad 2010, p. 370–380.

ISO 7730:2005, Ergonomics of the Thermal Environment—Analytical Determination and Interpretation of Thermal Comfort Using Calculation of the PMV and PPD Indices and Local Thermal Comfort Criteria.

ISO 7726:1998, Ergonomics of the Thermal Environment—Instruments for Measuring Physical Quantities.

Karlsson H., Hagentoft C.-E. 2005. Modelling of Long Wave radiation Exchange in Enclosures with Building Integrated Heating, Symposium on Building Physics in the Nordic Countries, Reykjavik, Iceland.

Karlsson H. 2006a. *Thermal system analysys of embedded building integrated heating.* Thesis for the degree of licentiate of engineering, Chalmers University of Technology, Göteborg, Sweden.

Karlsson H. 2006b. Building integrated heating: hybrid three—dimensional numerical model for thermal system analysis, *Journal of Building Physics.*

Khalifa A.J.N. 2001a. Natural convective heat transfer coefficient a review I. Isolated vertical and horizontal surfaces. *Energy Conversion and Management* 42: 491–504.

Khalifa A.J.N. 2001b. Natural convective heat transfer coefficient—a review: II. Surfaces in two—and three—dimensional enclosures, *Energy Conversion and Management* 42: 505–517.

Kilkis B., Eltez M., Sager S. 1995. A simplified model for the design of radiant in slab heating panels. *ASHRE Transactions*, vol. 101, part 1.

Kowalczyk A., Strzeszewski M. 1999. Comparison of calculation methods of heat flux density from radiant floor heating panel. *District Heating, Heating, Ventilation*, 3/99 (in Polish).

Kowalczyk A., Rosiński M. 2007. Comparative analysis of the european metod for dimensioning of massive floor radiators with empirically verified reference numerical method. *Archives of Civil Engineering*, 2, pp. 357–386.

PN-EN 1264-2:2008 Water based surface embedded heating and cooling systems—Part 2 Floor heating: Prove methods for the determination of the thermal output using calculation and test methods.

PN-EN 1264-5:2008 Water based surface embedded heating and cooling systems—Part 5: Heating and cooling surfaces embedded in floors, ceilings and walls—Determination of the thermal output.

Rahimi M., Sabernaeemi A. 2011. Experimental study of radiation and free convection in an enclosure with under-floor heating system. *Energy Conversion and Management* 52: 2752–2757.

Rosiński M., Cholewa T. 2010. The analysis of parameters influencing the calorific effect of floor radiators in aspect of heating installations designing. In *Proceedings of the XIII International Symposium on Heat Transfer and Renewable Sources of Energy* HTRSE 2010, Szczecin-Międzyzdroje, p. 543–550.

Wall G. 2013. Exergy, Life and Sustainable Development. *Problems of Sustainable Development* 8 (1): 27–41.

Environmental Engineering IV – Pawłowski, Dudzińska & Pawłowski (eds)
© 2013 Taylor & Francis Group, London, ISBN 978-0-415-64338-2

Study of different parameters impact on heat consumption in education buildings

R. Sekret & A. Jachura

Faculty of Environmental Protection and Engineering, Department of Heating, Ventilation and Air Conditioning, Czestochowa University of Technology, Czestochowa, Poland

ABSTRACT: The article presents the analysis of heat consumption variation and demand depending on the examined technical and constructional parameters of 5 schools, selected from the group of 50 educational buildings, located in the area of the city of Czestochowa. The purpose of the study was to determine the characteristic factors which have an impact on the discrepancies between the heat demand and the heat consumption for a selected group of education buildings. The derivation of such relationships helped to create a basis for developing indicators to correct the theoretical heat demand for the population of school buildings under the study.

Keywords: heat demand, heat consumption, education buildings, schools

1 INTRODUCTION

The rational use of energy is driven by two basic factors, namely economic criteria, and regulations in energy management. The economic criteria mean striving for reducing costs, which, in the case of energy consumption, are understood as operating costs. Regulations constitute a set of imposed prescriptions intended to achieve specific objectives in rational energy management (Jachura et al. 2010). The main legal instrument in the European Union for improving the efficiency and energy performance of buildings is European Directive 2010/31/EU (EPBD) (Directive 2010/31/UE of the European Parliament and of the Council of 19 May 2010 on the energy performance of buildings, 2010), which puts emphasis on the requirements regarding the energy performance, explains and simplifies some of its provisions to reduce the large differences between the practices of the Member States. It sets a challenging goal, whereby all new buildings shall have a near-zero energy characteristic by the 31st of December 2020, while public utility buildings shall be near-zero energy after the 31st of December 2018 (Dascalakia & Sermpetzogloub 2011). Figure 1 shows the structure of energy consumption in public buildings.

It should be borne in mind that the nature of public buildings can be very diverse. Thus, for example, in specific embodiments, the proportion of energy for lighting canincrease up to about 30%. However, the data presented in Figure 1 shows that the main component of the energy balance of the heat for the buildings, more than 50%, is used for heating and ventilation. In the municipal

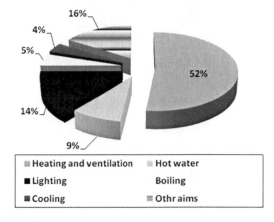

Figure 1. The structure of energy consumption in public buildings.

energy supply, among the strategic energy users there are education buildings. Among the most important organizational units in the education system there are schools, where so large discrepancies are found that it is hard to estimate even an approximate structure of energy needs. In the majority of countries, education facilities have similar constructions, operation principles and maintenance characteristics. The two most noteworthy similarities in those types of buildings are the high consumption of energy and the need to modernize many buildings in this sector (Erhorn et al. 2008). The factors that influence the level of heat consumption in public utility buildings are shown in Figure 2.

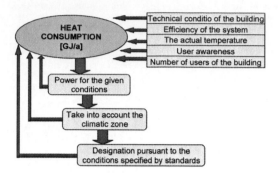

Figure 2. Factors influencing the heat consumption level in public utility buildings.

The energy consumption in school buildings, due to their large number in the country, contributes to the considerable overall energy consumption by public utility buildings. Moreover, school buildings are not normally used during the night, at the weekends, on holidays and in the summer, being left empty for up to the three fourths of the total number of hours in the year. Therefore, on the one hand, the study of alternative solutions towards the reduction of energy consumption in school buildings is advisable and necessary. On the other hand, the teaching function of school buildings requires the proper control of parameters that influence the internal conditions in classrooms, such as lack of thermal comfort and unsatisfactory air quality conditions, which lower the learning capacities of the students. Thus, the energy efficiency of school buildings has a twofold purpose, namely the saving of energy and better internal conditions in the classrooms (Dimoudi & Kostarela, 2009).

2 MATERIAL AND METHODS

The subject of the study is a set of 50 primary schools located in the area of Czestochowa. Figure 3 shows a map in which education buildings under the study are indicated.

As shown in Figure 3 a group of school buildings is distributed throughout the city. Cubic test heated buildings range from 2090 to 40903 m³. Almost 70% of the analyzed schools is supplied with district heating. These are buildings built in years 1913–1999, of which over 77% was established before 1969. Buildings constructed during that period significantly differ from the currently accepted standards of heat. Most of the educational buildings, about 80%, were not subjected to thermal process and the improvements, that have been made in a few schools, were obliged to replace windows and to repair central heating.

2.1 Defining the criteria for the selection of representative buildings

It was preliminarily proposed to divide education buildings into 5 groups, namely according to the year of construction, the heated cubage, the shape factor, the type of the heat source intended for central heating purposes, and the method of preparing hot water. Then, each group was divided into several subgroups. Table 1 presents 5 groups together with their respective subgroups obtained from the preliminary division of primary school buildings.

The main criterion for the selection of representative buildings was the determined g coefficient [GJ/m³] that represents the actual amount of heat supplied to a building and converted into a standard season, in relation to the number of degree days, in [GJ] by the heated cubage of that building in m³. Next, the coefficient was corrected with a temperature conversion factor:

$$g_t = g_{ave,y} \cdot \frac{t_{ave,c}}{t_{ave,y}} \left[\frac{GJ}{m^3} \right] \tag{1}$$

where: $g_{ave,y}$ = the average coefficient value for a given calculation year, $[GJ/m^3]$; $t_{ave,c}$ = the average temperature value for the entire calculation period, [°C]; $t_{ave,y}$ = the average temperature value for a given calculation year, [°C];
and the conversion factor for the number of degree-days:

$$g_{DD} = g_{ave,y} \cdot \frac{DD_{ave,c}}{DD_{ave,y}} \left[\frac{GJ}{m^3} \right] \tag{2}$$

where: $DD_{ave,c}$ = the average value of degree-day numbers for the entire calculation period, [days]; $DD_{ave,y}$ = the average value of degree-day numbers for a given calculation year, (days).

The analysis of changes in outside air temperature, the length of the heating season and the number of degree days in Czestochowa, was made on the basis of a set of data from the years 1951–2010, as shown in the work (Sekret & Wilczyński 2011). The mean values of temperature and the number of degrees were calculated from the period of 58 years, on the basis of which the following ratios were established:

- Temperature coefficient as the ratio of the average outdoor temperature from the period of 58 years and the average external temperature of the heating season of the year concerned.
- Degree ratio as the ratio of the average number of degrees from the period of 58 years and the average number of degree days in the year concerned the heating season.

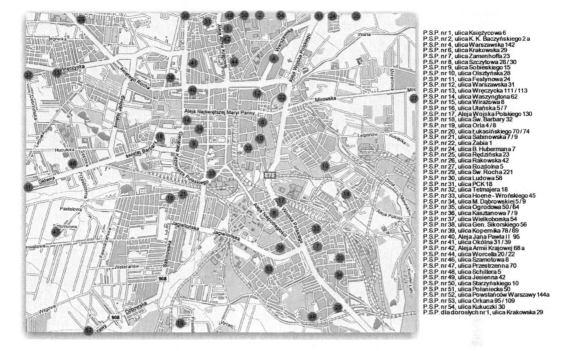

Figure 3.   A list and location of education buildings selected to analysis of heat demand and consumption in Czestochowa.

Table 1.   Division of education buildings according to the adopted selection criteria.

Group number	Year of construction	The heated cubage of a building [m³]	The shape factor of a building A/V [1/m]	The type of the heat source intended for heating system purposes	The method of preparing hot water
1	1913–1929	To 6000	0,16–0,18	District heat	District heat
2	1930–1959	6001–13000	0,19–0,21	Electricity	Electricity
3	1960–1969	13001–25000	0,22–0,24	Natural gas	Natural gas
4	1970–1999	Above 25000	0,25–0,27	Oil-fired boiler	Oil-fired boiler
5	–	–	0,28–0,32	Pot to a solid fuel	Pot to a solid fuel

After performing the above operations, 5 representative buildings were obtained. Table 2 gives characteristic quantities and g coefficient values, on the basis of which reference buildings were selected.

Five groups of quantities influencing the theoretical heat demand and the actual heat consumption were preliminarily proposed:

1. Climatic data (analytical external and internal air temperatures).
2. Technical and service parameters (the shape factor and service time of a building).
3. Overall heat-transfer coefficient (the thermal conductivity, the thickness of individual partition layers, and the thermal resistance of the partition).

4. Ventilation air flux (the air exchange rate and the number of people residing in the room).
5. Operability of the technical systems (the efficiency of the system individual elements).
6. Determination of the heat demand for school building heating.

The demand for heating and cooling constitutes an important component of the general energy characteristics of buildings. Many indicators based on the energy demand provide a basis for comparison of architectural concepts and estimation of the future operation costs of facilities, and even for evaluation of the environmental impact of buildings. In general, the balance of the energy demand of a public utility building encompasses (Turner,

399

Table 2. Values of g coefficients and the quantities characterizing the reference buildings.

No. school	Year of construction	The heated cubage of a building [m³]	A/V [1/m]	The heat source for the purposes c.o.	The heat source for the purposes c.w.u.	g [GJ/m³]	g_t [GJ/m³]	g_{DD} [GJ/m³]
1	1960	13184	0,26	CS	EE	0,12	0,12	0,16
2	1964	14522	0,22	G	EE	0,12	0,12	0,15
3	1968	15437	0,21	CS	EE	0,12	0,12	0,15
4	1980	39003	0,22	CS	CS	0,11	0,11	0,13
5	1982	40903	0,24	CS	CS	0,11	0,11	0,13

A/V—the shape factor of a building, c.o.—heating system, c.w.u.—hot utility water, CS—district heat, EE—electricity, G—natural gas.

2005) two groups of needs. On the one hand, there is the first group which concerns the assurance of thermal comfort to the user of a given facility, which requires a specified amount of heat or refrigeration to be supplied to that facility. On the other hand, the second group includes auxiliary needs, such as lighting, hot water preparation, etc. In this study, the first group of needs is examined.

2.2 Determination of the theoretical thermal load

The theoretical thermal load for the education buildings under the study was determined on the basis of standard PN–EN 12831 "Heating systems in buildings. A method for calculating the design thermal load". The total design heat loss by the heated space was calculated with the use of the following Formula (3):

$$\Phi_{(i)} = \left(\Phi_{(T,i)} + \Phi_{(V,i)} \right) \cdot f_{\Delta\theta,i} \left[W \right] \tag{3}$$

where: $\Phi_{(T,i)}$ = design heat loss by the heated space (i) through heat transmission [W]; $\Phi_{(V,i)}$ = design ventilation heat loss by the heated space (i) [W]; $f_{\Delta\theta,i}$ = correction temperature coefficient allowing for additional heat losses by rooms heated to a temperature higher than that of the adjacent heated rooms.

2.3 Determination of the seasonal heat demand

The time variability of climatic conditions, in which a building is situated, results in a variable energy demand. The form of climatic data in the mathematical description of this energy demand determines the approach to the analysis of the building energy needs (Sekret, Wilczyński, 2011), i.e. either statistical or dynamical. In the statistical approach, calculations are made assuming the heat transfer through the building partitions under steady conditions, i.e. with the heat flux and temperature being constant within the examined time

intervals (monthly), and with averaged external climate parameters. It causes some simplifications (in contrast to dynamic modelling that is characterized by a complex mathematical description) but the availability of climatic data in the form of average monthly values and the ease of use, caused the statistical approach to have been used for the determination of the building energy demand.

The seasonal heat demand for the examined education buildings was determined on the basis of standard PN–EN ISO 13790 "The energy service properties of buildings. The calculation of energy consumption for the purposes of heating and cooling". For determining the level of heat demand for heating a building, the analysis of the building thermal balance is necessary. This thermal balance includes, on the one hand, heat losses and on the other hand, heat gains occurring in the building. The value of the monthly heat demand for heating and ventilation of the building, $Q_{H,nd,n}$, was calculated on the basis of (4):

$$Q_{H,nd,n} = Q_{H,ht} - \eta_{H,gn} \cdot Q_{H,gn} \left[\frac{kWh}{month} \right] \tag{4}$$

where: $Q_{H,ht}$ = the heat losses in a monthly period, [kWh/month]; $Q_{H,gn}$ = the heat gains in a monthly period, [kWh/month]; $\eta_{H,gn}$ = the coefficient of heat gain utilization efficiency in the heating mode [–].

2.4 Final energy demand

The annual heat demand, $Q_{H,nd}$, is determined by the method of monthly calculations for consecutive months. The $Q_{H,dn}$ is the sum of the heat demands for the building heating and ventilation in individual months, $Q_{H,nd,n}$, in which the heat demand level has a positive value. The months from January to May and from September to December inclusive, are considered. The monthly balance results calculated for the months from January to May and from September to December are summed up to obtain the annual heating energy demand. The annual final energy demand, $Q_{K,H}$, was calculated (5):

$$Q_{K,H} = \frac{Q_{H,nd}}{\eta_{H,tot}} \left[\frac{kWh}{a} \right] \qquad (5)$$

where: $\eta_{H,tot}$ = the average seasonal overall efficiency of the building heating system, which has been adopted in accordance with the specifications in the Polish law regulations.

Data on the thermal parameters of the tested objects characterizing their full and transparent partitions, body and internal system parameters are based on construction documents, index cards and visits. Actual heat consumption for the reference buildings were measured at the source, using a heat meter (for power network) and based on the amount of fuel consumed (for their own source of heat).

To assess the impact of parameters on the differences in the use of real and theoretical heat used in the difference between the theoretical heat demand and the actual heat consumption:

$$Q_{t-r} = (Q_t - Q_r) \left[\frac{GJ}{year} \right] \qquad (6)$$

where: Q_t = the theoretical heat demand, $[GJ/year]$; Q_r = the actual heat consumption, $[GJ/year]$.

3 RESULTS AND DISCUSSION

On the basis of the monthly heat balances of the buildings (the theoretical heat demand) and the annual balances relating to the actual heat consumption by the representative buildings, comparison of the facilities under analysis is made in Figure 4.

It can be seen from the data shown in Figure 4 that the difference between the actual energy demand and the theoretical energy demand for the examined buildings ranges from 61% to 71%. To determine the causes of so large discrepancies between the actual and the theoretical heat consumptions, the influence of the following technical and constructional parameters was examined:

• The air exchange rate—the n_{50} factor [h^{-1}].
• The minimum external air exchange rate—n_{min} [h^{-1}].
• The internal heat gain load—q_{int} [W/m^2].
• The internal temperature – θ_{int} [°C]. and
• The heat transfer coefficient—U [W/m^2 K].

The results of the performed analyses are represented in the diagrams below (Figs. 5–9). Figure 5 illustrates the effect of the air exchange rate on the

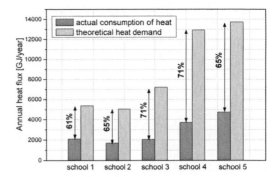

Figure 4. Diagram comparing the theoretical heat demand with the actual heat consumption in [GJ/year] and showing the differences between the theoretical heat demand and the actual heat consumption in [%] (for the reference buildings).

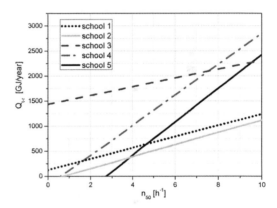

Figure 5. Effect of the air exchange ratio, n_{50}, on the difference between the theoretical heat demand and the actual heat consumption.

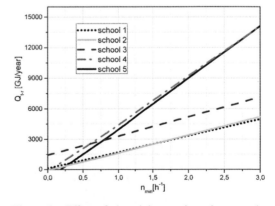

Figure 6. Effect of the minimum air exchange ratio, n_{min}, on the difference between the theoretical heat demand and the actual heat consumption.

401

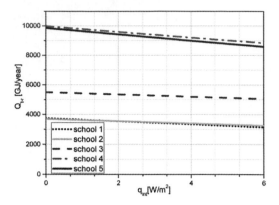

Figure 7. Effect of the value of the internal heat gain load factor, q_{int}, on the difference between the theoretical heat demand and the actual heat consumption.

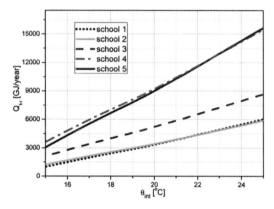

Figure 8. Effect of the building internal temperature, θ_{int}, on the difference between the theoretical heat demand and the actual heat consumption.

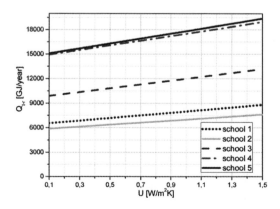

Figure 9. Effect of the heat transfer coefficient, U, on the difference between the theoretical heat demand and the actual heat consumption.

difference between the theoretical heat demand and the actual heat consumption for the school buildings under the study. The air exchange rate is a parameter for building characteristic and, as demonstrated by the diagrams in Figure 5, has a significant influence on the heat consumption in buildings 1 & 2 and 4 & 5. It is of smaller importance in the case of building 3. According to the currently applicable standard PN–EN ISO 13790, the value of the n_{50} factor [h^{-1}] lies in the following ranges: $n_{50} < 2$ for a high air-tightness level; n_{50} from 2 to 5 for a medium air-tightness level; and $n_{50} > 5$ for a low air-tightness level. The examined buildings are characterized by a medium air-tightness level, and assuming values from the range prescribed by the standard PN–EN ISO 13790 for buildings 1, 2, 4 and 5 allows to obtain very close theoretical energy demand and actual heat consumption values. In contrast, building 3 considerably deviates from the preferred values assumed for the n_{50} factor. Exceptions for building no. 3 may be due to the poor condition of the internal heating system and the outer shell of the building (envelope both full and windows), and a ventilation system malfunction, and a much longer lifetime of the school building.

Figure 6 illustrates the effect of the minimum air exchange rate on the difference between the theoretical heat demand and the actual heat consumption for the education buildings under study.

The minimum external air exchange rate is a quantity influencing the ventilation system. This parameter also has a great impact on heat consumption. As indicated by the data in Figure 6, the increase in the value of the n_{min} factor increases the difference between the theoretical and the actual heat consumption values. According to the standard PN–EN ISO 13790, the value of the n_{min} factor contains in the range from 0.5 to 2 [h^{-1}]. Adopting the lower limit from this range for buildings 1, 2, 4 and 5, allows to reduce considerably the difference between the theoretical and the actual heat consumption. Similarly, as in the previous diagram, building 3 requires a significant adjustment of the examined factor.

Figure 7 illustrates the effect of the value of the internal heat gain load factor on the difference between the theoretical heat demand and the actual heat consumption for the school buildings under examination.

It can be noticed from the diagram in Figure 7 that with the increase in the q_{int} factor, the difference between the theoretical and the actual heat consumption decreases, though it is a very weak correlation. In an average school room, internal heat gains from people and equipment range from 1.5 to 4.7 W/m^2 (Building Law of 19 September 2007 (Official Journal of 18 October 2007 No. 191, item. 1373)). To eliminate the occurring differences in heat consumption, the internal heat gain load factor would have to take on a value of the order of several hundred W/m^2, which in reality is not possible to appear in buildings of this type.

Figure 8 shows the effect of the internal temperature value on the difference between the theoretical heat demand and the actual heat consumption for the school buildings under examination.

As indicated by the data in Figure 8, the difference between the theoretical heat demand and the actual heat consumption is directly proportional to the temperature inside the building. The lower the internal temperature value is, the smaller the discrepancies in the heat consumption become. It is a strong relationship for each of the examined buildings. In accordance with standard PN–82/B–02402 "Temperatures heated rooms in buildings, in force applicable to the temperatures of heated rooms", the θ_{int} factor value is taken as 20°C. By lowering this value by 4°C, the difference in the utilization of heat is reduced to about 50%.

Figure 9 shows the effect of the heat transfer coefficient value on the difference between the theoretical heat demand and the actual heat consumption for the school buildings under examination. From the diagram in Figure 9, a weak relationship between the heat transfer coefficient and the discrepancies in heat consumption are obtained. This relationship is directly proportional: the lower the U coefficient value is, the smaller the $(Q_t - Q_r)$ difference is. By reducing the value of this parameter by as much as up to 30%, no significant improvement in the difference between the theoretical and the actual heat consumption was achieved.

4 CONCLUSION

Many education buildings in Poland are old structures characterized by high energy intensity. It can be accepted that the thermal protection in buildings constructed before 1981 is insufficient, poor in buildings dated from the years 1982–1990, sufficient in buildings from the years 1991–1994, and good in ones constructed after 1995. The group of examined buildings were constructed in the period from 1960 to 1982. Buildings constructed in this period considerably deviate from the present standards, which might be due to various factors.

After examining the influence of 5 technical and constructional parameters, i.e. the air exchange rate, the minimum external air exchange rate, the internal heat gain load, building internal temperature and the heat transfer coefficient, it was found that a significant quantity was stated that the volumes that characterize the ventilation system

have a significant influence. The deviations in the case of building 3 might result from poor condition of the internal heating system and the outer shell of the building (envelope both full and windows), and a ventilation system malfunction, and a much longer lifetime of the school building.

The temperature that should hold inside a building also determines the amount of consumed heat to a considerable extent, which is indirectly influenced by transparent partitions. Too intensive airing and faulty window framing cause excessive infiltration of cold air, which results in a drop in room temperatures and the need for reheating the rooms to equalize the temperatures. It, in turn, increases the level of heat consumed by the building.

Both the internal heat gain load and the heat transfer coefficient showed a weak correlation with the difference between the theoretical heat demand and the actual heat consumption; although these should not be disregarded. Prolonged neglecting of building partitions combined with their insufficient thermal insulation may contribute to an increase in energy demand and emissions of pollutants to the natural environment by school buildings. Solid external partitions in all of the schools examined have never been subjected to a thermal insulation process, which would have allowed a reduction of the heat loss. As far as the internal heat gains are concerned, their effect on the consumed heat level is only a matter of a few per cent.

It has been observed that buildings of a similar heated cubage and constructed in a coinciding period have curves of a very much alike behaviour in all diagrams examined. It might be due to the use of very similar structures and building materials that prevailed in a given period.

ACKNOWLEDGEMENTS

This scientific research has been financed by the MNiSW (Ministry of Science and Higher Education in Poland) with resources allocated for scientific development in the years 2011–2012, under Research Project No. 7418/B/T02/2011/40.

REFERENCES

Building Law of 19 September 2007 (Official Journal of 18 October 2007 No. 191, item. 1373).
Dascalakia E.G., Sermpetzogloub V.G., 2011, Energy performance and indoor environmental quality in Hellenic schools, *Energy and Buildings* 43: 718–727.
Dimoudi A., Kostarela P., 2009, Energy monitoring and conservation potential in school buildings in the C climatic zone of Greece, Renewable Energy 34: 289–296.

Directive 2010/31/UE of the European Parliament and of the Council of 19 May 2010 on the energy performance of buildings, 2010.

Erhorn H., Mroz T., Mørck O., Schmidt F., Schoff L., Thomsen K.E., 2008, The Energy Concept Adviser—A tool to improve energy efficiency in educational buildings, *Energy and Buildings* 40: 419–428.

Jachura A., Turski M., Lis P., Sekret R., 2010, Possibilities for the improvement of the energy effectiveness of the construction and installation systems of an hospital facility. In: Proceeding 2 st International Scientific Conference CASSOTHERM CD, Jasna, ISBN-978-80-89385-03-4.

PN–EN 12831. Heating systems in buildings. A method for calculating the design thermal load.

PN–82/B–02402. Temperatures heated rooms in buildings.

PN–EN ISO 13790. The energy service properties of buildings. The calculation of energy consumption for the purposes of heating and cooling.

Sekret R., Wilczynski J., 2011, The effect of variations in ambient air temperature and heating season length on the number of degree-days on the example of the city of Czestochowa, *Rynek Energi* 4 (95): 58–63.

Turner W.C., 2005, Energy management handbook, New York, Lilburn: Fairmont Press.

Environmental Engineering IV – Pawłowski, Dudzińska & Pawłowski (eds)
© *2013 Taylor & Francis Group, London, ISBN 978-0-415-64338-2*

The actual and calculated thermal needs of educational buildings

P. Lis
Faculty of Environmental Protection and Engineering, Czestochowa University of Technology, Czestochowa, Poland

ABSTRACT: This paper presents the selected results of examinations connected with a seasonal heat consumption (Q) and thermal power (q) for heating. The presented analysis and its results concern the group including 46 of 50 educational buildings, which form a municipal group of this type of the buildings. The purpose of presented analysis was to examine the influence of possible occurrence and level of differences between the seasonal heat consumption (Q) and the seasonal heat demand (Q_q) for heating, calculated on the basis of q values. A modification in the method for determination of Q_q for rooms heating on the basis of available data on q was introduced. A linear function, describing the changes in (Q_q–Q) depending on the changes in q values, which was applied for that purpose, allowed to improve the consistency of obtained heat demand values in relation to measured consumption of heat for heating by 65.6%.

Keywords: educational buildings, heating, thermal power, heat demand, heat consumption

1 INTRODUCTION

The calculative methods, which are adopted in various fields of engineering, are usually a certain kind of theoretical approximation of reality. The main problem, which occurs here, is the degree of consistency of theoretical description of some phenomenon or process with the actual conditions of its course.

The deviations from a full consistency of actual conditions and theoretical assumptions occur also in case of building heating. The building is considered as a constructional and installation entirety and constitutes a set of many installation, architectonic-building, constructional-material and operational properties, which have a direct or indirect connection with its heating. The appearing discrepancies are visible even in quantitative characteristics of heating, i.e. the amount of heat theoretically needed and actually used for that purpose. Not so long ago, the thermal power q, which was calculated according to PN-B-03406:1994—"Calculation of heat demand rooms" and the seasonal heat demand Q_d in a standard heating season, which was calculated according to PN-B-02025:2001—"The calculation of the seasonal heat demand of residential and public buildings public" were one of the basic quantities calculated in the projects of heating systems. Several years ago, in the design studies, the value of q was replaced by, so-called, design heat load, which is calculated according to PN EN ISO 12831:2006—"Heating systems in

buildings. Method for calculation of the design heat load". However, there are no obstacles to use still q in various considerations. One should only remember to apply consistently only one of mentioned quantities. A change of methodology in the design calculations regards also Q_d which is calculated according to PN EN ISO 13790:2008—"Energy performance of buildings—Calculation of energy use for space heating and cooling". In spite of the differences occurring in physical interpretation and the method of calculation of these quantities, both of them can be considered as theoretical, in contrast with the seasonal heat consumption Q for buildings heating, the amount of which was determined by measurement and, in a sense, "describes the actual conditions of heating".

The value of Q is the effect of, *inter alia*, the duration of the heating season and the conditions inside and outside of a room, while the calculated q is based on the most unfavorable of mentioned conditions that may occur during the heating season and is a "momentary" value. However, it is not so significant for the purposes of considerations conducted because the discrepancies between the actual conditions and their theoretical representation still appear, irrespective of applied methodology.

There are many publications on heat consumption and heat demand in buildings. However, most of them concern the housing or community on which these buildings have a dominant influence. There is much less publication on a group of public buildings, including community-specific educational buildings. These objects are different from residential buildings, in particular:

- The specificity of use resulting from the function of the education buildings and the resulting periodicity of use.
- Average height of the rooms much higher than in residential buildings.
- Average percentage of glazing significantly higher than in residential buildings.

The above differences make direct use of research results and analysis of residential buildings, what is impossible or difficult in the case of educational buildings. Publications concerning the heat consumption or heat demand for heating the educational buildings relate to individual buildings or random groups numbering from a few buildings up to 320 objects in the world e.g. (Butala & Novak 1999, Corgnati et al. 2008, Desideri & Proietti 2002, Filippin 2000, Gallachoir et al. 2007, Noren & Pyrko 1998) and in Poland e.g. (Piotrowska-Woroniak & Woroniak 2005, Prętka 2004, Stachniewicz 2007). These groups were created as a result of the research. Only a few publications cover a group of educational buildings linked administratively and territorially: located in one of the regions in Italy (Corgnati et al. 2008, Desideri & Proietti 2002), Ireland (Gallachoir et al. 2007) and Sweden (Noren & Pyrko 1998). The main theme of these publications (except for Corgnati et al. 2008, Desideri & Proietti 2002) is different from the subject of this article, that is discrepancies between the thermal needs of educational buildings in actual and calculation conditions. The signal confirming the occurrence of these discrepancies analysis of heat consumption Q and thermal power q for educational buildings conducted by the author, whose results are presented in the Figure 1. In the relationship shown in the Figure 1, it was noticed that about 84% of changes in Q depends on the changes in q, while 16% does not depend on the changes of this quantity (determination of coefficient

value amounts to 0.84 approximately). This state can be caused by not very accurate consideration of the actual conditions of buildings heating in the methodology of q calculation. Moreover, it seems that if the problem in a macro scale is considered, i.e. if we are looking at the entire object, then the differences between the reality and theory occur but their significance undergoes a "certain dissipation"; whereas while compared to the unit of building cubic capacity the appearing discrepancies undergo "concentration". Thus, it can be presumed that above-mentioned differences will be more sensible in smaller buildings and less significant in bigger objects. Such reasoning finds also a partial confirmation in the publications while taking into consideration the issues of educational buildings heating (Corgnati et al. 2008).

Estimated and simplified calculations of the seasonal heat demand Q_q based on the known value of thermal power q are used quite often in the engineering practice. The heat demand in analyzed heating season for a given building calculated with the using of known value of thermal power, on the basis of equation (1):

$$Q_q = q \cdot Nd \cdot 24\,h \cdot \frac{(T_{ical} - T_{eav})}{(T_{ical} - T_{emin})} \cdot 0.0036 \qquad (1)$$

where:
Q_q heat demand for heating in heating season a, calculated with the using of known value of thermal power, GJ/a;
q   thermal power, kW;
Nd   number of heating days in considered heating season, days;
24 h   duration of a day, h;
T_{ical} calculative temperature of air inside of heated building assumed in considered case ($T_{ical.} = +20°C$), °C;
T_{eav} average temperature of air outside in the heating season for considered period and for determined area (town) ($T_{eav} = +2.9°C$), °C;
T_{emin} calculative outdoor temperature, ($T_{emin} = -20.0°C$), °C;
0.0036 conversion factor for values expressed in various physical units.

The usage of available base quantities for calculation of sought quantities, the physical interpretation of which is often different from the "base", is not a new phenomenon (Corgnati et al. 2008; Kasperkiewicz 2006). Despite of its disadvantages, it will be probably still applied. In spite of the simplifications introduced in such cases, the obtained results of calculations should correlate with the results of measurements. It should be the same in case of theoretical Q_q, calculated on the basis of q and the actual (measured directly or indirectly)

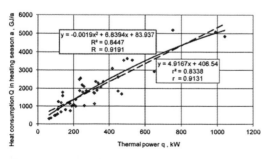

Figure 1. Dependence between heat consumption Q in heating season and thermal power q for heating.

seasonal heat consumption Q for heating. The connection between discussed quantities should be the stronger, the better are rendered the heating conditions and the specificity of an operated object.

The association between Q and the calculated q as well as the discrepancies appearing here (Fig. 1) for examined educational buildings and mentioned using of q values for estimating of the heat demand Q_q in the engineering practice were the main reason inducing to undertake the examinations and analyses. Bearing in mind the practice that the demand for heat needed for buildings heating is estimated on the basis of q as well as the conclusion formulated above (Fig. 1), the following important questions arise:

1. Will the relationships and discrepancies revealed in the analysis of graphs on Figure 1 be analogous in calculating of Q_q on the basis of q (it should be expected)?
2. What is a quantitative range of possible discrepancies?
3. What may be the reasons of such situation, other than not too accurate consideration of actual conditions of heating in the methodology of q value calculation?
4. Is it possible to eliminate, in a simplified manner, the above-mentioned discrepancies and thus reduce the distance between theory represented by Q_q, calculated on the basis of thermal power q and the reality in the form of heat consumption Q?

Examinations and analysis presented below are intended to enable the answers to these questions.

2 MATERIALS AND METHODS

An indirect target of the analysis was to establish if there are essential differences between the actual seasonal heat consumption Q for heating of examined buildings and the seasonal heat demand Q_q for heating. The realization of such formulated task should enable to achieve the direct target, i.e. proposing a modified version of method for seasonal heat demand calculation $Q1_q$. The modification should enable to reduce (which is mentioned in the title) differences between the thermal needs of educational buildings in actual and theoretical conditions, in case when these needs are estimated on the basis of known values of thermal power.

2.1 Examined educational buildings

The material presented in this work is a fragment of wider analysis and regards 46 of 50 educational buildings, which were constructed in years 1913–1992 (in the author's opinion data

for 4 objects were questionable). They form a full municipal complex of 50 objects, where the primary schools and junior high schools are located. The full statistical examinations included all the units of this complex and they were conducted by stages, in cooperation with the Municipal Office. The basic characteristic of this buildings is presented in the Table 1. This description (Table 1) includes statistical values (average value x_{av}, median Me, standard deviation $s(x)$ and variation $s^2(x)$), which describe the distribution of selected characteristics of the examined population of buildings. If the description of the distribution of selected characteristics in other populations of educational buildings is very similar to presented, this allows the use of the results included in this paper for those populations.

The statistical description of this group does not differ significantly from the description of the entire group of 50 buildings. Different from this analysis of heat consumption in educational buildings is also presented, inter alia, Butala & Novak (1999), Corgnati & Corrado & Filippi (2008), Desideri & Proietti (2002).

In the examined group, 23 educational buildings were provided with heat for heating by HPC (Heat Power Company), while 27 buildings had their own boiler-rooms. Data determining the thermal power (q) for individual objects were obtained from design and operating documentation and from own calculations. In case of buildings using the heat supplied from heat-generating plant the bills for thermal ("ordered") power and for consumed heat were an additional source of information.

The amount of heat consumed in a base heating season was determined in two manners, depending on the source supplying the heat to the central heating system. In the buildings equipped with remote systems of central heating the actual seasonal heat consumption Q was determined on the basis of readings from installed heat meters, with the measuring accuracy not lower than 2%. In case of own gas or coal-coke boiler-rooms, the amount of seasonal heat consumption was calculated on the basis of information on fuel consumption in the base heating season, kind of used fuel and its calorific value, average nominal efficiency and estimated average operational efficiency of the central heating boilers, kind of losses in the heat production and their average levels for various types of boilers.

2.2 External and internal conditions of examinations and analysis

The heating season used in the analyses was characterized by the average outdoor temperature of air for analyzed period and determined area (town)

407

Table 1. Selected measures of statistical description for the values characterizing 46 of 50 educational buildings forming the municipal group of objects of this type.

Value [x]	Selected measures of statistical description					
	Average value [x_{av}]	Median [Me]	Standard deviation [s(x)]	Variation [$s^2(x)$]	Limits of typicality [x_{typ}]	Coefficient of variation [$V_c(x)$, %]
Year of built, year	1959	1962	19.94	397.50	1939–1979	–
Volume of the building (heated) V, m³	14682.37	13300.00	9674.55	93596859.10	5007.82–24356.92	65.89
Usable area (heated) A_u, m²	3194.09	2864.50	2161.41	4671701.91	1032.67–5355.50	67.67
Number of classrooms N_s, rooms	21	19	12	156	9–33	58.48
Usable area of classroom A_c, m²	50.30	49.82	8.09	65.94	42.21–58.39	16.08
Number of pupil in classroom N_{pc}, pupil/classroom	32	32	10	87	22–42	29.62
Usable area of classroom per pupil $A_{cp} = (A_c/\text{pupil})$, m²	1.69	1.69	0.41	0.17	1.28–2.10	24.38
Relation of the building ext. partitions' area to the volume of building A/V, m⁻¹	0.40	0.38	0.09	0.01	0.31–0.50	23.43
Relation of windows' area to the area of façades A_w/A_f, –	0.25	0.25	0.05	0.003	0.19–0.30	21.73
Weighted average heat transfer coefficient for external partitions U_B, W/(m² K)	1.27	1.27	0.20	0.04	1.07–1.47	15.63
Heat losses through external partitions Q_{ep}, W/K	3152.23	3318.36	1568.34	2459701.1	1583.89–4720.57	49.75
Thermal power q, kW	323.38	254.07	235.15	5529658	88.23–558.54	72.72
Index q/V, W/(m³a)	21.93	22.84	5.11	26.15	16.81–27.04	23.32
Heat consumption Q, GJ/a	1996.52	1799.00	1266.14	1603106.42	730.38–3262.66	63.42
Index Q/V, GJ/(m³a)	138.36	129.29	39.26	1541.18	99.10–177.62	28.37

$T_{e\,śr} = +2.9°C$ and the duration $Ld = 230$ days. This season can be considered as typical for multiannual period in statistical respect (min. 30 years), which was confirmed by a positive result of testing a hypothesis on its statistical typicality (e.g. Sobczyk 2009).

The heat demand Q_q in analyzed heating season for a given building calculated with the using of known value of q, on the basis of equation (1). In the analyzed case this relationship will have the following form:

$$Q = q \cdot 230 \cdot 24 \text{ h} \cdot \frac{(20°C - 2.9°C)}{(20°C - (-20°C))} \cdot 0.0036 \qquad (2)$$

The equation (2) is the result of comparison of algorithms for calculation of heat demand for heating in the conditions of previously characterized heating season and thermal power necessary for fulfilling of these needs in extreme conditions. Obviously, it is an imperfect comparison due to the applied "conversion factor", which "eliminates" only the difference of temperatures outside of a heated building, included in considered algorithms.

The average temperature inside of heated educational buildings achieved the value $T_{iav} = +19.9°C$. The average grade of thermal comfort assessment made by the employees and students in examined objects amounted to 3.98 point in 7-point scale (from 1 to 7) (research methodology and results of thermal comfort can be found in publications (Filippin 2000, Fanger & Popiołek & Wargocki 2003/2004, Lis 2005, Olesen 2007) and is within, so-called, thermal comfort zone. The procedure described above was conducted in order to assure that there are no too big differences of temperatures in heated rooms between examined buildings, which could have a significant influence on analyzed values of Q.

2.3 Methods used for the analysis

A method of ad hoc statistical census was applied here, with the use of selected measures of descriptive statistics and correlation analysis. The statistical observation was carried out by correspondence and direct surveys, interviews, site inspections and personally conducted measurements. The research material obtained in this such a way is a primary material, collected especially for the purposes of conducted statistical examinations. The graphs of relationship, along with the coefficients of correlation and determination for linear function (Pearson's linear correlation coefficient—r, coefficient of determination—r^2) and for curvilinear functions (R and R^2 respectively) were applied,

inter alia, in the correlation analysis (included also in presented material), in order to determine the strength and direction of the relationship between analysed quantities (Sobczyk 2009). The author named the quantities R and R^2, used in the analysis, the coefficient of correlation for adjusted function of y variable in relation to x variable and the coefficient of determination R^2 for adjusted function, respectively. The differences of coefficients of determination R^2 and r^2 for adjusted and rectilinear functions, respectively, were used in order to assess the degree of regression curvilinearity. The smaller is the difference, the relationship between the variables is closer to linear.

For all analyzed relationships between selected characteristics of educational buildings the trend lines were drawn, in the form of linear functions and with specifying of determination coefficient values r^2. It is better to use a linear function than a function of a higher order in case when the task consists only in indicating the direction, the trend of relationship between the quantities, but not in an accurate graphical presentation and mathematical description of this relationship. In order to keep a sufficient legibility of the graphs, the points of data were not placed on them. The functional relationships describing the trend lines serve only for their more accurate identification and allow to determine the coefficient of linear function directivity. This coefficient is a number given at the x variable in the equation of linear function on the graph. It is equal to the tangent of the angle between a straight line and the plus part of X axis. The bigger the difference between the values of these coefficients is, the bigger the discrepancy between the trends of changes in quantities shown on a graph is. An additional index helpful in conducted analysis is also so coefficient, it allows e.g. a quantitative determination of convergence or divergence in the trends of changes in analyzed quantities.

The assessment of statistical significance of Pearson's linear correlation r, which is necessary to assess the significance of examined relationships, was made by applying a method of hypothesis testing, with using of a test built on t-Student statistics for a population with size $n < 122$ (Sobczyk 2009). This test is based on statistics, t—and has the following form (Sobczyk 2009):

$$t = \frac{r}{\sqrt{(1-r^2)}} \sqrt{(n-2)} \qquad (3)$$

where:

n	population size ($n = 46$ or $n = 50$);
r and r^2	Pearson's linear correlation coefficient and coefficient of determination.

2.4 *Analysis algorythm*

In order to answer previously mentioned questions there is a planned course of analysis made in accordance with the following algorithm:

- Calculation of the heat demand for heating of educational buildings Q_q on the basis of known values of thermal power q according to equation (2);
- Comparison of the heat consumption for heating Q and Q_q for the examined buildings;
- Determination of the quantitative discrepancy between the actual heat consumption Q for heating and the theoretical heat demand Q_q (marked as (Q–Q_q) i [(Q–Q_q)/Q]100%);
- Selecting characteristics of the educational buildings which may have significantly influence on the above discrepancies;
- Analysis of the influences of changes in selected characteristics of the educational buildings on the changes in discrepancies between Q and Q_q;
- Analysis of the influence of changes in thermal power q on the changes in discrepancies between the actual heat consumption Q and the theoretical heat demand Q_q;
- Mathematical description of the above dependence;
- Modification of the equation for calculation of the heat demand $Q1_q$ for heating calculated on the basis of thermal power. This modification includes the mathematical description of the influence of changes in q on the changes in discrepancies between Q and Q_q;
- Determination the effects of the above modification.

The analysis included the essential elements of statistical description that may affect the objectivity and transfer results to other population of educational buildings. In the same way, the use of statistical methods and the dependency graph seems to be ready for a general algorithm for use in other technical studies dealing with the discrepancy between theory and reality. Also the results of this analysis can be considered as a general for statistical population of the educational buildings. The statistical population with similar statistical distribution. In this case, the first cubic capacity of studied objects. From the standpoint of technical calculations and energy management in this type group of buildings it is quite sufficient.

3 RESULTS AND DISCUSSION

3.1 *Heat consumption Q and heat demand Q_q for heating*

The planned analysis calculated the heat demand for heating of educational buildings Q_q, compared

with the heat consumption for heating Q and the heat demand Q_q and determined the quantitative discrepancy between the actual Q and the theoretical Q_q (marked as (Q–Q_q) and [(Q–Q_q)/Q]100%). The results are shown in Table 2 and in graphical form in Figure 2. This comparison was the fundamental issue for analyses and interpretation of graphs presented in the next part.

The analysis (Table 1 and Fig. 2) reveals the differences between seasonal heat consumption Q and calculated seasonal heat demand Q_q. Occurrence of these differences confirms the divergence of trends for changes in analyzed quantities, which are shown on the graph (Fig. 2). The described differences could be considered as resulting only from the discrepancies between the calculative assumptions and the actual conditions of heating season.

However, it seems that their level (Fig. 2) in examined educational buildings and the course of trend line do not impose such a statement. Q_q is bigger by 41.6%, on the average, than Q. However, there are objects (7 of 46), in which the situation is opposite, i.e. Q > Q_q (Fig. 2). In the analyzed group of buildings the maximal discrepancy between the values of Q and Q_q amounted to 134% (Q_q > Q), while minimal discrepancy amounted to (–34)% (Q > Q_q).

A phenomenon, which was observed for several buildings and consists in a considerable diversity between the values of heat demand and heat consumption for heating of rooms in relation to its average level, in connection with presented results, may prove:

- an incorrect determination of thermal power q for the part of objects. Such errors may be a reason of insufficient heating of schools (Q > Q_q) or paying of excessive fixed duties for thermal power (q) (Q_q > Q);
- an improper operation of schools within the scope of appropriate ventilation of rooms, which is always connected with cubic capacity of examined buildings, mentioned here.

A percentage relation of the difference value (Q_q–Q) to the value of Q is diverse in individual objects. Does this diversity remain dependant on certain factors characterizing the selected objects, such as, *inter alia*, the architectonic shape of a building, energy consumption by heating and thermo-insulating power of external particles? It will be explained in the next section.

3.2 *Selected characteristics of educational buildings and the changes in discrepancies between heat consumption Q and heat demand Q_q*

In the next step of analysis selected characteristics of the educational buildings which may have

Table 2. Selected measures of statistical description for the values characterizing 46 of 50 educational buildings forming the municipal group of objects of this type.

Value [x]	Selected measures of statistical description					
	Average Value [x sr]	Median [Me]	Standard deviation [s(x)]	Variation [s²(x)]	Limits of typicality [x$_{typ}$]	Coefficient of variation [V$_c$(x), %]
Thermal power q, kW	323.38	254.07	235.15	5529658	88.23–558.54	72.72
Heat consumption Q, GJ/a	1996.52	1799.00	1266.14	1603106.42	730.38–3262.66	63.42
Heat demand Q$_q$ calculated on the basis of thermal power q Q$_q$, GJ/a	2747.24	2158.40	1997.68	3990741.87	749.56–4744.93	72.72
Modified heat demand Q1$_q$ calculated on the basis of thermal power q Q1$_q$, GJ/a	1996.67	1655.83	1156.31	1337051.45	840.36–3152.98	57.91
Calculated discrepancies (Q$_q$–Q), GJ/a	750.72	506.97	987.18	974520.57	–236.46–1737.90	–
Calculated discrepancies [(Q$_q$–Q)/Q]100%, %	41.60	43.92	40.72	1658.40	0.88–82.33	–
Calculated modified discrepancies (Q1$_q$–Q), GJ/a	0.15	117.52	516.12	266375.75	–515.96–516.27	–
Calculated modified discrepancies [(Q1$_q$–Q)/Q]100%, %	14.30	3.94	40.32	1625.41	–26.02–54.62	–

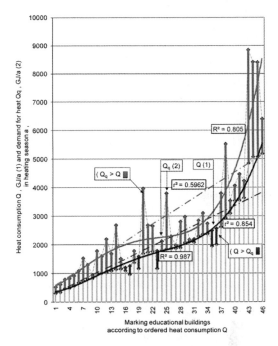

Figure 2. Heat consumption Q and heat demand Q$_q$ for heating in heating season a, in educational buildings according to ordered heat consumption Q.

significant influence on the discrepancies between the heat consumption Q and the heat demand Q$_q$ for heating calculated on the basis of thermal power q (marked as (Q–Q$_q$) and [(Q–Q$_q$)/Q]100%). These characteristics are described by the appropriate quantities.

The selection of quantities shown on the graphs on the axes X resulted from the need of direct or indirect take into account the main factors, which may have an influence on the analyzed quantities and on possible dependences revealed between them.

Thus:

- changes in above-mentioned quantity in the period 1913–92, when the examined objects were constructed (Fig. 3);
- size and shape of a building were taken into consideration by introducing of object cubic capacity V (Fig. 4) and superficial modulus A/V (Fig. 5) to the considered relationships;
- energy consumption by heating was included by the Q/V index (Fig. 6);
- thermo-insulating properties of external partitions were characterized by introducing the heat transfer coefficient for external partitions of educational building U$_B$ (Fig. 7) and heat losses through these particles Q$_{pz}$ (Fig. 8).

Taking into consideration the dependences graphs, it should be considered that these graphs

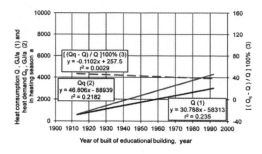

Figure 3. Dependence between Q, Q_q, $[(Q_q-Q)/Q]100\%$ and year of building.

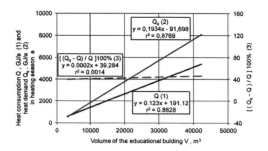

Figure 4. Dependence between Q, Q_q, $[(Q_q-Q)/Q]100\%$ and educational building capacity V.

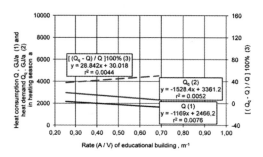

Figure 5. Dependence between Q, Q_q, $[(Q_q-Q)/Q]100\%$ and rate (A/V) of educational building.

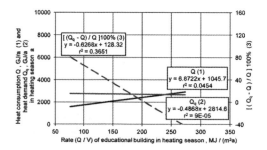

Figure 6. Dependence between Q, Q_q, $[(Q_q-Q)/Q]100\%$ and rate Q/V of educational building.

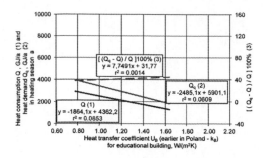

Figure 7. Dependence between Q, Q_q, $[(Q_q-Q)/Q]100\%$ and heat transfer coefficient U_B (earlier in Poland—k_B).

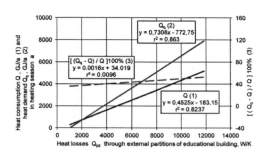

Figure 8. Dependence between Q, Q_q, $[(Q_q-Q)/Q]100\%$ and heat losses through external partitions of educational building Q_{pz}.

were drawn for the group of buildings diverse in respect of properties, which influence significantly the amount of heat consumed for heating. Being "invisible" on the graph, the mentioned properties have often a dominant impact on the course of examined relationships. Due to that fact, the reason of this course must be different from graphical presentation of a theoretical case (it could be stated that the relationships differ from the theory in this field), in which the influence of the changes in only one quantity on the values of other quantities, with simultaneous elimination of the influence of other significant factors, is analysed.

Moreover, the graphs illustrate certain phenomena, which can appear in conditions of real functioning of other groups of buildings of this type. In each of cases presented on Figures 2–8, the directions of trend lines for analyzed quantities are less or more divergent. It means that the increase of calculated Q_q value is quicker than the increase of the heat amount used for heating. Simultaneously, it causes the increase of the value of difference (Q_q-Q) for these objects.

As it results from the analysis of the graphs presented on Figures 3–8, it shows that the dependency between Q_q, Q and quantities characterising examined buildings occurs only in the case:

- changes in above-mentioned quantity in the period 1913–92, when the examined objects were constructed (Fig. 3). However, due to the calculated coefficient of determination $r^2 = 0.2182$ for heat demand Q_q and $r^2 = 0.235$ for heat consumption Q, this significance should be considered with a certain dose of a "statistical" caution;
- size of a building—volume of the building (Fig. 4);
- thermo-insulating properties of external partitions—heat losses through these partitions Q_{pz} (Fig. 8).

A quicker increase in the value of Q_q in relation to the increase of heat amount Q consumed for heating of buildings during the heating season, with a simultaneous increase of the value of $(Q_q–Q)$ difference for these objects, occurs along with the increase of the values mapped on the X axis of presented graphs, i.e. along with:

- increase in the amount of heat Q consumed in a season for buildings heating (Fig. 2);
- decrease of examined buildings age (Fig. 3);
- increase of size of heated educational objects (Fig. 4);
- increase of heat losses through external particles of heated buildings (Fig. 8).

The values specified above have a certain common property, which is a direct or indirect connection with the size of examined buildings. All these reasons incline to the interpretation that presented dependencies are, to a bigger or smaller extent, determined by the size of examined objects, which is described quantitatively by their cubic capacity V. Thus, the cubic capacity of examined objects has a significant and often dominant influence on analyzed dependencies. The justification of such reasoning is presented on the example of the graph showing the changes in Q_q and Q values depending on the changes in heat losses Q_{pz} through external particles (Fig. 8). The value of these losses is influenced mainly by the thermo-insulating power and the area of external particles. However, if the lack of significant connections between Q_q and Q and A/V and U_B was noticed previously, then it can be supposed that the diversity of external particles area, caused mainly by the size of a building, is a dominant reason for the appearing of relationship. The architectonic shape of the buildings (A/V index Fig. 5) and the thermo-insulating power of external particles (U_B coefficient—Fig. 7) are, in this case, a less significant reason of the occurrence and course of analyzed relationship. Taking that into consideration, it seems that the cubic capacity

of heated buildings (Fig. 6) and the factors connected with that have a dominant influence on the quantitative level of differences $(Q_q–Q)$.

As it results from the analysis of the graphs, it is presented that the dependency between percentage relation of the difference's value $(Q_q–Q)$ to the value of Q and quantities characterising the buildings shown in the graphs (Figs. 3, 4, 5, 7 and 8) does not occur (almost "flat" courses of the trend lines and the low values of determination coefficients). It indicates statistical insignificance of presented dependences. The statistical significance of the relation was found only for the Q/V index, which characterizes the energy consumption for heating of 1 m^3 of building cubic capacity (Fig. 6). However, due to the calculated coefficient of determination $r^2 = 0.3651$, this significance should be considered with a certain dose of a "statistical" caution.

3.3 Discrepancies between the actual heat consumption Q and theoretical heat demand Q_q and changes in thermal power q

Bearing in mind the results of previous analyses, in the next stage a trial to modify the relationship (1) was undertaken. The effect of such modification should be the determination of a method for calculating the values of seasonal heat demand for heating $Q1_q$, which would, on the average, differ less from the heat consumption Q in examined objects in relation to the difference occurring if the Q_q quantity is used. For this purpose, the analysis of the influence of changes in thermal power q on the changes in discrepancies Q and Q_q was conducted (Fig. 9). At this point it should be clear that the some dependencies shown in graphs, which are presented in this section, have a coefficient of determination $r^2 = 1$. It is a result of heat demand calculation with using of thermal power, and not a merit of an "ideal" adjustment of function to the points of data on a graph.

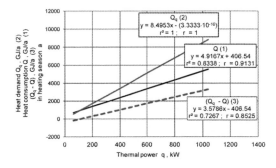

Figure 9. Dependence between heat demand Q_q, heat consumption Q, differences $(Q_q–Q)$ and thermal power q.

The course $(Q_q - Q)$ on the graph (Fig. 9) was described by the using of linear function:

$$y = 3.5786x - 406.54 \qquad (4)$$

After substituting for $x = q$ and $y = (Q_q - Q)$ was obtained:

$$(Q_q - Q) = 3.5786q - 406.54 \qquad (5)$$

In further part the mentioned linear function was used for quantitative determination of differences between the heat demand and the heat consumption for heating of educational buildings, including the changes in thermal power.

3.4 *Modification of the equation for calculation of the heat demand for heating calculated on the basis of thermal power q*

Equation (5) describes the difference (shown in Fig. 9) between calculated seasonal heat demand Q_q and the actual heat consumption Q for heating was used to create the relationship allowing for calculation of a modified seasonal heat demand for heating $Q1_q$ calculated on the basis of thermal power q:

$$Q1_q = Q_q - (Q_q - Q) \qquad (6)$$

By substituting the relationships (1) and (5) to the equation (6), was obtained:

$$Q1_q = q \cdot Nd \cdot 24\,h \cdot \frac{(T_{ical} - T_{eav})}{(T_{ical} - T_{emin})} \cdot 0.0036$$
$$- (3.5786q - 406.54) \qquad (7)$$

The symbols used in equation (7) are the same as in case of the equation (1) and they were described above. Similar, but quantitatively worse effect, can be achieved by replacing the mentioned functional notation with a multiplier, equal to an average difference between Q_q and Q, which in analyzed case amounted to 0.416. The values of Q_q and $Q1_q$ (Fig. 10), calculated with two various methods, were compared and differentiated in relation to the Q (Fig. 11), in order to determine if performed transformations allowed for reduction of the differences between the thermal needs of educational buildings in actual and theoretical conditions.

The method of $Q1_q$ determination allowed to make the difference $(Q1_q - Q)$ independent of the changes in thermal power q (Fig. 11). The comparison of graphs presented on Figures 12 and 13 and information with Table 2 allows to state that if the equation (7) is applied for calculating of heat demand $Q1_q$ on the basis of thermal power q,

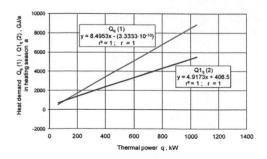

Figure 10. Dependence between Q_q, $Q1_q$ and thermal power q.

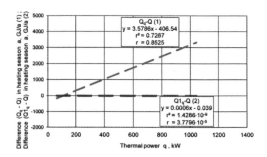

Figure 11. Dependence between differences $(Q_q - Q)$, $(Q1_q - Q)$ and thermal power q.

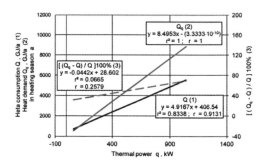

Figure 12. Dependence between Q, Q_q, $[(Q_q - Q)/Q]100\%$ and thermal power q.

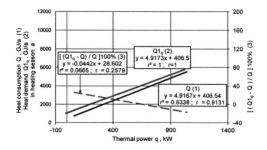

Figure 13. Dependence between Q, $Q1_q$, $[(Q1_q - Q)/Q]100\%$ and thermal power q.

then they obtained difference of theoretical values in relation to the heat consumption (Q) is smaller than it was in case of using of equation (1) and calculation of Q_q. An average level of mentioned differences amounted to 14.3% and 41.6%, respectively. The qualitative effect of described actions is presented on the Figures 12 and 13, in the form of graph.

4 CONCLUSIONS

Executed research and analysis provided responses to the questions posed in the introduction of this paper. To sum up, it can be stated that the occurrence of discrepancies between the seasonal heat consumption Q and the seasonal heat demand Q_q for schools heating, calculated in proposed manner, was found. The influences of the analysis of different characteristics of educational buildings on the heat demand, heat consumption and dependencies between these values incline to the interpretation that are, to a bigger or smaller degree, determined by the size of examined objects, which is described quantitatively by their cubic capacity V. Thus, the cubic capacity of these objects has a significant and often dominant influence on analyzed dependencies. Obviously, also the other factors have their influence, but it is less significant. This conclusion is more important for small educational buildings, consuming less heat for heating on this account. For large objects, it is less significant due to relation of the discrepancies value to heat consumption for heating. The percentage relation of the difference value (Q_q-Q) to the value of Q does not change on the analyzed graphs. The exception is a statistically significant decrease in the above relation with increasing ratio Q/V. It means that the discrepancies between the demand and consumption of heat will be much more important in educational buildings with a lower energy consumption for heating, that is, the smaller of ratio Q/V values. The above mentioned statements remain in a certain correlation with the results presented in the publication by Corgnati, Corrado and Filippi (2008).

Proposed the statistical and graphical methods applied in the analysis of the Q and Q_q values may be used in order to determine the scope of such incorrectness in the analysis of heating of statistically similar populations of educational buildings, as a help in providing other information connected with the specificity of such objects heating also as a tool in management of educational buildings that create the administrative, organizational and functional unit. The condition of these methods is the use of statistical similarity of the parameters basic (average value x_{av}, median Me, standard deviation $s(x)$ and variation $s^2(x)$)—describing the distribu-

tion of selected characteristics of others populations to the distribution of this urban population of educational buildings. In case of absence of similarity between the populations of buildings there can be an independent analysis conducted with the use of the analysis algorithm, which is designed and presented in this paper. A modification in the method for determination of heat demand for rooms heating on the basis of available data on thermal power was introduced. A linear function, describing the changes in (Q_q-Q) depending on the changes in thermal power q values, which was applied for that purpose, allowed to improve the consistency of obtained heat demand values in relation to measured consumption of heat for heating by 65.6%. Even better effect can be expected with the application of a function in the form of polynomial, which is better adjusted to data included on the graph. However, the selection of such a function is connected with complication of calculation method, which is "simplified" by assumption.

ACKNOWLEDGEMENTS

This examinations were conducted by stages, in cooperation with the Municipal Office in Czestochowa, Poland—formal assistance in data collection.

REFERENCES

Butala, V. & Novak, P. 1999. Energy consumption and potential energy savings in old school buildings. Energy and Buildings 29(1999): 241–246.

Corgnati, S.P. & Corrado, V, & Filippi, M. 2008. A method for heating consumption assessment in existing buildings: A field survey concerning 120 Italian schools. Energy and Buildings 40(2008): 801–809.

Desideri, U. & Proietti, S. 2002. Analysis of energy consumption in the high schools of a province in central Italy. Energy and Buildings 34(2002): 1003–1016.

Fanger, P.O. & Popiołek, Z. & Wargocki, P. 2003/2004. Indoor environment. Impact on health, comfort and productivity, Gliwice, Silesian University of Technology.

Filippin, C. 2000. Benchmarking the energy efficiency and greenhouse gases emissions of school buildings in central Argentina. Building and Environment 35(2000): 407–414.

Gallachoir, B. & Keane, M. & Morrissey, E. & O'Donnell, J. 2007. Using indicators to profile energy consumption and to inform energy policy in a university—A case study in Ireland. Energy and Buildings 39(2007): 913–922.

Kaspeskiewicz, K. 2006. Building heating needs assessment based on monitoring of the supplied energy. Cracow University of Technology Technical Journal 5B(2006): 251–258.

Lis, A. 2005. Analysis of factors affecting the thermal indoor climate components. Building Physics in Theory and Practice 1(2005): 256–264.

Noren, C. & Pyrko, J. 1998. Typical load shapes for Swedish schools and hotels. *Energy and Buildings* 28(1998): 145–157.

Olesen, B.W. 2007. The philosophy behind EN15251: Indoor environmental criteria for design and calculation of energy performance of buildings. *Energy and Buildings* 39(2007): 740–749.

Piotrowska-Woroniak, J. & Woroniak, G. 2005. The modernization of central heating and hot water in the modernized public buildings. District Heating, Heating, Ventilation 1(2005): 24–27.

Prętka, I. 2004. Analysis of the thermal efficiency of the education building. District Heating, Heating, Ventilation 1(2004): 10–14.

Sobczyk, M. 2009. Statistics, Warsaw, PWN.

Stachniewicz, R. 2007. Energy and environmental indicators for selected school buildings. Building Materials 1(2007): 50–51.

Environmental Engineering IV – Pawłowski, Dudzińska & Pawłowski (eds)
© *2013 Taylor & Francis Group, London, ISBN 978-0-415-64338-2*

Effect of personalized ventilation on individual employee performance and productivity

A. Bogdan, A. Łuczak & M. Zwolińska
Central Institute for Labour Protection, National Research Institute, Department of Ergonomics, Warsaw, Poland

M. Chludzińska
Warsaw University of Technology, Faculty of Environmental Engineering, Warsaw, Poland

ABSTRACT: The purpose of the research presented in this paper was to examine whether the use of personalized ventilation systems with the option of heating an occupant in winter will also affect his/her productivity, and what configuration of ambient temperature/PV air temperature flowing from a diffuser and what direction of the air supply will cause the highest work productivity. Based on the survey, it was observed that in all variants an improvement in subjective ratings of mood and fatigue occurred during the test conducted in a climatic chamber with the PV system turned on. The use of the PV system had also a positive effect on feelings about the subjective assessment of work productivity.

Keywords: personalized ventilation, productivity, effectiveness, HVAC

1 INTRODUCTION

The primary task of ventilation and air conditioning systems is to provide a room occupant with a sense of comfort, or satisfaction with the conditions prevailing in the room. The systems available now are based on mixing or displacement ventilation and are designed according to standards adopted for a large group of people, i.e. ASHRAE Standard 55–1992, EN ISO 7730:2006, and in Poland additionally PN-EN 15251:2007 and PN-B-03421:1978 and PN-B-03430:1983/Az3:2000. However, a Mixing Ventilation (MV) system, by diluting pollutants generated in the room and providing a uniform thermal environment, does not create conditions that would be acceptable to all room occupants. Classical mixing ventilation solutions allowing for air and temperature to be controlled by users do not translate into increased occupant comfort (Karjalainen 2007). The research conducted in 1991 (Grivel 1991) showed that individual differences in preferred values for ambient air temperatures can reach up to 10 K, and the value of air velocity chosen as the optimum can vary fourfold among occupants (Melikov 1994). A Displacement Ventilation (DV) system, in which air at the temperature of 2 to 4 K below the room temperature is supplied through supply air diffusers located near the floor level, may cause occupants feel a local thermal discomfort caused by drafts or a vertical temperature gradient. Only fully adjustable conditions of the immediate human environment can result in improved working conditions, and thus be reflected in the efficiency of work performed (Bako-Biro 2004, Chludzińska 2008, Melikov 2004). Development of systems that use individual air distribution for each user potentially offers independent control of thermodynamic air characteristics in the immediate human environment. This allows to adjust the immediate indoor environment to the needs at a given time: physical activity, metabolism, clothing insulation. The values of these characteristics can vary greatly from person to person—the value of human metabolism, depending on natural inclinations of an individual and physical activity, can range from 1 MET to 1.5 MET, and the clothing insulation worn can vary from about 0.4 clo to 1.2 clo (ASHRAE 2001). Factors such as human physiology, physical activity, psychological reactions, clothing insulation, and even individual's self-awareness of the capability to control ventilation parameters (Bauman 1998) are very important elements in the perception of environmental microclimate as an optimal and comfortable workplace (Fanger 2001). In the case of a personalized ventilation, thermal comfort is achieved by adjustment of the cooling effect of the human body by the air flow to suit occupant's personal preferences.

The cooling effect of an air flow depends on:

- the intensity of air turbulence in the flow (Melikov 1994),
- the difference in the temperature between the room air and the supply air (Faulkner 2002, Toftum 2000),
- the direction of air flowing on the human body (Toftum 2000).

Apart from ensuring appropriate thermal perception, a Personalized Ventilation (PV) system has a significant effect on the quality of the air breathed in by an occupant, as in this case there is no mixing of pollutants generated in the room, to be subsequently inhaled. A personalized ventilation system provides fresh, properly conditioned air directly to the breathing zone, thus the partly polluted air from the environment is being replaced by clean air, which is then inhaled by an occupant. Initially, a similar potential was perceived in displacement ventilation. However, measurements carried out in rooms with displacement diffusers showed that nearly 50% of people were dissatisfied with the prevailing room air quality (Melikov 2005, Naydenov 2002, Kaczmarczyk 2002, Zeng 2002, Gao 2005) and showed that personalized ventilation significantly improves the assessment of the air quality compared to the mixing ventilation for the same amount of fresh air provided by both systems.

2 PURPOSE OF RESEARCH

The purpose of the research presented in this paper was to examine whether the use of personalized ventilation systems with the option of heating an occupant in winter will also affect his/her productivity, and what configuration of ambient temperature/PV air temperature flowing from a diffuser, and what direction of the air supply will cause the highest work productivity. The research compared the test results obtained during the presence of volunteers in a laboratory room, in which constant air temperature of 22°C was maintained to results of tests carried out in a climatic chamber at a PV station. Productivity consists of a number of components, such as (Wyon 1993): occupational absenteeism, speed and accuracy of work performed, self-assessment of work performed, quality of work, etc. During the research, productivity may be assessed with the use of psychological tests examining employee's level of attention, as well as by subjective tests evaluating employee's well-being and his/her feelings about work productivity. In this research, productivity was assessed objectively (attention and perception test) and subjectively (Grandjean's scale). Subjective assessment of work productivity was also assessed.

3 RESEARCH METHODOLOGY

The tests were conducted in the Thermal Load Laboratory at Central Institute for Labour Protection—National Research Institute with the use of a climatic chamber and a set of microclimate meters. Because of the intention to see if the human body can be heated using PV systems, tests were conducted in the winter. During the tests, the ambient temperature in a climatic chamber was 20°C and 22°C, while the air supply temperature was higher by 1°C or 2°C. The air supply from the PV system was at the height of the face or ankles.

3.1 Test subjects

The tests were conducted with the participation of 20 male students. An average age was 22.4 years (sd = 1.6 years), height 1.81 m (sd = 0.046 m), weight 80.4 kg (sd = 10.7 kg). The subjects were asked to come to the tests at 9 am after a good night's sleep and breakfast. In addition they were asked not to drink alcohol 48 hours before the experiment. One volunteer would participate in 1 experiment (variant) on a given day. The volunteers were asked to adjust their clothes to thermal conditions of a given day. Before the tests were commenced, thermal insulation of clothes was assessed (on the basis of standards included in EN 7730).

Mean thermal insulation of clothing was equal to 0.76 clo (sd = 0.03 clo). The participants of the tests received payment. The methodology was granted the approval of the Committee for Research Ethics at CIOP-PIB.

3.2 The test stand and measurement equipment

The measuring station consisted of a purpose-designed desk with built-in diffusers directing conditioned air to the face or ankles of the occupant (Fig. 1). Diffusers were installed at the height

Figure 1. The desk designed for testing purposes with an integrated personalised ventilation system.

of the head and ankles of the person sitting in the central part of the desk. Ventilation louvers with moveable horizontal louver slats were used to allow changing the air supply direction in the horizontal plane within ± 45°. The louvers were selected assuming a speed of approximately 0.35 to 0.45 m/s to be achieved at the desk edge, i.e. in the area where the flow reaches the occupant. A set of air heating and cooling devices conditioning the air supplied by the desk was located in the double rear wall of the desk.

Delivered fresh air was directed to heating and cooling elements and then to an air damper directing the flow to the top or the bottom diffuser, as needed (Chludzińska 2009). The volume of the air flow supplied was measured using a sharp edge measuring orifice fabricated in accordance with the standard ISO 5221:1994. Air volume flow measurement was carried out throughout the experiments. During the tests, a constant amount of fresh air of 72 m³/h (20 l/s) was supplied through the personalized ventilation, which was then filtered and properly conditioned.

3.3 The test procedure

Every day, the volunteers upon arrival at the laboratory stayed in a laboratory room (at the temperature of 22°C) for 20 minutes to calm down the body. Then, they solved psychological tests, in accordance with the procedure outlined in Table 1. After the tests completion, volunteers entered the climate chamber, in which they stayed for 40 min and performed the same psychological tests again. In the climatic chamber, the following variables were applied: direction of supplied air, ambient temperature and the temperature of air supplied from PV. The number of variants and setting of particular air parameters are shown in Table 2.

The tests covered the following tasks:

A. Concentration and Perception tests (TUS): Concentration and perception tests (TUS) serve to examine how well a person is able to concentrate on a task. The test consists in crossing out a given number of symbols from amongst other similar ones in a 3-minute time. The TUS test is composed of four independent versions differing in symbols which should be crossed out. The symbols are as follows: "b" and "k", "3" and "8", "6" and "9" and "gw" (abbreviation from the Polish word meaning asterisks). Two versions of the TUS test were used in the experiment: "3" and "8", "6" and "9". The test indicators are the following:

- Speed SP—a number of all symbols examined in a 3-minute time
- Number of omissions LO—a number of symbols which were left out
- Number of mistakes LB—a number of incorrectly crossed out symbols.

Test psychometric parameters are satisfactory (Ciechanowicz 2006).

B. Grandjean's scale: Grandjean's scale is a subjective method for diagnosing mental load in terms of fatigue and mood (Costa 1993, Weber 1975, Baschera 1979). The scale consists of 14 subscales describing mood and fatigue; the two opposing concepts are put at opposite ends of a scale. The subject's task is to mark on each of the subscales the point which describes the most accurately the current (at the moment of making the assessment) intensification of a given feeling. A fatigue indicator is denoted by distance (measured in millimetres) from the beginning of the scale to the point marked by the subject on each of the following subscales: strong—weak, refreshed—tired, interested—bored, vigorous—exhausted, awake—sleepy, efficient—inefficient, attentive—distracted, able to concentrate—unable to concentrate; whereas in case of mood, the subscales are as follows: positive mood—negative mood, relaxed—tensed, happy—sad, energetic—lazy, exhilarated—angry, stimulated—sedated. Time for marking the scale is not limited but on average it takes about 3 minutes to make the assessment.

C. Scale for subjective assessment of work productivity: The subject's task is to mark on the scale the point which the most accurately describes current feeling, i.e. at the moment of making the assessment (Kaczmarczyk 2010). The scale

Table 1. Diagram of each test.

Test duration:	15 min	5 min	40 min	5 min
Test location:	Laboratory, 22°C (E1)		In the climatic chamber, air parameters in accordance with the test variant (E2)	
Test types:	Acclimatization 1. Attention and perception tests (TUS). 2. Grandjean's scale. 3. Scale of subjective assessment of work productivity.		Acclimatization 1. Attention and perception tests (TUS). 2. Grandjean's scale. 3. Scale of subjective assessment of work productivity.	

Table 2. Variants of conditions in a climatic chamber.

Variant	Direction	Ta	Tn
1	Face	20	21
2			22
3		22	23
4			24
5	Ankles	20	21
6			22
7		22	23
8			24

is described by the statement: "at this moment I can work" and the percentage from 0 to 100% is presented.

4 RESULTS AND DISCUSSION

The results of individual statistical analyses are presented below.

4.1 *Results of the type A analysis*

The results obtained for individual variants are presented in graphs 2–9. As a result of the statistical analysis for variant 1, only in 2 cases of the description of subjective feelings: 'refreshed—tired' and 'awake—sleepy', a statistically significant difference between the state in the laboratory and during the impact of the PV system was found. Volunteers claimed that during the operation of the PV system, they felt less tired and less sleepy. As a result of the statistical analysis for variant 2, in 10 tests statistically significant differences between the state in the laboratory room and during the effect of the PV system were identified: LB—the number of errors committed in the TUS test, work productivity, 'strong—weak', 'refreashed—tired', 'interesterd—bored', 'awake—sleepy', 'vigorous—exhausted', 'efficient—inefficient', 'able to concentrate—unable to concentrate'. The number of errors committed in the TUS test (concentration and perception test) was higher during the operation of the PV system. Volunteers, however, claimed that during the effect of the PV system, they were more productive at work and felt more strong, refreshed, interested, awake, vigorous, efficient and more able to concentrate than in the laboratory. As a result of the statistical analysis for variant 3, only in 8 cases of the description of subjective feelings the statistically significant differences between the state in laboratory room and during the effect of PV system were found: work productivity, 'interested—bored', 'vigorous—exhausted','awake—sleepy','efficient—inefficient',

'attentive—distracted', 'able to concentrate—unable to concentrate'. In each of these descriptions of perceptions, the volunteers claimed that during the effect of the PV system, they were more productive at work and felt more interested, awake, vigorous, efficient, attentive and more able to concentrate than in the laboratory. As a result of the statistical analysis for variant 4, only in 2 cases of the description of subjective feelings: 'vigorous—exhausted', 'awake—sleepy' a statistically significant difference was found. The volunteers claimed that during the effect of the PV system, they were more productive at work and felt mode awake and vigorous than in the laboratory. As a result of the statistical analysis for variant 5, only in 3 cases of the description of subjective feelings: 'refreshed—tired', 'awake—sleepy' and 'attentive—distracted' a statistically significant difference was found. The volunteers claimed that during the effect of the PV system, they experienced more refreshed, awake and attentive. As a result of the statistical analysis for variant 6, in 4 tests statistically significant differences were noted: LB—the number of errors committed in the TUS test, inefficient in action, distracted, unable to concentrate. The number of errors committed in the TUS test (concentration and perception test) was higher during the operation of the PV system. Volunteers, however, claimed that while staying in a climatic chamber and during the operation of the PV system, they felt more attentive and more able to concentrate than during operation of the PV system. As a result of the statistical analysis for variant 7, only in 3 cases of the description of subjective feelings: 'strong—weak', 'awake—sleepy' and 'interested—bored', a statistically significant difference between the state before the start of the test and during the effect of the PV system was found. However, the volunteers claimed that during the effect of the PV system they felt more strong, awake and interested. As a result of the statistical analysis for variant 8, no differences before testing and during the effect of the PV system were found.

The perceptiveness factor WS was analysed separately. As a result of the statistical analysis for WS, only in variant 8 a statistically significant difference between the state before the test and during the effect of PV system, i.e. during the effect of the PV system a reduction in the number of digits viewed in the TUS test was observed.

4.2 *Results of the type B analysis*

As a result of the statistical analysis of data for variants 1 and 5, no statistically significant differences between variants 1 and 5 were found. No effect was found of the "direction" of air supply towards the face—ankles ($t_a = 20°C$, $t_n = 21°C$) on the subjective

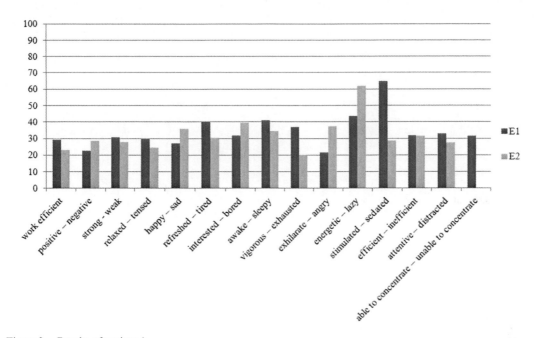

Figure 2. Results of variant 1.

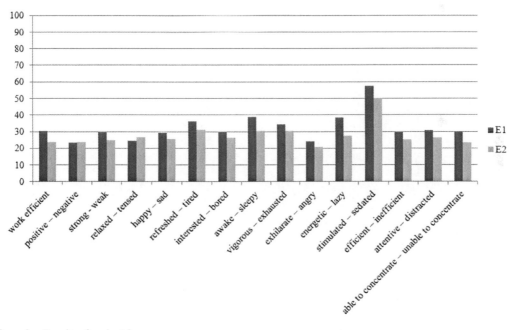

Figure 3. Results of variant 2.

evaluations of volunteers. As a result of the statistical analysis of data for variants 2 and 6, no statistically significant differences between variants 2 and 6 were found. No effect was found of the "direction" of air supply towards the face—ankles on the subjective evaluations of volunteers. As a result of the statistical analysis of data for variants 3 and 7, only 1 statistically significant difference in the number of errors (LB) committed in the TUS test was found. It was found that volunteers

421

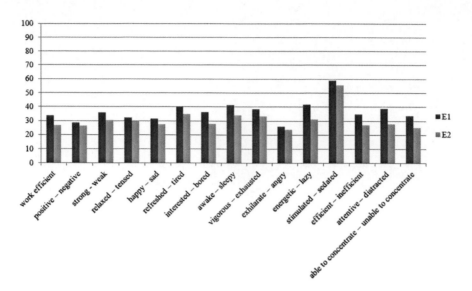

Figure 4. Results of variant 3.

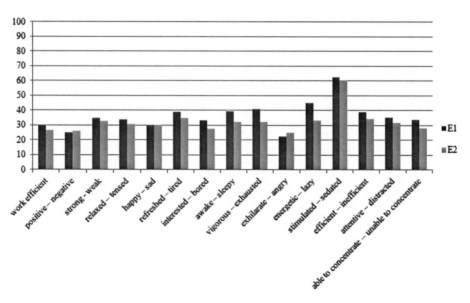

Figure 5. Results of variant 4.

commit significantly fewer errors after the air supply flowed towards the face. In other cases, there were no statistically significant differences between the variants. No effect was found of the "direction" of air supply towards the face—ankles on the subjective evaluations of volunteers. As a result of the statistical analysis of data for variants 4 and 8, only 1 statistically significant difference was found for 'vigorous—exhausted'. It was found that the volunteers felt less exhausted when the air

was directed to the face. In other cases, there were no statistically significant differences between the variants. No effect was found of the direction of air supply towards the face—ankles on the subjective evaluations of volunteers.

4.3 Results of the type C analysis

As a result of the statistical analysis of data for variants 1, 2, 5 and 6, no statistically significant

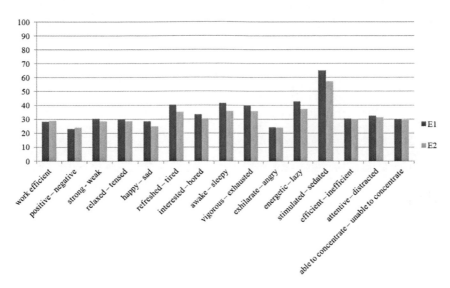

Figure 6.   Results of variant 5.

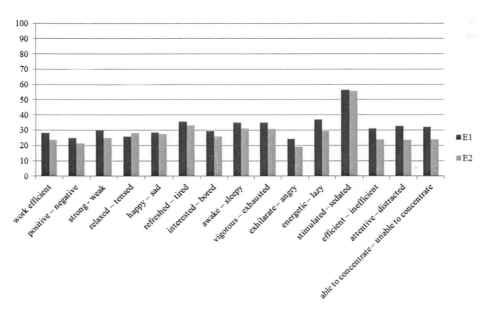

Figure 7.   Results of variant 6.

differences in this group were found. Neither the effect of air supply directed to the face and ankles, air supply temperature ($t_a = 20°C$, $t_n = 21°C$, $22°C$) nor the interaction t_nx direction on the subjective evaluations of volunteers were found. As a result of the statistical analysis of data for variants 3, 4, 7 and 8, statistically significant differences were found for 'interested—bored' and 'vigorous—exhausted'. It was found that the volunteers felt less

exhausted when the air was directed on the face and less depressed when the air temperature was 23°C. In all cases, no interaction was found between t_n and the direction of the air supply. In other cases, there was no significant statistical difference in the group. No effect of air supply directed to the face and ankles and the air supply temperature ($t_a = 22°C$, $t_n = 23°C$, $24°C$) on the subjective evaluations of volunteers was found

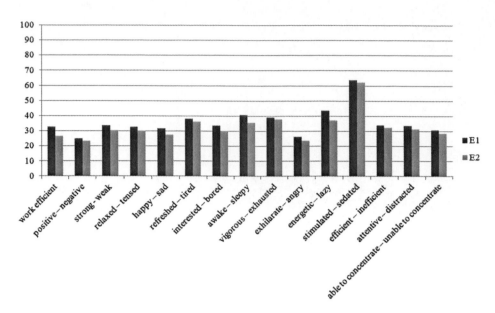

Figure 8. Results of variant 7.

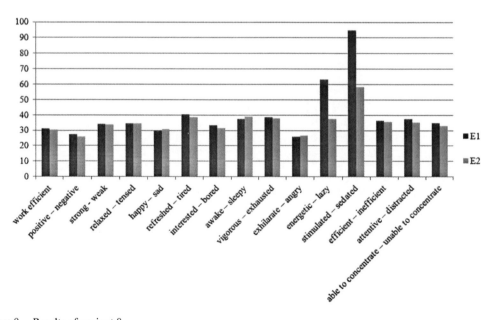

Figure 9. Results of variant 8.

5 CONCLUSIONS

In the survey, it was observed that in all variants volunteers reported more positive subjective ratings of mood and fatigue during the test conducted in a climatic chamber with the PV system turned on. At the same time, in a number of variants statistically significant differences with regard to mood and fatigue were found. In the variants with the air supply directed toward the face or ankles, the use of the PV system mainly reduced the feeling of fatigue. Thus, it can be assumed that directing warmer air supply results in stimulation of the occupant. The use of the PV system had also a positive effect on feelings about the subjective assessment of work productivity. In all cases, productivity increased, reaching a value above 70%. The verification, however, of which kind of air supply (face/ankles) contributed to a greater increase in favourable ratings—there were no significant differences, only the feeling of exhaustion, which was lower when the air supply was directed to the face. No significant differences were also observed by analyzing which type of air supply (variant) was the best in terms of the improvement of subjective assessments for a given ambient temperature. With respect to the concentration and perception test, there was no significant change in the perception ratio, only in variant 8. However, it was noted that the use of the PV system resulted in an average reduction of perception ratio, whether the air supply was directed to the face or ankles of a volunteer. Only in variant 1 and 5 (when the air supply temperature of the PV system was higher by 1°C from the ambient air temperature), an increase in the value of the perception ratio was noted. Therefore, a conclusion can be drawn that the air supply temperature should be slightly higher than the ambient air temperature.

The results of tests indicated a need for a PV system, which has a positive effect, in particular on the subjective feelings of users. At the same time, by selecting the air supply parameters and room air, it should be borne in mind, in particular, that an additional air supply in the immediate vicinity of a subject had no adverse effect on subject's level of focus and concentration.

ACKNOWLEDGMENTS

This research was part of the projects N N506 475034 and NR04-0018-10/2011 supported by the Polish national funds for science in 2008–2010 and The National Centre for Research and Development.

REFERENCES

ANSI/ASHRAE 55-2010 Standard 55-2010—Thermal Environmental Conditions for Human Occupancy (ANSI approved).

ASHRAE, 2001, ASHRAE Handbook of Fundamentals, Atlanta, American Society of Heating, Refrigerating and Air Conditioning Engineers.

Bako-Biro, Z., Wargocki, P., Weschler, C.J., Fanger, P.O. 2004. Effects of pollution from personal computers on perceived air quality, SBS symptoms and productivity in Office, *Indoor Air*, 14: 178–187.

Baschera, P., Grandjean, E.P. 1979. Effect of repetitive task with different degrees of difficulty on critical fusion frequency (CFF) and subjective state, *Ergonomics*, 22: 377–385.

Bauman, F.S., Carter, T.G., Baughman, A.V. et al. 1998. Field study of the impact of a desktop Task/Ambient Conditioning System in office buildings, *ASHRAE Transactions: Symposia*, 104: 1153–1171.

Chludzińska, M., Bogdan, A. 2008. Zastosowanie wentylacji indywidualnej w pomieszczeniach biurowych, Chłodnictwo i Klimatyzacja 11 (125), Warszawa (available in Polish).

Chludzińska, M., Bogdan, A., Mizieliński, B. 2009. Technique for measuring personalized ventilation comfort using thermal manikin, The 11th International Conference "Air Distribution in Rooms—Roomvent 2009", 24–27 May 2009, Busan, South Korea.

Ciechanowicz, A., Stańczak, J. 2006. Test Uwagi i Spostrzegawczości TUS (Concentration and Perception Tests). Warszawa: Pracownia Testów Psychologicznych 2006 (available in Polish).

Costa, G. 1993. Evaluation of workload in air traffic controllers, *Ergonomics*, 36: 1111–1120.

EN 15251 Indoor environmental input parameters for design and assessment of energy performance of buildings addressing indoor air quality, thermal environment, lighting and acoustics.

Fanger, P.O. 2001. Human requirements in future air-conditioned environments, *Journal of Refrigeration*, 24(2): 148–53.

Faulkner, D., Fisk, W.J., Sullivan, D.P., Lee, S.M. 2002. Ventilation efficiencies of a desk-edge-mounted task ventilation system, *Proceedings of Indoor Air*, 2002, 4: 1060–1065.

Gao, N., Niu, J. 2005. Modeling the Performance of Personalized Ventilation under Different Conditions of Room Air and Personalized Air, *HVAC&R Research*, 11, Issue 4.

Grivel, F., Candas, V. 1991. Ambient temperatures preferred by young European males and females at rest, *Ergonomics*, 34: 365–378.

ISO 5221:1984 Air distribution and air diffusion—Rules to methods of measuring air flow rate in an air handling duct.

ISO 7730:2005 Ergonomics of the thermal environment—Analytical determination and interpretation of thermal comfort using calculation of the PMV and PPD indices and local thermal comfort criteria.

Kaczmarczyk, J., Melikov, A., Sliva D. 2010. Effect of warm air supplied facially on occupants' Comfort, *Building and Environment*, 45(4): 848–855.

Kaczmarczyk, J., Zeng, Q., Melikov, A., Fanger, P.O. 2002. The effect of personalized ventilation system on perceived air quality and SBS symptoms, Proceedings Indoor Air 2002, Monterey, USA, 4: 1042–1047.

Karjalainen, S., Koistinen, O. 2007. Users problems with individual temperature controls in office, *Building and Environment*, 42: 2880–2887.

Melikov, A., Pitchurov, G., Naydenov, K., Langkilde, G. 2005. Field study on occupant comfort and the office thermal environment in rooms with displacement ventilation, *Indoor Air* 15, nr 3, pp. 205–214.

Melikov, A.K. 2004. Personalized ventilation, *Indoor Air* 2004, 14: 157–167.

Melikov, A.K., Arakelian, R.S., Halkjaer, L., Fanger, P.O. 1994. Spot cooling—part 2: Recommendations for design of spot cooling systems, *ASHRAE Transactions*, 100: 500–510.

Naydenov, K., Pitchurov, G., Langkilde, G., Melikov A. 2002. Performance of displacement ventilation in practice, Preceedings of Roomvent 2002, Copenhagen, T*echnical Univercity Denmark and DANVAK*, 483–486.

PN-B-03421:1978 Wentylacja i klimatyzacja—Parametry obliczeniowe powietrza wewnętrznego w pomieszczeniach przeznaczonych do stałego przebywania ludzi (available in Polish).

PN-B-03430:1983/Az3:2000 Wentylacja w budynkach mieszkalnych zamieszkania zbiorowego i użyteczności publicznej—Wymagania (available in Polish).

Toftum, J., Zhou, G., Melikov, A.K. 2000. Effect of airflow direction on human perception of draught, CLIMA 2000, 30 Brussels.

Weber, A., Jeremini, C., Grandjean, E.P. 1975. Relationship between objective and subjective assessment of experimentally induced fatigue. *Ergonomics*, 18: 151–156.

Wyon, D. 1993. Healthy Buildins and Their Impact on Productivity, *Proc. Indoor Air* '93, vol. 6, p.3–13, Helsinki, Finland.

Zeng, Q., Kaczmarczyk, J., Melikov, A. et al. 2002. Perceived air quality and thermal sensation with a personalized ventilation system, Proceedings of Roomvent 2002, Copenhagen, Denmark.

Environmental Engineering IV – Pawłowski, Dudzińska & Pawłowski (eds)
© 2013 Taylor & Francis Group, London, ISBN 978-0-415-64338-2

Application of the Background Oriented Schlieren (BOS) method for visualization of thermal plumes

J. Hendiger, P. Ziętek & M. Chludzińska
Warsaw University of Technology, Faculty of Environmental Engineering, Warsaw, Poland

ABSTRACT: Optical measurement technique called BOS (Background Oriented Schlieren) has the ability to visualize density gradients. This paper describes the possibility of BOS technique utilization for thermal plumes visualization. Tests, conducted in order to use the BOS technique for visualization of warm convective plumes over the objects of relatively low temperature, were performed with special background. The temperature difference between source of the thermal plume and ambient air was changed and the lowest analyzed value was about 10 K. The test results indicate that the BOS technique can be utilized for thermal plumes visualization, when the components of test stand are applied.

Keywords: Background Oriented Schlieren technique, thermal plumes, visualization

1 INTRODUCTION

One of the main tasks of today's HVAC systems is to provide their users with thermal comfort and remove pollutants generated in the rooms. To properly accomplish this task, it is necessary to obtain the relevant distribution of the airflows in the entire space of the room. Proper air circulation can be interrupted, among others by thermal plumes, forming over heat sources like persons or office equipment such as computers, printers or lamps.

Visualization of the real airflow allows to analyze its movement in the room, but due to the fact that the air is transparent, it is difficult to imagine the flow. Such visualizations can be pursued, for example using smoke generators. However, this method, although provides the results in real time, is tedious and time consuming. In addition, supply of the smoke of a different temperature than ambient air may interfere with the actual distribution of the airflows. Helpful in this case may be a method of visualization, based on the occurrence of the different degree phenomenon of deflection of light penetrating through the non-homogeneous mediums (fluids), called as the Background Oriented Schlieren (BOS). The difference in distance covered by the electromagnetic waves results from different values of light refractive index dependent on the density of the fluid, so it is possible to visualize the airflow of density other than the ambient air density.

1.1 *Schlieren technique history*

Although observations of thermal plumes with this method began in the 17th century, August

Toepler is considered to be the inventor of the Schlieren technique in the 19th century. The classic Schlieren technique is based on an optical system consisting of lenses or mirrors, light source, knife-edge and screen or camera. Although the image can be obtained using the systems of the different configurations of the optical elements, below on Figure 1, a simple Z-type Schlieren system based on parabolic mirrors is presented. Z-type system is one of the most popular arrangement using two oppositely-tilted, on-axis telescopic parabolas (Settles 2001). Both lens—and mirror-type systems require high quality optical elements, which are crucial to obtain a good sensitivity.

In the investigations of ventilation airflows, the classic Schlieren method has a very serious constraint. The size of the tested object is limited by the diameter of the available mirrors. What is more, concave mirrors of a large diameter are very expensive. Due to the relatively small size of the available mirrors (in practice to 1 m diameter), such phenomena as a discharge from the air

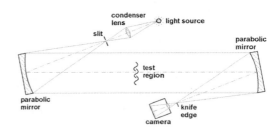

Figure 1. Example of the optical Z-type Schlieren system (Settles 2001).

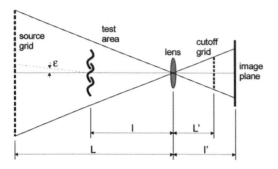

Figure 2. Diagram of a grid-type Schlieren system (Settles et al. 1995).

terminal devices, operation of heaters or thermal plumes forming over large heat sources, cannot be visualized. Therefore, Schlieren techniques enabling visualizations in large scale, were created. These techniques were developed and improved from the 1940s (Settles 2001). Basic configuration of the test standing for full scale visualization consisted of the background, made in the form of a grid of white and black lines, the optical system, the cutoff grid and a plane where the image is created. The cutoff grid is responsible for cutting off the light emitted from the source grid. Test space is located in the middle of the distance between the source grid and the lens. When the air of a different temperature and density appears in the test space, the refractive index of light changes. Refracted waves of light begin to reach the image plane, thus the image of non-isothermal air flow is created. In such systems, the size of the test area is limited only to the size of the source grid. In practice, it will also depends on the cutoff grid size, possible to execute, and the length of the test stand. The full scale Schlieren system based on the described principles was constructed in Penn State University and its source grid size is $4,9 \times 5,5$ m with black lines width 5 mm, whereas the test area has size $2,1 \times 2,7$ m (Settles, 1997). Images of ventilation airflows supplied from ventilation equipment obtained with such stand are very interesting. However, as in other optical Schlieren methods, the quality of the resulting images depends on the precision of the optical system implementation, especially the precision of the cutoff grid execution.

1.2 *Background Oriented Schlieren idea*

Background Oriented Schlieren technique was introduced as a simplification of the optical schlieren system and a shift to increased use of numerical image treatment by Meier (1999, 2002). The idea was also considered by Dalziel et al. (2000) who named it "qualitative synthetic schlieren" and

described it as one of the novel techniques for obtaining both high quality visualizations and accurate measurements of the density field.

The BOS technique is applied in many fields of science where flow visualization is needed. Some attempts to use the method in HVAC issues were made by Hargather and Settles (2010) where the results of flow visualization from air heater and air terminal device were presented. All of the Schlieren techniques use phenomenon of the light refractive index changes. The BOS technique is based exactly on the same phenomenon of light refraction occurred when the light is passing through the medium (fluid) of a different density and refractive index. However, the practical application of the BOS method is much easier and does not require complicated test stand or precise optical system. Moreover, resolution of the digital cameras matrixes is increasing rapidly, which allow to obtain higher resolution and better quality digital images. BOS compared to the traditional methods has several advantages, which are specified below.

The advantages of BOS:

• No sophisticated optical system is needed. Classic systems require the mirrors, lenses, cutoff grids, etc. In the BOS method, just the background, camera and of course software are required.

• No need for precise cutoff grid, which often determines the quality of obtained image. The result of visualization with BOS method is obtained on the basis of the comparison (correlation) of images.

• BOS method, similar to the full-scale methods, can be used to perform visualization of large space or area, limited only by the size of background.

• The lighting is not so crucial in the BOS method. In the classic test systems, the impact of additional light sources, which may disturb the measurement has to be reduced. The tests should be carried out in the premises where conditions are close to those in a darkroom. In the case of BOS method such a problem does not exist and tests may be performed even in natural light.

• Test stand for BOS analysis can be portable. In the classic methods all the elements of the optical system must be precisely set to one another. Therefore, moving of the system is in practice rather impossible. With BOS, setup of test stand is much faster and easier. The equipment can be easily moved to a place where, for example, examined ventilation elements are fitted.

• Construction of the BOS test stand is less expensive than traditional Schlieren systems, due to the simpler structure.

The BOS method is based on the computational analysis of variations in digital images of background occurred due to the refraction caused by the investigated flow of different density than surrounding. Performing BOS visualization requires special high-contrast, motionless background, which typically consists of different black and white patterns. High quality digital images of background are recorded by the high-resolution camera with and without refractive disturbances in the test area, which is between them. Pairs of digital images are then post-processed in the Particle Image Velocimetry (PIV) or Digital Image Correlation (DIC) software to reveal all, even marginal, distortions of the background.

Most of the DIC software measures the pixel shift between images analyzed in the horizontal and vertical direction. The shift is closely related to the spatial refractive index and can be used to obtain quantitative data (Goldhahn & Seume 2007).

BOS like other Schlieren techniques is based on the presence of variation of refractive index depending on the air density.

The relationship between the density and the refractive index can be described by the Gladstone-Dale equation:

$$\frac{n-1}{\rho} = G(\lambda) \tag{1}$$

where:
n—refractive index,
ρ—density [kg/m³],
G(λ)—Gladstone-Dale number.

Gladstone-Dale number G(λ) is defined by:

$$G(\lambda) = 2{,}2244 \times 10^{-4} \left[1 + \left(\frac{6{,}7132 \times 10^{-8}}{\lambda} \right)^2 \right] \tag{2}$$

where:
λ – length of wave.

The primary method used to analyze the received images is a digital image correlation. The image element is represented by a discrete function, accepting the values from 0 to 255 (the gray levels). The correlation calculation is made for origin consisted of a group of pixels. An initial image element before distortion described the discrete function f(x, y). After distortion and shift it is converted to the form f*(x*, y*).

The theoretical relationship between two discrete functions can be described as:

$$f^*\left(x^*, y^*\right) - f\left(x + u(x,y), y + v(x,y)\right) = 0 \tag{3}$$

where u(x,y) and v(x,y) represent the shift of the image element.

2 MATERIALS AND METHODS

In order to verify the possibility of using BOS to visualize thermal plumes generated in the premises, the investigations with the use of the objects of various size and temperature were conducted. Each test was performed in constant temperature of ambient air, therefore the analysis of the results depending on the temperature difference between object surface and surrounding was possible.

2.1 Experimental facilities

The test stand was consisted of background, camera, source of light and different sources of convective plum of various size and temperature.

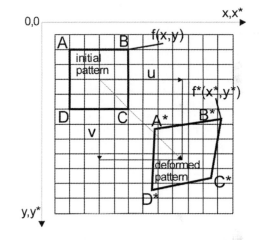

Figure 3. Principle of image correlation method (Mguil-Touchal et al. 1997).

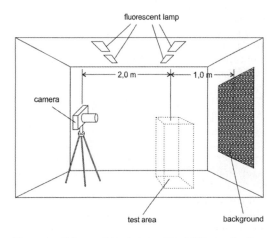

Figure 4. Scheme of the test stand for BOS visualization.

The background was prepared from randomly generated black and white dots pattern where the ratio of colors was 50/50%. Background was attached to the wall of the test room. The Digital Single-Lens-Reflex (DSLR) camera Canon 60D with 18 megapixel resolution CMOS matrix was utilized for recording of still digital images. Camera was placed on the stable tri-pod and equipped with Sigma DC 17-70 1:2,8-4 MACRO HSM zoom lens and remote switch. Lighting of the test area was provided by the 4 standard ceiling-mounted fluorescent lamps, which offered even illumination of the test room.

Examined source of convective flow was placed between background and camera with dimensions presented in Figure 4. Image correlation was performed with Vic 2D software.

3 RESULTS AND DISCUSSION

Test results are presented in the form of images obtained after image correlations. Warm convection plume formed over typical desk lamp is presented Figure 5. Temperature of metal lampshade was changing (increasing) during the test. The analysis was made on the basis of the photos taken in different time measured from turn-on of the lamp. Therefore, thermal plume visualization over identical source but in variable thermal conditions was possible.

Temperature of the lampshade surface was measured in one point at the top with contact temperature probe. The ambient air temperature was measured during the test and equaled 21 °C. Figure 5 shows that together with the increase of source temperature, the plume becomes more stable and the type of flow in the visible area changes from partially turbulent to fully laminar for the highest temperature. The best results

of plume visualization using the BOS technique were obtained for higher temperatures of the heat source and temperature difference between the air in the plume and ambient air. The density gradient between the plume and the ambient fluid was in those cases greater and the recorded refraction of light was more significant. The shape of thermal plume forming over the lampshade is similar to the plume from point heat source.

Other result of BOS visualization is presented in Figure 6, where thermal plume over a person was investigated. Body temperature measured on the forehead was 36,7 °C, whereas temperature of ambient air was 21 °C. Due to the irregular shape and temperature of the heat source, the results are less legible than for the concentrated source. However, convection plumes formed over the head are visible. The analysis of consecutive images recorded one after another can also be very useful in that case because the trend of changes in the shape of airflow can be observed. The last results in Figure 7, shows the visualization of

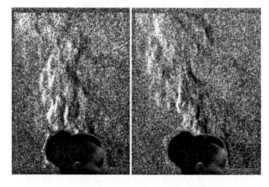

Figure 6.   BOS thermal plume visualization performed on a person (shift between recording time about 2 s).

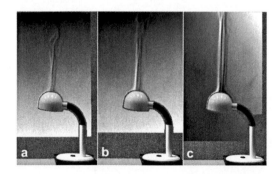

Figure 5.   BOS thermal plume visualization over the desk lamp of different surface temperature: a) time 80 s; temperature 30 °C, b) time 160 s, temperature 40,7 °C, c) time 420 s, temperature 66,3 °C.

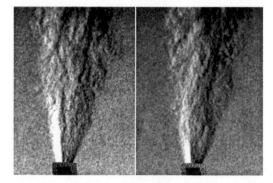

Figure 7.   Warm turbulent ventilation air flow supplied vertically from a circular outlet (shift between recording time about 3 s).

forced convection flow generated from the thermo-ventilation equipment with circular outlet. The axisymmetric turbulent flow of warm air is visualized on the images. Due to the turbulent nature of the flow, the uneven pattern of visible airflow can be observed. The air velocity measured in the outlet was about 7 m/s, whereas the temperature of supplied airflow was equal to 80 °C. On the basis of the results, the angle of airflow spread and its shape in the space room can be determined.

4 CONCLUSIONS

Background Oriented Schlieren technique is a very interesting alternative for classic methods of flow visualization. First of all, no additional marker, which could change the physical properties of fluid, is introduced in the examined airflow. The BOS method has a lot of advantages in comparison with other Schlieren techniques but its application range is limited and strongly impacted by the quality and resolution of equipment used during investigations. The results of investigations presented in this article and earlier ones confirm that airflows visualization of temperature higher than ambient air of about 10–15 K is possible. Such sensitivity is usually enough to visualize thermal plumes, even so complex one like that formed over a human body. In case of heat sources of higher temperature the effect of the BOS utilization is even more visible. Research on the use of the BOS technique in ventilation will be still developed in order to improve the sensitivity of method for small temperature difference typical for air conditioning.

REFERENCES

Dalziel S.B. & Hughes G.O. & Sutherland B.R. 2000. Whole-field density measurements by 'synthetic schlieren', *Experiments in Fluids* 28: 322–335.

Goldhahn E. & Seume J. 2007. The background oriented schlieren technique: sensitivity, accuracy, resolution and application to a three-dimensional density field, *Experiments in Fluids* 43: 241–249.

Hargather M.J. & Settles G.S. 2010. Background-Oriented Schlieren visualization of heating and ventilation flows: HVAC-BOS, *Proceedings of 14th International Symposium of Flow Visualization*, EXCO Daegu, Korea.

Meier G.E.A. 1999. Hintergrund-Schlierenverfahren, *Deutsche Patentanmeldung* DE 199 42 856 A1.

Meier G.E.A. 2002. Computerized background-oriented schlieren, *Experiments in Fluids* 33: 181–187.

Mguil-Touchal S. & Morestin F. & Brunet M. 1997. Various experimental applications of digital image correlation method, *CMEM 97*, Rhodes: 45–58.

Settles G.S. & Hackett E.B. & Miller J.D. &, Weinstein L.M. 1995. Full-scale Schlieren flow visualization, *Proceedings of the VII International Symposium on Flow Visualization*, Seattle, Washington, 11–14 September 1995.

Settles G.S. 1997. Visualizing full-scale ventilation airflows, *ASHRAE Journal*, July 1997: 19–26.

Settles G.S. 2001. Schlieren and shadowgraph techniques, Springer.

Environmental Engineering IV – Pawłowski, Dudzińska & Pawłowski (eds)
© *2013 Taylor & Francis Group, London, ISBN 978-0-415-64338-2*

Noninvasive moisture measurement in building materials

Z. Suchorab
Faculty of Environmental Engineering, Lublin University of Technology, Lublin, Poland

ABSTRACT: This article presents the noninvasive attempt to determine the moisture of building materials with the use of electric methods. In comparison to other measurement techniques like chemical or physical, electric methods enable quick moisture estimation and are suitable solution for moisture changes monitoring. Among the electric measurement devices the most popular are capacitance and resistance sensors. Since several years the attempts to determine moisture of building materials and barriers have been made with the use of the surface TDR (Time Domain Reflectometry) probes which, before the mentioned tests, had been invasive and improper for building materials moisture determination. This type of TDR probes enabled noninvasive measurements of moisture inside building materials and became useful for moisture changes monitoring in laboratory and in-situ conditions. For the experiment three model walls were built and exposed to moisture influence. Water uptake process was monitored by surface TDR sensor prototypes. The obtained results confirmed the functionality of described method.

Keywords: building materials, capacitance probe, TDR probe, moisture detection

1 INTRODUCTION

Water is one of the most important chemical compounds on earth. It determines functioning of all living organisms. However, its presence may lead to several problems, which may also be connected with building objects functioning. Water presence in building materials is a common phenomenon. It occurs not only in older buildings but also in modern objects as the consequence of improper building works. In building barriers water may occur in all three phases—solid, liquid and gaseous. Water presence in newly built objects is always connected with technology of materials production and building works. During their exploitation, buildings may run dry or get wet, depending on external and internal conditions but also on properties of the applied materials. It has been noticed that chemically bound water does not influence moisture properties of the building barriers. What is more, water vapor presence is less influencing, in comparison with the other types of water present in the building envelopes. The most meaningful is sorption and capillary water. The problem of water migration in building barriers is difficult to describe, mostly due to inhomogeneous structure of the materials and barriers and also because of the fact that particular elements like bricks or blocks are combined with mortar having completely different parameters from the main masonry material. All above mentioned problems cause serious difficulties for theoretical and experimental description of the mentioned phenomenon.

This paper is devoted to capillary rise phenomenon monitoring, which is considered to be a major cause of building destruction. The most popular in Polish building market materials are autoclaved aerated concrete, red ceramic brick and silicate brick. All these materials were used to build the model building barriers and investigated for capillary uptake susceptibility. The modified TDR (Time Domain Reflectometry) technique, which enabled noninvasive moisture measurement in porous materials, was applied to monitor the moisture changes in the model barriers.

2 MATERIAL AND METHODS

Laboratory research was conducted with the use of the TDR method. This method has been applied for about 30 years in moisture determination of porous materials. Firstly, the discussed methodology was developed for moisture determination of the soils (Topp et al. 1980, Noborio 2001, Walker et al. 2004) which was mainly connected with sensor construction (Fig. 1) and high demand for monitoring of soil moisture parameters important for plants cultivation. Probes presented in Figure 1 are designed for precise moisture measurement (together with salinity and temperature) in soft, porous materials like soils, etc.

Construction of the above presented probes is quite simple—they are the specific extensions of the concentric cable (Fig. 2). At the end of the cable the wires are separated into two measuring

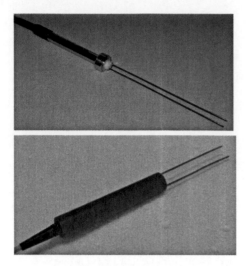

Figure 1. TDR probes for moisture measurement of soils (LP/ms and FP/mts, EasyTest, Lublin, Poland).

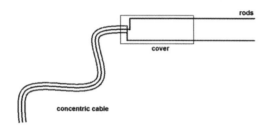

Figure 2. Idea of the typical TDR probe construction.

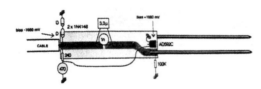

Figure 3. Construction of the typical FP/mts TDR probe (EasyTest, Lublin, Poland).

elements (steel rods). The construction is closed by a special cover, sometimes containing simple electronic elements (Fig. 3).

Furthermore, the literature shows other ideas of the TDR probes construction, for example, the coaxial TDR probe, being a direct extension of a concentric cable into a unit with a greater dimensions (Topp et al. 1980). Other idea is a three-rod sensor, where the braid goes into two external rods and axial cable into one, central rod (Nissen & Moldrup 1995).

The measurement is based on the determination of electromagnetic pulse propagation velocity along the measuring rods inserted into the measured material. It significantly depends on its dielectric properties and thus moisture (O'Connor & Dowding 1999). Reflectometric technique relies on the determination of signal velocity between the reflections at particular elements of the sensor and interval between the first and the second reflection peek (Fig. 4), which is the base for dielectric permittivity estimation according to the following equation:

$$\varepsilon = \left(\frac{ct_p}{2L}\right)^2 \tag{1}$$

where:

ε—relative dielectric permittivity of the porous material [–],

c—light velocity in vacuum [$3 \cdot 10^8$ m·s^{-1}],

t_p—time of signal propagation along the rods of the sensor [s],

L—length of the rods inserted into the measured material [m].

With measured dielectric permittivity, moisture can be recalculated using theoretical, physical models (Birchak et al. 1974, De Loor 1968, Tinga et al. 1973, Whalley 1993) or the empirical calibration formulas obtained by experimental examinations (Malicki et al. 1996, Noborio et al. 1999, Sobczuk & Suchorab 2005, Topp et al. 1980).

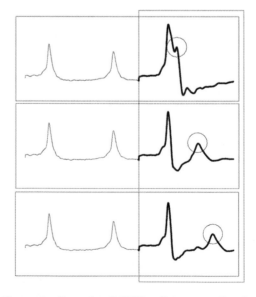

Figure 4. Example of TDR reflectorgram for dry, unsaturated and saturated material (thick line represents the signal propagation along the sensor burried into the material).

All above presented sensor solutions are improper for moisture measurements of building materials and envelopes which are harder from the soils.

This problems generated the idea of the surface TDR probes development, which allowed for noninvasive moisture measurements without steel rods inserted into the structure of the investigated material. First attempts on this technique were presented in the following papers Perrson & Berndtsson (1998) and Wraith et al. (2005), and the probe designed especially for building materials was presented by Sobczuk (2008). The new TDR sensor idea was developed at Lublin University of Technology. Within the experiments several constructions of the noninvasive TDR sensors have been developed there (Fig. 5).

Typical surface TDR probe consists of the dielectric cover (with handle), measuring element (flat or angle-bar), printed circuit and the BNC connector (Figs. 6 and 7).

Application of the surface TDR probes is comparable to the traditional sensors. It relies on the determination of electromagnetic pulse propagation velocity and, thus, its recalculation into dielectric permittivity and moisture.

It must be mentioned, that application of the surface sensor requires specific calibration for each sensor construction and type of material (Suchorab et al. 2009, Suchorab & Sobczuk 2009). It is mainly caused by the influence of other factors which are not present in the standard TDR probes solutions. Moreover, the readouts may be characterized by greater measurement errors caused by the influence of more parameters and shortened time of signal travel along the steel bars.

On the other hand, the application of the surface sensor does not require any special preparation of the samples, it does not destruct the structure of the material, which may by especially important in case of the old, historical objects. Surface measurements do not require drilling the holes or making any invasive procedures which, in case of the traditional sensors would change the material structure and thus influence the expected readouts.

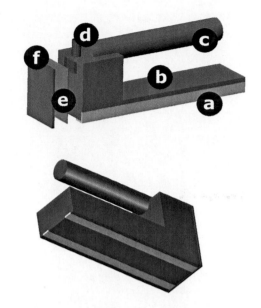

Figure 6. Model of the surface TDR probe, a—measuring element, b—dielectric, c—handle, d—BNC juction, e—printed circuit, f—cover (Suchorab et al. 2009).

Figure 7. Example of the surface TDR probe applied in the described research.

Capillary rise phenomenon was examined in the three porous media in the form of model walls (Fig. 8) made of autoclaved aerated concrete with bulk density 600 kg/m³, red ceramic brick with bulk density 1600 kg/m³ and finally silicate brick with bulk density of 1800 kg/m³. Particular elements of the masonry were combined with cement mortar 10 mm wide. Particular elements of ceramic and silicate bricks dimensions are the following: $510 \times 120 \times 290$ mm, for aerated concrete: $240 \times 240 \times 590$ mm and $120 \times 240 \times 590$ mm

Figure 5. Photo of surface TDR probes prototypes developed at Lublin University of Technology.

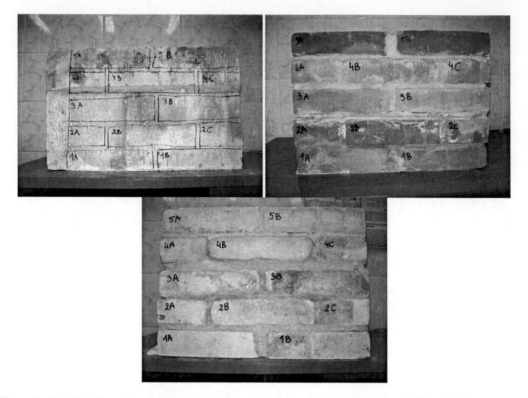

Figure 8. Model walls used in experiment built of aerated concrete, red ceramic brick and silicate brick.

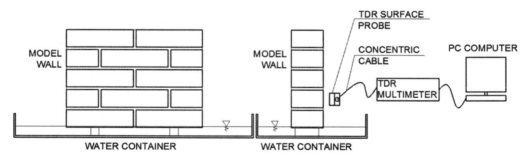

Figure 9. Experimental setup.

The modified TDR method, which allowed for noninvasive measurements of moisture processes inside the walls was applied in the experiment. The technique modification relies on substitution of classical invasive sensors with surface sensors (Figs. 5–7) that do not require sensor installation inside the measured material.

Measuring setup, presented in Figure 9 consisted of:

• three model walls placed in the water containers,

• TDR multimeter (LOM/EasyTest, Lublin, Poland),

• surface TDR probe (Lublin University of Technology, Poland),

• PC computer as control station.

In model walls prepared in such a way, particular bricks were signed with the following symbols 1A, 1B, 2A, 2B, 2C, 3A, 3B, 4A, 4B, 4C, 5A and 5B, where digit meant particular layer of the elements. In case of the aerated concrete, only two elements were used. However, in order to keep the dimension compatibility, they were divided into equivalent fragments to the brick masonries (Fig. 8).

Walls prepared in such a way were dried and put into the water container with constant water level 10 mm above bottom edge of the measured wall. Walls were monitored for moisture changes by the surface TDR probe for period of 30 days with 24 hour intervals. TDR instrumentation applied for the experiment enabled readouts of electromagnetic pulse propagation along the measuring elements of the surface probe, which were used for effective dielectric permittivity calculation (depending on surface sensor construction) and thus moisture. Furthermore, calibration experiments were presented in the following paper (Suchorab et al. 2009).

3 RESULTS AND DISCUSSION

On the basis of moisture readouts obtained by surface TDR probe, the dynamics of moisture increase in walls of aerated concrete, red ceramic brick and silicate brick, caused by capillary rise phenomenon were determined.

Data presented in Table 1 and Figure 10 depict average readouts of moisture at particular layers of the model wall made of aerated concrete.

Moisture readouts made on the model wall made of red ceramic brick are presented in Table 2 and Figure 11.

And finally Table 3 and Figure 12 present averaged readouts at particular level of the model wall made of silicate brick.

The greatest absorptivity of autoclaved aerated concrete, in comparison with the other materials, is the result of highly porous structure of the material. In the period of 30 days in the lowest parts of the model wall, volumetric moisture value was greater than $25\%_{vol}$. It must be underlined that maximum water content value for this material (about $35\%_{vol}$) was not reached. The dynamics of moisture increase in higher parts was noticeable but moisture increase occurred later and the maximum values were exceeding $20\%_{vol}$ in the second layer and $15\%_{vol}$ in the third layer. In layers 4 and 5 no significant moisture increase was observed.

In case of the red ceramic brick, average readouts at layers close to water table level, shown water content equaled about $25\%_{vol}$ (close to saturation), in the second level, volumetric water content was about $20\%_{vol}$ and no water content was observed in higher parts of the model wall.

According to the previous assumptions, the weakest absorptivity was observed for silicate brick. In the examined wall, moisture increase was

Table 1. Moisture values measured at each layer of the autoclaved aerated concrete masonry.

Day	Layer 1	Layer 2	Layer 3	Layer 4	Layer 5
0	0,012	0,011	0,015	0,006	0,005
1	0,058	0,011	0,015	0,005	0,005
2	0,106	0,011	0,015	0,006	0,005
3	0,138	0,039	0,015	0,005	0,007
4	0,165	0,090	0,015	0,004	0,005
5	0,188	0,132	0,015	0,005	0,005
6	0,205	0,140	0,015	0,005	0,005
7	0,224	0,147	0,015	0,006	0,004
8	0,240	0,147	0,032	0,005	0,005
9	0,242	0,159	0,032	0,005	0,005
10	0,245	0,162	0,054	0,004	0,003
11	0,245	0,161	0,078	0,005	0,005
12	0,248	0,166	0,088	0,004	0,005
13	0,252	0,172	0,088	0,011	0,006
14	0,251	0,171	0,088	0,010	0,005
15	0,256	0,170	0,100	0,013	0,005
16	0,257	0,171	0,112	0,013	0,006
17	0,253	0,171	0,116	0,012	0,005
18	0,261	0,171	0,127	0,012	0,005
19	0,259	0,171	0,127	0,013	0,003
20	0,264	0,177	0,127	0,013	0,006
21	0,263	0,174	0,127	0,013	0,005
22	0,260	0,179	0,129	0,013	0,005
23	0,263	0,184	0,145	0,012	0,007
24	0,263	0,188	0,151	0,013	0,011
25	0,263	0,193	0,155	0,012	0,011
26	0,259	0,193	0,155	0,013	0,011
27	0,257	0,193	0,155	0,012	0,011
28	0,261	0,194	0,155	0,013	0,011
29	0,263	0,197	0,154	0,013	0,011

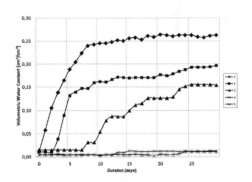

Figure 10. Moisture changes in time in particular levels of aerated concrete.

only observed in the lowest layer of the wall. In that part of the brick, moisture read value was about $15\%_{vol}$, so it was close to saturation. In higher levels of the wall no moisture increase was observed.

Table 2. Moisture values measured at each layer of the red ceramic brick.

Day	Layer 1	Layer 2	Layer 3	Layer 4	Layer 5
0	0,005	0,007	−0,006	0,005	0,011
1	0,141	0,020	−0,003	0,003	0,013
2	0,165	0,060	0,001	0,007	0,007
3	0,184	0,097	0,003	0,005	0,010
4	0,184	0,134	0,001	0,006	0,010
5	0,193	0,162	0,006	0,006	0,009
6	0,195	0,162	0,000	0,006	0,013
7	0,195	0,170	0,005	0,007	0,011
8	0,198	0,170	0,005	0,005	0,011
9	0,208	0,173	0,004	0,004	0,009
10	0,219	0,179	0,009	0,005	0,011
11	0,213	0,192	0,006	0,005	0,006
12	0,212	0,196	0,009	0,005	0,010
13	0,216	0,197	0,012	0,008	0,009
14	0,215	0,195	0,013	0,010	0,010
15	0,213	0,196	0,016	0,007	0,011
16	0,213	0,195	0,019	0,009	0,012
17	0,213	0,195	0,016	0,006	0,010
18	0,211	0,196	0,017	0,007	0,011
19	0,216	0,200	0,018	0,011	0,012
20	0,213	0,200	0,017	0,010	0,006
21	0,212	0,205	0,015	0,007	0,007
22	0,215	0,208	0,015	0,008	0,013
23	0,216	0,207	0,018	0,007	0,006
24	0,218	0,208	0,022	0,011	0,012
25	0,219	0,207	0,022	0,011	0,007
26	0,217	0,210	0,022	0,010	0,010
27	0,219	0,207	0,022	0,008	0,012
28	0,218	0,208	0,022	0,010	0,010
29	0,227	0,208	0,022	0,011	0,009

Table 3. Moisture values measured on each layer of the silicate brick.

Day	Layer 1	Layer 2	Layer 3	Layer 4	Layer 5
0	0,004	0,003	0,005	0,004	0,004
1	0,063	0,006	0,005	0,004	0,004
2	0,111	0,007	0,005	0,004	0,004
3	0,116	0,007	0,005	0,004	0,004
4	0,116	0,007	0,005	0,004	0,004
5	0,116	0,007	0,005	0,004	0,004
6	0,116	0,007	0,005	0,004	0,004
7	0,116	0,007	0,005	0,004	0,004
8	0,125	0,008	0,005	0,004	0,004
9	0,125	0,008	0,005	0,004	0,004
10	0,141	0,008	0,005	0,004	0,004
11	0,144	0,008	0,005	0,004	0,004
12	0,142	0,008	0,005	0,004	0,004
13	0,148	0,008	0,005	0,004	0,004
14	0,151	0,008	0,005	0,004	0,004
15	0,149	0,008	0,005	0,004	0,004
16	0,152	0,008	0,006	0,004	0,004
17	0,152	0,009	0,006	0,006	0,004
18	0,149	0,009	0,007	0,006	0,004
19	0,153	0,009	0,007	0,006	0,005
20	0,152	0,009	0,007	0,006	0,005
21	0,152	0,009	0,007	0,006	0,006
22	0,149	0,009	0,007	0,006	0,007
23	0,152	0,009	0,007	0,006	0,007
24	0,150	0,009	0,007	0,006	0,007
25	0,152	0,009	0,007	0,006	0,007
26	0,148	0,009	0,007	0,006	0,007
27	0,149	0,009	0,007	0,006	0,007
28	0,152	0,009	0,007	0,007	0,007
29	0,151	0,011	0,007	0,008	0,008

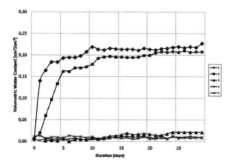

Figure 11. Moisture changes in time in particular levels of red ceramic brick.

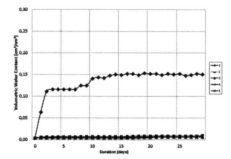

Figure 12. Moisture changes in time in particular levels of silicate brick.

4 CONCLUSION

Autoclaved aerated concrete is a very popular building material on Polish market. It is characterized by the highest ability of water uptake phenomenon. The process is very dynamic and moisture values read by the TDR equipment are higher than other materials.

Red ceramic brick is characterized by the lower water absorptivity than autoclaved aerated concrete, but is more prone to the discussed phenomenon than silicate brick. Silicate brick is the building material characterized by the lowest water absorptivity from all discussed materials. Moisture values read by the surface TDR equipment are about $10\%_{vol}$ less than aerated concrete and about $5\%_{vol}$ less than red ceramic brick. Also, the velocity of the phenomenon is strongly lower. All investigated materials significantly differ in structure which is highly inhomogeneous. In all materials several stages of moisture rise process can be highlighted. First, quick stage, when the smallest pores are filled with water. The second stage runs slower and is typical for the pores with greater dimensions. Cement mortar decreases the velocity of capillary rise phenomenon, therefore, model walls made of red ceramic brick and silicate brick showed moisture increase only in bottom layers. Modified TDR method has big measuring potential for research of capillary parameters of building barriers.

REFERENCES

Birchak, J.R., Gardner, C.G., Hipp, J.E. & Victor, J.M. 1974. High dielectric constant microwave probes for sensing soil moisture. *Proc. IEEE* 1(62): 93–98.

De Loor, G.P. 1968. Dielectric properties of heterogeneous mixtures containing water. *J. Microwave Power* 2(3): 67–73.

Malicki, M.A., Plagge, R. & Roth, C.H. 1996. Improving The Calibration Of Dielectric TDR Soil Moisture Determination Taking Into Account The Solid Soil. Eur. *J. Soil Sci.* 3(47): 357–366.

Nissen, H.H. & Moldrup, P. 1995. Theoretical background for the TDR methodology. Proceedings of the Symposium: Time Domain Reflectometry Applications in Soil Science held at the Research Centre Foulum, Danish Institute of Plant and Soil Science, P Report No. 11: 9–23.

Noborio, K., Horton, R. & Tan, C.S.. 1999. Time Domain Reflectometry Probe for Simultaneous Measurement of Soil Matric Potential and Water Content. *Soil Science Society of America Journal* 63: 1500–1505.

Noborio, K. 2001. Measurement of soil water content and electrical conductivity by time domain reflectometry: a review, Computers and Electronics in Agriculture 31: 213–237.

O'Connor, K.M. & Dowding, C.H. 1999. GeoMeasurements by Pulsing Cables and Probes, CRC Press.

Perrson, M. & Berndtsson, R. 1998. Noninvasive Water Content and Electrical Conductivity Laboratory Measurements using Time Domain Reflectometry. Soil Sci. Soc. *Am. J.* 62: 1471–1476.

Sobczuk, H. & Suchorab, Z. 2005. Calibration of TDR Instruments for Moisture Measurement of Aerated Concrete. Monitoring And Modelling the Properties of Soil as Porous Medium. Institute of Agrophysics, Polish Academy of Sciences, 13–16 February 2005: 156–165, Lublin, Poland.

Sobczuk, H., Sonda do pomiaru wilgotności ośrodków porowatych zwłaszcza materiałów budowlanych, Polish Patent P—198492 (30.06.2008).

Suchorab, Z., Sobczuk, H., Cerny, R., Pavlik, Z. & Plagge, R. 2009. Noninvasive moisture measurement of building materials using TDR method, Proceedings of the 8th International Conference on Electromagnetic Wave Interaction with Water and Moist Substances (ISEMA) 1–5 June 2009: 147–155. Espoo, Finland.

Suchorab, Z. & Sobczuk, H. 2009. Dielectric properties of building materials, Thermophysics, Conference Proceedings, 29–30 October 2009: 138–146. Valtice, Czech Republic.

Suchorab, Z., Widomski, M., Łagód, G. & Sobczuk, H. 2010. Capillary rise phenomenon in aerated concrete. Monitoring and simulations, Proceedings of ECOpole 4(2), 285–290.

Tinga, W.R., Voss, W.A.G. & Blossey, D.F. 1973. Generalized approach to multiphase dielectric mixture theorie. *J. Appl. Phys.* 44: 3897–3902.

Topp, G.C., Davis, J.L. & Annan, A.P. 1980. Electromagnetic determination of soil water content: Measurements in coaxial transmission lines. *Water Resour. Res.* 16: 574–582.

Walker, J.P., Willgoose, G.R. & Kalma, J.D. 2004. In situ measurement of soil moisture: a comparison of techniques, *Journal of Hydrology* 293: 85–99.

Whalley, W.R. 1993. Consideration on the use of Time Domain Reflectometry (TDR) for measuring soil water content. *J. Soil Sci.* 44: 1–9.

Wraith, J.M., Robinson, D.A., Jones, S.B. & Long, D.S. 2005. Spatially characterizing apparent electrical conductivity and water content of surface soils with time domain reflectometry. *Computers and Electronics in Agriculture* 46: 239–261.

Energy

Environmental Engineering IV – Pawłowski, Dudzińska & Pawłowski (eds)
© *2013 Taylor & Francis Group, London, ISBN 978-0-415-64338-2*

Fuel cell as part of clean technologies

A. Kacprzak, R. Włodarczyk, R. Kobyłecki, M. Ścisłowska & Z. Bis
Department of Energy Engineering, Faculty of Environmental Engineering and Biotechnology, Czestochowa University of Technology, Czestochowa, Poland

ABSTRACT: Carbon fuel cells directly fueled with coal is a technology for highly efficient conversion of chemical energy, through electrochemical reactions, into electrical energy without combustion. The development of coal technology of fuel cells due to high energy conversion efficiency can help to reduce emissions of pollutants such as NOx, SO_2 and fly ashes. The article presents the operating characteristics of the carbon fuel cell made of construction materials such as carbon steel, stainless steel or nickel and its alloys. Fuels used to fuel the cell were coal, biomass and graphite. The study also examines the impact of the actual working conditions of fuel cell on changes in cell cathode construction material: Ni-based Inconel® alloy 600.

Keywords: direct carbon fuel cells, clean coal technologies, coal, biocarbon

1 INTRODUCTION

Nowadays, electric power from coal is achieved by coal-fired power plants, however, it is well know that this method is not only a low-efficiency process, but also it emits substantial amounts of greenhouse gases and pollutants, such as NOx, SOx and fly ashes (Pieńkowski 2012, Shan 2012, Hoedel 2011). Therefore, it is logical to investigate a new and high-efficiency process for coal to be used for power generation. The Direct Carbon Fuel Cell (DCFC) is a power generation device in which the chemical energy of carbon is directly converted into electrical energy by electrochemical oxidation of carbon, without the combustion, gasification process or the moving machinery associated with conventional electric generators. This fuel cells uses solid carbon as fuel, which is different from the Molten Carbonate Fuel Cell (MCFC) and the Solid Oxide Fuel Cell (SOFC) operating on a gaseous fuel. DCFC has several unique attractive features. Firstly, DCFC offers great thermodynamic advantages over other fuel cell types. Secondly, it has high efficiency alternatively to the traditional coal fired electrical power plants, resulting in reduced carbon dioxide emissions per unit generated electricity. Thirdly, solid carbon fuel can be easily produced from many different resources, including coal, petroleum coke, biomass (*e.g.* grass, woods, nut shells, corn husks) and even organic garbage. The major motivation for recent work with DCFC has been of higher theoretical energy efficiency (100%) compared to thermal conversion processes (35–45%). DCFC releases lower emissions than

coal-firing power plants. Hence, DCFC may cut carbon emissions from coal by 50% and reduce off-gas volume by 10 times compared to conventional coal-burning power plants (Cao *et al.* 2007). The present work illustrates the construction and performance of a Direct Carbon Fuel Cell with hydroxide electrolyte. Molten alkaline hydroxides have many long-known advantages *e.g.* high ionic conductivity, higher electrochemical activity of carbon (higher anodic oxidation rate and lower overpotential), lower operating temperatures (450°C) and consequently allow for the usage of less expensive materials for cell fabrication. Cheaper materials such as carbon steel and stainless steel may be used to fabricate containers, anodes and cathodes, because of the lower corrosiveness at lower temperatures. To date, three generations of DCFC prototypes have been built and tested to demonstrate the technology—all using different fuels (biocarbon, coal and graphite rod) and construction materials.

2 MATERIAL AND METHODS

2.1 *Construction materials*

A number of construction materials may be used to produce the elements of DCFC (*e.g.* anode, cathode, container). Metals and alloys most frequently considered for use in carbon fuel cells with molten hydroxide electrolyte are carbon steel, stainless steel, nickel and high-nickel alloys. All these alloys have been tested directly in fuel cell to identify which material will be suitable for target device construction.

Alloys, with their nominal composition, used to build a fuel cell are given in Table 1.

Carbon steel is the predominant construction material in the first generations of DCFC (Jacques 1896), therefore it was used for building the first generation prototype of fuel cell (Kacprzak et al. 2009, 2010).

Austenitic stainless steel (i.e. 304, 316) series were chosen to construct the second prototype of DCFC (Kacprzak et al. 2011). Type 304L stainless steels are the most widely used of any stainless steel. Although they have a wide range of corrosion resistance they are not the most corrosion resistant of the austenitic stainless steels. The 304 series of stainless steels exhibit high temperature strength, oxidation resistance, ease of fabrication and weldability and good ductility (Schweitzer 2003). Type 316L stainless steel is low carbon version of type 316 and offers the additional feature of preventing excessive intergranular precipitation of chromium chlorides during welding and stress relieving (Schweitzer 2003). In general alloy is more corrosion resistant than type 304 stainless steels.

The third carbon fuel cell was made of nickel and high nickel alloys because of their ability to resist corrosive environments. Nickel is commonly used as the material for the hydroxides service (Paul et al. 1993, Davis 2000). However, nickel is rapidly oxidized in situ to NiO which is a p-type semiconductor. The good corrosion resistance of Ni-based alloy Inconel® 600 in molten NaOH was attributed by Tran et al. (Tran et al. 1995) to the formation of a protective passive film of Ni–Cr spinel-type oxide which prevents the leaching of Cr as chromite and chromate from the alloy. The corrosion resistance of the construction materials is listed in Table 2.

2.2 Direct carbon fuel cell designs

A photo of the constructed cell is shown in Figure 1. Main part of the cell was a crucible made of carbon steel (depth of 135 mm with a inside diameter of approximately 57 mm) in which molten electrolyte (NaOH) is contained. The steel pot acted also as the cathode. Air from a compressor (necessary for the electrochemical reaction) was distributed to the cathode by sparge pipe (outside diameter of approximately 5,7 mm and wall thickness 1 mm) perforated at the bottom— six holes with a diameter of about 2 mm. A metal basket current collector with particulate carbon fuel, positioned inside the cathode crucible was an anode. In order to maintain the electrolyte in the liquid phase the crucible was heated by an electric heater. Reduce heat loss was achieved by securing a prototype by mineral wool and seal off all in ceramic casing.

Table 1. Alloys used to construction of DCFC with molten hydroxide electrolyte.

| Alloy | Composition [wt%] | | | | | | | | |
	Ni	Cr	Fe	C	Mn	S	Si	P	Other
Carbon steel	–	–	Bal.	<0.24	<1.1	<0.05	0.10–0.35	<0.05	–
304L	8.0–12.0	18.0–20.0	Bal.	<0.03	<2.0	<0.03	<1.0	<0.045	–
316L	10.0–14.0	16.0–18.0	Bal.	<0.03	<2.0	<0.03	<1.0	<0.045	Mo = 2.0–3.0
316Ti	10.0–14.0	16.0–18.0	Bal.	<0.08	<2.0	<0.03	<0.75	<0.045	Mo = 2.0–3.0 Ti < 0.7
Nickel® 201	Min. 99	–	<0.4	<0.02	<0.35	<0.01	<0.2	–	Ti < 0.1 Cu < 0.25
Inconel® 600	Min. 72	14.0–17.0	6.0–10.0	<0.15	<1.0	<0.015	<0.5	–	Cu < 0.5

Table 2. Corrosion resistance of selected materials in molten NaOH in elevated temperatures.

| Materials | Corrosion rate | | Reference |
	Temp. [°C]	mm/yr	
Carbon steel	340	≈0.5	(Paul et al. 1993)
SS 304L	340	≈0.6	(Paul et al. 1993)
Nickel® 201	500	0.033	(Davis 2000)
Inconel® alloy 600	500	0.06	(Davis 2000)

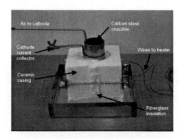

Figure 1. A photography of experimental prototype I of direct carbon fuel cell made of carbon steel.

Fuel cell prototype called as II is shown in Figure 2. All the cell steel parts were manufactured either from a corrosion-resistant stainless steel or Nickel® 201 alloy. The main cell container (inside diameter of 76 mm and with a height of about 142 mm) was manufactured from the Nickel® 201 in order to provide sufficient corrosion resistance to liquid sodium hydroxide. Container was covered by ceramic band heaters providing heat for melting the electrolyte and maintaining its temperature at 450°C during the experiments. The cathode was manufactured from sintered stainless steels 316Ti (bottom sparger, average pore size of 20 micrometers) and 316L (main air pipe, outer diameter of 10 mm and wall thickness of 2 mm), while the anode was made of 304L steel and formed as a specially-designed tube of 25 mm diameter. Since the design of some parts of the electrodes is subjected to be patented more details on the electrode design will be given after the patent is granted. Anode and cathode were mounted in the lid of the cell (the cathode is electrically isolated from the anode with a ceramic plug). Current collectors in the form of copper plates and wires were attached to the anode and cathode for collection of electrons.

A picture of the prototype III fuel cell is shown in Figure 3. That model was built only from nickel and its alloys. In addition, anodic and cathodic chambers were separated, making gases are not mixed with each other (CO_2 above the anode and excess air above the cathode). As in the case of the second prototype of fuel cell the main cell container (inside diameter of 83 mm and with a height of about 147 mm) was manufactured from the Nickel® 201. In that construction of cell anode was made from the Nickel® 201 and the cathode was made of Ni-based Inconel® alloy 600. The anode and cathode were specially designed constructions made of pipes with the outside diameters of 19,1 mm and 42 mm respectively. The components of the prototype III carbon fuel cell, in particular the anode and cathode materials, were subjected to simultaneous oxidation-lithiation process. Doping cation-defective p-type nickel oxide by lower valence cations (Li^+) makes them highly conductive, which is a requirement for a high performance cathode material. The lithiated NiO cathodes were made by *in situ* oxidizing and lithium-doping Ni-base cathode material.

Source of Li^+ was lithium hydroxide monohydrate ($LiOH \cdot H_2O$, $m_p = 470°C$). Shortly after the salt has been melted air was introduced into the system (0.2 dm^3/ min.) in order to accelerate the oxidation of nickel to NiO and lithium ions can incorporate into the surface and inner of NiO film. The process was carried out at 600°C for 24h. Afterwards anode and cathode materials were slowly cooled to room temperature and then were washed several times in distilled water and HCl

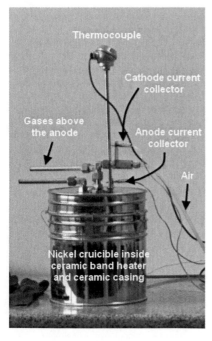

Figure 2.   A photography of experimental prototype II of direct carbon fuel cell made of stainless steels 300 series.

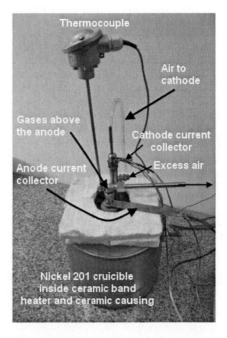

Figure 3.   A photography of experimental prototype III of direct carbon fuel cell made of nickel and high-nickel alloys.

solution to remove and neutralize the residue of lithium hydroxide. After washing, the cell components were dried in a convection oven for 2h. In order to reduce the dissolution of lithium ions in the electrolyte during operation of fuel cell, lithium hydroxide (10 mol%) was melted with NaOH. As in the case of prototype II more details on the electrodes design will be given after the patent is granted.

2.3 Characterization of the Inconel® alloy 600 after cell tests

The film and the reaction products of cathode material are characterized by using X-ray diffractometry. Phase analysis of the Inconel alloy 600® (after lithiation process, 64, 94 and 115 hours) was carried out by means of XRD Seifert 3003 T-T X-ray diffractometer with the use of $K_{\alpha Co}$ radiation (0.17902 nm). The corrosion rate was studied with the use of weight loss technique. Measurements on test material were conducted under the same conditions and the results were compared. Weights were determined with a precision of

0.0001 g on a RADWAG AS 220/X digital analytical balance.

2.4 Characterization of apparatus and experimental procedures

The experiments described in the present paper were conducted in a laboratory-scale facility shown schematically in Figure 4.

The electrolyte temperature was determined by a K-type thermocouple (NiCr-NiAl) and was maintained at the desired value by an electronic temperature controller. The data acquisition module Advantech USB-4711A was used for the measurement of the cell voltage and the decrease of the voltage on an external resistor. The module was connected to a Personal Computer (PC) where the data was displayed and stored. The Tektronix DMM4040 Digital Multimeter was used to measure the open circuit voltage of the fuel cell. In order to determine the cell current intensity at various loads an external resistance setup MDR-93/2-52 was used and connected to the cell circuit thus providing the possibility to adjust the

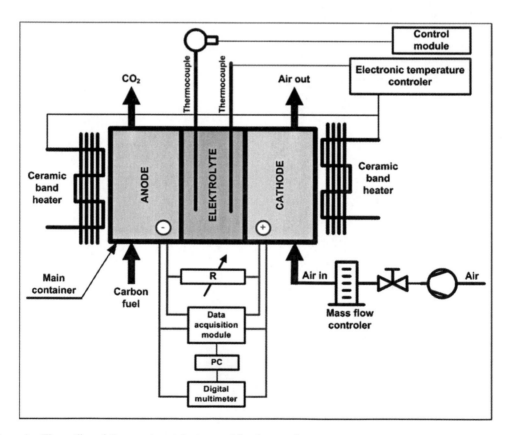

Figure 4.   The outline of the experimental setup used for the experiments.

electrical resistance of the external circuit (in range 0.1–10000Ω). The amount of air fed into the cell was controlled by a thermal mass flow controller (Brooks 4850) witch Local Operator Interface (LOI) to view, control and configure the control device. It was possible to adjust the gas flow rate from 0.1 dm³/min. to 2 dm³/min. In order to attenuate short-term surge suppression and eliminate the effects of power grid interferences on the recorded data, the emergency standby backup power device PowerCom UPS BNT-1500AP with a noise filter EMI/RFI, was also used during the experiments. The tool also acted as an 'emergency power supply device' for the data recording system in case of power failure. At the beginning of each test 500 g of sodium hydroxide was put into the main cell container (prototype I and II) or prepared eutectic mixture of NaOH (90 mol%) with LiOH (10 mol%) in case of prototype III and then heated up to the desired temperature. After the temperature level of 450°C was reached and the electrolyte was completely molten, both the cathode and the anode were slowly immersed into the electrolyte and the cell data (current intensity, voltage, temperature, *etc.*) were recorded.

After each test was finished the heating was turned off and the cell was 'shutdown'. The setup was then cooled down to room temperature and then all its parts were placed in special plastic container filled with roughly 25 liters of ionized water. All the elements were kept there for three hours

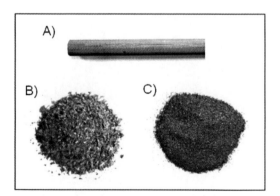

Figure 5.   The example fuel samples: A) graphite rod (d = 13 mm), B) biocarbon, C) hard coal.

in order to get the solidified electrolyte removed. A mechanical stirrer was used to improve the dissolution of the electrolyte. The water-electrolyte mixture was then removed and new 25 liters of ionized water were put into the container. The whole procedure was then repeated. Afterwards, the cell elements were removed from the container, cleaned with a soft sponge, and finally again rinsed with ionized water. All the elements were then dried for 3 hours in a drier.

2.5   *Characterization of carbon fuels*

Three kinds of carbon fuels with different characteristics were used as anode materials—graphite, biocarbon and coal. The example fuel samples are shown in Figure 5 while the main properties of the fuels are shown in Table 3.

These biocarbon and coal samples were ground and sieved on the laboratory vibration shaker to fraction of 0.18–0.25 mm granule size.

3   RESULTS AND DISCUSSION

3.1   *Effect of different electrode materials and carbon fuels on the cell performance*

3.1.1   *Prototype I*

Figure 6 shows the voltage and power density versus current density characteristics of prototype I direct carbon fuel cell at 450°C (air flow rate: 0.5 dm³/min).

The voltage-current density characteristics (I-V) show that a limiting current condition is reached nearly 80 A/m² with biocarbon as a fuel and above 40 A/m² for graphite. The figure shows that the maximum peak power densities achieved are 50 W/m² and 24 W/m² for biocarbon and graphite respectively. Biocarbon produced an Open Circuit Voltage (OCV) of up to 1,075 V is compared to graphite rod that produced an OCV of up to 0,85 V.

The overall simplicity of cell design showed in Figure 6 combined with the cheap construction materials makes it commercially attractive but high corrosion rates of carbon steel used in the cell at the operating temperatures lead to high levels of degradation particularly on the long term operation.

Table 3.   The main parameters of the fuels used during the experiments (air-dry state).

Fuel type	HHV [MJ/kg]	Volatile matter [%]	Ash [%]	Moisture [%]	Particle size [mm]
Biocarbon (granules)	29.9	18.4	2.6	4.5	0.18–0.25
Hard coal (granules)	26.7	36.6	11.5	4.5	0.18–0.25
Graphite (rod)	32.3	1.3	1.7	0.1	13 mm outside diameter

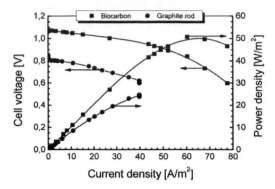

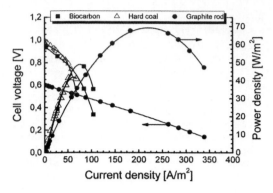

Figure 6. Voltage—current density and power density—current density characteristics of the prototype II fuel cell (electrolyte temperature 450°C, air flow rate 0.5 dm³/min).

Figure 7. Voltage—current density and power density—current density characteristics of the prototype II fuel cell (electrolyte temperature 450°C, air flow rate 0.5 dm³/min).

3.1.2 *Prototype II*

The materials of construction for this prototype were stainless steels: 304L, 316L and 316Ti. Figure 7 shows the polarization (I-V) curve and power obtained during the experiments on three different carbon materials: biocarbon, coal and graphite at 450°C. As can be seen that the DCFC prototype II produced an increased performance, with maximum power densities of >65 W/m² with graphite rod serves as a fuel. For biocarbon the maximum power densities were comparable with the first prototype and equaled to 48 W/m². For coal the cell achieves power density above 40 W/m².

Figure 7 shows current densities of above 300 A/m² achieved with graphite as the fuel and very low for biocarbon (104 A/m²) and coal (68 A/m²). The OCV of the prototype II with solid biocarbon and coal is higher than the one with graphite fuel—0,95 V, 0,99 V and 0,6 V respectively. After fuel cell tests at the temperature of 450°C, stainless steels 300 series were covered with an oxide layers (probably NiO, Fe_2O_3, Cr_3O_4, Cr_2O_5, and CrO). The protective oxide films formation of a dense film of corrosion products that firmly adheres to the steel surface and has a sufficiently high chemical resistance. However, after each test NaOH melt contained loose products of further interaction of oxides with the melt, which poorly adhere to the surface. Therefore, it was decided to build another prototype of the cell made of nickel and its alloys.

3.1.3 *Prototype III*

In Figure 8 the characteristics of the prototype III, manufactured from the Nickel® 201 and Inconel® alloy 600, operated with various fuel types are shown. The maximum current density for this cell

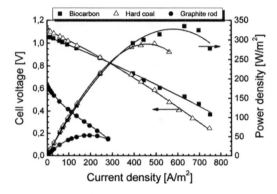

Figure 8. Voltage—current density and power density—current density characteristics of the prototype II fuel cell (electrolyte temperature 450°C, air flow rate 0.5 dm³/min).

was very high, more than 700 A/m² both for the biocarbon and coal as the fuels.

The observations on three samples of fuels showed a large variation in maximum peak power densities—335 W/m², 291 W/m² and 53 W/m² for biocarbon, coal and graphite, respectively. It was suggested that this variation was largely due to a significant differences in structures of those types of fuel samples and in the reactions kinetic on a new construction materials. Alloy 600 with high Ni content shows good catalytic activity to oxygen (cathode side) and new design of cell was employed to reduce the impact of ohmic losses on DCFC performance. The results presented in Figures 6–8 clearly indicate the effect of fuel type and the construction material on the operation performance of the fuel cell, and are quite promising with respect to the potential application of the DCFC technology for large-scale power generation since the data indicate that the DCFC may be easily supplied with granulated hard coal or carbonized and granulated biomass.

3.2 Inconel® alloy 600 (cathode material) characterization

Figure 9 shows the ring made of Inconel® alloy 600 mounted on the cathode cells during each run of the fuel cell. All runs were conducted at 450°C with air flow rate of 0.5 dm³/min. Electrolyte composition was NaOH-LiOH (90–10 mol%).

According to the XRD analysis, $Li_{1-x}Ni_xO$ existing in Inconel® alloy 600 and coating was in the form of NiO, $Li_{0.05}Ni_{0.95}O$ and $Li_{0.1}Ni_{0.9}O$.

The corrosion rate of Inconel® alloy 600 during direct carbon fuel cell operation at 450°C was studied with the use of the weight loss technique. Table 4 shows the weight changes of examined material during fuel cell operations. Figure 10 shows the XRD patterns of the Inconel® alloy 600 ring.

Table 4 shows that the weight gain of the Inconel® 600 increases during fuel cell operation time. In the first step the weight gain is related with the formation of NiO doped Li⁺ ions. During the fuel cell test, firstly it can observed the weight loss probably related to removal of $Li_{0.4}Ni_{0.6}O$

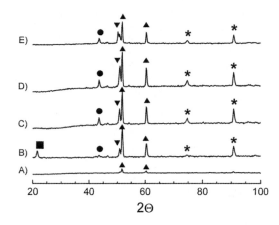

Figure 10. X-ray diffraction patterns for A) pure Inconel® alloy 600, B) after lithiation process, C) after 64 hours, D) after 94 hours, E) after 115 hours, (■) $Li_{0.4}Ni_{0.6}O$ peak, (●) $Li_{0.05}Ni_{0.95}O$ peak, (▲) Ni peak, (▼) NiO peak, (✳) $Li_{0.1}Ni_{0.9}O$ peak.

species (see Fig. 10 B). The overall weight gain after 20 and 40 hours of state corrosion test was found to be 0.6 and 7.2 mg, respectively. This result illustrated the corrosion protecting effect of the NiO coating very well (see Fig. 10 C-E).

4 CONCLUSIONS

Three DCFCs with different configurations and construction materials have been successfully tested in the laboratory scale. On the basis of a series of corrosion experiments with a number of different materials, it was decided to use a nickel (Nickel® 201) and high-nickel alloy (Inconel® alloy 600) to build the cell (anode, cathode, container, etc.). Various carbon fuels have been tested in different carbon fuels is still unclear. Apparently the disordered carbon is more reactive due to a preponderance of edges sites and defects. The physical and chemical properties of carbon fuels can be seen to highly influencing the electrochemical performance. However, the DCFC technology represent a simple way to convert carbon chemical energy to electricity efficiently and without forming by-products associated with conventional combustion (NO_x, SO_x, etc.). Power densities ranging from 24–335 W/m² were measured on a variety of carbon materials. The highest power densities were achieved with charred biomass and raw coal for prototype III fuel cell. These power densities are too low for a commercially viable system, however, with research and development in progress there is a potential for further improvements.

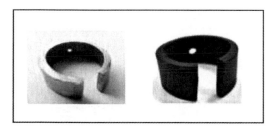

Figure 9. Inconel® alloy 600 ring before (left) and after (right) oxidation/lithiation process.

Table 4. Time weight changes measurement of Inconel® alloy 600 ring during fuel cell operation.

Exposure period (hours)	Weight change	
	[g]	[%]
24 (after oxidation/lithiation process)	+1.95	+6.1
10 (first test after lithiation)	−0.0004	−0.0008
20	+0.0006	+0.0038
40	+0.0072	+0.0723

Using conventional hard coal without pretreatment fuel cell had achieved power density greater than 290 W/m^2 and current density around 750 A/m^2 at 450°C (prototype III). The cell performance of direct carbon fuel cell with molten hydroxide electrolyte can be significantly improved by optimizing the cell design and the electrodes material. Inconel® alloy 600 was the good catalytic material for the cathode with the good corrosion resistance. There is also significant work to continue at a cell level on the optimization of operating temperature, air humidity, gas bubbling rate and electrolyte composition (single and mixed hydroxides of alkali metals such as Na, K, Li).

ACKNOWLEDGEMENTS

The present work was supported by the Polish Ministry of Education and Science under the grant No. N N513 396 736.

REFERENCES

Cao D., Sun Y., Wang G., 2007, Direct carbon fuel cell: fundamentals and recent developments, *J. Power Sources*, 167, 2, 2250–2257.

Davis J.R., 2000, ASM specialty handbook: nickel, cobalt, and their alloys, ASM International, ISBN 0871706857.

Hoedl E., 2011, Europe 2020 Strategy and European Recovery. Problemy Ekorozwoju/*Problems of Sustainable Development* 6(2), 11–18.

Jacques W.W., 1896, Method of converting potential energy of carbon into electrical energy, U.S. patent no. 555 511.

Kacprzak A., Kratofil M., Kobyłecki R., Bis Z., 2009, Characteristics of operation of direct carbon fuel cell, in: Proc. IX Conf. RDPE 2011, Research and Development in Power Engineering (in Polish).

Kacprzak A., Kobyłecki R., Bis Z., 2010, Clean electricity from the direct carbon fuel cell, Środowisko i rozwój, 2, 22, 87–100 (in Polish).

Kacprzak A., Kobyłecki R., Bis Z., 2011, Clean energy from a carbon fuel cell, *Archives of Thermodynamics*, 32, 3, 37–47.

Paul L.D., Barna J.L., Danielson M.J., Harper S.L., 1993, Corrosion-resistant tube materials for extended life of openings in recovery boilers, *Tappi Journal*, 76, 8, 73–77.

Pieńkowski D., 2012, The Jevons Effect and the Consumption of Energy in the European Union, *Problemy Ekorozwoju/Problems of Sustainable Development*, 7(1), 105–116.

Schweitzer P.A., 2003, Metallic Materials: Physical, Mechanical, and Corrosion Properties, CRC Press, ISBN: 978-0-8247-0878-8.

Shan S., Bi X., 2012, Low Carbon Development of China's Yangtze River Delta Region. *Problems of Sustainable Development* 7(2): 33–41.

Tran H., Katiforis N.A., Utigard T.A., et al., 1995, Recovery boiler air-port corrosion—part 3: corrosion of composite tubes in molten NaOH, Tappi Journal, 78, 9, 111–117.

Environmental Engineering IV – Pawłowski, Dudzińska & Pawłowski (eds)
© 2013 Taylor & Francis Group, London, ISBN 978-0-415-64338-2

Identification of heat transport processes in solar air collectors

M. Żukowski

Faculty of Civil and Environmental Engineering, Bialystok University of Technology, Bialystok, Poland

ABSTRACT: Solar air collectors have more and more applications in agriculture, food industry, and the building sector. In the paper, the most popular construction solutions are discussed. The author has proposed his own construction of solar air collector based on slot micro-jets. The article analyzes the heat transfer process inside these type of devices. Computer simulations based on algorithms of Computational Fluid Dynamics (CFD) were used as a research tool. Effectiveness of the traditional solution with a flat absorber was compared with that of the author's construction. Finally, results of the comparative analysis and findings are presented.

Keywords: solar air collectors, micro-jets, heat transfer, computational fluid dynamics

1 INTRODUCTION

Solar air collector is a device for conversion of solar radiation energy into heat. The basic principle is that the heat from the warming absorber plate is absorbed by air circulating around it. Heat exchange may occur via free or forced convection. The most important element of the device, determining its energy performance, is the plate that absorbs solar radiation. Its surface should have selective properties, i.e. it should have a high short-wave radiation absorption coefficient and low long-wave radiation emissivity. The air may pass over or/ and below the absorber plate, and also through it in a transpired design. Detailed review of such constructions was presented by Kumar & Rosen (2011). In order to increase the length of air flow and surface of heat exchange, collectors with flow on both sides of the absorber plate were developed (Soprian et al. 1999). A detailed characteristics of collectors with a transpired absorber plate was prepared by Augustus & Kumar (2007). This solution is getting more and more popular. Above all, it has wide application for initial heating of the ventilation air in single—on multiple-family houses.

Efficiency of solar air collectors may vary in a range from 25% to 75%. In order to increase the efficiency, different ways of increasing the heat exchange surface are implemented. One of these is profiling of the plate that absorbs solar radiation. Examples of individual solutions with v-groove profiled absorber plates can be found e.g. in the work of Tchinda (2009). The results of experimental studies on solar collectors with absorber plate formed as rectangular channels were presented by Youcef-Ali & Desmons (2006). Belusko et al. (2005)

determined thermal performance of collectors with absorber plate made of corrugated plate that was planned to be roof integrated.

Another way to increase the energy performance of solar air collectors is to attach fins into the plate that absorbs solar radiation. Such constructions were analyzed e.g. by Ho et al. (2011). According to the results of experimental studies, the increase of the heat exchange area gained with the use the method, above described, may result in the increase of efficiency of analyzed devices by over 20%, as compared to typical solutions with flat absorber plate.

The article analyzes the heat exchange process inside a solar air collector. Effectiveness of traditional solutions was compared with that of a construction designed by the author; the author's construction utilizes the micro-jet technique. Jets of fluid impinging on the absorber plate surface break the near-wall laminar sublayer, hence decreasing the heat transfer resistance in this area. This technique often uses axis-symmetrical jets (Belusko et al. 2008) and more rarely slot jets. The latter were used by the author in his solution (Fig. 1) for which patent application was submitted (No. WIPO ST 10/C PL398636).

The author chose a standard collector with a flat absorber plate to compare its thermal performance with the prototype presented in Figure 1. Wide review of solar air collector constructions presented above has shown a great variety of technical solutions. However, the great majority of mass-produced devices are based on plat absorber plates. It is due to, among others, a very simple construction, low cost, ease of cleaning and minimal air flow pressure drop. And the increase of efficiency

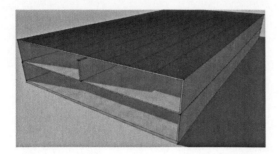

Figure 1. Scheme of a solar air collector using the slot jets.

is obtained via the increase of the air volume flow rate. Hence, the author chose this solution to perform the comparative analysis.

2 NUMERICAL MODEL

Computer simulations based on algorithms of computational fluid dynamics were used as a research tool. To simplify the numerical model construction, top glazed layer was not taken into consideration because it influences the thermal performance of both devices in an identical way.

Two three-dimensional models were developed; these represent the standard collector (model I—shown in Fig. 2) and the collector using the slot jets technique (model II—shown in Fig. 3).

To simulate the heat transfer by forced convection, a model of κ–ϖ turbulence with Shear Stress Transport option was used. It is most often recommended in literature (Heyerichs & Pollard 1996, Park et al. 2003) to describe heat-flow problems that concern the micro-jets. And for the standard collector, numerical calculations based on a κ–ε turbulence model were performed.

Heat loss on the top surface of the absorber plate was modeled using the third type boundary condition (Eq. 1). Solar Ray Tracing Algorithm (Tan & Chen 2010) was used to determine heat gain from the Sun's radiation.

$$\dot{q} = h_{c+r}\left(T_{abs} - T_a\right), \tag{1}$$

where h_{c+r} = total heat transfer coefficient, taking into account free convection and radiation, T_{abs} = absorber plate temperature, T_a = ambient temperature.

Geometry and calculation meshes were generated with the use of the GAMBIT preprocessor. The flow domain was divided using a mesh with the following parameters: number of control volumes—500 000 (model I) and 1 325 000 (model

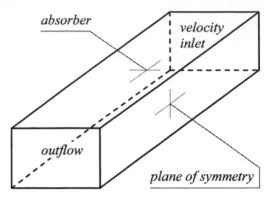

Figure 2. Model of the traditionally designed collector (model I).

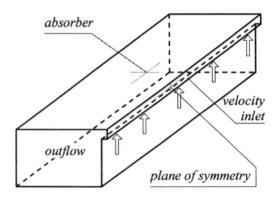

Figure 3. Model of the collector using the slot jet (model II).

II), number of surfaces—1 522 500 (model I) and 4 020 300 (model II), number of nodes—522 801 (model I) and 1 370 711 (model II). Calculations and processing of the results were performed using the latest version of a ANSYS Academic Research CFD software package, r. 14.

3 RESULTS AND DISCUSSION

Simulation calculations were performed with a constant value of air mass flow rate of 0,001505 kg/s. Air flow was not changed due to the fact that in most cases ventilators with a constant rotational speed are used. Constant temperature of the ventilated air, and hence constant ambient temperature of 20°C, was also assumed, due to the fact that the analyzed devices are designed mainly for operation in the summertime. The value that was changed, as it is in the real operating conditions, was the total rate of solar radiation. Four values were assumed, only with minimal differences for both devices; these are outlined in Table 1. The absorption coefficient of the absorber plate surface was assumed to be 0,9.

Analysis of the air velocity field in the cross-section of the standard collector shows that the highest value is present in the central part of the channel. It is not beneficial from the point of view of the amount of heat exchange. For the second collector, the high air velocity is observed near the absorber plate, which, of course, is beneficial for intensification of the heat transfer.

The velocity distribution discussed above has a significant influence on the temperature field in the flow channels of the collectors. In case of the traditional solution (Fig. 4), a phenomenon of thermal stratification is presented. The warm air zone is located directly below the absorber plate. This results in limitation of heat exchange.

A big advantage of micro-jet technique use is the introduction of a cold air directly onto the surface of the heated absorber plate (Fig. 5). Owing to the circular air movement, we observe a relatively uniform distribution of the temperature in almost the whole cross-section of the channel. Analysis of the absorber plate temperature field in the standard collector (Fig. 6) shows its significant gradient in the direction of the flow. It is connected with

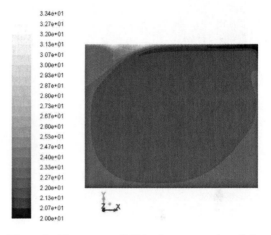

Figure 5. Temperature field in the cross-section of the micro-jet collector.

Table 1. Solar energy conversion efficiency for the analyzed collectors.

Collector type	G_{sol} [W/m²]	T_{out} [°C]	T_{abs} [°C]	η [%]
Standard collector	260.0	20.77	27.18	72.1
Microjet-based collector	260.7	20.90	22.87	83.8
Standard collector	561.3	21.66	35.56	71.4
Microjet-based collector	563.3	21.93	26.39	82.9
Standard collector	772.5	22.28	41.47	71.5
Microjet-based collector	775.3	22.64	28.81	82.5
Standard collector	1035.4	23.07	48.88	71.7
Microjet-based collector	1039.3	23.55	31.80	82.6

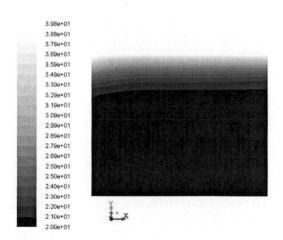

Figure 4. Temperature field in the cross-section of the standard collector.

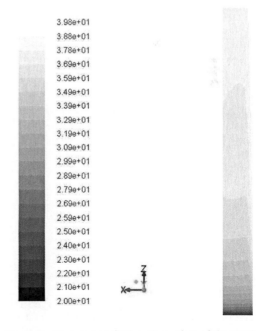

Figure 6. Temperature field on the surface of the standard collector.

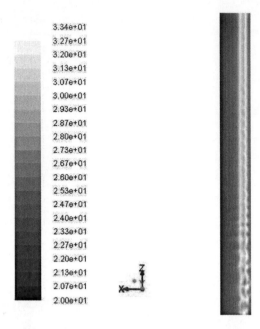

Figure 7. Temperature field on the surface of the micro-jet collector.

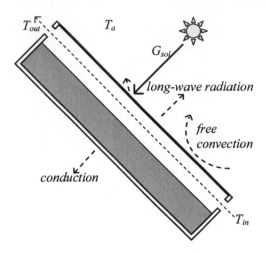

Figure 8. Solar air collector energy balance.

the increase of the temperature of the heated air, which results in the decrease of amount of the heat exchange. Absorber plate of the micro-jet collector (Fig. 7) has a gradient across the flow direction. The lowest temperature is observed at the point where the jet impinges on the plate. The highest value appears close to the channel wall. A characteristic lighter strip results from the decrease in heat exchange intensity in the area of air jet backing up.

4 COMPARISON OF THE ANALYZED SOLAR AIR COLLECTOR CONSTRUCTIONS

Air collector energy balance is presented schematically in Figure 8. Solar radiation falling onto the absorber plate is converted into heat, which results in the increase of the plate temperature T_{abs}. As a result of contact with the internal surface of the absorber plate, air flowing through the collector channels is heated from the initial temperature T_{in} (equal to the ambient temperature T_a) to the temperature T_{out} measured at the outflow. To calculate the effective amount of heat, the total amount of solar energy falling onto the collector must be decreased by energy loss. It includes loss of heat via conduction by the bottom casing of the device and also heat loss by the absorber plate and

optional top glazed shield of the collector, which results from the free convection and radiation. One must also take into account partial reflection of solar radiation due to the fact that the surface onto which the radiation falls is not a perfect black body.

Hence, heat efficiency of the analyzed collectors can be defined as a ratio of the rate of energy absorbed by the air flowing through the collector to the total rate of solar radiation falling onto the gross area of the collector.

$$\eta = \frac{Q_u/A_k}{G_{sol}}, \qquad (2)$$

where Q_u = useful energy rate, A_k = adsorption area (gross) of the collector, G_{sol} = total rate of the solar radiation, which is a sum of direct, diffuse and ground-reflected radiation. Amount of useful energy gained by the collector was determined with the use of the following formula:

$$Q_u = m_p c_p \left(T_{out} - T_{in}\right), \qquad (3)$$

where m_p = air mass flow rate, c_p = specific heat of air.

Results of efficiency calculations and basic operational parameters for both devices compared are summarized in Table 1. As it can be seen, the greatest differences occur in the temperature of the absorber surface. Their value increases in the range of 16%–35% along with the increase of solar radiation rate. It must be noted that the lower temperature of the plate, the more intensive heat transfer. Use of micro-jet technology allows to increase the

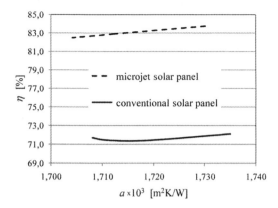

Figure 9. Dependency of the collector efficiency on parameter a.

efficiency of the solar collector significantly. In Figure 9, the η value is shown as a function of a characteristic parameter a, often used in the literature. The parameter is defined by Eq. 4. Increase of the efficiency, as compared to the traditional solution, is significant and on average equal to 13,6%.

$$a = \frac{\left(\dfrac{T_{in} + T_{out}}{2} - T_a \right)}{G_{sol}}. \tag{4}$$

5 CONCLUSION

The first of two types of solar air collectors has a construction most often used in mass-produced devices. When it comes to the second device, developed by the author, it is based on the micro-jet technique. In calculation of heat transfer, algorithms of the computational fluid dynamics were used.

Results of the analysis presented in the article are indicative of the fact that use of slot micro-jets impinging on the heat exchange surface may significantly improve energy efficiency. On the basis of calculations, higher efficiency of a micro-jet collector, on average by 13,6%, as compared to the device with the traditional construction, was found.

The article presents a report of the first stage of a project aiming at determination of energy efficiency of the proposed construction solution. The next stage of the project shall include building of a prototype micro-jet collector and performance of experiments in laboratory and real-life conditions.

ACKNOWLEDGEMENTS

The work was performed within the framework of a grant from the Ministry of Science and Higher Education N N523 615539—"Study of a heat exchanger based on microjets impinging on a heat exchange surface"—2010–2013.

REFERENCES

Augustus, M.L. & Kumar, S. 2007. Mathematical modeling and thermal performance analysis of unglazed transpired solar collectors. *Solar Energy* 81: 62–75.

Belusko, M. et al. 2005. Analysis of a roof integrated solar air collector. Proc. of the International Solar Energy Society World Congress, 6–12 August 2005, Orlando, Florida, USA.

Belusko, M. et al. 2008. Performance of jet impingement in unglazed air collectors. *Solar Energy* 82: 389–398.

Heyerichs, K. & Pollard, A. 1996. Heat transfer in separated and impinging turbulent flows. *International Journal of Heat and Mass Transfer* 39(12): 2385–2400.

Ho, C.-D. et al. 2011. Collector efficiency of upward-type double-pass solar air heaters with fins attached. *International Communications in Heat and Mass Transfer* 38: 49–56.

Kumar, R. & Rosen. M.A. 2011. A critical review of photovoltaic–thermal solar collectors for air heating. *Applied Energy* 88: 3603–3614.

Park, T.H. et al. 2003. Streamline upwind numerical simulation of two-dimensional confined impinging slot jets, *Heat and Mass Transfer* 46: 251–262.

Tan, T. & Chen, Y. 2010. Review of study on solid particle solar receivers. *Renewable and Sustainable Energy Reviews* 14: 265–276.

Tchinda, R. 2009. A review of the mathematical models for predicting solar air heaters systems. *Renewable and Sustainable Energy Reviews* 13: 1734–1759.

Youcef-Ali, S. & Desmons, J.Y. 2006. Numerical and experimental study of a solar equipped with offset rectangular plate fin absorber plate. *Renewable Energy* 31: 2063–2075.

Environmental Engineering IV – Pawłowski, Dudzińska & Pawłowski (eds)
© 2013 Taylor & Francis Group, London, ISBN 978-0-415-64338-2

Computer modeling of the mass transport influence on epilayer growth for photovoltaic applications

S. Gułkowski, J.M. Olchowik & K. Cieślak
Faculty of Environmental Engineering, Lublin University of Technology, Lublin, Poland

ABSTRACT: The advantage of the growth of Si layers from the Liquid Phase (LPE) is a possibility of using low temperature of the technological process. It leads to a lower concentration of the most unwanted impurities. With the use of ELO technique of silicon growth on substrate partially covered by dielectric mask, it is possible to obtain high quality Si epitaxial layers on silicon substrates with poor quality. For this reason and also due to using low-cost and simple apparatus, it seems to be a very promising method for photovoltaic applications. The main purpose in the method proposed in this paper is to determine technological parameters for which layers will be as wide and as thin as possible. It requires numerous technological experiments. Finite element analysis of the growth technique reduces number of experimental work and in consequence the cost of the optimizing process. Approach presented in this work is based on the assumption that growth is pure diffusion–controlled and mass transfer is the main process to reach thermodynamic equilibrium between the solid and liquid phase on the interface. On the basis of the diffusion of silicon in the solution the growth rate is calculated. Simulations have been carried out to obtain concentration profiles of the mass transport into grown interface of the layer for different conditions. To improve calculations near the border adaptive mesh method was used.

Keywords: computer simulation, mass transfer, epitaxial lateral overgrowth, liquid phase epitaxy

1 INTRODUCTION

In the last few years the cost of solar cell processing and module fabrication was considerably reduced. However, more than 80% of the current solar cells production requires cutting of large silicon crystal. Due to high material losses (about 50%) during this process, costs of sawing consist 29% of the total wafer production cost, and thus contribute considerably to the total module cost (Koch 2003). For this reason, the technique which could avoid the sawing step of the cell production can be a method for developing cheaper solar cells. Epitaxial Lateral Overgrowth (ELO) is one of the most promising techniques in this application due to the possibility of producing high quality Si epitaxial layers on silicon substrates with poor quality. Liquid Phase Epitaxy (LPE), which is preferred in equilibrium method for ELO use low-cost and simple apparatus which decrease costs of the crystal production.

ELO is a method of epitaxial growth on substrate partially covered by 100 nm thick dielectric mask. With the use of a conventional photolithography and etching mask-free seeding windows (opened windows) are created. It is the place where epitaxial growth begins. As soon as the growing crystal-solution interface exceeds the top layer of the mask, it proceeds in lateral and normal direc-

tion with different growth rates. If long enough time is given, a new epitaxial layer fully covers the masked substrate. The purpose of using dielectric mask deposited on growth substrate is to reduce the density defect in the new layer (Zytkiewicz 1999, Nishinaga 2002, Dobosz 2003). Growth rate depends on technological conditions of the experimental process such as cooling rate, temperature and geometry of the system. The main purpose in ELO method is to determine parameters for which ELO layers will be as wide and as thin as possible. It requires numerous technological experiments. For this reason, numerical analysis seems to be a promising way of finding parameters because it reduces a number of experimental work and, in consequence, the cost of the optimizing process. Computational calculations can also lead to better understanding of the growth mechanism in ELO from solution, so increasing number of simulation works can be observed in literature. However, there are not many works about growth of Si from Si-Sn solution. What is more, existing models are based on the assumption that growth is both diffusion and kinetic limited. It means that different boundary conditions are used on the two faces of layer growth (Liu 2005, Yan 2000, Liu 2004).

The approach presented in this work is based on the assumption that growth is pure diffusion–

controlled and mass transfer is the main process to reach thermodynamic equilibrium between the solid and liquid phase on the interface. The diffusion of silicon in the solution is the only factor which limits the growth rate of the layer (Moskvin 2007). In order to calculate the movement of the interface, the solute concentration in Si-Sn solution was determined after each time step. New interface position was obtained by calculating growth rate in normal to interface. The growth rate was computed from concentration gradients near the growth surface. To solve the mass transfer problem the finite element method was used. Solid-liquid front was tracked explicitly while diffusive transport was solved on a triangle grid. Simulations were carried out to obtain the evolution of the Si epilayer interface for given conditions. To improve the concentration profile near the border of grown layer adaptive mesh was used.

2 COMPUTATIONAL MODEL

A two-dimensional computational domain for ELO growth is shown in Figure 1. As it can be seen, it consists of two grids: triangle FEM mesh, which keeps information about Si concentration in the solution, and moving interface grid which is represented by the Lagrangian set of connected points. Interface nodes change of their position in time according to the calculated growth rates in the normal interface. Information about the position of moving grid on the one hand, and concentration field in the vicinity of the interface on the other, must be passed between these two grids.

Concentration field of solute in the computational domain is obtained by solving the 2D diffusion equations in time:

$$\frac{\partial C(\xi,t)}{\partial t} = D_{Si} \nabla^2 C(\xi,t), \qquad (1)$$

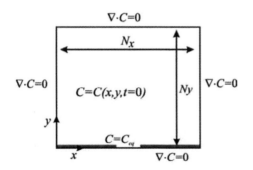

Figure 1. Schematic view of the system configuration used for the modeling of interface evolution during ELO growth.

where $C(\xi, t)$ is the mass fraction of the solute and D is the diffusion coefficient of solute, which can be taken from the Kimura's paper (1994). The initial concentration C_0 for all elements is set as equilibrium concentration at the starting temperature T_0 and it is obtained from the phase diagram of the solution. The phase diagram relation was obtained by fitting a polynomial into experimental points taken from work of (Mauk 2007). During the growth process, the temperature T decreases with time according to the formula $T = T_0 - c_r t$, where c_r is the constant cooling rate and t is the growth time. Therefore, equilibrium concentration must be calculated in each time step of the simulation. In order to solve equation (1) the finite element method was used.

Boundary conditions used in the numerical model are the following: For the two vertical walls ($x = 0$; $x = N_x$) Neumann boundary condition are used. No flux boundary conditions are used for calculations in the area between solution and oxide mask (2):

$$\left.\frac{\partial C}{\partial x}\right|_{\substack{x=0 \\ x=N_x}} = 0, \quad \left.\frac{\partial C}{\partial y}\right|_{y=N_y} = 0, \qquad (2)$$

Concentration at the upper liquid surface was set as C_o. For the grown layer interface following equation must be fulfilled:

$$\left. D_{Si} \frac{\partial C}{\partial n}\right|_L = v_n \cdot (C_{eq}^s - C_{eq}), \qquad (3)$$

where C_{eq}^s and $C_{eq}(T)$ are the equilibrium concentrations of components at the interface on the sides of the solid and liquid phases.

As it can be concluded from equation (3), growth rate v_n can be determined from the gradient of concentration in normal direction to the local curvature of the interface. It should be emphasized that the calculations of concentration field in the vicinity of the growing layer have to be very precise. For this reason adaptive mesh method was applied.

After the calculation of growth rate each point of the Lagrangian grid is moved to the new location according to the relation:

$$x^{t+1} = x^t + v_x \cdot \Delta t, \qquad (4)$$

$$y^{t+1} = y^t + v_y \cdot \Delta t, \qquad (5)$$

where v_x and v_y are x and y component of the growth rate, Δt—time step used in the calculation. As the interface moves, new points must be created or deleted to maintain given distance between two adjacent points of the interface grid. The idea of the calculation method presented in this paper has been taken from the basics of front tracking method. Detailed description and applications of the Front Tracking Method can be found in following papers: (C.Y. Li 2003, Y. Yang 2005, M. Muradoglu 2008).

3 RESULTS AND DISCUSSION

Computer simulations were carried out for Si ELO layer grown by LPE on Si partially masked substrate. Initial temperature of the system was selected as 920°C. Diffusion coefficient D came to $D = 5.0 \cdot 10^{-5}$ cm^2/s. Size of the window was 0.005 cm (50 μm). All calculations were made for the first few minutes of real time growth for cooling rate equals 0.5°C/min. Figures 2 and 3 present

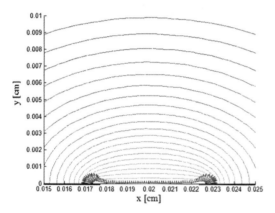

Figure 2. Iso-concentration lines of Si in Si-Sn solution, interface of the layer and growth rate after 60 s of growth.

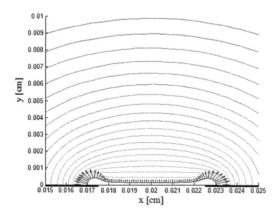

Figure 3. Iso-concentration lines, interface of the layer and growth rate after 150 s of growth.

Si concentration field, the position of interface and calculated growth rate near the layer in the two early stages of growth process. Figure 2 shows the concentration lines after 60 s of simulation with initial temperature $T_0 = 920$°C and cr = 0.5°C/min. Figure 3 presents the configuration of the system after 2.5 min of growth.

Due to the fact that the system temperature decreases with time, growth of the ELO layer can be observed. The concentration of Si species in the area of opened window decreases. The difference between concentration of Si near the interface and in faraway from its causes flux of species. It can be concluded from concentration contour map presented in Figures 3 and 4. It should be noted that the stream of solute is perpendicular to the iso-concentration lines. Due to no flux boundary condition for the mask region all Si species move from the mask area towards the region of the grown layer. It leads to higher gradient concentration in the region of layer edges. It can be visible very clearly in Figures 4 and 5 by the density of contour lines in this region. The system configuration with triangle mesh presented in Figures 5 and 6 proves high precision of calculation of the concentration profiles near the interface of the layer.

On the basis of concentration gradient, growth rate of the interface can be determined from Eq. (3). Higher concentration gradient at the edge of the layer leads to higher growth rate in that region in comparison with the planar one. It has been pointed out by arrows in above Figures. The growth rate in presented system is about six times larger for points located at the edge of the layer. Higher growth rate leads to the faster growth in this region.

Figure 6 shows the evolution of the interface of Si ELO layer calculated for 4.0 min of growth. Time distance between lines is 30 s. It can be seen that the higher growth rate leads to the faster growth in this region if we compare the distance between two adjacent lines. After 4 min of real time growth thickness of the layer in flat face is about 0.0002 cm (2 μm), whereas width is about 0.0007 cm (7 μm). It leads to aspect ratio equaled nearly 3 in this early stage of growth.

Shape of the layer presented in Figure 6 was calculated according to the pure diffusion—limited model of growth without the contribution of the Gibbs-Thomson effect. Lack of this effect leads to the creation of characteristic bump of Si at the edge of the Si window. It should be mentioned that creation of the bump is not observed in experimental observations at final stage of growth (for higher time than used in this paper). For this reason including influence of the interface curvature on the equilibrium in this model seems to be significant and it may determine final shape of the layer.

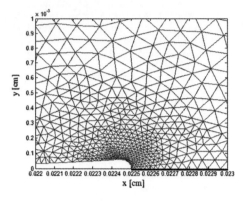
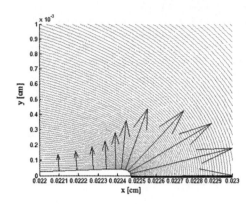

Figure 4.   Triangle mesh, iso-concentration lines of Si in Si-Sn solution and growth rate calculated for interface after 60 s of growth.

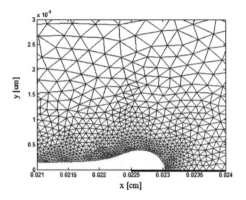
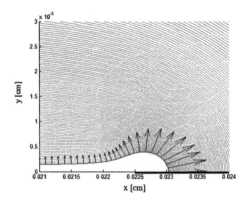

Figure 5.   Triangle mesh and iso-concentration lines of Si in Si-Sn solution and growth rate calculated for interface after 160 s of growth. Higher growth (bump) can be observed at the edge of the layer.

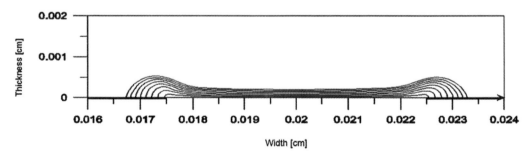

Figure 6.   Time evolution of the growth interface of the layer. Time interval between lines is equal 30 s. Initial temperature was selected as 920°C and cooling rate was equal 0.5°C/min.

4   CONCLUSION

Two dimensional computer simulations of ELO layer growth from liquid were carried on the basis of pure diffusion model without the contribution of the Gibbs-Thomson effect. Interface growth rate calculations were performed from the concentration gradient in the normal direction to the interface in the vicinity of the surface. Interface evolution during the growth was investigated and the results of calculation in the early stage of growth process were shown. It was pointed out that ELO layer started to growth laterally faster than in vertical direction due to geometry of the growth substrate, which means presence of the no flux mask regions on the substrate. Lack of the Gibbs-Thomson's effect in the model leads to the creation of characteristic bump of Si on the edge of the Si window, which is not observed in the experiments in the final stage of growth. It leads to the conclusion that the curvature of the interface may change the thermodynamic equilibrium in the area near the layer, and in consequence determine shape of the layer.

REFERENCES

Capper P., Mauk M. 2007. Liquid Phase Epitaxy of Electronic, Optical and Optoelectronic Materials, John Willey & Sons.

Dobosz D. et al. 2003. Epitaxial lateral overgrowth of GaSb layers by liquid phase epitaxy. Journal of Crystal Growth 253: 102–106.

Kimura M., Djilali N., Dost S. 1994. Convective transport and interface kinetics in liquid phase epitaxy. Journal of Crystal Growth 143: 334–348.

Koch. W. et al., 2003. Bulk Crystal Growth and Wafering for PV w Handbook of Photovoltaic Science and Enginieering (ed. Antonio Luque, Steven Hegedus) Wiley.

Li C.Y., Garimella S.V., Simpson J.E. 2003. Fixed-grid front-tracking algorithm for solidification problems, part I: method and validation. Numerical Heat Transfer, Part B, 43: 117–141.

Liu Y.C., Zytkiewicz Z.R., Dost S. 2005. Computational analysis of lateral overgrowth of GaAs by liquid-phase epitaxy. Journal of Crystal Growth 275: e953–e957.

Liu Y.C., Zytkiewicz Z.R., Dost S. 2004. A model for epitaxial lateral overgrowth of GaAs by liquid-phase electroepitaxy Journal of Crystal Growth 265: 341–350.

Moskvin P.P., Khodakovskii V.V. 2007. The Diffusion Kinetics in Isothermal Epitaxy of CdxHg1-xTe Solid Solutions from an Unlimited Liquid Phase Volume. Russian Journal of Physical Chemistry A Vol 81, 11: 1845–1850.

Muradoglu M., Tryggvason G. 2008. A front-tracking method for computation of interfacial flows with soluble surfactants Journal of Computational Physics 227: 2238–2262.

Nishinaga T. 2002. Microchannel epitaxy: an overview. Journal of Crystal Growth 237–239: 1410–1414.

Yan Z., Naritsuka S., Nishinaga T. 2000. Two-dimensional numerical calculation of solute diffusion in microchannel epitaxy of InP Journal of Crystal Growth 209: 1–7.

Yang Y., Udaykumar H.S. 2005. Sharp interface Cartesian grid method III. Solidification of pure materials and binary solutions. Journal of Computational Physics 210: 55–74.

Zytkiewicz Z.R., 1999. Epitaxial Lateral Overgrowth of GaAs: Principle and Growth Mechanism Cryst. Res. Technol. 34: 573–582.

Environmental Engineering IV – Pawłowski, Dudzińska & Pawłowski (eds)
© *2013 Taylor & Francis Group, London, ISBN 978-0-415-64338-2*

The investigation of electron transfer mechanism in dye-sensitized solar cells

A. Zdyb
Faculty of Environmental Engineering, Lublin University of Technology, Lublin, Poland

S. Krawczyk
Faculty of Mathematics, Physics and Computer Science, Maria Curie-Sklodowska University, Lublin, Poland

ABSTRACT: The electron transfer from sensitizer molecule to TiO_2 nanoparticle plays a very significant role in controlling the efficiency of Dye-Sensitized Solar Cells (DSSC) and its better understanding is essential for the improvement of DSSC performance. In this work several sensitizers namely: catechol, 9-anthracenecarboxylic acid (9-ACA), *all-trans* retinoic acid (ATRA) and *all-trans* bixin were studied in the context of application in Dye-Sensitized Solar Cells (DSSC). The authors analyzed absorption spectra at room temperature as well as absorption and electroabsorption spectra obtained by Stark spectroscopy at low temperatures. The main result of the spectroscopic method used is the dipole moment change which is the parameter providing quantitative information about the degree of instantaneous electron transfer that takes place in the course of the photon absorption. The highest value has been obtained for ATRA and it is about twice as much as that for other substances.

Keywords: TiO_2, charge-transfer, dye-sensitized solar cells, Stark effect

1 INTRODUCTION

Due to population growth, technology development and the urbanization of the third world power needs of our civilization remain at approximately 15 TW. Fossil fuels combustion cannot be a long-term solution because of the limited resources and serious environmental consequences. Undoubtedly, people need to conserve energy, seek environmentally clean alternative resources and develop sustainable energy conversion processes because by year 2050 we will need an additional 10–30 TW of power (Kamat 2000, Michałowski 2012, Pimentel 2012). Among many renewable energy sources, solar energy and its conversion to electric power in photovoltaic effect stands out as the most reliable one to meet our energy demand. The total solar energy that reaches the Earth surface in one day could power the planet for an entire year. The costs of photovoltaic technology continue to decrease while the average cost of grid-supplied electricity will continue to rise due to increased demand and fossil fuel prices.

Nowadays we can distinguish four generations of solar cells. The first generation—silicon based photovoltaic devices that accounts for 85% of the market, the second one—thin film devices, the third one—organic and DSSC and the fourth generation—Quantum Dot Sensitized Solar Cells (QDSSC). Crystalline Si and amorphous Si cells have the champion efficiency of 25% and 12% respectively (Nofuentes et al. 2011, Green et al. 2012, Greenpeace and EPIA Report 2011) but these values do not improve and remain stable for years. Nowadays the third generation solar devices based on new technologies and new materials, especially dye-sensitized solar cells (O'Regan & Grätzel 1991, Archer & Nozik 2010), are very promising. Recently DSSC has achieved 12.3% of efficiency under AM1.5 global sunlight (Yella et al. 2011) and this type of photocells has important advantages in consumer applications:

- flexibility,
- performing in bad light conditions,
- diversity of colors,
- interesting possibilities as Building Integrated Photovoltaics (BIPV),
- cheap materials and simple construction lowers manufacturing cost that is expected to reach 0.5 €/W by 2020.

Operating of DSSC is based on well-known sensitization process (Gerischer 1969) in which dye molecules absorb solar radiation in the range of visible light spectrum and electrons are injected from photoexcited dye into the conduction band of the semiconductor. Titanium dioxide is a favourable semiconductor in this type of photocell because it is cheap, environmentally friendly material with a large band gap of 3.1 eV. The main part of DSSC is an electrode covered by the layer of TiO_2 nanoparticles with photosensitizing dye molecules adsorbed to the surface. Upon visible light the following steps involving the Sensitizer (S) and the redox system in the adjacent electrolyte take place (Hara & Arakawa 2003):

– absorption of a photon

$$S + h\nu \rightarrow S^*$$

– injection from the dye excited state into semi-conductor conduction band

$$S^* \rightarrow S^+ + e^-$$

– regeneration of the dye

$$S^+ + e^- \rightarrow S$$

– reduction of I_3^- ion

$$I_3^- + 2e^- \rightarrow 3I^-.$$

The interfacial electron injection from the dye to the conduction band of TiO_2 is of great importance and has been studied extensively (Huber et al. 2002, Tae et al. 2005, Martini et al. 1998a,b). The electron transfer is the process directly influencing the efficiency of DSSC.

In this work we apply Stark effect spectroscopy (Liptay 1969, Bublitz & Boxer 1997, Nawrocka and Krawczyk 2008) as a method providing information about redistribution of electrons and electron transfer in the molecule-solid system. We have investigated the excited state of different sensitizers: catechol, which adsorbs to TiO_2 nanoparticles with the dihydroxy functions, as well as 9-anthracenecarboxylic acid (9-ACA), *all-trans* retinoic acid (ATRA) and *all-trans* bixin which bind with the carboxy group.

2 MATERIALS AND METHODS

Catechol, 9-ACA and ATRA were purchased from Sigma-Aldrich and their solutions were obtained in absolute (99.8%) ethanol. Bixin was extracted from annatto seeds in a few steps (Montenegro et al. 2004). First the seeds were boiled in ethyl acetate, then precipitated, filtered and subjected to TLC using chloroform:methanol 95:5 (v/v). This procedure gave *9'-cis* bixin that needed further treatment to obtain *all-trans* bixin. The extract was subject to photoisomerization in ethyl acetate in the presence of bengal rose dye, while cooling to 15 °C under irradiation with white light through a heat absorbing filter. Figure 1 presents molecular structures of the investigated molecules.

Coloidal TiO_2 nanoparticles in absolute ethanol were prepared by hydrolysis of titanium tetraiso-propoxide (Kamat et al. 1994).

The ethanolic solutions of the sensitizers were added to TiO_2 colloid to achieve adsorption. Stable absorption spectra in room temperature were obtained in a few minutes and they were recorded by Shimadzu UV-160 A spectrophotometer. The experimental setup used in Stark spectroscopy consists of:

– optical flow cryostat (Optistat, Oxford Instruments),
– grating monochromator (SPM-2, Zeiss) controlled by a computer,
– 150 W halogen or xenon lamp,
– voltage amplifier,
– digital lock-in amplifier (SR 830 DSP, Stanford Research Systems).

Sinusoidal electric voltage of 800–1200 V r.m.s. was applied to the samples that were built from two glass windows with conductive transparent layers and a thin layer (0.08 mm) of colloidal TiO_2 solution with adsorbed sensitizer frozen between them.

We have obtained absorption and electric field-induced electroabsorption (Stark effect) spectra for isolated molecules (without TiO_2) and molecules adsorbed on TiO_2 nanoparticles. The spectra were obtained at low temperatures 85–95 K. Experimentally obtained electroabsorption spectra present changes in absorbance ΔA which come from the change in electronic transition energy ΔE induced by the external electric field $\vec{F}$ applied to the sample:

$$\Delta E = \Delta E_e - \Delta E_g = -\Delta\vec{\mu} \cdot \vec{F} - \frac{1}{2}\left(\vec{F} \cdot \Delta\hat{\alpha} \cdot \vec{F}\right), \quad (1)$$

where $\Delta\mu = \mu_e - \mu_g$ is the permanent dipole moment change in the transition from the ground (g) to the excited (e) state, $\Delta\alpha = \alpha_e - \alpha_g$ is the polarizability difference, and μ_e, α_e—dipole moment and polarizability in the excited state, μ_g, α_g—parameters of the ground state. The values of $\Delta\mu$ and $\Delta\alpha$ characterize electron density redistribution between the ground and excited states in the system. The change of a given state energy (ground or excited) is described by the following equation:

$$\Delta E_{state} = -\vec{\mu} \cdot \vec{F} - \frac{1}{2}\left(\vec{F} \cdot \hat{\alpha} \cdot \vec{F}\right) \quad (2)$$

which has two components:

$$\Delta E_{state} = \Delta E' + \Delta E''$$

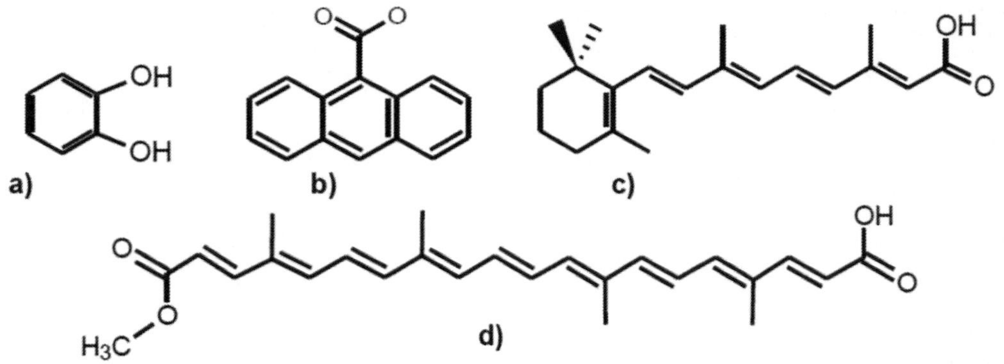

Figure 1. Molecular structures: a) catechol, b) 9-anthracenecarboxylic acid (9-ACA), c) all-trans retinoic acid (ATRA), d) all-trans bixin.

ΔE′ comes from the influence of electric field on the permanent dipole moment, ΔE″ is connected with induced dipole moment $\vec{\mu}_{ind}$. The changes of permanent dipole moment and of polarizability in the presence of an applied electric field cause the band shift effect that results in the change of absorbance ΔA measured in the experiment. Following the Stark effect theory, electroabsorption spectra were then fitted by the linear combination of the first and second derivatives of the absorption (Bublitz & Boxer 1997):

$$\Delta A(\tilde{\nu}) = a_1 \tilde{\nu} \frac{d\left(A(\tilde{\nu})/\tilde{\nu}\right)}{d\tilde{\nu}} + a_2 \tilde{\nu} \frac{d^2\left(A(\tilde{\nu})/\tilde{\nu}\right)}{d\tilde{\nu}^2} \qquad (3)$$

Coefficients a_1 and a_2 were then used to estimate Δμ and Δα values according to the following formula (Bublitz & Boxer, 1997):

$$a_1 = \frac{\Delta\alpha f^2 F^2}{10\sqrt{2}hc}\left[(3\cos^2\gamma - 1)\cos^2\chi + 2 - \cos^2\gamma\right]$$

$$a_2 = \frac{(f\Delta\mu)^2 F^2}{10\sqrt{2}h^2c^2}\left[(3\cos^2\delta - 1)\cos^2\chi + 2 - \cos^2\delta\right],$$

$$(4)$$

where γ is the angle between the polarizability axis and the transition moment vector, δ is the angle between the dipole moment change Δμ̄ and the transition moment vector, χ—the angle between the electric field vector of the light and the applied electric field vector, f—local field factor, f≈1,1–1,3 (Bublitz & Boxer 1997). Multiple scanning and signal averaging were employed in collecting the spectra. The values of electroabsorption parameters presented in results section are the mean values obtained for several samples and they are given with the standard deviation.

3 RESULTS AND DISCUSSION

All four sensitizers investigated in this work: catechol, 9-anthracenecarboxylic acid (9-ACA), all-trans retinoic acid (ATRA) and all-trans bixin bind to TiO₂ nanoparticles creating complexes which can be monitored by their characteristic absorption. Titanium dioxide absorbs light below 400 nm but the absorption of the investigated substances occurs in the visible spectrum, which is the fundamental requirement for the sensitization process to be effective. Upon binding to TiO₂ nanoparticles the red shift is observed in Figure 2. Absorption of catechol/TiO₂ complex is especially interesting because of the formation of a new absorption band at 400 nm which has a nature of Charge Transfer (CT) transition. In this case the electron injection pathway is direct from the ground state of the sensitizer into the TiO₂ conduction band. Photoexcitation of CT band result in a very fast electron injection (<1 ps) (Matylitsky et al. 2006, Gundlach et al. 2006, Ramakrishna et al. 2004) but it does not improve the overall photovoltaic performance of DSSC (Tae et al. 2005). The detailed analysis of sensitizer-TiO₂ complexes is made possible by using Stark spectroscopy—the method sensitive to the band structure of the absorption spectrum. Here we present absorption and electroabsorption spectra of ATRA and ATRA/TiO₂ complex as an example of our results (Krawczyk & Zdyb 2011). This sensitizer is characterized by the highest value of the dipole moment change that is favourable for the DSSC application. The absorption and electroabsorption spectra of ATRA in its free form (Fig. 3) serve as a reference in evaluation of the changes in electronic structure caused by its adsorption to nanoparticles. Panel A presents the absorption spectrum, B—the electroabsorption spectrum (squares are the experimental data

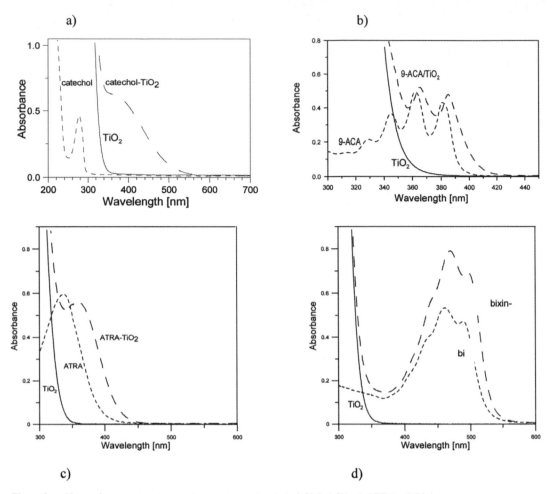

Figure 2.   Absorption spectra at room temperature: a) catechol, b) 9-ACA, c) ATRA, d) bixin.

and the line is the fitting curve), C—absorption derivatives that are components of the fit in panel B obtained according to Eq. (3, 4). The spectra of ATRA adsorbed on TiO_2 shown in Figure 4 are similar in shape to free ATRA but they are shifted about 1000 cm^{-1} to lower energies and no signs of the free pigment band can be seen, which points to the complete adsorption of ATRA. The authors observe prevailing role of the second derivative (panel C) which is due to the large value of $\Delta\mu$. The first derivative is not visible in Figure 4 since it is about 40 times smaller and as a consequence $\Delta\alpha$ cannot be reliably estimated in this case. Analogous analysis was performed on the basis of the experimental data obtained for other sensitizers. The main results of the present work, that is the values of the electrooptical parameters $\Delta\mu$ and $\Delta\alpha$—are

presented in Table 1. The dipole moment change values of catechol (Nawrocka et al. 2009), 9-ACA (Zdyb & Krawczyk 2010) and *all-trans* bixin are in the range of 7–11 D. Significantly larger $\Delta\mu$ characterize ATRA and its values are consistent with literature data for retinal which has similar molecular structure (Davidsson & Johansson 1984, Locknar & Peteanu 1998). However *all-trans* bixin which also belongs to carotenoids and is a close analogue of ATRA has the value of $\Delta\mu$ which is about two times smaller than that for ATRA (Table 1). The probable reason is slightly different molecule structure (Fig. 1) which is less symmetric in case of ATRA and promotes the electron transfer process. The computational DFT study (Ruiz-Anchondo et al. 2010) also shows the leading role of ATRA among the investigated carotenoids including bixin.

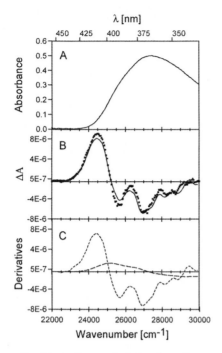

Figure 3. Absorption and electroabsorption spectra of ATRA in the free form. Temperature 85 K, voltage $1 \cdot 10^5$ V/cm, (Krawczyk & Zdyb 2011).

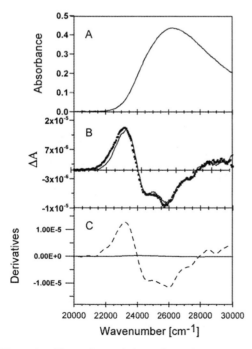

Figure 4. Absorption and electroabsorption spectra of ATRA/TiO$_2$. Temperature 85 K, voltage $1 \cdot 10^5$ V/cm.

Table 1. Electrooptical parameters of the investigated sensitizers.

Sample	$\tilde{\nu}_{(0-0)}$ [cm^{-1}], [nm]	$\Delta\alpha$ [Å^3]	$\Delta\mu$ [D]
Catechol/ TiO$_2$	24600, 407	25 ± 5	8 ± 1
9-ACA/ TiO$_2$	24900, 402	–	$10,5 \pm 0,4$
all-trans bixin/ TiO$_2$	19200, 521	1150 ± 150	$7,1 \pm 0,4$ (high energetic form) $11 \pm 0,6$ (low energetic form)
ATRA/ TiO$_2$	24650, 406	–	$22,5 \pm 1$ (high energetic form) ~30 D (low energetic form)

4 CONCLUSIONS

All the molecules examined: catechol, 9-anthracenecarboxylic acid (9-ACA), *all-trans* retinoic acid (ATRA) and *all-trans* bixin adsorb to TiO$_2$ nanoparticles creating complexes than can be useful in dye-sensitized solar cells.

Stark spectroscopy provides quantitative information about the degree of electron transfer and that makes this technique highly adequate in investigations of the sensitizers that are used in DSSC. The study was performed in low temperatures in which the increased spectral resolution could be observed.

Comparison of the values of dipole moment change for different sensitizers enables us to order the investigated sensitizers according to the lowering degree of electron transfer to TiO$_2$ as the following: ATRA > bixin > 9-ACA > catechol. The results suggest that ATRA is potentially the best sensibilizator among the dyes examined in this work, but its performance in DSSC may involve other processes of complex nature and depends also on other conditions that need further study.

REFERENCES

Archer, M.D., Nozik, A.J. 2010. Nanostructured and photoelectrochemical systems for solar photon conversion. London: Imperial College Press.
Bublitz, G.U., Boxer, S.G. 1997. Stark Spectroscopy: Applications in Chemistry. Biology and Materials Science. Annu. Rev. Phys. Chem. 48: 213–221.
Davidsson, A., Johansson, L.B.A. 1984. Electrochromism in viscous systems. Excited-state properties of all-trans-retinal. J. Phys. Chem. 88: 1094–1098.
Gerisher, H. 1969. Charge transfer processes at semiconductor-electrolyte interfaces in connection with problems of catalysis. Surf. Sci. 18: 97–122.

Green, M.A., Emery, K., Hishikawa, Y., Warta, W., Dunlop, E.D. 2012. Solar Cell Efficiency Tables. Prog. Photovolt: Res. Appl. 20: 12–20.

Greenpeace and EPIA Report, Solar Generation 6, 2011. Solar Photovoltaic Electricity Empowering the World, (www.greenpeace.org).

Gundlach, L., Ernstorfer, R., Willig, F. 2006. Escape dynamics of photoexcited electrons at catechol:TiO2. Phys. Rev. B 74: 035324.

Hara, K., Arakawa, H. 2003. Dye-sensitized solar cells. In A. Luque & S. Hegedus (eds), Handbook of Photovoltaic Science and Engineering: 663–696. London: Wiley.

Huber, R., Moser, J.E., Grätzel, M., Wachtveitl, J. 2002. Real-Time Observation of Photoinduced Adiabatic Electron Transfer in Strongly Coupled Dye/Semiconductor Colloidal Systems with a 6 fs Time Constant. J. Phys. Chem. B 106: 6494–6499.

Kamat, P. 2007. Meeting the Clean Energy Demand: Nanostructure Architectures for Solar Energy Conversion. J. Phys. Chem. C 111: 2834–2860.

Kamat, P.V., Bedja, I., Hotchandani, S. 1994. Photoinduced charge transfer between carbon and semiconductor clusters. One-electron reduction of C60 in colloidal TiO2 semiconductor suspensions. J. Phys. Chem. 98: 9137–9142.

Krawczyk, S., Zdyb, A., 2011. Electronic Excited States of Carotenoid Dyes Adsorbed on TiO2. J. Phys. Chem. C 115: 22328–22335.

Liptay, W. 1969. Electrochromism and solvatochromism, Angew. Chem. Internat. Edit. 8: 177–188.

Locknar, S.A., Peteanu, L.A. 1998. Investigation of the Relationship between Dipolar Properties of cis-trans Configuration in Retinal Polyenes: A Comparative Study Using Stark Spectroscopy and Semiempirical Calculations. J. Phys. Chem. B 102: 4240–4246.

Martini, I., Hodak, J.H., Hartland, G.V. 1998a. Effect of Structure on Electron Transfer Reactions between Anthracene Dyes and TiO2 Nanoparticles. J. Phys. Chem. B 102: 9508–9517.

Martini, I., Hodak, J.H., Hartland, G.V. 1998b. Effect of Water on the Electron Transfer Dynamics of 9-Anthracenecarboxylic Acid Bound to TiO2 Nanoparticles: Demonstration of the Marcus Inverted Region. J. Phys. Chem. B 102: 607–614.

Matylitsky, V.V., Lenz, M.O., Wachtveitl, J. 2006. Observation of pH-Dependent Back-Electron-Transfer Dynamics in Alizarin/TiO2 Adsorbates: Importance of Trap States. J. Phys. Chem. B 110: 8372–8379.

Michałowski, A. 2012. Ecosystem Services in the Light of a Sustainable Knowledge-Based Economy. Problemy Ekorozwoju/Problems of Sustainable Development 7: 97–106.

Montenegro, M.A., Rios, A.d.O., Mercadante, A.Z., Nazareno, M.A., Borsarelli, C.D. 2004. Model studies on the photosensitized isomerization of bixin. J. Agric. Food Chem. 52: 367–373.

Nawrocka, A., Krawczyk, S. 2008. Electronic excited state of alizarin dye adsorbed on TiO2 nanoparticles: A study by electroabsorption (Stark effect) spectroscopy. J. Phys. Chem. C 112: 10233–10241.

Nawrocka, A., Zdyb, A., Krawczyk, S. 2009. Stark Spectroscopy of Charge-Transfer Transitions in Catechol-Sensitized TiO2 Nanoparticles. Chem. Phys. Letters 475: 272–276.

Nofuentes, G., Munoz, J.V., Talavera, D.L., Aguilera, J., Terrados, J. 2011. Technical Handbook, in the framework of the PVs in Bloom Project, ISBN: 9788890231001.

O'Regan, B., Grätzel, M. 1991. A low-cost, high efficiency solar cell based on dye-sensitized colloidal TiO2 films. Nature 353: 737–740.

Piemental, D. 2012. Energy Production from Maize, Problemy Ekorozwoju/Problems of Sustainable Development 7: 15–22.

Ramakrishna, G., Singh, A.K., Palit, D.K., Ghosh, H.N. 2004. Dynamics of Interfacial Electron Transfer from Photoexcited Quinizarin (Qz) into the Conduction Band of TiO2 and Surface States of ZrO2 Nanoparticles. J. Phys. Chem. B 108: 4775–4783.

Ruiz-Anchondo, T., Flores-Holguín, N., Glossman-Mitnik, D. 2010. Natural Carotenoids as Nanomaterial Precursors for Molecular Photovoltaics: A Computational DFT Study. Molecules 15: 4490–4510.

Tae, E.L., Lee, S.H., Lee, J.K., Yoo, S.S., Kang, E.J., Yoon, K.B. 2005. A Strategy To Increase the Efficiency of the Dye-Sensitized TiO2 Solar Cells Operated by Photoexcitation of Dye-to-TiO2 Charge-Transfer Bands. J. Phys. Chem. B 109: 22513–22522.

Yella, A. et al., 2011. Porphyrin-Sensitized Solar Cells with Cobalt (II/III)-Based Redox Electrolyte Exceed 12 Percent Efficiency. Science 334: 629.

Zdyb, A., Krawczyk, S. 2010. Molecule-solid interaction: Electronic states of anthracene-9-carboxylic acid adsorbed on the surface of TiO2. Appl. Surf. Sc. 256: 4854–4858.

Environmental Engineering IV – Pawłowski, Dudzińska & Pawłowski (eds)
© 2013 Taylor & Francis Group, London, ISBN 978-0-415-64338-2

Thermal system diagnostics through signal modeling

D. Wójcicka-Migasiuk

Faculty of Fundamentals of Technology, Lublin University of Technology, Lublin, Poland

ABSTRACT: The paper presents the analysis of thermal system operation that is a part of a thermal power plant supplied from biofuel boilers and consists of volume heat exchangers. The measurements have been recorded in heating seasons, the models have been formulated on the real data. The presented research consists of the comparative analysis and verification of the models. The analysis has been performed by means of MathCad procedures and of the algorithms formulated individually for the purpose of the research. This diagnostics let us precise the control, minimise the lost of energy and the use of fuelsand through this reduce CO_2 emissions. The formulation of prognosis is advantageous from the point of environment and energy business. The possibility to increase the amount of energy production from biofuels generally known of medium efficiency can have an impact either on environment friendly technology or on public appreciation of biofuels.

Keywords: parametric identification models, thermal systems

1 INTRODUCTION

The rational use of energy and the sustainable development of technology belong to primary problems considered in the majority of research. The background of this research consists in the thermal—electric analogy which lets us take the advantage from both theories and apply similar models to both of them. This particular case uses electric modelling to the analysis of thermal and flow signals separately and simultaneously, and through the analysis it enables the real system diagnostics. The improved quality diagnostics in heat transport enables more efficient use of biofuels which follows more fluent response of the system temperature to demand requirements (Pieńkowski 2012). The combustion of biofuels is usually less efficient than the combustion of concentrated fuels such as e.g. gas, that is why, any improvement can benefit to the process. Moreover, broadly undertaken enterprises, research and other activities providing for the introduction and for the applications of the variety of clean development mechanisms, indicate the necessity of this type of research carried out in global dimension (Dasgupta, Taneja 2011, Shan, Bi 2012). The improvements in overall system diagnostics and control can lead to more advantageous use of this environment friendly technology of energy generation (Hoedl 2011). The general purpose for this research is included in widely understood reduction of CO_2 emissions. The paper contributes also to the reduction of global energy use and problems related to the processes of sustainable development (Wall 2013).

The analysis of real measurement results indicates the possibility to apply two basic models, i.e. a linear proportional model and a phase non-minimum model. Moreover, the classification identifies the signal as discrete, which means that it can be transformed on digital processors and this procedure can be performed with the use of Discrete Fourier Transform (DFT). This transformation is orthogonal in n-dimensional space and can be presented by means of the straight method and also by reverse equations (Worden 2011). The basic model consists of output signal and distortion signal but the input signal is recognised as non-determined. This approach let us mathematically anticipate inputs from outputs and noise, without the precise knowledge of the real input signal and its physics (Tchórzewski 2010). This is particularly important when signals are difficult to determine, for instance, such as nonlinear dynamic heat transfer, and are similar to non-measurable distortion $\eta(k)$, called color noise. There is visible auto-correlation of model error signal $\hat{\eta}(k)$ when such distortion influences the process. The following basic equation (eq. 1) describes the model initial in the analysis:

$$\hat{y}(k) = -a'_1 y(k-1) - \ldots - a'_n y(k-n) + c'_1 \hat{\eta}(k-1) + \ldots c'_n \hat{\eta}(k-n)$$

$$-\sum_{i=1}^{n} a'_i y(k-i) + \sum_{i=1}^{n} c'_i \hat{\eta}(k-1) = v(k)\theta$$

$$v(k) = -[-y(l-1), \ldots, -y(k-n), \hat{\eta}(k-1), \ldots, \hat{\eta}(k-n)],$$
$$\theta = [a'_1, \ldots, a'_n, c'_1, \ldots, c'_n]'$$

$$\tag{1}$$

where $\hat{\eta}(k)$ is signal estimation substituting unknown values of $\eta(k)/$. The dynamics of the process can be formulated by means of eq. 2 (Guinon 2010), (Škrjanc 2004).

$$y(k) = -a'_1 y(k-1) - \ldots - a_n y(k-n) + c'_1 \eta(k-1)$$
$$+ \ldots + c_n \eta(k-n) + \eta(k) = v(k)\theta + \eta(k) \qquad (2)$$

2 METHODS

The presented research shows the analysis carried out on the subject process model presented in Figure 1.

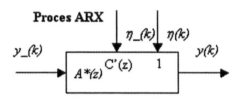

Figure 1. Simplified process model in a block diagram.

The sample of initial data to launch the model is presented in Figure 2 as a continuous series for a selected period.

The particular range of data presented, covers one month of the supply in the secondary circuit of the investigated volume exchanger. The analysis is performed on the fundament of real values received from continuous measurement and models formulated on this data basis. The system is a part of a thermal power plant supplied from biofuel boilers and consists of volume heat exchangers. The measurements have been recorded in heating seasons and collect several simultaneous parameters of operation. The graph indicates random drops caused by registering device switch offs, one switch off of the heat exchanger and regular up and downs caused by the standard control of the medium flow in the supply conduit. The temperature range is mostly between 65–75°C with one cycle visibly exceeding. The mean and often maintained temperature is 71.4°C.

The trend line determined by means of straight linear regression shows the dominant tendency for the signal. The regression analysis describes the

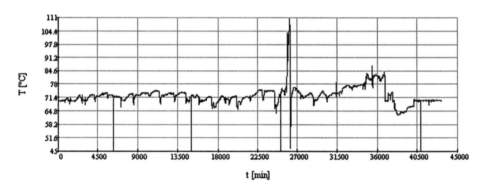

Figure 2. Data sample in selected period.

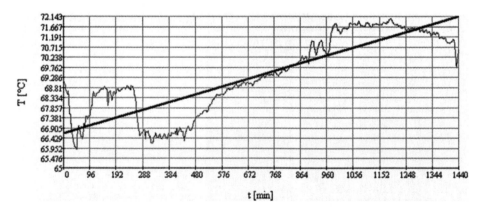

Figure 3. The trend line determined for one day.

470

dependence between the dependent variable and independent ones. The method of least squares is used to determine the trend function $y = ax + b$ where the selection of parameters for this trend is to meet the requirement that the sum of square values of experimental difference values of y_i and the calculated values of $ax_i + b$ is the least. Figure 3 shows the raising trend line determined for one day within the selected period. The definite technological explanation could only be given when other signals are analyzed, but there is no such need at this stage of research.

3 RESULTS AND DISCUSSION

It is justified from the point of view of this research to remove the trend line from signals. This procedure discovers the requested variability that carries the information on the process that is useful for further identification. The result of this procedure is presented in Figure 4.

The trend has been eliminated by means of Mathcad detrend(u) facility, which gives u-table with all trend lines removed where u is a real value (Guinon 2010), (Smith 2012). The remained set of data is then subjected to discrete Fourier transform and its reverse according to the following equation:

$$x(n) = \frac{1}{N} \sum_{k=0}^{N-1} X(k) e^{j\frac{2\pi}{N}kn} \qquad (3)$$
$$n = 0,1,2\ldots,N-1$$

The result of this computation is presented in Figure 5.

However, the physics of the signal sample had been unknown, one can guess from the DFT results that there is some over dimensioning

of the model and the confirmation for this is required. The spectrum resolution should be then increased again which is presented in Figure 6. Taking into account that sampling frequency should be higher than twice of the highest frequency component in the measured signal $f_s > 2f_{ms}$ and that the presented waveform is the image after *detrend* and DFT treatment, one can decide that *Nyquist*'s characteristic should also be performed.

The distribution range of DFT values for the presented sample of subsequent 60 measurements does not exceed 0.04 which also suggests that the over dimensioning of the model can be expected. The definite decision can be taken when the comparison between the modelled signal and the real one is available. However, it is possible that another model type (the non-minimal one) can be applied with the right dimensioning.

This solution can be obtained when polynomial coefficients are determined as presented in the further sequence of this procedure.

The polynomial coefficients let us formulate the exact equation presenting the signal waveform for the subject source, whatever it can be, without the necessity toanalyze the heat transfer model equations. This attempt to the diagnostic of thermal systems could be very advantageous.

First of all, usual practice of CA Design gives in result a fully controlled systems. However, in practice, such systems require additional control adjustments, although many of design solutions are standard and typical for their applications (e.g. heat exchanger control). The additional diagnostics through signal modeling is not cost consuming and requires only reliable procedures elaborated for separate modules of thermal systems. It can be applied such as simulation methods at every stage of construction, and would make possible

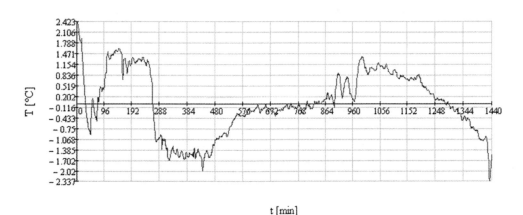

Figure 4.   The elimination of trend line from one day signal.

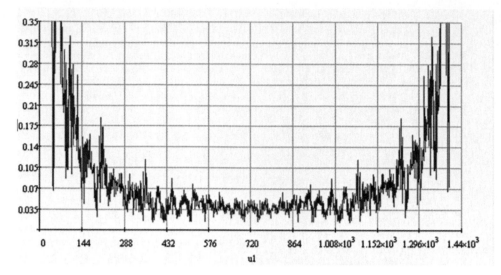

Figure 5. Reverse DFT sample results from a selected range (u1).

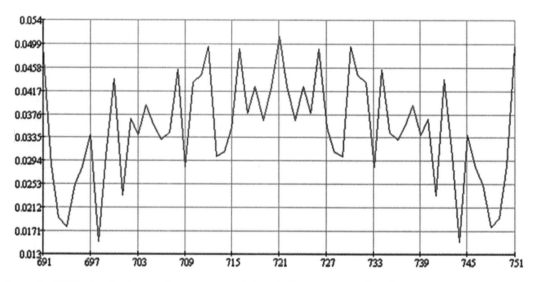

Figure 6. DFT values for the sample sequence consisting of 60 measurements (items).

to improve operational parameters of particular modules comprising the system giving the knowledge of expected output waveforms from obtained formulae based on polynomial coefficients (Worden 2011). The auto-regression model presented in Figure 1 is called ARX (AutoRegressive with eXogenous input) type MISO (Multiple Input, Single Output) and has the form of eq. 4.

$$A(q)y(t) = \sum_{i=1}^{nu} B_i(q)u_i(t - nk_i) + e(t) \qquad (4)$$

where:

$y(t)$—discrete input signal series,

$u(t)$—discrete output signal series,

nk—output—input delay, i.e. discrete step number, after which the discrete response $y(t+nk)$ is given for the discrete impulse $u(t)$,

nu—input signal number,

$e(t)$—white noise series,

$A(q) = 1 + a_1q^{-1} + \ldots + a_iq^{-i}$; $B(q) = b_0 + b_1q^{-1} + \ldots + b_jq^{-j}$ – polynomials of a_i, b_j parameters,

i, j – number of polynomial coefficients.

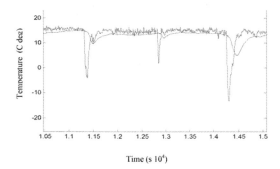

Time (s 10^4)

Figure 7. The result of temperature signal modeling (smooth line—model computations).

The mathematical tool properties of MathCAD and MatLAB have enabled to obtain the resulting formula in the form of a polynomial. The temperature model obtained in the result of the research procedure has the form of the following equation:

$$\begin{aligned}
y_1(t) = {} & 1{,}756y_1(t\text{-}1)\text{-}0{,}851y_1(t\text{-}2) + 0{,}087y_1(t\text{-}3) \\
& + 0{,}006y_1(t\text{-}4) + 0{,}23u_1(t\text{-}1) - 0{,}486u_1(t\text{-}2) \\
& + 0{,}305u_1(t\text{-}3)\text{-}0{,}486u_1(t\text{-}4) - 0{,}032u_2(t\text{-}1) \\
& + 0{,}072u_2(t\text{-}2)\text{-}0{,}041u_2(t\text{-}3) + 0{,}002u_2(t\text{-}4)
\end{aligned} \quad (5)$$

The computational model lets the author obtain the full graph of modeled signal and then, to compare it with the real one to make the assessment of the model coincidence and, thus, to verify the method. These have been presented in Figure 7 and allow the author to formulate some conclusions.

4 CONCLUSION

The conclusion at this stage of the research can be formulated on the basis of the resulting temperature graph (Fig. 7) and comments about it in relation to the physics of thermal processes occurring in heat exchanger. The suggested over-dimensioning of the model has not been confirmed but instead of this hypothesis, the necessity to formulate a more advanced non-minimum phase model has appeared. The real data (particularly considering Fig. 6) show distinctly the additional input signal forcing transient temperature changes. The authors' suggestion is that this input is caused by the continuous charging and discharging cycles, which in the case of thermal devices are related to their heating up and cooling down repetitions along the system operation during a heating period. In the research analysis it must be taken into consideration that the real data comes from a fully controlled thermal system, which neither comes to a full

discharge level nor to over-heating levels. There are undoubtedly the phases of delays, or better named, adjusting the temperature visible in the sample of 60 measurements. It is treated as the occurrence of the thermal hysteresis phenomenon. The presented signal modeling makes possible not only to confirm this but through the further application of more advanced non-minimum phase model should improve the system diagnostics. Moreover, as it is presented in Figure 7, the method of parametric modeling used for thermal flows, enables to obtain reliable temperature values along the whole heating period. This approach gives the method of prognosis for thermal system operation on the basis of selected data and then, in result, better control over the system effectiveness. These findings provide also for better use of fuel, in this case—an environment friendly biofuel.

REFERENCES

Dasgupta P., Taneja N. 2011. Low Carbon Growth: An Indian Perspective on Sustainability and Technology Transfer. *Problemy Ekorozowju/Problems of Sustainable Development* 6(1): 65–74.

Guinon J.L., Ortega E., García-Antón J., Pérez-Herranz V. 2010. Digital Filtering by means of the Fourier Transform Using Mathcad. *International Conference on Engineering Education* ICEE-2010. Gliwice. http://www.ineer.org/Events/ICEE2010/papers/W13E/Paper:959_1138.pdf.

Hoedl E. 2011. Europe 2020 Strategy and European Recovery. *Problemy Ekorozowju/Problems of Sustainable Development* 6(2): 11–18.

Pieńkowski D. 2012. The Jevons Effect and the Consumption of Energy in the European Union. *Problemy Ekorozowju/ Problems of Sustainable Development* 7(1): 105–116.

Shan S., Bi X. 2012. Low Carbon Development of China's Yangtze River Delta Region. *Problemy Ekorozowju, Problems of Sustainable Development* 7(2): 33–41.

Škrjanc I. et al. 2004. An approach to predictive control of multivariable time-delayed plant: Stability and design issues. *ISA Transactions* 43(4), Oct. 2004: 585–595.

Smith A., Luck R., Mago P.J. 2012. Integrated parameter estimation of multi-component thermal systems with demonstration on a combined heat and power system. *ISA Transactions* 51(4), Jul. 2012: 507–513.

Tchórzewski J. 2010. The Security of Polish Electrical Power System Development, Different Faces of Security from Knowledge to Management. *Institute for Security and Development Policy, Stockholm-Nacka.* Sweden 2010: 237–263.

Wall G. 2013. Exergy, Life and Sustainable Development. *Problemy Ekorozowju, Problems of Sustainable Development* 8(1): 27–41.

Worden K., Staszewski W.J., Hensman J.J. 2011. Natural computing for mechanical systems research: A tutorial overview, *Mechanical Systems and Signal Processing* 25(1), Jan. 2011: 4–111.

Environmental Engineering IV – Pawłowski, Dudzińska & Pawłowski (eds)
© 2013 Taylor & Francis Group, London, ISBN 978-0-415-64338-2

Influence of the liquid phase epitaxy conditions of growth on solar cells performance

K. Cieślak, J. Olchowik & S. Gułkowski

Faculty of Environmental Engineering, Lublin University of Science, Lublin, Poland

ABSTRACT: Liquid Phase Epitaxy (LPE) may be the only method of thin-film solar cell production that also enables the production of thin semiconductor layers on the growth substrate. This article presents results obtained by modification of the LPE method; thin Silicon (Si) layers were laterally grown on silicon substrates that were partially masked by dielectric SiO_2. This approach enabled changes in the specificity of crystallization. Moreover, a dielectric layer, present under the Si film, forms an inner mirror to photons that have not been absorbed in the active layer. The influence of the dimensions of the dielectric mirror on the short circuit current density was examined. For more reliable analysis, thin-film Si solar cells based on the LPE method, but without an inner dielectric mirror, were made and their performance compared with solar cells produced on substrates with the dielectric cover.

Keywords: thin film silicon solar cells, liquid phase epitaxy, epitaxial lateral overgrowth, inner mirror

1 INTRODUCTION

Solar cells are mainly based on semiconductor technologies. Due to economic considerations, the most popular material for solar cells is Silicon (Si). Despite the fact that Si is the second most common element in the Earth's crust, producing Si for electronic and photovoltaic applications, it is very expensive, and Liquid Phase Epitaxy (LPE) is the only method that enables economic production of thin active Si layers. Series of experiments were performed to obtain thin Si films on growth substrate that was partially masked in a specific way by dielectric. This work presents the influence of the LPE conditions and of growing substrate preparation on the performance of solar cells.

2 EXPERIMENT

As a growing substrate p+ <111> Si was used; the <111> surface orientation enables a mirror-like, homogenous layer to grow during the epitaxial process (Suzuki 1989). The p+ conductivity-type forms the back surface field (Gray 2003) in the solar cell structure. Epitaxial growth was performed on the growth substrate that was partially masked by a Silicon Dioxide (SiO_2) layer. Si growth windows (Fig. 1) were opened in the dielectric layer by a standard photolithography process (Jeager 2002). Four different photolithography masks were used to determine the influence of the dielectric surface coverage on photoconversion. These masks had

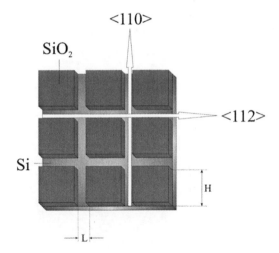

Figure 1. Schematically shown surface of a growing sample. L—width of an opened silicon growth window, H—width of a SiO_2 islands.

different surface coverage on the growing substrate: SC = 70, 75, 80 and 90%. To compare and establish the influence of the dielectric layer on photoconversion, LPE was also performed in the same thermodynamic conditions on a Si growth substrate without dielectric layer (i.e. SC = 0%). Dimensions of the opened silicon growth windows and dimensions of the SiO_2 islands are presented in Table 1.

This type of structure present on the surface of a growth substrate enables the production of

Table 1. Dimensions of the four different grids that were used on the growth substrates.

Surface coverage ratio of SiO_2 on the growth substrate [%]	Width of the opened silicon growth windows, L [μm]	Dimensions of the SiO_2 islands, H × H [μm × μm]
70	10	50 × 50
75	10	70 × 70
80	5,5	50 × 50
90	5,5	80 × 80

a)

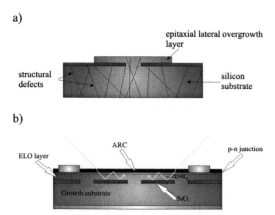

b)

Figure 2. Schematically shown growth of the lateral epitaxial layer (a), cross-section of the ELO-based solar cell with an inner mirror composed of SiO_2 (b).

a thin Si film characterized by better structural quality compared to the growth substrate alone. It is possible because the dielectric layer prevents crystallographic defects from the growth substrate propagating into the growing layer (Fig. 2a) (Nishinaga 2002, Zytkiewicz 2002). Moreover, SiO_2 trapped inside a solar cell structure forms an inner mirror, thus photons that are not absorbed in the epitaxial layer are reflected back from the SiO_2 to the active layer and so increasing absorption efficiency (Fig. 2b).

The LPE process was performed with the use of a horizontal system with a graphite cassette (Mauk 2007), with argon (Ar) as an ambient gas. A horizontal system seems to be the most reliable and is characterized by a small complexity of an apparatus for the LPE process.

The saturation temperature established was 1193 K and during growth temperature difference (ΔT) was 60 K. Tin (Sn) with a small addition (0.13% weight) of Aluminum (Al) was used as a metallic solvent. Al plays two roles during the epitaxial growth. First, it etches native oxides from

the surface of the open Si windows on the growing sample and enables growth to commence. Al is an acceptor in the Si, and so during epitaxial growth Al is introduced into the crystallographic structure of the lateral epitaxial layer and changes the material conductivity type to positive.

To examine the influence of epitaxial growth velocity on photoconversion, four different cooling rates (c.r.) were used during the series of experiments: 0.25, 0.5, 0.75 and 1 K/min. The same conditions were applied to obtain reference samples (standard LPE). The influence of the dielectric Surface Coverage ratio (SC) was also examined. SiO_2 deposited on the growth substrate forms an inner mirror inside solar cells (Figs. 1 and 2b) so coverage ratio parameter seems to have an important role in the analysis of solar cells performance. Samples obtained during the LPE process were used as a substrate for producing solar cells. All of the solar cells were produced with the same technological parameters. The standard procedure for solar cell production was:

1. Creating p–n junction in the epitaxial layer during phosphorous diffusion. Width of the n area was about 5 μm.
2. Deposition of the 80 nm SiNx anti-reflection coating on the upper surface of the samples.
3. Deposition of front and rear metallic contacts. For rear contacts aluminum was used and as for the front electrode Ti/Pl/Ag has been deposited.

After producing solar cells I-V characteristics were investigated to determine the main parameters characterizing solar cells, e.g. open circuit voltage (U_{OC}), short circuit current (I_{SC}) and efficiency. The solar cells produced had an average area of 1 cm².

3 RESULTS AND DISCUSSION

To establish the influence of the LPE technological parameters on photoconversion, a comparative parameter short circuit current density (J_{SC}) was chosen. No optimization procedure was carried out for metallic contacts and for quality of p–n junction, both of which have a strong influence on efficiency and U_{OC}. J_{SC} depends directly on the number of carriers generated by photons and separated by the p–n junction on the area unit. Comparison of this parameter seems to be the best way to establish quality of the produced solar cells and to determine the best growth conditions in the LPE process.

J_{SC} was shown to be dependent on surface coverage ratio, for particular cooling rates (Olchowik 2010, Cieslak 2010). Results presented are the summary of the previous work (Olchowik 2010, Cieslak 2010) along with J_{SC} obtained from solar cells produced in the same thermodynamic conditions on Si substrates without dielectric cover. It enabled the determination of any trend of changes in J_{SC}, depending on the epitaxial technological conditions. The J_{SC} dependence on the growing substrate SiO_2 coverage ratio is shown in Figure 3.

The maximal J_{SC} for cooling rates 0.25 and 0.5 K/min occurred for 80% coverage ratio and were 34.45 and 27.19 mA/cm², respectively. For cooling rates 0.75 and 1 K/min, the J_{SC} maximum

was for 70% c.r. (35.39 and 29.2 mA/cm², respectively). Taking into account the dimensions of the inner dielectric mirror, samples with a higher SC should have a higher J_{SC} than those with a lower SC. This dependence was not strong (Fig. 2). J_{SC} also depends on morphology of the growing epitaxial layer and the morphology depends on the distance from the Si windows on the sample. When dielectric surface coverage was 90%, the distance between Si windows (H in Fig. 1) was the highest, which may affect the Epitaxial Lateral Overgrowth (ELO) layer during growth. While propagating in

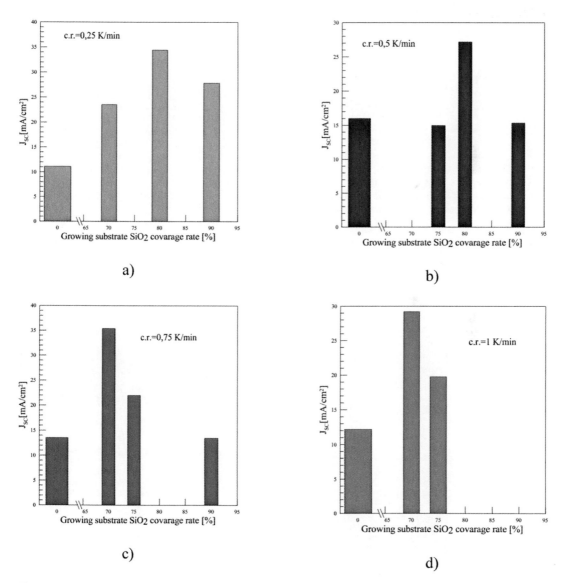

Figure 3.   Short circuit current density (J_{SC}) for different cooling rates and different sizes of the inner SiO_2 mirror; (a) c.r. = 0.25 K/min, (b) c.r. – 0.5 K/min, (c) c.r. = 0.75 K/min, (d) c.r. = 1 K/min.

a lateral direction, between ELO and dielectric layers, some voids may be formed (Kinoshita 1991), which would cause higher recombination in the ELO–dielectric interface—it can have a greater influence on samples with 90% SC resulting in reduction of J_{SC}.

Velocity of growth (c.r.) also had an important influence on performance of ELO-based solar cells. Higher c.r. caused higher defect densities in the growing layer and decreased the beneficial effect of inner reflectance (Fig. 3 c, d). It is worth noting that solar cells made from samples obtained in the classical LPE (no dielectric layer on the growing substrate) were characterized by lower J_{SC} compared with solar cells with the inner dielectric mirror (Fig. 3). The average thickness of the obtained epitaxial films was 21 μm.

4 CONCLUSIONS

The series of experiments was carried out to establish the best Si growth conditions in the LPE process. The specially prepared growth substrates enabled the production of an inner dielectric mirror inside the structure of the solar cells. Analysis of the results established the influence of the dielectric mirror on J_{SC}; samples with the inner mirror were compared with solar cells made on a base of epitaxial layers grown in the same thermodynamic conditions, on a Si substrate without dielectric coverage.

The highest J_{SC} for cooling rates of 0.25 and 0.5 K/min were obtained with samples with 80% of SC. For higher velocities of growth of 0.75 and 1 K/min, the best dielectric surface coverage was 70%. It is interesting that 90% SC was not best for photovoltaic conversion, as the morphology of a growing Si layer could be degraded by a higher density of voids formed in the ELO–dielectric interface. This effect can cause a higher recombination ratio and leads to decreased J_{SC}. Samples without the inner dielectric mirror were characterized by lower J_{SC} compared with samples with the

dielectric layer inside the structure. It leads to the conclusion that the dielectric layer on the growing substrate effectively increased optical way of photons inside the solar cell, resulting in an increase of J_{SC}.

ACKNOWLEDGEMENTS

The authors of this work would like to thank for all the help during solar cells production and examination process to dr Alain Fave from Institut National des Sciences Appliquées de Lyon and to dr Bachir Semmache from Irysolar Montpellier in France.

REFERENCES

Cieślak, K. 2010. Optimization of the liquid phase epitaxy process for photovoltaic applications; *Zeszyty Naukowe Politechniki Rzeszowskiej*, 271, z. 57: 83–86.

Gray, J. 2003. The physics of the solar cell, in: Handbook of Photovoltaic Science and Engineering (ed. A. Luque, S. Hegedus); West Sussex, Wiley,.

Jeager, R.C. 2002 *Introduction to Microelectronic Fabrication*; Prantice Hall, Upper Saddle River, New Jersey.

Kinoshita, S. 1991. Epitaxial lateral overgrowth of Si on non-planar substrate; *Journal of Crystal Growth* 115: 561–566.

Mauk, M. 2007. Equipment and instrumentation for liquid phase epitaxy, in: *Liquid Phase Epitaxy of Electronic, Optical and Optoelectronic Materials* (ed. P. Capper, M. Mauk); John Wiley and Sons Ltd., Chichester.

Nishinaga, T. 2002. Microchannel epitaxy: an overview; *Journal of Crystal Growth* 237–239: 1410–1417.

Olchowik, J.M. 2010. Analysis of internal reflectivity of silicon ELO PV cells obtained by LPE, *Proceedings of the 35th IEEE Photovoltaic Specialists Conference*, Honolulu, Hawaii, USA.

Suzuki, Y. 1989. Epitaxial lateral overgrowth of Si by LPE with Sn solution and its orientation dependence; *Japanese Journal of Applied Physics* 28: 440–445.

Zytkiewicz, Z.R. 2002. Laterally overgrown structures as substrates for lattice mismatched epitaxy; *Thin Solid Films*, 412: 64–67.

Environmental Engineering IV – Pawłowski, Dudzińska & Pawłowski (eds)
© 2013 Taylor & Francis Group, London, ISBN 978-0-415-64338-2

Life cycle assessment of solar hot water system for multifamily house

A. Żelazna & A. Pawłowski

Faculty of Environmental Engineering, Lublin University of Technology, Lublin, Poland

ABSTRACT: The renewable energy technologies are developed as a tool for reduction of the climate change as well as the dependence on fossil fuels. These technologies are generally considered as environmentally friendly, nevertheless the impacts of production, installation, operation and final disposal of devices needed for energy conversion cannot be neglected in the total environmental balance indicators. Considering the above mentioned, the use of Life Cycle Assessment method seems to be reasonable in the case of renewable energy technologies. In this paper, the LCA method was applied to assess resource use and other environmental burdens related to entire life cycle steps of solar hot water system for multifamily house. Based on Global Warming Potential method as impact assessment tool, the carbon dioxide payback time was calculated.

Keywords: solar hot water, Life Cycle Assessment, Global Warming Potential

1 INTRODUCTION

In the few past decades, Renewable Energy Sources (RES) became alternative for traditional energy production systems, mostly based on burning of coal. It is an important step towards the implementation of sustainable development concept since RES contribute to both diminishing of fossil fuels use, as well as the reduction of environmental emissions coming from energy sector (Pawlowski 2011). One of the examples of RES usage is thermal conversion of solar energy in Solar Hot Water systems (SHW). During the operation of such a system, electrical power is used for the supply of pumps and controller. The conventional energy amounts are relatively small (about 1%) in comparison to hot water needs. However, if we broaden the system boundaries to incorporate the usage of materials and energy for production, maintaining and final disposal of component devices, there are new environmental impacts emerging to be included in the total assessment.

In this research work, the Life Cycle Assessment method was used. According to European Environment Agency, "Life Cycle Assessment (LCA) is a process of evaluating the effects that a product has on the environment over the entire period of its life thereby increasing resource-use efficiency and decreasing liabilities". Such an assessment can be run both for the product or a system. LCA is treated as cradle to grave analysis and includes:

- the identification and quantification of pollutants introduced into the environment, as well as the energy and materials used;
- the assessment of potential impacts of the above mentioned;
- identification of possible reformations to diminish these impacts (Kowalski et al. 2007).

Life Cycle Assessment is based on the calculation of energy and materials use as well as pollutants and wastes emissions, which are connected with entire period of product's life. The life cycle should include pre-production (raw materials extraction) and production stage, operation and final disposal. The LCA analysis run for new technologies can support the decisions on their implementation to the market. However, it is necessary to emphasize that local conditions on specific market (ex. climatic data, materials availability) are necessary to be included in LCA.

The technological scheme of LCA is defined as consisted of four main phases (Fig. 1):

1. Goal and scope definition;
2. Inventory Analysis;
3. Impact Assessment;
4. Interpretation.

The aim of LCA is the assessment of potential dangers for environment through identification and calculation of amounts of materials and energy used and generated pollutants or wastes. Based on this, LCA can be used in marketing of production processes. Moreover, it can be useful in the estimation of the environmental costs minimization possibilities. As the environmental management technique, LCA is used in environmental certificates and indicators appointment (Zbiciński et al. 2006).

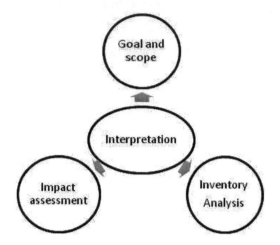

Figure 1. Technological scheme of LCA.

In the scientific literature of the last decade the new articles concerning Life Cycle Assessment of RES appeared, including these covering solar hot water systems. The chosen examples are summed up below. Ardente (Ardente et al. 2005) used LCA for the assessment of solar thermal collector (surface 2,13 m^2) production. Author established the primary energy consumption during the collector production as 11,5 GJ, and greenhouse gases emission as 721 kgCO$_{2eq}$. Energy used directly for collector manufacture in the plant accounts only 5% of total consumption; 6% is used for transportation. The remaining amount is used during the pre-production phase for raw materials extraction. Author stated that direct energy consumption is less important, than indirect energy usage.

Battisi and Corrado (Battisi, Corrado 2005) analysed solar thermal collector integrated with storage system through increased heat exchanger capacity. Primary energy consumption for production, distribution and final disposal of analyzed system was estimated as 3,1 GJ, and greenhouse gases emission as 219,4 kgCO$_{2eq}$. These influences are mostly connected with production phase (97,8%). In the analysis, the operation phase was neglected because of specific construction of a system without the additional medium, which excludes the necessity of pumping. In Kalogirou's papers (Kalogirou 2004, Kalogirou 2009) the information about LCA of solar flat collectors can be found. For the first one, the primary energy use was estimated as 2,7 GJ (system integrated with water storage). For traditional system author counted primary energy consumption as 3,5 GJ. The conducted analysis show that solar hot water systems are characterized by low greenhouse gases emission in comparison with traditional energy sources.

The results show that the payback time for production phase emissions is shorter than the total period of a system's life. However, it is necessary to underline that analysis were carried out for Mediterranean countries (Italy, Cyprus) so the solar radiation conditions were favorable. The aim of this paper is to conduct Life Cycle Assessment for solar hot water system operating in local climate conditions of Lublin area.

2 MATERIALS AND METHODS

Life Cycle Assessment of solar hot water system includes four phases described in ISO 14040:2009 standards.

2.1 Goal and scope definition

The goal of analysis was to estimate the environmental effects connected with the usage of solar hot water system. The installation is sourced both by solar thermal collectors (20 flat collectors, 2,13 m^2 each) and oil boiler as basic energy source. The oil boiler with atmospheric burning covers 48% of hot water needs, while the left 52% is covered by solar energy. The extent of analysis includes solar system construction (flat collectors, pipes and instrumentation, pump and steering device, medium—ethylene glycol, hot water storage bin, energy and fuels used for transport and building), the operation phase and final disposal. The functional unit is 1 kWh of energy produced during the operation of system in Lublin climate conditions.

2.2 Inventory analysis

The data used for analysis cover the materials and energy consumption in the mentioned life cycle stages. The Ecoinvent database was used, as well as producers and local data.

2.3 Impact assessment

For Life Cycle Impact Assessment, the Global Warming Potential method was used. It allows to appoint the greenhouse gases emissions in the mass unit—CO$_{2eq}$ (Cel et al. 2010). Carbon dioxide equivalent is a measure used to evaluate the emissions from diverse greenhouse gases based upon their global warming potential (http://stats.oecd. org). For example, 1 kg of methane is equivalent to 25 kg of carbon dioxide. As the complementary method, EcoIndicator'99 was used. It allows to simplify environmental burden categories into one indicator expressed in points Pt. Such a result enables to compare products or processes regardless of the material or amount differences.

2.4 *Interpretation*

The ultimate results were recalculated to the functional unit (1 kWh of produced heat energy). To ensure the right interpretation, the final result was compared to the traditional systems based on hard coal, lignite, coke, natural gas, light oil, wood.

3 RESULTS

The data characterizing greenhouse gases emissions counted by GWP100a method are stated on Figure 2.

Data describing the influence on natural environment according to EcoIndicator'99 method, divided into life cycle stages, are shown on Figure 3.

The results of GWP assessment for various sources of energy in analyzed system are presented

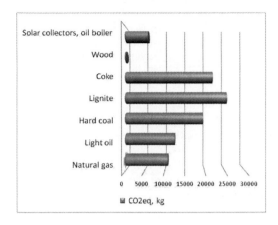

Figure 4. Greenhouse gases emissions from various systems, kg CO_{2eq}.

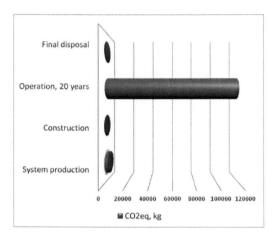

Figure 2. Greenhouse gases emissions from analyzed system, kg CO_{2eq}.

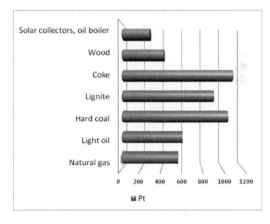

Figure 5. EcoIndicator'99 results for various systems, Pt.

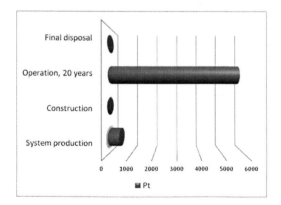

Figure 3. EcoIndicator results for life cycle stages of analyzed system.

on Figure 4. These data concern one year of operation, without the rest of life cycle stages.

The results of EcoIndicator assessment for various sources of energy in analyzed system are presented on Figure 5. The assumptions are equivalent to the above mentioned.

4 DISCUSSION

The results of conducted analysis show that the operation stage of system life cycle is the most meaningful. About 90% of the EcoIndicator'99 value is connected with this stage, mostly because of the use of fuel—light oil in boiler. Production phase has a share of 3% in the total greenhouse gases emission indicator and almost 10% in EcoIndicator'99. The stages of building and final disposal have light participation in the environmental indicators.

Due to polish climate conditions, the collectors alone are not able to secure the hot water delivery. It is necessary to install additional energy source. The analyzed solar system cooperates with oil boiler, therefore average winter and summer season emissions were included in the assessment. The carbon dioxide emission factor for the functional unit equals 0,167 $kgCO_2/kWh$ for such a configuration of system.

Based on Ecoinvent database, heat production processes from various energy sources were characterized. For the known amount of hot water needs, the one year emissions of greenhouse gases were estimated. The data from Figure 3 allow to show the favorable solutions from emissions minimizing perspective. These solutions include boilers for wood burning and analyzed system with solar thermal collectors and oil boiler. The wood is treated as less harmful because of the low nitrogen and sulphur oxides emissions. Carbon balance of biomass is close to zero for the reason that the plant needs to draw the CO_2 in the process of growing (Piemental 2012).

The results of analysis shown on Figure 4 allow to state that solar collectors in cooperation with oil boiler are favorable solution in comparison with other sources. The influence of operation stage on human health, ecosystem quality and resources is relatively low.

The research projects conducted on the process of wood burning showed that it is responsible for the emission of over 100 various chemical substances, some of them are toxic or carcinogenic (Łucki et al. 2011). This is the reason why the wood has the second position in the rank, with EcoIndicator'99 value insensibly lower than natural gas.

5 CONCLUSIONS

The use of life cycle assessment for energy systems in various technological configurations allows for the choice of the solutions which are the most favorable for natural environment.

The results of analysis consistently show that there is a possibility to limit the negative effects of hot water production by the use of solar thermal collectors. In the analyzed case, the low emission oil boiler was basic energy source. Nonetheless, annual greenhouse gases emissions decrease is high enough to state that, in the case of every alternative solutions, production emissions payback time does not exceed one year. The only exception is wood, treated as low-emission for greenhouse gases. The substitution of boiler for wood burning by the solar system with oil boiler would not bring the emissions reduction, but only the rise in installation use comfort by its unattended operation.

In the case of environmental burdens expressed by EcoIndicator, the payback time considered as time period needed for amortization production and construction emissions fluctuates from a few months for high-emission sources (hard coal, lignite, coke) to 3 years for wood.

The conducted analysis prove that the use of solar thermal collectors for hot water production is a highly favorable solution from the perspective of sustainable development concept.

REFERENCES

Ardente F., Beccali G., Cellura M., Lo Brano V. 2005. Life cycle assessment of solar thermal collector. Renewable Energy vol. 30, 1031–1054.

Battisti R., Corrado A. 2005. Environmental assessment of solar thermal collectors with integrated water storage. Journal of Cleaner Production vol. 13, 1295–1300.

Cel W., Pawłowski A., Cholewa T. 2010. Ślad węglowy jako miara zrównoważoności odnawialnych źródeł energii/Carbon footprint as the sustainability indicator for renewable energy sources. Prace Komisji Ekologii i Ochrony Środowiska Bydgoskiego Towarzystwa Naukowego, Tom IV, s. 15–22.

Kalogirou S. 2004. Environmental benefits of domestic solar energy systems. Energy Conversion and Management vol. 45, 3075–3092.

Kalogirou S. 2009. Thermal performance, economic and environment al life cycle analysis of thermosiphon solar water heaters. Solar Energy vol. 83, 39–48.

Kowalski Z., Kulczycka J., Góralczyk M. 2007. Ekologiczna ocena cyklu życia procesów wytwórczych (LCA)/ Ecological Life Cycle Assessment of manufacturing processes (LCA), PWN.

Łucki Z., Misiak W. 2011. Energetyka a społeczeństwo. Aspekty socjologiczne./ Energetics ans society. Social aspects. PWN.

Pawłowski A. 2011. Conditions of Polish Energy Security and Sustainable Development, Eco-Management For Sustainable Regional Development, 385–407.

Piementel D. 2012. Energy Production from Maize, Problemy Ekorozwoju/Problems of Sustainable Development vol. 7 no 2, 15–22.

Zbiciński I., Stavenuiter J., Kozłowska B., Van de Coevering H.P.M. 2006. Product Design and Life Cycle Assessment. The Baltic University Press.

http://stats.oecd.org (August 2012).

Environmental Engineering IV – Pawłowski, Dudzińska & Pawłowski (eds)
© 2013 Taylor & Francis Group, London, ISBN 978-0-415-64338-2

Effect of the gas absorption chiller application for the production of chilled water

J. Stefaniak & A. Pawłowski

Faculty of Environmental Engineering, Lublin University of Technology, Lublin, Poland

ABSTRACT: Due to fossil fuels resources depletion, there is a need to seek for new solutions that are based on the renewable energy sources or solutions that can be more energy efficient in order to ensure energy security in the future. In this paper, two types of devices used for chilled water production for air conditioning purpose were evaluated. Energetic and environmental performances of compressor chiller powered by mechanical energy and direct-fired absorption chiller powered by gas were analyzed and compared. Operating costs for devices were also estimated. Whereas, environmental impact of devices was calculated on the basis of Carbon Footprint. Although the absorption chiller is characterized by much smaller electricity demand, high coefficient of performance of compressor chiller makes it more competitive.

Keywords: compressor chiller, gas absorption chiller, Carbon footprint

1 INTRODUCTION

Energy is essential to satisfy the basic needs of human beings, therefore, ensuring its sustainable supply is one of the fundamental conditions for the implementation of sustainable development (Pawłowski 2009 a,b). In 1995, the word-wide energy consumption rate amounted to 2 kW per year per capita. In 2008, this consumption rate increased to 2.36 kW (Fay & Golomb 2012).

One of the most popular forms of energy is electric power, which is mainly used in households, services and industry. The world electric power sector produces an annual average of 1.4 TW of electrical power (Fay & Golomb 2012). It is based primarily on fossil fuels, which provide more than 65% of the world electricity demand. Two major energy carriers in this sector are coal and natural gas which supply 40% and 20%, respectively, of total electric power generation (Fig. 1). The remaining carriers are: nuclear fuels, renewables and oil.

1.1 *Characteristic of selected energy carriers*

1.1.1 *Coal*

Since the 18th century, coal has been used on a mass scale in the industrial processes and for heating purposes. Currently it is the second source of primary energy in the world after oil, however, in case of electricity generation it is the first one. But it should be noted, that the environment impact resulting from the use of this carrier is significant. GreenHouse Gases (GHG) emission, in particular

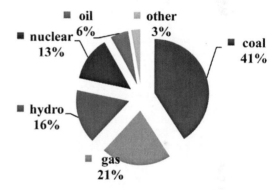

Figure 1. The world electric power generation by fuel (source: http://www.iea.org).

carbon dioxide, and tarry substances emission are threats to the environment and human health (IEA). Carbon footprint of electricity production from coal is the highest among other energy carriers. It is estimated, that generation of 1 kWh of electric power is connected with GHG emission at 0,4939 kg CO_2e (Guidelines to Defra 2009). Of course, this value depends on the applied technology (Weisser 2007).

Despite many efforts to build a modern, low-carbon and more efficient power plants and modernize existing ones, the current situation is not satisfactory. In this situation, the usage of Carbon Capture and Storage (CCS) technology seem to be the best solution. Unfortunately, this technology is not still developed enough. And it is an important challenge because coal consumption continues to increase, and according to some forecasts this trend will continue, despite the implementation of strict low regulations (IEA).

1.1.2 *Natural gas*
Technologies enabling the use of natural gas on a large scale were developed in the 20th century. In the aspect of the electricity generation, it is preferred to use this energy source. Combustion of gas is simple, because gas mixes with air quickly, and the devices for combustion like furnaces or boilers are characterized by small size. What is also important, gas plants are built in a relatively short time (about 2 years), and its control is not complicated. Natural gas combustion is low-risk (technically and financially) compared with coal or nuclear plant.

Gas plants can also cooperate with renewable energy sources, i.e. with wind power plants during windless weather (Fay & Golomb, IEA). The use of natural gas results in lower greenhouse gas (GHG) emissions compared to other fossil fuels. Carbon footprint of gas combustion amounts to 2.0322 kg CO_2e per 1 m^3 and 0.18521 kg CO_2e per 1 kWh of power from gas combustion (Guidelines to Defra 2009). Furthermore, the usage of gas combustion allows to avoid great energy losses due to energy conversion in conventional power plants (Fig. 2).

1.1.3 *Energy price*
Energy prices in selected European Union countries in years 2009, 2010 and 2011 are presented in Table 1. The prices include all taxes, therein VAT for industrial and business users.

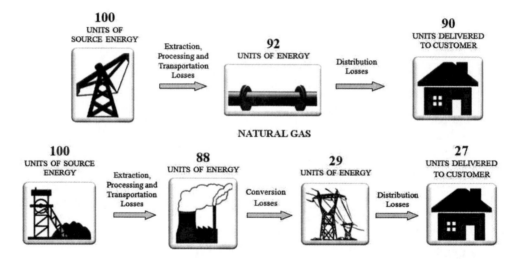

Figure 2. Energy production and transmission efficiency (source: http://www.energysolutionscenter.org).

Table 1. Energy prices in the European Union, the euro area and Poland (in Euro) (source: epp.eurostat.ec.europa.eu).

| | Electricity prices (per kWh) | | | | | | Gas prices (per kWh) | | | | | |
| | Households | | | Industry | | | Households | | | Industry | | |
	2009	2010	2011	2009	2010	2011	2009	2010	2011	2009	2010	2011
EU-27	0.164	0.173	0.184	0.103	0.105	0.112	0.053	0.057	0.064	0.030	0.034	0.038
Euro area	0.173	0.182	0.193	0.106	0.109	0.118	0.058	0.064	0.071	0.032	0.036	0.040
Czech R.	0.139	0.139	0.147	0.112	0.108	0.108	0.047	0.052	0.060	0.027	0.036	0.035
Germany	0.229	0.244	0.253	0.113	0.119	0.124	0.059	0.057	0.064	0.035	0.044	0.050
Ireland	0.186	0.188	0.209	0.118	0.113	0.129	0.055	0.053	0.062	0.026	0.032	0.040
Spain	0.168	0.185	0.209	0.112	0.109	0.116	0.054	0.054	0.054	0.027	0.029	0.033
France	0.121	0.135	0.142	0.065	0.072	0.081	0.058	0.058	0.065	0.032	0.035	0.038
Poland	0.129	0.138	0.135	0.093	0.099	0.094	0.046	0.051	0.050	0.030	0.033	0.032
Romania	0.098	0.105	0.109	0.083	0.081	0.080	0.027	0.028	0.028	0.021	0.022	0.025

Energy prices are different in different member countries because they depend on many factors, such as: geopolitical situation, energy dependence (necessity to import energy), energy demand, environmental costs, weather conditions and the amount of taxes and excise duties. However, it can be easily noticed that price of energy in most countries is rising year after year, with the largest increase observed for the price of electricity. In 2011, the price of 1 kWh of electricity sector reached 0.184 Euro cents in the household, while 1 kWh from gas cost 0.064 Euro cents.

1.1.4 Sustainable energy

The sustainable energy is energy that can persist into the future, has mineralized environmental impact and high energy efficiency. These requirements result from the need to ensure the future generations' unlimited access to the energy and clean environment. Unfortunately, it is not easy to meet these requirements. It happens due to several reasons. First of all, a strong dependance of society on the non-renewable energy, whose resources decrease rapidly, is observed. This may causes a serious problem of energy security in the future. In addition, currently used technologies are still not enough environmentally friendly (Pawłowski 2009a,b). Therefore, another dilemma appears—how to cope

with the increasing pollution of air, water and soil? Therefore, it is very important to carry out energy and environmental assessments of individual processes and the technologies, in order to improve energy management and decision-making process.

1.1.5 Air conditioning—cooling production

The growing importance of air conditioning systems is correlated with stronger need to provide clean indoor environment and microclimate that conduce good mental and physical condition (Grignon-Massé et al. 2011, Jones 2001). Two different types of devices can be used for chilled water production for air conditioning and ventilation purpose—compressor chiller and absorption chiller. Compressor chillers are widely used in Heating, Ventilation, Air Conditioning (HVAC) sector because of high cooling efficiency and simple control. In this kind of devices refrigerating cycle is realized through the use of mechanical compressor powered with electrical energy. An alternative for compressor chiller are absorption chillers. Absorption units use absorption effect to provide refrigeration cycle, in which one substance (refrigerant) is absorbed by other substance (working fluid). This process occurs due to high hygroscopic properties of working fluid. By the application of absorber and generator, so called thermal compressor, these

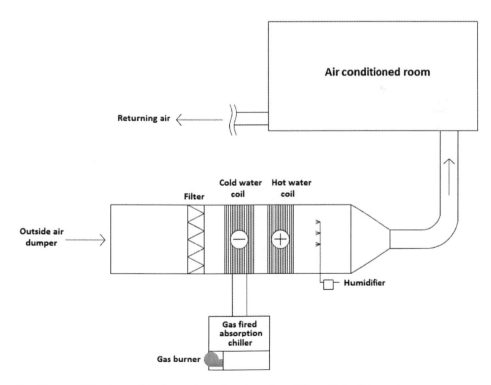

Figure 3.   Scheme of the air conditioning system using absorption chiller with built-in gas burner.

two substances are separated and recombined during the whole refrigeration cycle.

Absorption chillers have been used for more than 60 years. Over the years, these devices have been developed and upgraded. Nowadays they are used more often, especially in regions where the rapid increase of electricity demand is observed and where the access to the electricity is limited. They are also used in Combined Heat and Power (CHP) systems. Devices available in the market today, have smaller sizes and are more efficient relative to their antecedents. They can be also adapted to smaller objects like houses or offices (Jaruwongwittaya & Chen 2010, Martínez et al. 2003). Figure 3 shows the exemplary air conditioning system with absorption chiller powered by gas combustion.

The main advantage of absorption units is that they are powered by thermal energy and they are characterized by low electricity demand. They can be used in systems where waste heat or heat from renewable energy sources is available. They can be also powered by thermal energy which comes from gas combustion.

The Coefficient of Performance (COP) for the absorption chillers is much smaller than for compressor chiller (COP = 0.6–1.2 for absorption chillers, COP = 2 or more for compressor chillers). However, these devices have many different advantages. Their service life reaches 20–30 years (for compressor chillers is only about 5 years) due to a small number of moving parts. Thus, absorption chillers do not require frequent servicing and the service itself is facilitated (Rusowicz 2007).

Application of absorption chillers is particularly preferred in the following cases:

- total heat generated in the cogeneration system is not fully utilized or cogeneration system is at the design stage,
- an inexpensive heat source is available—e.g. solar or waste heat,
- there are power limits that restrict the use of compressor chillers.

2 METHODS

The aim of this study was to compare the performance of two chillers, compressor chiller and gas fired absorption chiller, applied in air conditioning system for chilled water production. Air conditioned object is a commercial building—a coffee house.

Evaluation of the performance involved estimation of energy demand (needed to achieve the required cooling capacity) and energy cost. Environmental assessment was made by calculating the total carbon dioxide emissions for the cooling process. This estimation was based on direct emission factors specified in the "Guidelines to Defra/DECC's Greenhouse Gas Conversion factors for the reporting of the Company in 2012".

2.1 Characteristic of air conditioned building

The cubage of coffee house is 480 cubic m^3. Cooling demand in the building occurs from May to September. In order to ensure proper indoor air parameters, it is necessary to supply fresh air of temperatures 20 °C in amount of 2.7 kg/s. The total annual cooling demand for the coffee house amounts to 25,359 kWh.

2.2 Used cooling devices

For the analysis purposes two different chillers were used—an absorption chiller with built-in gas burner and a regular compressor chiller. Absorption unit is powered by the heat comes from gas combustion. Gas consumption of the unit is 5.3 m3/h and the electric power demand (to power absorbent pump) is 1.98 kW. The Coefficient of Performance (COP) of the unit is 0.71. Compressor chiller, used in the research, is characterized by the cooling capacity of 35 kW and a Coefficient of Performance (COP) of 2.5.

3 RESULTS AND DISCUSSION

To achieve required cooling demand with gas absorption chiller 32,444 kWh of thermal energy from gas combustion and 3,273 kWh of electricity are required annually. Energy demand (only electricity) for compression chillers amounts to 10 150 kWh per year. Environmental effects related to the operational phase are shown in Figure 4. GHG emissions values for the chosen energy sources, i.e. electricity and thermal energy from natural gas combustion, are presented in the graph.

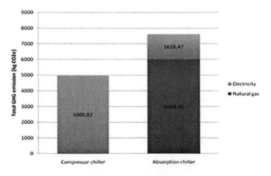

Figure 4. The total GHG emissions for chillers.

Total GHG emissions is greater for the absorption chiller (7,625.42 kg CO_2e) than for compression refrigerator (5,009.82 kg CO_2e), and the difference between them exceeds 2.5 tones CO_2e.

The percentage and quantitative share of individual greenhouse gases in total GHG emissions for both chillers is presented in Figures 5 and 6. In both cases carbon dioxide is responsible for more than 95% of the total emission. The remaining 5% is connected with emission of sulfur oxide (I) and methane.

The average prices of gas and electricity in the European Union in 2011 (Table 1) were used to calculate operation costs of used devices. The results of calculation are presented in Table 2. Application of compressor chiller in the air conditioning system generates lower costs. Application of gas absorption chiller increases annual operating costs by about 800 Euro.

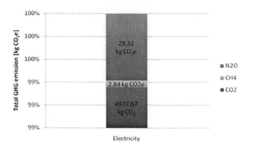

Figure 5. Share of individual gases in total GHG emission for compression chiller.

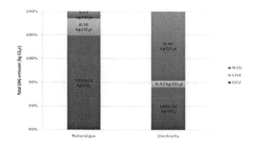

Figure 6. Share of individual gases in total GHG emission for gas absorption chiller.

Table 2. The annual cost of the energy needed to power chillers [Euro].

	Compressor chiller	Absorption chiller
Electricity	1,866.42	2,076.41
Gas	–	602.22
Total cost	1,866.42	2,678.63

4 CONCLUSION

This research aimed to evaluate and compare the energetic and environmental performance of two chillers—gas absorption chiller and compressor chiller—used for chilled water production for air conditioning purpose in coffee house building of total cubature 480 m^3 and annual cooling demand 25,359 kWh. For a particular cooling demand energy needed to power chillers was calculated. Energy demand of absorption chiller is the following: 32,444 kWh of thermal energy from gas combustion and 3,273 kWh of electric power. The same requirements can be achieved with compressor chiller and the energy needed to power refrigeration cycle is 10,150 kWh. It appears that compressor chiller is characterized by lower energy demand. The reason for that is higher value of COP. Despite the fact, that the cost of 1 kWh power from gas is lower than the cost of 1 kWh electric power, low value of COP of absorption chiller is a reason of its higher operation cost. The annual cost of chilled water production by absorption chiller is 2,678.63 Euro, while in case of compressor chiller is only Euro 1,866.42. The total GHG emissions from operation phase of gas absorption unit are 7,625.42 kg CO_2e. Compressor unit operating phase results with GHG emissions at 5,009.82 kg CO_2e. In both cases, carbon dioxide has the biggest share in total GHG emission—more than 99%.

Direct use of natural gas allows to avoid large energy losses resulting from energy conversion in conventional coal plants. However, in this particular case, better performance results of compression chiller were achieved due to its high-value of COP.

More research must be conducted to change this situation and improve the cooling efficiency of absorption units.

Another option is to use absorption chillers in systems, in which they can be powered by waste heat. Through the use of waste heat there is an opportunity to avoid emissions associated with the combustion of natural gas, which in the present case were significant.

REFERENCES

Dasgupta, P. & Taneja, N. 2011. Low Carbon Growth: An Indian Perspective on Sustainability and Technology Transfer. *Problemy Ekorozwoju/Problems of Sustainable Development* 6(1): 65–74.

Energy Solution Center: http://www.energysolutionscenter.org.

Eurostat: http://epp.eurostat.ec.europa.eu/.

Fay, J.A. & Golomb, D.S. 2012. Energy and the Environment. Scientific and technological Principles. New York: Oxford University Press.

Grignon-Massé, L., Rivière1, P. & Adnot, J. 2011. Strategies for reducing the environmental impacts of room airconditioners in Europe. *Energy Policy* 39(4): 2152–2164.

Guidelines to Defra/Decc's GHG Conversion Factors for Company Reporting 2009.

International Energy Agency (IEA): http://www.iea.org/.

Jaruwongwittaya, T. & Chen, G. 2010. A review: Renewable energy with absorption chillers in Thailand. Renewable and Sustainable Energy Reviews 14(5): 1437–1444.

Jones, W.P., 2001. *Klimatyzacja* (in Polish). Warszawa: Arkady.

Martínez, P.J., García, A. & Pinazo, J.M. 2003. Performance analysis of an air conditioning system driven by natural gas. *Energy and Buildings* 35(7): 669–674.

Rusowicz, A. 2007. Tendencje rozwojowe urządzeń chłodniczych absorpcyjnych (in Polish). In G. Krzyżaniak (ed.), **XXXIX Dni Chłodnictwa; Konferencja Naukowo-Techniczna**, Poznań, 14–15 November 2007. Poznań: Systherm Chłodnictwo i Klimatyzacja.

Shan, S. & Bi, X. 2012. Low Carbom Development of China's Yangtze River Delta Region. *Problemy Ekorozwoju/Problems of Sustainable Development* 7(2): 33–41.

Weisser, D. 2007. A guide to life-cycle greenhouse gas (GHG) emission from electricity supply technologies. *Energy* 32(9): 1543–1559.

Environmental Engineering IV – Pawłowski, Dudzińska & Pawłowski (eds)
© 2013 Taylor & Francis Group, London, ISBN 978-0-415-64338-2

Cost-effectiveness terms of a wind energy investment

W. Jarzyna & P. Filipek
Department of Electrical Drive Systems and Electrical Machines, Lublin University of Science, Lublin, Poland

A. Koziorowska
Rzeszow University, Institute of Technology, Rzeszow, Poland

ABSTRACT: In the paper factors that should be considered when planning Wind Power Station (WPS) investment were discussed. These factors affecting the final effect of investment are the average annual speed, roughness of the terrain, the Weibull distribution function, a wind turbulence amount and a level of grid infrastructure. Analysis of these factors help to find the final turbine selection which precise fundamental WPS parameters as: nominal power of turbine, rated speed of wind and tower height. However, decision about wind power investment requires multivariable economic analysis. In the paper Financial and Economic Balance results were computed for three variants of electrical energy prices raises. The received final results show the Return on Investment Time depending on average annual wind speed and energy prices. The discussed problems emphasize that the most rational solution should be found by verifying the computed financial results for specially selected different technical options.

Keywords: wind energy, terms of cost-effectiveness, wind power plant selection factors

1   INTRODUCTION

As a result of increasing demand for electricity and rising energy prices, the production of renewable energy from wind is an interesting opportunity for investors looking for an effective and safe capital investment in new rapidly growing market sectors. Wind power is an interesting form of fund investments due to high reliability and modernity of installed devices, relatively low operating costs and possible fast growing return on investment. Due to these factors, aeroenergetics investments are becoming more profitable business and are creating new competition for conventional methods of electricity production. What is more, wind power energy significantly contributes to creating distributed energy structure. When joining wind energy with energy storage systems and energy flow control, it promotes the development of modern smart grids, meets the local demands for electrical energy and has positive influence on economical growth of the region.

The development of the energy-efficient sector implies other positive effects, such as: reduction of CO_2 and other greenhouse gases emission, new jobs, revenues from property taxes for municipal funds, incomes for tenants (mostly farmers) for the lease of land for wind farms and high voltage connection lines.

Despite these advantages, the construction of new wind power stations has many obstacles that constipates planned actions implementation. At every stage of the project preparation, investor faces political, procedural, social, environmental and technical barriers. The choice of a localization is not affected only by the measurement results of wind speed, but also other important factors that determines the economical success of the project, such as: sale prices of electricity and green certificates, forecasts of electricity production, operational costs bared during the investment and incidental expenses on failures or downtimes. These factors coming with wind energy investments are causing the necessity of conducting preliminary analysis of the economical condition of its profitability. The crucial element that is needed in order to prepare the investment of wind power station construction is economical analysis that includes wide range of mentioned factors. After analyzing these factors, each investor that is planning to build wind power stations, should know how safely and efficiently operate on the competitive market (ENTSO-E 2012; Jarzyna, Lipnicki 2012). This paper was devoted to solve and present such problems. There were presented technical and local requirements including wind characteristics, results of economical analysis and exemplary cash flows of such investments.

2   METHODS

Methods determining effectiveness of Wind Power Station construction can be divided into three groups: localization condition, project and choice of construction type, economical conditions of the energy sale. Their characteristic is shown in the following sections of the paper.

2.1 Localization of wind power station

The choice of localization can be evaluated on account of the characteristics of the wind and the possibility of high voltage connection to the Grid Supply Point (GSP) (Jarzyna 2011).

Wind measurements are carried out for a minimum two—three years for the sake of a possibility of 20% variation of wind conditions in consecutive years. Wind characteristic is determined by measurements of meteorological conditions. The special measuring stations are equipped with two or three anemometers, as well as wind direction, air pressure, temperature and humidity sensors. The sketch of such installation is presented in Figure 1.

Anemometers used for measuring wind speed are mounted at different heights. Such localization allows to determine wind velocity gradient as a function of height. Thanks to that, the authors are able to estimate, so-called, roughness coefficient and plot a curve of wind speed changes. The curve determines wind speed above the highest measuring point of the audit tower and allows to limit the height of the towers. To eliminate caused interference near the surface, the first measuring point should be located at least 10 meters above the surface. Workflow of measuring system should be managed by a microcomputer, that is collecting and recording data, performing basic processing and transmitting the signal to the central station.

European energy audit standards (Wind Turbines Part 1 2005; Wind Turbines Part 12-1 2005) describe in details conditions of the measurement and positioning audit equipment in respect to constructional difficulties. Electronic system that is collecting, processing and conditioning data, enables to compute and determine average, maximum, minimum and standard deviation for wind speed and its direction, pressure, air density, temperature and humidity.

The average ten minutes winds are used to calculate the size of wind speed standards that are later on used to forecast annual power and energy produced by planned wind power station (Wind Turbines Part 12-1 2005). The turbulence coefficient is a crucial parameter that is based on audit wind measurements. It determines the size of the non-laminar wind flow, which is characterized by rapid changes in wind speed and direction. This wind is a serious threat for wind turbines. It causes sudden changes of mechanical stress and practically is not generating drive torque. This turbulence coefficient is recognized in the rating data and should be taken into account when selecting a wind power station.

A wind power plant location is valuated also due to availability of the power grid because the cost of building power line connection which can significantly increase the value of the investment. The costs of these lines are different and depend on country, characteristics and value of the land. In most cases 30kV overhead lines are less expensive alternative than the cable lines. However, recently the capital and operating costs of 110kV HV lines already became more expensive option than underground cable lines (2012 Polish data). Hence, the cable lines are preferable solution in most European countries.

2.2 Pricing influence on renewable energy investments profitability

The competitiveness of wind energy depends on the pricing policies carried out by particular country. This policy often exists in the form of subsidies for green energy production. In Poland, the basic average price per 1 MWh in 2012 was close to 49 Euro. But for producers of energy from renewable sources such value is unsatisfactory. The average return on investment could be amounted close to 20 years. Such a long period of time will not encourage investors to build wind power stations.

Therefore, in many European countries, additional financial support mechanisms are applied. In Poland, it is a subsidy made by government to produced renewable energy, commonly called "green certificates". In 2012 in Poland, the value of such payments for 1 MWh amounted to about 66 Euro. This significant payment changes completely payback terms. The results of the sample analysis are shown in Chapter 4.

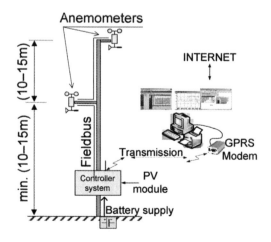

Figure 1.   The draft audit station for wind conditions.

In addition, there is a possibility to obtain a refund of expenses. Such funding may even be equal to 75% of their expenses and be financed from a special restructuring funds.

2.3 *Selection of the type of the wind power plant*

The purchase of a Wind Power Plant should be adapted to the local terrain and meteorological conditions. The most important conditions are as listed below:

• Average wind speed,
• Wind turbulence
• Distance between Wind Power Station and Grid Supply Point (GSP) of power grid and electrical parameters of this grid.

In the Wind Power Station market we can find stations, which have a different nominal power with the same blade diameter. For example, the Enercon E82 Wind Power Stations model has different types of rated power: 2MW, 2.3MW or 3MW. This variation comes from different rated speeds of a wind and different shape of power curve due to wind speed. It is easy to find, that the type of wind power plant having the biggest power has also the biggest rated wind speed. However, it rises a question, how can we know which type of turbine fits to the local wind conditions? To solve this problem we have to find the Weibull probability function, used to describe the distribution of wind speeds during an extended period of time. This statistical function tells us how often different speeds of the wind will be seen at a location with a certain average speed of the wind. Having this knowledge we are able to make a decision which wind turbine has the optimal rated speed.

In general, the average annual wind speed for lowlands is smaller than average annual wind speed for offshore. Taking into account Weibull distribution function, in most cases we accept that rated wind speed for lowlands should be around 12 m/s and for offshore territories around 17 m/s. The solution of this simple case clarifies that for different wind conditions manufacturer will offer different types of wind power plants (Kacejko, Pijarski 2009).

Therefore, the choice of the wind power plant should be legitimated by wind audit results, wind distribution function and electric parameters of power grid. It can be noticed that excessive rated power of wind power plant causes irrational increase of maximal power, to which should be fitted the all electric devices, converters, transformers, connection line and even legal contract with Transmission System Operator.

3 ECONOMIC ANALYSIS METHODS

While making the decision about wind energy investment, it is necessary to carry out economic analysis of the profitability of the project. The lack of such analysis can contribute to big losses (Paska, Klos 2010, Solinski, Jesionek 2004).

When analyzing the case, it has to be considered whether wind conditions are suitable to obtain profits within planned period. Moreover, type of wind power plant, which will maximize our incomes should be chosen. In order to conduct the analysis, the risk analysis should be parsed which includes assessment of market and financial profit abilities (Fig. 2).

Economic analysis of rentability of the investment enables to estimate basic static indicators

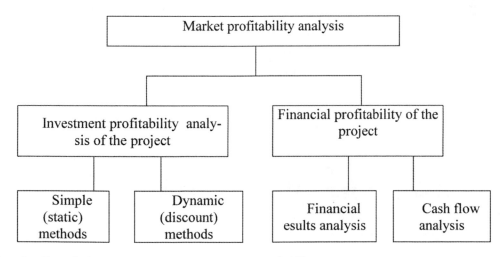

Figure 2. General scheme of the basic methods of market profitability.

like Return on Investments Time (RoIT), return on capital, break-even point and dynamic indicators: discountable costs, Net Present Value (NPV), Internal Rate of Return (IRR), efficiency of investment and time needed to receive return on investment. More accurate outcome will be received when using dynamical methods including whole period of time of the investment.

Cash flow analysis is one of the most used method of all financial analysis, especially when controlling the economic activity of the company. It is irreplaceable when synchronizing investment incomes, activating production and development of the company assets. The investor should be equipped with abundant financial means to cover costs of production, financial indebtedness, handling debt costs or taxes. The comparison of the cash flows is usually made yearly or monthly. The project is rated positively when there is favorable balance of the cash flows in all investigated periods (Solinski, Jesionek 2004).

4 EXEMPLARY PROFITABILITY RESULTS

During computations, input data, which are related to base value defined as a total investment value, were assumed.

- Investors own funds 48.8%
- Bank credit 39.4%
- Governmental grants 11.9%
- Annual development costs 1.5%
- Period of credit reimbursement 5 years

- Grace period of the credit 2 years
- Discount rate 8%
- Rate of depreciation 5%
- Working life 20 years
- Rated wind speed 12 m/s

On the basis of wind audit, average annual wind velocity 7 m/s has been assumed. Received total energy was determined as 27.6%, what describes the amount of total annual produced energy divided by the generated energy when power plant works continuously with rated power. The forecasting Financial and Economic Balance results were computed for three variants of electrical energy prices raises. During stable 1% annual growth of the prices, the Return on Investment Time is equal to 6.8 years, while when raise of the price is 4%, the RoIT is equal to 6 years. For variable price raise starting from 8% and decreasing during the first 10 years up to only 1% raise, the RoIT equals only 5,4 years. The histogram for those three variants of the Financial Balance is presented in Figure 3.

The shown in the Figure 3 the highest annual FBx is the Base Value of Serie 1–3. Series 1, 2 and 3 correspond to: Serie 1—1% stable price raise during 20 years, Serie 2—4% stable price raise during 20 years, Serie 3—variable price raise, where during the first 10 years there is 8% of annual price raise and in the next 10 years only 1% of the raise. Nevertheless, Financial and Economic Balances depend manly on annual average wind speed. For the considered project, where wind speed rated is equal to 12 m/s, an influence of average annual wind speed on foreseen RoIT illustrates Figure 4.

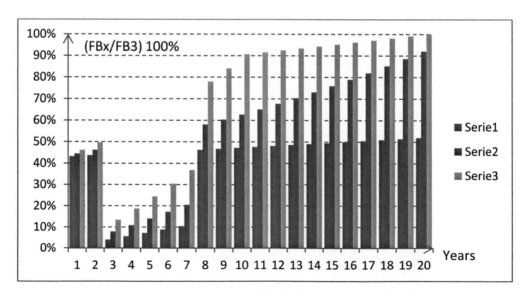

Figure 3. The Financial Balance (FB) computed for 3 variants of price raises and determined in relative values.

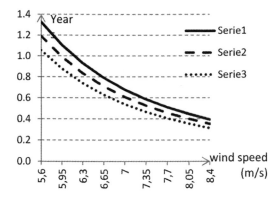

Figure 4. The Return on Investment Time in dependence on average annual wind speed for three variants of annual growth of energy prices, where Series 1–3 are explained in description of the Figure 3.

Assuming 7 m/s as the base value of Annual Wind Speed (AWS), reducing 20% of this value to 5.6 m/s results in nearly doubling the RoIT. However, 20% increase of annual wind speed up to 8.4 m/s, shortens time up to 60% of the based time.

5 CONCLUSION

Decision about wind power investment requires multivariable analysis made in accordance with a certain economic method. On the preliminary stage, this analysis can be made applying static and subsequent dynamic methods and then cash flow computations. As input parameters we have to introduce all the accompanying building costs. These costs vary depending on localization price, soil conditions, taxes and a distance to the Grid Supply Point (GSP). To these costs is should be added supplementary electrical devices as transformers and reactive power compensator, which include STATCOM converters.

The main factors affecting the final effect are obviously the average annual speed, roughness of the terrain, the Weibull distribution function and a wind turbulence level. All mentioned quantities can be defined on the basis of 2–3 years wind audit. Received results are fundamental to prepare the final turbine selection. These choices depend on the selected technical solution, whose fundamental parameters are: nominal power of turbine,

rated speed of wind and tower height depending on a terrain roughness. By verifying the computed financial results for different technical options, the most rational solution should be found.

Performing these computations, the probability of success can increase but the risk still exist because it is caused by the increase in energy prices that is hard to estimate, variable subsidies of green certificates and random changes of annual average wind speed.

REFERENCES

ENTSO-E. 2012. Draft Network Code for Requirements for Grid Connection applicable to all Generators. European Network of Transmission System Operators for Electricity, 24 January 2012. https://www.entsoe.eu, p.85.

Jarzyna W. 2011. Terms Of The Turbine and Generator Choice of Wind Power Stations. In Polish (Warunki wyboru turbin i generatorów elektrowni wiatrowych). Rynek Energii, vol.95, No 1, pp. 102–106.

Jarzyna W., Lipnicki P. 2012. The Comparison of Polish Grid Codes to Certain European Standards and Resultant Differences for WPP Requirements. Epe Joint Wind Energy and T&D Chapters Seminar, June 2012.

Kacejko P., Pijarski P. 2009. Connecting of wind farms—reasonable limitations instead of oversized investment. in Polish (przyłączanie farm wiatrowych—ograniczenia zamiast przewymiarowanych in-westycji). Rynek Energii,vol.80, No 1, pp. 10–15.

Kacejko P., Wydra M. 2011. Wind Energy in Poland—Analysis of Potential Power Systems Balance Limitations and Influence on Conventional Power units Operational Conditions, in Polish (Energetyka wiatrowa w Polsce, analiza potencjalnych ograniczeń bilansowych i oddzialywań na warunki pracy jednostek konwencjonalnych). Rynek Energii, vol.93, No 2, pp. 25–30.

Paska J., Klos M. 2010. Wind power plants in electric power system—connecting, influence on the system and economics. In Polish (Elektrownie wiatrowe w systemie elektroenergetycznym—przyłączanie, wpływ na system i ekonomika). Rynek Energii, vol 91, No 1.

Soliński I., Jesionek J. 2004. Financial risk analysis of proecological investments on example of wind energy production. Górnictwo i Geoinżynieria/Akademia Górniczo-Hutnicza im. Stanisława Staszica, Kraków, vol.28, No 2, pp. 27–38.

Wind Turbines, Part 1. 2005. Design Requirements. International Standard IEC 61400-1.

Wind turbines—Part 12-1. 2005. Power performance measurements of electricity producing wind turbines. *International Standard IEC 61400-12-1.*

Environmental Engineering IV – Pawłowski, Dudzińska & Pawłowski (eds)
© 2013 Taylor & Francis Group, London, ISBN 978-0-415-64338-2

Financial evaluation of heat and electrical energy cogeneration in Polish household

P.Z. Filipek & W. Jarzyna
Department of Electrical Drive Systems and Electrical Machines, Lublin University of Technology, Lublin, Poland

ABSTRACT: The article presents the financial analysis of a cogenerating system with a gas boiler with built-in Stirling engine, which both produces thermal energy for house and water heating, and generates electrical energy to supplement grid supply. A few models were developed: cogeneration system, thermal losses of the house, hot water requirement and climate conditions of Lublin region. The computer simulations were used to evaluate the efficiency of the system operation and to calculate the payout time of the system implementation.

Keywords: energy, cogeneration, renewable energy sources

1 INTRODUCTION

Cogeneration in the meaning of simultaneous production of heat and electrical energy is widely used in professional power plants to increase the efficiency of such systems and in consequence both to save fossil fuels and to decrease the emission of CO_2 to the atmosphere. (Polish Ministry of Environment 2003), (Pavlova-Marciniak 2006).

Cogeneration is also being applied more and more frequently in public buildings like hospitals and schools. Unlike industrial power-plants, the prime goal of such systems is producing thermal energy for house and water heating. Moreover, additionally generated electrical energy only supports the house needs. The specific character of such buildings, connected with continuous need for electrical energy and heat, makes such systems decrease the exploitation costs. Thus, common application of this technology in the country can save energetic sources and decrease emission of CO_2 in Poland. (Kalina 2003). There is a wide choice of gas boilers on the market which support cogeneration. Many of them are dedicated to household use. However, the usefulness and profitability of their implementation can be determined only after an analysis of the system functioning in a given house and for certain climatic localization. (Twidell & Weir 2007).

2 MODELLING OF THE SYSTEM

For the purpose of the research, several models in Matlab Simulink were designed: cogeneration gas boiler model, thermal model of the house, model of hot water demand and model of weather conditions for a given localization.

The computer simulations ware carried out for a typical detached house built in state-of-the-art technology with usable floor area of 120 square metres ($s_m = 120$ m^2) and cubature of 350 cubic metres ($V = 350$ m^3). Thermal demand of the heated building is the sum of heat loses (Klugmann-Radziemska 2009):

• heat transmission losses

$$Q_o = \frac{\lambda}{d}(t_i - t_e)A(1 + d_1 + d_2),\cdot W \qquad (1)$$

where: t_i and t_e—accordingly outer and inner temperature, λ—heat conductance coefficient of a building wall, d—thickness of heat insulation, A—total surface of outer walls, d_1 and d_2—factors which represent additional heat losses by heat

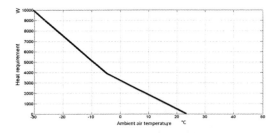

Figure 1. Correlation between the house heat demand and ambient temperature.

transmission accordingly for a low ambient temperature and sunny weather conditions,
• and ventilation heat losses:

$$Q_w = \left[0.34(t_i - t_e) - 9\right]V, \cdot W \qquad (2)$$

On the basis of the implemented Matlab-Simulink model, a building thermal power demand was estimated as the function of an ambient temperature.

For the four-person family the average consumption of 200 liters of hot water per day was used to estimate daily thermal energy consumption needed to heat running water.

The data mentioned above and climate conditions for Lublin district area were used to select a rated heat power of the gas boiler. Taking into account Polish Standard PN-83/B-03406 (less strict than equivalent European Standard) the boiler minimum heat power required for heating of the house was calculated as 9.24 kW. Eventually,

a gas boiler PPS24-ACLG5 manufactured by *WhisperGen™* [5] was chosen for the system which is indeed micro combined heat and power (micro-CHP) system. It has built-in Stirling motor fed by thermal energy generated during gas combustion in dedicated special gas burners. The motor drives en electric generator which produces electric energy whose amount depends on an operating mode of the boiler. Additionally, produced thermal power is used by the heating system.

The microCHP system has the following parameters: output rated heat power 8 kW (maximum to 12 kW), maximum output electric power to 1.2 kW with standard electric greed parameters ($U_n = 230$ V, $f_n = 50$ Hz); operating cycle from 1h to 24h per day; self-consumption of electric power: stand-by 30 W and operation 110 W; total efficiency 90%; feeding by natural gas or landfill gas, or biogas.

In the above model the input variables are heat demand of the building In1 and the caloric value of feeding gas In2, whose value can vary from 21 MJ/m³ (for landfill gas) to 38 MJ/m³

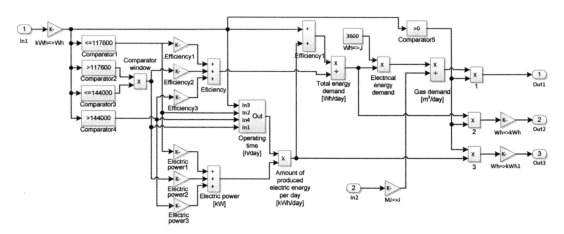

Figure 2. Model of a microCHP system applied in a gas boiler.

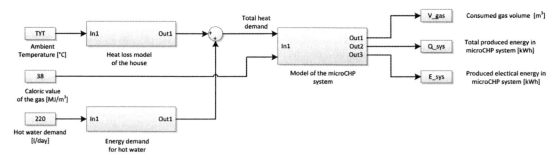

Figure 3. Complete model of the system.

(for natural gas). The model allows for system simulation in three operating modes defined by produced power threshold: I mode—from 0 to 4.9 kW produced heat and 0.4 kW electric power with total efficiency of $\eta = 98.3\%$; II mode—more than 6 kW of produced heat and 0.85 kW electric power with $\eta = 97.3\%$; III mode—more than 8 kW of produced heat and 1.2 kW electric power with $\eta = 96.4\%$;

Firstly, depending on a threshold of the produced power demand, one of four different input comparators is chosen and afterwards actual energetic efficiency, an amount of electric power and operating time of the boiler are estimated. Secondly, electrical energy and heat generated in the microCHP system are calculated. Output variables of the model are: Out3 (kWh/day)—the amount of generated electrical energy, and Out2 (kWh/day)—the amount of produced heat and Out1 (m³/day)—cubic volume of the consumed gas. The complete model of the microCHP system with previously defined units is shown in the following diagram (Fig. 3).

For computer calculations climatic conditions for localization of Lublin in S-E Poland (depicted on Fig. 4) were taken into account.

The results of computer simulations are shown in Table 1.

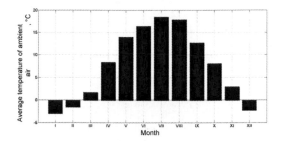

Figure 4. Average month temperatures for a typical year for Lublin, Poland.

3 FINANCIAL ANALYSIS

Financial analysis was made for a newly built heating system. Thus, only the cost difference between typical and cogeneration system was considered.

The reduction of a yearly maintenance cost ΔK_e is calculated as an income from exploiting of the

Table 1. Results of system simulations.

Month	Electric energy [kWh]		Thermal energy [kWh]		Gas consumption [m³]	
	Daily	Monthly	Daily	Monthly	Daily	Monthly
January	7.47	232	100.5	3116	9.523	295.21
February	6.77	190	91.1	2551	8.631	241.68
March	5.33	165	71.6	2221	6.789	210.45
April	2.2	66	29.6	889	2.806	84.19
May	0.73	23	9.8	305	0.934	28.95
June	0.73	22	9.8	296	0.934	28.02
July	0.73	23	9.8	306	0.934	28.95
August	0.73	23	9.8	306	0.934	28.95
September	0.73	22	9.8	296	0.934	28.02
October	2.34	72	31.5	977	2.985	92.53
November	4.72	141	63.5	1,905	6.016	180.48
December	7.15	222	96.1	2,980	9.108	282.31
Total per year	1,200		16,147		1,530	

Table 2. Comparison of purchase and plumbing labor, and maintenance cost for a conventional gas boiler and micro cogeneration system.

	Purchase (€)	Installation cost (€)	Inspection and maintenance service (€)
Gas boiler with cogeneration	3,500	750	150
Typical gas boiler	1,500	800	150
Cost difference	2,000	50	0

environment-friendly energy system in comparison with the cost of using another energy source. In the case in question cost lowering was computed assuming the energy cost of 1 kWh equal 0,15€. When total, yearly produced energy is 1200 kWh, the total savings can reach $\Delta K_e = 175€$ per year.

In order to estimate payback time, in which investment can be compensated by achieved savings, yearly cost reduction ΔK_e was mainly taken into account, and the total cost N_i was calculated as the difference between investment of microCHP system and a system with a typical boiler. Simple Payback Time (SPBT) was calculated with the use of the following equation:

$$SPBT = \frac{N_i}{\Delta K_e}, \text{ years} \qquad (3)$$

When energy costs are considered as constant, the total repayment will occur after 11 years of the system usage. When there is a 40% potential financial support from Polish National Fund for Environmental Protection and Water Management (the same as applied for solar systems) and gradual increase in fuel cost, the repayment can be achieved after 5 years.

4 CONCLUSIONS

The computer simulation research of the micro-CHP system applied for a typical Polish household proved system capability for both producing heat and generating electrical energy which can support the house needs. The total repayment of cost difference between purchasing classical gas boiler and microCHP can be seen after 11 years of the system usage in Polish conditions. This indicates that potential financial support from the Polish National Fund for Environmental Protection and Water Management will make this technology cost efficient and in consequence will lead to saving energy and reducing greenhouse gasses production on the country scale.

REFERENCES

Kalina, J. 2003. Small-scale cogeneration in Poland. 2003 market report. Gliwice. Silesian University of Technology.

Klugmann-Radziemska, E. 2009. *Renewable Energy Sources*. Examples of Calculation. Gdańsk.

Pavlova-Marciniak I. 2006: "Unconventional energy technologies and environmental preservation", *Przegląd Elektrotechniczny* 9'2006.

Polish Ministry of Environment. 2003. Strategies of Green-House Gases Reduction in Poland by 2020 year.

Twidell, J. & Weir, T. 2007. *Renewable Energy Resources*. CRC Press Taylor and Francis Group, second edition.

Environmental Engineering IV – Pawłowski, Dudzińska & Pawłowski (eds)
© *2013 Taylor & Francis Group, London, ISBN 978-0-415-64338-2*

The influence of leachate composition on anaerobic digestion stability and biogas yields

M. Lebiocka & A. Montusiewicz

Faculty of Environmental Engineering, Lublin University of Technology, Lublin, Poland

ABSTRACT: In this paper the leachate composition influence on anaerobic digestion stability and biogas production was presented. Co-digestion of Sewage Sludge (SS) and landfill Leachate (L) requires an adequate period of microorganisms acclimatization. The results show that addition of old leachate as a co-substrate to anaerobic digestion process requires longer acclimatization time about 75 days, while using leachate of medium age, 60-day acclimatization is adequate. The addition of medium leachate (landfill aged from 5 to 10 years) to anaerobic digestion process of sewage sludge in ratio SS:L = 20:1 and 10:1 provided the growth of biogas yield for about 50 and 15% respectively, compared with biomethanization process of only sewage sludge. In this case biogas yield was $1.55 \, m^3 \, kg^{-1}VS_{rem}$ (20:1) and $1.18 \, m^3 \, kg^{-1}VS_{rem}$ (10:1). Application of leachate obtained from long age landfill (>10 years) in ratio 20:1 and 10:1 resulted in biogas yield decrease for about 30%, despite achieving fermentation stable conditions.

Keywords: landfill leachate, mixed sewage sludge, anaerobic co-digestion, biogas production, microorganisms adaptation

1   INTRODUCTION

Commercial and industrial development, technological and technical rise and constantly growing living standards increase the volume of generated solid waste (Hoedl 2011, Michałowki 2012). Whereas the production of municipal solid waste grows in per capita and overall terms, thus landfilling is still an important issue of waste management system. Landfilling of municipal solid waste due to cheap exploitation and low operational costs is widely used in Europe and the rest of the world. During landfilling, solid waste undergoes a series of biological and physicochemical changes.

Consequently the degradation of organic fraction of municipal solid waste in connection with percolating rainwater leads to the generation of leachate. Leachate can potentially contaminate nearby ground and surface water (Wiszniowski et al. 2006). Leachate contain large amounts of organic matter (both biodegradable and refractory to biodegradation), ammonia nitrogen, heavy metals and toxic substances such as xenobiotic organic compounds. Leachate composition depends on waste type and their compaction, landfill hydrology, climate and landfill age. During landfilling leachate goes through aerobic, acetogenic and methanogenic stages of organic wastes stabilization. The characteristic of landfill leachate can be represented by basic parameters such

as COD, BOD, BOD/COD ratio, pH, ammonium nitrogen (Kjeldsen et al. 2001). The literature data show that the age of the landfill and the degree of organic fraction of municipal solid waste has a significant effect on leachate composition (Table 1).

The variability in landfill leachate quality and quantity make designing of a universal leachate treatment system complicated. Conventional treatment includes: i) leachate transfer, e.g. co-treatment with domestic sewage and leachate recirculation back through the tip; ii) biological aerobic or anaerobic treatment; and iii) physicochemical treatment that applies flotation, coagulation/flocculation, chemical precipitation, adsorption, chemical oxidation, ammonia stripping, ion exchange and electrochemical method. The biological method of landfill leachate treatment is the most effective to eliminate nitrogen from leachate. However, this method is inhibited by the toxic substances and by the presence of refractory organic compounds in leachate (Christensen et al. 2001, Welander et al. 1997, Li et al. 2007). Conventional treatment methods like air stripping and coagulation/flocculation are often expensive in terms of energy requirements and use of additional chemicals. Other, physicochemical methods such as reverse osmosis, active carbon adsorption, advances oxidation process for landfill leachate treatment of technical scale are not economically acceptable. Among many technological approaches to leachate treatment, co-digestion of leachate and sewage sludge seems to be

Table 1. Composition of landfill leachate with age (Renou S. et al., 2008, Kulikowska and Klimiuk, 2008).

Parameter	Young	Intermediate	Old
Age (years)	<5	5–10	>10
pH	<6.5	6.5–7.5	>7.5
COD (mg/L)	>10 000	4000–10 000	<4000
BOD/COD	0.5–1.0	0.1–0.5	<0.1
Organic compounds	80% VFA	5%–30% VFA + humic and fulvic acids	Humic and fulvic acids
NH_3-N (mg/L)	<400	N.A.	>400
Heavy metals	Low to medium	Low	Low
Biodegrability	High	Medium	Low

worth considering. Anaerobic treatment of sewage sludge leads both to sludge volume reduction and to biogas production. However, operational data have indicated possible reserves of the digesters capacity, frequently as much as 30% (Braun 2002). What is more, it is possible to introduce additional components to the co-digestion process. Considering the general problems related to one-source waste fermentation, co-digestion seems to be a promising solution. Benefits of co-digestion consist in (Cecchi et al. 1996):

– dilution of toxic substances coming from any of the substrates involved,
– improved nutrient balance,
– higher biogas yield and an increased load of biodegradable organic matter,
– synergetic effects on microorganisms,
– high digestion rate,
– possible removal of some xenobiotic organic compounds.

Using landfill leachate as a co-substrate to the co-digestion process seems to be a promising solution due to a high concentration of leachate contaminants, particularly soluble organic matter (expressed as soluble COD, TOC and VFA). However, there are some issues that have to be considered. They include mostly the landfill age and the corresponding phase of refuse decomposition, commonly known as major determinants of leachate quality (El-Fadel et al. 2002). In the present study the influence of leachate composition on anaerobic digestion stability and digest quality was presented.

2 MATERIALS AND METHODS

2.1 Material characteristics

Sewage sludge (primary and secondary) was obtained from Pulawy municipal Wastewater Treatment Plant (WWTP). Sludge originating

from gravity thickener (i.e. primary thickened sludge) and from a mechanical belt thickener (i.e. waste thickened sludge) was used as material for the present study. The characteristics of mixed sludge from Puławy WWTP are shown in Table 2.

Leachate was achieved from Rokitno municipal solid waste landfill, from intermediate (medium) and mature (old) basins. Leachate composition is presented in Table 3.

2.2 Sample preparation procedure

Sludge was sampled once a week in Puławy municipal wastewater treatment plant and then provided immediately to the laboratory of the Lublin University of Technology (Poland). Primary sludge and waste sludge were transported in separate containers. Under laboratory conditions, sludge was mixed at volume ratio of 60:40 (primary: waste sludge), then homogenized, manually screened through a 3 mm screen and partitioned. The sludge samples were frozen at −25°C in laboratory freezer and thawed daily for 12 h at 20°C in the indoor air. Sludge prepared in this manner was considered as mixed sludge, which fed the reactor. The preliminary experiments showed that the mixed sludge frozen completely at −25°C and at least 12 h were required to completely thaw the samples at the room temperature.

Leachate was sampled once as averaged collected sample taken from leachate storage tank with the capacity of 25 m³ and transported, as soon as possible, to the laboratory. To ensure the same experiment conditions for co-digestion components, leachate samples were prepared and stored similarly to the sludge ones.

2.3 Laboratory installation for co-digestion process

The laboratory installation consisted of an anaerobic, completely mixed, hermetic reactor equipped

Table 2. Characteristics of mixed sludge from Puławy WWTP during experiments.

Parameter	Unit	Average values
COD	mg/L	42519
TOC[a]	mg/L	720
VFA[a]	mg/L	1710
TS	g/kg	39,0
VS	g/kg	28,7
pH	–	6,36
Alkalanity	mg/L	956
N-NH$_4$$^{+a}$	mg/L	150,8
P-PO$_4$$^{3-a}$	mg/L	176,2

[a]concentration determined in supernatant.

Table 3. Leachate composition.

Parameter	Unit	Intermediate landfill leachate	Old landfill leachate
COD	mg/L	1990	6615
BOD	mg/L	290	355
BOD/COD	–	0,15	0,05
TOC	mg/L	761	2495
pH	–	7,84	8,12
Alkalanity	mg/L	8500	15050
VFA	mg/L	423	426
TS	g/kg	8,55	12,05
VS	g/kg	2,5	2,24
N-NH$_4$$^+$	mg/L	1040	1390
P-PO$_4$$^{3-}$	mg/L	4,9	6,2

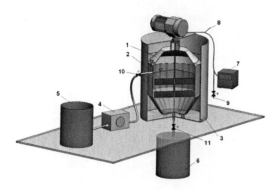

Figure 1. Laboratory installation for co-digestion process. 1—anaerobic reactor, 2—mechanical stirrer, 3—heating jacket, 4—influent peristaltic pump, 5—influent storage vessel, 6—effluent storage vessel, 7—drum gas meter, 8—gaseous installation and gas sampler with a rubber septum, 9—dewatering valve, 10—inlet valve, 11—outlet valve.

with gaseous installation, an influent peristaltic pump and storage vessels. Anaerobic reactor with a working volume of 40 dm^3 was inserted into the heating jacket at stable temperature of 35°C. Mixing was carried out using a mechanical stirrer with rotational speed of 50 min^{-1}. Influent was supplied to the upper part of digester, whereas effluent was wasted through the bottom by gravity. The biogas installation was attached at the headspace of the reactor. The gas system consisted of pipelines linked with the pressure equalization unit and the drum gas meter. This was equipped with gas valves, a dewatering valve and a gas sampler with a rubber septum, which enabled insertion of a syringe with pressure lock. The laboratory installation is shown in Figure 1.

2.4 Operational set-up

An inoculum for the laboratory reactor was taken from Puławy wastewater treatment plant as a collected digest from a mesophilic anaerobic digester operating at 35–37°C with a volume of 2500 m^3

and a Hydraulic Retention Time (HRT) about 25 d. The adaptation of the digester biomass was achieved after 30 days. The study was carried out in the reactor operating at a controlled mesophilic temperature of 35°C and in a semi-flow mode (digester was supplied regularly once a day). Two phases with three runs were conducted and each experiment lasted over 60 days.

2.5 The first phase of experiment

In the first run (1.R1) the reactor was fed daily with 2 L of mixed sludge. Hydraulic retention time reached 20 days and hydraulic loading rate was 0.05 d^{-1}. Organic Loading Rate (OLR) was time-dependent with the average value of 1.47 kg VS m^{-3}d^{-1}. The second run (1.R2) was conducted according to previous schedule, however, reactor was fed using sludge with medium leachate addition in a volumetric ratio of 20:1 (the influent consisted of the sludge and leachate mixture in the proportion of 2 L and 100 mL, respectively). HRT reached 19 days and hydraulic loading rate was 0.053 d^{-1}. Average value of OLR was 1.51 kg VS m^{-3}d^{-1}. In the third run, the arrangement was the same as in 1.R2 but a volumetric ratio was assumed at 10:1 (influent consisted of the sludge and leachate mixture in the proportion of 2 L and 200 mL, respectively). HRT reached 18 days and hydraulic loading rate was 0.055 d^{-1}. Average value of OLR was 1.36 kg VS m^{-3}d^{-1}.

2.6 The second phase of experiment

In the first run off the second phase (2.R1) the reactor was fed daily with 2 L of mixed sludge.

Hydraulic retention time reached 20 days and hydraulic loading rate was 0,05 d^{-1}. Organic Loading Rate (OLR) was time-dependent and ranged from 1,50 to 1,55 kg VS m^{-3}d^{-1}. During second run (2.R2) reactor was fed with 2,1 L using sludge with old leachate addition in a volumetric ratio of 20:1. HRT reached 19 days and hydraulic loading rate was 0,053 d^{-1}. OLR ranged from 1,38 to 1,48 kg VS m^{-3}d^{-1}. In the third run (2.R3) arrangement was the same as in R2, but a volumetric ratio was assumed at 10:1 thus feed was 2,2 L. HRT reached 18 days and hydraulic loading rate was 0,055 d^{-1}. OLR ranged from 1,61 to 1,69 kg VS m^{-3}d^{-1}.

2.7 Analytical methods

In mixed sludge the following parameters were analysed once a week: Volatile Fatty Acids (VFA), Volatile Solids (VS), alkalinity and pH level. The same schedule was used for determining the values of parameters that characterized supernatant (sludge liquid phase) before digestion: Total Organic Carbon (TOC), ammonia nitrogen (N-NH$_4^+$), nitrite and nitrate nitrogen (N-NO$_x^-$) and ortho-phosphate phosphorus (P-PO$_4^{3-}$). The supernatant samples were obtained by centrifuging the sludge at 4000 r/min for 30 min. Leachate composition was determined once, after providing it to the laboratory. The following parameters were analyzed: BOD$_5$, COD, TOC, TS, VS, alkalinity, pH, N-NH$_4^+$, N-NO$_x^-$, P-PO$_4^{3-}$. In digested sludge specified parameters (VFA, Alkalinity, pH, VS) were determined three times a week.

Most analyses were carried out in accordance with the procedures in Polish Standard Methods. The anaerobic digestion efficiency was controlled by the daily evaluation of the biogas yield. Biogas production was determined with the use of Drum-type Gasmeter TG-Series (Ritter, Germany).

3 RESULTS AND DISCUSSION

Many authors suggested volatile fatty acids as control parameter of the anaerobic digestion condition. These acids are indicative of the activity of the methanogenic consortia (Münch et al. 1998, Strik et al. 2006, Madsen et al. 2011). In case of VFA accumulation during biomethanization process this can suggested either inhibition of the methanogenic microorganisms or organic overloading. Variability of VFA concentrations in reactor during two phases of experiment is shown in Figures 2 and 3.

The VFA concentration in R1 at the end of the run was about 208 mg/L, while in the supernatant from WWTP in Pulawy was on average 940 mg/L. Addition of medium age landfill leachate, in

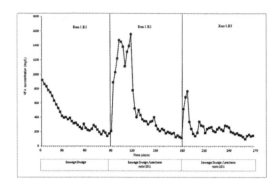

Figure 2. Variability of the VFA concentrations during the first phase of experiment.

dose 20:1 (sewage sludge: landfill leachate) as a co-substrate in anaerobic digestion process of sewage sludge resulted in a significant increase of VFA concentration in digester effluent. The high level of VFA concentration maintained for 30 days. Then gradually decreased and after 60 days the VFA concentration reached a value indicating stable process conditions (Fig. 2). The increase of the share of leachate in the mixture feeding the reactor (dose 10:1) also caused an increase of VFA concentration in the digest but it was smaller than R1 and R2. Stable conditions were obtained after 30 days of the experiment. Addition of old landfill leachate (age > 10 years) in dose 20:1 also resulted in a significant increase of VFA in the digester effluent. Acclimatization time during this phase of experiment was longer and lasted about 75 days (Fig. 3). Increasing the dose of leachate (10:1) resulted in increase of the VFA concentration in the digest but stable process conditions were obtained after 30 days. The results show that addition of old leachate as a co-substrate to anaerobic digestion process requires longer acclimatization time about 75 days, while during co-digestion of sewage sludge and medium age leachate microorganisms require 60-day acclimatization.

According to Laloui-Carpentier et al. (2006) adaptation of the microorganisms in anaerobic digestion reactor was probably supported by an increase in share of the microbial community in the medium landfill leachate. This thesis was confirmed by Vavlilin et al. (2006), who demonstrated an advantages effect of methanogenic leachate recirculation on the degradation of solid waste on landfills. There is no data showing the influence of the old landfill leachate on the microorganisms adaptation during anaerobic digestion process.

Another parameter preferable to monitor the anaerobic digestion process in alkalinity. This parameter allows the timely detection of changes of

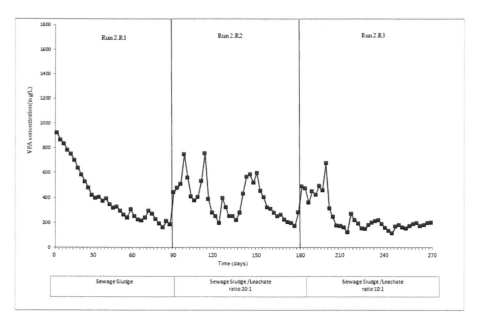

Figure 3. Variability of the VFA concentrations during the second phase of experiment.

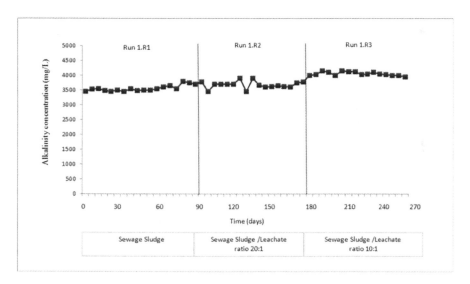

Figure 4. Variability of the alkalinity during co-digestion of sewage sludge and medium landfill leachate.

the process buffer capacity and, as a consequence, a more accurate information about the risk of process inhibition by the VFA. Variability of alkalinity concentrations in reactor during two phases of experiment is shown in Figures 4 and 5.

The present results indicate that alkalinity concentration during two phases of experiment was at comparable levels. The highest differences in alkalinity concentrations were observed during co-digestion of sewage sludge and old landfill leachate, when leachate dose was 10:1 (2.R3). The alkalinity concentration in Run 2.R3 was higher by about 25% than in 2.R1 and 18% than 2.R2. Méndez-Acosta et al. (2009) and Şentürk et al. (2010) emphasized the need for balance between VFA concentration and alkalinity (VFA/Alkalinity). They suggested that such ratio should be below 0.35. When VFA/A ratio exceed 0.35 it is believed

503

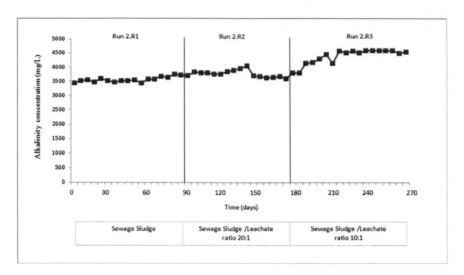

Figure 5. Variability of the alkalinity during co-digestion of sewage sludge and old landfill leachate.

that it has an inhibitive effect on biogas production and can lead to the collapse of the anaerobic digestion process. The VFA/Alkalinity ratios for all runs during two phases of experiment are shown in Table 4.

Taking into account the achieved values of the VFA/Alkalinity ratio it can be pointed out that process (in both phases of experiment) was carried out in stable conditions. Anaerobic bacteria, specially the methanogens, are sensitive to the acid concentration in the digester and their growth can be inhibited by acidic conditions. It has been determined that an optimum pH value for microorganisms leading anaerobic digestion process is between 5.5 and 8.5. The average pH values in the digester influent and effluent are presented in Table 5.

The obtained results indicate that anaerobic digestion in laboratory conditions was performed in stable conditions. Addition of leachate to the mixture feeding the reactor resulted in insignificant reduction of the pH values. According to Chen et al. (2008) methane production is stabilized, the pH level in digest stays between 7.2 and 8.2. The pH values in effluent in both phases of experiment were at a comparable level, and did not exceed the value of 8.2.

Taking into account all the obtained results it can be stated that anaerobic co-digestion of landfill leachate and sewage sludge was carried out in stable process conditions. Biogas production during the experiment was shown in Figure 6.

The addition of medium leachate (landfill aged 5–10 years) to anaerobic digestion process of sewage sludge in ratio SS:L = 20:1 and 10:1 provided the growth of biogas yield for about 50 and 15%,

Table 4. The VFA/Alkalinity ratios during experiment.

	VFA/Alkalinity [–]
First phase of experiment	
Run 1.R1	0.07
Run 1.R2	0.156
Run 1.R2	0.06
Second phase of experiment	
Run 2.R1	0.07
Run 2.R2	0.1
Run 2.R3	0.05

Table 5. The average pH values in the digester influent and effluent.

	Influent	Effluent
First phase of experiment		
Run 1.R1	6.58	7.78
Run 1.R2	6.21	7.80
Run 1.R3	6.50	7.87
Second phase of experiment		
Run 2.R1	6.58	7.78
Run 2.R2	6.20	7.73
Run 2.R3	6.40	7.81

respectively, compared with biomethanization process of only sewage sludge. In this case biogas yield was 1.55 m^3 kg^{-1}VS$_{rem}$ (20:1) and 1.18 m^3 kg^{-1}VS$_{rem}$ (10:1). Application of old leachate (>10 years) in ratio 20:1 and 10:1 resulted

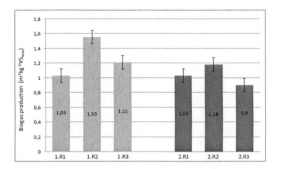

Figure 6. Biogas production during the different runs in the experiment (error bars present confidence level, $\alpha = 0.05$).

in biogas yield decrease for about 30%, despite achieving fermentation stable conditions.

4 CONCLUSIONS

Co-digestion of sewage sludge and landfill leachate requires an adequate period of microorganisms acclimatization. The results show that addition of old leachate as a co-substrate to anaerobic digestion process requires longer acclimatization time about 75 days, while during co-digestion of sewage sludge and medium age leachate microorganisms require 60-day acclimatization. The efficiency of co-digestion in stabilized process conditions was dependent on the leachate composition and its share in the reactor feedstock (in the mixture feeding the reactor). The addition of medium leachate (in ratio 20:1 and 10:1) provided the growth of biogas yield for about 50 and 15%, respectively, compared with biomethanization process of only sewage sludge. Application of leachate obtained from old landfill in ratio 20:1 and 10:1 resulted in biogas yield decrease compared with co-digestion process of sewage sludge and medium landfill leachate. Differences in biogas production during experiment resulted probably from different landfill leachate composition.

REFERENCES

Braun, R. 2002. Potential of co-digestion: limits and merits. Report. IEA Bioenergy Task 37, 3 (http://www.novaenergie.ch/iea-bioenergy-task37/publicationsreports.htm).
Cecchi F., Pavan P., Mata-Alvarez J. 1996. Anaerobic co-digestion of sewage sludge: application to the macroalgae from the Venice lagoon, Resources. *Conservation and Recycling*, 17, 57–66.
Chen Y., Cheng J.J, Creamer K.S. 2008. Inhibition of anaerobic digestion process: A review. *Bioresource Technology*, 99, 4044–4064.

Christensen T.H., Kjeldesen P., Bjerg P.L., Jensen D.L., Christensen J.B., Baun A., Albrechtsen H.J., Heron G. 2001. Biogeochemistry of landfill leachate plumes. *Applied Geochemistry*, 16, 659–718.
El-Fadel M., Bou-Zeid E., Chahine W., Alayli B. 2002. Temporal variation of leachate quality from pre-sorted and baled municipal solid waste with high organic and moisture content. *Waste Management*, 22, 269–282.
Hoedl E. 2011. Europe 2020 Strategy and European Recovery. *Problems of sustainable development*, 6(2), 11–18.
Kjeldsen P., Barlaz M.A., Rooker A.P., Baun A., Ledin A., Christensen T. 2002. Present and long-term composition of MSW landfill leachate: a review. *Critical Review in Environmental Science of Technology*, 32(4), 297–336.
Kulikowska D., Klimiuk E. 2008. The effect of landfill age on municipal leachate composition. *Bioresource Technology*, 99, 5981–5985.
Laloui-Carpentier W., Li, T., Vigneron V., Mazéas L., Bouchez T. 2006. Methanogenic diversity and activity in municipal solid waste landfil leachates. *Antonie van Leeuwenhoek*, 89, 423–434.
Li Z.Z., Zhou S., Qiu J. 2007. Combined treatment of landfill leachate by biological and membrane filtration technology. *Environmental Engineering Science*, 24(9), 1245–1256.
Madsen M., Holm-Nielsen J.B., Esbensen K.H. 2011. Monitoring of anaerobic digestion process: A review perspective. *Renewable and Sustainable Energy Reviews*, 15, 3141–3155.
Méndez-Acosta H.O., Palacios-Ruiz B., Alcaraz-González V., González-Álvarez V. 2009. Improving the operational stability in continuous anaerobic digestion process, Congreso Anual 2009 de la Asociación de México de Control Automático. Zacatecas, México.
Michałowski A. 2012. Ecosystem services in the light of a sustainable knowledge-based economy. *Problems of Sustainable Development*, 7(2), 97–106.
Münch E., Greenfield P.F. 1998. Estimating VFA concentrations in prefermenters by measuring pH. *Water Research*, 32(8), 2431–2441.
Renou S., Givaudan J.G., Poulain S., Dirassouyan F., Moulin P. 2008. Landfill leachate treatment: Review and opportunity. *Journal of Hazardous Materials*, 150, 468–493.
Şentürk E., İnce M., Onkal Engin G. 2010. Treatment efficiency and VFA composition of a thermophilic anaerobic contact reactor treating food industry wastewater. *Journal of Hazardous Materials*, 176, 843–848.
Vavilin V.A., Jonsson S., Ejlertsson J., Svensson B.H. 2006. Modelling MSW decomposition under landfill conditions considering hydrolytic and methanogenic inhibition. *Biodegradation*, 17, 389–402.
Welander U., Henryson T., Welander T. 1997. Nitrification of landfill leachate using suspended-carrier biofilm technology. *Water Research*, 31, 2351–2355
Wiszniowski J., Robert D., Surmacz-Górska J., Miksch K., Weber J.V. 2006. Landfill leachate treatment methods: A review. *Environmental Chemical Letters*, 4, 51–61.

Biology and technology

Environmental Engineering IV – Pawłowski, Dudzińska & Pawłowski (eds)
© *2013 Taylor & Francis Group, London, ISBN 978-0-415-64338-2*

Effect of overburden management in open cast limestone mining on soil quality in reclamation of post-extraction pits

M. Kacprzak & M. Szewczyk

Faculty of Environmental Engineering and Biotechnology, Czestochowa University of Technology, Czestochowa, Poland

ABSTRACT: Open cast limestone mining is concentrated mainly in the southern part of Poland. The largest mine of the Opole province in terms of carbonate extraction is the Górażdże Limestone Mine. The effects of the extensive limestone extraction in this region are minimized by rational management of overburden. Overburden are used to form a top layer for future tree planting. Properly prepared layers of overburden affect primary physicochemical properties of soils and determine viability of planted seedlings and success in biological reclamation of these areas. In terms of physicochemical properties of initial soils studied in the post-calcareous pit, such as: accumulation of organic carbon, total nitrogen, calcium carbonate, available phosphorus, pH and conductivity in the subsurface layer (20 cm), the analyzed soils exhibit properties similar to those observed in natural zonal soils. Overburden soils are therefore a convenient substrate for planting *Pinus sylvestris* pine trees in a reclaimed pit of mine.

Keywords: limestone mine, reclamation, overburden, initial soil

1 INTRODUCTION

Mining of carbonate raw materials occupies a high position in Poland and it is supposed that the increase in production in this sector will be maintained in the coming years due to the development of engineering, residential and industrial construction (Dulewski & Wtorek 2000, Pietrzyk-Sokulska 2003). The province of Opole is in the forefront in terms of extraction of carbonate raw materials for cement and lime industry. This region is particularly rich in limestone, marls and dolomite (Kusza 2007, Kacprzak & Bruchal 2011). Apart from the rich deposits of carbonate raw materials in the Opole province and the positive aspects of their widespread use, the method of extraction poses a major problem. Opencast extraction of limestone results in serious transformations in the natural environment of the deposit, which has the most noticeable effect on the landscape in the form of post-calcareous pits (Dulewski & Wtorek 2000, Pietrzyk-Sokulska 2003, Hopper & Bonnier 2004, Siemek 2013). The harmful effect of mining activity that causes land transformation can be limited by means of a comprehensive reclamation of post-extraction areas (Maciejewska & Kwiatkowska 2002, Kusza 2007, Fijałkowski et al. 2008, Kacprzak & Bruchal 2011). An essential factor in successful reclamation of post-extraction areas is rational management of overburden, which consists in selective removal and tipping of rock material lying above the deposit (Mikłaszewski 1972). The selective method of tipping overburden material improves the efficiency of land reclamation, positively affects grain composition and basic physical and chemical properties of post-mining areas (Stachowski 2005). Tipped overburden is used to form humus layer, which is a soil substrate and determines the viability of planting trees and shrubs during the first years of cultivation in post-extraction areas of limestone mines.

The aim of the present study was to demonstrate the usefulness of overburden layers used for forest reclamation in post-calcareous pits by analysis of changes in physicochemical properties of the soil formed in sill pit of limestone mine reclaimed after 1 year and 5 years.

2 METHODS

The Górażdże deposit of Triassic limestone discussed in the study is located in Opole Silesia, Poland. Under the current system of administrative division in Poland, the Górażdże deposit belongs in more than 90% to the municipality of Gogolin. Only a small portion of this deposit in its northern part is located in the municipality of Tarnów Opolski. The immediate surroundings of the site are covered by forests and belong to the Forest District of Strzelce Opolskie and areas of the town of Górażdże. The extraction of the

Triassic limestone in Górażdże is conducted with the division into two pits, west and east one, which results from presence of a central protecting pillar in the deposit for the nature reserve Kamień Śląski. The Górażdże Limestone Mine is the largest mine in terms of the surface area and annual production in the region, where, apart from extraction of raw materials, reclamation of post-calcareous pits is also carried out. Management strategies for post-mining areas in Górażdże Mine are oriented towards forest reclamation.

Two research plots with a surface area of approximately 40 m² each were designated in the western pit of the Górażdże Limestone Mine, reclaimed after 1 year and 5 years. The areas designated for experimental research are currently overgrown with *Pinus sylvestris* introduced within biological reclamation procedures. Nine samples were collected from the surface layer (20 cm) of two plots in the spring of 2011 to represent the whole surface and mixed to represent the average sample. The collected soil samples were analyzed by means of physicochemical methods commonly used in soil science to determine the quality of soil formed from overburden layers used for reclamation of post-calcareous pits.

Laboratory tests were used to determine:

– granulometric composition: sieve method according to PN-ISO 11277:2005;
– soil pH: potentiometry method according to PN-ISO 10390:1997;
– conductivity: conductometric method according to PN-ISO 11 265 + ACI:1997;
– calcium carbonate content: Scheibler volumetric method according to PN-ISO 10693:2007;
– organic carbon content: Tiurin modified method according to PN-ISO 14235:2003;
– total nitrogen content: Kjeldahl method according to PN-ISO 11261:2002;
– content of available phosphorus according to PN-R-04 023:1996.

3 RESULTS AND DISCUSSION

In the case of opencast extraction of carbonate raw materials, overburden remains within the range from several centimeters to several meters and is entirely used for reclamation of post-calcareous pits. It usually consists of clay, silt, sand, and sometimes rock waste of varied thickness (Mikłaszewski 1972). Granulometric composition of formations that constitute overburden is an important factor which significantly affects the grain size of soils formed in post-mining areas and viability of the introduced reclamation plantings (Kusza 2007, Janecka & Sobik Szołtysek 2009). In the Górażdże

Mine, overburden material is formed by Quaternary formations: sand, clay and degraded rubble of variable thickness, from 0.3 m to 13.0 m (Dreszer 2005, Dulewska & Kusza, 2008). It was found, on the basis of the examination of granulometric composition in the surface layer with thickness of 20 cm in both experimental plots designated in the Górażdże Mine pit, that the dominant formation in the surface layer of both one-year and five-year plots is sandy formations with a predominance of sandy fractions of 0.25 mm. Overburden material deposited on the plots is a mixture of loose soil, characterized by a significant looseness, which is the most noticeable in the 20 cm layer of the one-year plot.

Similar to soils formed from overburden, pH in overburden layers is, apart from grain size, the second major factor in soil productivity in the post-mining lands (Stachowski 2005, Kusza 2007, Kacprzak & Bruchal 2012). In the opencast limestone mining, pH in rock material from overburden ranges between 7.3 and 7.6, i.e. at the level of alkaline reaction (Mikłaszewki 1972). The pH of the soil that forms the top layer in the surface studied is maintained at the alkaline levels of 7.8 and 7.6 for one-year and 5-year plot, respectively. Over the subsequent years of biological reclamation, the soils formed in the post-calcareous pits showed a small decrease in pH in the surface layer, which is confirmed by the results obtained for pH and studies carried out by Kusza (2007) in two dumping grounds in the area of a former limestone mine, forested 45 years ago. The initial anthropogenic soils present in this site exhibited constant alkaline pH at the level of from 7.3 to 7.6 in the Kamień Śląski Mine and 7.3 to 7.9 in Strzelce Opolskie Mine.

The low acidification rate in surface layers of initial soils in post-mining lands is caused by significant contents of calcium carbonates (Spychalski & Gilweska 2008, Kacprzak & Bruchal 2012), which reached the levels of 7.45% and 6.46% in the layer with thickness of 20 cm in 1-year plot and 5-year plot, respectively. According to Stachowski (2005), this significant content of calcium carbonate results from genetic properties of the overburden material used for forming the pit surface.

Furthermore, soils formed from overburden material showed low salinity in surface layers, which, in both 1-year and 5-year plot studied, did not exceed 100 μs/cm⁻¹ (Table 1). Over the subsequent years after biological reclamation, the levels of soil conductivity in the post-calcareous pits remained at an average of 100–280 μs cm⁻¹ in a study carried out by Kusza (2007) in Opole Limestone Mines and 275 μs cm⁻¹ in a mine in Spain examined by Jordan et al. (2009). Enrichment of subsurface layers with organic matter over the years

Table 1. Physical and chemical properties of surface layers of reclaimed plots in the sill limestone mine.

Parameters	Layer 0–20 cm	
	1-year plot	5-year plot
pH (KCL)	7,8 ± 0,07	7,6 ± 0,08
Electrical conductivity [µS cm^{-1}]	56 ± 9,21	89,2 ± 18,27
Organic carbon [%]	0,19 ± 0,05	0,56 ± 0,13
Total nitrogen [%]	0,028 ± 0,00	0,056 ± 0,01
Calcium carbonate [%]	7,45 ± 1,07	6,46 ± 1,22
Available phosphorus [mgP$_2$O$_5$ kg^{-1} soil]	10,09 ± 5,42	12,17 ± 5,90

of forestation is an important factor in the activation of soil forming processes in post-mining areas (Gilewska & Otremba 2004, Wójcik & Kowalik 2006, Kacprzak & Bruchal 2012). The studied areas of forestation (1-year and 5-year plants) show significant changes in the dynamics of organic matter content in the subsurface layer. In the 5-year surface, the contents of nitrogen and carbon in the layer with thickness of 20 cm amounts to 0.056% of total nitrogen, 0.56% of organic carbon and is twice higher compared to the 1-year surface (Table 1). The noticeable increase in the content of organic matter in the subsequent years of forestation in reclaimed areas was also demonstrated in a study by Kusza (2007). The dumping grounds in the two limestone mines examined by the author and results of the studies showed that the content of organic matter in the surface layer was at the level of 0.72% of organic carbon and 0.018% of total nitrogen in the Kamień Śląski Mine and 0.61% organic carbon and 0.020% total nitrogen in the Strzelce Opolskie Mine. However, these levels of accumulated organic matter were small considering that these areas were forested 45 years before. The results obtained by the author showed that the post-mining soil remained over the years the ecosystem which accumulated and immobilized nitrogen (Bender & Gilewska 2004, Spychalski & Gilewska 2008, Kacprzak & Bruchal 2012).

Apart from the accumulation of organic matter and weathering processes in overburden material, the effect of the changes observed in the post-extraction pit is the increase in abundance of the top layers of mineral nutrients, especially phosphorus (Wójcik & Kowalik 2006). A moderate increase in abundance of available phosphorus in the top 20-cm layer of forming soil is observed in the plots over the years of the study. This effect is caused by progressive accumulation of organic matter and changes in pH levels, which promotes absorption of nutrients. The content of phosphorus measured in the top layer of the analyzed 1-year

and 5-year plots was low, at the level of 10.08 and 12.17 mg P$_2$O$_5$ kg^{-1} soil, respectively. On the basis of the results obtained in 1-year and 5-year plots designated in the sill in the Górażdże Limestone Mine, it can be concluded that overburden material used for reclamation of the pits is a convenient soil substrate, in which particular changes in physicochemical properties occur over the subsequent years towards progressing activation of soil-forming processes. The advanced soil-forming processes that occur in the surface layer are confirmed by the fact that the analyzed soils begin to exhibit properties similar to non-degraded soils located beyond the effect of the mining area. The redominant type of soils in the region of the mine are rendzinas i.e. soils formed during the process of limestone rock weathering. Rendzinas with different degree of development can be found in the area of the mine, from initial soils to humus rendzinas. These are soils with pH close to neutral, rich in the forms of nitrogen which are bioavailable to plants, with a relatively high agricultural utility (Dulewska & Kusza 2008).

4 CONCLUSIONS

The results of the studies conducted in two experimental plots in a sill of a limestone mine, obtained after one year and five years of reclamation, lead to the following conclusions:

1. Physical and chemical analyzes carried out in the layer of soil with the thickness of 20 cm confirm a beneficial direction of soil forming processes stimulated by particular overburden management.
2. Significant trends of changes concerned mainly the accumulation of organic substances such as nitrogen and carbon in the surface layer studied.
3. The overburden material used for reclamation is a potentially suitable soil substrate, which, over the subsequent years, is transformed into the initial soil with properties similar to the surrounding non-degraded soils.

REFERENCES

Bender, J. & Gilewska, M. 2004. Rekultywacja w świetle badań i wdrożeń. *Roczniki gleboznawcze* LV (2): 29–46.

Dulewski, J. & Wtorek, L. 2000. Problemy przywracania wartości użytkowych gruntom zdegradowanym działalnością górniczą. *Inżynieria ekologiczna* 1: 14–22.

Fijałkowski, K.; Janecka, B.; Kacprzak, M. & Bień, J. 2008. Określenie wpływu różnych dodatków organicznych na uzyskanie pokrywy roślinnej zwałowiska odpadów cynkowo-ołowianych. *Inżynieria i ochrona środowiska* 11(2): 141–148.

Hopper, B.D. & Bonnier, F.J. 2004. *Reclamation of the Irish Cove Limestone Quarry.* Report of Activities 2003, Nova Scotia Department of Natural Resources: 15–24.

Janecka, B. & Sobik-Szołtysek, J. 2009. Badania przydatności wybranych technik remediacji terenów zdegradowanych działalnością przemysłu cynkowo-ołowiowego. *Inżynieria i ochrona środowiska* 12(4): 281–294.

Jordán, M.M.; García-Sánchez, E.; Almendro-Candel, M.B.; Navarro-Pedreño, J.; Gómez-Lucas, I. & Melendez, I. 2009. Geological and environmental implications in the reclamation of limestone quarries in Sierra de Callosa (Alicante, Spain). *Environmental Earth Science* 59: 687–694.

Kacprzak, M. & Bruchal, M. 2011. Procesy rekultywacji terenów pogórniczych na przykładzie Kopalni Wapienia Górażdże. *Inżynieria i ochrona środowiska* 14(1): 49–58.

Kacprzak, M. & Bruchal, M. 2012. Zmiany właściwości fizykochemicznych gleb w trakcie prowadzenia procesu rekultywacji w kierunku leśnego zagospodarowania terenów kopalni wapienia [w:] *Rekultywacja i rewitalizacja terenów zdegradowanych.* Monografia.

Kusza G. 2007. Zmiany wybranych właściwości fizyko-chemicznych zwałowisk poeksploatacyjnych kopalni wapieni Strzelce Opolskie i Kamień Śląski, pod wpływem roślinności wysokiej [w:] *Ochrona środowiska na uniwersyteckich studiach przyrodniczych.* Monografia. Opole.

Maciejewska, A. & Kwiatkowska, J. 2002. Przydatność preparatów z węgla brunatnego do rekultywacji gruntów pogórniczych. *Inżynieria i ochrona środowiska* 5(1): 55–66.

Mikłaszewski, A. 1972. Rekultywacja w górnictwie skalnym. *Górnictwo odkrywkowe* 11–12: 372–378.

Pietrzyk-Sokulska, E. 2003. Eksploatacja surowców skalnych—problem nieużytków pogórniczych. *WUG* 5: 33–34.

PN-ISO 11277:2005. Jakość gleby. Oznaczanie składu granulometrycznego w mineralnym materiale glebowym. Metoda sitowa i sedymentacyjna.

PN-ISO 10390:1997. Jakość gleby. Oznaczanie pH.

PN-ISO 11265+ACI:1997. Jakość gleby. Oznaczanie przewodności elektrolitycznej właściwej.

PN-ISO 10693:2007. Jakość gleby. Oznaczanie zawartości węglanów. Metoda objętościowa.

PN-ISO 14235:2003. Jakość gleby. Oznaczanie zawartości węgla organicznego przez utlenianie dwuchromianem(VI) w środowisku kwasu siarkowego (VI).

PN-ISO 11261:2002. Jakość gleby. Oznaczanie azotu ogólnego. Zmodyfikowana metoda Kjeldahla.

PN-R-04023:1996. Analiza chemiczno-rolnicza gleby. Oznaczanie zawartości przyswajalnego fosforu w glebach mineralnych.

Rosik-Dulewska, Cz. & Kusza, G. 2008. Projekt rekultywacji wyrobisk powstałych w wyniku eksploatacji złoża Górażdże w Górażdżach, Opole.

Siemek, J. Nagy, S. Siemek, P. 2013, Challenges for Sustainable Development: The case of shale gas exploitation in Poland. *Problemy Ekorozwoju—Problems of Sustainable Development*, 8(1): 91–104.

Stachowski, P. 2005. Wpływ zabiegów rekultywacyjnych na właściwości gruntów pogórniczych. VII Ogólnopolska Konferencja Naukowa. Koszalin.

Spychalski, W. & Gilewska, M. 2008. Wybrane właściwości chemiczne gleby wytworzonej z osadów pogórniczych. *Roczniki gleboznawcze* LIX (2): 207–214.

Wójcik, J. & Kowalik, S. 2006. Kształtowanie się wybranych właściwości inicjalnej gleby na zrekultywowanej w kierunku leśnym hałdzie górnictwa miedzi. *Inżynieria środowiska* 11(1): 87–99.

Environmental Engineering IV – Pawłowski, Dudzińska & Pawłowski (eds)
© 2013 Taylor & Francis Group, London, ISBN 978-0-415-64338-2

Application of non-ionic surfactant in the remediation of PAHs contaminated soil

K. Bułkowska & Z.M. Gusiatin
Faculty of Environmental Sciences, University of Warmia and Mazury in Olsztyn, Olsztyn, Poland

ABSTRACT: In the present study, the efficiency of phenanthrene removal from artificially contaminated soil by nonionic Rokanol NL8 was investigated. The experiments were performed under static conditions, at constant soil to solution ratio (m/V = 1/5), and under dynamic conditions, using a column reactor at flow rate of 2 mL/min. Three series at 0.1%, 0.5% and 1% Rokanol NL8 concentration were conducted. At concentration 0.1%, the efficiency of phenanthrene removal did not exceed 2% in both conditions. In static conditions, at 0.5% and 1% Rokanol NL8, the process efficiency was about 41.72% and 70.46%, respectively. In dynamic conditions, at 0.5% and 1% of surfactant concentration, the phenanthrene removal from soil increased through the time and reached 93.30% regardless of surfactant concentration.

Keywords: non-ionic surfactant, phenanthrene, CMC, soil washing

1 INTRODUCTION

Anthropogenic sources of PAHs are industrial processes (gasification and coking, petroleum processing), power and heat generation, waste incineration and traffic. Nadal et al. (2004) estimated the sum of 16 PAHs concentrations in soils near chemical industries in Catalonia (Spain) it was 1002 µg/kg, whereas in residential zone—736 µg/kg. Gong et al. (2006) investigated two soils from a former manufactured gas plant site polluted for several decades in Berlin (Germany). The content of 13 PAH concentrations ranged from 724 mg/kg to 4721 mg/kg. In Poland, the sum of 9 PAHs (2–6 rings) in soil from coking plant area was 280 mg/kg (Czaplicka 2004). As soil contamination, Polycyclic Aromatic Hydrocarbons (PAHs) show many threats to environment as they have been classified as carcinogens, mutagens and immunosuppressants (Prithard 2006). In term of soil remediation, low volatility, low water solubility and poor biodegradability of PAHs creates technological problems. For that reason, initially thermal methods were applied (McGowan et al. 1996). However, presently they are not recommended due to high cost and destroying of soil structure. On the contrary, thermal desorption enables separation of organic compounds from soils without the combustion of the media or contaminants (Gitipour et al. 2011). However, it is also a cost-ineffective process (on average $200 US/t of soil) (Khan et al. 2004). Presently, soil washing is extensively used in Europe (López-Vizcaíno et al. 2012) for rapid remediation of soils highly contaminated by heavy metals and/or organic pollutants (Li et al. 2011). However, the process efficiency depends on many factors: a) soil type, b) washing agent type, and c) operational conditions.

In general, soils with low cation exchange capacity (50–100 meq/kg) and particle sizes of 0.25–2 mm can be most effectively cleaned by soil washing (Mulligan et al. 2001). However, the soil rich in organic matter diminish the process efficiency because PAHs tend to be strongly sorbed on soil organic matter (Rulkens & Honders 1996; Chang et al. 2000). For that reason, the selection of washing agents is especially important. It should be based both on their ability to solubilize PAHs and environmental and health effects. Among them, primarily water was proposed. However, owing to low PAHs solubility and their high affinity for surfaces, the process efficiency is negligible. For example, the mixture of 7 PAHs in soil from former manufactured gas plant area was removed by water in column reactor with efficiency on the level of 0.3% (Enell et al. 2004). The higher efficiencies are reported for organic solvents. Khodadoust et al. (2000) used solvent mixture (5% 1-pentanol, 10% water and 85% ethanol) at soil/solvent ratio of 1/4 for PAHs removal from soil. The efficiency clearly depended on their structure. For 2-rings compounds, the efficiency was 90%, whereas for 6-rings PAHs it decreased to 65%.

Recently, the application of surfactants in remediation of soils contaminated by PAHs is also observed. Among them, mainly anionic, nonionic (Chang et al. 2000; Zheng & Obbard 2002; Peng et al. 2011), their mixtures (Zhou & Zhu 2008) and biosurfactants (Urum et al. 2006; Kobayashi et al. 2012; An et al. 2011) were tested. Surfactants enhance the solubility of hydrophobic organic compounds by partitioning them into the hydrophobic cores of surfactant micelles (Peng et al. 2011) that decrease the interfacial tension between PAH and water (Ahn et al. 2008). Among PAHs, the phenanthrene is commonly used as model contaminants in soil washing. Until now, most experiments using surfactants were conducted under static conditions. Enell et al. (2004) pointed some disadvantages of using stirred aqueous static reactor like the grinding effect, which can increase the desorption rate of contaminants and to obtain sufficient elution concentrations for analysis of the hydrophobic organic contaminants. The column leaching method is considered as more realistic in simulating the leaching processes occurring in the field. However, from literature review, there is a lack of consistent data which conditions are the most suitable.

Therefore, in this study, the efficiency of phenanthrene removal from loamy sand in static and dynamic conditions by nonionic surfactant was investigated. The research was aimed at determination of surface properties for Rokanol NL8 and its sorption to soil, evaluation of process efficiency depending on surfactant concentration under both conditions within time, and comparison of process efficiency in static and dynamic conditions based on the volume of surfactant used in the experiment.

2 MATERIALS AND METHODS

2.1 Soil

As model soil, loamy sand collected from 0-surface layer (30 cm depth) in Baranowo (north-east Poland) was used. The soil was air-dried indoors at ambient temperature ($22 \pm 2°C$), screened through a 1 mm sieve and mixed to ensure homogeneity. The characteristic of selected physicochemical parameters of the soil is shown in Table 1.

Table 1. Physicochemical properties of loamy sand.

Parameter	Value
Organic matter (%)	2.04
Organic carbon (%)	0.81
Cation exchange capacity (cmol/kg)	18.5
pH	7.6

2.1.1 Soil contamination

The soil was artificially contaminated with phenanthrene (Sigma-Aldrich, Germany) by mixing 50 mg of the PAH dissolved in 250 ml dichloromethane with 500 g of soil (24 h at 120 rpm) in amber glass container. The total initial concentration of phenanthrene in soil was 107.6 ± 1.74 mg/kg.

2.2 Surfactant

Nonionic Rokanol NL8 obtained from chemical plant ROKITA S.A. in Brzeg Dolny (Poland) was tested without further purification. Chemically, it is oxyethylenated synthetic fatty alcohol with formula $(R\text{-}O(C_2H_4O)_nH$, where $R = C_{9+11} H_{2(9+11)+1}$; $n_{av.} = 8)$. This product was 99 wt% aqueous solutions (paste-like) with molar mass of 510 g/mol. It is characterized by relatively high biodegradation rate ($k = 0.19$ d^{-1}).

2.3 Material preparation

2.3.1 Determination of CMC and surfactant sorption to soil

Surface tension was used to estimate the Critical Micelle Concentration (CMC) of Rokanol NL8 in fresh aqueous surfactant solutions at concentration between 0.005 and 10 g/L. For the evaluation of its sorption to soil, surfactant solutions at the same concentration range were shaken with soil at designed soil/solution ratio (1/5, w/v) on rotary shaker at 150 rpm by 24 h. Then, the supernatants were centrifuged at 8000 rpm for 1 h, filtered and submitted to measurements of surface tension. The values of surface tension were plotted vs. the logarithm of the surfactant concentration. The point of intersection of two regression lines made on the basis of experimental data refers to CMC. Triplicate measurements were performed for each sample. The amount of surfactant sorbed to soil at critical micelle concentration was calculated according to the equation (Liu et al. 1992):

$$Q_{surf} = (CMC_{SS} - CMC_{FS}) \cdot (V/m) \qquad (1)$$

where Q_{surf} is the amount of surfactant sorbed to soil (mg/g), CMC_{SS}—critical micelle concentration of soil/surfactant supernatant (mg/L), CMC_{FS}—critical micelle concentration of fresh surfactant solution (mg/L), V—volume of surfactant solution (L), m—weight of soil (g).

On the basis of CMC_{SS} value, the range of Rokanol NL8 concentrations for the PAH removal under static and dynamic conditions was adopted.

2.3.2 Phenanthrene removal from soil sample in static conditions

In static experiment, Rokanol NL8 concentrations were 0.5, 2.5 and 5 times the CMC_{SS}. The soil washing was performed in polyethylene tubes at m/V ratio equal 1/5 (8 g of soil + 40 ml of surfactant solution). The samples were shaken (24 h at 150 rpm), centrifuged (8000 rpm for 1 h) and filtered. Next, the phenanthrene from washing solution was extracted with the use of a Solid Phase Extraction (SPE) method and its concentration was measured by High Performance Liquid Chromatography (HPLC).

2.3.3 *Phenanthrene removal from soil samples in dynamic conditions*

In dynamic conditions, the column reactor was used (Fig. 1), at which contaminated soil (50 g) was placed between the layers of gravel and debris. Rokanol NL8 solution, at the same concentrations as in static conditions, was pumped from the bottom to the top of reactor by peristaltic pomp at the flow rate equal 2 mL/min. The eluate (80 ml) was collected at upper part of reactor at every hour to polyethylene containers.

2.4 *Analytical methods*

2.4.1 *Surface tension of surfactant*

Surface tension was measured with a Krüss K100 tensiometer (Germany) employing a Wilhelmy plate method.

2.4.2 *Total phenanthrene concentration in soil*

The total concentration of phenanthrene in loamy sand was determined on the basis of extraction in MARS microwave oven (CEM Corporation, USA). Soil samples of 5 g were weighted to PTFE vessels, and 30 ml of mixture containing hexane (Eurochem BGD—Poland) and acetone (P.P.H. Stanlab—Poland) at 1:1 ratio was added. The extraction was performed at the following conditions: 10 min., 115°C, 75 Pa. After extraction, supernatants were filtered and analyzed by HPLC.

2.4.3 *Phenanthrene concentration in washing solutions*

The surfactant solutions after washing contaminated soil were passed through Solid Phase

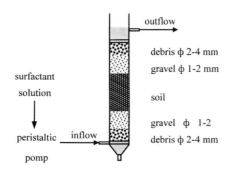

Figure 1. Scheme of the column reactor system.

Extraction (SPE) system using Discovery DSC-18 columns (Supelco) with 500 mg of sorbent. The SPE stationary phases were conditioned with acetonitrile (3 ml), water (3 ml). Next the aqueous samples (1 ml) were passed through the column. Cartridges were cleaned by water (3 ml) and then dried under vacuum for 5 min. The phenanthrene from DSC-18 columns was eluted using dichloromethane (2 × 0.8 mL) to Teflon-lined screw capped amber-glass vials. After dichloromethane evaporation in N_2 stream, 1 mL of acetonitrile was added to the vials. The samples were analyzed by HPLC.

2.4.4 *HPLC instrument*

The high performance liquid chromatograph (Varian, Australia) was equipped with autosampler and Supelcosil LC-PAH column (15 cm × 4.6 mm, 5 μm) with precolumn (LC-18, 2 cm) (Supelco). The mixture of water and acetonitrile (22:78) was used as the solvent at a flow rate of 0.8 mL/min. Phenanthrene was determined by UV detector at wavelength of 254 nm.

2.5 *Statistical analysis*

Data were statistically evaluated with the use of STATISTICA 9.0 (StatSoft, Inc.). Significantly different means ($p < 0.05$) were assessed by a post hoc Tukey's Least Significant Difference (LSD) test.

3 RESULTS AND DISCUSSION

3.1 *Critical micelle concentration of surfactant and its sorption on soil*

In the present study, the changes of surface tension in fresh and supernatant solutions depending on Rokanol NL8 concentration are shown in Figure 2.

Regardless of sample type (with or without soil), Rokanol NL8 quite effectively reduced surface tension to 29.65 mN/m on average. The critical micelle concentration determined in aqueous solution (CMC_{FS}) of Rokanol NL8 was 293 mg/L but in supernatant of soil/surfactant system (CMC_{SS}) it was 7.4 time higher (2177 mg/L). According to Wang & Mulligan (2004), effective physicochemical properties of surfactants for soil washing include low CMC (to 2000 mg/L) and low surface tension (below 30 mN/m).

The increase of CMC in solutions containing soil is an evidence of surfactant sorption. The amount of Rokanol NL8 (8 oxyethyl groups) adsorbed to loamy sand, calculated from equation 1, was 9.4 mg/g. Similar results were obtained by Rao & He (2006). The adsorbed

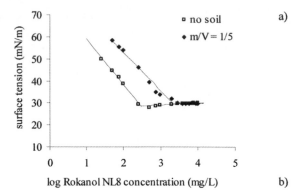

a)

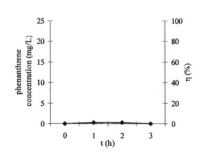

b)

Figure 2. Plot of surface tension versus the logarithm of Rokanol NL8 concentration in aqueous and soil/surfactant system (m/V).

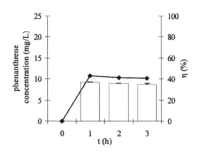

c)

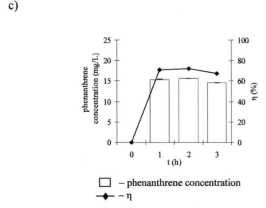

☐ – phenanthrene concentration
◆ – η

Figure 3. The phenanthrene concentration in supernatant and the removal efficiency (η).

amount of nonionic alcohol ethoxylates with 9 oxyethyl groups to silty loam (at soil/solution ratio equal 1/10) was 8.5 mg/g. Nevskaia et al. (1996) investigated the Triton's sorption characterized by different amount of ethylene oxides n = 7; 9.5; 16; 30, respectively, on kaolin. The data results that Triton X-114 with the lowest amount of ethylene oxides (n = 7) revealed the highest adsorption on kaolin—17.26 mg/g. The amount of adsorbed others surfactants (n = 9.5–30) were about 7 mg/g. Zhou & Zhu (2007) stated that the amount of adsorbed Triton X-114 was 8.88 mg/g on soil with organic carbon content—1.93%. In our studies, concentration of organic carbon in soil was twice lower (0.81%) but the amount of Rokanol NL8, which had the same amount of oxyethyl groups as Triton X-114, adsorbed on soil was similar.

3.2 Effectiveness of phenanthrene removal from soil under static conditions

The phenanthrene removal from loamy sand at 0.1, 0.5 and 1% Rokanol NL8 was performed. The experimental data are shown in Figure 3. The efficiency of phenanthrene removal from loamy sand at Rokanol NL8 concentration of 0.1% (below the CMC_{SS}) did not exceed 1.2% (Fig. 3a). With increase of surfactant concentration in the washing solution to 0.5% and 1% (above the CMC), the efficiency of phenanthrene removal after the 1st hour of experiment visibly increase to 43.1%, and 71.3%, respectively (Fig. 3b, c).

The differences of phenanthrene removal depending on Rokanol NL8 concentration were probably affected by various mechanisms included in PAHs desorption. Based on literature, the mechanism of hydrophobic compounds removal from soil by surfactants below and above the CMC concentration is different. Below the CMC, surfactant monomers are responsible for the soil roll up mechanism and above the CMC—their solubilization into the hydrophobic core of surfactant micelles (Deshpande et al. 1999). The PAH solubilization increase among the increase of number of micelles in solution. The surfactant concentrations above the CMC should be used for obtain high soil washing efficiency.

Deshpande et al. (1999) studied surfactant-enhanced soil washing of petroleum contaminated soils by anionic and nonionic surfactants. The surfactants concentrations below and above the CMC (0.25, 4 and 25 times the CMC) were used.

The authors stated that at surfactant concentration below CMC, the hydrophobic contaminants removal from soil may be relevant only when the surfactants were characterized by the low soil adsorption. The authors preferred nonionic surfactants instead of anionic for soil washing because 10–100 times lower concentration is needed to obtain the same results.

At the same time, if nonionic surfactant is significantly sorbed on soil, it negates the soil washing efficiency, so the differences may be reduced or even reversed for that system.

Above the CMC, surfactant micelles begin competing for PAH molecules in soil-solution system, therefore the increase of PAHs removal from soil can be observed. Zhou & Zhu (2007) investigated the phenanthrene desorption from artificially contaminated soil by nonionic surfactants (Triton X-114, Triton X-100, Triton X-305). At the surfactant concentration to CMC_{FS} ratio of 28.6 for TX-114, 19.4 for TX-100 and 5.2 for TX-305, the phenanthrene removal from soil was 80.3%, 73.5% and 63.7%, respectively. Below the CMC, phenanthrene removal did not reach 10% for all surfactants. In our investigation at 1% of Rokanol NL8 concentration, for which the surfactant concentration/CMC_{FS} ratio was 34.1, the soil washing efficiency was the highest and reached 71.3% (after 1 hour of experiment) (Fig. 3c). In static conditions, the investigations on the phenanthrene removal from loamy sand was conducted for 3 hours. The increase of soil washing time was not influenced on process efficiency regardless of surfactant concentration (Fig. 3). Similarly, Peng et al. (2011) investigated the PAHs removal from soil by nonionic surfactants (Tween 80 and Triton X-100). The process efficiency increased relatively quickly, within 30 minutes, but after 60 minutes it was approaching a stable level. Longer time was needed for Total Petroleum Hydrocarbon (TPH) removal from sandy loam by biosurfactants (Lai et al. 2009). The authors stated that irrespective of the biosurfactant type and its concentration or TPH content in soil, the time of 1 day was optimal for TPH removal. Soil washing efficiency depends on surfactant concentration. Grasso et al. (2001) investigated desorption of PAHs from aged soil (former manufactured gas plant site) using nonionic alkil ethoxylate surfactant (Alfonic 1412-7).

At surfactant concentration 2 times lower than its CMC, the PAH desorption was negligible despite long reaction time (about 30 days). However, at concentration 5 times higher than CMC, the authors obtained higher desorption from about 8–18% within 3–7 days.

Another investigation conducted by Paterson et al. (1999) revealed that the steady state of phenanthrene removal from a coal tar contaminated soil by surfactants was obtained after 50 hours for four of the five tested surfactants. The phenanthrene removal was between 13% and 47% for different types of surfactants.

To sum up, the surfactant concentration influenced more phenanthrene removal from soil than reaction time. As optimum time, 1 h seems to be enough to obtain the maximum efficiency of soil washing process under static conditions.

3.3 Effectiveness of phenanthrene removal from loamy sand under dynamic conditions

The investigations on phenanthrene removal from loamy sand in column reactor were conducted at the same Rokanol NL8 concentrations as under the static conditions. In Figure 4 the mass and the efficiency of phenanthrene removal from soil within time was shown. The mass of removed phenanthrene from soil ($M_{PHE,E}$) (Fig. 4a, c, e) was calculated from the equation:

$$M_{PHE,E} = V_{E,i} \cdot C_{PHE,Ei} \tag{2}$$

where $V_{E,i}$—the volume of the eluate in time (i) of the experiment (L), $C_{PHE,Ei}$—the concentration of phenanthrene in eluate (mg/L).

The sum of removed phenanthrene mass ($M_{PHE,S}$) in time (Fig. 4b, d, f) was calculated as:

$$M_{PHE,S} = \Sigma M_{PHE,Ei} \tag{3}$$

The efficiency of phenanthrene removal (%) was calculated from following equation (Fig. 4b, d, f):

$$\eta = \frac{\Sigma M_{PHE,Si}}{M_{PHE,0}} \cdot 100\% \tag{4}$$

where $M_{PHE,0}$—the mass of phenanthrene at the beginning of the experiment (mg), $M_{PHE,Si}$—the sum of removed phenanthrene mass in following hours (i) (mg). From the data obtained, it can be seen that similarly as it was under static conditions, Rokanol NL8 at 0.1% concentration was not effective during the whole experiment. After the 3rd h, the mass of removed phenanthrene was 0.1 mg which responded to 1.9% process efficiency. The increase of the surfactant concentration influenced phenanthrene removal. After the 1st hour of experiment, at 0.5% Rokanol NL8, 1.96 mg of phenanthrene was removed. During the same time at 1% Rokanol NL8 it was 1.7-times higher. In contrast to static conditions, the increase of washing time enhanced process efficiency. At 0.5% Rokanol NL8

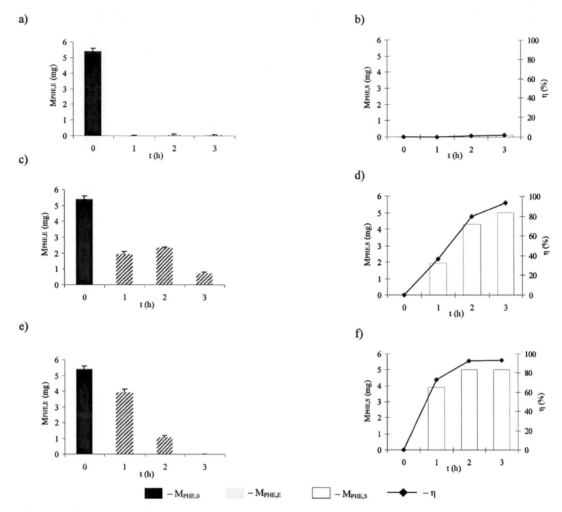

a)

b)

c)

d)

e)

f)

■ – $M_{PHE,0}$ ▨ – $M_{PHE,E}$ ▢ – $M_{PHE,S}$ ◆ – η

Figure 4.   The mass and efficiency of phenanthrene removal from loamy sand in column reactor at surfactant concentrations: a, b) 0.1%, c, d) 0.5%, e, f) 1%. Error bars represent standard deviation (n = 3).

the efficiency of phenanthrene removal increased from 36.5% (1 h) to 93.3% (3 h). For comparison, at the highest surfactant concentration it changed from 73.05% to 93.3% (Fig. 4f).

Zhou & Zhu (2008) investigated the phenanthrene removal from contaminated soils by Triton X-100 in column reactor. According to the authors at the early stage of column flushing, the water is responsible for the solubilization of phenanthrene, because surfactant was strongly sorbed into soil.

When the surfactant sorption approached saturation, the concentration of phenanthrene in effluent sharply increased with the increase of surfactant concentration, ascribing the solubilization of phenanthrene to surfactant micelles.

When the surfactant sorption approached saturation, the concentration of phenanthrene

in effluent sharply increased with the increase of surfactant concentration, ascribing the solubilization of phenanthrene to surfactant micelles. In our study, the increase of Rokanol NL8 concentration to 0.5% and 1% visible rise of the phenanthrene removal efficiency were already observed after the 1st hour of experiment. It was probably caused by lower sorption compared to 0.1% concentration and higher surfactant availability for phenanthrene.

3.4 Comparison of soil washing under the static and dynamic conditions

The comparison of phenanthrene removal from loamy sand by Rokanol NL8 under the static and dynamic conditions was performed on the basis of

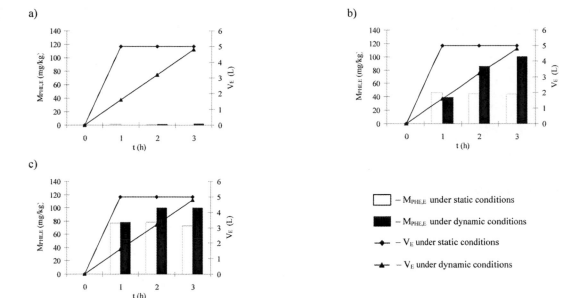

a)

b)

c)

- ☐ – $M_{PHE,E}$ under static conditions

- ■ – $M_{PHE,E}$ under dynamic conditions

- ◆ – V_E under static conditions

- ▲ – V_E under dynamic conditions

Figure 5. The mass of removed phenanthrene and volume of surfactant solution per kg of soil sample at different Rokanol NL8 concentrations: a) 0.1%, b) 0.5%, c) 1%.

the mass of removed phenanthrene ($M_{PHE,E}$) and the volume of surfactant solution ($V_{E,i}$) used in relation to 1 kg of soil sample within time (Fig. 5).

The volume of used surfactant solution per kg of soil within time under static conditions was constant and amounted to 5 L. In column reactor the volume increased from 1.6 L (1 h) to 4.8 L (3 h) that was comparable with static conditions. At 0.1% Rokanol NL8 concentration, the soil washing efficiency was insignificant under both conditions (Fig. 5a).

At the concentration of 0.5% the mass of removed phenanthrene after 1 h was higher in static than dynamic conditions (p > 0.05). After the 3rd hour of experiment, in spite of comparable volume of used surfactant, the mass of removed phenanthrene was 2.3 times higher under dynamic than static conditions (Fig. 5b).

Analogous tendency was observed at 1% surfactant concentration, however, the differences under both conditions were lower than at 0.5% concentration (Fig. 5c). It means that besides surfactant concentration also solution, supplying the fresh surfactant, had the influence on the process efficiency under the dynamic conditions.

remediation of loamy sand contaminated with phenanthrene. The advantage of tested surfactant was low CMC_{SS} being in the range considered as optimal, high ability to reduce the surface tension and low soil adsorption.

The efficiency of phenanthrene removal strongly depended on surfactant concentration and mode of experiment. At 0.1% of Rokanol NL8 removed phenanthrene did not exceed 2% under both static and dynamic conditions.

At 0.5% of Rokanol NL8 and after the 3rd hour of soil washing higher efficiency was obtained under dynamic than static conditions at comparable volume of used surfactant. The similar results were obtained at 1% of surfactant concentration but the differences of removed phenanthrene mass were lower.

To sum up, after 3 hours of experiment under the dynamic conditions, the same results were obtained both at 0.5% and 1% of surfactant concentration. Therefore, from economical point of view, the phenanthrene removal from soil should be conducted in column reactor at 0.5% of Rokanol NL8.

4 CONCLUSION

The investigations showed that nonionic Rokanol NL8 can be used as washing solution in

REFERENCES

An, C.J., Huang, G.H., Wei, J., Yu, H. 2011. Effect of short-chain organic acids on the enhanced desorption of phenanthrene by rhamnolipid biosurfactant in soil-water environment. *Water Research* 45(17): 5501–5510.

Chang, M.Ch., Huang, CH.R., Shu, H.Y. 2000. Effects of surfactants on extraction of phenanthrene in spiked sand. *Chemosphere* 41: 1295–1300.

Czaplicka, M. 2004. Ocena stopnia zanieczyszczenia wybranych terenów przemysłowych wielopierścieniowymi węglowodorami aromatycznymi. W: Lemański J.F. & Zabawa S., *Zanieczyszczenie środowiska produktami naftowymi ich monitoring i usuwanie.* X Jubileuszowa Konferencja Naukowo-Techniczna. Poznań—Ustronie Morskie.

Deshpande, S., Shiau, B.J., Wade, D., Sabatini, D.A., Harwell, J.H. 1999. Surfactant selection for enhancing ex situ soil washing. *Water Research* 33: 351–360.

Enell, A., Reichenberg, F., Warfvinge, P., Ewald, G. 2004. A column method for determination of leaching of polycyclic aromatic hydrocarbons from aged contaminated soil. *Chemosphere* 54: 707–715.

Gitipour, S., Mohebi, M., Taheri, E. 2011. Evaluation of Carcinogenic Risk Due to Accidental Ingestion of PAHs in Contaminated Soils. *Clean—Soil, Air, Water* 39(9): 820–826.

Gong, Z., Wilke, B.M., Alef, K., Li, P., Zhou, Q. 2006. Removal of polycyclic aromatic hydrocarbons from manufactured gas plant-contaminated soils using sunflower oil: Laboratory column experiments. *Chemosphere* 62: 780–787.

Grasso, D., Subramaniam, K., Pignatello, J.J., Yang, Y., Ratté, D. 2001. Micellar desorption of polynuclear aromatic hydrocarbons from contaminated soil. *Colloids and Surfaces A: Physicochemical and Engineering Aspects* 194: 65–74.

Khan, F., Husain, T., Hejazi, R. 2004. An overview and analysis of site remediation technologies. *Journal of Environmental Management* 71: 95–122.

Khodadoust, A.P., Bagchi, R., Suidan, M.T., Brenner, R.C., Sellers, N.G. 2000. Removal of PAHs from highly contaminated soils found at prior manufactured gas operations. *Journal of Hazardous Materials* 80(1–3): 159–174.

Kobayashi, T., Kaminaga, H., Navarro, R.R., Iimura, Y. 2012. Application of aqueous saponin on the remediation of polycyclic aromatic hydrocarbons-contaminated soil. *Journal of Environmental Science and Health—Part A Toxic/Hazardous Substances and Environmental Engineering* 47(8): 1138–1145.

Lai, C.C., Huang, Y.C., Wei, Y.H., Chang, J.S. 2009. Biosurfactant-enhanced removal of total petroleum hydrocarbons from contaminated soil. *Journal of Hazardous Materials* 167(1–3): 609–614.

Li, Y.S., Hu, X.J., Sun, T.H., Hou, Y.X., Song, X.Y., Yang, J.S., Chen, H.L. 2011. Soil washing/flushing of contaminated soil: A review. *Chinese Journal of Ecology* 30(3): 596–602.

Liu, Z., Edwards, D.A., Luthy, R.G. 1992. Sorption of non-ionic surfactants onto soil. *Water Research* 26: 1337–1345.

López-Vizcaíno, R., Sáez, C., Cañizares, P., Rodrigo, M.A. 2012. The use of a combined process of surfactant-aided soil washing and coagulation for PAH-contaminated soils treatment. *Separation and Purification Technology* 88: 46–51.

McGowan, T.F., Greer, B.A., Lawless, M. 1996. Thermal treatment and non-thermal technologies for remediation of manufactured gas plant sites. *Waste Management* 16: 691–698.

Mulligan, C.N., Yong, R.N., Gibbs, B.F. 2001. Surfactant-enhanced remediation of contaminated soil: a review. *Engineering Geology* 60: 371–380.

Nadal, M., Schuhmacher, M., Domingo, J.L. 2004. Levels of PAHs in soil and vegetation samples from Tarragona County, Spain. *Environmental Pollution* 132: 1–11.

Nevskaia, D.M., Guerrero-Ruíz, A., López-González, J.D.D. 1996. Adsorption of polyoxyethylenic surfactants on quartz, kaolin and dolomite: A correlation between surfactant structure and solid surface nature. *Journal of Colloid and Interface Science* 181: 571–580.

Paterson, I.F., Chowdhry, B.Z., Leharne, S.A. 1999. Polycyclic aromatic hydrocarbon extraction from a coal tar-contaminated soil using aqueous solutions of nonionic surfactants. *Chemosphere* 38(13): 3095–3107.

Peng, S., Wu, W., Chen, J. 2011. Removal of PAHs with surfactant-enhanced soil washing: Influencing factors and removal effectiveness. *Chemosphere* 82(8): 1173–1177.

PN-C-04645:2001. Woda i ścieki. Badanie biodegradacji "częściowej" anionowych i niejonowych substancji powierzchniowo-czynnych. Test wstępny.

PN-EN ISO 9408:2005. Jakość wody. Oznaczanie całkowitej biodegradacji tlenowej związków organicznych w środowisku wodnym przez oznaczanie zapotrzebowania tlenu w zamkniętym respirometrze.

Rao, P. & He, M. 2006. Adsorption of anionic and nonionic surfactant mixtures from synthetic detergents on soils. *Chemosphere* 63: 1214–1221.

Richard, H., Jones-Meehan, J., Nestler, C., Hansen, L.D., Straube, W., Joes, W., Hind, J., Talley, J.W. 2006. Polycyclic aromatic hydrocarbons (PAHs): improved land treatment with bioaugmentation. *Bioremediation of Recalcitrant Compounds*, J.W. Talley: 215–300.

Rulkens, W.H. & Honders, A. 1996. Clean-up of contaminated sites: experiences in The Netherlands. *Water Science and Technology* 34: 293–301.

Urum, K., Grigson, S., Pekdemir, T., McMenamy, S. 2006. A comparison of the efficiency of different surfactants for removal of crude oil from contaminated soils. *Chemosphere* 62: 1403–1410.

Wang, S. & Mulligan, C.N. 2004. An evaluation of surfactant foam technology in remediation of contaminated soil. *Chemosphere* 57(9): 1079–1089.

Zheng, Z., Obbard, J.P. 2002. Evaluation of an elevated non-ionic surfactant critical micelle concentration in a soil/aqueous system. *Water Research* 36: 2667–2672.

Zhou, W. & Zhu, L. 2007. Efficiency of surfactant-enhanced desorption for contaminated soils depending on the component characteristics of soil-surfactant—PAHs system. *Environmental Pollution* 147: 66–73.

Zhou, W. & Zhu, L. 2008. Influence of surfactant sorption on the removal of phenanthrene from contaminated soils. *Environmental Pollution* 152: 99–105.

Environmental Engineering IV – Pawłowski, Dudzińska & Pawłowski (eds)
© 2013 Taylor & Francis Group, London, ISBN 978-0-415-64338-2

Effectiveness of the household hybrid wastewater treatment plant in removing mesophilic, psychrophilic and *Escherichia coli* bacteria

T. Bergier & A. Włodyka-Bergier

AGH University of Science and Technology, Faculty of Mining Surveying and Environmental Engineering, Department of Environmental Protection and Management, Cracow, Poland

ABSTRACT: The article presents the results of research on the domestic wastewater treatment effectiveness by the Enkosystem household wastewater treatment plant—a hybrid of the conventional biological technology and constructed wetlands using a hydroponic bed. The study was conducted from November 2010 to June 2011, samples were taken every month. The microbiological parameters as well as the additional selected physico-chemical parameters have been measured. Relatively high effectiveness of Enkosystem has been observed in removing bacteria from domestic sewage, namely the highest removal effectiveness (about 95%) has been observed for psychrophilic bacteria, 90% for mesophilic bacteria and 89% for *Escherichia coli*.

Keywords: constructed wetlands, microorganisms removal, psychrophilic bacteria, mesophilic bacteria, *Escherichia coli*

1 INTRODUCTION

Recently, municipal wastewater management in Poland has significantly developed, higher ratio of wastewater from the urban agglomerations is collected by centralized systems and is treated with highly efficient biological processes (GUS 2011). However, several aspects still need to be substantially improved, namely the most urgent and crucial among them is unsatisfactory quality of Polish rivers (GUS 2011), which is mainly caused by the bacteriological pollution and nutrients (Bergier 2010). One of the major challenges in Poland is to develop the sustainable wastewater management model for areas of scattered housing. One of valid and rational options to manage domestic sewage from such areas is to employ the decentralized local systems, based on household wastewater treatment plants, which are not popular and rarely used in Poland (Bergier 2010). The interesting examples of this kind of option are Constructed Wetlands (CWs), especially subsurface flow ones, which are extensively used world-wide (Kadlec 2009, Masi et al. 2008) and slowly getting popular also in Poland (Obarska et al. 2010). Apart from several advantages, they request relatively large area for location, which sometimes hinders their usage. The interesting alternative for CWs is a hybrid technology, which combines the conventional biological treatment chambers with the CWs in a form of hydroponic bed. Such a solution has been developed and currently offered in Poland by the Enko company as Enkosystem.

As it was mentioned above, microbiological pollution is a key issue for Polish water management but it is especially important in the case of individual household systems. Sewage from such utilities is released to local water and soil, which has a positive effect on the small water cycle and local hydrology as long as there is no healthy and/or environmental risk. CWs ability to efficiently remove bacteria from sewage is commonly reported (Axler et al. 2001, Decamp & Warren 1998), and explained by the complex interaction of several biochemical and physical processes, which include mechanical filtration and sedimentation, oxidation, influence of plants exudates, sorption on organic matter, antagonistic activities of CWs microorganisms (Pundsack et al. 2001, Morales et al. 1996, Decamp & Warren 2001, Gersberg 1989).

The goal of this article is to examine the bacteria removal effectiveness with the use of the hybrid technology (constructed wetlands and biological treatment). The study was conducted on the Enkosystem Wastewater Treatment Plant (WWTP), which was installed in July 2010 to treat domestic sewage from the Town Hall of Zawoja. Zawoja is a municipality (in southern part of Malopolska), in which the issue of sewage management has not been solved and many problems have occurred with employing the centralized sewage system on the area of very scattered housing. The Enkosystem WWTP was installed as the pilot project to test possibilities to use decentralized solutions in that area.

The presented research was conducted from November 2010 to June 2011, which are the first eight months of WWTP functioning. In the article, the following microbiological parameters have been presented: psychrophilic bacteria, mesophilic bacteria, *Escherichia coli*, which are the main subject of this work. However, in order to fully understand and interpret these results selected physico-chemical parameters are presented, which had already been published (Bergier & Włodyka-Bergier 2012).

2 MATERIALS AND METHODS

The Enkosystem WWTP is located directly next to the Town Hall in Zawoja, behind the parking lot (Fig. 1). The WWTP treats the entire amount of sewage from the Town Hall. The average sewage flow is 0.65 m³/d. Prior to the treatment, sewage is stored in a reinforced concrete tank (10 m³), which played the role of septic tank before the WWTP installation. Sewage from this tank flows by gravity to the Enkosystem WWTP, which consists of two underground cylindrical plastic tanks made of high thermo isolation material. Both tanks have the same diameter 2050 mm, the first one is 1828 mm high, and second—2220 mm. Inside these tanks, there is a complex system of compartments, whose role is to direct sewage flow and to create zones of different conditions to effectively remove pollutants. In the first tank, the mechanical pre-treatment takes place, ammonification and dephosphatation processes are also initialized. After the initial treatment, sewage is dosed from the first tank to the second one with a pump controlled by the electronic system. The main sewage treatment is realized in the second tank, on top of which the multi-species hydroponic bed is sited and inside of which the

Figure 1. The main tank of Enkosystem WWTP for the Zawoja Town Hall (photo T. Bergier).

different oxidation level zones function. A pump installed in the second tank oxides sewage and controls its flow and recirculation. Thanks to which the proper conditions for microorganisms to grow are created on the plants roots and the hydroponic bed. Treated sewage is released to Skawica river flowing in a distance about 100 m from the WWTP. The outlet from the second tank is equipped with a valve, which enhances other forms of effluent utilization and/or its sampling.

The presented study was conducted in an initial period of the Enkosystem WWTP functioning. The sewage samples were taken from November 2010 to June 2011, every month approximately. Every time, three kinds of samples were taken, that is raw sewage—from the tank prior the Enkosystem, pre-treated sewage—from the first tank of the Enkosystem WWTP, effluent—treated sewage from the WWTP outlet. Moreover, water meter in the Town Hall was checked to determine the amount of water consumed and to calculate the average sewage flow through the WWTP. The air temperature was also measured during sewage sampling.

In sewage samples microbiological pollution was measured representing the total numbers of: psychrophilic bacteria, mesophilic bacteria, bacteria from coli group, thermotolerant coli bacteria and *Escherichia coli*. In accordance with the Polish PN-EN ISO standard 6222:2004, the sowing method was employed for culturable psychrophilic microorganisms (72 h incubation at 22°C) and mesophilic ones (48 h incubation at 36°C). The membrane method, according to Polish PN-EN ISO standard 9308.1:2004, was employed for bacteria from coli group, thermotolerant coli bacteria and *Escherichia coli*. Measurements for the late three kinds of bacteria provided the same results in all experimental series, for this reason only results for *Escherichia coli* are presented and discussed in the remaining part of the article. During observations, the following physico-chemical parameters have also been measured in sewage, namely temperature, pH, total suspended solids, Total Organic Carbon (TOC), 5-day Biochemical Oxygen Demand (BOD₅). They were analyzed accordingly to the adequate Polish standards—the detailed description of the methodology can be found in (Bergier & Włodyka-Bergier 2012). In the same article, there are also the results for more physico-chemical parameters.

3 RESULTS AND DISCUSSION

The concentration of the physico-chemical parameters of sewage as well air temperature have been presented in Table 1. As it can be observed from the numbers, the Enkosystem WWTP treats sewage in stable and efficient way. As it can be

Table 1. Physico-chemical parameters of sewage from Enkosystem, 2010–2011 (Bergier & Włodyka-Bergier 2012).

Sewage sample	Date of sampling							
	16 XI	14 XII	18 I	08 II	07 III	04 IV	05 V	08 VI
Air temperature, °C								
	4.7	–10.5	0.3	2.2	2.4	6.5	4.3	15.6
Temperature, °C								
Raw sewage	10.5	–	5.0	6.0	7.0	13.0	10.0	16.0
Pre-treated sewage	9.0	–	2.0	5.0	6.0	12.0	10.0	14.0
Effluent	5.5	–	3.0	3.5	4.5	11.5	8.0	15.0
pH								
Raw sewage	7.52	7.55	7.34	7.40	7.16	6.98	7.15	7.22
Pre-treated sewage	7.48	7.56	7.65	7.70	7.51	7.63	7.52	7.04
Effluent	7.14	7.72	7.28	7.60	7.64	7.97	7.47	7.28
Total suspended solids, mg/dm³								
Raw sewage	130	80	152	200	128	303	500	268
Pre-treated sewage	90	78	128	172	102	158	156	154
Effluent	21	10	40	10	57	44	48	21
Total organic carbon, mg/dm³								
Raw sewage	126.0	140.4	140.2	165.2	200.3	203.9	82.2	76.9
Pre-treated sewage	109.5	45.4	111.6	123.3	125.9	153.7	74.0	73.4
Effluent	39.5	30.1	32.1	30.7	50.3	62.6	25.6	54.0
BOD₅, mg O₂/dm³								
Raw sewage	–	261	499	565	314	405	437	251
Pre-treated sewage	–	226	267	290	274	223	227	213
Effluent	–	13	6	16	20	23	19	17

observed in the table, for all presented physico-chemical parameters, the concentrations measured in treated sewage comply with the Polish standards for wastewater release to surface water or soil (the Regulation of the Minister of Envir. 2006). However, the main reason for presenting this data is to complement and supplement these bacteriologic parameters, and to enable their full interpretation and understanding.

The results for *Escherichia coli* bacteria have been presented in Figure 2. The average number of bacteria was $5.5 \cdot 10^6$ cfu/100 ml ($2.0 \cdot 10^6$–$10.8 \cdot 10^6$ cfu/100 ml) in influent to WWTP, and $0.5 \cdot 10^6$ cfu/100 ml ($0.1 \cdot 10^6$–$1.1 \cdot 10^6$ cfu/100 ml) in effluent. The average removal effectiveness was 90.3% (78.6%–98.40%). The number of *Escherichia coli* in pre-treated sewage averaged $6.4 \cdot 10^6$ cfu/100 ml ($3.5 \cdot 10^6$–$11.5 \cdot 10^6$ cfu/100 ml) and was generally higher than in influent to WWTP.

The results for mesophilic bacteria have been presented in Figure 3. The average number of bacteria was $1.49 \cdot 10^6$ cfu/ml ($0.38 \cdot 10^6$–$2.82 \cdot 10^6$ cfu/ml) in influent to WWTP, and $0.07 \cdot 10^6$ cfu/ml ($0.01 \cdot 10^6$–$0.15 \cdot 10^6$ cfu/ml) in effluent. The removal effectiveness was on average 95.6%, and it remained stable in an experimental period (90.3%–97.6%).

The average number of mesophilic bacteria in pre-treated sewage was averagely $0.93 \cdot 10^6$ cfu/ml ($0.16 \cdot 10^6$–$2.39 \cdot 10^6$ cfu/ml) and was lower than in influent to WWTP, opposite than in a case of *Escherichia coli*.

The results for psychrophilic bacteria have been presented in Figure 4. The average number of bacteria was $5.05 \cdot 10^6$ cfu/ml ($2.57 \cdot 10^6$–$8.93 \cdot 10^6$ cfu/ml) in influent to WWTP, and $0.24 \cdot 10^6$ cfu/ml ($0.05 \cdot 10^6$–$0.65 \cdot 10^6$ cfu/ml) in effluent. The average removal effectiveness was 93.8% (79.6%–98.7%). The average number of psychrophilic bacteria in pre-treated sewage averaged $4.20 \cdot 10^6$ cfu/ml ($1.05 \cdot 10^6$–$6.97 \cdot 10^6$ cfu/ml) and was slightly lower than in influent to WWTP.

There were no significant changes observed in removal effectiveness in winter months, which was also confirmed by the statistical analysis that showed no statistically significant correlations between the bacteria removal effectiveness and sewage temperature (air temperature either). Only for mesophiles, the minor influence (correlation coefficient: −0.77) of sewage temperature on their removal effectiveness was observed ($R^2 = 0.59$, $p = 0.04$). This effect is evidently caused by their more intensive reproduction in higher temperatures.

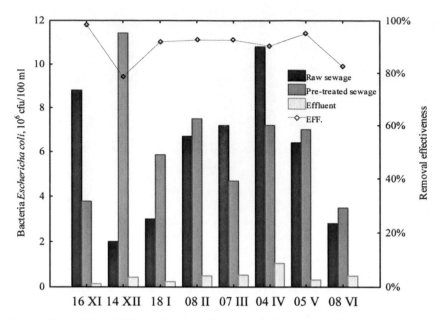

Figure 2.   Bacteria *Escherichia coli* (10^6 cfu/100 ml) and its removal effectiveness (EFF) by Enkosystem in Zawoja.

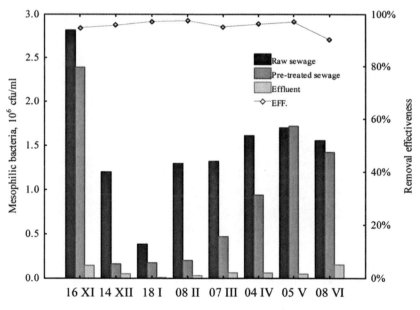

Figure 3.   Mesophilic bacteria (10^6 cfu/ml) and its removal effectiveness (EFF) by Enkosystem in Zawoja.

For *Escherichia coli* the correlation analyses indicated relatively strong (correlation coefficient: 0.67) and statistically significant ($R^2 = 0.45$, $P < 0.07$) influence of its initial concentration (in raw sewage) on the removal effectiveness. The similar effect was not observed for two remaining (considered) groups of bacteria. The similarly significant correlation was observed between the bacteria removal effectiveness and BOD_5 in raw sewage for both *Escherichia coli* ($R^2 = 0.54$, $P = 0.06$, correlation coefficient: 0.73) and mesophiles ($R^2 = 0.56$, $P = 0.05$, correlation coefficient: 0.746). This effect

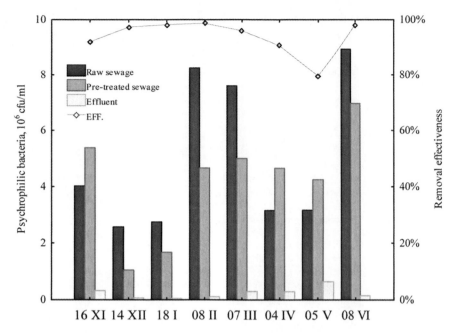

Figure 4. Psychrophilic bacteria (10^6 cfu/ml) and its removal effectiveness (EFF) by Enkosystem in Zawoja.

was not found for psychrophiles. Statistical analyses for an influence of other physico-chemical parameters on the bacteria removal rate were of no significance. As it was mentioned in the introduction, constructed wetlands are considered as the technology that removes bacteria from domestic sewage in an efficient way (Khatiwada & Polprasert 1999, Włodyka-Bergier et al. 2010). However, when considering the values of bacteria removal efficiency, relatively wide range of numbers can be found in research reports. For *Escherichia coli* the removal rates of 99% and higher are commonly reported (Ottova et al. 1997, Gersberg et al. 1989, Khatiwada & Polprasert 1999).

Nevertheless, significantly lower rates were reported in several articles (Karathanasis et al. 2003, Frazer—Williams et al. 2008, Steer et al. 2002), e.g. in the British study (Decamp & Warren 1998) the bacteria removal was observed on the level of 67% for CWs used in the secondary treatment, and 15%–39% for ones used in tertiary treatment.

The bacteria removal effectiveness observed in the presented study on the Enkosystem WWTP was relatively high, although it did not achieve the highest levels reported for conventional CWs, what could be explained by the relatively short time of functioning (processes of microcosms maturing probably did not finished) and much shorter retention time (1–2 days for Enkosystem and usually 6–7 days for CWs).

CONCLUSIONS

The Enkosystem WWTP has showed the relatively high bacteria removal effectiveness. It was removing *Escherichia coli* with the average removal rate of 90.3% (78.6%–98.40%), mesophilic bacteria with 95.6% (90.3%–97.6%) and psychrophilic bacteria with 93.8% (79.6%–98.7%). The number of bacteria was comparable to those reported in other articles considering the issues of domestic sewage (Ottova et al. 1997, Decamp & Warren 1998). For raw wastewater it was: $5.5 \cdot 10^6$ cfu/100 ml ($2.0 \cdot 10^6$–$10.8 \cdot 10^6$ cfu/100 ml) for *Escherichia coli*, $1.49 \cdot 10^6$ cfu/ml ($0.38 \cdot 10^6$–$2.82 \cdot 10^6$ cfu/ml) for mesophilic bacteria and $5.05 \cdot 10^6$ cfu/ml ($2.57 \cdot 10^6$–$8.93 \cdot 10^6$ cfu/ml) for psychrophilic bacteria. The sewage detention (approximately 15-day) in an initial, septic tank prior the Enkosystem WWTP did not noticeably remove bacteria, for *Escherichia coli* even the slight elevation of bacteria number was observed in pre-treated sewage. As it was mentioned above, the number of bacteria in treated sewage was considerably lower than in influent—it was: $0.5 \cdot 10^6$ cfu/100 ml ($0.1 \cdot 10^6$–$1.1 \cdot 10^6$ cfu/100 ml) for *Escherichia coli*, $0.07 \cdot 10^6$ cfu/ml ($0.01 \cdot 10^6$–$0.15 \cdot 10^6$ cfu/ml) for mesophilic bacteria and $0.24 \cdot 10^6$ cfu/ml ($0.05 \cdot 10^6$–$0.65 \cdot 10^6$ cfu/ml) for psychrophilic bacteria. There was no significant decrease observed in bacteria removal effectiveness in winter months. For mesophilic bacteria, an opposite effect was observed, that is their removal rate was higher in lower temperature. The similarly significant correlation was observed between the bacteria removal effectiveness and BOD_5 in raw sewage, for both *Escherichia coli* ($R^2 = 0.54$, $P = 0.06$, correlation coefficient: 0.73) and mesophiles ($R^2 = 0.56$, $p = 0.05$, correlation coefficient: 0.746). This effect did not find its application for psychrophiles. Statistical analyses for an influence of other physico-chemical parameters on the bacteria removal rate have been of no significance.

As it was described in the introduction, there is an urgent and substantial need in Poland to find the effective and sustainable methods and techniques of wastewater management for areas with scattered housing, which would enable creation of modern decentralized sewage systems, wherever the conventional centralized solutions cannot be adopted for technical, economical and/or environmental reasons. Enkosystem is one of the most valuable options, which should be considered when planning and designing such a decentralized system. Apart from the good results in removing physico-chemical pollutants from domestic sewage, it also removes bacteria with a relatively high rate, as it was presented the research. However, due to still high concentration of bacteria in treated sewage, it would be recommended to further improve this technology in that matter or supplement it with a disinfection unit, before employing it on the wide scale, especially in applications when water reuse is considered.

ACKNOWLEDGEMENTS

The work has been financed by the AGH statutory research of the Department of Environmental Protection and Management (Katedra Kształtowania i Ochrony Środowiska).

REFERENCES

Axler, R., Henneck, J. & Mccarthy, B. 2001. Residential subsurface flow treatment wetlands in northern Minnesota, *Wetland Systems for Water Pollution Control*, Vol. 44 (11–12), pp. 345–352.

Bergier, T. 2010. Municipal management. In Kronenberg, J. & Bergier, T. (ed.) *Challenges of Sustainable Development in Poland*. Krakow: The Sendzimir Foundation.

Bergier, T. & Włodyka-Bergier A. 2012. Efektywność oczyszczania ścieków w przydomowej hybrydowej oczyszczalni hydrofitowo-biologicznej. *Woda-Środowisko-Obszary Wiejskie* 12 (1): 97–109.

Decamp, O. & Warren, A. 2001. Abundance, biomass and viability of bacteria in wastewaters: impact of treatment in horizontal subsurface flow constructed wetlands. *Water Research* 35 (14): 3496–3501.

Decamp, O. & Warren, A. 1998. Bacteriovory in ciliates isolated from constructed wetlands (reed beds) used for wastewater treatment. *Water Research* 32: 1989–1996.

Frazer-Williams, R., Avery, L., Winward, G., Jeffrey, P., Shirley-Smith, Ch., Liu, S., Memon, F. & Jefferson B. 2008. Constructed wetlands for urban grey water recycling. *International Journal of Environment and Pollution* 33 (1): 93–109.

Gersberg, R.M., Gearhart, R.A. & Yves, M. 1989. Pathogen removal in constructed wetlands. In: Hammer D.A. (ed.) *Constructed Wetlands for Wastewater Treatment; Municipal, Industrial and Agricultural*. Chelsea, Michigan: Lewis Publisher.

GUS 2011. Ochrona środowiska/Environment 2011. Warszawa: GUS.

Kadlec, R.H. 2009. Comparison of free water and horizontal subsurface treatment wetlands. *Ecological Engineering* Vol. 35: 159–174.

Karathanasis, A.D., Potter, C.L. & Coyne, M.S. 2003. Vegetation effects on fecal bacteria, BOD, and suspended solid removal in constructed wetlands treating domestic wastewater. *Ecological Engineering* 20: 157–169.

Khatiwada, N.R. & Polprasert, C. 1999. Kinetics of fecal coliform removal in constructed wetlands. *Water Science and Technology* 40 (3): 109–116.

Masi, F., Conte, G. & Martinuzzi, N. 2008. Sustainable Sanitation by Constructed Wetlands in the Mediterranean Countries: Experiences in Small/Medium-Size Communities and Tourism Facilities. In Al Baz, I., Otterpohl, R. & Wendland, C. (ed.) *Efficient Management of Wastewater, Its Treatment and Reuse in Water Scarce Countries*. New York: Springer.

Morales, A., Garland, J.L. & Lim, D.V. 1996. Survival of potentially pathogenic human-associated bacteria in the rhizosphere of hydroponically brown wheat. *FEMS Microbiology Ecology* 20: 155–167.

Obarska-Pempkowiak, H., Gajewska, M. & Wojciechowska E. 2010. *Hydrofitowe oczyszczanie wód i ścieków*. Warszawa: PWN.

Ottova, V., Balcarova, J. & Vymazal J. 1997. Microbial characteristics of constructed wetlands. *Water Science and Technology* 30 (5): 117–124.

Pundsack, J., Axler, R., Hicks, R., Henneck, J., Nordmann, D. & Mccarthy, B. 2001. Seasonal pathogen removal by alternative on-site wastewater treatment systems. *Water Environment Research* 73: 204–212.

Regulation of the Minister of Environment (Rozporządzenie Ministra Środowiska) z dnia 24 lipca 2006 roku w sprawie warunków, jakie należy spełnić przy wprowadzaniu ścieków do wód lub do ziemi, oraz w sprawie substancji szczególnie szkodliwych dla środowiska wodnego. Dz.U. 2006 nr 137 poz. 984.

Steer, D., Fraser, D.L., Boddy, J. & Seibert B. 2002. Efficiency of small constructed wetlands for subsurface treatment of single family domestic effluent, *Ecological Engineering* 18: 429–440.

Włodyka-Bergier, A., Dziugieł, M. & Bergier T. 2010. The Possibilities of Using Constructed Wetlands to Disinfect Water. *Geomatics and Environmental Engineering* 4 (3): 87–93.

Environmental Engineering IV – Pawłowski, Dudzińska & Pawłowski (eds)
© 2013 Taylor & Francis Group, London, ISBN 978-0-415-64338-2

The effect of ion exchange substrate on plant root system development during restoration of sandy mine spoil

M. Chomczyńska, M. Wągrowska & A. Grzywa
Faculty of Environmental Engineering, Lublin University of Technology, Lublin, Poland

ABSTRACT: The paper presents the studies on effects of an ion exchange substrate (nutrient carrier) on the root systems development of *Dactylis glomerata* L. and *Lotus corniculatus* L. in pot experiments. During the experiments plants were grown on sand (model of degraded soil), and on sand with 2% addition of the ion exchange substrate (Biona-112). The number of roots and total length of roots (from 100-cm^2 cross-section of medium) were measured during plant vegetative growth. When the experiments were terminated, the wet and dry biomass of stems and roots were determined. The addition of Biona-112 to sand significantly intensified the root system development of *D. glomerata* and *L. corniculatus*. The observed intensification of root system development would favor better stabilization and anti-erosive functions of plants in the biological restoration of degraded sandy soil.

Keywords: ion exchange substrate, grass, legume, erosion, root system

1 INTRODUCTION

Sand mining produces excavations that are covered with sandy spoil materials with weak physical and chemical properties and, similarly to other degraded areas, require technical and biological reclamation (Baran & Turski 1996, Maciak 1996, Vankatesh 2012). Biological restoration of such lands involves cultivation of plants, which have a dual role in such reclamation. Plant growth on degraded soils supplies an organic material necessary for the formation of humus; in addition, plant root systems bind loose soil material, thus stabilizing the ground and preventing wind and water erosion (Krzaklewski 1990, Maciak 1996).

Intensive plant growth on degraded lands can be supported by mineral and organic fertilization. For mineral fertilization of degraded grounds, ion exchange substrates can be used. The ion exchange substrates (trade name Biona®) are mixtures of cation and anion exchangers saturated with macronutrient and micronutrient ions in appropriate ratios (Soldatov et al. 1968) and differ from each other in characteristics such as nutrient content and exchanger composition. These materials are produced in limited amounts at the experimental plant of the Institute of Physical Organic Chemistry of the Belarus National Academy of Sciences (BNAS). The ion exchange substrate differs from conventional fertilizers in several aspects. The ions of nutrients are bound to the polymer matrices of cation or anion exchanger and tend not to be washed away by water. However, they can be readily exchanged with ionic metabolites of a plant and absorbed by the root. The high concentration of the nutrient ions in the ion exchange substrate (≈8% of mass) does not cause osmotic shock, and many biological experiments have shown that plants grow successfully in the pure ion exchange substrate. Ion exchangers play the same role as the exchange complex of natural soils, retaining ions of nutrients and delivering them to the plant root in response to the requirements of the plant, which releases ionic metabolites (H^+ and HCO_3^-) into the rhizosphere. Due to an inherent equivalence of ion exchange, overdosing of nutrients is not possible. Ion exchange substrates contain all nutrient elements (including micronutrients) in an optimum ratio, which allows for the ion concentration of the equilibrium solution (in this case the 'soil solution') to be maintained at a level identical to that of Hewitt's nutrient solution, which is recognized as one of the best for plant nutrition (Soldatov et al. 1978, 1985, Soldatov 1988).

The advantages of ion exchange substrates led to the idea of using them in biological soil restoration. Lublin University of Technology, Poland, has commenced modeling studies with different kinds of ion exchange substrates (Biona) as fertilizing additions to sand. The results showed that the addition of 1% (v/v) of Biona-111 to sand led to the fertility of this mixture similar to that of a commercial garden soil (Soldatov et al. 1998). It was also found that a 0.3–0.4% addition of the ion exchange substrate Biona-312 increased the productivity of sand to that of cultivated soil (podzolic soil). Moreover, the studies showed that various spatial distribution of Biona-111 on the height of cultivated layer (of 10 cm depth) did not affect plant yield (Soldatov et al. 1998, Wasąg et al. 2000). In these studies, the main objective was to maximize plant biomass but the effect of substrate addition on the development of plant root systems, which is important for the stabilizing function of plants in soil reclamation, has not been studied.

Thus, the objective of this study was to determine the influence of a small addition of the ion exchange substrate Biona-112 on the development of root systems of orchard grass (*Dactylis glomerata* L.) and bird's foot trefoil (*Lotus corniculatus* L.), species recommended as components of plant restoration mixtures.

2 MATERIALS AND METHODS

The ion exchange substrate Biona-112 and sand (as a model of degraded soil) were used. Sand consisted of the following fractions of particle diameters: <0.10 mm, 52%; 0.10–0.25 mm, 37%; and 0.25–0.50 mm, 11%. The pH of the sand in distilled H_2O equaled 6.2. The ion exchange substrate Biona-112 was prepared at the Institute of Physical Organic Chemistry of BNAS in Minsk with a base of the strong cation exchanger KU-2 and polyfunctional anion exchanger EDE-10P mixed in a 1:1.5 mass ratio. The total ion exchange capacity of these ion exchangers was 5 mmolg^{-1} (cation exchanger) and 11 mmolg^{-1} (anion exchanger), respectively. The content of nutrients in Biona-112 was (mgg^{-1}): K, 7.43; Ca, 20.40; Mg, 4.86; Fe, 1.30; N-NH$_4^+$, 0.90; N-NO$_3^-$, 28.51; P, 10.44; and S, 17.93.

Two series of media were prepared for the plant experiments: the control series with sand; and the test series with mixture of sand and 2% addition (v/v) of Biona-112 substrate. There were 10 pots in each series (Table 1). The 2050 cm^3 pots were equipped with one transparent wall that enabled root development to be observed (Bohm 1985). After media preparation, seedlings of orchard grass and birdsfoot trefoil were planted in specific pots. The experiments were carried out in a phytotron

with 12/12 h light/dark. During plant vegetative growth, the air temperature and relative air humidity were monitored with the use of a TZ-18 theromohygrograph ('Zootechnika', Krakow, Poland). The range in daytime air temperature (between 7 a.m. and 7 p.m.) was 25–28°C. The night-time air temperature (between 7 p.m. and 7 a.m.) was 17–18°C. The daytime air humidity range was 27–71%, and night-time air humidity was 28–74%. The vegetative growth period lasted 35 and 39 days (d) for orchard grass and bird's foot trefoil, respectively. During the plant vegetative growth, the root systems of the tested species were traced on foils placed on the transparent pot walls every 2–4 days, starting from the fourth day of growth for bird's foot trefoil and from the seventh day for orchard grass. When the experiments were terminated, the aboveground shoots of plants were cut and roots were separated. The wet and dry (dried at 105°C) biomass of shoots and roots was measured. The number of plant roots was determined with the use of the drawings from the media cross-sections, in the areas of squares: 10 cm × 10 cm (at depth of 3–13 cm). In the same areas, the root lengths were measured with the use of a curvimeter (Silena Sweden AB) and were summed to obtain the total values per pot. The wet and dry biomass of stems and roots, total root length and root number were used to calculate arithmetic mean values for both experimental series.

The statistical significance of differences between mean values was assessed using Student's t-test or the Aspin–Welch v-test at the significance level $\alpha = 0.05$. Student's t-test was used when variances for compared mean values did not differ significantly (the Fisher–Secendor F-test). When variances differed significantly, the Aspin–Welch v-test was applied (Czermiński et al. 1992, Zgirski & Gondko 1998, Zieliński & Zieliński 1990).

3 RESULTS AND DISCUSSION

The mean values of vegetation growth parameters for orchard grass in pots with Biona-112 addition (S+B) were significantly higher than for controls (S) (Figs. 1 and 2). Wet and dry stem biomass of

Table 1. Characteristics of the experimental series.

Series	Sand amount [g per pot] [cm³ per pot]	Substrate addition [g per pot] [cm³ per pot]	Pot number for orchard grass	Pot number for bird's foot trefoil
Sand (S)	2800 1840	–	5	5
Sand+Biona-112 (S+B)	2744 1803	23 37	5	5

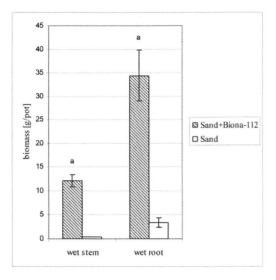

Figure 1. Mean wet stem and root biomass of orchard grass. a—mean values of test series (S+B) were significantly higher ($\alpha = 0.05$) than mean values of controls (S), I—standard deviation.

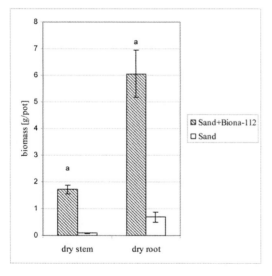

Figure 2. Mean dry stem and root biomass of orchard grass. a—mean values of test series (S+B) were significantly higher ($\alpha = 0.05$) than mean values of controls (S), I—standard deviation.

orchard grass in the pots with Biona-112 were 29 and 19 times higher, respectively, than on sand alone.

The Biona-112 addition to sand caused more than ten-fold increase in wet root biomass and almost nine-fold increase in dry root biomass of orchard grass. Mean root number and total root

length of *D. glomreata* in the sand–Biona mix were higher than the controls during the entire observation period (Table 2).

At the end of the experiment, mean root number in the 100-cm² cross-section of sand–Biona mix was 303% greater than in sand. The mean total root length obtained for the cross-section through the sand–Biona mix was 237% greater than in sand alone (Table 2).

Almost all mean values of vegetation growth parameters for bird's foot trefoil in the sand–Biona mix were significantly greater than those in controls (Figs. 3 and 4). Wet and dry stem biomass of bird's foot trefoil growing in the sand–Biona mix were higher than those in sand alone by eleven and almost nine times, respectively. Sand enrichment with a 2% Biona dose increased wet and dry root biomass almost five and over five times, respectively. However, the difference in dry root biomass between the two growth media was not statistically significant. During the observation period, the mean root number and total root length of *L. corniculatus* in sand–Biona mix were,

Table 2. Root number and total root length of orchard grass.

Vegetative growth period (d)	Sand+Biona-112 (S+B)	Sand (S)	Ratio S+B/S
Mean root number [per 100 cm²]			
7	3.2 ± 0.4	2 ± 0.8^a	1.60
11	8.2 ± 2.3	6.8 ± 2.4	1.21
13	15.6 ± 3.1	12.4 ± 3.8	1.26
15	38 ± 6	17.8 ± 3.9^a	2.13
19	100.2 ± 16.5	41.2 ± 10.8^a	2.43
22	150.2 ± 15.2	48.6 ± 9.5^a	3.09
25	215.8 ± 37.3	58.8 ± 11.8^a	3.67
27	248.6 ± 44.3	65.8 ± 10.1^a	3.78
29	279.8 ± 49.7	72.4 ± 10^a	3.86
32	317 ± 36.9	78.6 ± 10.7^a	4.03
Mean total root length [cm/100 cm²]			
7	2.3 ± 0.68	1.67 ± 0.06	1.38
11	15.58 ± 2.5	7.35 ± 3.04^a	2.12
13	34.62 ± 7.69	14.37 ± 3.79^a	2.41
15	60.31 ± 12.27	22.04 ± 5.1^a	2.74
19	137.49 ± 23.75	46.64 ± 7.94^a	2.95
22	208.63 ± 36.52	60.89 ± 8.18^a	3.43
25	293.97 ± 40.02	80 ± 11.02^a	3.67
27	322.75 ± 33.59	96.38 ± 13.74^a	3.35
29	367.19 ± 20.37	107.39 ± 12.28^a	3.42
32	399.84 ± 15.27	118.7 ± 11.57^a	3.37

[a]Mean values of test series (S+B) were significantly higher ($\alpha = 0.05$) than mean values of controls (S), ± standard deviation.

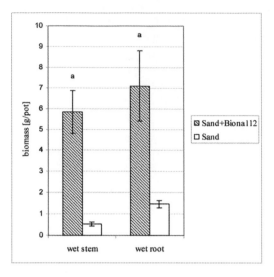

Figure 3. Mean wet stem and root biomass of bird's foot trefoil. a—mean values of test series (S+B) were significantly higher ($\alpha = 0.05$) than mean values of controls (S), I—standard deviation.

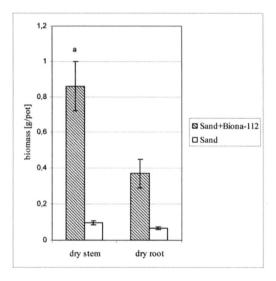

Figure 4. Mean dry stem and root biomass of bird's foot trefoil. a—mean value of test series (S+B) was significantly higher ($\alpha = 0.05$) than mean value of controls (S), I—standard deviation.

Table 3. Root number and total root length of bird's foot trefoil.

Vegetative growth period (d)	Sand+Biona-112 (S+B)	Sand (S)	Ratio S+B/S
Mean root number [per 100 cm²]			
4	4.2 ± 1.6	3.3 ± 1.7	1.27
6	7 ± 2.2	7.3 ± 2.2	0.96
10	17 ± 6	15.2 ± 2.8	1.12
13	27 ± 9.7	20.8 ± 7.9	1.30
16	52 ± 15.5	33 ± 10.2	1.58
18	67.6 ± 13.4	39.6 ± 13.3[a]	1.71
20	76.8 ± 14.2	44 ± 13.6[a]	1.75
23	94.6 ± 19.5	50.6 ± 15.2[a]	1.87
25	101.2 ± 17.8	54.6 ± 15.4[a]	1.85
27	108 ± 18.4	59.6 ± 15[a]	1.81
30	118.2 ± 21.8	61.8 ± 14.2[a]	1.91
32	129.2 ± 20.8	66.2 ± 13.8[a]	1.95
34	140.8 ± 20.1	69.8 ± 14.1[a]	2.02
37	155.8 ± 23.7	72.8 ± 12.6[a]	2.14
Mean total root length [cm/100 cm²]			
4	4.36 ± 0.87	3.83 ± 1.68	1.14
6	9.96 ± 3.88	9.56 ± 4.63	1.04
10	34.5 ± 7.76	36.99 ± 11.1	0.93
13	56.93 ± 9.21	60.13 ± 17.91	0.95
16	87.8 ± 15.83	82.72 ± 20.97	1.06
18	103 ± 15.02	95.01 ± 22.52	1.08
20	111.04 ± 15.13	102.18 ± 20.79	1.09
23	125.26 ± 18.2	110.45 ± 20.24	1.13
25	128.8 ± 18.3	114.19 ± 19.9	1.13
27	132.98 ± 17.73	116.75 ± 19.34	1.14
30	146.93 ± 24.99	119.05 ± 18.50	1.23
32	158.33 ± 24.13	122.47 ± 18.97[a]	1.29
34	166.28 ± 23.78	124.35 ± 19.29[a]	1.34
37	177.71 ± 25.81	125.96 ± 19.19[a]	1.41

[a]Mean values of test series (S+B) were significantly higher ($\alpha = 0.05$) than mean values of controls (S), ± standard deviation.

in general, higher than those in controls (Table 3). At the end of the experiment, mean root number in the 100-cm² cross-section through sand–Biona mix was 114% greater than for sand alone.

The mean total root length of bird's foot trefoil found for the cross-section of sand–Biona mix was 41% higher than for sand-only medium. The

increases in root parameter values for bird's foot trefoil were lower than those for orchard-grass. It is probably due to interspecific differences between orchard grass (a grass) and bird's foot trefoil (a legume) in nutrient requirements, for example, which could be met by Biona-112 addition to different extents.

The increase in dry root biomass of orchard grass and bird's foot trefoil in these experiments was higher than that in the previous studies using the same amounts of different substrates. Wasąg et al. (2000) showed that Biona-111 in sand caused a six-fold increase in dry root biomass of orchard grass after six weeks of vegetative growth. The enrichment of the same sand with Biona-312 substrate increased dry root biomass of *D. glomerata* over seven times. A similar increase in dry root biomass of orchard grass after Biona-312 application was found by Chomczyńska & Pawłowski (2003), whereas there was almost five-fold increase in dry root biomass of *L. corniculatus* in the same trials. The differences in increases of root biomass observed in the present experiments, and those in previous studies, could result from different amounts of macronutrients in ion exchange substrates as well as from different experimental conditions (e.g. air temperature and light duration).

Increases in values of root morphological parameters after enriching sand with Biona-112 indicated intensified development of root systems, which was partly due to the formation of more lateral roots. The enhanced production of root branches in the presence of the ion exchange substrate as nutrient carriers (containing phosphorus at rate of 4.5 mM) was not consistent with reports on uniform phosphorus supply (Linkohor et al. 2002, Liu et al. 2004, Lopez-Bucio et al. 2002). These studies showed that low doses of P (0.001 mM P) and even complete phosphate deficiency (0 mM KH_2PO_4) increased the density of lateral roots in the tested plants compared with higher P doses: 2.5 mM P (Linkohor et al. 2002), 1 mM P (Lopez-Bucio et al. 2002) and 0.25 mM KH_2PO_4 (Liu et al. 2004). In the present study the ion exchange substrate Biona-112 supplied plants with P together with other nutrients. Therefore, the formation of lateral roots could depend not only on P level but also on the presence and activity of other elements, among which nitrogen, iron and sulfur can also alter postembryonic root development (Lopez-Bucio et al. 2003).

4 CONCLUSIONS

The results of the study indicate that sand enrichment with Biona-112 had a positive effect on the vegetation of test species (*D. glomerata* L. and *L. corniculatus* L.), significantly increasing wet and dry biomass of stems and roots. The findings confirm that the 2% addition of ion exchange substrate into sand (as a model of degraded sandy mine spoil) intensified development of the root systems of test species. Hence, the addition of 2% Biona-112 could increase the stabilization and anti-erosive functions of plants during the biological restoration of sandy mine spoil.

ACKNOWLEDGMENTS

Presented results are the effect of the cooperation between Lublin University of Technology and Belarus National Academy of Sciences.

REFERENCES

Bohm, W. 1985. *Methods of studying root systems*. Warszawa: PWRiL (in Polish).

Chomczyńska, M. & Pawłowski, L. 2003. Utilization of spent ion exchange resins for soil reclamation. *Environ. Eng. Sci.* 20 (4): 301–306.

Czermiński, J.B., Iwaszewicz, A., Paszek, Z. & Sikorski, A. 1992. *Statistical methods for chemists*. Warszawa: PWN (in Polish).

Krzaklewski, W. 1990. Restoration of mine lands. In S. Kozłowski (ed.), *Rules for the protection and development of the natural environment in the areas of mineral exploitation*: 158–69. Warszawa: Wydawnictwo SGGW (in Polish).

Linkohor, B.I., Williamson, L.C., Fitter, A.H. &. Ottoline Leyser, H.M. 2002. Nitrate and phosphate availability and distribution have different effects on root system architecture of Arabidopsis. *Plant J.* 29(6): 751–60.

Liu, Y., Mi, G., Chen, F., Zhang, J. & Zhang, F. 2004. Rhizosphere effect and growth of two maize (Zea mays L.) genotypes with contrasting P efficiency at low P availability. *Plant Sci.* 167: 217–23.

Lopez-Bucio, J., Hernandez-Abreu, J., Sanches-Calderon, L., Nieto-Jacobo, M.F., Simpson, J. & Herrera-Estrella, L. 2002. Phosphate availability alters architecture and causes changes in hormone sensitivity in the Arabidopsis root system. *Plant Physiol.* 129: 244–56.

Lopez-Bucio, J., Cruz-Ramires, A. & Herrera-Estrella, L. 2003. The role of nutrient availability in regulating root architecture. *Curr. Opin. Plant Biol.* 6: 280–86.

Maciak, F. 1996. *Environment protection and restoration*. Warszawa: Wydawnictwo SGGW (in Polish).

Soldatov, V.S., Terent'ev, V.M. &. Peryškina, N.G. 1968. Artificial soils on the basis of ion exchange materials. *Dokl. Akad. Nauk BSSR* 12: 357–59 (in Russian).

Soldatov, V.S., Peryškina, H.G. & Horoško, R.P. 1978. *Ion exchanger soils*. Minsk: Nauka i Technika (in Russian).

Soldatov, V.S., & Periškina, H.G. 1985. *Artificial soils for plants*. Minsk: Nauka i Technika (in Russian).

Soldatov, V.S. 1988. Ion exchanger mixtures used as artificial nutrient media for plants. In M. Streat (ed.), *Ion exchange for industry*: 652–58. London: E. Horwood.

Soldatov, V.S., Pawłowski, L., Szymańska, M., Matusevich, V., Chomczyńska, M. & Kloc, E. 1998. Ion exchange substrate Biona-111 as an efficient measure of barren grounds fertilization and soils improvement. *Zesz. Probl. Post. Nauk Rol.* 461: 425–35.

Venkatesh, G. 2012, Future prospects of industrial ecology as a set of tools for sustainable development, *Problemy Ekorozwoju/Problems of Sustainable Development*, 7(1) 77:80.

Wasąg, H., Pawłowski, L., Soldatov, V.S., Szymańska, M., Chomczyńska, M. Kołodyńska, M., Ostrowski, J., Rut, B., Skwarek, A. & Młodawska, G. 2000. *Restoration of degraded soils using ion exchange resins*. Research project KBN No 3 T09 C 105 14. Lublin: Politechnika Lubelska (in Polish).

Zgirski, A. & Gondko, R. 1998. *Biochemical calculations*. Warszawa: PWN (in Polish).

Zieliński, R. & Zieliński, W. 1990. *Statistical arryies*. Warszawa: PWN (in Polish).

Environmental Engineering IV – Pawłowski, Dudzińska & Pawłowski (eds)
© 2013 Taylor & Francis Group, London, ISBN 978-0-415-64338-2

Equilibrium and studies on sorption of heavy metals from solutions by algae

M. Rajfur, A. Kłos & M. Wacławek

Department of Biotechnology and Molecular Biology, Opole University, Opole, Poland

ABSTRACT: The kinetics of heavy-metal sorption process, by saltwater algae *Palmaria palmate*, was investigated under the laboratory conditions. The sorption process was carried out under the static conditions. Salts of the following heavy metals were used in the study: Mn, Cu, Zn and Cd. Sorption kinetics studies of those metals has shown that the equilibrium was attained after approximately 50 min. Under the experiment conduction conditions, 60–70% of metals were sorbed in the first 10 min. The sorption equilibria were approximated with the Langmuir isotherm model. It was established that the algae sorb heavy metals proportionally to the metals concentration in solution. The studies confirmed that 50 min of exposure to algae slightly contaminated with heavy metals results in an increase in concentration of these analytes. The results of the conducted studies were compared with the results of studies carried out on the freshwater algae *Spirogyra* sp.

Keywords: heavy metals, algae *Palmaria palmate*, algae *Spirogyra* sp., sorption kinetics, Langmuir isotherm

1 INTRODUCTION

Algae are a pioneering species which populate new environments partly because they occur in very diverse ecosystems but also because of their high resistance to physicochemical conditions. Due to their abundance in very diverse environmental conditions, they become more and more popular biomonitors (Boubonari 2008, Hill 2010, Kaonga 2008, Melville 2007, Rajfur 2011a).

The studies conducted using the algal biomass may be divided into two basic groups, namely studies aiming at the assessment of the algal thalli ability to accumulate analytes (Feng 2004, Gupta 2001, Gupta 2008b, Gupta 2008c, Nuhoglu 2002) and studies using algae as bioindicators of surface water contamination by, for instance, heavy metals (Giusti 2001, Karez 1994, Nguyen 2005, Topcuoğlu 2004, Topcuoğlu 2010). The conducted laboratory studies, using different species of algae, are aiming, among others, at the assessment of their sorption characteristics, and also at their ability of desorption of heavy metals accumulated in their thalli.

The differences in the sorption characteristics of different genera and species of algae are indicated. Those differences are connected, among others, with the physiological and morphological structure of algal thalli, with the habitat from which the biomass was collected, and with the method of preparation of algae for the analysis (Skowroński 2002). The most frequently described is the kinetics of the

process and the most frequently determined are the parameters of the equilibrium state and algae sorption capacity, e.g. (Gupta 2006).

The studies of kinetics of heavy metal sorption by algae suggest that the time necessary to attain the dynamic equilibrium depends, among others, on the type of algae and the fineness of the biomass. Marine algal biomass of *Ecklonia maxima*, with the thalli size of 1.2 mm, attained the equilibrium during Cu, Pb and Cd sorption after approximately 60 min, whereas when the grinded thalli size amounted to 0.075 mm, the equilibrium was attained after approximately 10 min (Feng 2004). The studies confirmed that the cosmopolitan algae *Chlorella vullgaris* and freshwater algae *Scenedesmus quadricauda* sorb 90–95% of Cu in the first 15 min of the experiment conduction (equilibrium was attained after approximately 2 h) (Harris 1990), marine algae *Padina* sp. sorb 90% of Cd in 35 min (equilibrium attained after approximately 60 min) (Kaewsarn 2001). Marine algae *Ulva fasciata* and *Sargassum* sp. in 30 min sorb approximately 90% of Cu^{2+} ions from the solution (equilibrium was attained after approximately 2 h) (Karthikeyan 2007). Biosorbents (DP95Ca and ER95Ca) prepared on the basis of marine algae *Durvillaea potatorum* and *Ecklonia radiata* sorb 90% of Cu^{2+} and Pb^{2+} ions in 30 min (equilibrium attained after approximately 60 min) (Matheickal 1999), and marine algae *Ulva lactuca* sorb 80% of Cr^{6+} ions in the first 20 min of the process (equilibrium was attained after approximately 40 min) (El-Sikaily 2007).

Various models are used to describe the kinetics of sorption, among others, pseudo-first-order model, determined by means of Lagergren equation (Apiratikul 2008, Pavasant 2006, Qaiser 2009), pseudo-second-order model (Apiratikul 2008, Pavasant 2006, Qaiser 2009), and Weber and Morris model (Apiratikul 2008, Pavasant 2006). Kinetics of sorption process, when expressing concentrations, converted to unit ion charge $c* = z \cdot c$ [mol/L]; z—ion electric charge) can be described in detail by means of second-order reaction equation (Kłos 2006), while it is necessary to trace changes in concentration of ions as well sorbed as desorbed from algal thalli. In the case of algae *Palmaria palmate*, this model is difficult to apply, due to the lack of data regarding the quality of ions desorbed from the algae in the process of ion exchange, and the influence of competitive H+ ions absorption (Rajfur 2012). Various models are used to describe the equilibrium of the heavy metal sorption process, including, among others, Freundlich (Dönmez 1999, Kumar 2008, Tien 2002), Langmuir (Dönmez 1999, Kumar 2008, Lee 2004), Redlich-Peterson (El-Sikaily 2007) and Koble-Corrigan isotherm models (El-Sikaily 2007). On the basis of those models, one may calculate sorption capacity of algae, and metal affinity series to algae thalli. The best correlations between test data and calculated data are obtained very often with the use of Langmuir isotherm model (Gupta 2001, Gupta 2002, Gupta 2006, Gupta 2008a, Gupta 2008b, Saeed 2005). The laboratory studies indicate that the maximal sorption capacity of, i.e. *Cladophora glomerata* algae amounts to: 15.0 mg/g dry mass (d.m.) and 22.5 mg/g d.m. for Cu and Pb respectively (Yaçin 2008), *Ecklonia radiata, Ecklonia maxima, Laminaria japonica* and *Laminaria hyperbola* algae, amounts to: 1.0 to 1.6 mmol/g d.m. for Pb; 1.0 to 1.2 mmol/g d.m. for Cu and 0.8 to 1.2 mmol/g d.m for Cd (Yu 1999), whereas the sorption capacity of algal-bacterial biomass amounts to 8.5 ± 0.4 mg/g d.m for Cu (Muñoz 2006). Two-year studies confirmed that biosorbents (DP95Ca and ER95Ca), prepared on the basis of marine algae *Durvillaea patatorum* and *Ecklonia radiata,* are characterised by sorption capacity, comparable to the capacity of ion-exchange synthetic resins, and much greater than capacity of natural zeolites or powdered activated carbon (DP95Ca—1.3 mmol/g for Cu, 1.6 mmol/g for Pb; ER95Ca—1.1 mmol/g for Cu, 1.3 mmol/g for Pb) (Matheickal 1999). The presented results indicate significant differences concerning the sorption capacity of algae, depending on their type. Unfortunately, the authors do not give the uncertainties of the measurements results.

The aim of the studies was the assessment of the kinetics, as well as the description of the sorption equilibrium of the following heavy metals: Mn, Cu, Zn and Cd, by marine algae *Palmaria palmate*. The sorption equilibria were approximated with the Langmuir isotherm model. The attempt to compare marine algae sorption and the freshwater algae *Spirogyra* sp characteristics was undertaken. The determined correlations between the analyte concentrations in marine algae, and in the solution in which they were immersed, may be used in the future to prepare a simple biosensor to measure heavy metal concentrations in marine and freshwater reservoirs.

2 MATERIAL AND METHODS

Marine algae *Palmaria palmate*, purchased from BogutynMłyn company from Radzyń Podlaski (PL), were used in the studies. The algae samples for the analysis were rinsed in demineralised water (conductivity $\kappa = 0.5$ μS/cm) and dried at the temperature of 323 K (24 h). Algae prepared in such a way were stored in sealed polyethylene containers. The concentrations of the metals naturally accumulated in the algae: $c_{(a,0)}$, amounted to: $c_{Mn(a,0)} = 0.033 \pm 0.005$ mg/g d.m.; $c_{Cu(a,0)} = 0.0033 \pm 0.0008$ mg/g d.m.; $c_{Zn(a,0)} = 0.021 \pm 0.002$ mg/g d.m.; $c_{Cd(a,0)} = 0.0016 \pm 0.0002$ mg/g d.m.

2.1 Study of the kinetics of sorption

The algae samples, with a mass of approximately 0.5 g d.m., were placed in perforated containers, the volume of which amounted to approximately 15 mL, and they were immersed with the container in the salt of a specific heavy metal (Mn, Cu, Zn and Cd). The volume of those metals amounted to 200 mL. Before the experiment, the dried algae were immersed in demineralised water for 30 min (Rajfur 2012). The solution was intensely stirred with the use of the magnetic stirrer. The solution was periodically sucked in, directly from the container in which the study was carried out, in order to determine the concentration of copper (AAS). The process lasted 70 min. The conductometric and pH-metric measurements were carried out during the process of heavy metal sorption by the algal biomass.

2.2 Determination of the equilibrium parameters in static system

The experiments were carried out analogously in accordance with the studies of the kinetics of the process. Only the initial concentrations of Mn^{2+}, Cu^{2+}, Zn^{2+} and Cd^{2+} ions in the solution were changed. In order to determine the concentration of the selected metal, the samples of the solution were collected at the beginning and at the end of the sorption process, which lasted 50 min.

2.3 Analytical instruments and chemicals

Heavy metals were determined with an atomic absorption spectrometer SOLAAR 969 from UNI-CAM, produced by *Thermo Electron* Corporation, USA. ANALYTIKA Ltd. (CZ) standards were used to calibrate the apparatus. The values of the most concentrated calibration standard (2 mg/L for Cd and 5 mg/L for other metals) were assumed to be the upper limit of the linear relation between the signal and the concentration. The conductivity and pH measurements of the solution in which the algae were immersed, were carried out using the instruments from Elmetron general partnership from Zabrze (PL); CP551 pH-meter and CC551 conductometer, which absolute indication errors

Table 1. The limits of detection and the limits of quantification for atomic absorption spectrometer UNICAM-SOLAAR 969 (Manual 1997).

Metal	Instrumental detection limit IDL (mg/L)	Instrumental quantification limit IQL (mg/L)
Mn	0.0016	0.029
Cu	0.0045	0.041
Zn	0.0033	0.013
Cd	0.0028	0.032

amount to $\Delta pH = 0.02$ and $\Delta\kappa = 0.1$ µS/cm respectively. The solutions were prepared with the use of the chemicals from MERCK.

2.4 Quality control

Table 1 presents the detection limit and determination limit for heavy metals, which are characteristic for the Atomic Absorption Spectrometer (AAS) used in our studies (Manual 1997).

The quality control of measurements was assured by test analyses of the BCR 414 *plankton* and BCR-482 *lichen* reference materials from the Institute for Reference Materials and Measurements in Belgium. The obtained results are summarized in Table 2.

2.5 Langmuir isotherm model

The Langmuir isotherms model is valid for monolayer adsorption on to surface containing finite number of identical sorption sites which is described by the following equation (Saeed 2005):

$$c_{(a,1)} = (c_{(a,max)} \cdot K \cdot c_{(s,1)}) \cdot (1 + K \cdot c_{(s,1)})^{-1} \qquad (1)$$

where: $c_{(a,1)}$—equilibrium concentration of a given metal in algae (mg/g), $c_{(s,1)}$—equilibrium concentration of this metal ions in solution (mg/L), $c_{(a,max)}$—sorption capacity of algae (mg/g d.m.), K—constant.

The Langmuir equation can be rearranged to linear form for the convenience of plotting and determining the Langmuir constants as below:

$$(c_{(a,1)})^{-1} = (c_{(a,max)} \cdot K \cdot c_{(s,1)})^{-1} + (c_{(a,max)})^{-1} \qquad (2)$$

Table 2. Measured and certified values of heavy metals concentration in the BCR 414 *plankton* and BCR 482 *lichen* reference material.

Metal	Certified value (mg/kg d.m.)	±Uncertainty (mg/kg d.m.)	AAS		D* (%)
			Mean (mg/kg d.m.)	±SD (mg/kg d.m.)	
BCR 414 plankton					
Mn	299	12	276	15	−7.7
Cu	29.5	1.3	27.8	1.9	−5.8
Zn	112	3	103	4	−8.0
Cd	0.383	0.014	n.d.	n.d.	n.d.
			Mean (mg/kg d.m.)	Mean (mg/kg d.m.)	
BCR 482 lichen					
Mn	33.0	0.5	31.2	0.8	−5.5
Cu	7.03	0.19	6.54	0.18	−7.0
Zn	100.6	2.2	93.9	2.5	−6.7
Cd	0.56	0.02	0.52	0.04	−7.1

*Deviation: the relative difference between measured by AAS and certified concentrations in (%).

3 RESULTS AND DISCUSSION

3.1 Kinetics of heavy metal ions sorption by palmaria palmate algae

The diagram in Figure 1 presents the kinetics of the process of heavy metals sorption in algal biomass of *Palmaria palmate*. During the process, the concentrations of metals in the $c_{(s)}$, solution, in which the algae were immersed, were measured periodically. On the basis of those measurements, the temporary changes in metal concentrations in 1 g of dry mass of algae were determined; $c_{(a)}$. The concentrations of heavy metals in initial solutions amounted to $2 \cdot 10^{-5}$ mol/L.

The result of the conducted studies indicates that dynamic equilibrium during the heavy metal sorption is attained by *Palmaria palmeta* algae, after approximately 50 min. During this period, the following amounts of ions from the initial solution were accumulated in the algae sample: 87% of Mn^{2+}, 95% of Cu^{2+}, 96% of Zn^{2+} and 97% of Cd^{2+}.

In the first 10 min of the process, approximately 60–70% of determined heavy metal ions are sorbed from the solution to the biomass, with reference to their concentration accumulated in algae in an equilibrium state (for Cu—0.56 mg/g d.m.). On the contrary, the dynamic equilibrium during the sorption process of, among others, Mn, Cu, Zn and Cd by *Spirogyra* sp. is attained after approximately 30 min. During this period, the following amounts of ions from the initial solution were sorbed by the freshwater algae: 80% of Mn^{2+}, 94% of Cu^{2+}, 92% of Zn^{2+} and 94% of Cd^{2+}. The diagrams in Figures 2 and 3 present respectively: the changes of conductivity and of H^+ ions concentration, during the process of sorption of Mn, Cu, Zn and Cd in algal biomass of *Palmatia palmate*.

The increase of conductivity was observed during the conduction of the heavy metal sorption process in algae *Palmaria palmate* (Fig. 2). The same effect was found during the results analysis of

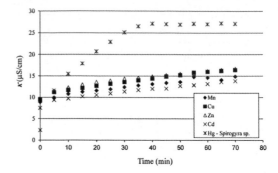

Figure 2. Changes in the solution conductivity during the heavy metal ions sorption in the algal biomass.

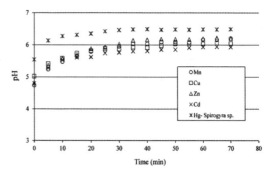

Figure 3. Changes in pH of the solution during the heavy metal ions sorption in algal biomass.

studies of the Hg^{2+} ions sorption in lyophilized biomass of alga *Spirogyra* sp. (Rajfur 2011b), which is presented in Figure 3. It may be assumed that the abovementioned effect results from the irreversible changes, developing in time in the cell membrane structure, which lead to a leakage of ionic substance from algal cells into the solution. That effect was observed during studies on the influence of copper on lichen thalli. The potassium ions were released into the solution, as a result of the destruction of cell membranes (Cabral 2003). The diagram presented in Figure 3 leads to the conclusion that the process of Mn^{2+}, Cu^{2+}, Zn^{2+} and Cd^{2+} ions sorption is accompanied by the sorption of H^+ ions. This conclusion is confirmed by the studies of heavy metal sorption kinetics from solutions with pH = 4.0. The 0.1 M hydrochloric acid was added, in order to lower the pH of metal salts' solutions. The results of the analyses were presented in Figure 4.

It should be assumed that the greater the concentration of H^+ ions in the solution, in which the algae were placed, the greater their competition towards the heavy metal ions. It was also affirmed

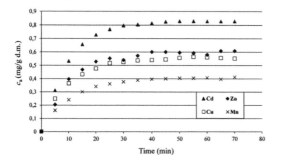

Figure 1. Kinetics of heavy metal sorption ions in algae *Palmaria palmate*.

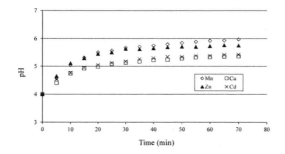

Figure 4. The changes in the concentration of H⁺ ions in the solutions with pH = 4.0 during the heavy metal ions sorption in algal biomass.

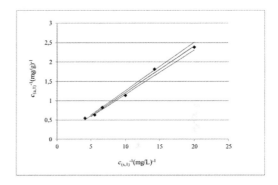

Figure 5. Langmuir isotherm, describing the sorption equilibrium of Mn ($t_{sorption}$ = 50 min).

that less Cu^{2+} ions (0. 53 mg/g d.m.) were sorbed from the solution with initial pH amounting to 4.0 and metal concentration amounting to c_{Cu2+} = 1.47 mg/L, than from the solution with the same copper concentration and pH = 4.82 (0.56 mg/g d.m.). This conclusion confirms the results of the studies performed by other authors (Feng 2004, Gupta 2006, Hamdy 2000, Herrero 2005).

3.2 Study of equilibria parameters during the heavy metals sorption by palmaria palmate algae

The next stage included the equilibrium studies during the process of sorption of Mn^{2+}, Cu^{2+}, Zn^{2+} and Cd^{2+} ions from solutions of their salts.

On the basis of the results of the measurements of heavy metal concentration in the $c_{(s)}$ solution, before (0) and after (1) sorption process, the concentrations of metals $c_{(a)}$ were determined and calculated per 1 g d.m. In order to maintain the experiment conditions (0.5 g d.m. of algae were introduced into 200 L of metal salt solution); it should be assumed that sorption in 1 g d.m. of algae takes place from the solution, the volume of which amounts to 400 mL. The change of one of the parameters (algae sample mass, solution volume) has an impact on the equilibrium of the process.

The results approximated with the Langmuir model (dependence 2) are presented in Figures 5–8. The dashed lines were marked on the diagrams, and they indicate the uncertainties of measurements of heavy metal concentration in algae, indicated by standard deviation $\pm SD_a$ determined for the a parameter of the function: $y = a \cdot x + b$ (Table 3).

The isotherms parameters were tabulated in Table 3 and compared with the parameters of the isotherms determined for the algae *Spirogyra* sp (Rajfur 2010).

Sorption capacities $c_{(a,max)}$, determined from Langmuir isotherms, are only approximate in

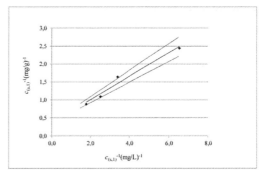

Figure 6. Langmuir isotherm, describing the sorption equilibrium of Cu ($t_{sorption}$ = 50 min).

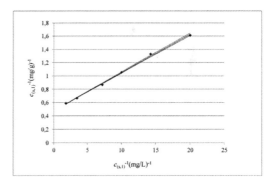

Figure 7. Langmuir isotherm, describing the sorption equilibrium of Zn ($t_{sorption}$ = 50 min).

character, due to the fact that they are characterised by large uncertainty indicated by means of $\pm SD_b$. The results presented in Figures 5–8 indicate the possibility of using algae for active biomonitoring of surface waters. It should be assumed that the exposition of the marine algae *Palmaria palamate*, transferred from waters with

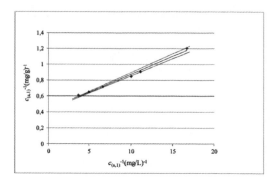

Figure 8. Langmuir isotherm, describing the sorption equilibrium of Cd ($t_{sorption}$ = 50 min).

Table 3. The parameters of $y = a \cdot x + b$ function presented in Figures 3–7; linear regression parameters: $\pm SD_a$—standard deviation of a parameter, $\pm SD_b$—standard deviation of b parameter, R^2—Correlation coefficient and sorption capacity determined from Langmuir isotherm: $c_{(a,max)}$.

Metal	a	b	$\pm SD_a$	$\pm SD_b$	R^2	$c_{(a,max)}/$ (mg/g d.m.)
Palmaria palamate						
Mn	0.328	0.344	0.042	0.168	0.968	2.91
Cu	0.120	0.006	0.005	0.057	0.993	167
Zn	0.058	0.473	0.001	0.015	0.998	2.11
Cd	0.046	0.419	0.002	0.022	0.991	2.39
Spirogyra sp.						
Mn	0.528	0.229	0.072	0.456	0.988	4.4
Cu	0.198	0.115	0.0051	0.101	0.994	8.7
Zn	0.210	0.222	0.050	0.072	0.999	4.5
Cd	0.216	0.084	0.0039	2.93	0.998	11.9

low content of heavy metals, or of cultivated algae, by means of proportional sorption in contaminated marine and freshwater reservoirs, shall facilitate the quality assessment of those waters.

Determined on the basis of the slope value of Langmuir isotherms: $a = (c_{(a,max)} \cdot K)^{-1}$, heavy metal affinity to thalli of *Palmaria palamate* and *Spirogyra* sp. algae increases in the following order: $Mn^{2+} < Cu^{2+} < Zn^{2+} < Cd^{2+}$ and $Mn^{2+} < Cd^{2+} \approx Zn^{2+} \approx Cu^{2+}$. Due to the large uncertainty of the results, no attempts were undertaken to indicate which species is a better sorbent. It should be assumed that sorption of H^+ ions, accompanying the accumulation of heavy metals in algal thalli, shall also have an impact on heavy metal sorption in freshwater as well as in marine algae.

4 CONCLUSIONS

The conclusion reached is in the conducted studies is that marine algae *Palmaria palamate,* as well as freshwater algae *Spirogyra* sp., are very good heavy-metal sorbents. Algal biomass accumulates heavy metals in its thallus proportionally to their concentration in the solution. The assumption that algae may be used as biosensors of the contamination of aqueous ecosystems by analytes may be drawn, on the basis of the above mentioned conclusions.

The mechanisms of sorption and dynamic equilibria in the structure of algae thalli, including the biological half-life of the contaminant, are still insufficiently identified. It is a result of a multidimensionality of interactions: alga—environment. In the future, the determined correlations between concentrations of analytes in algae and in solution (or *in situ*), in which they were immersed, as well as the identification of abiotic factors, influencing the sorption process, may be used to measure, in a simple way, heavy metals concentrations in surface waters, e.g. to draw up the surface water classification method, in which the algae shall perform the function of biosensors of the quality of surface waters.

ACKNOWLEDGEMENTS

The Project received financial assistance from the funds of the National Science Centre granted by force of the decision no. DEC-2011/03/D/NZ9/00051.

REFERENCES

Apiratikul, R., Pavasant, P. 2008. Batch and column studies of biosorption of heavy metals by Caulerpa lentillifera. *Bioresource Technology* 99: 2766–2777.

Boubonari, T., Malea, P., Koyro, H.W., Kevrekidis, T. 2008. The red macroalga Gracilaria bursa-pastoris as a bioindicator of metals (Fe, Zn, Cu, Pb, Cd) in oligohaline coastal environments. *Fresenius Environmental Bulletin* 17(12): 2207–2216.

Cabral, J.P. 2003. Copper toxicity to five Parmalia lichens in vitro. *Environmental and Experimental Botany* 49: 237–250.

Dönmez, G.C., Aksu, Z., Öztürk, A., Kutsal, T. 1999. A comparative study on heavy metal biosorption characteristics of some algae. *Process Biochemistry* 34: 885–892.

El-Sikaily, A., El Nemr, A., Khaled, A., Abdelwehab, O. 2007. Removal of toxic chromium from wastewater using green alga Ulva lactuca and its activated carbon. *Journal of Hazardous Materials* 148: 216–228.

Feng, D., Aldrich, C. 2004. Adsorption of heavy metals by biomaterials derived from the marine alga Ecklonia maxima. *Hydrometallurgy* 73: 1–10.

Giusti, L. 2001. Heavy metal contamination of brown seaweed and sediments from the UK coastline between the Wear river and the Tees river. *Environment International* 26(4): 275–286.

Gupta, V.K., Shrivastava, A.K., Neeraj, J. 2001. Biosorption of chromium(VI) from aqueous solutions by green algae Spirogyra species. *Water Research* 35(17): 4079–4085.

Gupta, V.K., Rastogi, A., Nayak, A. 2002. Biosorption of nickel onto treated alga (Oedogonium hatei): Application of isotherm and kinetic models. *Journal of Colloid and Interface Science* 342: 533–539.

Gupta, V.K., Rastogi, A., Saini, V.K., Jain, N. 2006. Biosorption of copper(II) from aqueous solutions by Spirogyra species. *Journal of Colloid and Interface Science* 296: 59–63.

Gupta, V.K., Rostogi, A. 2008a. Biosorption of lead(II) from aqueous solutions by non-living algal biomass Oedogonium sp. and Nostoc sp.—A comparative study. *Colloids and Surfaces B* 64: 170–178.

Gupta, V.K., Rostogi, A. 2008b. Biosorption of lead from aqueous solutions by green algae Spirogyra species: Kinetics and equilibrium studies. *Journal of Hazardous Materials* 152: 407–414.

Gupta, V.K., Rostogi, A. 2008c. Equilibrium and kinetic modelling of cadmium(II) biosorption by nonliving algal biomass Oedogonium sp. from aqueous phase. *Journal of Hazardous Materials* 153: 759–766.

Hamdy, A.A. 2000. Biosorption of heavy metals by marine algae. *Current Microbiology* 41: 232–238.

Harris, P.O., Ramelow, G.J. 1990. Binding of metal ions by particulate biomass derived from Chlorella vulgaris and Scenedesmus quadricauda. *Environmental Science and Technology* 24: 220–228.

Herrero, R., Lodeiro, P., Rey-Castro, C., Vilariño, T., Sastre de Vicente, E.M. 2005. Removal of inorganic mercury from aqueous solutions by biomass of the marine macroalga Cystoseira baccata. *Water Research* 39: 3199–3210.

Hill, W.R., Ryon, M.G., Smith, J.G., Adams, S.M., Boston, H.L., Stewart, A.J. 2010. The role of periphyton in mediating the effects of pollution in a stream ecosystem. *Environmental Management* 45(3): 563–576.

Kaewsarn, P., Yu, Q. 2001. Cadmium(II) removal from aqueous solutions by pre-treated biomass of marine alga Padina sp. *Environmental Pollution* 112: 209–213.

Kaonga, C.C., Chiotha, S.S., Monjerezi, M., Fabiano, E., Henry, E.M. 2008. Levels of cadmium, manganese and lead in water and algae Spirogyra aequinoctialis. *International Journal of Environmental Science and Technology* 5(4): 471–478.

Karez, C.S., Amado Filho, G.M., Moll, D.M., Pfeiffer, W.C. 1994. Metal concentrations in benthic marine algae in 3 regions of the state of Rio de Janeiro. *Anais da Academia Brasileira de Ciencias* 66(2): 205–211.

Karthikeyan, S., Balasubramanian, R., Iyer, C.S.P. 2007. Evaluation of the marine algae Ulva fasciata and Sargassum sp. for the biosorption of Cu(II) from aqueous solutions. *Bioresource Technology* 98:, 452–455.

Kłos, A., Rajfur, M., Wacławek, M., Wacławek, W. 2005. Ion exchange kinetics in lichen environment. *Ecological Chemistry and Engineering A* 12(12): 1353–1365.

Kumar, D., Singh, A., Gaur, J.P. 2008. Mono-component versus binary isotherm models for Cu(II) and Pb(II) sorption from binary metal solution by the green alga Pithophora oedogonia. *Bioresource Technology* 99: 8280–8287.

Lee, H.S., Suh, J.H., Kim, I.B., Yoon, T. 2004. Effect of aluminum in two-metal biosorption by an algal biosorbent. *Minerals Engineering* 17: 487–493.

Manual of spectrometer AAS SOLAR 969 UNICAM 1997. Spectro-Lab, Warsaw, Poland.

Matheickal, J.T., Yu, Q. 1999. Biosorption of lead(II) and copper(II) from aqueous solutions by pre-treated biomass of Australian marine algae. *Bioresource Technology* 69: 223–229.

Melville, F., Pulkownik, A. 2007. Investigation of mangrove macroalgae as biomonitors of estuarine metal contamination. *Science of the Total Environment* 387; 301–309.

Muñoz, R., Alvarez, M.T., Muñoz, A., Terrazas, E., Guieysse, B., Mattiasson, B. 2006. Sequential removal of heavy metals ions and organic pollutants using an algal-bacterial consortium. *Chemosphere* 63: 903–911.

Nguyen, L.H., Leermakers, M., Elskens, M., Ridder, D.F., Doan, H.T., Baeyens, W. 2005. Correletions, partitioning and bioaccumulation of heavy metals between different compartments of Lake Balaton. *Science of the Total Environment* 341: 211–226.

Nuhoglu, Y., Malkoc, E., Gürses, A., Canpolat, N. 2002. The removal of Cu(II) from aqueous solutions by Ulothrix zonata. *Bioresource Technology* 85: 331–333.

Pavasant, P., Apiratikul, R., Sungkhum, V., Suthiparinyanont, P., Wattanachira, S., Marhaba, T.F. 2006. Biosorption of Cu2+, Cd2+, Pb2+ and Zn2+ using dried marine green macroalga Caulerpa lentillifera. *Bioresource Technology* 97: 2321–2329.

Qaiser, S., Saleemi, A.R., Umar, M. 2009. Biosorption of lead(II) and chromium(VI) on groundnut hull: Equilibrium, kinetics and thermodynamics study. *Electronic Journal of Biotechnology* 12(4): 1–17.

Rajfur, M., Kłos, A., Wacławek, M. 2010. Sorption properties of algae Spirogyra sp. and their use for determination of heavy metal ions concentrations in surface water. *Bioelectrochemistry* 80: 81–86.

Rajfur, M., Kłos, A., Wacławek, M. 2011a. Application of alga in biomonitoring of the Large Turawa Lake. International *Journal of Environmental Science and Health* 46: 1401–1408.

Rajfur, M., Kłos, A., Wacławek, M. 2011b. Kinetics of Hg2+ ions sorptio on algae Spirogyra sp. *Proceedings of ECOpole* 5(2): 589–594.

Rajfur, M., Kłos, A., Wacławek, M. 2012. Sorption of copper(II) ions in the biomass of alga Spirogyra sp. *Bioelectrochemistry* 87: 65–70.

Saeed, A., Iqbal, M., Akhtar, M.W. 2005. Removal and recovery of lead(II) from single and multimetal (Cd, Cu, Ni, Zn) solutions by crop milling waste (black gram husk). *Journal of Hazardous Materials B* 117:, 65–73.

Skowroński, T., Kalinowska, R., Pawlik-Skowrońska, B. 2002. Glony środowisk zanieczyszczonych metalami ciężkimi [The algae in environments contaminated with heavy metals], *Kosmos* 2: 165–173.

Tien, C.J. 2002. Biosorption of metal ions by freshwater algae with different surface characteristics. *Process Biochemistry* 38: 605–613.

Topcuoğlu, S., Kırbaşoğlu, Ç., Yılmaz, Y.Z. 2004. Heavy metal levels in biota ve sediments in the northern coast of The Marmara Sea. *Environmental Monitoring and Assessment* 96: 183–189.

Topcuoğlu, S., Kılıç, Ö., Belivermiş, M., Ergül, H.A., Kalaycı, G. 2010. Use of marine algae as biological indicator of heavy metal pollution in Turkish marine environment. *Journal of the Black Sea/Mediterranean Environment* 16(1): 43–52.

Yalçın, E., Çavuşoğlu, K., Maraş, M., Bıyıkoğlu, M. 2008. Biosorption of lead (II) and copper (II) metal ions on Cladophora glomerata (L.) Kütz. (Chlorophyta) Algae: Effect of algal surface modification. *Acta Chimica Slovenica* 55: 228–232.

Yu, Q., Matheickal, J.T., Yin, P., Kaewsarn, P. 1999. Heavy metal uptake capacities of common marine macro algal biomass. *Water Research* 33(6): 1534–1537.

Environmental Engineering IV – Pawłowski, Dudzińska & Pawłowski (eds)
© *2013 Taylor & Francis Group, London, ISBN 978-0-415-64338-2*

Use of plants for radiobiomonitoring purposes

A. Dołhańczuk-Śródka, Z. Ziembik & M. Wacławek
Independent Chair of Biotechnology and Molecular Biology, Opole University, Opole, Poland

ABSTRACT: The study includes a comparison of ^{137}Cs content in 3 species of plants from 3 separate classes, which were collected in the same areas, as well as an assessment of the correspondence of the obtained results. Samples of the plants were collected in the north-west part of Poland. Using a single-factor ANOVA, the authors assessed the influence of the plant species on accumulation of ^{137}Cs. Good correlations between the transformed ^{137}Cs content values in particular plants collected in the same area were observed. This enables a reliable comparison of contamination in the areas, where different species of the plants grow.

Keywords: ^{137}Cs, biomonitoring, plants

1 INTRODUCTION

The nuclear weapon tests carried out in the mid-20th century are the main source of synthetic radioactive isotopes to be found in the natural environment. The radioactive substances released during those tests are still present in the ecosystems. Today, nuclear weapon tests are not carried out any more; yet, the radioactive materials may still be a hazardous. Wide use of radioactive materials in various areas of economy results in concerns related to the possible uncontrolled release of such materials into the natural environment, and subsequently into our systems. Despite the advanced technology and appropriate procedures to follow when handling radioactive materials, there is a risk of unrestricted spreading of such materials. The issue may be relevant to the entire world as the range of released radioactive materials can be global.

Release incidents, as a result of which significant amounts of radioactive substances are released into the natural environment, are rare. Throughout the last 30 years, only two such incidents took place. The first one in 1986, when the failure of the Chernobyl Nuclear Power Plant resulted in a release of a significant amount of radioactive material into the atmosphere. The other one took place in Japan in 2011. The tsunami wave, which was created by an extremely powerful earthquake, flooded some parts of Fukushima Dai-ichi Nuclear Power Plant, causing failure of the nuclear reactor cooling system. Due to the failure, significant amounts of material containing radioactive isotopes were released to the surrounding environment.

Monitoring or radioactive contamination is one of the basic nuclear safeguards. It facilitates identification of hazard sources, which allows undertaking measures aiming at minimising the consequences of a spread of hazardous radioactive substances. Various organisms, particularly plants, can be used to assess the existing radioactive contamination as well as the trends in terms of its movement. In the land environment, virtually all contaminants, sooner or later, penetrate into the soil. It is an extremely complex system of dead, inorganic material, organic material as well as living organisms. The substances found in the land environment are accumulated in the soil and are subject to various types of chemical and physical processes. ^{137}Cs is one of the longest-lasting synthetic radioactive isotopes to be found in the natural environment. From the soil, the radionuclide can be extracted by plants or absorbed by microorganisms and, by the same token, it is further transferred to other links of the food chain. The isotope is absorbed by plants through their roots as well as their overground parts.

The mechanism for absorption of elements, including ^{137}Cs, through the roots from the soil is a result of the following processes: cationic exchange through cell membranes, intracellular transport as well as rhizosphere processes. Two fundamental mechanisms of absorption of elements through the root can be distinguished: passive and active. The passive mechanism is based on the diffusion of ions from the soil solution into the endodermal root cells and subsequently into the overground parts of the plant along with the transpiration water flow. Whereas the active transport takes place against the concentration gradient through the special channels in the cell membrane using the metabolic energy (Kabata-Pendias & Pendias 1999). Plants are also capable of absorbing elements and chemical compounds through their overground parts, particularly their leaves. The absorption mechanism is based mainly on penetration through the cuticle into the mesophyll layer as well as xyleme bundles, which may be used for transport to various parts of the plant. Participation of the above-mentioned mechanisms in various periods after a release of radionuclides into the natural environment may vary. During the fallout, absorption through the overground parts of the plants is decisive. The radionuclides, which due to deposition find their way onto the surface of leaves or needles, are absorbed by their stomatal apparatuses.

It was found that caesium contained in soil particles, stirred in the form of dust, can be deposited on the overground parts of the plants (Kłos et al. 2007). The other mechanism involved the diffusion of metal cations from the soil through water wetting the plant, what was demonstrated for moss (Kłos et al. 2012). Transfer of caesium from the soil and into the plants depends on its characteristics. Among 30 species of plants, the discrepancies in terms of accumulation of caesium reached 10,000% (Broadley & Willey 1997). A similar variability for 6 different species of plants cultivated on the same soil was observed by Skarlou (Skarlou et al. 1996). It was also noticed that the type of soil significantly affects uptake of caesium through the plant root. The phenomenon was particularly evident in the case of sandy soils characterised by relatively minor affinity with caesium (Guivarch et al. 1999; Massas et al. 2010). It was proved that volume-wise, the dependency between accumulation of ^{137}Cs and the plant species can be illustrated with the following sequence: fern > heather > berries > grass (Pietrzak-Flis 1996). The above-mentioned plants, particularly ferns and berries, are sued to assess radioactive contamination in the examined areas, also time-wise. But, in contrast to these results, negligible influence of plant type on quantity of absorbed ^{137}Cs was also observed. Similar concentration of activities was determined in epigeic lichen and epiphytic moss collected in neighbour sites (Kłos et al. 2009).

Fungi are good biosorbents for caesium (e.g. Nimis 1990, Nimis et al. 1986, Nimis and Cebulez 1989, Haselwandtner 1988). It is assumed that fungi are the most important biological factors for transfer of radioactive caesium in the soil. The mycelium can accumulate substantial amounts of radionuclides from large forest areas.

Mosses and lichens are also good accumulators of contaminants in the forest ecosystems. These plants, which have no root system, accumulate substantial amounts of contaminants, including radionuclides, which can be absorbed directly from the atmospheric aerosol. In a humid environment, these organisms additionally absorb substances from the substratum,

on which they grow. In the case of mosses, it is the major direction of transfer, and the mineral salts from the substratum are absorbed through the rhizoids. Lichens accumulate contaminants due to the ion exchange, which takes place between the surrounding (water solution, lichens come into contact with) and the cation-active layer of the lichens (extracellular structure). In a water environment, the process, due to the ambient acidification, increases solubility of selected substances contained in the substratum (Kłos et al. 2007). Due to the good absorption qualities of mosses and lichens, they are widely used as biomonitors for the assessment of environment contamination. Trees are also effective "catchers" of aerosols (Yamagata et al. 1969, Bunzl & Kracke, 1988, Desmet and Myttenaere 1988, Sokolov 1990). Therefore, forests are extremely efficient filters for collecting dry fallout. Use of plants for ^{137}Cs biomonitoring may be restricted due to the lack of representation of a given species in different parts of the examined area. Even when using popular species, one must take into account the fact that in certain locations samples of the required plant will not be available. It is vital to assess, whether it is possible to replace one species with another and compare the results obtained for particular regions of the examined area.

The objective of the research was to compare ^{137}Cs content in 3 species of plants from 3 separate classes, which were collected in the same areas and assess the correspondence between the obtained results. The research included mosses (*Pleurozium schreberi*), lichens (*Hypogymnia physodes*) as well as roots and shoots with leaves of Common Bilberry (*Vaccinium myrtillus L.*). For statistical computations, the R language was utilized (R Development Core Team 2009). R is a free software environment for statistical computing and graphics. In our computations, the centered logratio (clr) transformation was used (van den Boogaart & Tolosana-Delgado 2006; K. Gerald van den Boogaart & Tolosana-Delgado 2008; K.G. van den Boogaart et al. 2011).

2 MATERIAL AND METHODS

Samples of the plants (mosses, lichens, berries root, leaves and stalk) were taken at 21 places in forests around Opole (south-western Poland) in an area of Bory Stobrawskie limited approximately by a 40 km × 20 km rectangle.

This region was particularly contaminated by radioactive materials released to atmosphere after Chernobyl catastrophe, and elevated ^{137}Cs activity concentrations in soil are still determined in this area. In Figure 1 positions of places in which the samples were collected are shown. The places were selected mainly randomly, although presence of

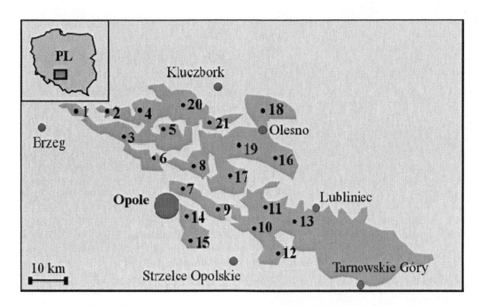

Figure 1.   Location of places from which the plants samples were taken. Each place is indicated by its number.

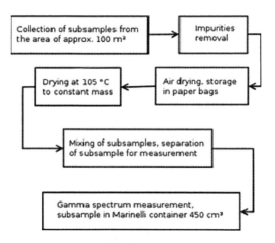

Figure 2.   Scheme of consecutive steps of sample preparation for measurement.

moss, lichens and berries growing on the common area of approx. 100 m² was required.

The plants were taken manually from forest glades at the places located not less than 200 m from roads. The samples were stored in paper bags in which air circulation was enabled. The scheme in Figure 2 shows consecutive sample preparation steps, starting from material collection to measurement.

The measurement of radionuclide activity in plants samples was carried out by means of a gamma-spectrometer with a germanium detector HPGe (Canberra) of high resolution: 1.29 keV (FWHM)

at 662 keV and 1.70 keV (FWHM) at 1332 keV. Relative efficiency: 21.7%. Energy and efficiency calibration of the gamma spectrometer was performed with the standard solutions type MBSS 2 (Czech Metrological Institute, Praha), which covers an energy range from 59.54 keV to 1836.06 keV. The geometry of the calibration source was a Marinelli container (447.7 ± 4.48 cm³), with density 0.985 ± 0.01 g/cm³, containing ^{241}Am, ^{109}Cd, ^{139}Ce, ^{57}Co, ^{60}Co, ^{137}Cs, ^{113}Sn, ^{85}Sr, ^{88}Y and ^{203}Hg. The geometry of sample container was a similar Marinelli of 450 cm³. Time of measurement was 24 h for all of moss samples. Measuring process and analysis of spectra were computer controlled with the use of software GENIE 2000. The results were corrected to the same date of measurement.

The following markings were assigned to the plant samples: *ms*—mosses, *ln*—lichens, *bb*—Common Bilberry root and *bg*—Common Bilberry shoots and leaves.

3   RESULTS AND DISCUSSIONS

The results of ^{137}Cs activity concentrations in the plants samples in the Bory Stobrawskie area are given in Table 1. In this table Min is the lowest value in data, Q1 is lower quartile, Median is median, Mean is arithmetic mean, Q3 is upper quartile, Max is the highest value, CV is coefficient of variability.

The ^{137}Cs radioactivity concentrations found in the present study ranged from 12.7

Table 1. Characteristics of ^{137}Cs activity concentrations in dry mass (d.m.) of the plants samples collected in the Bory Stobrawskie area.

| | | ^{137}Cs radioactivity concentration [Bq/kg d.m.] | | | | | |
	Min	Q1	Median	Mean	Q3	Max	CV
bb	12.7	35.3	61.0	166	152	974	1.43
bg	22.4	62.6	132	279	320	1821	1.49
ms	20.3	41.4	81.5	225	382	1234	1.30
ln	0.00	19.8	36.5	183	258	959	1.49

to 974 Bq/kg d.m. (the mean is 166 Bq/kg d.m.) for berries root, from 22.4 to 1821 Bq/kg d.m. (the mean is 279 Bq/kg d.m.) for berries leaves and stalk, from 20.3 to 959 Bq/kg d.m. (the mean is 225 Bq/kg. d.m.) for mosses and from 0.0 to 959 Bq/kg d.m. (the mean is 183 Bq/kg d.m.) for lichens.

Activity values are always nonnegative and must not exceed the set upper limit value. The limit value can be determined by assuming that the sample is composed exclusively of the examined isotope. Such a method for identification of the limit value is effective, provided that single radioisotope content is being determined. A more universal method for determining the upper limit value of results is based on conversion of the specific activity of the isotope(s) into mass concentration. Measurement instruments are usually calibrated in such a manner, as to make sure the result is displayed in relation to the unit mass. Therefore, it should be concluded that the total mass of all ingredients must not exceed the set reference mass.

Activities of ^{137}Cs can be easily recalculated to their mass concentrations from the formula 1:

$$m = \frac{t_{1/2}M}{\ln(2)N_A} a \qquad (1)$$

where $t_{1/2}$ is the half life time of the radionuclide, M is its molar mass and N_A is Avogadro constant. Because the activities are referred to 1 kg of sample mass, the radionuclide mass m expressed in g unit corresponds to its concentration expressed in g kg^{-1} unit. The quantities with nonnegative values and the sum not exceeding the set constant are usually referred to as compositional or rational data. Such data displays certain peculiar qualities. First of all, particular variables are not independent of one another. The sum of all the variables describing composition of the examined object is constant. An increase in the value of one variable must be related with a decrease in the value of another variable or other variables. Due to the peculiar qualities of the compositional data, the use of standard statistical methods for the purpose of the analysis

of such data may lead to erroneous conclusions (Pawlowsky-Glahn & Buccianti 2011, Pawlowsky-Glahn & Egozcue 2006). Aitchison (2010) presented methods, which should be applied for the purpose of a statistical analysis of compositional data. The methods mentioned were also expanded and elaborated on by other authors (Pawlowsky-Glahn & Buccianti 2011).

It has been proven that the geometric mean is a useful tool for describing the central tendency of compositional data. Table 2 includes the average ^{137}Cs content in the plant samples as well as the average content calculated jointly for all the plants. Geometric mean values were provided as well as, for comparison purposes, arithmetic mean values.

It can be noted that the geometric means are approximately 2 times smaller than the arithmetic means. As the geometric mean, better than the arithmetic mean, illustrates the central tendency of the data, it may be concluded that the assessment of ^{137}Cs content using the arithmetic mean leads to substantially overestimated values.

One of the available solutions to the above-mentioned issue of the upper and lower limit of the variables is their transformation into such a space, in which their values are not limited (Filzmoser et al. 2010, Filzmoser et al. 2009). Such a transformation enables one to use standard statistical methods for the purpose of data analysis.

For a vector c of nonzero compositions the *clr* transformation is defined by the relationship 2:

$$clr(c) = \ln\left(\frac{c}{g(c)}\right) \qquad (2)$$

When distinguishing only one of the constituents of the system characterised by concentration c, one can notice that the values of all the other constituents c_r can be expressed using a simple relation: $c_r = 1000 - c$, where c and c_r are expressed in g/kg. In such a case, when $c \ll 1000$, the relation 2 is reduced to the following form:

$$clr(c) = \frac{1}{2} \ln \frac{c}{1000} \qquad (3)$$

Table 2. Arithmetic and geometric mean values describing ^{137}Cs content in the plants, which were calculated jointly and by particular plant species.

	Total	ln	ms	bg	bb
Geometric mean · 10^{11} g/kg	3.02	2.15	3.34	4.38	2.63
Arithmetic mean · 10^{11} g/kg	6.72	5.99	7.00	8.69	5.16

Figure 3 presents box plots reflecting activity of ^{137}Cs as well as transformed concentrations ct of the isotope in the examined plant material.

The left-hand column in Figure 3 illustrates box plots of activity of ^{137}Cs in the examined plants. For the purpose of comparison, the right-hand column of Figure 3, using the same type of plots, illustrates distribution of transformed concentrations of ^{137}Cs. One can notice outstanding measurement points on the plot of activity of ^{137}Cs, whereas in the plot illustrating the transformed values using the clr transformation, no such points are visible. Whereas the distribution of activity clearly deviates from the standard one, the distribution of transformed ^{137}Cs concentrations is quite symmetrical. Normality of the transformed concentrations distributions was verified using the following tests: Anderson-Darling test, Cramer-von Mises test, Shapiro-Francia test and Lilliefors test (Lehman & Romano 2005; Gross 2012). Each of the above-mentioned tests verifies another quality of the experimental distribution by comparing it with the relevant quality of the normal distribution. The obtained values of the confidence levels enable one to accept the hypothesis regarding the normal distribution of the transformed concentrations of ^{137}Cs in all of the examined plants.

Normality of the distributions enables one to use ANOVA variance analysis to assess the hypothesis regarding the lack of influence of the type of plant on the average values of transformed concentrations of ^{137}Cs (Weisberg 2005; Everitt & Hothorn 2010; Samuels et al. 2012). The obtained F-statistics value amounted to 1.1. For 3 and 74 degrees of freedom, the significance level of the F-statistics amounts to 0.36. There are no reasons to reject the hypothesis, which assumes no influence of the plant species on the average transformed concentrations of ^{137}Cs. Tukey's HSD test (Montgomery & Runger 2003; Everitt & Hothorn

2010) was used to assess the significance of the differences between the average values for all pairs of the plants. All of the obtained values of the significance level exceed 0.3. This means that there are no statistically significant differences between the average transformed concentrations compared in pairs of different plants. Figure 4 illustrates the values of average differences and their 95% confidence intervals calculated for all pairs of the plants. One can notice that all the differences in terms of the average values are located close to the 0 point and their confidence intervals cover it.

Table 3 illustrates the calculated coefficients of the correlation between the transformed variables.

When analysing the data in Table 3, once can notice good correlations between the transformed ^{137}Cs content values in different plants collected in the same area. This enables a reliable comparison of contamination in the areas, where separate species of the plants grow. The highest correlation coefficient was obtained for the transformed concentrations of ^{137}Cs in mosses and lichens. It might be suggested that both plants could be used interchangeably in biomonitoring of ^{137}Cs. Concentrations of ^{137}Cs in stems and leaves as well as the root of Common Bilberry are also well correlated. The lowest value of the correlation coefficient was obtained in the case of lichens and roots of Common Bilberry. Nevertheless, the correlations between concentrations in the other

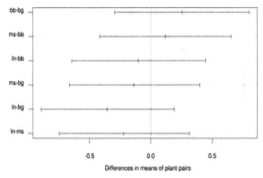

Figure 4. Values of differences between the average values and their 95% confidence intervals calculated for all pairs of the plants.

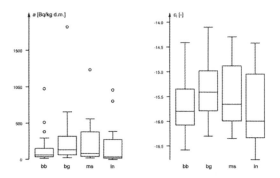

Figure 3. Box plots reflecting activity of ^{137}Cs as well as transformed concentrations of the isotope in the examined plant material.

Table 3. Coefficients of the correlation between the transformed variables.

	ms	*ln*	*bb*
ln	0.914	1.000	
bb	0.653	0.530	1.000
bg	0.674	0.611	0.886

plants remain fairly high. Therefore, one can compare concentrations of ^{137}Cs in various species of plants, although one must take into consideration the uncertainty of such an estimate as well as the reduced sensitivity in terms of detection of changes of concentration.

4 CONCLUSIONS

In order to assess the contamination of natural environment with ^{137}Cs, one can use whole plants or their parts. Due to the type of data, a correct assessment of the average content of radiocaesium in a plant requires use of appropriate calculation methods, encompassing, among others, the correct transformation of the raw measurement data.

It was concluded that the average values of content of ^{137}Cs in various plants as well as various parts of Common Bilberry are statistically indistinguishable. What is more, values of radiocaesium content in different plants collected in the same regions are well correlated. This enables one to compare contamination of particular regions of the examined area using different biomonitoring plants.

REFERENCES

Aitchison, J. 2010. A Concise Guide to Compositional Data Analysis, http://www.leg.ufpr.br/lib/exe/fetch.php/pessoais:abtmartins: a_concise_guide_to_compositional_data_analysis.pdf, accessed June 10th, 2010.

Broadley, M.R. & Willey, N.J. 1997. Differences in root uptake of radiocaesium by 30 plant taxa. *Environmental Pollution* 97: 11–15.

Bunzl, K. & Kracke, W. 1988. Cumulative deposition of Cesium-137, Plutonium-238, Plutonium-239, Plutonium-240 and Americium-241 from global fallout in soils from forest, grassland and arable land in Bavaria, West Germany. *J. Environ, Radioact.* 8(1): 1–14.

Desmet, G.M. & Myttenaere, C. 1988. Considerations on the role of natural ecosystem in the eventual contamination of man and his environment. *J. Environ. Radioact.* 6: 197–202.

Everitt, B.S. & Hothorn, T. 2010. A Handbook of Statistical Analyses Using R, CRC Press Taylor & Francis Croup.

Filzmoser, P., Hron, K., Reimann, C. 2009. Univariate statistical analysis of environmental (compositional) data: Problems and possibilities. *Sci. Total Environ.* 407: 6100–6108.

Filzmoser, P., Hron, K. & Reimann, C. 2010. The bivariate statistical analysis of environmental (compositional) data. Sci. Total Environ. 408: 4230–4238.

Gross, J. 2012. nortest: Tests for Normality. R package version 1.0.

Guivarch, A., Hinsinger, P., Staunton, S. 1999. Root uptake and distribution of radiocaesium from contaminated soils and the enhancement of Cs adsorption in the rhizosphere. *Plant and Soil* 211: 131–138.

Haselwandtner, K., Berreck, M. & Brunner P. 1988. Fungi as bioindicators of radiocaesium contamination: pre- and post-Chernobyl activities. *Trans. Brit. Mycol. Soc.* 90(2): 171–174.

Kabata-Pendias, A. & Pendias H. 1999. Biogeochemia pierwiastków śladowych. PWN, Warszawa.

Kłos A, Rajfur M. & Wacławek M. 2007. Application of lichens for the determination of precipitation pH by the exposure method, Environ. Engin. (eds. Pawłowski L., Dudzińska M. & Pawłowski A), pp. 505–512.

Kłos, A., Czora, M., Rajfur, M. & Wacławek, M. 2012. Mechanisms for Translocation of Heavy Metals from Soil to Epigeal Mosses. *Water, Air, & Soil Pollution* 223: 1829–1836.

Kłos, A., Rajfur, M., Waclawek M. & Waclawek W. 2009. ^{137}Cs transfer from local particulate matter to lichens and mosses. *Nukleonika* 54(4): 297–303.

Lehman, E.L. & Romano, J.P. 2005. Testing statistical hypotheses, Springer.

Massas, I, Skarlou, V., Haidouti, C., Giannakopoulou, F. 2010. ^{134}Cs uptake by four plant species and Cs–K relations in the soil–plant system as affected by Ca(OH)2 application to an acid soil. *Journal of Environmental Radioactivity* 101: 250–257.

Montgomery, D.C. & Runger G.C. 2003. Applied Statistics and Probability for Engineers 3. wyd., USA: John Wiley & Sons, Inc.

Nimis, P.L. & Cebulez, E. 1989. I macromiceti quali indicatori di contaminazion da cesio radioattivo nel friulli-Venezia Giulia. *Inform. Bot. Ital.* 21(1–3): 181–188.

Nimis, P.L., Giovani, C. & Padavani P. 1986. La contaminazione da Cs-137 nei macromiceti del Friuli-Venezia Gilia nel 1986. *Stud. Geobot.* 6: 3–121.

Nimis, P.L. 1990. Air quality indicator and indices. The use of plants as bioindicators and biomonitors of air pollution. Proc., Workshop on Indicator and Indices, JRC Ispra. EUR 13060 EN, 93.

Pawlowsky-Glahn, V. & Buccianti, A. (eds.) 2011. Compositional Data Analysis. Theory and applications, John Wiley & Sons, Ltd.

Pawlowsky-Glahn, V., Egozcue, J.J. 2006. Compositional data and their analysis: an introduction. *Geological Society, London, Special Publications* 264: 1–10.

Pietrzak-Flis, Z. 1996. Obieg cezu promieniotwórczego w środowisku leśnym, ss. 150–166, Materiały konferencyjne XVI Szkoły Jesiennej: Czarnobyl—10 lat później: skażenia środowiska i żywności, Zakopane 14–18.10.1996.

R Development Core Team, 2009. R: A language and environment for statistical computing. R Foundation for Statistical Computing, Vienna, Austria, Available at: http://www.R-project.org.

Samuels M.L., Witmer J.A. & Schaffner A.A., 2012. Statistics for the life sciences 4. Ed. Boston Columbus Indianapolis: Prentice Hall.

Skarlou, V., Papanicolaou, E.P., Nobeli, A. 1996. Soil to plant transfer of radioactive cesium and its relation to soil and plant properties. *Geoderma* 72: 53–63.

Sokolov, V.E. 1990. Ecological and genetic consequences of the Chernobyl-Atomic power plant accident, ICSU, SCOPERADPATH, Lancaster Radpath Meting, 26–30 March 1990, University of Leicester, UK.

van den Boogaart, K.G., Tolosana, R. & Bren, M. 2011. compositions: Compositional Data Analysis. R package version 1.10–2. Available at: http://CRAN.R-project.org/package=compositions.

van den Boogaart, K.G. & Tolosana-Delgado, R. 2008. "compositions": A unified R package to analyze compositional data. *Computers & Geosciences* 34: 320–338.

van den Boogaart, K.G. & Tolosana-Delgado, R. 2006. Compositional data analysis with 'R' and the package 'compositions'. Geological Society, London, Special Publications 264(1): 119–127.

Weisberg, S. 2005, Applied Linear Regression, Wiley.

Yamagata, N., Matsuda, S. & Chiba, M. 1969. Radiology of ^{137}Cs and ^{90}Sr in a forest. *J. Radiat. Res.* 10: 107–112.

Environmental Engineering IV – Pawłowski, Dudzińska & Pawłowski (eds)
© *2013 Taylor & Francis Group, London, ISBN 978-0-415-64338-2*

Attachment of silica and alumina nanoparticles as studied by QCM

A. Hänel, J. Nalaskowski, P.R. Satyavolu & M. Krishnan
IBM T.J. Watson Research Center, Yorktown Heights, New York, USA

ABSTRACT: Silica and alumina abrasives are commonly used in Chemical Mechanical Planarization (CMP) processes. Unfortunately, post-CMP attachment of abrasive particles to polished surfaces is a significant problem, which adversely affects the device yield. These particles have to be efficiently removed in post-CMP cleaning process. In the present research, the authors applied quartz crystal microbalance to investigate the silica and alumina abrasive attachment to copper and silica surfaces in the presence of surface active CMP slurry components. For copper surfaces with adsorbed Benzotriazole (BTA), alumina and silica abrasives attached as separate particles regardless of the zeta potential sign. At SiO_2 surfaces, in the presence of adsorbed CTAB structure silica particles preferentially attached.

Keywords: particle attachment, post-CMP, quartz crystal microbalance

1 INTRODUCTION

Chemical Mechanical Planarization (CMP) is one of the significant steps during the production of microelectronic devices. It comprises the removal of material overburden by the synergetic interaction of chemistry and abrasive particles. The main goal of CMP is to achieve highly planar and smooth surfaces with no surface defects caused by scratching, corrosion, erosion, or dishing. The slurry is usually composed of oxidizer, pH adjusters, surfactants, complexing agents, water-soluble polymers and abrasive particles. Alumina and silica abrasives are commonly used in CMP processes. However, the particular slurry composition depends on the material, which has to be removed.

One issue is that abrasives, metallic ions and other chemical components attach to the surface during the CMP and have to be removed in the post-CMP cleaning since any surface contamination influences the device yield and reliability. Thus, it is essential to find a method to study the particle-surface attachment and to understand the mechanism of surface contamination.

Planarized structures consist of metal lines, barrier layers and interlayer dielectrics. A typical currently employed dielectric is silicon dioxide. Surfactants like Cetyl Trimethyl Ammonium Bromide (CTAB) have been used to optimize the planarization efficiency and material removal selectivity since it adsorbs on metal and dielectric surfaces (Hong et al. 2005). The adsorption of CTAB to SiO_2 surfaces was the object of various researches (Parida et al. 2006, Tyrode et al. 2008).

The CTAB adsorption depends on the pH. With increasing pH more CTAB adsorbs, which is attributed to the increase in surface charge sites (Fleming et al. 2001). The adsorption is interplay of electrostatic and hydrophobic interactions. At the beginning, the head group of the CTA+ cations adsorb electrostatically on the SiO_2 surface, which makes the SiO_2 surface hydrophobic. At higher concentrations, CTAB adsorbs via hydrophobic interaction reversing the charge of the surface (Wang et al. 2004). This means that with increasing CTAB concentration the zeta potential increases.

Copper is the metal of choice as interconnect material because of its lower resistance, superior resistance to electromigration and the reduction of resistance-capacitance time delay (Hong et al. 2004). Both alumina and silica abrasives are used for copper CMP. Benzotriazole (BTA) is known to strongly adsorb on copper and thus it is most widely used in CMP slurries for removal rate control, as well as corrosion inhibitor in post-CMP cleaning. It is a heterocyclic compound with a triazole ring. The exact BTA adsorption mechanism is still under discussion (Finšgar & Milošev 2010). The formation of Cu(I)-BTA surface complex involving strong Cu-N bonds plays a major role. Moreover, BTA forms intermolecular aggregates, such as [BTA-Cu]$_n$ polymeric complex, which contributes to the stability of the inhibitor film on the Cu-surface (Kokalj et al. 2010). One consequence of the BTA adsorption is that the Cu(I)-BTA complex makes the copper surface hydrophobic. Kim and Oh (2008) report contact angles between 50° and 65° and Chen et al. (2004) determined contact angles of about 90°. It is believed that the hydrophobicity of the BTA-Cu surface promotes the attachment of the abrasives and may lead to increased particulate contamination during post-CMP cleaning. Chen et al. (2004) propose that van der Waals attraction accounts for the attachment but the hydrophobic forces related to the instability of water film can be also responsible.

For the improvement of post-CMP cleaning process, it is essential to understand the attachment of abrasive particle to surfaces. In this paper, we employed Quartz Crystal Microbalance (QCM) to quantitatively measure abrasive attachment to copper and silicon oxide surfaces.

2 MATERIALS AND METHODS

All chemicals were used without further purification. The used water was degassed (by sonication) ultra-pure water with a conductivity of 0.1 µS/cm. The investigated nanoparticles were obtained as slurries. The two kinds of colloidal silica slurries (30H50 and S50) as well as alumina slurry (iCue600) were provided by IBM. The zeta potential and particle size was determined using the Colloidal Dynamics AcustoSizer II. AFM images were taken using tapping mode with Dimension V AFM (Veeco). For image evaluation WSxM 3.0 and Gwyddion 2.22 were used (Horcas et al. 2007).

Nanoparticle attachment onto Cu and SiO_2 surface was examined using a QCM200 from Stanford Research Systems, Inc. The polished gold/chromium QCM crystals were coated with 700 Å of Cu or 1000 Å of SiO_2. Before Cu or SiO_2 were deposited onto the crystal surface a 50 Å TaN and 50 Å Ta intermediate layer were deposited onto the surface. In order to minimize Cu oxidation, the crystals were kept under nitrogen atmosphere before the experiments. The 5 MHz quartz crystals were mounted in a flow cell, through which the solution/slurry was pumped with a constant flow rate of 0.2 ml/min. The flow cell was immersed in a water bath providing a constant temperature of 25°C. At first ultra-pure water was pumped through the flow cell for all experiments. For the experiments with the Cu-coated crystals 5 mM BTA solution was pumped through the flow cell followed by abrasives (silica 30H50 and alumina iCue600) suspended in 5 mM BTA solution. The SiO_2-coated crystal was overflowed by 500 µM CTAB followed by water and 0.1 wt% S50 silica abrasive in CTAB solution. Before the crystal was taken out, the flow cell was flushed with ultra-pure water. After each experiment tubes and connectors were replaced and non-exchangeable parts were cleaned with isopropyl alcohol and rinsed with water.

3 RESULTS AND DISCUSSION

Table 1 summarizes the characteristics of the used slurries. For the attachment of silica and alumina abrasives to BTA coated Cu surface, the pH was ~3.6. The used silica abrasive has a negative zeta potential at this pH, whereas alumina abrasive exhibits a positive zeta potential.

For the attachment of silica abrasives to CTAB coated SiO_2 surface, different silica abrasive was used to compare abrasive attachment to BTA coated Cu surface. This colloidal silica slurry was weakly acidic and exhibited a negative zeta potential. At first, the results of abrasive attachment onto BTA-Cu surface are discussed.

3.1 Silica and alumina abrasive attachment to BTA coated Cu surfaces

Figure 1 shows the topographic AFM image of the copper deposited onto the quartz crystal. Cu grains with a diameter between 60 and 270 nm are observed, which have an average height of 10.3 nm.

Initially Cu-coated quartz crystals are exposed to ultra-pure water. Figure 2 shows the mass change over time of the Cu-coated quartz crystal. The decrease in mass corresponds to the dissolution of copper. The average rate of copper dissolution in water for all experiments was determined to be 1.65E-4 µg/cm^2 s by the equation: rate = $-\partial m/\partial t$.

Figure 3 illustrates the mass change over the time, when the crystal was exposed to 0.1 wt% alumina abrasives. The dissolution rate was determined to be 6.05E-4 µg/cm^2 s. The dissolution rate by the slurry is larger than by water, since the pH of the 0.1 wt% alumina abrasive slurry was 3.7. The predominant dissolution species of copper in acidic environment is Cu^{2+}. In near neutral region CuO and Cu_2O are formed (Tamilmani et al. 2002). No abrasive attachment to the copper surface was observed using the QCM and AFM measurements (Fig. 4). The highest elevations on the AFM image are 8.8 nm high.

When the crystal was exposed to 5 mM BTA solution, the BTA adsorbed on the copper and suppressed the copper dissolution as can be seen in Figure 5. Multilayer of BTA and BTA-complex are formed which are reported to build up layers of several thousand Å in thickness (Al Kharafi et al. 2007, Finšgar & Milošev 2010). The structure and the orientation of the adsorbed BTA molecule is not fully understood. It was proposed that the BTA molecule and Cu(I)-BTA complex are orientated perpendicular, tilted or parallel to the Cu surface, whereas parallel orientation—especially for the first layer—is more likely (Finšgar & Milošev 2010). Assuming a molecular area of 31.4 Å^2 in

Table 1. pH, zeta potential and size of silica and alumina abrasives.

Slurry		Concentration (wt%)	pH (–)	Smoluchowski zeta potential (mV)	Abrasive size (nm)	Experiment (–)
Name	Compound					
30H50	SiO₂	0.1	3.5	−73.9	50	BTA—Cu surface
iCue600	Al₂O₃	0.1	3.7	+102.1	50	BTA—Cu surface
S50	SiO₂	0.1	5.4	−48.5	50	CTab—SiO₂ surface

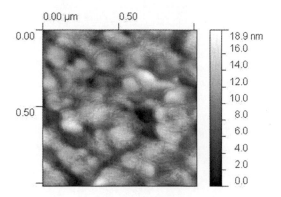

Figure 1. 1 μm × 1 μm AFM image of the Cu-coated quartz crystal before the QCM experiments.

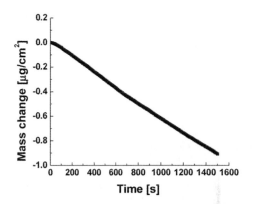

Figure 3. Mass change of Cu-quartz crystal in the flow of 0.1 wt% alumina abrasive.

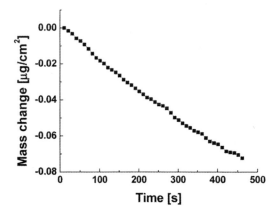

Figure 2. Mass change of the Cu-quartz crystal in the flow of water.

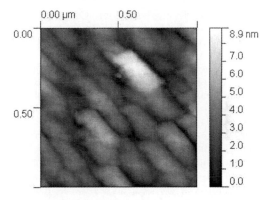

Figure 4. 1 μm × 1 μm AFM image of the Cu coated quartz crystal after the exposure to 0.1 wt% alumina abrasive suspension.

parallel orientation, a monolayer should correspond to a mass increase of 0.063 μg/cm². The adsorption rate of BTA varies from crystal to crystal, although the temperature, flow rate and concentration were kept constant. A possible explanation can be that the slight variations in the Cu surface oxidation during the storage contribute significantly to this discrepancy. The BTA adsorption rate is higher for oxidized Cu than on oxide free surface (Finšgar & Milošev 2010). The topographical AFM image (Fig. 6) of the BTA layer shows streak artifacts, which may be caused by the adsorbed BTA. The topography of the underlying Cu surface is still visible.

After exposing the Cu quartz crystal to 5 mM BTA solution, the solution was changed to 0.1 wt% abrasive slurry in 5 mM BTA. It was observed that both silica and alumina abrasives attached to the

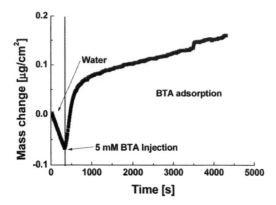

Figure 5. Mass change before and after pumping 5 mM BTA solution. The line marks the point, when the solution was changed from water to BTA solution.

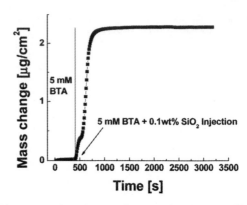

Figure 7. Mass change after pumping 0.1 wt% silica abrasives with 5 mM BTA. Before the abrasive BTA solution was injected, 5 mM BTA was pumped through the flow cell.

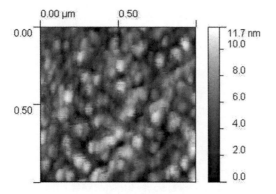

Figure 6. 1 μm × 1 μm AFM image of the Cu coated quartz crystal after initial Cu dissolution in water followed by 5 mM BTA injection.

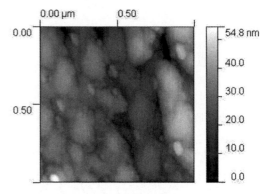

Figure 8. 1 μm × 1 μm AFM image of the Cu coated crystal with adsorbed BTA on the surface after the injection of 0.1 wt% silica abrasive in 5 mM BTA suspension.

surface, independent from the positive or negative zeta potential of the abrasive.

0.24 μg/cm² BTA was adsorbed during 5440s for the experiment of silica abrasive attachment. After the injection of 0.1 wt% silica abrasive, the mass increase was determined to be 2.2 μg/cm² in 500 s, which corresponds to a surface coverage of 3.0% (Fig. 7).

The AFM image presented in Figure 8 shows the silica abrasives attached to the surface, where the highest elevation is 54.8 nm high and where the average elevation is 27.3 nm. The abrasives are between 50 to 60 nm wide and 90 to 130 nm long. The silica abrasives seem to be attached separately, which can be caused by electrostatic repulsion between the silica particles. After the silica attached to the BTA covered Cu-surface, a mass increase of 0.028 μg/cm² in 2100s was observed. The authors propose that after the

abrasive attachment, only BTA adsorbs to the BTA-Cu surface because attached silica prevents further attachment of abrasives by electrostatic repulsion.

The experiments with alumina abrasives were conducted for different concentrations (Tab. 2). For 0.1 wt% alumina, the attachment was with 0.82 μg/cm² larger than for 0.01 and 1.0 wt%, which could be caused by contamination. Also the AFM image showed agglomerated alumina abrasives (image not shown).

The alumina abrasive attachment for 0.01 and 1.0 wt% was in the same range between 0.18 and 0.25 μg/cm². Thus, we conclude that the abrasive concentration has no effect on the amount of attachment. However, a well formed hydrophobic BTA layer seems to increase the attachment. After abrasive particle attachment a continuously mass increase was observed, which can be explained

Table 2. Alumina abrasive attachment in dependence of abrasive concentration and BTA adsorption.

Alumina concentration (wt%)	BTA adsorbed on the Cu surface ($\mu g/cm^2$)	Abrasive attachment ($\mu g/cm^2$)	Surface coverage (%)
0.01	0.2	0.18	0.14
0.1	0.45	0.82	0.67
1.0	0.52	0.22	0.18
1.0	0.66	0.25	0.20

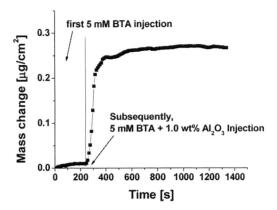

Figure 9. Mass change after pumping 1.0 wt% alumina abrasives with 5 mM BTA.

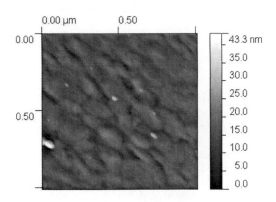

Figure 10. 1 μm × 1 μm AFM image of the Cu coated crystal with adsorbed BTA on the surface after the injection of 1.0 wt% alumina abrasive 5 mM BTA suspension.

by further BTA adsorption on the formed BTA layer (see Fig. 9). Figure 10 shows the AFM image after 1.0 wt% alumina abrasive injection. The alumina abrasives are separately attached, which can be caused by electrostatic repulsion between the abrasive particles. The experiments with 0.01 wt% alumina abrasives showed also separately attached alumina abrasives (AFM image not shown). The attached abrasives have a diameter between 40 and 60 nm and are up to 43 nm high. In order to investigate if BTA adsorbs onto the abrasive surface and causes agglomeration, zeta potential and particle size of 5 wt% abrasive solution was measured in function of BTA concentration using Accustosizer. No change in zeta potential and particle size of silica or alumina slurry was observed up to 9.5 mM BTA concentration.

The authors propose that electrostatic interactions are not responsible for the deposition phenomena since silica and alumina nanoparticles attached to the BTA-Cu surface regardless of their zeta potential. Furthermore, van der Waals attraction is not able to fully describe the attachment, since the geometries of the surface and particles are not ideal and also shearing forces are acting on the abrasives (Adamczyk & Weronski 1999, Cooper et al. 2001). Thus, it can be assumed that non-DLVO hydrophobic forces can be responsible for abrasive attachment to hydrophobic surfaces.

3.2 Silica abrasive attachment onto CTAB coated SiO$_2$ surface

The deposited SiO$_2$ on the crystal has grains with an average length of 360 nm and an average width of 175 nm (see Fig. 11). The average high of the grains is 6.5 nm with a maximum height of 17.83 nm. When the SiO$_2$ coated crystal was overflowed by water, no change of the mass was detected.

The critical micelle concentration (cmc) of CTAB is 0.9 mM in pure water (Pagac et al. 1998). During the experiments the concentration was chosen below the cmc. The SiO$_2$ coated crystal was overflown by 500 μM CTAB solution for 100 min.

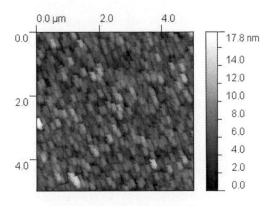

Figure 11. 5 μm × 5 μm AFM image of the SiO$_2$ coated crystal before the QCM experiments.

Subsequently, the solution was changed to pure water for at least 2000s, since the silica abrasives agglomerated in 500 μM CTAB solution. After flushing the flow cell with water, the SiO$_2$ surface was covered by an approximately 0.28 μg/cm^2 CTAB layer. The structure of the CTAB layer, especially for layer much below the cmc, is still under discussion. Formation of aggregates with semi-structured, globular appearance on the surface is reported for concentrations of 0.06 cmc (Fleming et al. 2001). Liu & Ducker (1999) find that aggregates form at concentrations of about 0.3 to 0.5 cmc. At 0.9 cmc, the structure seems to exist of spheres and short rods (Velegol et al. 2000). Above the cmc the aggregates are reported to be either spherical (Liu & Ducker 1999) or rod-like (Velegol et al. 2000). However, it was found that independent of concentration the surface is incompletely covered by CTAB and that the surfactant layer is between 2.8 and 3.4 nm thick (Tyrode et al. 2008). In the first experiment, 0.1 wt% silica in 55 μM CTAB solution was injected, which led to the attachment of 4.17 μg/cm^2 of silica abrasives in 450 s to the CTAB coated surface (Fig. 12). It corresponds to a surface coverage of 5.7%. When 0.1 wt% silica in 100 μM CTAB solution was injected, 4.79 μg/cm^2 of silica abrasives attached in 470 s to the CTAB coated surface and covered 6.6% of the surface (Fig. 13). For the 100 μM CTAB solution, more silica abrasives attached to the surface, since more CTAB was adsorbed on the abrasive surface, which decreased the electrostatic repulsion between the silica abrasives and the CTAB-SiO$_2$ surface. The adsorbed CTAB on the silica abrasives also decreases the electrostatic repulsion between the abrasives, what can be seen on the AFM image (Fig. 14). Adsorbed abrasives formed agglomerates on the surface, which are up to 1 μm long, 300 to 400 nm wide and 47 nm high.

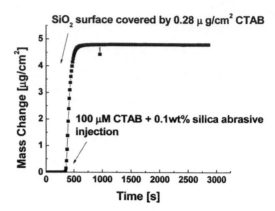

Figure 13. Mass change after pumping 0.1 wt% silica abrasive in 100 μM CTAB solution.

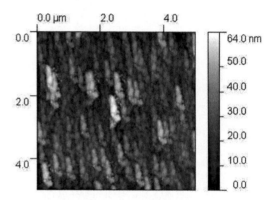

Figure 14. 5 μm × 5 μm AFM image of the SiO$_2$ coated crystal after the injection of 0.1 wt% silica abrasive in 55 μM CTAB solution.

The SiO$_2$ surface is not totally covered by CTAB, since grains of the deposited SiO$_2$ with the same dimensions like in Figure 11 can be observed. Thus, the abrasives attached to sites on the SiO$_2$ surface where CTAB was adsorbed.

Adsorbed on the abrasive surface, which decreased the electrostatic repulsion between the silica abrasives and the CTAB-SiO$_2$ surface. The adsorbed CTAB on the silica abrasives also decreases the electrostatic repulsion between the abrasives, what can be seen on the AFM image (Fig. 14). Adsorbed abrasives formed agglomerates on the surface, which are up to 1 μm long, 300 to 400 nm wide and 47 nm high. The SiO$_2$ surface is not totally covered by CTAB, since grains of the deposited SiO$_2$ with the same dimensions like in Figure 11 can be observed. Thus, the abrasives attached to sites on the SiO$_2$ surface where CTAB was adsorbed.

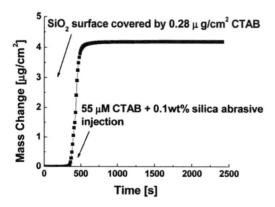

Figure 12. Mass change after pumping 0.1 wt% silica abrasive in 55 μM CTAB solution.

4 CONCLUSIONS

The object of this research was to investigate the attachment of silica and alumina abrasives onto hydrophobic Cu and SiO_2 surfaces. It was accomplished by quartz crystal microbalance measurements. The attachment of both kinds of abrasives was observed and confirmed by AFM images taken after the measurements. Making Cu surface hydrophobic by the adsorption of BTA promoted the abrasive attachment, which did not depend on the sign of the zeta potential. Both silica and alumina particles attached as separate particles to the BTA covered Cu surface. Change of alumina abrasive concentration did not influence the attachment. The authors propose that non-DLVO forces, like hydrophobic force, may be responsible for the abrasive attachment.

When CTAB was previously adsorbed to the SiO_2 surface, the silica abrasive attachment was enhanced. The CTAB incompletely covered the SiO_2 surface and made it more hydrophobic. The silica particles attached favorably to the CTAB covered areas.

REFERENCES

Adamczyk, Z. & Weronski, P. 1999. Application of the DLVO theory for particle deposition problems. Advances in Colloid and Interface Science 83(1–3): 137–226.

Al Kharafi, F.M., Abdullah, A.M. & Ateya, B.G. 2007. A quartz crystal microbalance study of the kinetics of interaction of benzotriazole with copper. Journal of Applied Electrochemistry 37(10): 1177–1182.

Chen, P.L., Chen, J.H., Tsai, M.S., Dai, B.T. & Yeh, C.F. 2004. Post-Cu CMP cleaning for colloidal silica abrasive removal. Microelectronic Engineering 75(4): 352–360.

Cooper, K., Gupta, A. & Beaudoin, S. 2001. Simulation of Particle Adhesion: Implications in Chemical Mechanical Polishing and Post Chemical Mechanical Polishing Cleaning. Journal of the Electrochemical Society 148: G662–G667.

Finšgar, M. & Milošev, I. 2010. Inhibition of copper corrosion by 1, 2, 3-benzotriazole. A review. Corrosion Science 52: 2737–2749.

Fleming, B., Biggs, S. & Wanless, E. 2001. Slow organization of cationic surfactant adsorbed to silica from solutions far below the CMC. Journal of Physical Chemistry B 105(39): 9537–9540.

Hong, Y., Eom, D., Lee, S., Kim, T., Park, J. & Busnaina, A. 2004. The effect of additives in post-Cu CMP cleaning on particle adhesion and removal. Journal of the Electrochemical Society 151(11): G756–G761.

Hong, Y., Patri, U., Ramakrishnan, S. & Babu, S. 2005. Novel Use of Surfactants in Copper Chemical Mechanical Polishing (CMP). Chemical-Mechanical Planarization-Integration, Technology and Reliability 867: 41–46.

Horcas, I., Fernandez, R., Gomez-Rodriguez, J.M., Colchero, J., Gomez-Herrero, J. & Baro, A.M. 2007. WSXM: A software for scanning probe microscopy and a tool for nanotechnology. Review of Scientific Instruments 78(1): 013705–013713.

Kim, K. & Oh, S. 2008. Analysis on Delamination Phenomena for Nonpatterned Wafers during Abrasive-Free CMP by Finite Element Simulation. Journal of the Electrochemical Society 155(10): H791–H796.

Kokalj, A., Peljhan, S., Finšgar, M. & Milošev, I. 2010. What Determines the Inhibition Effectiveness of ATA, BTAH, and BTAOH Corrosion Inhibitors on Copper? Journal of American Chemical Society 132(46): 16657–16668.

Liu, J.F. & Ducker, W.A. 1999. Surface-induced Phase Behavior of Alkyltrimethylammonium Bromide Surfactants Adsorbed to Mica, Silica, and Graphite. The Journal of Physical Chemistry B 103(40): 8558–8567.

Pagac, E.S., Prieve, D.C. & Tilton, R.D. 1998. Kinetics and Mechanism of Cationic Surfactant Adsorption and Coadsorption with Cationic Polyelectrolytes at the Silica—Water Interface. Langmuir 14(9): 2333–2342.

Parida, S.K., Dash, S., Patel, S. & Mishra, B.K. 2006. Adsorption of organic molecules on silica surface. Advances in Colloid and Interface Science 121(1–3): 77–110.

Tamilmani, S., Huang, W., Raghavan, S. & Small, R. 2002. Potential-pH diagrams of interest to chemical mechanical planarization of copper. Journal of the Electrochemical Society 149(12): G638–G642.

Tyrode, E., Rutland, M.W. & Bain, C.D. 2008. Adsorption of CTAB on Hydrophilic Silica Studied by Linear and Nonlinear Optical Spectroscopy. Journal of American Chemical Society 130(51): 17434–17445.

Velegol, S.B., Fleming, B.D., Biggs, S., Wanless, E.J. & Tilton, R.D. 2000. Counterion Effects on Hexadecyltrimethylammonium Surfactant Adsorption and Self-assembly on Silica. Langmuir 16(6): 2548–2556.

Wang, W., Gu, B., Liang, L. & Hamilton, W.A. 2004. Adsorption and Structural Arrangement of Cetyltrimethylammonium Cations at the Silica Nanoparticle-Water Interface. The Journal of Physical Chemistry B 108(45): 17477–17483.

Environmental Engineering IV – Pawłowski, Dudzińska & Pawłowski (eds)
© 2013 Taylor & Francis Group, London, ISBN 978-0-415-64338-2

Thermal treatment for the removal of mercury from solid fuels

M. Wichliński, R. Kobyłecki, M. Ścisłowska & Z. Bis
Department of Energy Engineering, Czestochowa University of Technology, Częstochowa, Poland

ABSTRACT: This paper presents the results of mercury release during thermal treatment of hard coal and lignite from Polish coal mines. The thermal treatment of the samples was conducted at temperatures 170–410 °C, and in three different atmospheres: air, nitrogen and carbon dioxide. The results indicated that it was possible to remove almost 100% of mercury from the coal samples at temperatures <410 °C i.e. below the ignition temperature. It was also found that the efficiency of mercury removal was depended on the type of the gas that surrounded the sample during its thermal treatment.

Keywords: mercury, coal, thermal treatment

1 INTRODUCTION

Poland has quite a unique structure of energy carriers, since over 90% of electricity and heat in the country is produced due to the combustion of solid fuels, mainly hard coal and lignite (Grudziński, 2010). This situation is difficult to change and despite the current trend to replace the electricity produced from 'dirty' coal by biomass and other renewable energy sources the combustion of fossil fuels will still remain the primary source of electricity in Poland for a long time (Grudziński 2010, Pieńkowski 2012). The combustion of hard coal and lignite is, however, often responsible for the emission of many harmful substances, such as particulate matter, mercury, lead or cadmium, and several relevant regulations limiting the emission of those substances into the atmosphere have been implemented so far. After the introduction of the emission standards for sulfur dioxide, nitrogen oxides, and particulate matter the current activities focus on the reduction of carbon dioxide and mercury. The emission of anthropogenic mercury to the atmosphere has recently been estimated at around 1000–6000 t/year (Yang et al., 2007). According to the United Nations Environmental Programme (2002) about half of that amount is the result of the combustion of solid fuels at power plants.

The mercury emitted into the atmosphere may be transported over long distances and then deposited in terrestrial and aquatic ecosystems where it is transformated into methylmercury. The methylmercury compounds enter then the living organisms and cannot be excreted. The whole food chain causes thus the accumulation of methylmercury in the consumers of higher orders—in human bodies

those compounds may cause several diseases, neurological problems, increase the risk of heart attack, or lower the IQ of unborn babies (Sundseth et al. 2010, United Nations Environmental Programme 2002).

Although the mercury content in coal is relatively low (approximately 100 ng/g) the annual emission of mercury from coal-fired power plants may be significant due to large amount of the coal burned each year. Głodek et al. (2009) estimated the emission in Poland at roughly 57.5 kg Hg/year. The United States was the first country were the regulations of the mercury emission from the combustion of solid fuels were introduced. Currently, the European Union and Japan are also making affords to implement similar regulations (Pavlish et al. 2010). As reported by (Kobyłecki et al. 2007, Wichliński et al. 2013) Polish coals contain relatively low amount of mercury (12–250 ng/g). The average value for the samples investigated by those authors was 80 ng/g and their data were consistent with the results obtained by other authors who reported that the average mercury content in Polish coals was roughly 100 ng/g for hard coals, and 250 ng/g for lignites (Wojnar et al. 2006). Just for comparison the average mercury content in US coals was approximately 220 ng/g, with the lowest reported values at roughly 80 ng/g (Toole-O`Neil et al. 1999). The research conducted by (Wichliński et al. 2011) indicated that the mercury in Polish coals was mainly in the form of HgO, HgO$_2$, Hg$_3$(SO$_4$) O$_2$, Hg$_2$SO$_4$, or HgCl$_2$. The majority of those compounds decomposes below 400 °C, except HgO, that decomposes at >500 °C (Lopez-Anton et al. 2010).

In order to remove the mercury during the combustion of coal two main possibilities are currently being taken into consideration: The first one is the removal of mercury from the fuel after the combustion process ('post-combustion') by eg. injection of the activated carbon (usually impregnated with iodine, sulfur or bromine) into the fuel gas duct before the Electrostatic Precipitator (ESP), while the other possibility is focused on thermal treatment of coal before the combustion process ('pre-combustion') in order to provide the optimum conditions for mercury release. The temperature of the thermal treatment of coal has to be sufficiently high to release the mercury but low enough to minimize the loss of volatile matter. Accordingly, the coal is usually treated at temperatures roughly 500 °C (Wang et al. 2000) or at 200–300 °C (Guffy et al. 2004, Chmielniak et al. 2010). The effectiveness of the 'post combustion' methods is relatively high, even over 95% (Sloss 1995, Wade et al. 2012) but the technology it is quite sophisticated and expensive. The effectiveness of the 'pre-combustion' methods may also exceed 95%, but those methods are usually much cheaper and furthermore offer the possibility to remove not only mercury but also other compounds, such as eg. H_2O. The 'pre-combustion' methods seems thus to be especially suitable for eg. fuels containing large amount of moisture, such as lignite, coal slurry, refuse coal, etc.

Due to the lack of information on the effectiveness of the 'pre-combustion' methods application for mercury removal from Polish coals the present paper is intended to feel the gap and report the result of some experimental investigation focused on the thermal treatment of some selected Polish hard coal and lignite samples in order to get rid of mercury at moderate temperatures, i.e. before the volatiles are released.

2 RESEARCH METHODOLOGY

The sketch of experimental setup designed for the present study is shown in Figure 1. It was dedicated to conduct the experiments in a fluidized bed apparatus since it provides excellent conditions for uniform mass and heat transfer. The experimental setup consisted of an electrically heated quartz glass column, 0,36 m high and 0,026 m i.d. The bottom of the column was closed with a ceramic grate and coupled to the gas distributor. The fluidized bed consisted of 10 g of ceramic beads (particle size: 2 mm) and 40 g of silica sand (particle size: 250–500 µm). During each of the experiments the bed was fluidized with various gases (air, N_2 or CO_2) and heated up to an assumed temperature. Then roughly 0.5 g of coal (particle diameter 500–1000 µm) of known ultimate and proximate analysis and mercury content was fed into the bed.

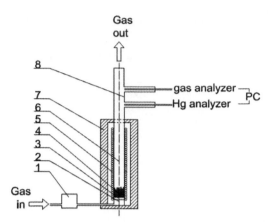

Figure 1. The schematic of the experimental apparatus: 1-mass/gas flow controller, 2-gas distributor, 3-grate, 4-fluidized bed, 5-electric heater, 6-quartz tube, 7-thermal insulation, 8-measurement zone.

The gas flow rate was maintained constant at roughly 10 l/min by a mass flow controlling device. The bed temperature was measured by a K-type thermocouple.

After the introduction of the coal sample into the column the fuel particles were heated up and the volatiles and mercury compounds were decomposed and evaporated. During the experiment the concentration of CO, CO_2, O_2, NO, NO_2 and SO_2 in the flue gas was continuously measured by the gas analyzer MRU Delta 65, while the mercury concentration in the gas was determined *on-line* by spectrometer Lumex RA-915+. Both devices were connected to a PC data storage system. After the end of each test the column heater was turned off and the whole setup was cooled in a nitrogen atmosphere. The coal sample was then removed from the column and burnt in a pyrolysis snap-RP-91c connected to the spectrometer RA-915+. The combination of the snap-RP-91c and the spectrometer allowed to determine the remaining amount of mercury in the fuel sample. Knowing the amount of mercury in the gas phase and the remaining mercury in the coal the relative amount of released mercury could be calculated from the following equation:

$$RR = \frac{m_{Hg}^{gas}}{m_{Hg}^{gas} + m_{Hg}^{char}} \cdot 100\% \qquad (1)$$

where:
RR is the relative amount of mercury released from the sample [%], m_{Hg}^{gas} is the amount of mercury released to the gas phase [ng/g], and m_{Hg}^{char} is the amount of mercury in the remaining coal [ng/g].

Table 1.   Proximate and ultimate analysis of the coal samples (air-dry).

| Symbol | Unit | Hard coal | | Lignite |
		Coal A	Coal B	Coal C
External moisture, W_p (as received)	%	8.4	6.1	10.3
Internal moisture, W_h	%	3.1	7.1	15.9
Ash, A^a	%	24.8	24.8	5.6
Volatile matter, VM^a	%	26.2	20.2	39.6
Fixed carbon, FC^a	%	45.6	47.2	38.7
Sulfur, S^a	%	1.04	0.58	0.5
Carbon, C^a	%	51,5	53.0	53.3
Total mercury, Hg_{total}	ng/g	131	79	231
HHV, C_{sp}^a	kJ/kg	26246	24800	21522
LHV, W_{op}^a	kJ/kg	25295	23794	20181

3   ANALYSIS OF THE COALS

Three coal samples were selected for the present experiments: two hard coal samples, called *Coal A*, and *Coal B*, and lignite sample, *Coal C*. The main results of the proximate and ultimate analysis of the samples are shown in Table 1. The proximate analysis of coals indicated that all samples contained rather low amount of external moisture, below 10%. The hard coal samples *Coal B* and *Coal C* contained roughly 25% of ash and 20–25% of volatile matter, while the lignite (*Coal C*) contained much less ash (5.6%) and much more volatile matter (39.6%). All coals contained rather small amount of sulfur (0.5–1.04%). The mercury content in the samples was between 79 ng/g (*Coal B*), and 231 ng/g (*Coal C*).

4   RESULTS AND DISCUSSION

The results of thermal treatment of the samples in the fluidized bed apparatus in the air atmosphere are shown in Figure 2 where the relative amount of mercury released from the coal samples is plotted versus the bed temperature. The results indicate that in the air atmosphere the release of Hg from both hard coals *A* and *B* is very similar regardless of the temperature. In both cases, the amount of released mercury was proportional to the temperature up to roughly 350 °C. Above that temperature level the change of the amount of released mercury was almost negligible. Quite different results were obtained for lignite (*Coal C*): the mercury release increased rapidly up to roughly 320 °C, while above that level the amount of released mercury was similar regardless of the temperature of the thermal treatment. The amount of mercury released from *Coal C* due to its heating over 320 °C was also roughly 90% of the initial amount of

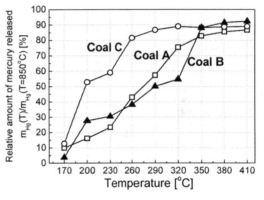

Figure 2.   The release of mercury versus the temperature of coal thermal treatment. Fluidizing gas: air.

mercury in the coal sample. The results in Figure 2 clearly indicate that the thermal treatment of all the samples at temperatures above 320 °C was sufficient to release the majority of coal mercury. The remaining amount of Hg (roughly 10% of the initial amount) could not be released from the coal samples since it is tied up in components such as HgO, which decompose at much higher temperatures (Wichlinski et al., 2011).

In the case of any potential commercial application of the investigated process the increase of the bed temperature above 410 °C is not recommended due to the danger of explosion or loss of the chemical energy of the coal sample. In the case of thermal treatment in air those issues are associated with eg. ignition and combustion of the fuel. In order to run the process at higher temperatures it is thus necessary to replace air by an inert gas such as nitrogen or carbon dioxide. The corresponding results are shown in Figures 3 and 4. In the case of coals A and B the results of its thermal treatment in nitrogen atmosphere were similar to

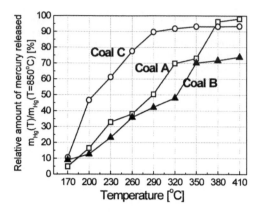

Figure 3. The release of mercury versus the temperature of coal thermal treatment. Fluidizing gas: nitrogen.

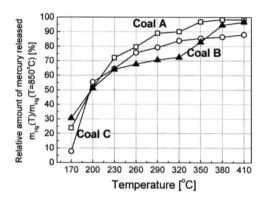

Figure 4. The release of mercury versus the temperature of coal thermal treatment. Fluidizing gas: CO_2.

the data shown in Figure 2 but the relative amount of released mercury was higher, particularly at temperatures of 380 °C and 410 °C. In the case of coal B the amount of released mercury was also significant (>70%).

The results of the thermal treatment of the coal samples in carbon dioxide atmosphere are shown in Figure 4. The results are very promising since the amount of mercury released was much higher than in the two previous cases (Figs. 2 and 3). The release of Hg from coals A and B was almost 100% at temperatures over 380 °C. In the case of lignite (coal C) the maximum amount of released mercury was also high (roughly 90%).

Apart from the efficiency of mercury release an important aspect that also has to be taken into consideration, particularly before any commercial application of the thermal treatment technology, is the loss of chemical energy of the fuel and decrease of its HHV and LHV due to the devolatilization during fuel processing. Optimally, the thermal

treatment should be conducted at temperature high enough to provide the condition for mercury release but also low enough to prevent any significant 'loss' of the volatile matter. Taking those 'requirements' into consideration the proposed temperature of the thermal treatment (roughly <400 °C—Figs. 2–4) seems to be acceptable. The results reported by (Edgar, 1983) indicated that the thermal treatment of coal at temperatures <410 °C brought about very low loss of the chemical energy of the fuels since at those temperatures the amount of released volatiles was not high and, furthermore, the major components of the released gas were practically only carbon dioxide and carbon monoxide.

The experimental results obtained by the authors of the present paper confirm the results of (Edgar, 1983). Some chosen results are shown in Figures 5 and 6. The loss of the chemical energy of the fuel due to its thermal treatment was determined directly from the analysis of the HHV of the

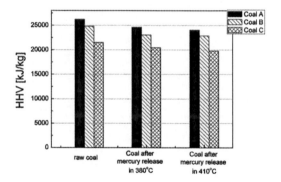

Figure 5. The High Heating Values (HHV) of some chosen coal samples after thermal treatment in nitrogen atmosphere.

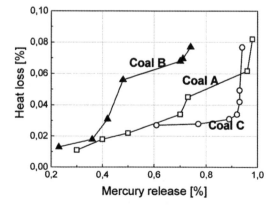

Figure 6. The heat loss versus mercury release for some coal samples. Fluidizing gas: nitrogen.

samples and furthermore independently calculated by integrating the CO and CO_2 concentration curves in the flue gases and assuming that the carbon in those two compounds was formed as the result of the oxidation of the carbon in the coal sample during its thermal treatment. The results in Figures 5 and 6 indicate that the maximum loss of HHV is less than 8%. The results directly indicate that the majority of mercury may be removed from coal with relatively low loss of the chemical energy of the fuel.

5 CONCLUSIONS

The results of the investigation, reported in the present paper may be briefly formulated as follows:

1. The thermal treatment of coals at temperatures below their ignition (roughly <410 °C) enables to get rid of up to 95% of mercury from the coal samples.
2. At similar temperature of the thermal treatment of the samples the relative amount of mercury released from lignite is much higher than from the hard coal.
3. The heating and thermal treatment of the coal samples at temperatures < 400 °C provide the conditions for the removal of the majority (>90%) of mercury and low loss of the chemical energy of the fuel (<8%).
4. The relative amount of released mercury is affected by the type of the fluidizing gas at temperatures of the thermal treatment above 350 °C.

REFERENCES

Chmielniak T., 2011; Reduction of mercury emission to the atmosphere from coal combustion process using low-temperature pyrolysis—a concept of process implementation on a commercial scale., *Rynek Energii,* 2 (93), 76–181.

Edgar T.F., Coal Processing and Pollution Control, Gulf Publishing Co, Houston 1983.

Głodek A., Pacyna J.M., 2009; Mercury emission from coal-fired power plants in Poland, *Atmospheric Environment,* 43, 5568–5673.

Grudziński Z., 2010; Konkurencyjność wytwarzania energii elektrycznej z węgla brunatnego i kamiennego, *Polityka Energetyczna,* 13, (2) 157–170.

Guffy F., Bland A., 2004; Thermal pretreatment of low-ranked coal for control of mercury emissions, *Fuel Processing Technology,* 85, 521–531.

Kobyłecki R., Wichliński M., Bis Z., 2007; Emisja rtęci z polskich węgli energetycznych. Współczesne Technologie i Urządzenia Energetyczne, Kraków, 269–274.

Lopez-Anton M.A., Yuan Y., Perry R., Maroto-Valer M.M., 2010; Analysis of mercury species present during coal combustion by thermal desorption, *Fuel,* 89, 629–634.

Pavlish J.H., Hamre L.L., Ye Z., 2010; Mercury control technologies for coal combustion and gasification systems, *Fuel,* 89, 838–847.

Pieńkowski, D., 2012. The Jevons effects and the consumption of energy in the European Union, *Problemy Ekorozwoju, Problems of Sustainable Development,* 7 (1): 105–116.

Sloss L., 1995; Mercury emissions and effects-the role of coal, IEA Coal Research, 19, 34–39.

Sundseth K., Pacyna J.M., Pacyna E.G., Munthe J., Belhaj M., Astrom S., 2010; Economic benefits from decreased mercury emissions: Projections for 2020, *Journal of Cleaner Production,* 18, 386–394.

Toole-O`Neil B., Tewalt S.J., Finkelman R.B., Akers D.J., 1999; Mercury concentration in coal—unraveling the puzzle, *Fuel,* 78, 47–54.

United Nations Environmental Programme, Global Mercury Assessment, UNEP Chemicals, Geneva 2002.

Wade C.B., Thurman C., Freas W., Student J., Matty D., Mohanty D.K., 2012; Preparation and characterization of high efficiency modified activated carbon for capture of mercury from flue gas in coal-fired power plants, *Fuel Processing Technology*, 97, 107–117.

Wichliński M., Kobyłecki R., Bis Z., 2011; Emisja rtęci podczas termicznej obróbki paliw, *Polityka Energetyczna,* 11, (2).191–202.

Wichliński M., Kobyłecki R., Bis Z., 2013; The investigation of the mercury contents in Polish coal samples, *Archives of Environmental Protection* (submitted for publication).

Wojnar K., Wisz J., 2006; Rtęć w polskiej energetyce, *Energetyka,* 4 (59), 280–283.

Yang H., Xu Z., Fan M., Bland A.E., Judkins R.R., 2007; Adsorbents for capturing mercury in coal-fired boiler flue gas, *Journal of Hazardous Materials,* 146, 1–11.

Environmental Engineering IV – Pawłowski, Dudzińska & Pawłowski (eds)
© 2013 Taylor & Francis Group, London, ISBN 978-0-415-64338-2

Author index